Biotechnology for Biofuel Production and Optimization

Biotechnology for Biofuel Production and Optimization

Edited By

Carrie A. Eckert
Cong T. Trinh

ELSEVIER
AMSTERDAM • BOSTON • HEIDELBERG • LONDON • NEW YORK • OXFORD
PARIS • SAN DIEGO • SAN FRANCISCO • SINGAPORE • SYDNEY • TOKYO

Elsevier
Radarweg 29, PO Box 211, 1000 AE Amsterdam, Netherlands
The Boulevard, Langford Lane, Kidlington, Oxford OX5 1GB, UK
50 Hampshire Street, 5th Floor, Cambridge, MA 02139, USA

Library of Congress Cataloging-in-Publication Data
A catalog record for this book is available from the Library of Congress

British Library Cataloguing in Publication Data
A catalogue record for this book is available from the British Library

For information on all Elsevier publications
visit our website at http://store.elsevier.com/

ISBN: 978-0-444-63475-7

Contents

Contributors

H.S. Alper McKetta Department of Chemical Engineering, and Institute for Cellular and Molecular Biology, The University of Texas at Austin, Austin, Texas, USA

M.C. Bassalo Department of Molecular, Cellular, and Developmental Biology, University of Colorado Boulder, Boulder, Colorado, USA

A.E. Beck Center for Biofilm Engineering, and Department of Microbiology and Immunology, Montana State University, Bozeman, Montana, USA

G.J. Bentley Division of Biological & Biomedical Sciences, Washington University in St. Louis, Saint Louis, Missouri, USA

H.C. Bernstein Chemical and Biological Signature Science, Pacific Northwest National Laboratory, Richland, Washington, USA

Abhijeet P. Borole Biosciences Division, Oak Ridge National Laboratory, Oak Ridge, Tennessee, USA; Energy Science and Engineering, Bredesen Center for Interdisciplinary Research and Graduate Education; Chemical and Biomolecular Engineering Department, University of Tennessee, Knoxville, Tennessee, USA

Nanette R. Boyle Chemical and Biological Engineering, Colorado School of Mines, Golden, Colorado, USA

Christopher J. Brigham Department of Bioengineering, University of Massachusetts Dartmouth, North Dartmouth, Massachusetts, USA

Katherine A. Brown National Renewable Energy Laboratory, Biosciences Center, Golden, Colorado, USA

Cody L. Brown Department of Agricultural and Biosystems Engineering, The University of Arizona, Tucson, Arizona, USA

Cassandra R. Burke Departments of Chemical Engineering and Bioengineering, Molecular Engineering & Sciences Institute, Center for Synthetic Biology, University of Washington, Seattle, Washington, USA

R.P. Carlson Center for Biofilm Engineering, and Department of Chemical and Biological Engineering, Montana State University, Bozeman, Montana, USA

James M. Carothers Departments of Chemical Engineering and Bioengineering, Molecular Engineering & Sciences Institute, Center for Synthetic Biology, University of Washington, Seattle, Washington, USA

Yingxi Chen Department of Chemical and Biological Engineering, Iowa State University, Ames, Iowa, USA
K. Chu Department of Energy, Environmental and Chemical Engineering, Washington University in St. Louis, Saint Louis, Missouri, USA
Peter Ciesielski National Renewable Energy Laboratory, Biosciences Center, Golden, Colorado, USA
Cameron Cotten Department of Chemical and Biological Engineering, and Great Lakes Bioenergy Research Center, University of Wisconsin-Madison, Madison, Wisconsin, USA
Joel L. Cuello Department of Agricultural and Biosystems Engineering, The University of Arizona, Tucson, Arizona, USA
Y. Deng National Engineering Laboratory for Cereal Fermentation Technology (NELCF), Jiangnan University, Wuxi, China
Qiang Fei National Bioenergy Center, National Renewable Energy Laboratory (NREL), Golden, Colorado, USA
S.S. Fong Department of Chemical and Life Science Engineering, Virginia Commonwealth University, Richmond, Virginia, USA
Meirong Gao Department of Chemical and Biological Engineering, The Center for Biorenewable Chemicals, Iowa State University, Ames, Iowa, USA
R.T. Gill Department of Chemical and Biological Engineering, and Renewable and Sustainable Energy Institute, University of Colorado Boulder, Boulder, Colorado, USA
Sanjay Guleria Centre for Biotechnology and Interdisciplinary Studies, Department of Chemical and Biological Engineering, Rensselaer Polytechnic Institute, Troy, New York, USA
A.L. Halweg-Edwards Department of Chemical and Biological Engineering, University of Colorado Boulder, Boulder, Colorado, USA
W.R. Henson Department of Energy, Environmental and Chemical Engineering, Washington University, Saint Louis, Missouri, USA
A. Hill McKetta Department of Chemical Engineering, The University of Texas at Austin, Austin, Texas, USA
Takanori Hoshino Department of Agricultural and Biosystems Engineering, The University of Arizona, Tucson, Arizona, USA
David Humbird DWH Process Consulting, Centennial, Colorado, USA
K.A. Hunt Center for Biofilm Engineering, and Department of Chemical and Biological Engineering, Montana State University, Bozeman, Montana, USA
C.M. Immethun Department of Energy, Environmental and Chemical Engineering, Washington University, Saint Louis, Missouri, USA
Laura R. Jarboe Department of Chemical and Biological Engineering, Iowa State University, Ames, Iowa, USA
Tao Jin Department of Chemical and Biological Engineering, Iowa State University, Ames, Iowa, USA
M.G. Kalyuzhnaya Biology Department, Viral Information Institute, San Diego State University, San Diego, California, USA

Aram Kang Joint BioEnergy Institute, Emeryville, and Physical Biosciences Division, Lawrence Berkeley National Laboratory, Berkeley, California, USA

Omar Khalidi Centre for Biotechnology and Interdisciplinary Studies, Department of Chemical and Biological Engineering, Rensselaer Polytechnic Institute, Troy, New York, USA

Paul W. King National Renewable Energy Laboratory, Biosciences Center, Golden, Colorado, USA

Mattheos A.G. Koffas Centre for Biotechnology and Interdisciplinary Studies, Department of Chemical and Biological Engineering, Rensselaer Polytechnic Institute, Troy, New York, USA

Sara Kuwahara Department of Agricultural and Biosystems Engineering, The University of Arizona, Tucson, Arizona, USA

Taek Soon Lee Joint BioEnergy Institute, Emeryville, and Physical Biosciences Division, Lawrence Berkeley National Laboratory, Berkeley, California, USA

Sophia Li Department of Biological Engineering, Massachusetts Institute of Technology, Cambridge, Massachusetts, USA

D. Liu Department of Energy, Environmental and Chemical Engineering, Washington University in St. Louis, Saint Louis, Missouri, USA

Jingnan Lu Department of Chemistry, Massachusetts Institute of Technology, Cambridge, Massachusetts, USA

K. Merklein Department of Biological and Agricultural Engineering, Kansas State University-Olathe, Olathe, Kansas, USA

S.D. Minteer Departments of Chemistry and Materials Science & Engineering, University of Utah, Salt Lake City, Utah, USA

T.S. Moon Department of Energy, Environmental and Chemical Engineering, Washington University, Saint Louis, Missouri, USA

D.R. Nielsen Chemical Engineering, School for Engineering of Matter, Transport, and Energy, Arizona State University, Tempe, Arizona, USA

Jennifer L. Reed Department of Chemical and Biological Engineering, and Great Lakes Bioenergy Research Center, University of Wisconsin-Madison, Madison, Wisconsin, USA

Zengyi Shao Department of Chemical and Biological Engineering, The Center for Biorenewable Chemicals, Iowa State University, Ames, Iowa, USA

Whitney E. Sinclair Chemical and Biological Engineering, Colorado School of Mines, Golden, Colorado, USA

Anthony J. Sinskey Department of Biology, and Engineering Systems Division, Massachusetts Institute of Technology, Cambridge, Massachusetts, USA

Miguel Suástegui Department of Chemical and Biological Engineering, The Center for Biorenewable Chemicals, Iowa State University, Ames, Iowa, USA

X. Wang School of Life Sciences, Arizona State University, Tempe, Arizona, USA

J.D. Winkler Department of Chemical and Biological Engineering, University of Colorado Boulder, Boulder, Colorado, USA

F. Zhang Department of Energy, Environmental and Chemical Engineering, and Division of Biological & Biomedical Sciences, Washington University in St. Louis, Saint Louis, Missouri, USA

Preface

Energy is the driving force of our modern society, and thus there is a great appetite for it. Current sources of energy are primarily derived from fossil-based resources that are not renewable, sustainable, or environmentally friendly for extraction and use, posing an immense challenge to meet the growing energy demands of the twenty-first century and beyond. To address this challenge, there is a great interest in the development of new technologies to harness alternative energy sources that are renewable, sustainable, and beneficial to the environment. This book presents state-of-the-art biotechnologies that employ biological systems for renewable biofuels production with the potential to replace fossil-based resources.

This book compiles a collection of 20 chapters that cover the entire process of biofuels production from manipulation of genes and pathways to platform organisms and renewable feedstocks for efficient biofuel production, as well as different cultivation techniques and process scale-up considerations. It is structured into six main topics: (1) harnessing unique metabolic pathways for advanced biofuel production; (2) pathway and strain design; (3) addressing complex phenotypes for improvement of biofuel production; (4) choice of microorganism and substrate utilization for production of targeted fuel molecules; (5) biohybrid methods for harvesting bioenergy; and (6) bioreactor design and optimization of scale up.

Topic 1 focuses on recent developments in the engineering of central and secondary metabolism pathways to make alcohols, gasolines, biodiesels, and jet fuels that resemble those derived from petroleum-based feedstocks. These include fermentative pathways, non-fermentative amino-acid derived pathways, reverse β-oxidation pathways, isoprenoid pathways, and *de novo* fatty acid biosynthesis pathways.

Topic 2 emphasizes the importance of pathway and strain design to optimize the production of targeted fuel molecules. Since native metabolic pathways utilized for target fuels are diverse, modular pathway engineering is important to quickly assemble and optimize production by using standardized parts such as promoters, ribosomal binding sites, regulatory elements, target genes, and plasmid copy number in a predictable and systematic fashion. In addition, dynamic pathway control is also critical to balance intracellular fluxes and hence avoid accumulation of intermediate metabolites that are toxic to biocatalysts and their production capabilities. The latest synthetic biology tools developed for parts prediction, pathway assembly, and rational design of dynamic pathway controllers are also presented.

Topic 3 will address complex cellular phenotypes for improvement of biofuel production. Biofuel molecules such as alcohols, biodiesels, and hydrocarbons are well known to be toxic to cells by interfering with intracellular and membrane processes. This inhibition severely affects cell growth and biofuel production. Since biofuel tolerance is a complex phenotype governed by multiple genes, effective system-wide engineering approaches—evolutionary engineering and combinatorial genome engineering—will be presented. Furthermore, the role of constraint-based metabolic network modeling will also be highlighted for the systematic search of genetic modifications across large-scale metabolic networks for enhanced production of fuel molecules.

Topic 4 will discuss the important consideration of microorganism platforms and substrate utilization for optimal production of targeted molecules. The economical biorefinery for biofuels production requires robust microbial biocatalysts that can use complex substrates such as lignocellulosic biomass, carbon dioxide, and other natural gases. Unique non-conventional microbes that can be harnessed for conversion of complex substrate into transportation liquid biofuels will be presented, including consolidated bioprocessing microbes, chemolithoautotrophs, phototrophs, methanotrophs, and mixed cell populations.

Topic 5 will discuss the latest biohybrid methods for harvesting bioenergy. Despite the vast array of enzymes and pathways that can be utilized from nature, understanding their complex chemistry within the cellular context is often limiting for optimal productivity. It is of great interest to study reactions under cell-free conditions using a blend of biological and material science to better understand and optimize these systems. This section will discuss unique biohybrid methods for the production of energy and biofuels including cell-free biotransformation, microbial electrochemical fuel cells, and PhotoBioHybrid Solar Conversion.

Topic 6 will cover bioreactor design and optimization of scale up. With the advances made at the laboratory scale in the production and titers of a variety of biofuels, the need to understand what factors are limiting for scale-up to industrial processes are of utmost importance. This section will address bioreactor design strategies for optimal production and collection as well as methods to optimize scale up of biofuel production.

Carrie A. Eckert
Cong T. Trinh

CHAPTER 1

Engineering Central Metabolism for Production of Higher Alcohol-based Biofuels

C.M. Immethun*, W.R. Henson*, X. Wang†, D.R. Nielsen‡, T.S. Moon*

*Department of Energy, Environmental and Chemical Engineering, Washington University, Saint Louis, Missouri, USA †School of Life Sciences, Arizona State University, Tempe, Arizona, USA ‡Chemical Engineering, School for Engineering of Matter, Transport, and Energy, Arizona State University, Tempe, Arizona, USA

1.1 Introduction: Longer Chain Bioalcohols as Gasoline Alternatives

In the United States in 2012, nearly 120 billion gallons of gasoline were consumed for an average of 329 million gallons each day.[1] Of this total, 40% of the crude oil used for gasoline production was imported from foreign sources,[2] exposing the domestic supply chain and national security to possible disruptions. The combustion of non-renewable petroleum-based fuels also causes harm to the environment, releasing previously entombed carbon to the atmosphere in the form of carbon dioxide, a greenhouse gas, along with other nitrogenous greenhouse gases, particulate matter, and additional organic species (e.g., polyaromatic hydrocarbons).[3] As energy consumption continues to soar both domestically and abroad, changes are now emerging in the atmosphere and global climate.[4] Sustainable biofuels could provide a solution to many problems that stem from the dependence on petroleum-based fuels.

First generation biofuels utilize agricultural crops as their feedstock, with the most suitable crops being those that thrive in precise geographic settings: for example, corn in the United States, sugarcane in Brazil, palm oil in Malaysia, and rapeseed in Germany. In all cases, however, the limited availability of arable land, water, and other resources spurs a vehement debate over the competition of crop use for food versus fuel. As a result, second generation biofuels, which are produced using lignocellulosic agricultural waste products such as corn stalks, sugarcane bagasse, or the hardy perennial switchgrass, are emerging as more sustainable options. However, these cheap feedstocks require energy intensive pretreatment to separate the sugar polymers (i.e., cellulose and hemicellulose) from the lignin. Cellulose, for example, must then be further depolymerized using cellulases[5] and acid hydrolysis[6] to render the sugars suitable as fermentation feedstocks. To date, however, these expensive feedstock pretreatment steps are largely responsible for limiting the economic viability of second generation biofuels.[7]

Biotechnology for Biofuel Production and Optimization. http://dx.doi.org/10.1016/B978-0-444-63475-7.00001-7

Ethanol remains the most technologically mature biofuel. Ethanol can be produced by yeast, fermenting the sugars derived from corn or sugar cane at high yield (90%), titer (10-14 w%), and productivity (2.5 g/L/h).[8] As a fuel, its high octane number prevents premature detonation (i.e., knocking) that causes engine damage and a loss of fuel economy. However, ethanol suffers from problems that prevent its full adoption by consumers and industry. For instance, its energy density is only two-thirds of that of gasoline (Table 1.1). Furthermore, the hygroscopic nature of ethanol both increases separation costs and can expose engines and other infrastructure (e.g., pipelines) to residual and damaging water.

To address the physical and thermodynamic limitations of ethanol, recent interest has been directed towards the microbial production of a series of longer chain alcohols, including the products of both natural and engineered pathways. Researchers are applying the tools and strategies of metabolic engineering to create "cellular factories" suited for the task. Butanol, a four carbon alcohol, can be synthesized in both *n*- (naturally occurring) and iso- (engineered) forms. Relative to ethanol, both forms possess an energy density much closer to that of gasoline (Table 1.1). This trend furthermore continues as the alkyl chain length increases, with the energy content of *n*-hexanol reaching 93% of that of gasoline.[9] Since longer chain alcohols are less hygroscopic, they are also more easily separated and display greater compatibility with conventional engines and other related infrastructure. Many recent examples illustrate how a range of emerging tools and strategies are being applied to engineer robust microbial production of higher alcohols.

While engineering microbes to produce a compound of interest is not a new topic, the emergence of systems biology and synthetic biology makes the practice of metabolic engineering even more powerful. Equipped with systems biology tools, including bioinformatics, genomics, transcriptomics, and proteomics, metabolic engineers have expanded understanding of elusive reactions in the central metabolism. Many organisms have well-defined central pathways of energy metabolism, glycolysis, the citric acid cycle, electron transport, oxidative phosphorylation, oxidation of amino acids and fatty acids, steroid and fatty acid biosynthesis, and gluconeogenesis.[10,11] Such well-characterized central pathways provide engineers with starting points for re-directing a native organism's carbon and energy towards biofuel production while native regulation can complicate such efforts. Fortunately, the

Table 1.1: Comparison of fuel properties[9]

	Gasoline	Ethanol	*n*-Butanol	Isobutanol
Chemical structure	C_4 to C_{12} mixture of hydrocarbons	CH_3CH_2OH	$CH_3(CH_2)_2CH_2OH$	$(CH_3)_2CHCH_2OH$
Fuel material	Crude oil	Sugars	Sugars	Sugars
Energy density (gasoline equivalent)	100%	67%	83%	83%
Blended octane number	82–92	99	91	98

building blocks used by organisms all originate from precursor metabolites whose biosynthesis is highly conserved across species.[12] Thus, synthetic biologists can build new pathways from heterologous parts, avoiding some of the regulation present in a host. Metabolic engineering enabled by systems biology and synthetic biology has produced alcohols that range from three to ten carbons in length by utilizing the host's central metabolism. Some of the pathways used to create longer chain alcohols can be found in nature while other non-native pathways have been created by researchers. All of these alcohol-producing pathways can be divided into two main categories: the pathways that function in the absence of oxygen and the pathways that require oxygen to function as discussed in Section 1.2.

1.2 Producing Longer Chain Alcohols Through the Central Metabolism

A wide variety of longer chain alcohols have been produced microbially through their central metabolism. Straight chain alcohols with up to eight carbons,[13] as well as isobutanol,[14] have been produced anaerobically, while a diversity of branched and straight chain alcohols have been produced through pathways that require oxygen.[15] Fermentation offers advantages and disadvantages over aerobic production. Aeration control can be challenging at the industrial scale.[16] Production costs are further reduced since the risk of contamination by unwanted microbes is diminished when cultures are grown anaerobically.[16] For low value biofuels, more than half the cost is associated with the carbon source.[17] The engineered microbe uses this carbon source for both growth and biofuel production. During fermentation, only 5-15% of the carbon is used for growth.[18] The addition of oxygen in a semi-aerobic fermentation increases the carbon used for growth up to 30%,[18,19] decreasing the carbon available to make biofuels.[20] Reduction of the carbohydrate carbon source to biofuels requires a balanced supply of reducing equivalents. In anaerobic conditions, the microbes regenerate NAD^+ natively by producing reduced metabolites such as lactate and alcohols, decreasing the reducing equivalents without causing a metabolic imbalance.[21,22] While reducing equivalents are regenerated under aerobic conditions, this process creates ATP that the cells use for growth, rather than it produces biofuels.[14] NADH is the major reducing equivalent in *E. coli* under anaerobic conditions, limiting the heterologous enzymes for biofuel production without changing their native cofactor specificity (from NADPH to NADH).[20] With advantages and disadvantages for both modes of growth, there has been extensive work towards the production of longer chain alcohols anaerobically as well as aerobically which is discussed in this section.

1.2.1 Fermentative Pathways

With a history of its fermentative production that extends over a century, *n*-butanol is the best characterized, naturally occurring higher alcohol. An endogenous product of numerous *Clostridium* species, most notably the type strain *Clostridium acetobutylicum*

ATCC 824, *n*-butanol is synthesized together with acetone and ethanol by the so-called ABE (acetone-butanol-ethanol) fermentation. Through the middle of the twentieth century, *Clostridium* was employed by industry to produce acetone and butanol for use as solvents.[23] As efficiencies in the petrochemical processes improved and the price of the preferred substrate (molasses) climbed, ABE fermentation was phased out.[24] Since the 1980's, there has been renewed interest in using the *n*-butanol from ABE fermentation as a biofuel. More recently, many efforts have been made to overcome several key and inherent limitations of the *n*-butanol pathway in *Clostridium*[25] (discussed in detail below), as well as to produce other higher alcohol products[26] and to functionally reconstruct the pathway in other more tractable hosts.

ABE Pathway in Clostridium Acetobutylicum

Despite its long history and well-understood physiology, *n*-butanol production by native *Clostridium* sp. suffers several inherent limitations from an industrial standpoint. First, native *n*-butanol fermentation involves two distinct growth stages, namely acidogenesis and solventogenesis. The organic acids acetate and butyrate are the principal fermentation products during exponential growth, the first stage. When the cells stop dividing during stationary phase, the organism shifts to solventogenesis followed by sporulation.[23] The acids produced during acidogenesis are re-assimilated and channeled back to acetyl-CoA and butyryl-CoA before being converted to ethanol, acetone, and butanol.[25] Both solventogenesis and sporulation have been shown to be initiated in part by the same regulator Spo0A although sporulation is not required for the production of the solvents.[25]

However, not only does the need for a two stage culture complicate efforts related to the development of continuous bioprocesses,[27] but the co-production of acetone and ethanol along with butanol detracts from competitive *n*-butanol yields. An enzyme involved in the production of both acetone and butanol, thiolase condenses two acetyl-CoA molecules to form acetoacetyl-CoA (Figure 1.1). Thiolase competes with aldehyde dehydrogenase for the acetyl-CoA as the aldehyde dehydrogenase funnels some of the carbon towards ethanol production. The acetoacetyl-CoA can either proceed to acetone formation through acetoacetyl-CoA:acyl-CoA transferase or to *n*-butanol via 3-hydroxybutyryl-CoA dehydrogenase and five subsequent steps.[25] Figure 1.1 illustrates *Clostridium* fermentative CoA-dependent pathway.

In native *C. acetobutylicum*, the ABE solvent products primarily accumulate at a mass ratio of 3:6:1.[28] Whereas the theoretical maximum yield of butanol on glucose is 0.41 g/g, butanol and total solvent yields of about 0.15-0.25 and 0.28-0.35 g/g, respectively, are typical.[23,28,29] For example, butanol and total solvent yields of 0.23 and 0.32 g/g, respectively, were reported by Li et al. using *C. acetobutylicum* ATCC 824.[30] Significant efforts have concentrated on altering the product mixture in *C. acetobutylicum*, primarily focusing on increasing the total alcohol content by reducing or even eliminating acid and acetone formation. Knocking

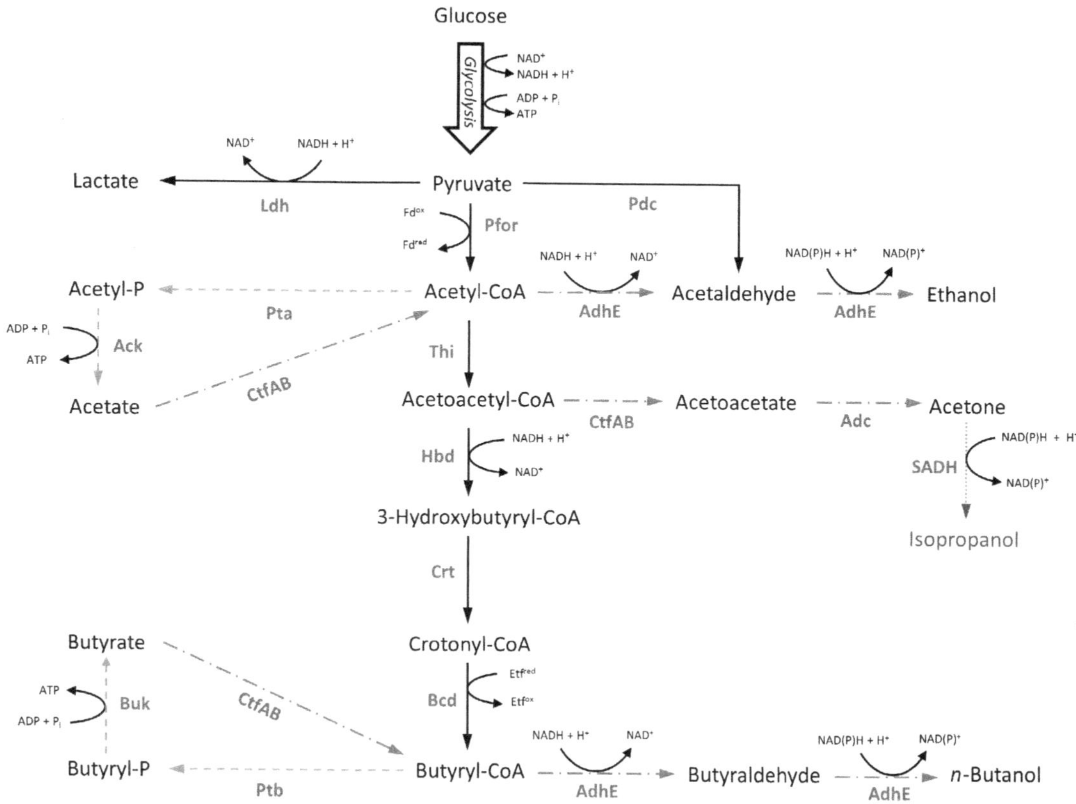

Figure 1.1

Acetone-butanol-ethanol fermentation in native *Clostridium acetobutylicum*. Reactions active just during acidogenesis are green dashed lines. Reactions active just during solventogenesis are blue dot-dashed lines. Reactions active during both acidogenesis and solventogenesis are black solid lines. Exogenous pathways are purple two parallel dashed lines and their products in purple. Enzyme names are in bold red: Ack, acetate kinase; Adc, acetoacetate decarboxylase; AdhE, aldehyde/alcohol dehydrogenase; Bcd, butyryl-CoA dehydrogenase; Buk, butyrate kinase; Crt, crotonase; CtfAB, acetoacetyl-CoA:acyl-CoA transferase; Fd, ferredoxin; Etf, electron transfer flavoprotein; Hbd, 3-hydroxybutyryl-CoA dehydrogenase; Ldh, lactate dehydrogenase; Pdc, pyruvate decarboxylase; Pfor, pyruvate:ferredoxin oxidoreductase; Pta, phosphotransacetylase; Ptb, phosphotransbutyrylase; SADH, secondary alcohol dehydrogenase; and Thi, thiolase. *Adapted from Lutke-Eversloh and Bahl (2011)*[25].

out, down-regulating, and over-expressing various genes in the highly interconnected and regulated ABE pathway often produce unwanted effects.[31–34] During acidogenesis, ATP is produced along with acetate and butyrate.[25] Knocking out acid production to improve flux towards the solvents therefore limits energy available to the cell.[31] Knocking out[32,34] or knocking down[35] genes in the acetone pathway often increased acid and ethanol concentrations, but not the concentration of *n*-butanol. A recent approach to minimizing acetone production in *C. acetobutylicum* converted the acetone to isopropanol, which can be used as a fuel additive, by introducing *Clostridium beijerinckii*'s secondary alcohol dehydrogenase.[36] This strategy creates a mixture of alcohols: ethanol, isopropanol, and *n*-butanol. While more alcohols are produced, the butanol titer remained unchanged.[28] Higher alcohol production using the ABE pathway in its native *C. acetobutylicum* remains tangled with additional products, increasing separation costs and reducing the yield of the longer chain alcohols.

In addition, product toxicity remains a key factor that limits the industrial prospects of ABE fermentation. For example, all three solvent products contribute to the destabilization and increased fluidity of the cytoplasmic membrane,[37] the generally accepted mechanism of solvent toxicity in *Clostridium* sp.[38] Furthermore, solvent stresses have also been implicated in spore formation and deterioration of the solvent forming phenotype (by the shedding of pSOL1, a mega-plasmid encoding several genes essential to the solvent biosynthesis pathways).[39] *n*-Butanol is the greatest contributor to the solvent toxicity, fully inhibiting growth and production in wild type cultures at concentrations as low as 13-20 g/L.[23] Furthermore, the economic recovery of *n*-butanol is challenged at lower titers due to its lower vapor pressure relative to water (0.0109 atm versus 0.0312 atm at 25°C). As a consequence, *n*-butanol recovery typically requires multi-stage distillation.[40] Conservative estimates, meanwhile, suggest that *n*-butanol titers as high as 20-30 g/L might be required to truly recoup the energy input needed for its distillative recovery.[41]

ABE Pathway in Heterologous Hosts

To overcome several of the inherent, undesirable issues of *n*-butanol production via the ABE fermentation in *Clostridium*, the *n*-butanol pathway (beginning with thiolase) has been functionally reconstructed and studied in several heterologous hosts. These hosts have been chosen for their genetic tractability, solvent tolerance, and ability to grow on a diversity of sustainable substrates, including *Escherichia coli*,[13,21,22,26,42–44] *Pseudomonas putida*,[44] *Bacillus subtilis*,[44] *Lactobacillus brevis*,[45] *Thermoanaerobacterium saccharolyticum*,[46] *Saccharomyces cerevisiae*,[47,48] and *Synechococcus elongatus* PCC 7942.[49] The clostridial genes have been expressed under synthetic control either as an operon[42–44] or individually[22,44,47] to optimize their expression. Enzyme homologs from other genetic sources have been substituted for clostridial pathway enzymes to improve activity,[21,45–47] introduce irreversibility,[13,50,51] and/or alter reducing equivalent requirements.[21,22,47,51]

One key challenge in these efforts has been to effectively balance the intracellular redox state of heterologous hosts with the demands of the *n*-butanol pathway. For example, the most common hosts, *E. coli* and *S. cerevisiae*, suffer from insufficient reducing equivalents for the pathway.[43,47] To address this problem, native pathways that compete for the reducing equivalents have been knocked out,[21,43] and heterologous enzymes that regenerate the co-factors have been introduced.[44] Meanwhile, metabolite flux through the pathway has also been promoted by knocking out native regulation[43] and overproducing required precursors.[44,48] For example, by knocking out *E. coli* lactate dehydrogenase (Ldh), alcohol/acetaldehyde dehydrogenase (AdhE), and fumarate reductase (FrdBC), which reduces competition for NADH and acetyl-CoA, *n*-butanol production doubled; however, acetate production also increased slightly.[43] Phosphate acetyltransferase was then deleted to decrease the acetate concentration, but the concurrent deletion of the anaerobic regulator, fumarate and nitrate reductase (Fnr), was required to improve *n*-butanol production to three times that of the wild type strain. Deleting *fnr* removed native regulation of pyruvate dehydrogenase complex, allowing the complex to produce acetyl-CoA and NADH in anaerobic conditions.[43] Through continuing work to improve performance, fermentative *n*-butanol production via the clostridial pathway is moving closer to commercial viability.

Extending the ABE Pathway

The *n*-butanol pathway of *Clostridium* has recently been extended in *E. coli* to produce *n*-hexanol and even *n*-octanol by leveraging promiscuity of different enzyme homologs at the first committed pathway step (thiolase). More specifically, by replacing *C. acetobutylicum* thiolase with β-ketothiolase from *Ralstonia eutropha* (used in the biosynthesis of polyhydroxyalkanoates), acetyl-CoA could be condensed with butyryl-CoA to produce 3-ketohexanoly-CoA.[13] This 6-carbon intermediate then proceeded analogously through the butanol pathway to ultimately yield *n*-hexanol. Initial strains were capable of producing 47 mg/L of *n*-hexanol and 5.1 g/L of *n*-butanol in 48 h. The bifunctional acetaldehyde-CoA/alcohol dehydrogenase of *C. acetobutylicum* is not substrate specific, and *n*-butanol production outcompeted *n*-hexanol.[13,26] To further increase the production of six carbon or longer chain alcohols, Machado et al. evolved pathway enzymes to prefer longer substrates.[13] First, starting with the work of Shen et al. to create a NADH driving force,[21] the mixed acid fermentation pathways for lactate, succinate, and ethanol production were knocked out to create a build-up of NADH in anaerobic conditions and a cessation of growth. Second, the electron transfer flavoprotein-dependent butyryl-CoA dehydrogenase was replaced with *Treponema denticola*'s NADH-dependent *trans*-enoyl-CoA reductase (Ter). As a result, the cells could recycle enough NADH for growth by utilizing the heterologous pathway.[21] Machado et al. then substituted an acyl-CoA thioesterase from fatty acid biosynthesis that prefers six carbon acyl-CoA substrates for the clostridial alcohol dehydrogenase, tying the evolution of upstream enzymes' preference for longer chain compounds to recycling reducing equivalents. *C. acetobutylicum*'s 3-hydroxybutyryl-CoA dehydrogenase was also replaced with *R. eutropha*'s PaaH1 since PaaH1 had shown superior activity towards longer substrates. Error prone PCR of *paaH1* followed by screening of the clones for improved growth in anaerobic conditions resulted in an increase of hexanoic acid and a decrease in butanoic acid. The acyl-CoA thioesterase was then replaced with the original alcohol dehydrogenase, producing approximately 450 mg/L *n*-hexanol and 50 mg/L *n*-octanol as well as 1-1.5 g/L *n*-butanol and 200-500 mg/L ethanol.[13] These works show promise for the continued development of novel fermentative routes to higher alcohols with superior properties as biofuels.

1.2.2 Non-Fermentative Pathways

Longer chain alcohols can also be produced through non-fermentative pathways, the most popular of which is the modified Ehrlich pathway. The modified Ehrlich pathway uses as precursors a variety of keto acids, intermediates in amino acid biosynthesis, which are converted to alcohols by the combined action of a keto acid decarboxylase and an alcohol dehydrogenase. The broad substrate specificity of these two enzyme classes has enabled production of straight and branched chain alcohols through this non-natural combination of pathways. Two additional non-fermentative routes for producing longer chain alcohols have also been built from natural catabolic processes. In one case, organisms can use the β-oxidation cycle to release energy stored in lipids by removing two carbons per cycle as acetyl-CoA.[52] This ubiquitous process can be reversed to elongate carbon chains two carbons per cycle. Reverse β-oxidation coupled with endogenous acetaldehyde dehydrogenase and *n*-butanol dehydrogenase has produced straight chain alcohols.[53] Lastly, longer chain alcohols have also been produced via natural catabolism in a process that begins with the oxidation of amino acids.[54] Pyruvate and keto acids are the products of amino acid oxidation, which can be converted to a variety of alcohols through the non-natural coupling with the Ehrlich pathway. All three of the non-fermentative pathways discussed begin with metabolic processes common to all organisms, making them versatile for longer chain biofuel production.

Amino Acid Biosynthesis/Ehrlich Pathway

Linking the Ehrlich pathway with amino acid biosynthesis has created a large variety of branched and straight chain alcohols. Pyruvate produced during glycolysis is the main branch point in the pathways used to produce longer chain alcohols from different keto acids (Figure 1.2). Following the valine biosynthetic pathway, decarboxylation and reduction of 2-ketoisovalerate yields isobutanol. 2-ketoisocaproate from the leucine biosynthetic pathway has been used for the production of 3-methyl-1-butanol while 2-keto-3-methylvalerate from the isoleucine biosynthetic pathway yields 2-methyl-1-butanol. Straight chain alcohols have also been produced through the same strategy. *n*-Propanol has been derived from 2-ketobutyrate while *n*-butanol has been produced from 2-ketovalerate through the norvaline pathway (Figure 1.2). Further engineering of these starting pathways has also yielded longer chain, non-natural alcohols. By altering the binding pocket of 2-isopropylmalate synthase, LeuA, straight chain alcohols with three to eight carbons[55] and non-natural branched chain alcohols with five to eight carbons[15] have also been produced (Figure 1.2). To date, different pathways have been constructed for higher alcohol production in a variety of hosts,

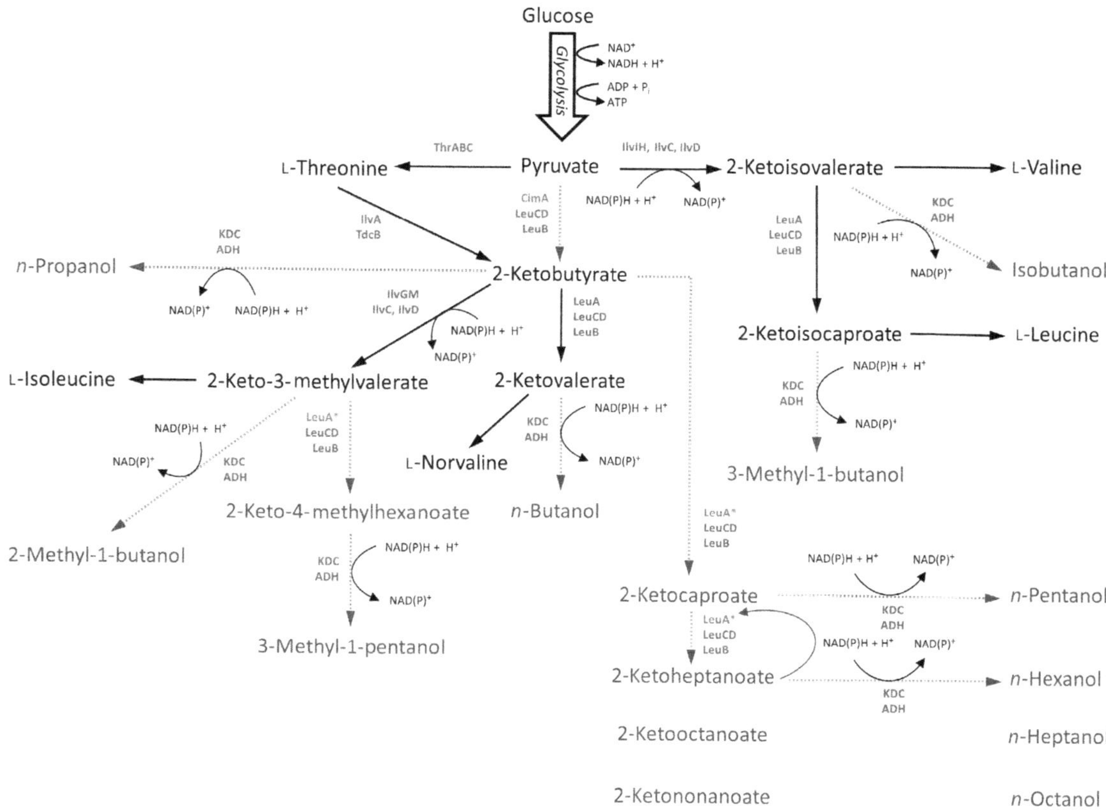

Figure 1.2

Amino acid biosynthesis/Ehrlich pathway. Decarboxylation of keto acids, followed by reduction, yields different alcohols: valine pathway, isobutanol; leucine pathway, 3-methyl-1-butanol; isoleucine pathway, 2-methyl-1-butanol; and norvaline pathway, 1-butanol. Further engineering of these starting pathways has yielded longer chain, "non-natural" alcohols: 3-methyl-1-pentanol, *n*-pentanol, *n*-hexanol, *n*-heptanol, and *n*-octanol. Introduced pathways are purple two parallel dashed lines and their products in purple. Enzyme names are in bold red while engineered enzyme names are in gray: ADH, alcohol dehydrogenase; IlvA, threonine deaminase; IlvC, acetohydroxy acid isomeroreductase; IlvD, dihydroxy acid dehydratase; IlvGM, acetohydroxybutanoate synthase/acetolactate synthase; IlvIH, acetolactate synthase/acetohydroxybutanoate synthase; KDC, 2-keto-acid decarboxylase; LeuA, 2-isopropylmalate synthase; LeuB, 3-isopropylmalate dehydrogenase; LeuCD, isopropylmalate isomerase; TdcB, catabolic threonine dehydratase; and ThrABC, aspartate kinase/homoserine dehydrogenase/homoserine kinase/threonine synthase. The reaction arrows between 2-ketooctanoate and *n*-heptanol as well as between 2-ketononanoate and *n*-octanol are not shown for clarity.

including *E. coli*,[14,15,20,54–63] *S. cerevisiae*,[64–70] *Corynebacterium glutamicum*,[71,72] *Clostridium celluloyticum*,[73] *B subtilis*,[74,75] *S. elongatus* PCC7942,[76] *R. eutropha*,[77] *Pseudomonas* sp strain VLB120,[78] and *Synechocystis* sp. PCC 6803.[79] The non-natural coupling of amino acid biosynthesis and the Ehrlich pathway creates an enormous diversity of pathways in a variety of hosts for the production of longer-chain alcohol biofuels.

Reverse β-Oxidation Pathway

The catabolic β-oxidation cycle can be reversed to elongate carbon chains two carbons at a time.[52] Cycle reversal (Figure 1.3) was achieved through several mutations of the genes involved in the fatty acid metabolism in addition to growing the host in media lacking fatty acids, but containing glucose. The first set of mutations allowed the *fad* regulon and *ato* operon to be expressed constitutively, enabling the organism to break down both short and long chain fatty acids at the same time.[80] To relieve carbon catabolite repression of steps in the β-oxidation cycle, the native *crp* gene was replaced with a cAMP independent version. Repression by ArcA was also removed by deleting its gene. Finally, the endogenous

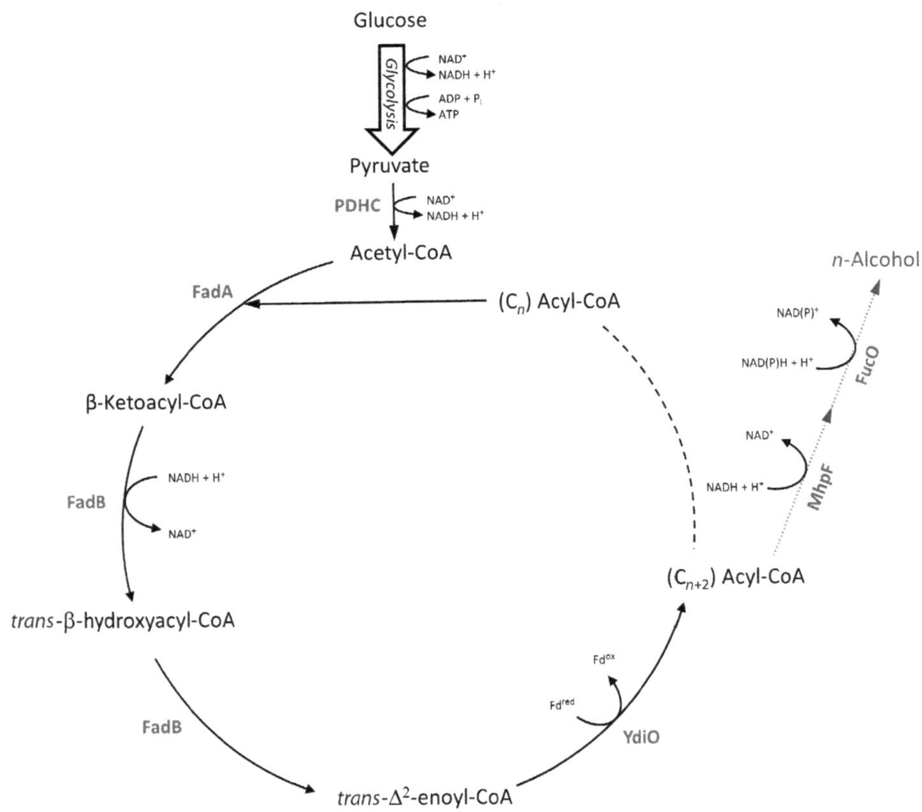

Figure 1.3

Reverse β-oxidation pathway. *E. coli* catabolic β-oxidation cycle is reversed to elongate carbon chains two carbons at a time. Introduced pathways are purple two parallel dashed lines and their products in purple. Enzyme names are in bold red: FadA, 3-ketoacyl-CoA thiolase; FadB, 3-hydroxyacyl-CoA dehydrogenase; Fd, ferredoxin; FucO, L-1,2-propanediol oxidoreductase acting as a *n*-butanol dehygrogenase; MhpF, acetaldehyde dehydrogenase; PDHC, pyruvate dehydrogenase; and YdiO, enoyl-CoA reductase. *Adapted from Dellomonaco et al. (2011)*[53].

acetaldehyde dehydrogenase, MhpF, and *n*-butanol dehydrogenase, FucO,[81] were introduced to convert the C_{n+2}-acyl-CoA to straight chain alcohols.[52,53] After proceeding through the pathway, *n*-butanol was produced. Multiple turns of the engineered reverse β-oxidation cycle were possible when the first gene in the cycle, encoding 3-ketoacyl-CoA thiolase (*fadA*), was overexpressed. FadA has broad substrate specificity, permitting the non-natural longer substrates that led to six, eight, and ten carbon straight chain alcohols.[53] By increasing the chain length two carbons at a time, this novel strategy could be an important pathway for biofuel production.

Amino Acid Oxidation/Ehrlich Pathway

Just as keto acids produced from amino acid biosynthesis pathways can be converted to alcohols using the modified Ehrlich pathway, keto acids produced through amino acid oxidation can also be converted to alcohols in the same manner. *S. cerevisiae* produces small quantities of higher alcohols naturally through this mechanism.[67,82] For example, the amino acid valine is produced in the mitochondria, oxidized in the cytosol to 2-ketoisovalerate by the branched-chain amino-acid aminotransferase (BAT2), and subsequently converted to isobutanol by pyruvate decarboxylases and alcohol dehydrogenases (Figure 1.4). However, the transportation of the pathway intermediates between different compartments limits production. This challenge has been addressed by overexpressing genes for production of pathway intermediates[65–67] and by linking isobutanol production to valine biosynthesis either in the cytosol or in the mitochondria. The first approach expressed the genes in the valine biosynthesis pathway, *ILV2*, *ILV5*, and *ILV3*, in the cytosol instead of the mitochondria.[68,70] Avalos et al. instead expressed the pyruvate decarboxylases, PDC6,5,1, and the alcohol dehydrogenase, ADH, in the mitochondria.[66] Both approaches simplified the steps required to synthesize isobutanol in *S. cerevisiae*.

Biofuels have also been produced in *E. coli* using keto acids available from amino acid degradation as the substrates for the Ehrlich pathway.[54] To increase the supply of keto acids, Huo et al. created an amino acid degradation driving force.[54] The amino acids threonine, glycine, serine, asparagine, arginine, proline, and glutamine are transaminated, transferring the amino group from one amino acid to synthesize glutamate, aspartate or one of the branched chain amino acids as a nitrogen storage mechanism. Except for aspartate, these subsequent amino acids can then be degraded through deamination, releasing ammonia.[83] To release the ammonia from amino acids not naturally deaminated, three exogenous cycles were introduced to transaminate and subsequently deaminate aspartate, alanine, leucine, isoleucine, and valine by overexpressing *Thermoactinomyces intermedius leuDH* as well as *E. coli avtA*, *ilvA*, *ilvE*, and *sdaB*.[54] Normally, the ammonia released by deamination can be taken back up by the cell, converting 2-ketoglutarate to glutamate and glutamine with the ammonia assimilation enzymes GlnA and GdhA.[84] Huo et al. knocked out these two genes, making more keto acids available

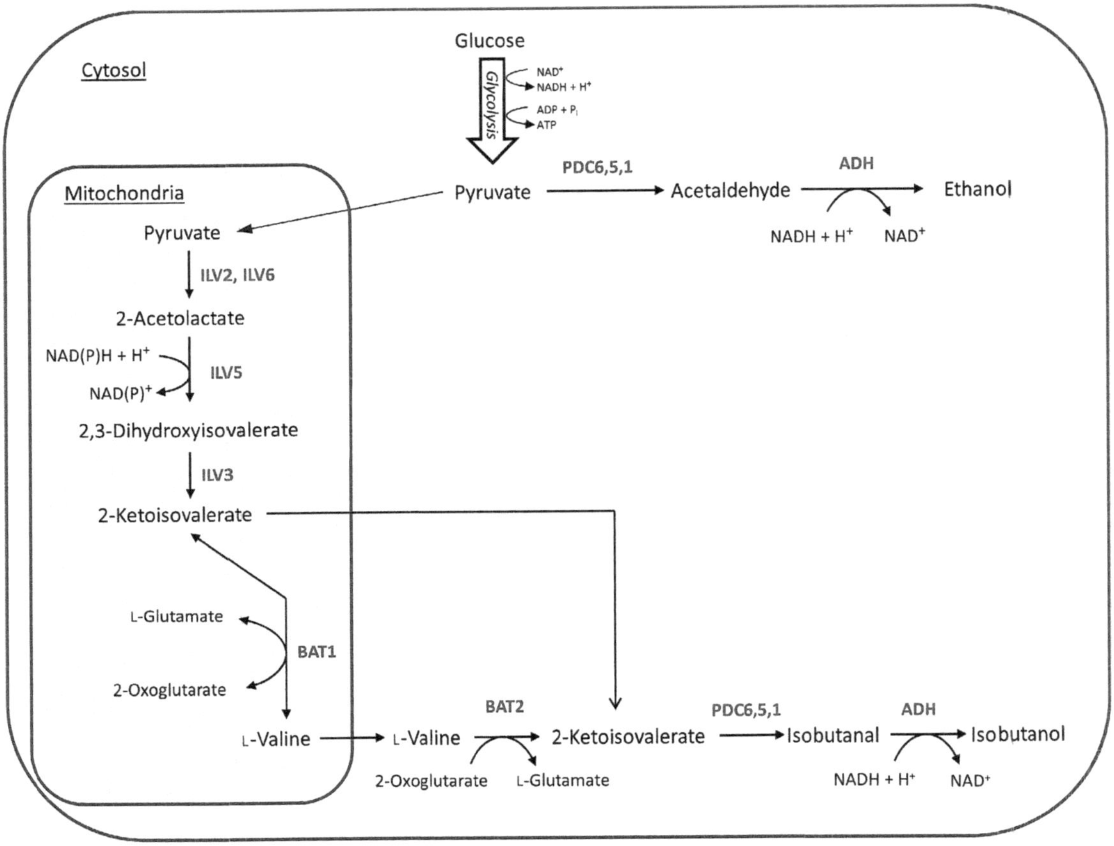

Figure 1.4

Native isobutanol pathway in *Saccharomyces cerevisiae*. Enzyme names are in bold red: ADH, alcohol dehydrogenase; BAT1, mitochondrial branched-chain amino acid aminotransferase; BAT2, cytosolic branched-chain amino acid aminotransferase; ILV2, catalytic subunit of acetolactate synthase; ILV3, dihydroxyacid dehydratase; ILV5, acetohydroxyacid reductoisomerase; ILV6, regulatory subunit of acetolactate synthase; and PDC6,5,1, pyruvate decarboxylases. *Adapted from Chen et al. (2011)*[67].

for biofuel production by heterologous keto acid decarboxylase and alcohol dehydrogenase.[54] This unique pathway can utilize protein by-products from bio-refining, including the protein by-products from genetically modified organisms which cannot be recycled into fertilizers or animal feed without additional treatment.

1.2.3 Opportunities for Further Improvement

With the aid of *de novo* engineered or heterologous natural pathways and extensive re-engineering of their central metabolism, microorganisms have been successfully engineered to produce higher alcohol biofuels up to ten carbons in length. In nearly all cases, however, product titer, yield, and overall productivity require significant improvement before commercially viable production metrics can be met. Accordingly, this challenge is being tackled through a variety of rational and combinatorial approaches intended to overcome key inherent and technological roadblocks.

Pathway function, for example, is contingent upon its composite enzyme function. Enzyme choice, therefore, can significantly alter the yield, titer, and productivity of the product, as well as affect the health of the host. The vast number of available protein sequences, easily found by homology, allows the possible isozymes to be available for each step in a pathway.[85] The enzyme's performance, such as activity and specificity, can be further optimized through rational engineering or random mutagenesis. Not only are the characteristics of individual enzymes important, but how the enzymes function together also greatly affects the host and product. Issues in pathway engineering, such as thermodynamic preference, kinetic properties, cellular ATP yield, and cofactor imbalance, also need to be addressed for favorable performance. Host selection and engineering, meanwhile, are similarly vital to successful biofuel production. As genetic tools are developed for new organisms, the number of choices for production hosts expands. Issues beyond genetic tractability, including nutrition requirements, growth rate, product tolerance, and substrate utilization, should also be considered. The host response to production conditions as well as tolerance to the desired product can limit the applicability of an optimized and balanced pathway. In particular, many production hosts need to be engineered to be tolerant to their target fuel product(s) at high concentrations. Engineering of the host for product tolerance can be either rational or through evolution. While commercial viability has yet to be attained for microbially produced biofuels, significant work has improved the titer, yield, and productivity of longer chain alcohols produced through the central metabolism, as discussed in Section 1.3.

1.3 Strategies for Improving Higher Alcohol Production

1.3.1 Approaches to Enhance Enzyme and Pathway Function

The function of both native *n*-butanol and engineered higher alcohol biosynthesis pathways has been enhanced through the application of alternative enzymes and expression level tuning to systematically optimize metabolite flux and overcome other limitations. In *Clostridium*, the impacts of said effects have yielded only marginal returns due in part to the highly regulated nature of the ABE pathway and its intimate connection to host fitness.[25] For example, knocking out acid production to improve flux towards the solvents limits ATP generation and thus energy availability to the cell.[31] Other approaches include the selective overexpression of *n*-butanol pathway enzymes or elimination of competing effects to enhance net activity and flux towards *n*-butanol. The success of these, however, has been limited due to diminished precursor availability.[33] Promoter replacement to enable the expression of aldehyde-alcohol dehydrogenase (normally only expressed during solventogenesis) during acidogenesis resulted in earlier alcohol production and less acid accumulation but not elevated butanol titers.[33]

Knocking out,[34] or knocking down with antisense RNA,[35] enzymes in the acetone pathway often increased acid and ethanol concentration[34] as did overexpressing thiolase which had been engineered to be less sensitive to its inhibitor, CoA-SH.[86]

In heterologous hosts, modulations to the *n*-butanol pathway enzymes have provided more substantial impacts. For example, thiolase has been replaced with *E. coli*'s native acetyl-CoA acetyltransferase (*atoB*) to take advantage of its five-time higher specific activity, while in *L. brevis* both native thiolase and aldehyde-alcohol dehydrogenase have been substituted.[45] Pathway function, meanwhile, benefits from modulated expression of the constituent enzymes. For example, it was found that by expressing each pathway gene individually under control of its own *T7lac* promoter in *E. coli*, *n*-butanol titers improved almost 6-fold relative to their polycistronic expression.[44]

Enzyme choice and expression strategies are equally important in the case of Ehrlich pathway higher alcohols. Choosing pathway enzymes to increase the supply of substrates to the pathways is critical. For example, as isobutanol is derived from 2-ketoisovalerate, overexpression of the genes responsible for converting pyruvate and 2-ketoisovalerate has been found to be essential for boosting flux and isobutanol yields while decreasing titers of other byproduct alcohols (produced as a result of enzyme promiscuity).[57] Several studies have collectively shown that this is best achieved via the combined overexpression of *B. subtilis* acetolactate synthase (*alsS*)[54,57,59,60,71,73,74,76] along with acetohydroxyacid isomeroreductase (*ilvC*) and dihydroxy acid dehydratase (*ilvD*)[14,54,59,60,73,76] of *E. coli*. While *E. coli* possesses three acetolactate synthase isozymes, AlsS from *B. subtilis* boasts a stronger affinity for pyruvate[57] and does not suffer from allosteric inhibition (by valine).[14] Similarly, to produce 2-methyl-1-butanol, *ilvGM* from *Salmonella typhimurium* and *ilvA* from *C. glutamicum* are both overexpressed to enhance the availability of isoleucine pathway precursors.[63]

The notable versatility of the Ehrlich pathway arises from the promiscuity of its two enzymes responsible for converting the 2-ketoacid precursors to their respective alcohol products.[57] Several 2-ketoacid decarboxylases have been tested, but the most common choice for all of the prokaryotic studies is Kivd from *Lactococcus lactis*.[21,54,55,57,59,60,64,68,72–79] In contrast to 2-ketoacid decarboxylase, the performance of the alcohol dehydrogenase chosen for the last step of the Ehrlich pathway varies widely from study to study, with the native gene often providing the best performance.[14,15,54,64–67,69,70,72,77] In other cases, improvements have been made to enhance or alter enzyme performance in heterologous hosts. For example, Bastian et al. improved the catalytic efficiency of *L. lactis* AdhA in *E. coli* by 40-fold through random mutagenesis[20] while others have used protein engineering to modify enzyme specificity towards both straight chain (with three to eight carbons)[55] and branched chain alcohols (with five to eight carbons).[15] Rational approaches to the design of enzymes are emerging and powerful tools that metabolic engineers can use to expand and tune what can be produced microbially, and even to include compounds not made by any organism today.

1.3.2 Engineering Microbial Hosts to Enhance Cofactor and Precursor Availability

Optimization of kinetic properties and expression levels of enzymes in pathways can increase the carbon flow to the target product. Another engineering approach is to optimize ATP yields and cofactor amounts.[87] They are the driving forces to alter metabolic flux and to maximize the final products. For example, the native clostridial CoA-dependent *n*-butanol pathway does not have an intrinsic, strong driving force to direct carbon flux from acetyl-CoA to *n*-butanol.[21] The crotonyl-CoA reduction step, catalyzed by the *C. acetobutylicum* butyryl-CoA dehydrogenase complex (Bcd-EtfAB), uses both NADH and reduced ferredoxin (Bcd-EtfAB complex) as reducing cofactors (Figure 1.1).[88] NADH is a preferred reducing cofactor for fermentative production due to its relative abundance as a result of glycolysis.[50,88,89] The crotonyl-CoA reduction can also be catalyzed by *trans*-enoyl-coenzyme A reductase (Ter), and this enzyme lacks catalytic activity for the reverse reaction, thus creating a kinetic trap to enhance the formation of butyryl-CoA from crotonyl-CoA reduction.[21,50] In addition, the deletion of the major acetyl-CoA-consuming enzyme phosphate acetyltransferase further increased *n*-butanol production to reach a high-titer (30 g/liter) and high-yield (70 to 88% of the theoretical values).[21] In the native clostridial CoA-dependent *n*-butanol pathway, besides the undesired crotonyl-CoA reduction step catalyzed by Bcd-EtfAb, the thiolase-mediated condensation of two acetyl-CoA molecules to produce acetoacetyl-CoA is reversible but thermodynamically unfavorable (Figure 1.1).[49] ATP-driven malonyl-CoA synthesis and the following decarboxylative condensation to form acetoacetyl-CoA create a thermodynamically favorable, synthetic alternative route to replace the original condensation step to form acetoacetyl-CoA.[49] This synthetic pathway successfully increased *n*-butanol production in *S. elongatus* PCC 7942.[49]

The yeast Ehrlich pathway has been constructed in *E. coli* for isobutanol production with 2-keto-isovalerate serving as a precursor (Figure 1.2).[57] Both ketol-acid reductoisomerase and alcohol dehydrogenase in the native pathway prefer NADPH as their reducing substrate. Under anaerobic fermentation conditions, NADPH is rather limited compared to NADH generated by glycolysis.[89,90] Bastian et al. used directed evolution to alter the substrate preference of these two enzymes for NADH.[20] The synthetic isobutanol pathway using these modified enzymes significantly improved the product yield in *E. coli* under anaerobic conditions.[20] Another strategy to address cofactor imbalance is improving NADPH supply.[20,91] One effective approach overexpresses pyridine nucleotide transhydrogenase PntAB to reduce $NADP^+$ to NADPH by oxidizing NADH.[20,91] In *E. coli*, PntAB is a membrane bound transhydrogenase, coupling reversible $NADP^+$ reduction by NADH with proton translocation across the inner membrane and into the cytosol.[92] The activity of PntAB provides the major source of NADPH in *E. coli* under aerobic, glucose conditions.[93] The inter-conversion between NADH and NADPH catalyzed by PntAB has been used to address cofactor imbalance issues in different metabolic engineering scenarios.[20,72,91,94,95] It was reported that overexpression of the genes encoding both NAD kinase NadK and PntAB enhanced isobutanol production in *E. coli*.[91]

NAD kinase catalyzes the ATP-dependent phosphorylation of NAD^+ to form $NADP^+$, which may increase the intracellular pool of $NADP^+$.[91] However, both reduction of $NADP^+$ by NADH and phosphorylation of NAD^+ require cellular energy input, thus limiting the potential of these approaches for anaerobic fermentative production of isotubanol. Rational design for pathway engineering should include an optimization strategy of kinetic and thermodynamic driving forces to maximize the carbon flow into the targeted pathway.

1.3.3 Engineering Microbial Hosts to Address Product Toxicity

Understanding Product Toxicity

Current fermentation technologies for the commercial production of higher alcohols face clear economic hurdles.[96,97] One critical challenge stems from the fact that as the alcohol's carbon chain length increases, so does the toxicity to the microorganism.[98] Exposure to higher alcohols negatively affects membrane fluidity,[99–101] and even sub-lethal concentrations can change the membrane lipid composition and the membrane permeability, leading to poor fermentation performance.[102–105] Besides membrane damage, a variety of negative effects including DNA damage and protein denaturation were observed in the presence of higher alcohols.[106,107] Relative toxicity is positively correlated with the hydrophobicity of the alcohol, determined by comparing the logarithm of the octanol-water partition coefficient (LogP). Organic solvents with a higher LogP value have the higher preferential partitioning into the plasma membrane, leading to more severe membrane damage.[106,107] LogP values of *n*-butanol and isobutanol are 0.88 and 0.68, respectively, much higher than that of ethanol (–0.29).[98] As a consequence, native butanol-producing bacterial strains such as *C. acetobutylicum* rarely tolerate more than 2% (v/v) butanol while native ethanol producing microbes such as *S. cerevisiae* are able to tolerate 10 to 20% (v/v) ethanol.[96,98,108]

Cellular toxicity caused by higher alcohols is a complex phenotype that has been difficult to elucidate and to mitigate. Many efforts have been made, however, to identify the genetic traits associated with toxicity of and tolerance to higher alcohols in different organisms. Microarray analysis of *C. acetobutylicum* ATCC 824 and its derived strains showed an *n*-butanol dose-dependent increase in expression of almost all major stress related genes, all major solvent formation genes, and major fatty acid synthesis genes.[109,110] Genome-wide transcriptome and metabolome analyses of exogenous *n*-butanol and isobutanol stress in *E. coli* strains show very complex responses to various factors including heat shock, cell envelope stress, and oxidative stress as well as changes in metabolite transport, respiration, and biosynthesis.[111–113] The increase of reactive oxygen species and quinone/quinol malfunction were also proposed as the important causes for cellular toxicity associated with high alcohols.[111,113] Genome-wide analyses of exogenous butanol stress in *Synechocystis* sp. PCC 6803[114] and *Staphylococcus warneri* SG1[115] similarly show an increased expression of genes associated with heat-shock response, oxidative stress response, energy metabolism, and biosynthesis.

Adaptive Evolutionary Approaches to Overcoming Product Toxicity

Adaptive laboratory evolution has been successfully used to improve *n*-butanol or isobutanol tolerance in *E. coli*,[116–119] *S. cerevisiae*,[120,121] and *C. acetobutylicum*.[122] For example, *n*-butanol tolerance limits have been significantly raised through adaptive evolution, wherein cells are exposed to and cultivated in the presence of gradually increasing concentrations of exogenous *n*-butanol, resulting in the generation of spontaneous mutants with improved tolerance characteristics.[108,123–125] However, improved tolerance to exogenous *n*-butanol typically has minimal bearing on increased production capacity. For example, although after just eight rounds of evolutionary adaptation, Liu et al. were able to isolate a strain (T64) of *C. acetobutylicum* capable of tolerating up to 40 g/L *n*-butanol, its maximum achievable product titers were raised only from 12.2 to 15.3 g/L.[122] Interestingly, different evolution strategies seem to yield different genetic mechanisms to improve alcohol tolerance. For example, an engineered isobutanol-producing *E. coli* strain has evolved to be more resistant to isobutanol by sequential transfers in Luria-Bertani broth containing isobutanol.[116] The evolved isobutanol-tolerant strain also shows resistance to *n*-butanol and 2-methyl-*n*-butanol.[116] The increased tolerance is largely due to the inactivation of genes *acrA*, *gatY*, *tnaA*, *yhbJ*, and *marCRAB*, elucidated by whole genome sequencing and allele replacement.[116] *acrA* encodes a component of the AcrAB-TolC multidrug efflux system, and inactivation of the AcrAB-TolC system may potentially reduce the isobutanol-induced quinone depletion.[116] *yhbJ* encodes a predicted ATPase, and deletion of *yhbJ* potentially changes lipopolysaccharide constituents of the outer membrane through activation of L-glutamine:D-fructose-6-phosphate aminotransferase. The *marCRAB* gene is known to mediate several different types of stress response, and its inactivation results in the decreased expression of the AcrAB-TolC system.[116] The isobutanol tolerant *E. coli* strain does not exhibit increased isobutanol production, suggesting the toxicity is probably not the engineering bottleneck for its production. However, it should be noted that toxicity induced by exogenous alcohols may not represent intracellular toxicity caused by metabolic production inside the cells. An *E. coli* K12 derivative strain has been adaptively evolved for approximately 500 generations in minimum medium supplemented with isobutanol.[117] Mutations in *marC*, *hfq*, *mdh*, *acrAB*, *gatYZABCD*, and *rph* genes seem to be responsible for the improved isobutanol tolerance phenotype.[117] Inactivation of genes *marC*, *acrAB*, and *gatYZABCD* as a tolerance mechanism during laboratory evolution[117] overlaps with the previous discoveries,[116] suggesting the presence of a common tolerance mechanism. Analysis of genetic determinants of *n*-butanol tolerance in the improved *E. coli* K12 strain by adaptive evolution suggests that iron-related and osmotic stress-related genes are able to confer *n*-butanol tolerance.[118,119] Adaptive evolution of *S. cerevisiae* under *n*-butanol stress also generates mutant strains with high tolerance to *n*-butanol. The exact tolerance mechanisms remain largely uncharacterized; however, proteomics analysis and whole genome sequencing link the tolerance to mitochondria constituents and protein degradation.[120,121]

Targeted Approaches to Overcoming Product Toxicity

The complexity of cellular responses to alcohol stress continues to present a major challenge, with respect to identifying primary toxicity mechanisms and rational approaches for improving tolerance. Nevertheless, new insights gained so far (discussed in the earlier sections) have inspired a number of effective solutions. For example, one approach involves overexpression of the chaperone system that endogenously functions as part of the general stress response system in *C. acetobutylicum*. Among the major chaperones that are up-regulated in response to both heat and solvent stresses,[109] GroESL overexpression in *C. acetobutylicum* ATCC 824 has been found to prolong and enhance growth while improving metabolism and solvent production up to 40% higher than the wild type.[110,124] In all cases, however, the ability to improve *n*-butanol production by addressing the tolerance limitation alone has not proven to be the key to enhancing *n*-butanol production. In related efforts, overexpression of heat shock proteins in *E. coli*, *Lactobacillus paracasei*, and others has also increased their tolerance to higher alcohols.[126–129] This common approach towards enhancing tolerance further suggests that besides membrane damage, alcohol toxicity is at least partially due to protein misfolding and denaturation. However, optimization of heat shock protein expression only has a limited effect on alcohol tolerance, suggesting the presence of other toxicity factors.[124,129]

Transcription factor engineering is an approach for reprogramming transcription of multiple genes for the improvement of complex phenotypes including alcohol tolerance.[130,131] For example, a variant of the RNA polymerase α subunit, identified by a growth-based screening of *rpoA* mutant libraries generated by error-prone PCR, conferred *n*-butanol tolerance to *E. coli*.[131] Mutant libraries for global transcriptional factors, such as *E. coli* cyclic AMP receptor protein (CRP)[132] and heterologous global regulator IrrE from *Deinococcus radiodurans*,[133] were screened in *E. coli* for improved tolerance to long-chain alcohols. Mutants conferring tolerance were identified, with the improved tolerance linked to the decreased level of intracellular reactive oxygen species and the up-regulation of stress related genes.[132,133] Similarly, a mutant library of novel artificial transcription factors composed of zinc finger DNA-binding proteins fused to *E. coli* CRP was screened to find the variant conferring *n*-butanol tolerance in *E. coli* (up to 1.5%).[134]

Bacterial efflux pumps have also been engineered to improve solvent tolerance.[134–136] The AcrAB-TolC efflux system is regarded as a suitable protein engineering target for solvent tolerance. *E. coli* AcrB protein is an inner membrane efflux pump and is responsible for substrate recognition in AcrAB-TolC system.[137] AcrB variants conferring *n*-butanol tolerance were identified by growth-based selection and these variants also conferred improved tolerance to isobutanol and straight chain alcohols up to heptanol.[135] However, the overexpression of membrane proteins is commonly associated with cellular toxicity, and pumping molecules outside of cells requires energy input, both limiting the potential of this strategy.[138]

An effective approach to identify alcohol tolerance genetic traits is to elucidate the genetic basis of strains already with improved alcohol tolerance as a result of screening of genomic libraries or adaptive laboratory evolution. A *C. acetobutylicum* ATCC 824 genomic library was constructed, and genetic elements conferring increased tolerance to *n*-butanol were isolated by growth-based selection under *n*-butanol stress.[39] Expression of two transcriptional regulators CAC0003 and CAC1869 in *C. acetobutylicum* ATCC 824 resulted in 13% and 81% increases in *n*-butanol tolerance, respectively, relative to the strain with empty plasmid.[39] *C. acetobutylicum* ATCC 824 with a plasmid containing CAC1869 showed increased growth and prolonged metabolic activity, while CAC1869 shares homology with the xenobiotic-responsive element (XRE) family of regulatory proteins. A similar genomic library enrichment strategy has been used to successfully identify beneficial genetic traits in *E. coli*[139] and *S. cerevisiae*.[140] Eleven beneficial genes identified in *E. coli* are able to confer butanol tolerance by overexpression from a plasmid.[139] The genes found to confer the strongest alcohol tolerance from plasmid-based overexpression are *entC* and *feoA*, which are related to iron metabolism and transport.[139] Overexpression of *entC* and *feoA* increased the butanol tolerance by 32.8% and 49.1%, respectively.[139] Four native *S. cerevisiae* genes related to isobutanol tolerance have also been identified by the genomic plasmid library enrichment approach.[140] Overexpression of the genes encoding INO1 (inositol-1-phosphate synthase), DOG1 (2-deoxyglucose-6-phosphatase), HAL1 (involved in halotolerance), or a truncated form of MSN2 (a stress-responsive transcriptional activator) resulted in increased tolerance to high concentrations of isobutanol.[140] The molecular mechanisms of alcohol tolerance conferred by overexpression of these genes remain to be elucidated. A complementary approach of genomic plasmid library enrichment is to screen the non-essential gene knockout collections for the gene deletions conferring alcohol tolerance. Collections of gene deletion strains are available for some microorganisms such as *S. cerevisiae*,[141] *E. coli*,[142] and *B. subtilis*.[143] Some of these strain collections have been used to screen for gene deletions conferring sensitivity and tolerance to higher alcohols.[120,139] In *E. coli*, three gene deletions, especially deletion of *astE* encoding succinylglutamate desuccinylase, enhanced *n*-butanol tolerance.[139] *S. cerevisiae* strains with deletions of genes involved in the ubiquitin-proteasome system and in vacuolar degradation of damaged proteins showed hypersensitivity to *n*-butanol.[120]

Isolating and Engineering More Tolerant Hosts

It is challenging to integrate multiple beneficial genetic traits into one strain for synergistic tolerance increase because most epistatic interactions between genetic traits are not predictable and searching for the optimal combination of multiple genes is time consuming and labor intensive.[144] Due to complex toxicity mechanisms of higher alcohols, increasing the tolerance by genetic engineering and laboratory adaptation has been limited. To our knowledge, *E. coli* strains modified by either random mutation or rational genetic engineering still cannot tolerate over 2% (vol/vol) butanol.[94,111,116–119,126,132–134,137,139] To overcome the toxicity of higher alcohols, others have sought to identify and isolate new hosts with greater inherent tolerance as more robust chassis for higher alcohol production.[145] Searches for naturally solvent-tolerant strains from environmental samples and other sources have identified a variety of microbes including *P. putida*,[146] *Lactobacillus* strains,[147,148] *B. subtilis*,[149] and *Enterococcus faecium*.[150] Some of these strains show strong tolerance to different alcohols after laboratory adaptation under alcohol stress. For example, an adapted *P. putida* strain was able to grow in the presence of up to 6% (v/v) butanol.[146] The native clostridial CoA-dependent *n*-butanol pathway has been successfully implemented in heterologous microorganisms such as *P. putida* and *B. subtilis*.[44] Despite the potential of these new hosts as higher alcohol producers, titers have yet to approach these elevated tolerance limits.

1.4 Successes and Challenges on the Path Towards Commercialization

1.4.1 Examples of Successful Commercialization

Production of *n*-butanol from biomass was first developed in 1912 in the UK.[97] The most well-known route is ABE fermentation (Figure 1.1) developed by Dr. Chaim Weizmann, and this biological route was replaced with a petrochemical process due to its high operational cost.[151] Recent technology development in biobutanol production has stimulated re-commercialization efforts using ABE fermentation. It was reported that over $200 million has recently been invested in China to install 0.21 million tons per annum of solvent capacity using ABE fermentation.[97] There are six major plants with production capacity of 30,000 tons of butanol per annum from corn starch.[97] In Brazil, a new plant operated by HC Sucroquimica was built with capacity of 8000 tons of solvent per annum from sugarcane.[97,152] Production cost is still the main challenge for the current ABE fermentation route. Isobutanol is a promising biofuel product, and it has the potential to be produced at a higher yield than *n*-butanol produced by ABE fermentation. The most commonly used metabolic routes are modified Ehrlich pathways, and many efforts have been made to engineer different organisms including *S. cerevisiae* for isobutanol production. Isobutanol bioproduction is currently being commercialized by a few companies including Gevo Inc., Butamax™ Advanced Biofuels (a joint venture between BP and DuPont), and Butalco.[97,153] These new processes are at the developing phase, and their operational cost should be minimized to be competitive at commercial scales.

1.4.2 Product Separation

One of the major problems for the commercialization of biobutanol is economical product separation. In general, there are two main separation methods: alcohol recovery from the fermentation broth and alcohol dehydration to remove water and purify the product.[154] Many separation techniques function as both, while others require coupling a recovery process with a dehydration process.[154] New technologies for product separation must improve over current separation technologies in terms of energy usage, product purity, or total alcohol recovery. While older technologies for ethanol production typically allow the fermentation to complete before harvesting the product, newer technologies need continuous removal of product to minimize product toxicity. The main technologies for bio-alcohol separation are distillation, flash distillation, gas stripping, adsorption, and liquid-liquid extraction.[97,154,155]

The major benchmark for alcohol recovery is distillation, the most common alcohol recovery process in ethanol plants. It leads to high total alcohol recovery and has good economies of scale, but it requires large amounts of energy for separation of low concentration feeds which are common for biological processes.[156] The respective water-alcohol azeotropes can limit the product purity, but this can be overcome using additives or multiple stages.[157,158] Because of its use in other chemical processes, heat and energy recovery can be integrated with other processes more easily than newer technologies.[154] A variant of distillation is flash distillation. Flash distillation uses a pressure drop to rapidly vaporize the alcohol phase in a continuous fashion.[159] Companies such as Gevo Inc. have filed patents to combine flash distillation with a phase separator to continuously distill and purify isobutanol.[160]

Gas stripping uses a gas stream to recover the volatile component (i.e., alcohols) from the fermentation. This concept is attractive because it can use inert gas to carry the alcohols, it can operate at fermenter temperatures, and it does not require cleaning for regeneration, compared to a process using a membrane or an adsorber.[161] While many systems use a condenser to remove the alcohols from the vapor phase, other systems have used membranes as well to maximize energy savings.[162] The gas used for stripping can be recycled to recover the remaining alcohols and lower the amount of gas needed. It has been commonly used for recovery of both ABE products and isobutanol to high yields at lab scale,[163,164] but it has not been widely adopted at an industrial scale.[165] The energy required for separation depends on the type of gas stream and the method for condensation from the vapor phase, but it typically requires less energy than distillation.

Adsorption uses a solid material with selective affinity for the target product in the fermentation broth. After adsorption, the product is released by desorption to make a high concentration of bioalcohol while the solid material is regenerated for re-use. Typical materials used for adsorption include silicates, activated carbon, resins, and zeolite.[166] The materials are non-toxic to cells, but biofilm formation can foul the material and hinder separation. Adsorption requires less energy for separation compared to gas stripping and distillation, but cleaning and regenerating the adsorbent can hinder continuous processes and increase operating costs.[155]

Liquid-liquid extraction is another method that can remove the alcohols from the fermentation broth. A selective solvent contacts the fermentation broth and removes the alcohol products while leaving cells and nutrients unaffected. Ideal solvents have a high partitioning coefficient for butanol and other alcohols and a low partitioning coefficient for nutrients in the broth. However, most solvents that have high partitioning coefficients for the alcohols are toxic to cells, while most non-toxic solvents have low partitioning coefficients for the alcohols. A mixture of decane or decanol with oleyl alcohol has been proposed as extractants for butanol production.[167] Other systems, such as patented systems by Butamax, use a two-step extraction process with mixtures of C12 to C22 fatty alcohols, esters, and aldehydes to possibly eliminate the problems of toxicity and low partitioning coefficients for the alcohols.[160]

1.4.3 Commercially Viable Hosts

While demonstration of high yields and productivities is achievable at the lab scale, producing bioalcohols at a large scale requires the use of a commercially viable host. Commercial scale production of bioalcohols needs high titers, but even low concentrations of alcohols often inhibit growth.[168] For butanol and isobutanol, *E. coli* strains have natural tolerances below 2%, and *Clostridium* species have natural tolerances near 2.5%.[169] Increasing the concentration of alcohols in the fermentation broth would greatly reduce downstream costs for separation. Thus, increasing tolerance of the organism while retaining productivity is a major priority of strain engineering efforts. In addition, lab-scale conditions have optimal nutrient concentrations, aeration, and mixing, which are not easily achieved in industrial fermentations. Native alcohol producers such as *Clostridium* species as well as recombinant bacterial and eukaryotic hosts have been investigated for bioalcohol production at pilot and industrial scales.[165] Each host has its own advantages and disadvantages, and the choice of a particular host is typically driven by the economics of the process and previous experience with the organism.

Clostridium species naturally produce *n*-butanol at high concentrations and have been the first choice for higher alcohol production.[31] Since the beginning of the twentieth century, *Clostridium* species have been isolated and tested for ABE fermentation. Common strains for large scale production include *C. beijerinckii* BA101, *C. tyrobutyricum*, and *C. acetobutylicum*. To directly convert cellulose into ABE products, *C. celluloyticum* has also been used.[73,170] However, attempts to increase production of ABE products or increase tolerance to ABE products have been limited due to the difficulty in genetic modification of this genus.[169] Recent developments of *Clostridium* genetic tools such as plasmids for overexpression and gene knockout methods have enabled metabolic engineering of these strains.[124,171]

Other more genetically tractable hosts such as *E. coli* have also been used for producing high titers of alcohols. By engineering new pathways for C4 alcohol production, titers of 50 g/L isobutanol and 30 g/L *n*-butanol have been achieved with gas stripping.[62,172] To attain such high titers, the bioalcohol production had to be paired with a removal process to avoid product toxicity (e.g., ~2% butanol for *E. coli*).[59] In addition, introducing biosynthetic pathways into new organisms often requires large-scale rewiring of carbon flux and optimizing enzyme kinetics. Using recombinant organisms in large-scale fermentations can suffer from genetic instability (e.g., plasmid loss and mutation).[173] There has been discussion of whether to engineer biofuel producing pathways in heterologous hosts or to use natural producers as industrial strains for biofuel production.[174]

Due to *E. coli*'s low tolerance to *n*-butanol and isobutanol, solvent-tolerant strains with some genetic tools, such as *P. putida*, *C. glutamaticum*, *Lactococcus brevis*, and various *Bacillus* strains, have also been investigated for bioalcohol production. *Corynebacterium* has been engineered to produce a titer of 12.6 g/L isobutanol while titers of 6.2 g/L and 0.122 g/L of isobutanol have been achieved in *B. subtilis* and *P. putida*, respectively.[44,45,72,74,175,176] Titer improvements for each organism have been made using traditional metabolic engineering techniques such as gene knockout and overexpression. Higher tolerance to alcohols and other stresses could increase the yield and productivity in a large scale fermentation while no large scale process that uses these hosts is known. Better characterization of metabolic pathways in these strains and further strain engineering could yield increased titers in each of these organisms. Some organisms, such as *C. glutamaticum*, have already been utilized in industrial-scale processes for amino acid production and could be easily applied to bioalcohol production.[177]

Cyanobacteria have been investigated for bioalcohol production due to their ability to fix CO_2 and use energy directly from light to split water and generate NADPH. Two strains have been engineered for bioalcohol production: *S. elongatus* PCC 7942 and *Synechocystis* sp. PCC 6803. While incorporation and engineering of the isobutanol pathway has been done in both strains, it was difficult to engineer the *n*-butanol pathway in cyanobacteria because of oxygen sensitivity of the clostridial pathway along with fewer genetic tools and the requirement of stronger driving forces for production.[49] An interesting approach for bioalcohol production is to engineer pathways for production of a precursor of the alcohol with a lower toxicity. For example, isobutyraldehyde was selected as a target for engineering, and it was produced with a titer of 1.1 g/L, compared to 0.45 g/L of isobutanol.[76] One challenge with cyanobacteria is to obtain high cell density cultures that may prevent sufficient light penetration.[178] The limited genetic tools also hinder development of these photosynthetic bacteria as alcohol producers.

S. cerevisiae is an attractive choice for industrial scale production because of its long history with ethanol production. Yeasts can be used in current infrastructure, they have demonstrated higher tolerance to inhibitors and low pH than bacterial hosts, and they are not affected by phage contamination like bacterial hosts.[82] However, the goals to increase titers and productivities of longer chain alcohols have been difficult to achieve.[47] One difficulty with yeasts is the internal compartmentalization of the TCA cycle in the mitochondria, which requires transport of molecules through the barrier surrounding the organelle, in contrast to prokaryotic cells where all reactions occur in the cytoplasm.[68] By comparison, glycolysis and ethanol production occur in the cytoplasm while pathways for isobutanol production which use amino acids are compartmentalized (Figure 1.4). By using metabolic engineering techniques, 0.243 g/L of *n*-butanol has been produced in yeast,[69] while Gevo has filed patents for titers up to 18.6 g/L isobutanol in complex media.[179]

1.4.4 Alternative Feedstocks

As competitiveness drives cost reductions, the choice of feedstock will become more important. While most laboratory experiments use purified sugars such as glucose, large scale fermentations must use cheaper sugar sources such as sugar cane and beet sugar. Still, as more sustainable feedstocks that do not compete with the food chain are required for truly scalable solutions, researchers are now exploring alternative options. Utilization of lignocellulose has been the focus of engineering efforts because of its abundance and high sugar content. The three main components of lignocellulose are cellulose, hemicellulose, and lignin. Cellulose and hemicellulose are polymers of hexose and pentose sugars, and many research efforts have focused on releasing these sugars for conversion to bioalcohols. However, pretreatment of lignocellulose leads to inhibitor formation in the form of furans from sugar breakdown and aromatics from lignin breakdown. These inhibitors are toxic and lower growth rates and overall conversion of sugars to products, but efflux pumps have been applied to remove pretreatment impurities or solvents.[134,180] To avoid complications with pretreatment of lignocellulose, some have approached this problem by using bacteria that can degrade cellulose. Some organisms contain a cellulosome, or a cellulose degrading catabolic cluster that can attack and break down cellulose into glucose. After the release of glucose, it can be fermented by the same bacteria into bioalcohols. This process, called consolidated bioprocessing, would lower the overall costs associated with using lignocellulose as a feedstock and would increase the cost-competitiveness of producing bioalcohols. Some *Clostridium* strains that can degrade cellulose have been engineered to produce up to 0.66 g/L isobutanol using this approach.[73] However, the authors used crystalline cellulose, not raw pretreated biomass which can contain many different polymeric forms of cellulose that are more difficult to degrade than crystalline cellulose.[181] Some microbial consortia have also been used for producing bioalcohols, in which one organism can degrade the cellulose and another organism can produce the biofuel.[182,183] While these systems show promise, they have not been demonstrated at an industrial scale.

Another carbon source that can be used for bioalcohol production is the protein waste from food processing and other bacterial fermentations. Using mutagenesis, Huo et al. created *E. coli* strains that can utilize 13 amino acids compared to four by the parent strain.[54] These engineered *E. coli* strains produced a titer of 4.04 g/L of isobutanol from amino acids as carbon sources. This interesting work demonstrates that proteinaceous microalgae could be potential carbon sources that could fix CO_2 and then could be digested by *E. coli* to produce biofuels.

Syngas is another potential carbon source that can be utilized by microorganisms for conversion to bioalcohols. Syngas is a mixture of CO, CO_2, and H_2 generated by digestion of biomass and different waste materials. CO and CO_2 can be used for ethanol and butanol production using native pathways in methanotrophic species.[170] A microbial consortium was shown to convert syngas to butanol by using *Butyribacterium methylotropicum* to convert syngas to butyrate and then *C. acetobutylicum* to convert butyrate to butanol.[184] *Clostridium ljungdahlii* can utilize syngas for growth and naturally produces ethanol, but it lacks the clostridial *n*-butanol pathway.[170] *C. ljundahlii* was engineered by introducing this pathway and produced small amounts of butanol in a co-culture with *C. ragsdelei*.[185,186] More recent work showed the production of ethanol, propanol, and butanol from a microbial consortium of three different strains.[186] Titers produced were 8 g/L ethanol, 6 g/L propanol, and 1 g/L butanol. It was also shown that an engineered *Clostridium* strain could produce 21 g/L isobutanol using a mixture of CO_2 and H_2 using a specially designed bioreactor.[186] While syngas shows promise as a carbon source, it requires specialized bioreactors for fermentation and separation, and new technologies must be developed before implementation at an industrial scale.

CO_2 has also been discussed as a possible carbon source for bioalcohol production. CO_2 is abundant in the atmosphere, and it is a major pollutant from burning fossil fuels. However, carbon fixation is an energy intensive process, and it is typically coupled with light reactions to overcome the energetic limitations associated with carbon fixation. Isobutyraldehyde, the precursor for isobutanol, was produced in a cyanobacterium strain up to a titer of 1.1 g/L.[76] Other approaches have focused on finding ways to utilize CO_2 without using photosynthesis. One approach was to use electricity to overcome the inherent energetic limitations of CO_2 fixation to convert it to formate, which could be used by cells for isobutanol production and energy for growth.[187] This process used *R. eutropha*, and it appears to be applicable to other organisms. Combining electricity with fermentation had previously been shown to inhibit cell growth, and it was hypothesized that this inhibition was due to production of hydrogen peroxide, reactive oxygen species, and nitric oxide. They used intracellular sensors to show that nitric oxide and reactive oxygen species were the main causes that inhibited growth. To overcome this limitation, they used a ceramic cup to shield the anode where these chemicals were produced. Titers of 0.846 g/L isobutanol and 0.57 g/L 3-methyl-*n*-butanol were achieved using this system.[187] Another approach for CO_2 fixation is to use the bioalcohol product as an alternative metabolic sink.[188] This concept uses the product as an alternative storage compound by knocking out a vital storage compound, glycogen in this case. A titer of 0.5 g/L isobutanol was achieved using this process.[188]

1.5 Outlook: Overcoming Roadblocks

Moving from the laboratory scale to the industrial scale poses challenges for many bioconversion processes. For bioalcohol production, the major hurdles include relatively high substrate costs, low product tolerance and titer, by-product formation (especially for ABE processes), and the high energy costs for distillation.[155,97] First, alternative feedstocks have been explored as a means to reduce costs, but utilization of pretreated lignocellulose and other feedstocks has only recently begun to be investigated at an industrial scale for production of longer chain alcohols. To achieve this goal, future research can focus on development of new strains that can utilize a variety of feedstocks and/or consolidation of multiple pathways into a single organism. Alternatively, some strains could work to depolymerize feedstocks (e.g., cellulose, hemicellulose, and lignin), and others could produce biofuels from the depolymerized monomer substrates to divide the labor. Second, increasing product tolerance and titers will lower separation costs, and different mechanisms for increasing tolerance have been investigated. Tolerance itself is a complex phenotype, so multiple mechanisms should be investigated at system levels, including efflux pumps,[135] stress responses,[129] and membrane composition changes.[189] Last, current industrial scale processes for *n*-butanol production are mostly based on the traditional ABE fermentation because of the demonstrated high yield and titer.[97] However, process improvements are still needed to minimize energy consumption during the process.[154] New materials for adsorption and desorption as well as new pervaporation methods could lead to lower costs for production.[154]

New technologies hold promise for commercial host development for bioalcohol production. The development of synthetic biology tools has led to faster development times for new hosts and pathways. The next decade will benefit especially from both next generation sequencing and DNA synthesis technologies that have increased the rate of discovery for new enzymes. For example, the decreased costs and increased speeds of DNA sequencing allow new genetic parts to be identified via gene mining and metagenomics. Recent examples of the use of gene mining or metagenomics include increased tolerance to inhibitors found in pretreated biomass by screening metagenomic libraries,[190] identification of efflux pumps for various biofuels,[134] and use of bioinformatics libraries to identify different genes for methyl halide synthesis.[191] Assisted by systems and synthetic biology, commercialization of higher alcohol-based biofuels will be the reality in the near future.

Acknowledgments

Development of this manuscript was supported by funding from a National Science Foundation Graduate Research Fellowship to Cheryl M. Immethun, an ASU start-up fund to Xuan Wang, the National Science Foundation grants (CBET-1159200 and CBET-1067684) to David R. Nielsen, and the National Science Foundation and the Department of Energy grants (MCB-1331194, CBET-1350498, and DE-SC0012705) to Tae Seok Moon.

References

1. Choi DG, Kreikebaum F, Thomas VM, Divan D. Coordinated EV adoption: double-digit reductions in emissions and fuel use for $40/vehicle-year. *Environ Sci Technol* 2013;**47**:10703–7.
2. Administration USEI. United States energy information administration: frequently asked questions. In: Administration USEI, editor. *Independent statistics & analysis*. Washington, D.C: United States Department of Energy; 2014. http://www.eia.gov/forecasts/aeo/er/early_production.cfm.

3. Aakko-Saksa PT, Rantanen-Kolehmainen L, Skyttä E. Ethanol, isobutanol, and biohydrocarbons as gasoline components in relation to gaseous emissions and particulate matter. *Environ Sci Technol* 2014;**48**:10489–96.
4. Karl TR, Trenberth KE. Modern global climate change. *Science* 2003;**302**:1719–23.
5. Sun Y, Cheng J. Hydrolysis of lignocellulosic materials for ethanol production: a review. *Bioresour Technol* 2002;**83**:1–11.
6. Lee RA, Lavoie J. From first- to third-generation biofuels: challenges of producing a commodity from a biomass of increasing complexity. *Anim Front* 2013;**3**:6–11.
7. Sims REH, Mabee W, Saddler JN, Taylor M. An overview of second generation biofuel technologies. *Bioresour Technol* 2010;**101**:1570–80.
8. Houghton J, Weatherwax S, Ferrell J. Breaking the biological barriers to cellulosic ethanol: a joint research agenda. In: Energy USDo, editor. *A research roadmap resulting from the biomass to biofuels*. Office of Science and Office of Energy Efficiency and Renewable Energy, Department of Energy Genomes to Life; 2005.
9. Wallner T, Ickes A, Lawyer K. Analytical assessment of C2-C8 alcohols as Spark-Ignition engine fuels. In: *Proceedings of the FISITA 2012 world automotive congress*, 3; 2013. p. 15–26.
10. Mathews CK, van Holde KE, Appling DR, Anthony-Cahill SJ. *Biochemistry*. 4th ed. Prentice Hall; 2012.
11. Liao JC, Hou SY, Chao YP. Pathway analysis, engineering, and physiological considerations for redirecting central metabolism. *Biotechnol Bioeng* 1996;**52**:129–40.
12. Nielsen J. Transcriptional control of metabolic fluxes. *Mol Syst Biol* 2011;**7**:478.
13. Machado HB, Dekishima Y, Luo H, Lan EI, Liao JC. A selection platform for carbon chain elongation using the CoA-dependent pathway to produce linear higher alcohols. *Metab Eng* 2012;**14**:504–11.
14. Trinh CT, Li J, Blanch HW, Clark DS. Redesigning Escherichia coli metabolism for anaerobic production of isobutanol. *Appl Environ Microbiol* 2011;**77**:4894–904.
15. Zhang K, Sawaya MR, Eisenberg DS, Liao JC. Expanding metabolism for biosynthesis of nonnatural alcohols. *Proc Natl Acad Sci U S A* 2008;**105**:20653–8.
16. Blanch HW. Bioprocessing for biofuels. *Curr Opin Biotechnol* 2012;**23**:390–5.
17. Zhang YH. Production of biocommodities and bioelectricity by cell-free synthetic enzymatic pathway biotransformations: challenges and opportunities. *Biotechnol Bioeng* 2010;**105**:663–77.
18. Huang W-D, Percival Zhang YH. Analysis of biofuels production from sugar based on three criteria: thermodynamics, bioenergetics, and product separation. *Energy Environ Sci* 2011;**4**:784–92.
19. Unden G, Bongaerts J. Alternative respiratory pathways of Escherichia coli: energetics and transcriptional regulation in response to electron acceptors. *Biochim Biophys Acta* 1997;**1320**:217–34.
20. Bastian S, Liu X, Meyerowitz JT, Snow CD, Chen MM, Arnold FH. Engineered ketol-acid reductoisomerase and alcohol dehydrogenase enable anaerobic 2-methylpropan-1-ol production at theoretical yield in Escherichia coli. *Metab Eng* 2011;**13**:345–52.
21. Shen CR, Lan EI, Dekishima Y, Baez A, Cho KM, Liao JC. Driving forces enable high-titer anaerobic 1-butanol synthesis in Escherichia coli. *Appl Environ Microbiol* 2011;**77**:2905–15.
22. Lim JH, Seo SW, Kim SY, Jung GY. Model-driven rebalancing of the intracellular redox state for optimization of a heterologous n-butanol pathway in Escherichia coli. *Metab Eng* 2013;**20**:56–62.
23. Jones DT, Woods DR. Acetone-butanol fermentation revisited. *Microbiol Rev* 1986;**50**:484–524.
24. Yang S-T, El Enshasy Hesham, Thongchul Nuttha. *Bioprocessing technologies in biorefinery for sustainable production of fuels, chemicals, and polymers*. Hoboken, New Jersey: Wiley; 2013.
25. Lütke-Eversloh T, Bahl H. Metabolic engineering of Clostridium acetobutylicum: recent advances to improve butanol production. *Curr Opin Biotechnol* 2011;**22**:634–47.
26. Dekishima Y, Lan EI, Shen CR, Cho KM, Liao JC. Extending carbon chain length of 1-butanol pathway for 1-hexanol synthesis from glucose by engineered Escherichia coli. *J Am Chem Soc* 2011;**133**:11399–401.
27. Antoni D, Zverlov VV, Schwarz WH. Biofuels from microbes. *Appl Microbiol Biotechnol* 2007;**77**:23–35.
28. Lee J, Jang YS, Choi SJ, Im JA, Song H, Cho JH, et al. Metabolic engineering of Clostridium acetobutylicum ATCC 824 for isopropanol-butanol-ethanol fermentation. *Appl Environ Microbiol* 2012;**78**:1416–23.

29. Dusséaux S, Croux C, Soucaille P, Meynial-Salles I. Metabolic engineering of Clostridium acetobutylicum ATCC 824 for the high-yield production of a biofuel composed of an isopropanol/butanol/ethanol mixture. *Metab Eng* 2013;**18**:1–8.
30. Li SY, Srivastava R, Suib SL, Li Y, Parnas RS. Performance of batch, fed-batch, and continuous A-B-E fermentation with pH-control. *Bioresour Technol* 2011;**102**:4241–50.
31. Jang YS, Lee JY, Lee J, Park JH, Im JA, Eom MH, et al. Enhanced butanol production obtained by reinforcing the direct butanol-forming route in Clostridium acetobutylicum. *mBio* 2012;**3**:e00314-12.
32. Lee JY, Jang YS, Lee J, Papoutsakis ET, Lee SY. Metabolic engineering of Clostridium acetobutylicum M5 for highly selective butanol production. *Biotechnol J* 2009;**4**:1432–40.
33. Sillers R, Al-Hinai MA, Papoutsakis ET. Aldehyde-alcohol dehydrogenase and/or thiolase overexpression coupled with CoA transferase downregulation lead to higher alcohol titers and selectivity in Clostridium acetobutylicum fermentations. *Biotechnol Bioeng* 2009;**102**:38–49.
34. Jiang Y, Xu C, Dong F, Yang Y, Jiang W, Yang S. Disruption of the acetoacetate decarboxylase gene in solvent-producing Clostridium acetobutylicum increases the butanol ratio. *Metab Eng* 2009;**11**:284–91.
35. Tummala SB, Junne SG, Papoutsakis ET. Antisense RNA downregulation of coenzyme A transferase combined with alcohol-aldehyde dehydrogenase overexpression leads to predominantly alcohologenic Clostridium acetobutylicum fermentations. *J Bacteriol* 2003;**185**:3644–53.
36. Hanai T, Atsumi S, Liao JC. Environ Microbiol Appl. Engineered synthetic pathway for isopropanol production in Escherichia coli. *Appl Environ Microbiol* 2007;**73**:7814–8.
37. Osborne SJ, Leaver J, Turner MK, Dunnill P. Correlation of biocatalytic activity in an organic aqueous 2-liquid phase system with solvent concentration in the cell-membrane. *Enzym Microb Technol* 1990;**12**:281–91.
38. Isken S, de Bont JAM. Bacteria tolerant to organic solvents. *Extremophiles* 1998;**2**:229–38.
39. Borden JR, Papoutsakis ET. Dynamics of genomic-library enrichment and identification of solvent tolerance genes for Clostfidium acetobutylicum. *Appl Environ Microbiol* 2007;**73**:3061–8.
40. Luyben WL. Control of the heterogeneous azeotropic n-butanol/water distillation system. *Energy Fuel* 2008;**22**:4249–58.
41. Flickinger MC, Drew SW. *Encyclopedia of bioprocess technology: fermentation, biocatalysis, and bioseparation*. New York: John Wiley; 1999.
42. Inui M, Suda M, Kimura S, Yasuda K, Suzuki H, Toda H, et al. Expression of Clostridium acetobutylicum butanol synthetic genes in Escherichia coli. *Appl Microbiol Biotechnol* 2008;**77**:1305–16.
43. Atsumi S, Cann AF, Connor MR, Shen CR, Smith KM, Brynildsen MP, et al. Metabolic engineering of Escherichia coli for 1-butanol production. *Metab Eng* 2008;**10**:305–11.
44. Nielsen DR, Leonard E, Yoon SH, Tseng HC, Yuan C, Prather KL. Engineering alternative butanol production platforms in heterologous bacteria. *Metab Eng* 2009;**11**:262–73.
45. Berezina OV, Zakharova NV, Brandt A, Yarotsky SV, Schwarz WH, Zverlov VV. Reconstructing the clostridial n-butanol metabolic pathway in Lactobacillus brevis. *Appl Microbiol Biotechnol* 2010;**87**:635–46.
46. Bhandiwad A, Shaw AJ, Guss A, Guseva A, Bahl H, Lynd LR. Metabolic engineering of Thermoanaerobacterium saccharolyticum for n-butanol production. *Metab Eng* 2014;**21**:17–25.
47. Steen EJ, Chan R, Prasad N, Myers S, Petzold CJ, Redding A, et al. Metabolic engineering of Saccharomyces cerevisiae for the production of n-butanol. *Microb Cell Fact* 2008;**7**:36.
48. Krivoruchko A, Serrano-Amatriain C, Chen Y, Siewers V, Nielsen J. Improving biobutanol production in engineered Saccharomyces cerevisiae by manipulation of acetyl-CoA metabolism. *J Ind Microbiol Biotechnol* 2013;**40**:1051–6.
49. Lan EI, Liao JC. ATP drives direct photosynthetic production of 1-butanol in cyanobacteria. *Proc Natl Acad Sci U S A* 2012;**109**:6018–23.
50. Bond-Watts BB, Bellerose RJ, Chang MC. Enzyme mechanism as a kinetic control element for designing synthetic biofuel pathways. *Nat Chem Biol* 2011;**7**:222–7.
51. Griese M, Lange C, Soppa J. Ploidy in cyanobacteria. *FEMS Microbiol Lett* 2011;**323**:124–31.
52. Gulevich AIu, Skorokhodova AIu, Morzhakova AA, Antonova SV, Sukhozhenko AV, Shakulov RS, et al. 1-Butanol synthesis by Escherichia coli cells through butyryl-CoA formation by heterologous enzymes of clostridia and native enzymes of fatty acid beta-oxidation. *Appl Biochem Microbiol* 2012;**48**:344–9.

53. Dellomonaco C, Clomburg JM, Miller EN, Gonzalez R. Engineered reversal of the beta-oxidation cycle for the synthesis of fuels and chemicals. *Nature* 2011;**476**:355–9.
54. Huo YX, Cho KM, Rivera JG, Monte E, Shen CR, Yan Y, et al. Conversion of proteins into biofuels by engineering nitrogen flux. *Nat Biotechnol* 2011;**29**:346–51.
55. Marcheschi RJ, Li H, Zhang K, Noey EL, Kim S, Chaubey A, et al. A synthetic recursive "+1" pathway for carbon chain elongation. *ACS Chem Biol* 2012;**7**:689–97.
56. Shen CR, Liao JC. Metabolic engineering of Escherichia coli for 1-butanol and 1-propanol production via the keto-acid pathways. *Metab Eng* 2008;**10**:312–20.
57. Atsumi S, Hanai T, Liao JC. Non-fermentative pathways for synthesis of branched-chain higher alcohols as biofuels. *Nature* 2008;**451**:86–9.
58. Atsumi S, Wu TY, Eckl EM, Hawkins SD, Buelter T, Liao JC. Engineering the isobutanol biosynthetic pathway in Escherichia coli by comparison of three aldehyde reductase/alcohol dehydrogenase genes. *Appl Microbiol Biotechnol* 2010;**85**:651–7.
59. Baez A, Cho KM, Liao JC. High-flux isobutanol production using engineered Escherichia coli: a bioreactor study with in situ product removal. *Appl Microbiol Biotechnol* 2011;**90**:1681–90.
60. Connor MR, Liao JC. Engineering of an Escherichia coli strain for the production of 3-methyl-1-butanol. *Appl Environ Microbiol* 2008;**74**:5769–75.
61. Savrasova EA, Kivero AD, Shakulov RS, Stoynova NV. Use of the valine biosynthetic pathway to convert glucose into isobutanol. *J Ind Microbiol Biotechnol* 2011;**38**:1287–94.
62. Atsumi S, Liao JC. Directed evolution of methanococcus jannaschii citramalate synthase for biosynthesis of 1-propanol and 1-butanol by Escherichia coli. *Appl Environ Microbiol* 2008;**74**:7802–8.
63. Cann AF, Liao JC. Production of 2-methyl-1-butanol in engineered Escherichia coli. *Appl Microbiol Biotechnol* 2008;**81**:89–98.
64. Matsuda F, Ishii J, Kondo T, Ida K, Tezuka H, Kondo A. Increased isobutanol production in Saccharomyces cerevisiae by eliminating competing pathways and resolving cofactor imbalance. *Microb Cell Factories* 2013;**12**:119.
65. Kondo T, Tezuka H, Ishii J, Matsuda F, Ogino C, Kondo A. Genetic engineering to enhance the Ehrlich pathway and alter carbon flux for increased isobutanol production from glucose by Saccharomyces cerevisiae. *J Biotechnol* 2012;**159**:32–7.
66. Avalos JL, Fink GR, Stephanopoulos G. Compartmentalization of metabolic pathways in yeast mitochondria improves the production of branched-chain alcohols. *Nat Biotechnol* 2013;**31**:335–41.
67. Chen X, Nielsen KF, Borodina I, Kielland-Brandt MC, Karhumaa K. Increased isobutanol production in Saccharomyces cerevisiae by overexpression of genes in valine metabolism. *Biotechnol Biofuels* 2011;**4**:21.
68. Lee WH, Seo SO, Bae YH, Nan H, Jin YS, Seo JH. Isobutanol production in engineered Saccharomyces cerevisiae by overexpression of 2-ketoisovalerate decarboxylase and valine biosynthetic enzymes. *Bioprocess Biosyst Eng* 2012;**35**:1467–75.
69. Si T, Luo Y, Xiao H, Zhao H. Utilizing an endogenous pathway for 1-butanol production in Saccharomyces cerevisiae. *Metab Eng* 2014;**22**:60–8.
70. Brat D, Weber C, Lorenzen W, Bode HB, Boles E. Cytosolic re-localization and optimization of valine synthesis and catabolism enables inseased isobutanol production with the yeast Saccharomyces cerevisiae. *Biotechnol Biofuels* 2012;**5**:65.
71. Smith KM, Cho KM, Liao JC. Engineering Corynebacterium glutamicum for isobutanol production. *Appl Microbiol Biotechnol* 2010;**87**:1045–55.
72. Blombach B, Riester T, Wieschalka S, Ziert C, Youn JW, Wendisch VF, et al. Corynebacterium glutamicum tailored for efficient isobutanol production. *Appl Environ Microbiol* 2011;**77**:3300–10.
73. Higashide W, Li Y, Yang Y, Liao JC. Metabolic engineering of clostridium cellulolyticum for production of isobutanol from cellulose. *Appl Environ Microbiol* 2011;**77**:2727–33.
74. Li S, Wen J, Jia X. Engineering Bacillus subtilis for isobutanol production by heterologous Ehrlich pathway construction and the biosynthetic 2-ketoisovalerate precursor pathway overexpression. *Appl Microbiol Biotechnol* 2011;**91**:577–89.

75. Jia X, Li S, Xie S, Wen J. Engineering a metabolic pathway for isobutanol biosynthesis in Bacillus subtilis. *Appl Biochem Biotechnol* 2012;**168**:1–9.
76. Atsumi S, Higashide W, Liao JC. Direct photosynthetic recycling of carbon dioxide to isobutyraldehyde. *Nat Biotechnol* 2009;**27**:1177–80.
77. Lu J, Brigham CJ, Gai CS, Sinskey AJ. Studies on the production of branched-chain alcohols in engineered Ralstonia eutropha. *Appl Microbiol Biotechnol* 2012;**96**:283–97.
78. Lang K, Zierow J, Buehler K, Schmid A. Metabolic engineering of Pseudomonas sp strain VLB120 as platform biocatalyst for the production of isobutyric acid and other secondary metabolites. *Microb Cell Factories* 2014;**13**:2.
79. Varman AM, Xiao Y, Pakrasi HB, Tang YJ. Metabolic engineering of Synechocystis sp. strain PCC 6803 for isobutanol production. *Appl Environ Microbiol* 2013;**79**:908–14.
80. Dellomonaco C, Rivera C, Campbell P, Gonzalez R. Engineered respiro-fermentative metabolism for the production of biofuels and biochemicals from fatty acid-rich feedstocks. *Appl Environ Microbiol* 2010;**76**:5067–78.
81. Rodriguez GM, Atsumi S. Isobutyraldehyde production from Escherichia coli by removing aldehyde reductase activity. *Microb Cell Fact* 2012;**11**:90.
82. Branduardi P, Longo V, Berterame NM, Rossi G, Porro D. A novel pathway to produce butanol and isobutanol in Saccharomyces cerevisiae. *Biotechnol Biofuels* 2013;**6**:68.
83. Prüss BM, Nelms JM, Park C, Wolfe AJ. Mutations in NADH:ubiquinone oxidoreductase of Escherichia coli affect growth on mixed amino acids. *J Bacteriol* 1994;**176**:2143–50.
84. Atkinson MR, Blauwkamp TA, Bondarenko V, Studitsky V, Ninfa AJ. Activation of the glnA, glnK, and nac promoters as Escherichia coli undergoes the transition from nitrogen excess growth to nitrogen starvation. *J Bacteriol* 2002;**184**:5358–63.
85. Lukk T, Sakai A, Kalyanaraman C, Brown SD, Imker HJ, Song L, et al. Homology models guide discovery of diverse enzyme specificities among dipeptide epimerases in the enolase superfamily. *Proc Natl Acad Sci U S A* 2012;**109**:4122–7.
86. Mann MS, Lutke-Eversloh T. Thiolase engineering for enhanced butanol production in Clostridium acetobutylicum. *Biotechnol Bioeng* 2013;**110**:887–97.
87. Lan EI, Liao JC. Microbial synthesis of n-butanol, isobutanol, and other higher alcohols from diverse resources. *Bioresour Technol* 2013;**135**:339–49.
88. Zheng YN, Li LZ, Xian M, Ma YJ, Yang JM, Xu X, et al. Problems with the microbial production of butanol. *J Ind Microbiol Biotechnol* 2009;**36**:1127–38.
89. Jarboe LR, Zhang X, Wang X, Moore JC, Shanmugam KT, Ingram LO. Metabolic engineering for production of biorenewable fuels and chemicals: contributions of synthetic biology. *J Biomed Biotechnol* 2010;**2010**:761042.
90. Grabowska D, Chelstowska A. The ALD6 gene product is indispensable for providing NADPH in yeast cells lacking glucose-6-phosphate dehydrogenase activity. *J Biol Chem* 2003;**278**:13984–8.
91. Shi A, Zhu X, Lu J, Zhang X, Ma Y. Activating transhydrogenase and NAD kinase in combination for improving isobutanol production. *Metab Eng* 2013;**16**:1–10.
92. Clarke DM, Bragg PD. Purification and properties of reconstitutively active nicotinamide nucleotide transhydrogenase of Escherichia coli. *Eur J Biochem* 1985;**149**:517–23.
93. Sauer U, Canonaco F, Heri S, Perrenoud A, Fischer E. The soluble and membrane-bound transhydrogenases UdhA and PntAB have divergent functions in NADPH metabolism of Escherichia coli. *J Biol Chem* 2004;**279**:6613–9.
94. Chin JW, Khankal R, Monroe CA, Maranas CD, Cirino PC. Analysis of NADPH supply during xylitol production by engineered Escherichia coli. *Biotechnol Bioeng* 2009;**102**:209–20.
95. Miller EN, Jarboe LR, Turner PC, Pharkya P, Yomano LP, York SW, et al. Furfural inhibits growth by limiting sulfur assimilation in ethanologenic Escherichia coli strain LY180. *Appl Environ Microbiol* 2009;**75**:6132–41.
96. Durre P. Fermentative production of butanol—the academic perspective. *Curr Opin Biotechnol* 2011;**22**:331–6.

97. Green EM. Fermentative production of butanol—the industrial perspective. *Curr Opin Biotechnol* 2011;**22**:337–43.
98. Liu S, Qureshi N. How microbes tolerate ethanol and butanol. *Nat Biotechnol* 2009;**26**:117–21.
99. Grisham CM, Barnett RE. The effects of long-chain alcohols on membrane lipids and the (Na++K+)-ATPase. *Biochim Biophys Acta* 1973;**311**:417–22.
100. Hui FK, Barton PG. Mesomorphic behaviour of some phospholipids with aliphatic alcohols and other non-ionic substances. *Biochim Biophys Acta* 1973;**296**:510–7.
101. Paterson SJ, Butler KW, Huang P, Labelle J, Smith IC, Schneider H. The effects of alcohols on lipid bilayers: a spin label study. *Biochim Biophys Acta* 1972;**266**:597–602.
102. Osman YA, Ingram LO. Mechanism of ethanol inhibition of fermentation in Zymomonas mobilis CP4. *J Bacteriol* 1985;**164**:173–80.
103. Ingram LO. Changes in lipid composition of Escherichia coli resulting from growth with organic solvents and with food additives. *Appl Environ Microbiol* 1977;**33**:1233–6.
104. Ingram LO. Adaptation of membrane lipids to alcohols. *J Bacteriol* 1976;**125**:670–8.
105. Berger B, Carty CE, Ingram LO. Alcohol-induced changes in the phospholipid molecular species of Escherichia coli. *J Bacteriol* 1980;**142**:1040–4.
106. Ramos JL, Duque E, Gallegos MT, Godoy P, Ramos-Gonzalez MI, Rojas A, et al. Mechanisms of solvent tolerance in gram-negative bacteria. *Annu Rev Microbiol* 2002;**56**:743–68.
107. Zingaro KA, Nicolaou SA, Papoutsakis ET. Dissecting the assays to assess microbial tolerance to toxic chemicals in bioprocessing. *Trends Biotechnol* 2013;**31**:643–53.
108. Lin YL, Blaschek HP. Butanol production by a butanol-tolerant strain of Clostridium acetobutylicum in extruded corn broth. *Appl Environ Microbiol* 1983;**45**:966–73.
109. Alsaker KV, Paredes C, Papoutsakis ET. Metabolite stress and tolerance in the production of biofuels and chemicals: gene-expression-based systems analysis of butanol, butyrate, and acetate stresses in the anaerobe Clostridium acetobutylicum. *Biotechnol Bioeng* 2010;**105**:1131–47.
110. Tomas CA, Beamish J, Papoutsakis ET. Transcriptional analysis of butanol stress and tolerance in Clostridium acetobutylicum. *J Bacteriol* 2004;**186**:2006–18.
111. Rutherford BJ, Dahl RH, Price RE, Szmidt HL, Benke PI, Mukhopadhyay A, et al. Functional genomic study of exogenous n-butanol stress in Escherichia coli. *Appl Environ Microbiol* 2010;**76**:1935–45.
112. Wang J, Chen L, Tian X, Gao L, Niu X, Shi M, et al. Global metabolomic and network analysis of Escherichia coli responses to exogenous biofuels. *J Proteome Res* 2013;**12**:5302–12.
113. Brynildsen MP, Liao JC. An integrated network approach identifies the isobutanol response network of Escherichia coli. *Mol Syst Biol* 2009;**5**:277.
114. Zhu H, Ren X, Wang J, Song Z, Shi M, Qiao J, et al. Integrated OMICS guided engineering of biofuel butanol-tolerance in photosynthetic Synechocystis sp PCC 6803. *Biotechnol Biofuels* 2013;**6**:106.
115. Fu F, Cheng VW, Wu Y, Tang Y, Weiner JH, Li L. Comparative proteomic and metabolomic analysis of Staphylococcus warneri SG1 cultured in the presence and absence of butanol. *J Proteome Res* 2013;**12**:4478–89.
116. Atsumi S, Wu TY, Machado IM, Huang WC, Chen PY, Pellegrini M, et al. Evolution, genomic analysis, and reconstruction of isobutanol tolerance in Escherichia coli. *Mol Syst Biol* 2010;**6**:449.
117. Minty JJ, Lesnefsky AA, Lin F, Chen Y, Zaroff TA, Veloso AB, et al. Evolution combined with genomic study elucidates genetic bases of isobutanol tolerance in Escherichia coli. *Microb Cell Fact* 2011;**10**:18.
118. Reyes LH, Almario MP, Winkler J, Orozco MM, Kao KC. Visualizing evolution in real time to determine the molecular mechanisms of n-butanol tolerance in Escherichia coli. *Metab Eng* 2012;**14**:579–90.
119. Reyes LH, Abdelaal AS, Kao KC. Genetic determinants for n-butanol tolerance in evolved Escherichia coli mutants: cross adaptation and antagonistic pleiotropy between n-butanol and other stressors. *Appl Environ Microbiol* 2013;**79**:5313–20.
120. González-Ramos D, van den Broek M, van Maris AJ, Pronk JT, Daran JM.Genome-scale analyses of butanol tolerance in Saccharomyces cerevisiae reveal an essential role of protein degradation. *Biotechnol Biofuels* 2013;**6**:48.

121. Ghiaci P, Norbeck J, Larsson C. Physiological adaptations of Saccharomyces cerevisiae evolved for improved butanol tolerance. *Biotechnol Biofuels* 2013;**6**:101.
122. Liu XB, Gu QY, Yu XB. Repetitive domestication to enhance butanol tolerance and production in Clostridium acetobutylicum through artificial simulation of bio-evolution. *Bioresour Technol* 2013;**130**:638–43.
123. Baer SH, Blaschek HP, Smith TL. Effect of butanol challenge and temperature on lipid-composition and membrane fluidity of butanol-tolerant *Clostridium acetobutylicum*. *Appl Environ Microbiol* 1987;**53**:2854–61.
124. Tomas CA, Welker NE, Papoutsakis ET. Overexpression of *groESL* in *Clostridium acetobutylicum* results in increased solvent production and tolerance, prolonged metabolism, and changes in the cell's transcriptional program. *Appl Environ Microbiol* 2003;**69**:4951–65.
125. Wang F, Kashket S, Kashket ER. Maintenance of Delta pH by a butanol-tolerant mutant of *Clostridium beijerinckii*. *Microbiology* 2005;**151**:607–13.
126. Luan G, Dong H, Zhang T, Lin Z, Zhang Y, Li Y, et al. Engineering cellular robustness of microbes by introducing the GroESL chaperonins from extremophilic bacteria. *J Biotechnol* 2014;**178**:38–40.
127. Desmond C, Fitzgerald GF, Stanton C, Ross RP. Improved stress tolerance of GroESL-overproducing Lactococcus lactis and orobiotic Lactobacillus paracasei NFBC 338. *Appl Environ Microbiol* 2004;**70**:5929–36.
128. Mann MS, Dragovic Z, Schirrmacher G, Lütke-Eversloh T. Over-expression of stress protein-encoding genes helps Clostridium acetobutylicum to rapidly adapt to butanol stress. *Biotechnol Lett* 2012;**34**:1643–9.
129. Zingaro KA, Papoutsakis ET. Toward a semisynthetic stress response system to engineer microbial solvent tolerance. *mBio* 2012;**3**:e00308-12.
130. Alper H, Moxley J, Nevoigt E, Fink GR, Stephanopoulos G. Engineering yeast transcription machinery for improved ethanol tolerance and production. *Science* 2006;**314**:1565–8.
131. Klein-Marcuschamer D, Santos CN, Yu H, Stephanopoulos G. Mutagenesis of the bacterial RNA polymerase alpha subunit for improvement of complex phenotypes. *Appl Environ Microbiol* 2009;**75**:2705–11.
132. Chong H, Geng H, Zhang H, Song H, Huang L, Jiang R. Enhancing E. coli isobutanol tolerance through engineering its global transcription factor camp receptor protein (CRP). *Biotechnol Bioeng* 2014;**111**:700–8.
133. Chen T, Wang J, Yang R, Li J, Lin M, Lin Z. Laboratory-evolved mutants of an exogenous global regulator, IrrE from deinococcus radiodurans, enhance stress tolerances of Escherichia coli. *PLoS One* 2011;**6**:e16228.
134. Dunlop MJ, Dossani ZY, Szmidt HL, Chu HC, Lee TS, Keasling JD, et al. Engineering microbial biofuel tolerance and export using efflux pumps. *Mol Syst Biol* 2011;**7**:487.
135. Fisher MA, Boyarskiy S, Yamada MR, Kong N, Bauer S, Tullman-Ercek D. Enhancing tolerance to short-chain alcohols by engineering the Escherichia coil AcrB efflux pump to secrete the non-native substrate n-butanol. *ACS Synth Biol* 2014;**3**:30–40.
136. Foo JL, Leong SSJ. Directed evolution of an E. coli inner membrane transporter for improved efflux of biofuel molecules. *Biotechnol Biofuels* 2013;**6**:81.
137. Bohnert JA, Schuster S, Seeger MA, Fähnrich E, Pos KM, Kern WV. Site-directed mutagenesis reveals putative substrate binding residues in the Escherichia coli RND efflux pump AcrB. *J Bacteriol* 2008;**190**:8225–9.
138. Dunlop MJ. Engineering microbes for tolerance to next-generation biofuels. *Biotechnol Biofuels* 2011;**4**:32.
139. Reyes LH, Almario MP, Kao KC. Genomic library screens for genes involved in n-butanol tolerance in Escherichia coli. *PLoS One* 2011;**6**:e17678.
140. Hong ME, Lee KS, Yu BJ, Sung YJ, Park SM, Koo HM, et al. Identification of gene targets eliciting improved alcohol tolerance in Saccharomyces cerevisiae through inverse metabolic engineering. *J Biotechnol* 2010;**149**:52–9.
141. Giaever G, Chu AM, Ni L, Connelly C, Riles L, Véronneau S, et al. Functional profiling of the Saccharomyces cerevisiae genome. *Nature* 2002;**418**:387–91.
142. Baba T, Ara T, Hasegawa M, Takai Y, Okumura Y, Baba M, et al. Construction of Escherichia coli K-12 in-frame, single-gene knockout mutants: the Keio collection. *Mol Syst Biol* 2006;**2**:2006.0008.
143. Kobayashi K, Ehrlich SD, Albertini A, Amati G, Andersen KK, Arnaud M, et al. Essential Bacillus subtilis genes. *Proc Natl Acad Sci U S A* 2003;**100**:4678–83.

144. Sandoval NR, Kim JY, Glebes TY, Reeder PJ, Aucoin HR, Warner JR, et al. Strategy for directing combinatorial genome engineering in Escherichia coli. *Proc Natl Acad Sci U S A* 2012;**109**:10540–5.
145. Kanno M, Katayama T, Tamaki H, Mitani Y, Meng XY, Hori T, et al. Isolation of butanol- and isobutanol-tolerant bacteria and physiological characterization of their butanol tolerance. *Appl Environ Microbiol* 2013;**79**:6998–7005.
146. Rühl J, Schmid A, Blank LM. Selected pseudomonas putida strains able to grow in the presence of high butanol concentrations. *Appl Environ Microbiol* 2009;**75**:4653–6.
147. Li J, Zhao JB, Zhao M, Yang YL, Jiang WH, Yang S. Screening and characterization of butanol-tolerant micro-organisms. *Lett Appl Microbiol* 2010;**50**:373–9.
148. Knoshaug EP, Zhang M. Butanol tolerance in a selection of microorganisms. *Appl Biochem Biotechnol* 2009;**153**:13–20.
149. Kataoka N, Tajima T, Kato J, Rachadech W, Vangnai AS. Development of butanol-tolerant Bacillus subtilis strain GRSW2-B1 as a potential bioproduction host. *AMB Express* 2011;**1**:10.
150. Ting CN, Wu J, Takahashi K, Endo A, Zhao H. Screened butanol-tolerant Enterococcus faecium capable of butanol production. *Appl Biochem Biotechnol* 2012;**168**:1672–80.
151. Awang GM, Jones GA, Ingledew WM. The acetone-butanol-ethanol fermentation. *Crit Rev Microbiol* 1988;**15**(Suppl 1):S33–67.
152. Mariano AP, Dias MO, Junqueira TL, Cunha MP, Bonomi A, Filho RM. Butanol production in a first-generation Brazilian sugarcane biorefinery: technical aspects and economics of greenfield projects. *Bioresour Technol* 2013;**135**:316–23.
153. Buijs NA, Siewers V, Nielsen J. Advanced biofuel production by the yeast Saccharomyces cerevisiae. *Curr Opin Chem Biol* 2013;**17**:480–8.
154. Vane LM. Separation technologies for the recovery and dehydration of alcohols from fermentation broths. *Biofuels Bioprod Biorefin* 2008;**2**:553–88.
155. Bankar SB, Survase SA, Ojamoa H, Granström T. Biobutanol: the outlook of an academic and industrialist. *RSC Adv* 2013;**3**:24734–57.
156. Collura MA, Luyben WL. Energy-saving distillation designs in ethanol production. *Ind Eng Chem Res* 1988;**27**:1686–96.
157. Barba D, Brandani V, di Giacomo G. Hyperazeotropic ethanol salted-out by extractive distillation. Theoretical evaluation and experimental check. *Chem Eng Sci* 1985;**40**:2287–92.
158. Seiler M, Jork C, Kavarnou A, Arlt W, Hirsch R. Separation of azeotropic mixtures using hyperbranched polymers or ionic liquids. *AIChE J* 2004;**50**:2439–54.
159. Shi Z, Zhang C, Chen J, Mao Z. Performance evaluation of acetone-butanol continuous flash extractive fermentation process. *Bioprocess Biosyst Eng* 2005;**27**:175–83.
160. Grady MC, Jahic M, Patnaik R. *A method for producing butanol using two-phase extractive fermentation.* (A2, W., Ed.), USA 2009.
161. Ezeji TC, Karcher PM, Qureshi N, Blaschek HP. Improving performance of a gas stripping-based recovery system to remove butanol from Clostridium beijerinckii fermentation. *Bioprocess Biosyst Eng* 2005;**27**:207–14.
162. Vane LM, Alvarez FR, Rosenblum L, Govindaswamy S. Hybrid vapor stripping-vapor permeation process for recovery and dehydration of 1-butanol and acetone/butanol/ethanol from dilute aqueous solutions. Part 2. Experimental validation with simple mixtures and actual fermentation broth. *J Chem Technol Biotechnol* 2013;**88**:1448–58.
163. Liao YC, Lu KM1, Li SY. Process parameters for operating 1-butanol gas stripping in a fermentor. *J Biosci Bioeng* 2014;**118**(5):558–64.
164. Inokuma K, Liao JC, Okamoto M, Hanai T. Improvement of isopropanol production by metabolically engineered Escherichia coli using gas stripping. *J Biosci Bioeng* 2010;**110**:696–701.
165. Nimcevic D, Gapes JR. The acetone-butanol fermentation in pilot plant and pre-industrial scale. *J Mol Microbiol Biotechnol* 2000;**2**:15–20.
166. Groot WJ, Luyben KCAM. In situ product recovery by adsorption in the butanol/isopropanol batch fermentation. *Appl Microbiol Biotechnol* 1986;**25**:29–31.

167. Evans PJ, Wang HY. Enhancement of butanol formation by clostridium acetobutylicum in the presence of Decanol-Oleyl alcohol mixed extractants. *Appl Environ Microbiol* 1988;**54**:1662–7.
168. Nicolaou SA, Gaida SM, Papoutsakis ET. A comparative view of metabolite and substrate stress and tolerance in microbial bioprocessing: from biofuels and chemicals, to biocatalysis and bioremediation. *Metab Eng* 2010;**12**:307–31.
169. Lee SY, Park JH, Jang SH, Nielsen LK, Kim J, Jung KS. Fermentative butanol production by clostridia. *Biotechnol Bioeng* 2008;**101**:209–28.
170. Köpke M, Held C, Hujer S, Liesegang H, Wiezer A, Wollherr A, et al. Clostridium ljungdahlii represents a microbial production platform based on syngas. *Proc Natl Acad Sci* 2010;**107**:13087–92.
171. Heap JT, Pennington OJ, Cartman ST, Carter GP, Minton NP. The ClosTron: a universal gene knock-out system for the genus Clostridium. *J Microbiol Methods* 2007;**70**:452–64.
172. Shen CR, Liao JC. A synthetic iterative pathway for ketoacid elongation. *Methods Enzymol* 2011;**497**:469–81.
173. Chen CK, Blaschek HP. Acetate enhances solvent production and prevents degeneration in Clostridium beijerinckii BA101. *Appl Microbiol Biotechnol* 1999;**52**:170–3.
174. Alper H, Stephanopoulos G. Engineering for biofuels: exploiting innate microbial capacity or importing biosynthetic potential? *Nat Rev Microbiol* 2009;**7**:715–23.
175. Yamamoto S, Suda M, Niimi S, Inui M, Yukawa H. Strain optimization for efficient isobutanol production using Corynebacterium glutamicum under oxygen deprivation. *Biotechnol Bioeng* 2013;**110**:2938–48.
176. Qi H, Li S, Zhao S, Huang D, Xia M, Wen J. Model-driven redox pathway manipulation for improved isobutanol production in Bacillus subtilis complemented with experimental validation and metabolic profiling analysis. *PLoS One* 2014;**9**:e93815.
177. Hermann T. Industrial production of amino acids by coryneform bacteria. *J Biotechnol* 2003;**104**:155–72.
178. Kumar K, Dasgupta CN, Nayak B, Lindblad P, Das D. Development of suitable photobioreactors for CO2 sequestration addressing global warming using green algae and cyanobacteria. *Bioresour Technol* 2011;**102**:4945–53.
179 Lies DP-SS, Kelly J, Dundon CA, Aristidou A, Hawkins A. Reduced by-product accumulation for improved production of isobutanol. (Office, U. S. P., Ed.), USA; 2012.
180. Bokinsky G, Peralta-Yahya PP, George A, Holmes BM, Steen EJ, Dietrich J, et al. Synthesis of three advanced biofuels from ionic liquid-pretreated switchgrass using engineered Escherichia coli. *Proc Natl Acad Sci U S A* 2011;**108**:19949–54.
181. O'Sullivan A. Cellulose: the structure slowly unravels. *Cellulose* 1997;**4**:173–207.
182. Minty JJ, Singer ME, Scholz SA, Bae CH, Ahn JH, Foster CE, et al. Design and characterization of synthetic fungal-bacterial consortia for direct production of isobutanol from cellulosic biomass. *Proc Natl Acad Sci* 2013;**110**:14592–7.
183. Brethauer S, Studer MH. Consolidated bioprocessing of lignocellulose by a microbial consortium. *Energy Environ Sci* 2014;**7**:1446–53.
184. Worden RM, Grethlein AJ, Jain MK, Datta R. Production of butanol and ethanol from synthesis gas via fermentation. *Fuel* 1991;**70**:615–9.
185. Perez JM, Richter H, Loftus SE, Angenent LT. Biocatalytic reduction of short-chain carboxylic acids into their corresponding alcohols with syngas fermentation. *Biotechnol Bioeng* 2013;**110**:1066–77.
186. Gak E, Tyurin M, Kiriukhin M. Genome tailoring powered production of isobutanol in continuous CO2/H2 blend fermentation using engineered acetogen biocatalyst. *J Ind Microbiol Biotechnol* 2014;**41**:763–81.
187. Li H, Opgenorth PH, Wernick DG, Rogers S, Wu TY, Higashide W, et al. Integrated Electromicrobial Conversion of CO2 to Higher Alcohols. *Science* 2012;**335**:1596.
188. Li X, Shen CR, Liao JC. Isobutanol production as an alternative metabolic sink to rescue the growth deficiency of the glycogen mutant of Synechococcus elongatus PCC 7942. *Photosynth Res* 2014;**120**:301–10.
189. Beney L, Gervais P. Influence of the fluidity of the membrane on the response of microorganisms to environmental stresses. *Appl Microbiol Biotechnol* 2001;**57**:34–42.
190. Sommer MO, Church GM, Dantas G. A functional metagenomic approach for expanding the synthetic biology toolbox for biomass conversion. *Mol Syst Biol* 2010;**6**:360.
191. Bayer TS, Widmaier DM, Temme K, Mirsky EA, Santi DV, Voigt CA.Synthesis of methyl halides from biomass using engineered microbes. *J Am Chem Soc* 2009;**131**:6508–15.

CHAPTER 2

Secondary Metabolism for Isoprenoid-based Biofuels

Aram Kang, Taek Soon Lee

Joint BioEnergy Institute, Emeryville, California, USA; Physical Biosciences Division, Lawrence Berkeley National Laboratory, Berkeley, California, USA

2.1 Introduction

Isoprenoids are one of the most diverse groups of natural products in existence, comprising of over 50,000 compounds. Isoprenoids are components of primary metabolites such as sterols, carotenoids, quinones, hormones as well as secondary metabolites, which often play a part in defensive mechanisms of plants or of microbes.

Isoprenoids biosynthesis is initiated by the condensation of two universal precursors, isopentenyl diphosphate (IPP) and its isomer, dimethylallyl diphosphate (DMAPP). These two precursors share a branched C_5 hydrocarbon backbone with one double bond, but differ in the position of the double bond. Prenyltransferase enzymes condense the isoprene unit of IPP to various lengths of prenyl diphosphate molecules such as DMAPP (C_5), geranyl diphosphate (C_{10}, GPP), farnesyl diphosphate (C_{15}, FPP), geranylgeranyl diphosphate (C_{20}, GGPP), and other less abundant prenyl diphosphate molecules.[1] The number of iterative condensation reactions of IPP to one unit of DMAPP is determined by the specificity of the relevant prenyltransferase enzymes.[2] Isoprenoids are split up into a number of different classes based on the number of the condensed isoprene unit: monoterpenes (C_{10}, IPP+DMAPP), sesquiterpenes (C_{15}, 2IPP+DMAPP), and diterpenes (C_{20}, 3IPP+DMAPP) amongst others. Prenyl diphosphates at various lengths are the substrates of terpene synthase enzymes that catalyze a divalent metal-aided removal of a diphosphate group to form a carbocation intermediate. This carbocation intermediate can either be cyclized to generate various ring structures or lose a proton, adding another double bond in the molecule and generating more diversity.[3] The methyl branches and the ring structures of isoprenoids make them promising alternatives to diesels and jet fuels by conferring high-energy density and a lower freezing point.[4]

Isoprenoids serve various biological roles across diverse species such as bacteria, plants, and mammals (Figure 2.1a), but are most plentiful in plants. In plants, isoprenoids play key roles in essential cellular functions such as respiration (e.g., ubiquinone), photosynthesis

Biotechnology for Biofuel Production and Optimization. http://dx.doi.org/10.1016/B978-0-444-63475-7.00002-9

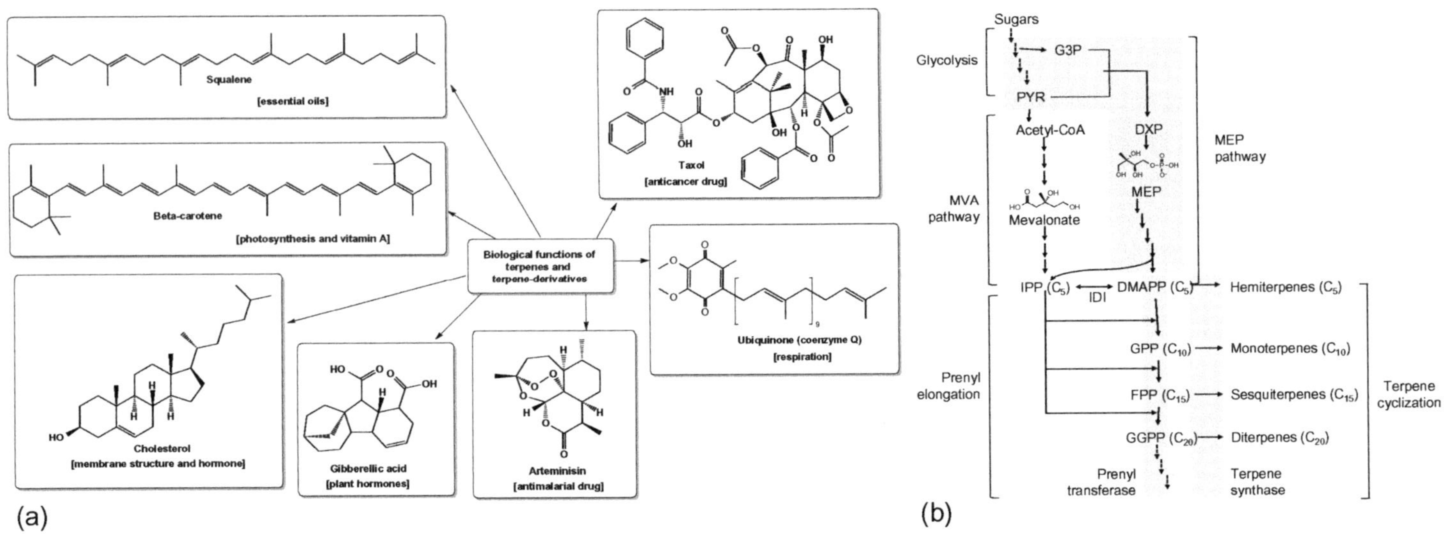

Figure 2.1

Isoprenoid natural products and the pathways. (a) Examples of isoprenoid natural products and their biological functions. (b) Overview of the isoprenoid pathways. Two pathways are found in nature [mevalonate (MVA) pathway and methylerythritol (MEP) pathway]. Two universal precursors of isoprenoids, IPP and DMAPP, are isomers and IPP isomerase (IDI) catalyzes the isomerization reaction. IPP is condensed to DMAPP by prenyl transferases. Prenyl diphosphates are cyclized to various terpenes by terpene synthases. Abbreviations: PYR, pyruvate; G3P, glyceraldehyde-3-phosphate; DXP, deoxyxylulose-5-phosphate; IPP, isopentenyl diphosphate; DMAPP, dimethylallyl diphosphate; GPP, geranyl diphosphate; FPP, farnesyl diphosphate; GGPP, geranylgeranyl diphosphate.

(e.g., carotenoids, chlorophylls, and tocopherols), and membrane architecture (e.g., sterols).[5] Isoprenoids are also found in insects and animals. In insects, isoprenoids (mostly acyclic sesquiterpenes) play a role as juvenile hormones that regulate development and reproduction.[6] In addition, isoprenoids are components of insect pheromones, which induce orchestrated behaviors of a group of insects such as aggregation of pest insects on their host trees.[7] In animals, isoprenoids are the source of vitamin A, K, and E, and are abundant in animal oils (e.g., squalene).[8,9] In addition to these essential roles, isoprenoids serve as secondary metabolites, which are used in the defensive mechanisms of plants. For example, plants produce isoprenoids to inhibit insect hormones, to repel predators, and to seal wounds (e.g., polyisoprene and rubber). These various biological roles of isoprenoids have facilitated their potential applications in medicine, fragrance, and flavors. Consequently, the efficient production of isoprenoids has been of great interests to chemical and medical industries.

Even though isoprenoids are most abundant and diverse in plants, it is not economically feasible to extract isoprenoids from plants in general. Plant cell cultures have been tested as a platform to produce pharmaceutically relevant isoprenoids.[10] More efficient production systems are needed, however, for large-scale production of the isoprenoids, especially when a significantly large scale is required as is the care in biofuel production. As a result, microbes, particularly *Escherichia coli* and yeast, have been exploited as hosts for isoprenoids biosynthesis. Engineering isoprenoid biosynthetic pathways in microbial systems has been supported and facilitated by an understanding of the isoprenoid biosynthetic pathways present in various species.

In this chapter, we present current progress in isoprenoid-based biofuel production with emphasis on secondary metabolism pathways for isoprenoid biosynthesis. In Section 2.2, two distinct biosynthetic pathways for production of the universal precursors of isoprenoids are discussed: first, the mevalonate (MVA) pathway (Section 2.2.1) and its modified archaeal version (Section 2.2.2), and second, the methylerythritol phosphate (MEP) pathway (Section 2.2.3). After the biochemical properties of the relevant enzyme reactions are briefly explored, regulation of the enzyme activity (Section 2.2.4) and stoichiometry of two pathways (Section 2.2.5) are reviewed. The prenyl diphosphate synthases and the terpene synthases that produce specific isoprenoids are reviewed in Section 2.2.6. In Section 2.3, we will discuss the development of isoprenoid pathways in microbial hosts. The recent progress of isoprenoid-based biofuel production and engineering strategies to optimize the isoprenoid pathways are explored in Section 2.4.

2.2 Biosynthesis of Isoprenoids

Figure 2.1b provides an overview of isoprenoid biosynthetic pathways. The substrates for the MVA pathway (acetyl-CoA) and the MEP pathway [pyruvate (PYR) and glyceraldehyde-3-phosphate (G3P)] are supplied through glycolysis pathway. These C_5 prenyl diphosphate molecules are elongated by serial tail-to-head condensation of IPP to DMAPP, and the elongated prenyl diphosphate molecules (e.g., GPP, FPP, and GGPP) are transformed to various isoprenoids by a terpene synthase. Therefore, synthesis of IPP and DMAPP is the first committed step of isoprenoids biosynthetic pathways. Until the late 1980s, the MVA pathway was considered to be the sole biosynthetic pathway that produces IPP and DMAPP, but Rohmer and Arigoni's research group proposed an alternative isoprenoid biosynthetic pathway (i.e., the MEP pathway) by showing that glyceraldehyde-3-phosophate (G3P) was incorporated to isoprenoids.[11,12]

The MVA and the MEP pathways consist of completely different sets of enzymes, catalytic reactions, and metabolites. For example, the MVA pathway produces MVA as an important intermediate while the MEP pathway, the so-called nonmevalonate pathway or deoxyxylulose-5-phosphate (DXP) pathway, does not produce MVA. Also, these two pathways require different precursors for the condensation reactions they catalyze and produce IPP and DMAPP at different ratios. The MVA pathway uses three molecules of acetyl-CoA to produce one molecule of IPP, which can be further isomerized to DMAPP by the isomerase (IDI). In contrast, the MEP pathway produces IPP and DMAPP at 5-6:1 ratio *in vitro* via the condensation of PYR and G3P.[13,14] Since the MVA pathway and the MEP pathway do not share even a single intermediate, most enzymes in both pathways catalyze unique chemical reactions on structurally distinctive substrates. The major differences between the MVA pathway and the MEP pathway are summarized in Table 2.1. In the following three sections, we introduce the MVA pathway (Section 2.2.1), the modified MVA pathway in archaeal species (Section 2.2.2), and the MEP pathway (Section 2.2.3) with focus on enzymes and their reactions.

2.2.1 The MVA Pathway

In the first step of the MVA pathway (Figure 2.2), two molecules of acetyl-CoA (**1**) are condensed to acetoacetyl-CoA (**2**) by an acetoacetyl-CoA thiolase (AACT; Ec-AtoB, Sc-ERG10). The acetoacetyl-CoA is further condensed with another molecule of acetyl-CoA

Table 2.1: Comparison of the MVA pathway and the MEP pathway

	MVA pathway	MEP pathway
Starting substrates	Acetyl-CoA	Pyruvate and glyceraldehyde-3-P
Initial isoprenyl precursor produced	IPP	IPP and DMAPP
Number of reactions	6	7
Organisms	Eukaryote and archaea	Bacteria* and plants
Required cofactors	2 NAD(P)H and 3 ATP	3 NADPH and 2 ATP
Stoichiometric efficiency	0.56 Cmol/Cmol glucose	0.83 Cmol/Cmol glucose
Redox balance	NADH accumulating	Balanced

*There are some bacteria that have the MVA pathway instead of the MEP pathway.

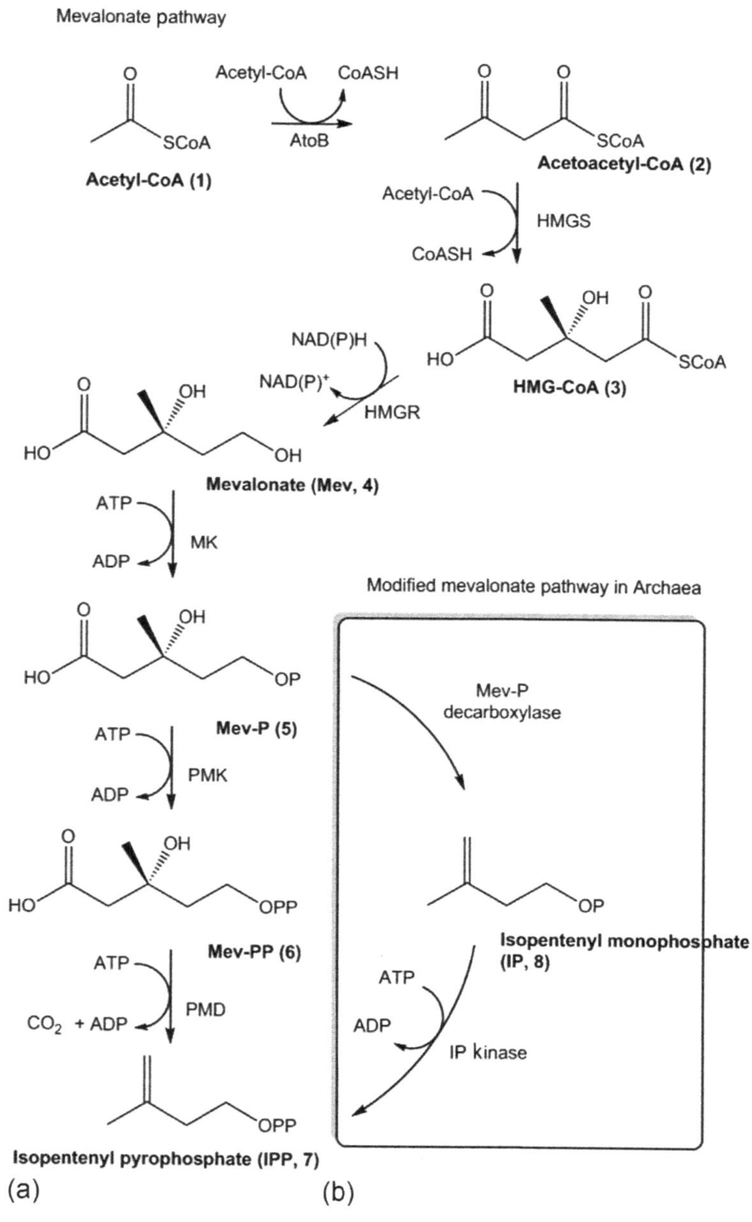

Figure 2.2
The MVA pathways. (a) The original MVA pathway and (b) the modified MVA pathway in Archaea.

by 3-hydroxy-3-methylglutaryl-CoA synthase (HMGS), which produces (S)-3-hydroxy-3-methylglutaryl-CoA (HMG-CoA) (**3**). In these two sequential condensation reactions, total three molecules of acetyl-CoA are required to produce C_6-hydrocarbon backbone with one methyl branch. MVA is then synthesized by HMG-CoA reductase (HMGR), which oxidizes two molecules of NADH or NADPH and reductively substitutes the CoA-group to a hydroxyl group.

The first enzyme, AACT which belongs to the class of biosynthetic thiolases (EC 2.3.1.9), catalyzes a reaction characterized as Claisen-like condensation,[15] which results in carbon-carbon bond formation. The Claisen-like condensation is widely observed in other metabolic pathways such as fatty acids and polyketide biosynthesis.[16] *E. coli* lacks the MVA pathway, but harbors a homologous enzyme of AACT (encoded by the gene, *atoB*), which catalyzes a reaction in the acetoacetate degradation pathway in *E. coli.*[17] The catalytic reaction mechanism of the AACT is summarized in Figure 2.3a: first, one molecule of acetyl-CoA is transferred to the enzyme to form an enzyme-bound acetyl group (acetyl-S-enzyme) followed by release of free CoA-SH. Second, the deprotonated C5 carbanion of the other acetyl-CoA attacks C2 of acetyl-S-enzyme, and forms a carbon-carbon bond between C2 and C5 of acetoacetyl-CoA. Another known mechanism to produce acetoacetyl-CoA is the decarboxylative condensation of acetyl-CoA and malonyl-CoA.[18] In the MVA pathway gene cluster of *Streptomyces* sp., a homologous enzyme of the AACT was not found, but an open reading frame that encodes a ketoacyl-(acyl carrier protein (ACP)) synthase-like enzyme, NhpT7, was identified. The co-expression of NhpT7 with HMGS and HMGR in *Streptomyces albus* increased the MVA concentration 3.5-fold (380 ± 60 mg of MVA per gram of dry cells), which presumably resulted from an increased intracellular concentration of acetoacetyl-CoA by decarboxylative condensation of acetyl-CoA and malonyl-CoA.[18]

The second enzyme of the MVA pathway, HMG-CoA synthase (HMGS), catalyzes the condensation reaction of acetyl-CoA to acetoacetyl-CoA (Figure 2.3b). The first step of the HMGS-catalyzed condensation resembles that of the AACT-mediated condensation, in which the acetyl group in acetyl-CoA is transferred to the cysteine residue of the active site, forming an acetyl-S-enzyme complex. The condensation reaction of HMGS also creates a carbon-carbon bond, but unlike the AACT, C1 carbanion of the activated acetyl-S-enzyme attacks C6 of acetoacetyl-CoA to form the carbon-carbon bond between C1 and C6 (Figure 2.3b). The ketone group at C6 is reduced to a hydroxyl group and the reaction is completed by release of HMG-CoA from the enzyme through hydrolysis of the thioester bond. The distinct catalytic mechanisms of AACT and HMGS have been deciphered by mutating signature residues in the active sites, and it was revealed that the position of base residues in the catalytic triads (HMGS has a conserved Cys-His-Glu triad and AACT has a Cys-His-Cys triad) were critical to distinguish two mechanisms.[15] The condensation by HMGS is also reversible, but the reverse reaction is much slower than the forward reaction, unlike the thiolase-mediated condensation reaction.[19] Eukaryotes have isoforms of HMGS in both mitochondria and cytoplasm which are involved in different metabolic pathways: the mitochondrial isoform is involved in the ketogenic pathway for acetoacetate biosynthesis and the cytosolic isoform is involved in the MVA pathways.[15] It is not clear whether these two isoforms have different enzyme kinetics or not.

The third enzyme of the MVA pathway, HMG-CoA reductase (HMGR), performs two-step reduction consuming either two NADH or two NADPH in the process (Figure 2.3c). First, HMGR

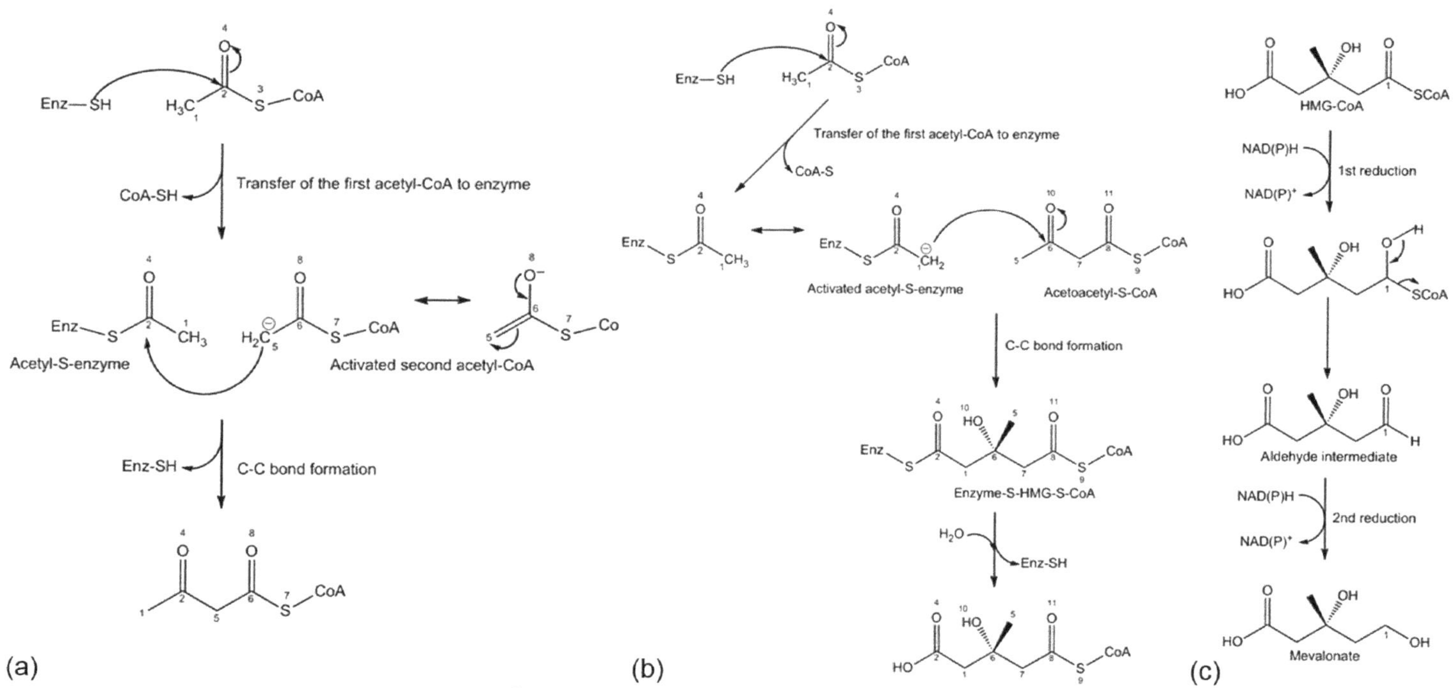

Figure 2.3
Reactions in the top portion of the MVA pathway. (a) Schematic reaction diagram of nondecarboxylative condensation: thiolytic cleavage and Claisen condensation reaction of AACT. (b) Schematic reaction diagram of condensation reaction of HMG-CoA synthase (HMGS). (c) Two-step reduction by HMG-CoA reductase (HMGR).

reduces the C1-ketone group of HMG-CoA to a C1-hydroxyl group (OH), which assists the formation of the aldehyde produced following the departure of the CoA group from C1. Then, the second reduction step converts this aldehyde to an alcohol, which is MVA. Among the enzymes involved in the MVA pathways, HMGR has been most extensively studied as a target for statin drugs, which inhibit the activity of HMGR and downregulate cholesterol biosynthesis.[15] HMGRs are categorized into two subclasses: class I consists of eukaryotic HMGRs, which are generally bound to the membrane of the endoplasmic reticulum, while class II comprises bacterial HMGRs, which are in cytoplasm. Although the sequence and structural homology between class I and II HMGRs are low, catalytically important amino acid residues are well conserved in the active sites across both classes. However, class I HMGRs and class II HMGRs show different inhibitor sensitivities because the enzymes in the two different classes form distinct active site pockets where inhibitors bind. So far, many studies on HMGR have focused on drug development and disease treatment, but little work has been done characterizing enzyme kinetics for engineering purposes. Pitera and colleagues[20] reported bacteriocidal or bacteriostatic effects of accumulated HMG-CoA, suggesting that HMGR could be a limiting enzyme of the MVA pathway. Recently, Ma and colleagues[21] compared the cofactor preference of HMGR from various species and measured the forward and reverse reaction rates *in vitro*. In vivo production of amorphadiene was improved by substituting NADPH-dependent HMGR of *Saccharomyces cerevisiae* to NADH-dependent HMGR of *Delftia acidovorans*.

Following the production of MVA from three molecules of acetyl-CoA, three enzymes convert MVA to IPP (Figure 2.2a). First, the hydroxyl group of the MVA is phosphorylated by two different kinases. Mevalonate kinase (MK) produces mevalonate-5-phosphate (Mev-P), which is subsequently phosphorylated to mevalonate-5-diphosphate (Mev-PP) by phosphomevalonate kinase (PMK). Finally, Mev-PP is decarboxylated to IPP by diphosphomevalonate decarboxylase (PMD). Each reaction by MK, PMK, and PMD requires one ATP. Sequence and structure analysis of these enzymes revealed that they all share a Galactokinase/Homoserine kinase/Mevalonate kinase/Phosphomevalonate kinase (GHMP kinase) motif. The major postulated function of the GHMP kinase motif is to position ATP by placing the adenine moiety and the phosphate group in the active site with appropriate orientation to transfer γ-phosphate to its recipient molecule.[22]

MK activity is regulated by a feedback inhibition of several downstream intermediates such as FPP and GPP. Animal MK is more sensitive (effective inhibition in the nM range) to feedback inhibition compared to bacterial MK (effective inhibition in the μM range). The structural basis and the mechanism of the feedback inhibition was investigated in human and rat MKs.[23] In this study, the MK crystal structure with a farnesyl thiodiphosphate (FSP, an analog of FPP) bound to the active site was compared to the structures of the ATP-bound enzyme and it suggested that FPP would compete with ATP for binding to the active site.[23]

PMK transfers the γ-phosphate group of ATP to Mev-P yielding Mev-PP and ADP. The enzyme kinetics and structure have been investigated in two representative nonorthologous PMKs from *Homo sapiens* and *Streptococcus pneumoniae*, respectively. These two PMKs have different structural motifs: invertebrate PMKs have been characterized as nucleoside monophosphate kinases (NMP kinases) while bacterial, fungal, and plant PMKs belong to the GHMP kinase family. The kinetics of *S. cerevisiae* PMK was recently reported.[24] *S. cerevisiae* PMK is not orthologous to *H. sapiens* PMK, but is orthologous to the bacterial and fungal PMKs. The K_m of *S. cerevisiae* PMK for ATP and Mev-P is 74.3 μM and 880 μM at 37 °C, respectively. It is noteworthy that K_m of *S. cerevisiae* PMK for Mev-P is much higher than that of other PMKs such as those from *S. pneumoniae* (4.2 μM)[25] and *H. sapiens* (34 μM).[26] On the other hand, the V_{max} values of *S. cerevisiae* PMK (4.5 μmol/min/mg enzyme and 5.3 μmol/min/mg enzyme at 30 °C and 37 °C, respectively) were similar to that of *S. pneumoniae* (5.5 μmol/min/mg enzyme), but tenfold lower than that of *H. sapiens* (46 μmol/min/mg enzyme).

There are few reports on reverse reaction of MK, but the reversibility of the PMK reaction was demonstrated in various PMKs.[25–27] The K_m values of *S. pneumoniae* PMK for the substrate (Mev-P) are slightly lower than that of the product (Mev-PP) (4.2 μM for Mev-P and 12 μM for Mev-PP),[25] but the k_{cat} of the forward reaction ($3.4\ s^{-1}$) is similar to that of the reverse reaction ($3.9\ s^{-1}$). Therefore, the reaction slightly favors product (Mev-PP) formation ($K_{eq} = 1.7$), though it was reported that the product could competitively bind to the active site of PMK.

PMD catalyzes an ATP-dependent decarboxylation reaction, where divalent metals (such as Mg^{2+}) play a role as cofactors. PMD also has a GHMP kinase motif because the decarboxylation is mediated by the formation of a 3-phospho-mevalonate-5-diphosphate intermediate where a γ-phosphate of ATP is transferred to the 3-phosphate group. In MK and PMK enzymes, the γ-phosphate of ATP is located in-line with the C5-hydroxyl group or the phosphate ester group of MVA or Mev-P, respectively. However, in PMD, the γ-phosphate must be coordinated to be in-line with the C3-hydroxyl group of Mev-PP while the diphosphate group of Mev-PP is stabilized by hydrogen bonds to multiple serine residues in the active site. Structural analysis revealed that positively charged amino acid residues such as arginine and lysine might play important roles in substrate binding as well as catalysis. In the proposed catalytic mechanisms, the lysine residues neutralize negative charges of the diphosphate group, and arginine residues interact with the carboxyl group, tying the angle of the C3-OH group and also positioning the C3-methyl group. Upon binding of the two substrates in the appropriate orientation, the C3-hydroxyl group of Mev-PP is phosphorylated, and ADP is released from the active site. Subsequently, the C3-phosphate is eliminated to produce a C3-carbocation intermediate, which becomes an electron sink that drives decarboxylation of the C1-carboxyl group. During decarboxylation, one carbon is lost as carbon dioxide (CO_2) and IPP is produced.

In summary, the MVA pathway produces one molecule of IPP (C_5) from three molecules of acetyl-CoA (C_2) in consumption of three molecules of ATP and two molecules of NAD(P)H. The net reaction stoichiometry is given as follows:

$$3\text{Acetyl}-\text{CoA}+2\text{NAD}(\text{P})\text{H}+3\text{ATP}\rightarrow\text{IPP}+\text{CO}_2+2\text{NAD}(\text{P})^{+}+3\text{ADP}$$

2.2.2 Variations of MVA Pathway—Example of Archaeal MVA Pathway

Recently, it has been elucidated that archaeal species have a modified MVA pathway (Figure 2.2b). Understanding of the archaeal MVA pathway is of particular interest because isoprenoids are major components of membrane lipids in many archaeal species, and it was demonstrated that the isoprenoids in the membrane were synthesized from both acetate and MVA in archaeal species,[28] which implies that the archaeal MVA pathway be part of archaeal primary metabolism, supplying isoprenoids for essential cellular functions. However, researchers were unable to identify any enzymes homologous to either PMD or PMK from any of the known archaeal genome sequences,[28] suggesting the presence of a modified MVA pathway in archaeal species. Supporting the hypothesis of a modified MVA pathway, isopentenyl phosphate kinase (IPK) was identified.[28] In this study, researchers demonstrated that the identified IPK could phosphorylate isopentenyl monophosphate (IP) to IPP. More recently, Mev-P decarboxylase was also identified in *Haloferax volcanii*, completing an alternative route of the archaeal MVA pathway.[29] In this study, researchers attempted to identify amino acid residues in the active site of archaeal PMD that determine the substrate specificity for Mev-P. It is still not clear why archaeal species have different route for the MVA pathway, but one possible explanation is that this observation stems from the promiscuous activity of enzymes in the MVA pathway.[29]

2.2.3 MEP Pathway

In the more recently elucidated MEP pathway (Figure 2.4a), pyruvate (PYR, **10**) and glyceraldehyde-3-phosphate (G3P, **11**) are condensed to 1-deoxyxylulose-5-phosphate (DXP, **12**) by DXP synthase (DXS). Both PYR and G3P are 3-carbon metabolites, and one carbon is lost as CO_2 in this first step resulting in one molecule of DXP. DXP is then converted to 2-C-methyl-D-erythritol 4-phosphate (MEP, **13**) by DXP reductoisomerase (DXR) consuming one NADPH in the process. The phosphate group of MEP is then replaced by CDP in the subsequent reaction by 4-diphosphocytidyl-2C-methyl-D-erythritol synthase (IspD), after which the 2-hydroxyl group is phosphorylated by 4-diphosphocytidyl-2-C-methylerythritol kinase (IspE), producing 2-phospho-4-(cytidine 5′-diphospho)-2-C-methyl-D-erythritol (MEP-CDP, **15**). 2-C-methyl-D-erythritol-2,4-cyclodiphosphate (MEcPP, **16**) synthase (IspF) subsequently forms an ester between the phosphate group of CDP and the 2-phosphate group of MEP resulting in MEcPP and CMP. MEcPP is then catalyzed by 1-hydroxy-2-methyl-2-(E)-butenyl 4-diphosphate

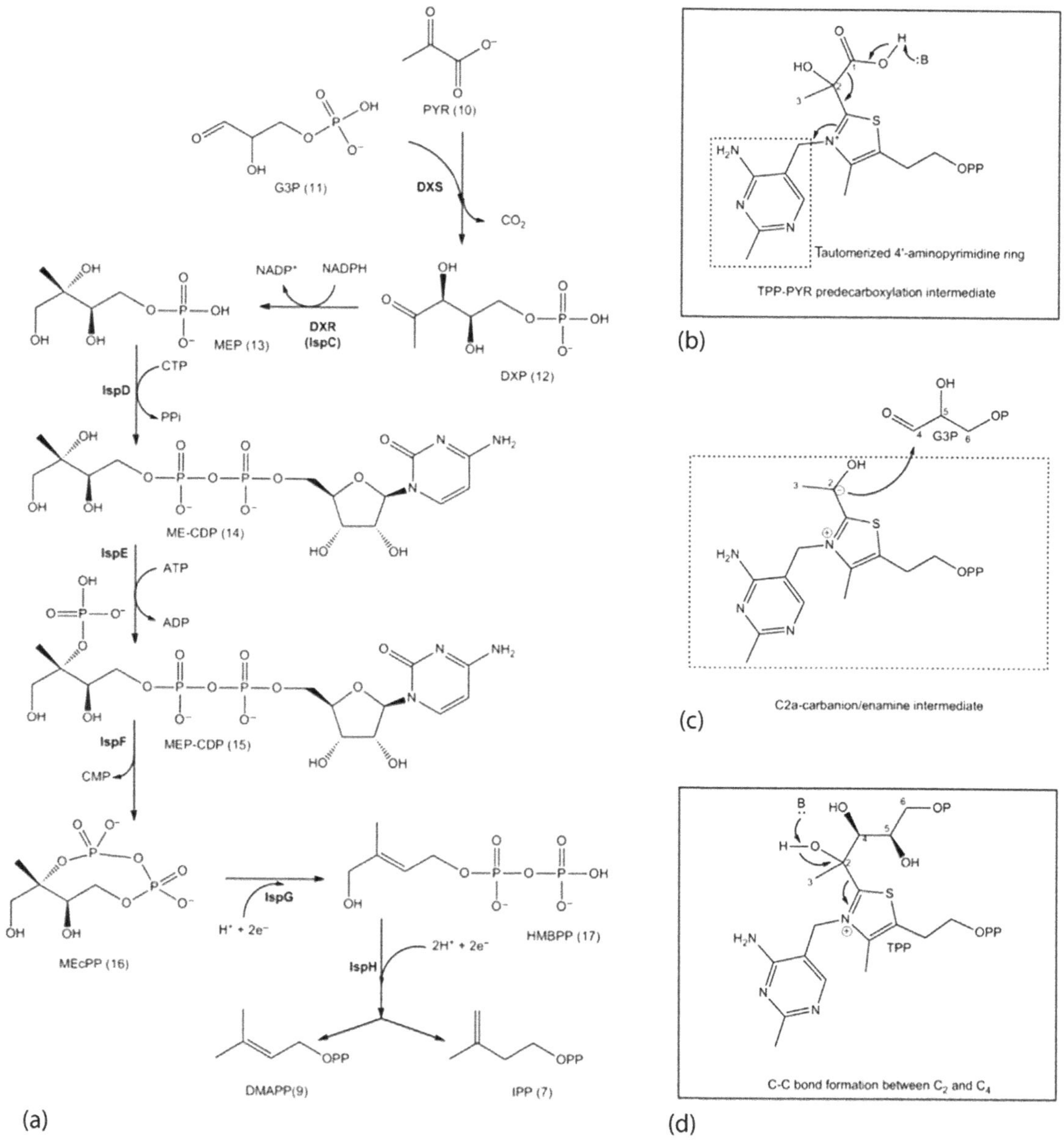

Figure 2.4
The MEP pathway. (a) Overall pathway summary and (b) TPP-dependent reaction mechanism of DXP synthesis.

(HMBPP, **17**) synthase (IspG), finally the C1-hydroxy group of HMBPP is reductively removed by IPP oxidoreductase (ispH), producing IPP (**7**) or DMAPP (**9**).

The MEP pathway has been extensively studied since its discovery in 1990s.[30] However, the reaction mechanisms of the enzymes involved in the MEP pathway still remain obscure due to unusual reaction chemistry and difficulties in obtaining crystal structures of the relevant enzymes. The first reaction is the condensation of PYR and G3P catalyzed by DXS, which requires thiamine diphosphate (TPP) as a cofactor. Most TPP-dependent enzymes are activated via ionization and tautomerization of the 4′-aminopyrimidine ring of TPP.[31] In particular, DXS becomes catalytically active to G3P after the ionized TPP forms a predecarboxylation intermediate with PYR (Figure 2.4b).[32] Subsequently, the PYR-TPP complex is decarboxylated to form a nucleophilic C2α-carbanion intermediate (Figure 2.4c), which then forms a carbon-carbon bond with G3P (Figure 2.4d). DXP is then released from the DXP-TPP complex. Thus far, the substrate binding mechanism of DXS remains obscure, and three different mechanisms have been proposed, each with conflicting supporting evidence. Initially, DXS was thought to exhibit a ping-pong mechanism,[33] as it is the case of most TPP-dependent enzymes. On the other hand, an ordered sequential binding mechanism was also suggested in a study where the PYR is irreversibly bound to the enzyme releasing CO_2.[34] Recently, rapid equilibrium and random sequential binding model[32] was proposed in light of new evidence inconsistent with the ordered binding mechanism.

The second enzyme of the MEP pathway is DXP reductoisomerase (DXR), and this enzyme was recently covered in a review.[35] This enzyme catalyzes the intramolecular rearrangement of a three-carbon fragment by switching C1-methyl and C2-ketone to C2-hydroxyl and methyl of MEP (**13**). A retroaldol/aldol mechanism was suggested to support this reaction, producing two putative intermediates glycolaldehyde phosphate and the enediolate of hydroxyacetone in the process. Subsequently, the C1-ketone group of the aldehyde intermediate 2-C-methyl-D-erythrose 4-phosphate (MEsP) is reduced to a C1-hydroxyl group yielding MEP. While the isomerization of DXP (**12**) to MEP (**13**) is a reversible reaction, DXR favors the formation of MEP.[36]

Three enzymes, IspD, IspE, and IspF, sequentially catalyze the formation of MEcPP (**16**) from MEP (**13**). IspD and IspE require CTP and ATP as cofactors, respectively. In the IspD active site, two substrates (MEP and CTP) and a divalent metal ion (Mg^{2+}) are coordinated in a way that enables in-line transfer of a CMP moiety to phosphate of MEP. The nucleophilic attack of the MEP phosphate on the α-phosphate of CTP results in formation of the product, ME-CDP (**14**).[37] A similar mechanism governs catalytic activity of IspE, which results in the transfer of a phosphate from ATP to a C2-hydroxyl group. Structure of IspE remains unclear. It was reported that IspE is a symmetric homodimer in which two monomers are connected by salt bridges and hydrogen bonds.[38] In this crystal structure, interface of two monomers are relatively small, and the homodimer forms a solvent-filled channel where two active sites are located. However, another study suggested that homodimerization of IspE could be an artifact of crystallization based on the fact that the active sites of each monomer are present independently.[39] IspD is also a symmetric homodimer in which two monomers have large interacting surface. IspF forms a symmetric homotrimer and catalyzes an intramolecular phosphate-transfer reaction. The activated, nucleophilic 2-phosphate group of MEP-CDP (**15**) attacks the β-phosphate of CDP forming a new P-O bond between the two phosphate groups. As a result, a cyclized, unusual structure of MEcPP (**16**) is formed.[40] It is also known that IspF produces 2C-methyl-D-erythritol 3,4-cyclophosphate and 2-phospho-2C-methyl-D-erythritol 3,4-cyclophosphate from ME-CDP (**14**) and MEP-CDP (**15**), respectively.[41] Interestingly, some organisms harbor a fused form of IspD and IspF, even though reactions catalyzed by two enzymes are not consecutive.

The last two enzymes, IspG and IspH, are iron-sulfur cluster (Fe-S)-dependent enzymes. Understanding of IspG and IspH has remained limited compared to the other enzymes of the MEP pathway, primarily due to unsuccessful crystallization of these enzymes with Fe-S clusters. IspG is a ring-opening enzyme that cleaves one of the phosphate ester bonds of MEcPP (**16**). IspH produces IPP and DMAPP, reducing the hydroxyl group and removing the diphosphate group.

Therefore, the MEP pathway provides one molecule of IPP or DMAPP (C_5) from one molecule of PYR (C_3) and one molecule of G3P (C_3). The net reaction stoichiometry is given as follows:

$$\mathrm{G3P} + \mathrm{PYR} + 3\mathrm{NAD(P)H} + 2\mathrm{ATP} \rightarrow \mathrm{IPP} / \mathrm{DMAPP} + \mathrm{CO_2} + 3\mathrm{NAD(P)}^{+} + 2\mathrm{ADP}$$

2.2.4 Regulations of the Isoprenoid Pathways

The regulatory mechanisms of the isoprenoid production pathways have been most extensively investigated in eukaryotes, more specifically in plants. Transcriptional regulations of the bacterial MEP pathway have not been well characterized. In general, relevant genes for a specific function are clustered together and transcribed simultaneously as an operon in bacterial genome. However, the genes annotated as a part of the MEP pathway are scattered in *E. coli* genome. For example, only *xseB-ispA-dxs* and *ispD-ispF* are clustered in the predicted operons, while other genes (*dxr*, *ispE*, *ispG*, and *ispH*) are located apart from each other. In addition, no transcription factor has been found to induce or to repress the MEP pathway genes. Although transcriptional regulation of the MEP pathway is still obscure in prokaryotes, the regulation mechanisms of the MVA pathway have been reported in eukaryotes at the transcriptional, translational, and posttranslational levels. In humans, the MVA pathway is an essential metabolic pathway for sterol biosynthesis, and regulation of HMGR activity was found to be important to control the level of sterols, especially cholesterol. The transcription and the degradation of HMGR is regulated by: (1) the intracellular level of sterols, (2) the sterol regulatory element binding protein (SREBP), (3) the SREBP cleavage-activating protein (Scap), and (4) Insig proteins.[42] Depending on the concentration of sterols, SREBP is either activated or deactivated, which subsequently turns on or off the HMGR transcription, respectively. The AMP-activated protein kinase and protein phosphatase 2A also regulate the catalytic activity of HMGR by phosphorylating the serine residue in the active site when the AMP:ATP ratio is high.[42] On the other hand, the high level of geranylgeraniol was shown to activate the degradation pathway of HMGR.[42] In *S. cerevisiae*, Hap1p, the transcriptional activator of HMGR1, is activated by aerobic growth and the synthesis of heme. HMGR1 is also regulated at the translational level by negative feedback regulation, and MVA has been suggested to serve as a signaling molecule that directly control the translation of HMGR1.[42,43] On the other hand, the activity of HMGR2 is regulated by degradation pathway responding to two signals, GGPP[44] and sterols.[45]

Plants exhibit both the MEP pathway and the MVA pathway specific to cellular compartments, and the regulation of the MEP pathway has been well investigated in plants. For example, the second enzyme of the MEP pathway (DXR) has an active site serine residue whose phosphorylation may regulate DXR activity, and it was supported by experiments demonstrating that mutations of this serine to aspartate or glutamate abolished the normal activity of DXR.[46] IspG and IspH may also have a significant role in the posttranslational regulation of the MEP pathway due to their dependence on Fe-S clusters. IspG and IspH require Fe-S clusters for the activity, and the biosynthesis of Fe-S clusters is controlled by the feedback regulation of IscR and by environmental stimuli such as iron starvation and the presence of reactive oxygen species.[47] While Fe-S clusters serve as electron donors for the substrate, they require redox partner proteins that reduce Fe-S clusters back for the catalytic cycle. Most of these Fe-S cluster-reducing proteins, however, are not well characterized, which makes the study of IspG and IspH enzymes difficult. Given that Fe-S clusters are sensitive to oxidative stress,[48] inactivation of IspG under certain conditions might result in the accumulation of MEcPP (**16**), which has regulatory roles in the disruption of the histone-DNA interactions, thereby affecting the chromosomal structures in *Chlamydia trachomatis*.[49] In addition to the regulatory role of MEcPP, the potential involvement of IspH in stringent responses has been suggested in relation to regulating the level of guanosine 3′,5′-bispyrophosphate ((p)ppGpp).[50,51]

Overall, there has been an extensive search for regulatory mechanisms involved in the MEP pathway and the MVA pathway as a part of efforts to gain a better understanding and ability to engineer the pathways, but further investigation is required. From an engineering standpoint, researchers have concerned that the MEP pathway is under more stringent regulation than the MVA pathway, and as a result, engineering a heterologous MVA pathway has been the preferred strategy for developing isoprenoid production strains that unaffected by the regulatory mechanisms in the native hosts.

2.2.5 Stoichiometry of the MVA and the MEP Pathways and Choice of Pathways for Engineering

The stoichiometry of the two major isoprenoid biosynthesis pathways has been previously analyzed.[52] The overall stoichiometry of the MVA pathway and the MEP pathway are given by Equations (2.1) and (2.2), respectively.

$$3AcCoA + 2NADPH + 3ATP \rightarrow IPP + CO_2 + 2NADP^+ + 3ADP \quad (2.1)$$

$$GAP + PYR + 3NADPH + 2ATP \rightarrow IPP + CO_2 + 3NADP^+ + 2ADP \quad (2.2)$$

Equations (2.1) and (2.2) give the same carbon yield (5/6 = 0.83) since both the MVA and MEP pathways lose one carbon in the form of CO_2 for each molecule of IPP produced. However, when the stoichiometry of substrates produced from glucose was taken into consideration, the two pathways resulted in different carbon yields. The stoichiometry equations were derived for the MEP pathway and the MVA pathway in *E. coli* and yeast, respectively, in consideration of the energy and redox requirement for IPP synthesis.[52] Assuming that glucose (GLC) is the sole carbon source to produce IPP, the following two equations (Equation (2.3) for the MEP pathway in *E. coli* and Equation (2.4) for the MVA pathway in yeast) were obtained:

$$GLC + 3NADPH + 2ATP + NAD^+ \rightarrow IPP + CO_2 + 3NADP^+ + 2ADP + NADH \quad (2.3)$$

$$1.5GLC + 1NADP^+ + 3ATP + 3NAD^+ \rightarrow IPP + 4CO_2 + 1NADPH + 3AMP + 3NADH \quad (2.4)$$

It is noteworthy that the differences of two pathways are apparent when overall cellular metabolisms are considered. Obviously, the MVA pathway is not as efficient as the MEP pathway. The MVA pathway consumes more energy (3 ATPs vs. 2 ATPs), more glucose (1.5 GLC vs. 1 GLC) than the MEP pathway, and loses 4 CO_2 to produce one IPP (0.56 Cmol/Cmol vs. 0.83 Cmol/Cmol of the MEP pathway). Also, the MVA pathway produces two more NADH and is less redox balanced than the MEP pathway.

The stoichiometry analysis suggests that sugar sources and sugar utilization pathways also affects the efficiency of the pathways. Different glycolytic pathways have been analyzed for isoprene production. Liu and colleagues optimized the MEP pathway to produce isoprene by implementing four different glycolytic pathways, which is the Embden-Meyerhof-Parnas (EMP) pathway, Entner-Doudoroff pathway (EDP), Pentose Phosphate Pathway (PPP), and the Dahms pathway together with the MEP pathway, and showed that isoprene production could be increased when the MEP pathway was combined with the EDP.[53] This report showed that energy and redox balance of the MEP pathway could be further optimized by redistributing sugar fluxes of the central metabolism to generate two precursors, PYR and G3P. A similar strategy may be applied to the MVA pathway by supplying PYR from different glycolytic pathways although the MVA pathway uses only single substrate, acetyl-CoA, which is mostly supplied from PYR.

Although the stoichiometry analysis considering the heterologous pathway and the background cellular metabolisms suggest that the MEP pathway is energetically more efficient and more redox balanced than the MVA pathway, the MVA pathway has been more robust in many engineering attempts for isoprenoid-based fuel productions. To more precisely evaluate two pathways in terms of efficiencies and energetics, additional parameters must be considered. These parameters include energy landscape of metabolites (i.e., how energetically favored the reactions are), cellular cost for enzyme translation and degradation, enzyme kinetics, cellular cost for biosynthesis of specific cofactors and efficiency of regenerating cofactors, and enzyme promiscuity and specificity. So far, it has been reasoned that the MEP pathway is more difficult to engineer due to uncharacterized regulatory mechanisms, but more efforts to engineer the MEP pathway are ongoing and it will show some progresses in future.

2.2.6 Prenyl Diphosphate Synthase and Terpene Synthase

IPP and DMAPP are used for carbon chain elongation reactions to produce longer prenyl diphosphate precursors. The last enzyme of the MEP pathway, IspH, produces both IPP and DMAPP, but the MVA pathway needs an additional enzyme to convert IPP to DMAPP for chain elongation reaction. Therefore, most organisms, which are solely dependent of the MVA pathway for the isoprenoids pathways, have IPP-DMAPP isomerase (IDI) enzymes. Some MEP pathway-dependent species have a copy of IDI, and the function of IDI is to maintain the ratio of IPP and DMAPP.[12]

Based on the number of carbons in the hydrocarbon backbone of precursor prenyl diphosphates, isoprenoids (terpenoids) are classified to hemiterpenoids (C_5, one isoprene unit), monoterpenoids (C_{10}, two isoprene units), sesquiterpenoids (C_{15}, three isoprene units), diterpenoids (C_{20}, four isoprene units), and so on (Figure 2.1b). All the prenyl diphosphates are synthesized and elongated by prenyltransferase enzymes, which condense 1 IPP to various allyl diphosphates (e.g., DMAPP, GPP, and FPP). Some prenyltransferase enzymes catalyze the condensation iteratively. *E. coli* FPP synthase (coded in *ispA*), for example, condenses one molecule of IPP with DMAPP to form GPP first, and then another molecule of IPP to GPP to produce FPP. However, GGPP is produced by GGPP synthase (coded in *IspB*), which condenses one IPP molecule only to FPP. In general, prenyltransferases show different substrate specificity for the hydrocarbon chain length of the product, and prenyltransferases with different specificities have been identified in various organisms. For example, bifunctional FPP/GGPP synthase was found in *Pyrococcus furiosus* and *Methanothermobacter thermautotrophicus*, and the dedicated prenyltransferases such as GGPP synthase (C_{20}), hexaprenyl diphosphate synthase (C_{30}), and octaprenyl diphosphate synthase (C_{40}) in other organisms.[54] It has been suggested that the depth of the enzyme active site determines hydrocarbon chain length of the prenyl diphosphates. For example, *E. coli* FPP synthase (IspA) could produce GGPP as a major product when the active site was mutated to deform the depth of the active site.[55]

It is remarkable that a large number of (more than 30,000) structurally distinct terpenoids are produced from such a small number of precursors, and a family of terpene synthases drives this diversity of terpenes from their acyclic precursors.[3] Although the specific cyclization mechanism of prenyl diphosphates differs from one terpene synthase to another, there are mechanistic aspects that most terpene synthases have in common. To initiate the reaction, terpene synthases either (1) generate an initial carbocation triggered by metal-aided diphosphate group removal (ionization dependent) or (2) protonate an epoxide ring or a carbon-carbon double bond (protonation dependent). This initiation step often accompanies an electrophilic attack of one of the π-bonds to generate diverse carbocation intermediates.[56,57] Monoterpene synthases, sesquiterpene synthases, and diterpene synthases, which are important for the production of many useful chemicals and isoprenoid-based fuels, are all ionization-dependent terpene synthases, and possible carbocation intermediates of monoterpenes and sesquiterpenes are shown in Figure 2.5. The structural and mechanistic features of different terpene synthases are out of the scope of this chapter and we do not further discuss it here. Complex cyclization and rearrangement reactions of various terpene synthases have been extensively reviewed.[58]

It should be noted that plants are an important resource of these useful terpene synthases. Since plant-based enzymes generally do not express well in bacterial hosts, and terpene synthases have been one of the limiting enzymes in heterologously engineered terpene production pathways, it has been an important task to find and engineer soluble and functional terpene synthases when the isoprenoid pathways were reconstructed in microbial hosts. In the next section, we will review how the engineering of isoprenoid pathways

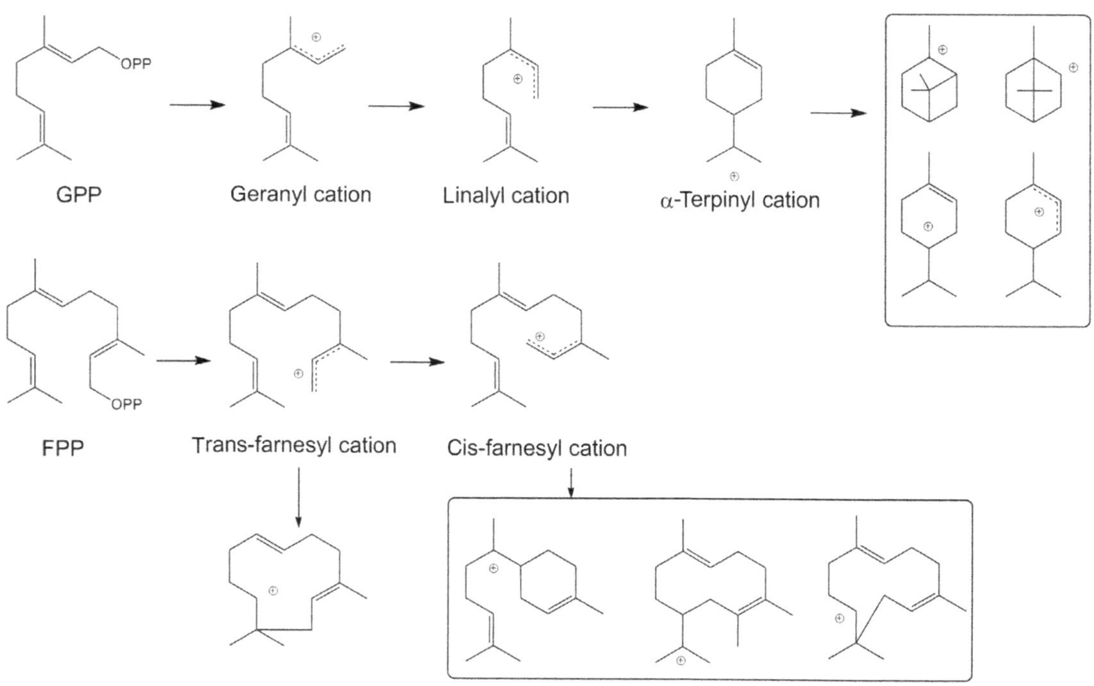

Figure 2.5
Isoprenyl carbocation intermediates of GPP and FPP.

in combination with different terpene synthases has established a strong foundation for microbial production of isoprenoids secondary metabolites.

2.3 Engineering of the Isoprenoid Pathways for Biofuels Production

To effectively produce isoprenoid-based biofuels, isoprenoid production pathways have been extensively engineered in microbial hosts. In Section 2.3.1, we will summarize the limitations of isoprenoid-based biofuel production from natural sources such as plants. In Section 2.3.2, we will discuss development of the isoprenoids pathways for an antimalarial precursor amorphadiene production as a model platform to review engineering efforts in *E. coli* and yeast. This work has laid the groundwork for the development of other terpene production pathways, and the strategies demonstrated in engineering of amorphadiene biosynthetic pathway can be categorized to the three levels (Figure 2.6a). First, the target pathways are reconstructed by assembling individual enzymes from various sources. At the second level, the overall pathway is characterized by identifying bottleneck enzymes and toxic metabolites, and enzyme expression level is balanced. Simultaneously, the host strains are improved by modulating genetic backgrounds and the native metabolisms. Lastly, the fermentation process is improved by optimizing operational parameters.

2.3.1 Limitations of Plants Platform for Isoprenoid Production

Isoprenoids are naturally produced from many species, and plants are the most abundant source for diverse isoprenoids and the corresponding terpene synthases. However, plants may not be an ideal isoprenoid production host due to the following limitations: first, most of the naturally occurring isoprenoids from plants are produced and recovered in very small quantities. For example, it has been estimated that the bark of 2000-3000 *Taxus brevifola* trees are required to produce 1 kg of an anticancer compound, Taxol.[59] Also, extracting isoprenoids from plants is not economical due to the complexity of plants metabolites and the high density of woody biomass, which requires extensive purification and pretreatment processes. There have been efforts to engineer isoprenoids biosynthetic pathways of plants to produce terpenes in large scale from plants or plant cells.[60] However, plants generally grow slow and this lengthens the period of engineering cycle where a genetic modification needs to be verified by acquisition of the desired phenotype. To meet industrial needs for fragrance, drugs, chemical, and biofuel, a large-scale production of isoprenoids is desired, and microbial platforms for isoprenoids have been developed in the context that requires a better host for isoprenoids production to overcome limitations of the natural hosts (i.e., plants). The engineered production of amorphadiene, a precursor of antimalarial artemisinin, is a representative microbial engineering effort in this field. Given that development of isoprenoids-based biofuels pathways can share the same strategy, we will discuss the engineering efforts in this section by reviewing advances made for amorphadiene production.

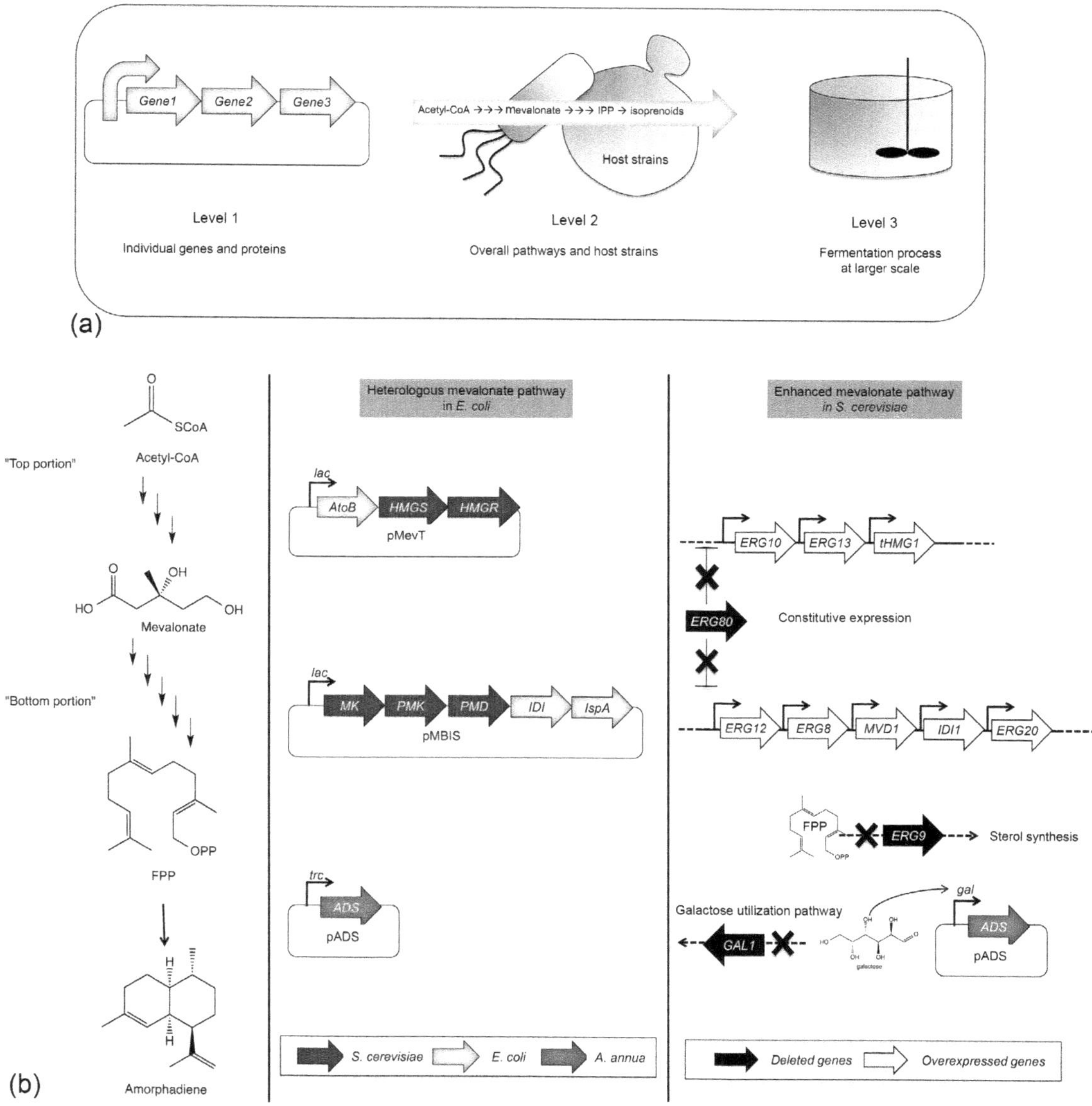

Figure 2.6
(a) General engineering strategies at three different levels and (b) the MVA pathway engineering for amorphadiene production in *E. coli* and yeast.

2.3.2 Development of Isoprenoids Pathways in Microbial Platforms

Engineering of the MEP pathway

Initial engineering strategies for isoprenoid production in *E. coli* focused on increasing availability of the precursors (IPP and DMAPP) by improving the endogenous MEP pathway. Microbial production of lycopene was extensively pursued via engineering of the MEP pathway because of lycopene's commercial and health values as antioxidant, anticancer, and antiinflammatory agents.[61] Since the early demonstration of lycopene production in *E. coli* by introducing the *crtEBI* operon of *Erwinia euridova*,[62] studies to improve lycopene production in *E. coli* employed various engineering strategies.[63] First, the rate-limiting enzymes of the MEP pathways were expressed to facilitate lycopene production.[64–66] Central metabolisms and global regulatory factors were also investigated and perturbed by knockout or overexpression of the relevant genes.[65,67,68] The target genes were identified by stoichiometric modeling,[69] *in silico* flux analysis,[70] and the identified genes have been tested by combinatorial deletion or overexpression.[63] Also, the *E. coli* genome was subjected to multiplex automated genome engineering (MAGE).[71] Similar strategies were used to enhance the production of other terpenoids such as diterpene taxadiene, the precursor of anticancer taxol, in *E. coli*.[72,73]

Although the MEP pathway has been extensively exploited for production of various isoprenoids, the MEP pathway often turned out to be not as efficient as the MVA pathway to achieve high titers. For example, a significant improvement in lycopene production has been made only by introducing the heterologous MVA pathway in *E. coli*.[74] Similarly, the highest titer of isoprene has also been achieved by employing the MVA pathway rather than the MEP pathway. Isoprene production was reported at 60 g/L titer in *E. coli*[75] by heterologously expressing the MVA pathway genes from *Enterococcus faecalis*, *Methanosarcina mazei*, *S. cerevisiae*, and *Populus alba*. The MEP pathway has also been engineered for isoprene production in *E. coli*, but the titer and yield were much lower than those using the MVA pathway.[76]

It was initially suspected that engineering of the MEP pathway in native hosts was impeded by unknown endogenous regulatory mechanisms that block high-level production of isoprenoids.[41,77] Later, the lower metabolic activity of the MEP pathway was attributed to accumulation of the indole that was result of imbalanced pathway expression,[72] and it was also suggested that much weaker catalytic strength of IspG and IspH might be another factor that limits the overall activity of the MEP pathway.[41]

Engineering of the MVA pathway for amorphadiene production in E. coli

Engineering of the MVA pathway in a heterologous host *E. coli* has been successfully demonstrated by reconstituting heterologous MVA pathway genes from *S. cerevisiae* for amorphadiene production.[77] Production of amorphadiene is the first committed step to produce antimalarial artemisinin, a primary component of antimalarial combination therapies (ACTs). To produce IPP and DMAPP at high levels, the heterologous MVA pathway was reconstituted in *E. coli* to bypass the innate regulation of the MEP pathway in the host. The "top portion" enzymes (Figure 2.6b), which condense three acetyl-CoAs into one MVA, were cloned from *E. coli* and *S. cerevisiae* and formed as an operon under the control of an IPTG-inducible promoter, resulting in a plasmid, pMevT. The second plasmid, pMBIS, delivered the "bottom portion" (Figure 2.6b) which catalyzes multistep conversion of MVA to FPP.

The pMBIS was made up of *ERG8 (mk)*, *ERG12 (pmk)*, and *MVD1 (pmd)* from *S. cerevisiae* and *idi* and *ispA* from *E. coli* under IPTG-inducible *lac* promoter. The amorphadiene synthase (ADS) was cloned from *Artemisia annua* and codon optimized for the expression in *E. coli* to produce amorphadiene from FPP. Co-expression of all the nine genes resulted in production of 3.1 μg caryophyllene-equivalent amorphadiene/mL/OD_{600}, which was a 36-fold higher titer than the titer produced by the endogenous MEP pathway.[77] This foundational work led to the development of a robust pipeline to produce various isoprenoids in high titers and also at a large scale.

The early studies on the MVA pathway for the amorphadiene production helped to identify the limiting enzymes, the toxic metabolites, and the inhibitory regulation of the intermediates and products in the MVA pathway. First, Martin and colleagues[77] have suggested that accumulation of FPP and IPP were toxic to the host cells. Later, the accumulation of HMG-CoA was shown to also be toxic[20] probably due to the inhibition of the fatty acid biosynthesis.[78] It has also been suggested that expression level of the individual genes of the MVA pathway need to be balanced to minimize accumulation of the toxic intermediates. The expression level of the top portion genes (*atoB*, *HMGS*, and *HMGR*) was randomly altered by introducing tunable intergenic regions (TIGRs), and the MVA concentration was unexpectedly increased when the expression level of HMGS and HMGR were decreased.[79] The entire amorphadiene pathway was also balanced by optimizing codon usage, promoter strength, and plasmid copy number, which resulted in sevenfold titer increase.[80] In addition, metabolite fluxes through the top portion was further investigated, and a modular control of the metabolite fluxes by synthetic protein scaffolds could increase the MVA yield up to 77-fold.[81] On the other hand, as a bottom-up approach, targeted proteomics approach was applied to quantify enzyme expression level of the MVA pathway.[82] In this study, two enzymes in the bottom portion of the MVA pathway (MK and PMK) were identified as potential limiting enzymes due to very low protein expression levels in the producing strains. Subsequently, an additional promoter was introduced at the upstream of *mk* and *pmk* genes to increase the expression of MK and PMK protein, which resulted in a threefold titer improvement.[82] Another approach to balance protein expression level was demonstrated by a dynamic regulation of gene expression via FPP-responsive promoters, P_{gadE} and P_{rstA}, which were identified by microarray studies of the *E. coli* under FPP stress.[83] In this study, a dynamically controlled MVA pathway was constructed using the FPP-repressive promoter (P_{gadE}) for FPP-producing enzymes and the FPP-inducible promoter (P_{rstA}) for the terpene synthase, and the resulting pathway maintained the FPP concentration below its toxic threshold and increased the amorphadiene titer twofold up to 1.5 g/L.[83] Improvement of isoprenoid production was further achieved by optimizing fermentation conditions. Organic overlay also increased the titer by minimizing product loss and pulling the pathway forward.[84] Also, optimized carbon source and feed rates of nutrients yield further improvement in titer to 27.4 g/L of amorphadiene production in *E. coli*.[85]

Engineering of the MVA pathway for amorphadiene production in yeast

Yeast natively expresses the MVA pathway (Figure 2.7), and the primary strategies for increased isoprenoid production in yeast were to overexpress the MVA pathway genes and to reinforce the carbon flux to FPP by deleting nonessential competing pathways. FPP production was improved by overexpression of UPC2, a transcriptional regulator of sterols, and downregulation of squalene synthase gene (ERG9) which competes with ADS for FPP.[86] A galactose-inducible promoter regulated the transcription of the MVA pathway genes and terpene synthase gene. This galactose-induction system was further optimized by deleting GAL1, encoding the enzyme that utilizes galactose as a carbon source. Furthermore, the deletion of GAL80 allowed the MVA pathway and ADS to be constitutively expressed under

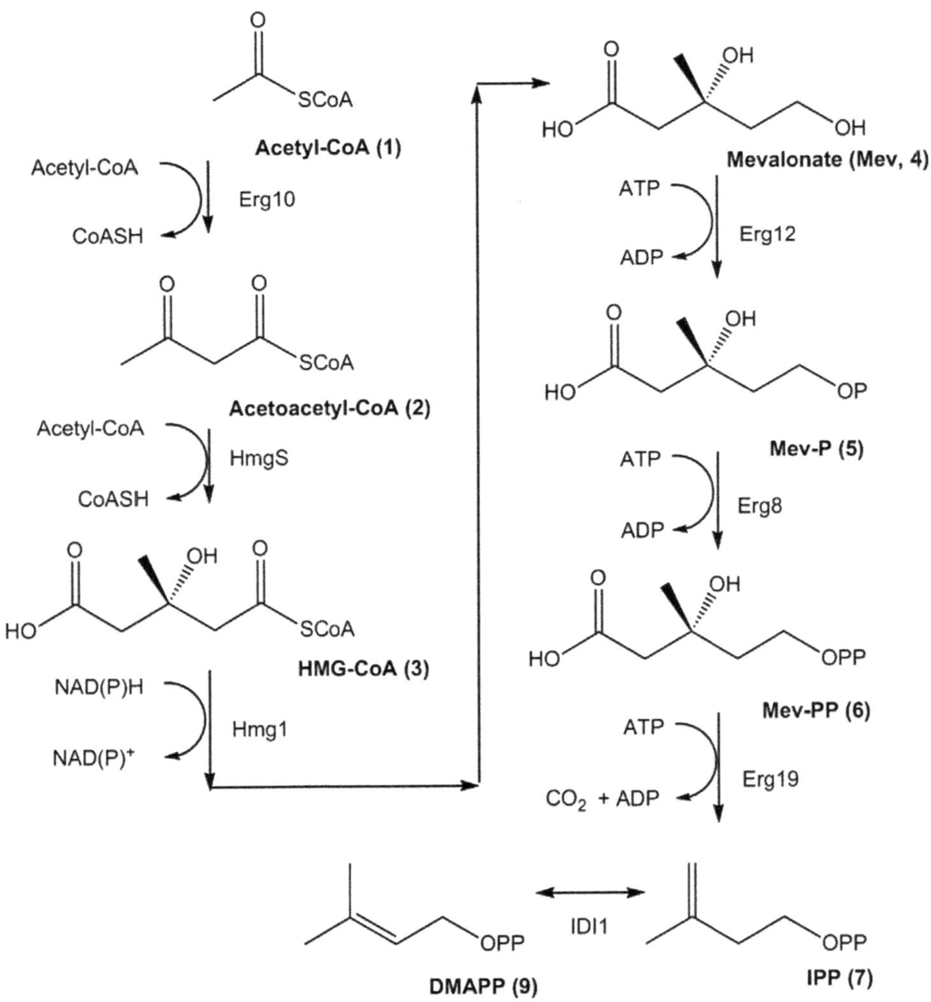

Figure 2.7
The MVA pathway of yeast.

GAL1 promoter in the absence of galactose.[87] Upon overexpression of the MVA pathway genes and ADS in *S. cerevisiae* CEN.PK2 strain, the highest amorphadiene titer of 41 g/L was achieved with all modifications made in host together.[88]

Efforts to engineer the MEP pathway in yeast

Since yeast natively utilizes the MVA pathway for isoprenoid production, there have been efforts to engineer yeast with a heterologous MEP pathway to improve isoprenoid production. As we have seen in Section 2.2, the MEP pathway has several advantages over the MVA pathway: a better theoretical yield, energy efficiency, and redox balance. In addition, it is believed that the regulatory mechanisms that can limit flux through the native MVA pathway in yeast may not affect the heterologous MEP pathway. But in spite of the advantages, engineering the MEP pathway in a heterologous host has shown only limited success so far, especially due to the difficulty in functional expression of IspG and IspH, two Fe-S cluster enzymes.[89] A fully functional expression of the MEP pathway in yeast would be an important milestone in improving isoprenoids titers.

2.4 Engineering of the MVA Pathway for Isoprenoid-based Biofuels

The MVA pathway platforms for production of the antimalarial drug precursor amorphadiene have led to the development of the isoprenoid-based biofuels pathways. In the past decade, diverse classes of terpenoids have been investigated with their fuel-like properties and engineered for microbial production. In this section, several terpene compounds will be discussed with emphasis on their fuel properties and the metabolic engineering efforts to produce them.

2.4.1 Sesquiterpene Biofuels (C_{15})

Sesquiterpenes are a class of terpenes derived from FPP (C_{15}). A large number of natural terpenes (>7000 compounds) belong to sesquiterpenoids,[90] and they have a wide range of physiological roles such as antimicrobial agents and alarm pheromones.[59] The catalytic diversity of the terpene synthases determines the structure of the terpene products, and sesquiterpenes can take on acyclic, monocyclic, bicyclic, or tricyclic structures. Among the naturally existing sesquiterpenes, farnesene, bisabolene, and farnesol have been suggested as biofuel precursors and their productions have been extensively investigated in both *E. coli* and yeast (Figure 2.8).

Farnesene

Farnesene is a collective name that refers to four stereoisomers of α-farnesene (3,7,11-trimethyl-1,3,6,10-dodecatetraene) and two stereoisomers of β-farnesene (7,11-dimethyl-3-methylene-1,6,10-dodecatriene) (Figure 2.8). All of these isomers have

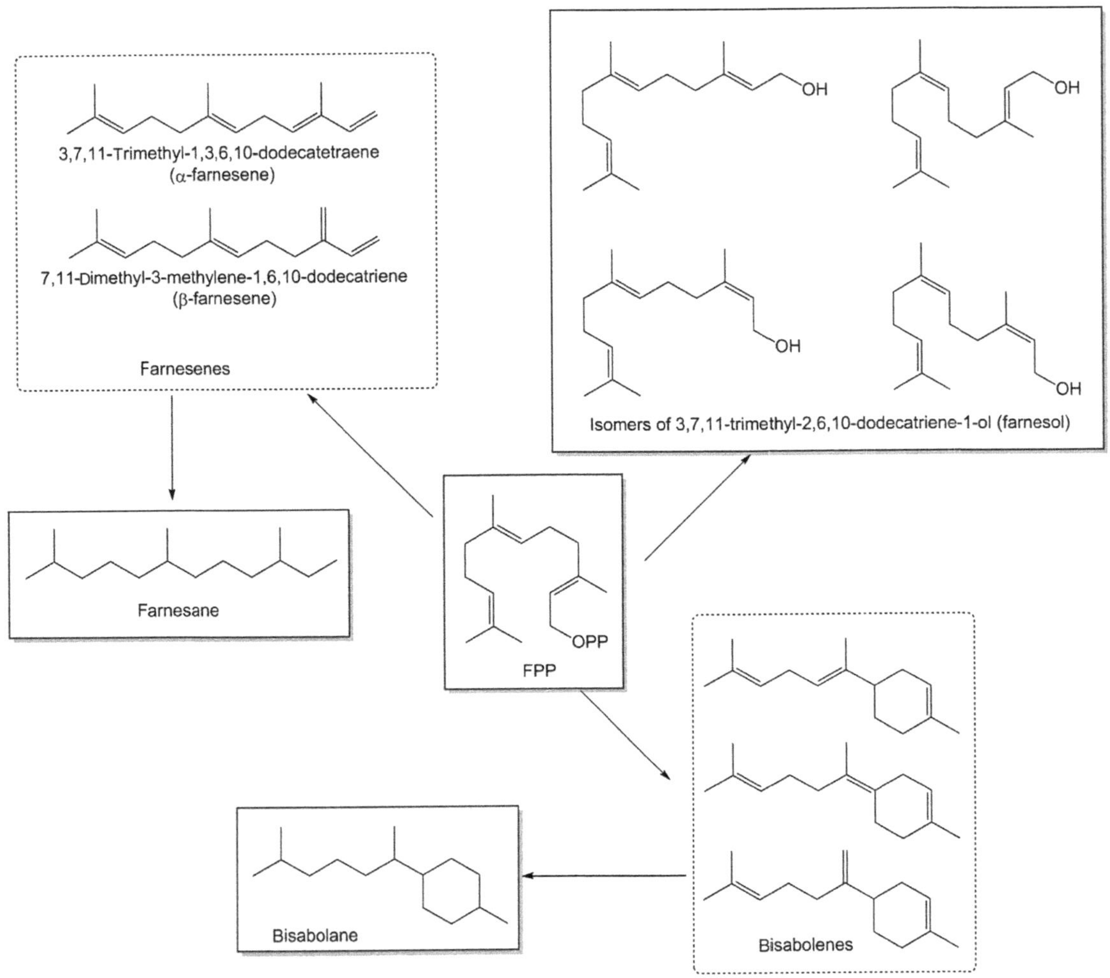

Figure 2.8
Sesquiterpene-based biofuels.

a branched hydrocarbon chain backbone and four double bonds in common, but they are differentiated by location of the double bonds and chirality of carbon-carbon (C–C) bonds. Only some of these farnesene isomers can be found in nature. For example, α-farnesene is a chemical component of fragrance or scent in various organisms, in which these strong smells play a role as attractant or repellent to other organisms, assists pollination, protects the hosts from their predators, or serves as signaling molecules such as pheromones in a group of insects.[91] In addition to their roles as fragrances, farnesene is also used in solvents, emollients, food supplements (e.g., vitamins), cosmetics (e.g., squalene),[92] and more recently precursors of high-performance polymers for tire industry.[93] Finally, the fully hydrogenated compound, farnesane, has been evaluated as an additive in diesels produced from renewable biomaterials.[94] Farnesane has an appropriate cetane number (~58), which qualifies it to be a potential replacement of diesels (cetane number range of D2 diesels is from 40 to 60),[4] and better cold properties with the lower cloud point of −78 °C than that of D2 diesels (−3 °C).

Farnesene is biologically produced from FPP by farnesene synthase, of which homologous enzymes have been identified in various organisms such as *Mentha piperita*,[95] *A. annua*,[96] *Picea abies*,[97] *Zea mays*,[98] or *Citrus junos*.[99] The FPP-overproducing *E. coli* and yeast strains, which have been previously developed for amorphadiene production, could be used for the farnesene production by incorporating adequate farnesene synthase to the platform.[94,100] This fast development of the production platform led to smooth transition to commercialization, and one of the farnesene isomers, trans-β-farnesene, is now produced under the product name, "Biofene®," by Amyris, a biotechnology company known for microbial antimalarial drug production. The initial microbial strains produced 1.1 g/L of farnesene after 120 h of *E. coli* fermentation and 728 mg/L farnesene after 72 h of yeast fermentation.[94] Amyris has continuously evolved the host strain, *S. cerevisiae* PE-2, by extensive mutagenesis and selection and has achieved more than 50% of its theoretical yield (23.8 g farnesene per 100 g glucose).[91] In the report published by Amyris in 2010, 104.3 g/L of farnesene was produced with a productivity of 16.9 g/L/day and a recovery of 95%.[101]

Bisabolene

Bisabolenes are another group of sesquiterpenes identified as potential diesel precursors. Bisabolenes are naturally found in essential oils of plants, and their derivatives are also found in insect pheromones.[102] The fuel properties and the microbial production of bisabolene was demonstrated recently.[103] The chemical hydrogenation of bisabolene produced a biodiesel compound bisabolane, and it showed a cetane number of 53, which qualifies as a potential replacement of diesels. Bisabolane also showed competitive cold properties as farnesane with low cloud point of −78 °C, and its density was a little higher (0.88 g/mL) than that of farnesane (0.77 g/mL) due to its ring structure.

Bisabolene is produced by carbocationic cyclization of FPP, which is catalyzed by bisabolene synthase. The homologous enzymes of the bisabolene synthase have been identified in several plants such as *Abies grandis*, *Arabidopsis thaliana*, *P. abies*, and *Pseudotsuga meseisi*. Using the FPP-producing platform from the amorphadiene work,[82] about 400 mg/L of bisabolene was produced with *E. coli* codon-optimized version of *A. grandis* bisabolene synthase. Further engineering such as codon optimization of pathway genes and the introduction of an additional promoter at the upstream of *MK* resulted in more than twofold of titer increase by increasing protein expression level of bottom pathway enzymes, especially MK and PMK. A similar titer was also reported in yeast by overexpressing the same bisabolene synthase in the FPP-accumulating yeast strain developed in a previous study.[86] Gene deletions in yeast CEN.PK2 that improved the isoprenoid pathways were identified by utilizing carotenoid-based screening methods.[104] In this study, three genes [two genes with un-characterized functions and a gene coding a transcriptional regulator (Rox1) that negatively regulate the genes involved in the MVA and sterol biosynthesis pathways] from the yeast chromosome were deleted, and the engineered yeast strain was able to produce 5.2 g/L bisabolene titer at fed-batch fermentation condition when combining all engineering efforts to improve the host strain and the pathway.[104]

Farnesol

Terpene-derived alcohols have also been proposed as fuel molecules.[4,105] Farnesol (3,7,11-trimethyl-2,6,10-dodecatriene-1-ol, Figure 2.8) is a branched-chain alcohol derived from FPP by hydrolysis of the diphosphate group of FPP. Potential use of farnesol as an antimicrobial agent,[106] as a precursor of anticancer drugs,[107,108] and as pesticides[105,109] have been reported. Recently its properties as alternative diesels or jet fuels were also reported.[4,105] Microbial production of farnesol has been enabled by engineering the MVA pathway in *E. coli* and yeast, and its progress has been recently reviewed.[105] The strategies used for developing yeast strains were to maximize the metabolites flux to FPP, and a farnesol titer of 3.05 g/L was obtained by fed-batch fermentation of the strain with HMGR (Hmg1 and Hmg2) overexpression and the squalene synthase (ERG9) deletion to prevent from losing FPP into sterol biosynthesis.

In *E. coli*, heterologous expression of the MVA pathway and overexpression of FPP synthase (IspA) resulted in 135 mg/L farnesol production without any additional phosphatase expression.[110] Even though endogenous phosphatases, whose deletion decreased farnesol production in *E. coli*, were identified, overexpression of these genes did not improve farnesol production, which implies either the hydrolysis of FPP may not be the rate-limiting step or the diphosphate group of FPP may be hydrolyzed by promiscuous activities of multiple endogenous phosphatases. The phosphatase may not be the best engineering target for high-titer production if hydrolysis of the FPP-pyrophosphate group is not the rate-limiting step. However, the identification and engineering of the FPP-specific phosphatases would still be important to improve the kinetics and the titer of farnesol production when the reaction relies only on the promiscuous activity of endogenous enzymes.

2.4.2 Monoterpene Biofuels (C_{10})

Monoterpenes (and monoterpenoids) are a class of isoprenoids produced from geranyl diphosphate (GPP, C_{10}) by various monoterpene synthases. Monoterpenes are valuable compounds as ingredients in cosmetics, food flavorings, cleaning products, and drugs. Some monoterpenes are acyclic hydrocarbons (such as myrcene and ocimene) or their derivatives (such as linalool and geraniol), but cyclic monoterpenes and their derivatives (such as limonene, pinenes, camphor, alpha-terpineol, perillyl alcohol, carveol, carvone, and menthols) are more abundant in nature and have more industrial applications. Among these monoterpenes, the hydrogenated form of myrcene and ocimene (2,6-dimethyloctane) and the hydroxylated forms such as linalool and geraniol have been suggested as potential biofuel precursors (Figure 2.9).[111,112] Recently, microbial platforms of two promising monoterpenes, limonene and pinene, have been reported.[113,114] These microbial platforms were slightly modified from the amorphadiene production platform by replacing FPP synthase (IspA) with GPP synthase, and by expressing heterologous monoterpene synthases from plants for cyclization of GPP to desired monoterpenes.

Early work toward microbial production of monoterpenes was reported using the endogenous MEP pathway in *E. coli.*[115] In this work, the heterologously expressed monoterpene synthases produced very low level of the corresponding monoterpenes, and it was reasoned that the low titer was primarily resulted by the limited availability of GPP supplied by the native MEP pathway, considering that GPP can be further condensed to FPP by the endogenous FPP synthase. By employing the MVA pathway, the high-titer production of limonene was achieved in *E. coli.*[113] Limonene is abundant in orange peels and traditionally obtained as a byproduct of orange juice industry. Limonene is a precursor of various useful terpenoids such as perillyl alcohol, carveol, carvone, and menthol in fragrance and health industry,[116] and the hydrogenation product of limonene shows good chemical properties as fuels.[112] A step-by-step optimization of limonene production was performed in *E. coli* strains with the heterologous MVA pathway, GPP synthase from *A. thaliana*, and limonene synthase from *Mentha spicata*, and the titer improved up to 450 mg/L. This titer is over 100-fold higher than the previously reported titer using the MEP pathway.[115]

Microbial production of pinene was recently demonstrated by heterologously expressing GPPS2 from *A. grandis* and pinene synthase (PS) Pt30 of *Pinus taeda* in *E. coli* with the titer of 5.44 mg/L and 0.97 g/L in flasks and in fed-batch fermentor, respectively.[117] Pinene production was subsequently improved by combinatorial expression of GPPS and PS from three organisms.[114] The higher titer (27.9 mg/L) was achieved in flask by co-expression of GPPS and PS from *A. grandis*, and the titer was increased further to 32 mg/L by expressing a fused GPPS-PS enzyme.[114] Significantly lower titer of pinene than that of limonene suggested that PS activity is probably the limiting factor considering both pathways use the same MVA pathway. Geraniol, a hydrolyzed product of GPP, was also produced in *E. coli* in titer to 182.5 mg/L.[118]

The production of monoterpenes close to the theoretical yields or at high titers would be detrimental to the microbial hosts due to the toxicity of monoterpenes,[119–121] and it is important to develop resistant strains that can tolerate the high titers required for competitive production of monoterpenes at industrial scale. One strategy is an export of the biofuel to the extracellular medium, and in a recent study, efflux pumps were identified and expressed in biofuel-producing *E. coli* strains.[122] Efflux pumps serve as a cellular defense mechanism, and actively transports toxic compounds out of the cells. Many efflux pumps have been identified in gram-negative bacteria, which has solvent-pumping transporters that belong to the resistance-nodulation-division (RND) family.[123] Expression of these RND pumps could confer the tolerance to monoterpene-based biofuels such as pinene and limonene probably by

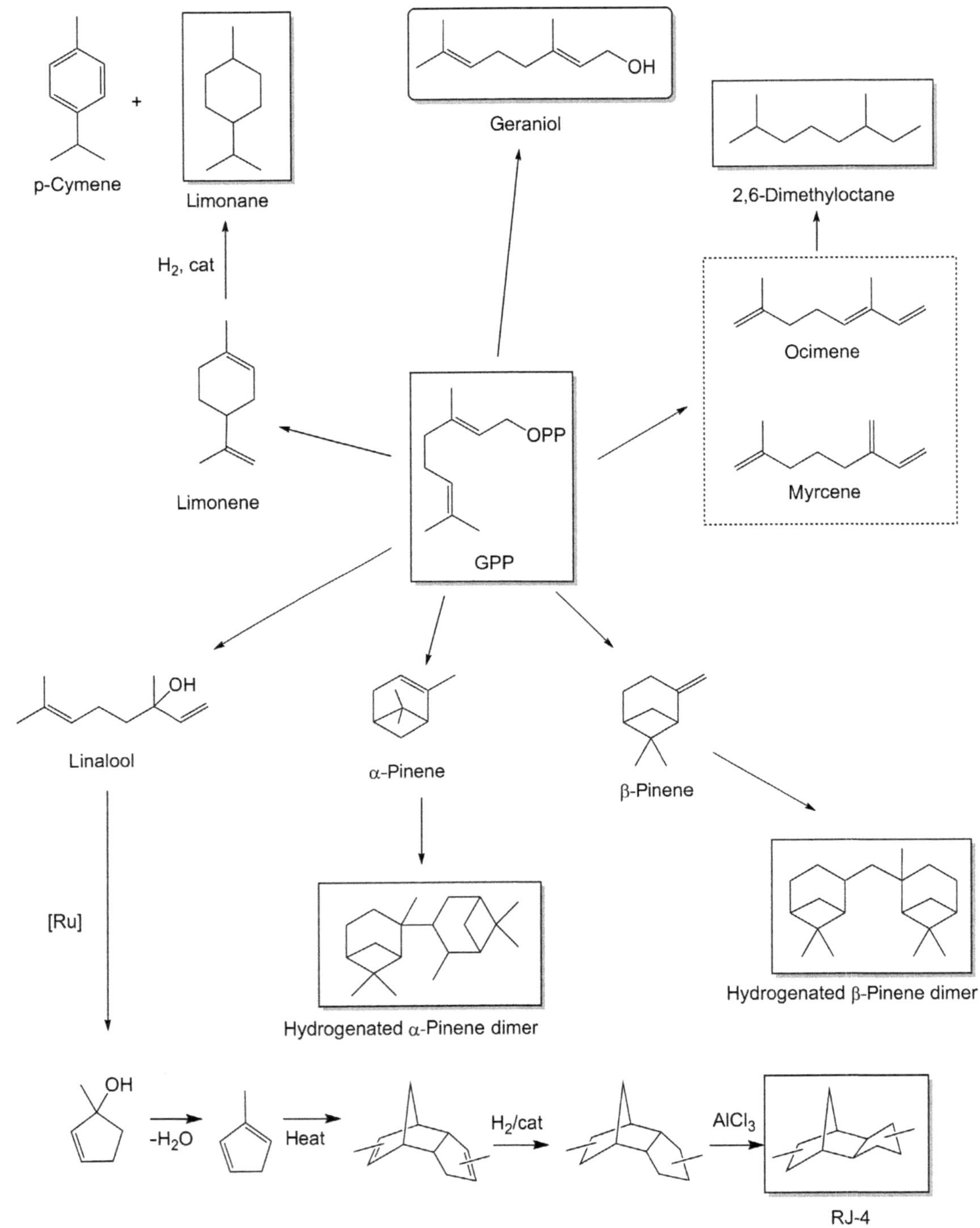

Figure 2.9
Monoterpene-based biofuels.

lowering the intracellular concentration of the fuel molecules.[122] Although the utilization of efflux pumps were shown to be effective, more research is needed to control the expression of these efflux pumps since unregulated overexpression of membrane proteins can impede viability of cells by trafficking membrane translocation machineries.[124]

In addition to the efflux pump, there have been more approaches to improve strain robustness against toxic biofuels such as the adaptation of the host strains under stress-imposed conditions, random engineering of the tolerance by mutagenesis and genome shuffling[125] or targeted recombineering methodologies.[71] Moreover, more stress-resistant species such as *Pseudomonas* have been considered as alternative hosts for engineering biofuel production pathways.[126–128]

2.4.3 Hemiterpene Biofuels (C_5)

Hemiterpenes and hemiterpenoids are derived from C_5-precursors IPP and DMAPP. The previous studies on hemiterpenes have focused on production of isoprene (Figure 2.10), the monomer of natural rubber. Isopentanol (isoamyl alcohol or 3-methyl-1-butanol, Figure 2.10) is a branched short-chain alcohol which can be produced either by using a 2-keto acid decarboxylase pathway from re-factored leucine biosynthesis pathway[129] or by hydrogenation of 3-methyl-3-butenol and 3-methyl-2-butenol (isopentenols) from the isoprenoid pathways.[130] All these C_5 alcohols have a higher energy content than that of ethanol[131] and high octane numbers that allow them to meet the requirements to serve as gasoline replacement[132] and as antiknocking additives.[133]

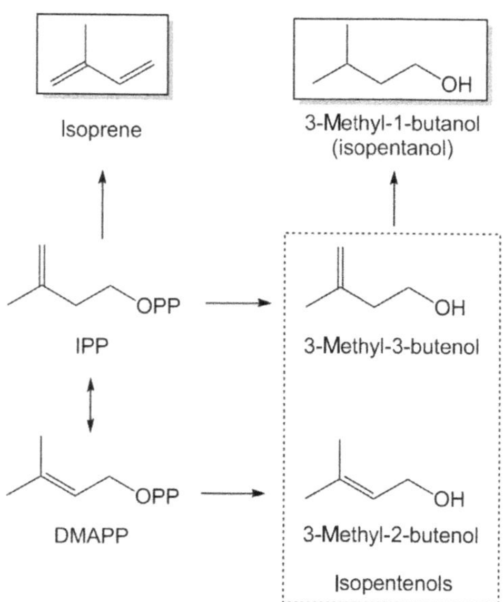

Figure 2.10
Hemiterpene-based biofuels.

Isopentenol was produced in the engineered *E. coli* using heterologous MVA pathway. While searching for an enzyme that hydrolyzes IPP and DMAPP to produce respective isopentenols, Withers and colleagues found that the expression of *nudF* gene from *Bacillus subtilis* could relieve the toxic effects of IPP and DMAPP.[134] Later, Chou and colleagues[130] identified *E. coli* NudB as the more efficient IPP-dephosphorylating enzyme by screening promiscuous activities of the enzymes in haloacid dehydrogenase (HAD) and Nudix hydrolase superfamilies in *E. coli.* In the same study, NemA from Old Yellow Enzyme (OYE) family in *E. coli* showed a promiscuous reductase activity that reduces 3-methyl-2-butenol into 3-methyl-1-butanol (isopentanol) while consuming NADPH. To improve the isopentenol titer, Zheng and colleagues replaced two genes in the top portion of the MVA pathway from *S. cerevisiae* to those from the *E. faecalis*, of which expression resulted in higher MVA titer. Finally, when co-expressed with NudF from *B. subtilis*, the engineered MVA pathway increased the titer to 1.3 g/L from 3% glucose (12% theoretical yield).[135] The more systematic approach using proteomics and metabolomics led to a higher yield and titer of isopentenol.[136] George and colleagues rationally constructed variants of the precursor pathway by employing a combinatorial approach to vary "top" and "bottom" modules of the MVA pathway (Figure 2.6b), and quantitatively analyzed the omics data of these strains to explain the behavior of the pathways. Based on the conceptual pathway model constructed by this approach, they engineered the pathway to achieve the isopentenol titer of 1.5 g/L from 1% glucose (46% theoretical yield).[136]

2.5 Concluding Remarks

In this chapter, we have discussed the secondary metabolism pathways for isoprenoids production and the current status of isoprenoid-based biofuel production. Isoprenoid-based biofuels have promising properties as replacements for or additives to existing petroleum-based fuels, and many researchers have demonstrated the feasibility of microbial platforms to produce isoprenoid-based fuels. Extensive engineering efforts have allowed researchers to overcome the low yield from plants by either improving native or implementing heterologous isoprenoid pathways in microbial hosts. Particularly, balancing the isoprenoid pathways by modulating protein expression level have been a major theme to minimize the accumulation of toxic intermediates or funneling of intermediates toward production of unwanted products.

It is noteworthy that understanding biochemical properties of enzymes and their catalytic reactions has facilitated the engineering of isoprenoid pathways in microbial hosts. The MVA pathway and the MEP pathway have been elucidated in many organisms, and reconstructed in heterologous microbial hosts. Studies of reaction mechanism, enzyme kinetics, and protein structure have provided the foundation to modulate enzyme activities and to balance protein expression level. In addition, several limiting enzymes have been identified and overexpressed to increase the metabolite fluxes toward the isoprenoid-based biofuels. Moreover, elucidating inhibitory effects of pathway intermediates or products allowed researchers to improve the pathway by removing or decreasing the inhibitory intermediates.

Among currently developed isoprenoid-based fuels, farnesane has started to serve as a transportation fuel in Brazil by a biotech company, Amyris, and other isoprenoid-based biofuels such as bisabolane and isopentanol are also produced at high titer in microbial hosts. We envision that further optimization of the isoprenoid pathways and production hosts along with enhanced fermentation processes will enable the isoprenoid-based biofuels to be economically feasible and fungible transportation fuels.

Acknowledgments

This work conducted by the Joint BioEnergy Institute was supported by the Office of Science, Office of Biological and Environmental Research, of the U.S. Department of Energy under Contract No. DE-AC02-05CH11231.

References

1. Rodríguez-concepción M. Plant isoprenoids. *Methods Mol Biol* 2014;**1153**:1–5.
2. Wang K, Ohnuma S. Chain-length determination mechanism of isoprenyl diphosphate synthases and implications for molecular evolution. *Trends Biochem Sci* 1999;**24**:445–51.
3. Degenhardt J, Köllner TG, Gershenzon J. Monoterpene and sesquiterpene synthases and the origin of terpene skeletal diversity in plants. *Phytochemistry* 2009;**70**:1621–37.
4. Peralta-Yahya PP, Zhang F, del Cardayre SB, Keasling JD. Microbial engineering for the production of advanced biofuels. *Nature* 2012;**488**:320–8.
5. Pulido P, Perello C, Rodriguez-Concepcion M. New insights into plant isoprenoid metabolism. *Mol Plant* 2012;**5**:964–7.
6. Mayoral JG, Nouzova M, Navare A, Noriega FG. $NADP^+$-dependent farnesol dehydrogenase, a corpora allata enzyme involved in juvenile hormone synthesis. *Proc Natl Acad Sci U S A* 2009;**106**:21091–6.
7. Keeling CI, Chiu CC, Aw T, Li M, Henderson H, Tittiger C, et al. Frontalin pheromone biosynthesis in the mountain pine beetle, *Dendroctonus ponderosae*, and the role of isoprenyl diphosphate synthases. *Proc Natl Acad Sci U S A* 2013;**110**:18838–43.
8. Mata-Gómez LC, Montañez JC, Méndez-Zavala A, Aguilar CN. Biotechnological production of carotenoids by yeasts: an overview. *Microb Cell Fact* 2014;**13**:12.
9. Ačimovič J, Rozman D. Steroidal triterpenes of cholesterol synthesis. *Molecules* 2013;**18**:4002–17.
10. Roberts SC. Production and engineering of terpenoids in plant cell culture. *Nat Chem Biol* 2007;**3**:387–95.
11. Rohmer M, Knani M, Simonin P, Sutter B, Sahm H. Isoprenoid biosynthesis in bacteria: a novel pathway for the early steps leading to isopentenyl diphosphate. *Biochem J* 1993;**295**(Pt 2):517–24.
12. Gräwert T, Groll M, Rohdich F, Bacher A, Eisenreich W. Biochemistry of the non-mevalonate isoprenoid pathway. *Cell Mol Life Sci* 2011;**68**:3797–814.
13. Rohdich F, Zepeck F, Adam P, Hecht S, Kaiser J, Laupitz R, et al. The deoxyxylulose phosphate pathway of isoprenoid biosynthesis: studies on the mechanisms of the reactions catalyzed by IspG and IspH protein. *Proc Natl Acad Sci U S A* 2003;**100**:1586–91.
14. Rohdich F, Hecht S, Gärtner K, Adam P, Krieger C, et al. Studies on the nonmevalonate terpene biosynthetic pathway: metabolic role of IspH (LytB) protein. *Proc Natl Acad Sci U S A* 2002;**99**:1158–63.
15. Miziorko HM. Enzymes of the mevalonate pathway of isoprenoid biosynthesis. *Arch Biochem Biophys* 2011;**505**:131–43.
16. Heath RJ, Rock CO. The claisen condensation in biology. *Nat Prod Rep* 2002;**19**:581–96.
17. Duncombe GR, Frerman FE. Molecular and catalytic properties of the acetoacetyl-coenzyme a thiolase of *Escherichia coli*. *Arch Biochem Biophys* 1976;**176**:159–70.
18. Okamura E, Tomita T, Sawa R, Nishiyama M, Kuzuyama T. Unprecedented acetoacetyl-coenzyme a synthesizing enzyme of the thiolase superfamily involved in the mevalonate pathway. *Proc Natl Acad Sci U S A* 2010;**107**:11265–70.

19. Theisen MJ, Misra I, Saadat D, Campobasso N, Miziorko HM, Harrison DHT. 3-Hydroxy-3-methylglutaryl-CoA synthase intermediate complex observed in "real-time". *Proc Natl Acad Sci U S A* 2004;**101**:16442–7.
20. Pitera DJ, Paddon CJ, Newman JD, Keasling JD. Balancing a heterologous mevalonate pathway for improved isoprenoid production in *Escherichia coli*. *Metab Eng* 2007;**9**:193–207.
21. Ma SM, Garcia DE, Redding-Johanson AM, Friedland GD, Chan R, Batth TS, et al. Optimization of a heterologous mevalonate pathway through the use of variant HMG-CoA reductases. *Metab Eng* 2011;**13**:588–97.
22. Andreassi JL, Leyh TS. Molecular functions of conserved aspects of the GHMP kinase family. *Biochemistry* 2004;**43**:14594–601.
23. Fu Z, Voynova NE, Herdendorf TJ, Miziorko HM, Kim J-JP. Biochemical and structural basis for feedback inhibition of mevalonate kinase and isoprenoid metabolism. *Biochemistry* 2008;**47**:3715–24.
24. Garcia DE, Keasling JD. Kinetics of phosphomevalonate kinase from *Saccharomyces cerevisiae*. *PLoS ONE* 2014;**9**:e87112.
25. Pilloff D, Dabovic K, Romanowski MJ, Bonanno JB, Doherty M, Burley SK, et al. The kinetic mechanism of phosphomevalonate kinase. *J Biol Chem* 2003;**278**:4510–5.
26. Herdendorf TJ, Miziorko HM. Phosphomevalonate kinase: functional investigation of the recombinant human enzyme. *Biochemistry* 2006;**45**:3235–42.
27. Eyzaguirre J, Valdebenito D, Cardemil E. Pig liver phosphomevalonate kinase: kinetic mechanism. *Arch Biochem Biophys* 2006;**454**:189–96.
28. Grochowski LL, Xu H, White RH. *Methanocaldococcus jannaschii* uses a modified mevalonate pathway for biosynthesis of isopentenyl diphosphate. *J Bacteriol* 2006;**188**:3192–8.
29. Vannice JC, Skaff DA, Keightley A, Addo JK, Wyckoff GJ, Miziorko HM. Identification in *Haloferax volcanii* of phosphomevalonate decarboxylase and isopentenyl phosphate kinase as catalysts of the terminal enzyme reactions in an archaeal alternate mevalonate pathway. *J Bacteriol* 2014;**196**:1055–63.
30. Chang W, Song H, Liu H, Liu P. Current development in isoprenoid precursor biosynthesis and regulation. *Curr Opin Chem Biol* 2013;**17**:571–9.
31. Nemeria NS, Chakraborty S, Balakrishnan A, Jordan F. Reaction mechanisms of thiamin diphosphate enzymes: defining states of ionization and tautomerization of the cofactor at individual steps. *FEBS J* 2009;**276**:2432–46.
32. Brammer LA, Smith JM, Wade H, Meyers CF. 1-Deoxy-D-xylulose 5-phosphate synthase catalyzes a novel random sequential mechanism. *J Biol Chem* 2011;**286**:36522–31.
33. Frank RAW, Leeper FJ, Luisi BF. Structure, mechanism and catalytic duality of thiamine-dependent enzymes. *Cell Mol Life Sci* 2007;**64**:892–905.
34. Eubanks LM, Poulter CD. *Rhodobacter capsulatus* 1-deoxy-D-xylulose 5-phosphate synthase: steady-state kinetics and substrate binding. *Biochemistry* 2003;**42**:1140–9.
35. Murkin AS, Manning KA, Kholodar SA. Mechanism and inhibition of 1-deoxy-D-xylulose-5-phosphate reductoisomerase. *Bioorg Chem* 2014;**57**:171–85.
36. Hoeffler J-F, Tritsch D, Grosdemange-Billiard C, Rohmer M. Isoprenoid biosynthesis via the methylerythritol phosphate pathway. *Eur J Biochem* 2002;**269**:4446–57.
37. Richard SB, Lillo AM, Tetzlaff CN, Bowman ME, Noel JP, Cane DE. Kinetic analysis of *Escherichia coli* 2-c-methyl-D-erythritol-4-phosphate cytidyltransferase, wild type and mutants, reveals roles of active site amino acids. *Biochemistry* 2004;**43**:12189–97.
38. Miallau L, Alphey MS, Kemp LE, Leonard GA, McSweeney SM, Hecht S, et al. Biosynthesis of isoprenoids: crystal structure of 4-diphosphocytidyl-2c-methyl-D-erythritol kinase. *Proc Natl Acad Sci U S A* 2003;**100**:9173–8.
39. Kalinowska-Tłuścik J, Miallau L, Gabrielsen M, Leonard GA, McSweeney SM, Hunter WN. A triclinic crystal form of *Escherichia coli* 4-diphosphocytidyl-2c-methyl-D-erythritol kinase and reassessment of the quaternary structure. *Acta Crystallogr Sect F: Struct Biol Cryst Commun* 2010;**66**(Pt 3):237–41.
40. Richard SB, Ferrer JL, Bowman ME, Lillo AM, Tetzlaff CN, Cane DE, et al. Structure and mechanism of 2-c-methyl-D-erythritol 2,4-cyclodiphosphate synthase. An enzyme in the mevalonate-independent isoprenoid biosynthetic pathway. *J Biol Chem* 2002;**277**:8667–72.
41. Zhao L, Chang W, Xiao Y, Liu H, Liu P. Methylerythritol phosphate pathway of isoprenoid biosynthesis. *Annu Rev Biochem* 2013;**82**:497–530.

42. Burg JS, Espenshade PJ. Regulation of HMG-CoA reductase in mammals and yeast. *Prog Lipid Res* 2011;**50**:403–10.
43. Dimster-Denk D, Thorsness MK, Rine J. Feedback regulation of 3-hydroxy-3-methylglutaryl coenzyme a reductase in *Saccharomyces cerevisiae*. *Mol Biol Cell* 1994;**5**:655–65.
44. Garza RM, Tran PN, Hampton RY. Geranylgeranyl pyrophosphate is a potent regulator of HRD-dependent 3-hydroxy-3-methylglutaryl-CoA reductase degradation in yeast. *J Biol Chem* 2009;**284**:35368–80.
45. Gardner RG, Shan H, Matsuda SPT, Hampton RY. An oxysterol-derived positive signal for 3-hydroxy-3-methylglutaryl-CoA reductase degradation in yeast. *J Biol Chem* 2001;**276**:8681–94.
46. Jawaid S, Seidle H, Zhou W, Abdirahman H, Abadeer M, Hix JH, et al. Kinetic characterization and phosphoregulation of the *Francisella tularensis* 1-deoxy-D-xylulose 5-phosphate reductoisomerase (MEP synthase). *PLoS ONE* 2009;**4**:e8288.
47. Johnson DC, Dean DR, Smith AD, Johnson MK. Structure, function, and formation of biological iron-sulfur clusters. *Annu Rev Biochem* 2005;**74**:247–81.
48. Roche B, Aussel L, Ezraty B, Mandin P, Py B, Barras F. Iron/sulfur proteins biogenesis in prokaryotes: formation, regulation and diversity. *Biochim Biophys Acta* 2013;**1827**:455–69.
49. Grieshaber NA, Fischer ER, Mead DJ, Dooley CA, Hackstadt T. Chlamydial histone-DNA interactions are disrupted by a metabolite in the methylerythritol phosphate pathway of isoprenoid biosynthesis. *Proc Natl Acad Sci U S A* 2004;**101**:7451–6.
50. Kanjee U, Ogata K, Houry WA. Direct binding targets of the stringent response alarmone (p)ppGpp. *Mol Microbiol* 2012;**85**:1029–43.
51. Traxler MF, Summers SM, Nguyen HT, Zacharia VM, Hightower GA, Smith JT, et al. The global, ppGpp-mediated stringent response to amino acid starvation in *Escherichia coli*. *Mol Microbiol* 2008;**68**:1128–48.
52. Gruchattka E, Hädicke O, Klamt S, Schütz V, Kayser O. *In silico* profiling of *Escherichia coli* and *Saccharomyces cerevisiae* as terpenoid factories. *Microb Cell Fact* 2013;**12**:84.
53. Liu H, Sun Y, Ramos KRM, Nisola GM, Valdehuesa KNG, Lee WK, et al. Combination of Entner-Doudoroff pathway with MEP increases isoprene production in engineered *Escherichia coli*. *PLoS ONE* 2013;**8**:e83290.
54. Chang TH, Guo RT, Ko TP, Wang AHJ, Liang PH. Crystal structure of type-III geranylgeranyl pyrophosphate synthase from *Saccharomyces cerevisiae* and the mechanism of product chain length determination. *J Biol Chem* 2006;**281**:14991–5000.
55. Lee PC, Petri R, Mijts BN, Watts KT, Schmidt-Dannert C. Directed evolution of *Escherichia coli* farnesyl diphosphate synthase (IspA) reveals novel structural determinants of chain length specificity. *Metab Eng* 2005;**7**:18–26.
56. Li R, Chou WKW, Himmelberger JA, Litwin KM, Harris GG, Cane DE, et al. Reprogramming the chemodiversity of terpenoid cyclization by remolding the active site contour of epi-isozizaene synthase. *Biochemistry* 2014;**53**:1155–68.
57. Christianson DW. Structural biology and chemistry of the terpenoid cyclases. *Chem Rev* 2006;**106**: 3412–42.
58. Gao Y, Honzatko RB, Peters RJ. Terpenoid synthase structures: a so far incomplete view of complex catalysis. *Nat Prod Rep* 2012;**29**:1153–75.
59. Kirby J, Keasling JD. Biosynthesis of plant isoprenoids: perspectives for microbial engineering. *Annu Rev Plant Biol* 2009;**60**:335–55.
60. Moses T, Pollier J, Thevelein JM, Goossens A. Bioengineering of plant (tri)terpenoids: from metabolic engineering of plants to synthetic biology in vivo and in vitro. *New Phytol* 2013;**200**:27–43.
61. Zhou Y, Nambou K, Wei L, Cao J, Imanaka T, Hua Q. Lycopene production in recombinant strains of *Escherichia coli* is improved by knockout of the central carbon metabolism gene coding for glucose-6-phosphate dehydrogenase. *Biotechnol Lett* 2013;**35**:2137–45.
62. Misawa N, Shimada H. Metabolic engineering for the production of carotenoids in non-carotenogenic bacteria and yeasts. *J Biotechnol* 1998;**59**:169–81.
63. Jin YS, Stephanopoulos G. Multi-dimensional gene target search for improving lycopene biosynthesis in *Escherichia coli*. *Metab Eng* 2007;**9**:337–47.

64. Kim SW, Keasling JD. Metabolic engineering of the nonmevalonate isopentenyl diphosphate synthesis pathway in *Escherichia coli* enhances lycopene production. *Biotechnol Bioeng* 2001;**72**:408–15.
65. Kang MJ, Lee YM, Yoon SH, Kim JH, Ock SW, Jung KH, et al. Identification of genes affecting lycopene accumulation in *Escherichia coli* using a shot-gun method. *Biotechnol Bioeng* 2005;**91**:636–42.
66. Yuan LZ, Rouvière PE, Larossa RA, Suh W. Chromosomal promoter replacement of the isoprenoid pathway for enhancing carotenoid production in *E. coli*. *Metab Eng* 2006;**8**:79–90.
67. Farmer WR, Liao JC. Improving lycopene production in *Escherichia coli* by engineering metabolic control. *Nat Biotechnol* 2000;**18**:533–7.
68. Alper H, Stephanopoulos G. Uncovering the gene knockout landscape for improved lycopene production in *E. coli*. *Appl Microbiol Biotechnol* 2008;**78**:801–10.
69. Alper H, Jin YS, Moxley JF, Stephanopoulos G. Identifying gene targets for the metabolic engineering of lycopene biosynthesis in *Escherichia coli*. *Metab Eng* 2005;**7**:155–64.
70. Choi HS, Lee SY, Kim TY, Woo HM. In silico identification of gene amplification targets for improvement of lycopene production. *Appl Environ Microbiol* 2010;**76**:3097–105.
71. Wang HH, Isaacs FJ, Carr PA, Sun ZZ, Xu G, Forest CR, et al. Programming cells by multiplex genome engineering and accelerated evolution. *Nature* 2009;**460**:894–8.
72. Ajikumar PK, Xiao WH, Tyo KEJ, Wang Y, Simeon F, Leonard E, et al. Isoprenoid pathway optimization for taxol precursor overproduction in *Escherichia coli*. *Science* 2010;**330**:70–4.
73. Jiang M, Stephanopoulos G, Pfeifer BA. Downstream reactions and engineering in the microbially reconstituted pathway for taxol. *Appl Microbiol Biotechnol* 2012;**94**:841–9.
74. Zahiri HS, Yoon SH, Keasling JD, Lee SH, Kim SW, Yoon SC, et al. Coenzyme Q10 production in recombinant *Escherichia coli* strains engineered with a heterologous decaprenyl diphosphate synthase gene and foreign mevalonate pathway. *Metab Eng* 2006;**8**:406–16.
75. Cervin MA, Whited GM, Chotani GK, Valle F, Fioresi C, Sanford KJ, et al. Compositions and methods for producing isoprene. US 20090203102 a1; 2009.
76. Zhao Y, Yang J, Qin B, Li Y, Sun Y, Su S, et al. Biosynthesis of isoprene in *Escherichia coli* via methylerythritol phosphate (MEP) pathway. *Appl Microbiol Biotechnol* 2011;**90**:1915–22.
77. Martin VJJ, Pitera DJ, Withers ST, Newman JD, Keasling JD. Engineering a mevalonate pathway in *Escherichia coli* for production of terpenoids. *Nat Biotechnol* 2003;**21**:796–802.
78. Kizer L, Pitera DJ, Pfleger BF, Keasling JD. Application of functional genomics to pathway optimization for increased isoprenoid production. *Appl Environ Microbiol* 2008;**74**:3229–41.
79. Pfleger BF, Pitera DJ, Smolke CD, Keasling JD. Combinatorial engineering of intergenic regions in operons tunes expression of multiple genes. *Nat Biotechnol* 2006;**24**:1027–32.
80. Anthony JR, Anthony LC, Nowroozi F, Kwon G, Newman JD, Keasling JD. Optimization of the mevalonate-based isoprenoid biosynthetic pathway in *Escherichia coli* for production of the anti-malarial drug precursor amorpha-4,11-diene. *Metab Eng* 2009;**11**:13–9.
81. Dueber JE, Wu GC, Malmirchegini GR, Moon TS, Petzold CJ, Ullal AV, et al. Synthetic protein scaffolds provide modular control over metabolic flux. *Nat Biotechnol* 2009;**27**:753–9.
82. Redding-Johanson AM, Batth TS, Chan R, Krupa R, Szmidt HL, Adams PD, et al. Targeted proteomics for metabolic pathway optimization: application to terpene production. *Metab Eng* 2011;**13**:194–203.
83. Dahl RH, Zhang F, Alonso-Gutierrez J, Baidoo E, Batth TS, Redding-Johanson AM, et al. Engineering dynamic pathway regulation using stress-response promoters. *Nat Biotechnol* 2013;**31**:1039–46.
84. Newman JD, Marshall J, Chang M, Nowroozi F, Paradise E, Pitera D, et al. High-level production of amorpha-4,11-diene in a two-phase partitioning bioreactor of metabolically engineered *Escherichia coli*. *Biotechnol Bioeng* 2006;**95**:684–91.
85. Tsuruta H, Paddon CJ, Eng D, Lenihan JR, Horning T, Anthony LC, et al. High-level production of amorpha-4,11-diene, a precursor of the antimalarial agent artemisinin, in *Escherichia coli*. *PLoS ONE* 2009;**4**:e4489.
86. Ro DK, Paradise EM, Ouellet M, Fisher KJ, Newman KL, Ndungu JM, et al. Production of the antimalarial drug precursor artemisinic acid in engineered yeast. *Nature* 2006;**440**:940–3.

87. Westfall PJ. Production of amorphadiene in yeast, and its conversion to dihydroartemisinic acid, precursor to the antimalarial agent artemisinin. *Proc Natl Acad Sci U S A* 2012;**109**:E111–8.
88. Paddon CJ, Keasling JD. Semi-synthetic artemisinin: a model for the use of synthetic biology in pharmaceutical development. *Nat Rev Microbiol* 2014;**12**:355–67.
89. Maury J, Asadollahi MA, Møller K, Schalk M, Clark A, Formenti LR, et al. Reconstruction of a bacterial isoprenoid biosynthetic pathway in *Saccharomyces cerevisiae*. *FEBS Lett* 2008;**582**:4032–8.
90. Misawa N. Pathway engineering for functional isoprenoids. *Curr Opin Biotechnol* 2011;**22**:627–33.
91. Chandran SS, Kealey JT, Reeves CD. Microbial production of isoprenoids. *Process Biochem* 2011;**46**:1703–10.
92. Amyris. *Amyris-news-amyris and soliance partner to commercialize renewable bio-sourced cosmetics*; 2010. http://www.amyris.com/News/145/Amyris-and-Soliance-Partner-to-Commercialize-Renewable-Bio-Sourced-Cosmetics#downloads.
93. Amyris. *Amyris-news-Amyris and Kuraray expand collaboration in farnesene-based polymers*; 2014. http://www.amyris.com/News/368/Amyris-and-Kuraray-Expand-Collaboration-in-Farnesene-Based-Polymers.
94. Renninger N, McPhee D. Petroleum component, fuel additive and microorganism produced isoprenoids; conversion of simple sugars, polysaccharides and/or nonfermentable carbon sources; diesel fuel, jet fuel, kerosene or gasoline replacements. US 20080098645; 2008.
95. Crock J, Wildung M, Croteau R. Isolation and bacterial expression of a sesquiterpene synthase cDNA clone from peppermint (*Mentha×piperita*, L.) that produces the aphid alarm pheromone (E)-beta-farnesene. *Proc Natl Acad Sci U S A* 1997;**94**:12833–8.
96. Picaud S, Brodelius M, Brodelius PE. Expression, purification and characterization of recombinant (E)-beta-farnesene synthase from *Artemisia annua*. *Phytochemistry* 2005;**66**:961–7.
97. Martin D, Fäldt J, Bohlmann J. Functional characterization of nine Norway spruce TPS genes and evolution of gymnosperm terpene synthases of the TPS-d subfamily. *Plant Physiol* 2004;**135**:1908–27.
98. Köllner TG, Gershenzon J, Degenhardt J. Molecular and biochemical evolution of maize terpene synthase 10, an enzyme of indirect defense. *Phytochemistry* 2009;**70**:1139–45.
99. Maruyama T, Ito M, Honda G. Molecular cloning, functional expression and characterization of (E)-beta farnesene synthase from *Citrus junos*. *Biol Pharm Bull* 2001;**24**:1171–5.
100. Wang C, Yoon S-H, Jang H-J, Chung Y-R, Kim J-Y, Choi E-S, et al. Metabolic engineering of *Escherichia coli* for α-farnesene production. *Metab Eng* 2011;**13**:648–55.
101. Pray T. *Technical advisory committee: drop-in fuels panel director of program operations*; 2010. http://www.biomassboard.gov/pdfs/biomass_tac_todd_pray_09_29_2010.pdf.
102. Lu F, Teal PEA. Sex pheromone components in oral secretions and crop of male Caribbean fruit flies, *Anastrepha suspensa* (loew). *Arch Insect Biochem Physiol* 2001;**48**:144–54.
103. Peralta-Yahya PP, Ouellet M, Chan R, Mukhopadhyay A, Keasling JD, Lee TS. Identification and microbial production of a terpene-based advanced biofuel. *Nat Commun* 2011;**2**:483.
104. Özaydın B, Burd H, Lee TS, Keasling JD. Carotenoid-based phenotypic screen of the yeast deletion collection reveals new genes with roles in isoprenoid production. *Metab Eng* 2013;**15**:174–83.
105. Wang C, Kim J, Choi E, Kim S. Microbial production of farnesol (FOH): current states and beyond. *Process Biochem* 2011;**46**:1221–9.
106. Inoue Y, Shiraishi A, Hada T, Hirose K, Hamashima H, Shimada J. The antibacterial effects of terpene alcohols on *Staphylococcus aureus* and their mode of action. *FEMS Microbiol Lett* 2004;**237**:325–31.
107. Goldberg L, Haklai R, Bauer V, Heiss A, Kloog Y. New derivatives of farnesylthiosalicylic acid (salirasib) for cancer treatment: farnesylthiosalicylamide inhibits tumor growth in nude mice models. *J Med Chem* 2009;**52**:197–205.
108. Wiseman D, Werner S, Crowell P. Cell cycle arrest by the isoprenoids perillyl alcohol, geraniol, and farnesol is mediated by p21Cip1 and p27Kip1 in human pancreatic adenocarcinoma cells. *J Pharmacol Exp Ther* 2007;**320**:1163–70.
109. Grace MH. Chemical composition and biological activity of the volatiles of *Anthemis melampodina* and *Pluchea dioscoridis*. *Phytother Res* 2002;**16**:183–5.

110. Wang C, Yoon SH, Shah AA, Chung YR, Kim JY, Choi ES, et al. Farnesol production from *Escherichia coli* by harnessing the exogenous mevalonate pathway. *Biotechnol Bioeng* 2010;**107**:421–9.
111. Meylemans HA, Quintana RL, Goldsmith BR, Harvey BG. Solvent-free conversion of linalool to methylcyclopentadiene dimers: a route to renewable high-density fuels. *ChemSusChem* 2011;**4**:465–9.
112. Tracy NI, Chen D, Crunkleton DW, Price GL. Hydrogenated monoterpenes as diesel fuel additives. *Fuel* 2009;**88**:2238–40.
113. Alonso-Gutierrez J, Chan R, Batth TS, Adams PD, Keasling JD, Petzold CJ, et al. Metabolic engineering of *Escherichia coli* for limonene and perillyl alcohol production. *Metab Eng* 2013;**19**:33–41.
114. Sarria S, Wong B, García Martín H, Keasling JD, Peralta-Yahya PP. Microbial synthesis of pinene. *ACS Synth Biol* 2014;**3**:466–75.
115. Carter OA, Peters RJ, Croteau R. Monoterpene biosynthesis pathway construction in *Escherichia coli*. *Phytochemistry* 2003;**64**:425–33.
116. Duetz WA, Bouwmeester H, van Beilen JB, Witholt B. Biotransformation of limonene by bacteria, fungi, yeasts, and plants. *Appl Microbiol Biotechnol* 2003;**61**:269–77.
117. Yang J, Nie Q, Ren M, Feng H, Jiang X, Zheng Y, et al. Metabolic engineering of *Escherichia coli* for the biosynthesis of alpha-pinene. *Biotechnol Biofuels* 2013;**6**:1.
118. Zhou J, Wang C, Yoon SH, Jang HJ, Choi ES, Kim SW. Engineering *Escherichia coli* for selective geraniol production with minimized endogenous dehydrogenation. *J Biotechnol* 2014;**169**:42–50.
119. Trombetta D, Castelli F, Sarpietro MG, Venuti V, Cristani M, Daniele C, et al. Mechanisms of antibacterial action of three monoterpenes. *Antimicrob Agents Chemother* 2005;**49**:2474–8.
120. Souza AB, de Souza MGM, Moreira MA, Moreira MR, Furtado NAJC, Martins CHG, et al. Antimicrobial evaluation of diterpenes from *Copaifera langsdorffii* oleoresin against periodontal anaerobic bacteria. *Molecules* 2011;**16**:9611–9.
121. Fonseca AP, Estrela FT, Moraes TS, Carneiro LJ, Bastos JK, dos Santos RA, et al. *In vitro* antimicrobial activity of plant-derived diterpenes against bovine mastitis bacteria. *Molecules* 2013;**18**:7865–72.
122. Dunlop MJ, Dossani ZY, Szmidt HL, Chu HC, Lee TS, Keasling JD, et al. Engineering microbial biofuel tolerance and export using efflux pumps. *Mol Syst Biol* 2011;**7**:487.
123. Putman M, van Veen HW, Konings WN. Molecular properties of bacterial multidrug transporters. *Microbiol Mol Biol Rev* 2000;**64**:672–93.
124. Wagner S, Klepsch MM, Schlegel S, Appel A, Draheim R, Tarry M, et al. Tuning *Escherichia coli* for membrane protein overexpression. *Proc Natl Acad Sci U S A* 2008;**105**:14371–6.
125. Patnaik R, Louie S, Gavrilovic V, Perry K, Stemmer WPC, Ryan CM, et al. Genome shuffling of lactobacillus for improved acid tolerance. *Nat Biotechnol* 2002;**20**:707–12.
126. Dunlop MJ. Engineering microbes for tolerance to next-generation biofuels. *Biotechnol Biofuels* 2011;**4**:32.
127. Mirata AM, Heerd D, Schrader J. Integrated bioprocess for the oxidation of limonene to perillic acid with *Pseudomonas putida* DSM 12264. *Process Biochem* 2009;**44**:8.
128. Rühl J, Schmid A, Blank LM. Selected *Pseudomonas putida* strains able to grow in the presence of high butanol concentrations. *Appl Environ Microbiol* 2009;**75**:4653–6.
129. Atsumi S, Hanai T, Liao JC. Non-fermentative pathways for synthesis of branched-chain higher alcohols as biofuels. *Nature* 2008;**451**:86–9.
130. Chou HH, Keasling JD. Synthetic pathway for production of five-carbon alcohols from isopentenyl diphosphate. *Appl Environ Microbiol* 2012;**78**:7849–55.
131. Cann AF, Liao JC. Pentanol isomer synthesis in engineered microorganisms. *Appl Microbiol Biotechnol* 2010;**85**:893–9.
132. Hull A, Golubkov I, Kronberg B, Marandzheva T, van Stam J. An alternative fuel for spark ignition engines. *Int J Engine Res* 2006;**7**:203–14.
133. Mack JH, Rapp VH, Broeckelmann M, Lee TS, Dibble RW. Investigation of biofuels from microorganism metabolism for use as anti-knock additives. *Fuel* 2014;**117**:939–43.

134. Withers ST, Gottlieb SS, Lieu B, Newman JD, Keasling JD. Identification of isopentenol biosynthetic genes from *Bacillus subtilis* by a screening method based on isoprenoid precursor toxicity. *Appl Environ Microbiol* 2007;**73**:6277–83.
135. Zheng Y, Liu Q, Li L, Qin W, Yang J, Zhang H, et al. Metabolic engineering of *Escherichia coli* for high-specificity production of isoprenol and prenol as next generation of biofuels. *Biotechnol Biofuels* 2013;**6**:57.
136. George KW, Chen A, Jain A, Batth TS, Baidoo EEK, Wang G, et al. Correlation analysis of targeted proteins and metabolites to assess and engineer microbial isopentenol production. *Biotechnol Bioeng* 2014;**111**:1648–58.

CHAPTER 3

Metabolic Engineering for Fatty Acid and Biodiesel Production

A. Hill*, H.S. Alper*,†

*McKetta Department of Chemical Engineering, The University of Texas at Austin, Austin, Texas, USA

†Institute for Cellular and Molecular Biology, The University of Texas at Austin, Austin, Texas, USA

3.1 Introduction

Petroleum has a footprint on everything in our day to day lives including transportation fuels, chemicals, and pharmaceuticals. However, finite global petroleum reserves and increasing costs spur the development of renewable processes for functionally equivalent (or ideally, superior) oleochemicals and fuels. While many advances have been made to source alternative energy, in 2013, less than 10% of the energy consumed in the United States originated from renewable sources.[1] Furthermore, adoption of bio-based alternatives in the transportation and energy sector has been slow, especially due to the need to be cost-competitive. For this reason, alternative processes for fuels and oleochemicals remain scientifically complex and economically challenging.

Finding an alternative to natural petroleum has taken two predominant routes: drop-in substitutes that are chemically distinct from petroleum and petroleum mimics (such as hydrocarbons). Natural petroleum was formed over long time frames from biomass exposed to extreme temperatures and pressures. Synthetic, biological routes to mimic this process in industrial scale time frames are nonetheless challenging. However, one such solution has been to engineer organisms for the production of fatty acids, natural hydrocarbons, which can be readily converted into biodiesel through chemical conversion. Biodiesel is a fatty acid ester that is produced by trans-esterifying cellular lipids with alcohols, most commonly methanol or ethanol. The resulting product is similar to conventional petroleum-derived diesel and can be used as a promising drop-in fuel alternative.

Unfortunately, current biodiesel production primarily utilizes oils extracted from plant and animal sources, straining the global food supply and requiring long harvesting times in only limited climates. For this reason, there is a great interest in developing microbes as surrogate producers. This can be achieved through either direct cellular conversion of sugar into biodiesel, or chemical conversion of cellular lipids into biodiesel. Moreover, the potential to modify these fatty acids and lipids *in vivo* also raises the possibility to customize fuel

Biotechnology for Biofuel Production and Optimization. http://dx.doi.org/10.1016/B978-0-444-63475-7.00003-0

compositions and performance. Additionally, lipid metabolism provides a means to convert cellular fatty acids into other useful oleochemicals such as alkanes, wax esters, fatty alcohols, and modified fatty acids. Here, we describe recent advances in engineering lipid metabolism for fuels and chemicals production after first providing an overview of lipid biochemistry and metabolism.

3.2 Overview of Lipid Metabolism

Despite the complexity and diversity of lipids produced across organisms, many of these chemical species are quite similar in that they contain aliphatic chains that are linked. These aliphatic chains, also called acyl chains, are synthesized by a multidomain enzyme complex called the fatty acid synthase (FAS). The FAS complex synthesizes fatty acids via the cyclical and incremental addition of malonyl-CoA to a bound, growing fatty acid chain. This process (Figure 3.1) is highly energy-intensive requiring two NADPH per elongation cycle (as an example, 14 NADPH are required for the elongation to palmitic acid).

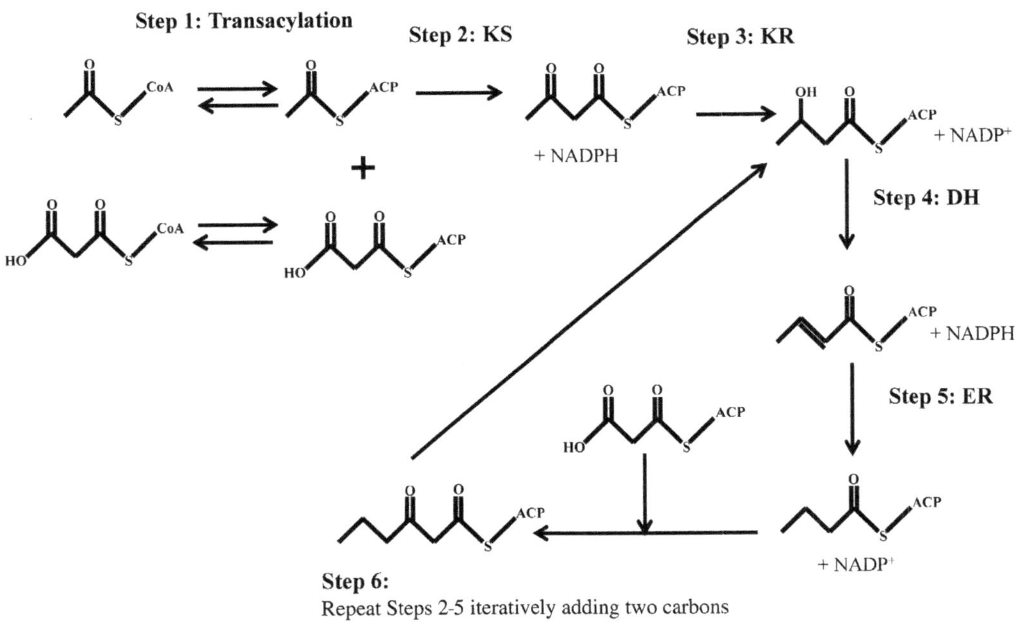

Figure 3.1

Reactive steps catalyzed by the FAS complex in fatty acid chain elongation. First, acetyl-CoA and malonyl-CoA are transacylated to the ACP domain of the complex (Step 1). Next, the acetyl-ACP and malonyl-ACP are condensed to form β-ketoacyl-ACP (Step 2). The β-ketone is then reduced to form β-hydroxyacyl-ACP (Step 3), which is then converted to enoyl-ACP by the dehydratase domain (Step 4). Finally, enoyl-ACP is reduced to form a saturated acyl chain two carbons longer than the original primer (2 carbons for acetyl-CoA) and can be subjected to further cycles to produce longer fatty acids.

The first step of the cycle involves the transacylation of the primer acetyl-CoA and extender malonyl-CoA to the complex's mobile acyl carrier protein (ACP) domain, forming acetyl-ACP and malonyl-ACP respectively. Next, the β-ketoacyl-ACP synthetase (KS) domain condenses the acetyl-ACP (or acyl chain of variable length) with the extender malonyl-ACP to form an extended β-ketoacyl-ACP chain that is two carbons longer. Subsequently, the β-ketoacyl-ACP reductase (KR) domain reduces the β-ketone group to a secondary alcohol. Next, the β-hydroxyacyl-ACP dehydratase (DH) domain removes water from the acyl chain to form enoyl-ACP. In the last step of the cyclical elongation process, the *trans*-Δ^2-enoyl-ACP reductase (ER) reduces the enoyl-ACP to produce a saturated acyl chain that can serve as the primer for additional rounds of elongation. This process may be repeated until chain termination occurs through the activity of an acyl-ACP thioesterase (TE) that releases a free fatty acid (FFA) from the FAS complex.

Once synthesized, fatty acids can have various fates including existing within the cell as FFAs, serving as a substrate for further downstream modification, or incorporating into other lipid species (esp. triacylglycerides). In the latter case, the FFAs are activated by CoA to form acyl-CoAs, a potent species that can enable acyl-chain transfer to form neutral lipids such as triacylglycerols (TAGs) or polar lipids such as phospholipids (PLs). Each of these lipid classes has its own biosynthesis pathway and multiple intermediates that play a role in native cellular function.

Neutral lipids such as sterol esters (SEs) and TAGs are synthesized by the activity of acyltransferases. In TAG formation, glycerol-3-phosphate (G3P) serves as the chemical backbone and sequential addition of acyl chains form lysophosphatidic acid (LPA), diacylglycerol (DAG), and finally the TAG (Figure 3.2). Since TAGs are primarily used as energy storage for cells, they are commonly stored for future use as aggregates in lipid bodies. Similarly, SEs are formed through the activities of enzymes such as an acyl-CoA:cholesterol acyltransferase (ARE1), which esterifies the hydroxyl group of the sterol with the aliphatic chain of acyl-CoA. Upon hydrolysis, free sterols can be incorporated into membranes to alter their stability, fluidity, and permeability.[2]

Another major class of lipids, called PLs, plays an important role in the formation of cellular membranes due to their amphiphilic nature. PLs are generally distinguished by modifications to their polar component and are commonly derived from phosphatidic acid (PA), an intermediate in TAG biosynthesis. During PL synthesis, PA is first converted into cytidine diphosphate (CDP) DAG through the activity of CDP-DAG synthase (CDS1), which attaches cytidine monophosphate (CMP) to the polar head of PA. CDP-DAG is then subjected to a series of modifications, as depicted in Figure 3.3, which culminates in the formation of phosphatidylcholine (PC).[3]

An alternative PC biosynthesis pathway has been identified that utilizes ethanolamine and choline as precursors. First, ethanolamine and choline are phosphorylated

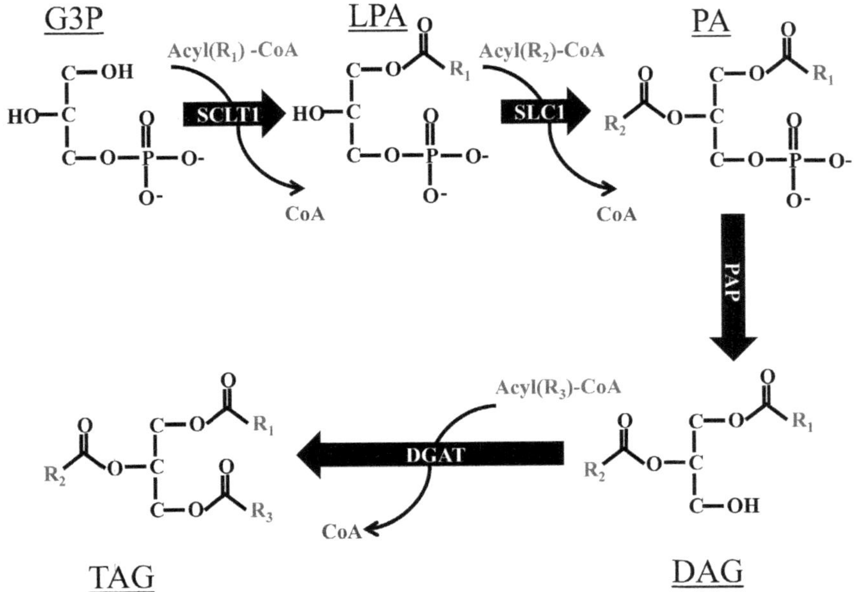

Figure 3.2
Triacylglycerol biosynthesis pathway. G3P serves as the backbone for glycerolipids and is first subjected to two acylations catalyzed by the G3P acyltransferase (SCT1) and 1-acyl-sn-glycerol-3-phosphate acyltransferase (SLC1) enzymes to form lysophosphatidic acid and PA, respectively. PA is then dephosphorylated by the phosphatidate phosphatase (PAP) to produce DAG that is further acylated into TAG.

by their respective kinases and then primed with cytidine triphosphate (CTP) by ethanolaminephosphate cytidylyltransferase and cholinephosphate cytidylyltransferase, respectively. Phosphoethanolamine and phosphocholine are then transferred to DAG by phosphotransferases to form phosphatidylethanolamine (PE) and PC, respectively. This pathway typically functions in an auxiliary manner, but is highly active in mutants unable to produce PC from CDP-DAG.[3]

In organisms, very long chain fatty acids (VLCFAs) of length greater than 20 carbons are produced via further elongation that is catalyzed by enzymes other than the FAS complex. Typically, the FAS complex produces cystolic fatty acids upward of 16-18 carbons in length for yeast species; however, yeasts also possess VLCFAs up to 26 carbons in length. In addition, systems for further elongation of fatty acids have been found in nature, such as the FAS-II system in *Mycobacterium tuberculosis*, which is capable of producing C_{48}-C_{56} fatty acid chains used in the synthesis of mycolic acids.[4,5] The production of these VLCFAs is catalyzed by a completely different set of enzymes located in the endoplasmic reticulum (ER) by membrane-bound enzymes that carry out chemically identical elongation reactions as those catalyzed by the FAS complex.[6] As an example, acyl-CoAs transferred to the ER can

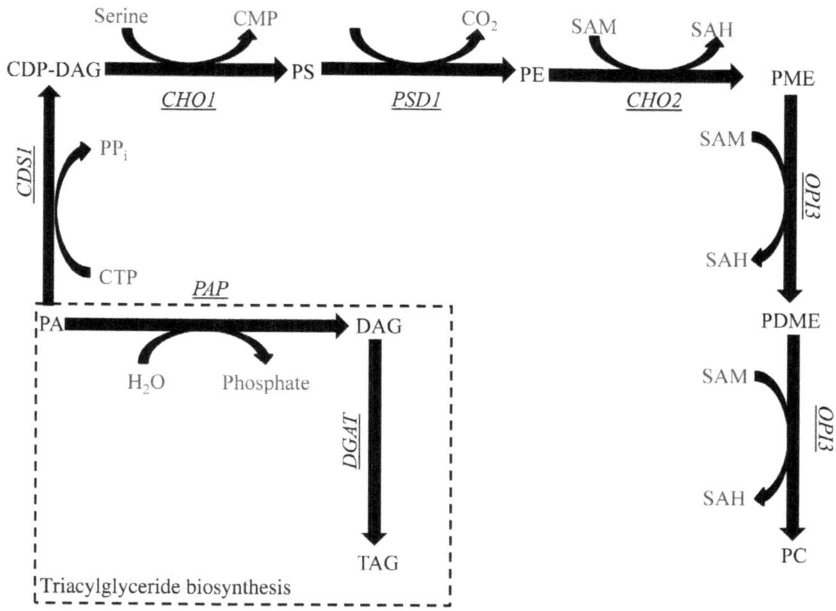

Figure 3.3

Phospholipid biosynthesis pathway. In phospholipid synthesis, PA that is not used for TAG synthesis (boxed) can be converted into various phospholipid species. First, PA is converted into CDP-DAG by the CDP-DAG synthase (CDS1). Next, the CMP head group is replaced with serine by phosphatidylserine (PS) synthase (CHO1) to form PS. PS is then decarboxylated by PS decarboxylase (PSD1) to form PE. PE is then methylated by phosphatidylethanolamine methyltransferase (CHO2) to form phosphatidylmonomethylethanolamine (PME), which is then subjected to two methylations that are catalyzed by methylene-fatty-acyl-phospholipid synthase (OPI3) to form phosphatidyldimethylethanolamine (PDME) and then phosphatidylcholine.

be modified by an elongase protein such as FEN1 or SUR4 to condense the acyl-chain with malonyl-CoA. It is important to note that while such VLCFAs are usually only present in small amounts, they are vital for cellular functions as demonstrated by the synthetic lethality of a *fen1*, *sur4* double mutant.[7]

While we focused a great deal on fatty acid and lipid anabolism, the catabolic processing of these molecules is also important in many organisms. The process responsible for degrading fatty acids, β-oxidation, mimics a reversed fatty acid synthesis process. In the first step of β-oxidation, an acyl-CoA oxidase oxidizes acyl-CoA into *trans*-Δ^2-enoyl-CoA which is then converted into β-ketoacyl-CoA through two enzymatic steps. The final step in each cycle of β-oxidation is catalyzed by a β-ketoacyl-CoA thiolase, which frees one molecule of acetyl-CoA from the acyl-CoA chain, which has been reduced in length by two carbons.[3] For each cycle of β-oxidation, one molecule of both $FADH_2$ and NADH are produced. This process can be repeated until the starting acyl-CoA chain has been reduced to either acetyl-CoA if the initial chain was even in length or propionyl-CoA if the initial number of carbons was odd.

Intracellular regulation of this pathway also plays a critical role in lipid accumulation. One such example is the feedback inhibition of the acetyl-CoA carboxylase (ACC) enzyme in *Escherichia coli* by acyl-ACP.[8] Since ACC activity is responsible for converting acetyl-CoA into malonyl-CoA, the building blocks of fatty acids, this regulation point represents an important challenge for engineering *E. coli* for the production of chemicals derived from fatty acids. Furthermore, it is recognized that the depletion of critical nutrients, particularly nitrogen, encourages lipid accumulation in oleaginous organisms. Under nitrogen starvation conditions, activity of an AMP deaminase (AMPD) cleaves AMP to replenish intracellular ammonium. The resulting loss of AMP inhibits isocitrate dehydrogenase and leads to the buildup of early TCA cycle intermediates, particularly citrate. Oleaginous organisms, such as *Yarrowia lipolytica*, possess an enzyme called ATP citrate lyase that converts the accumulated citrate into acetyl-CoA that can be used for fatty acid biosynthesis.[9] Additionally, evaluating the impact of leucine biosynthesis lipid accumulation in *Y. lipolytica* indicated that TOR-related regulation plays a role in lipogenesis.[10] These processes indicate that cellular regulation of fatty acid metabolism is highly important as well and should be addressed applying metabolic engineering to the production of fatty acids and their derivatives.

This section highlighted key elements of lipid and fatty acid anabolic and catabolic pathways (Figure 3.4), as well as some of the ways in which metabolic flux into fatty acid biosynthesis is regulated. It is important to note that various modifying enzymes will typically use one form of lipid rather than the others, whether that lipid is a FFA, acyl-CoA, acyl-ACP, or an acyl group linked to a PL. As a result, a basic understanding of lipid metabolism is important in generating strains for biodiesel and other oleochemicals. In the next sections, we highlight advances in the field involving the metabolic engineering of strains for efficient oleochemical production.

3.3 Modified Fatty Acid Species

Although fatty acids produced by the FAS complex are initially saturated, there are many modifying enzymes that alter and create fatty acid functional groups (Figure 3.5). While these modifications are important for maintaining membrane integrity and other essential biochemical functions, many of these modifications are also likewise important for the production of oleochemicals. The most common modification to fatty acids is the introduction of a double bond by a fatty acid desaturase (FAD), forming an unsaturated fatty acid. Typically, FADs introduce a *cis*-configuration double bond at a specific location within the chain. A common example of this process is the formation of oleic acid ($18{:}1^{\Delta 9}$) from stearic acid (18:0) through the activity of a Δ^9-desaturase, which introduces a *cis* double bond between the 9th and 10th carbons of stearic acid. The newly formed monounsaturated fatty acid may be further desaturated into a polyunsaturated fatty acid (PUFA) by additional desaturases; for instance, linoleic acid [$18{:}2^{\Delta 9,12}$] is formed from the desaturation of oleic

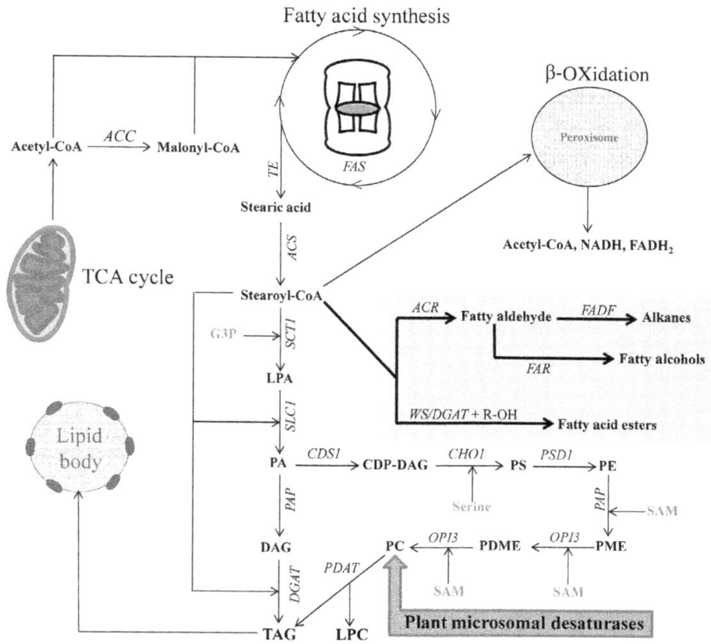

Figure 3.4

Overview of lipid metabolism. Acetyl-CoA produced through processes such as the TCA cycle is converted into malonyl-CoA, which is used by the FAS complex to elongate acetyl-CoA into fatty acids such as stearic acid. Once stearic acid is ligated to CoA, it can then be transferred to glycerol backbones by acyltransferases to form glycerolipids, converted into oleochemicals such as alkanes (highlighted), or degraded by peroxisomal β-oxidation. In glycerolipid synthesis, G3P is first acylated twice to form PA, which can be used to form phospholipids or TAGs. The ultimate product in phospholipid synthesis is PC, which can have its acyl chains modified by various plant FADs to form uncommon fatty acids. PA can also be dephosphorylated to form DAG, which can be converted into TAG by a third acylation catalyzed by either DGAT or PDAT. Once formed, TAGs can be stored in lipid bodies for later use. ACR, acyl-CoA reductase; FADF, fatty aldehyde deformylase; FAR, fatty aldehyde reductase; SAM, S-adenosylmethionine.

acid at the 12 carbon. This process can be repeated iteratively, with the double bonds typically being methylene separated, as can be seen in the case of important PUFAs such as Eicosapentaenoic acid [EPA, $20:5^{\Delta 5,8,11,14,17}$] docosahexaenoic acid [DHA, $22:6^{\Delta 4,7,10,13,16,19}$].

Although double bonds are typically spaced out by two carbon-carbon (C-C) bonds as described above, there are a number of natural fatty acids with a conjugated double-bond structure. Conjugated linoleic and linolenic acids (CLAs) are among the most well studied and have gained recognition for their health-related effects.[11] One such CLA is calendic acid [18:3(8E, 10E, 12Z)], a major lipid species in *Calendula officinalis*. Calendic acid is formed by the activity of a FAD-class enzyme that converts the *cis*-9 double bond of linoleic acid

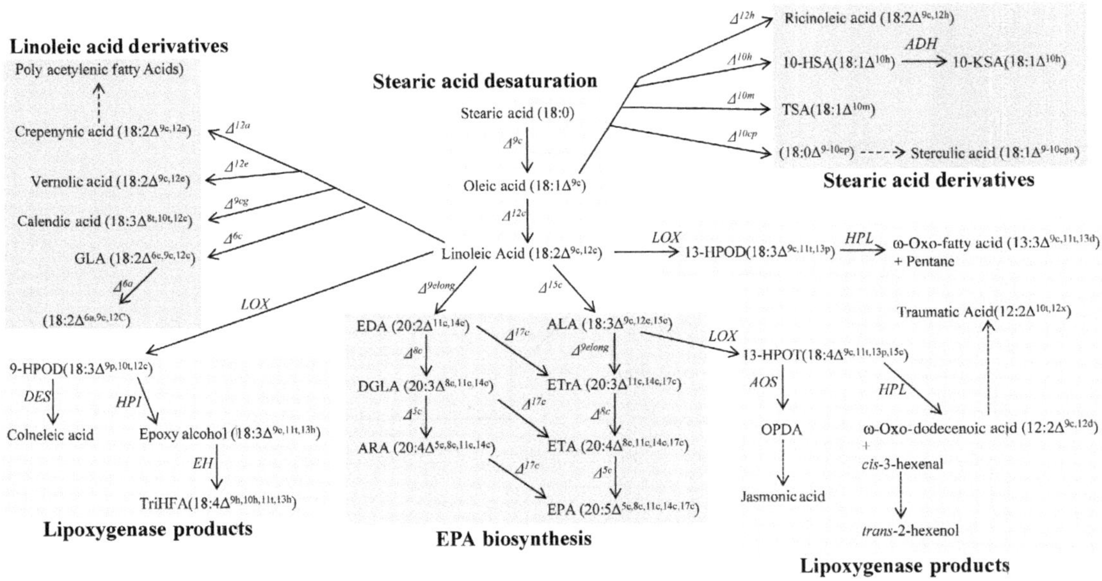

Figure 3.5

Overview of various biomodified fatty acids. The pathways for stearic acid desaturation into linoleic acid, EPA biosynthesis, lipoxygenase products, and the biomodification of oleic and linoleic acid are illustrated to illustrate the diversity of naturally occurring fatty acids. Fatty acid abbreviations are as follows: ALA, α-linolenic acid; EDA, eicosadienoic acid; ETrA, eicosatrienoic acid; ETA, eicosatetraenoic acid; DGLA, dihomo-γ-linolenic acid; ARA, arachidonic acid; HPOD, hydroperoxyoctadecadienoate; HPOT, hydroperoxyoctadecatrienoate; OPDA, 12-oxo-phytodienoic acid. Modifications introduced into the fatty acids are described in the form $\Delta^{\#x}$ where "#" is the site in the fatty acid chain and *x* is the modification where *x* is defined as follows: a, acetylene; c, *cis*; t, *trans*; h, hydroxy; e, epoxy; cp, cyclopropane; cpn, cyclopropene; m, methyl; p, hydroperoxy; d, aldehyde; o, oxo; b, carboxylic acid; cg, conjugase. This notation is also used to describe the enzymes that catalyze these conversions.

into conjugated *trans* double bonds.[12] Another CLA of interest is α-eleostearic acid [18:3(9Z, 11E, 13E)], which has potential to serve as a drying oil and is produced from linoleic acid in *Momordica charantia*.[13]

Although C-C double bonds are the most common functional group found in fatty acid chains, there are enzymes that can produce further desaturated C-C triple bonds. One such enzyme, which produces 6-ynoic acids, was isolated from *Ceratodon purpureus*. Through feeding assays, it was found that this enzyme exhibits substrate promiscuity and is able to introduce triple bonds into linoleic acid, γ-linolenic acid (GLA), and stearidonic acid (SDA).[14] Another acetylenic fatty acid is crepenynic acid [9-octadecen-12-ynoic acid], which is produced in *Crepis alpina* and accounts for roughly 70% of the plant's seed oil. Researchers isolated a FAD-like enzyme further desaturates the 12Z double bond of linoleic acid to form crepenynic acid.[15] Crepenynic acid also feeds into the production of numerous acetylenic products including diverse polyacetylene derivatives[16] that are of economic interest due to their antitumor, antibacterial, antimicrobial, antifungal, and immunosuppressive properties.[17]

Beyond saturation level, fatty acids may be modified through additional functional groups. For instance, castor oil is primarily comprised of the hydroxylated octadecanoic acid called ricinoleic acid, which has unique properties as a result of a higher polarity over traditional fatty acids due to the secondary alcohol.[18] It has been determined that ricinoleic acid in *Ricinus communis* is formed by introducing a hydroxyl moiety into oleic acid.[19] These hydroxy fatty acids can be useful as an intermediate to other chemicals as seen with the oleaginous yeast *Y. lipolytica* that can convert ricinoleic acid into the fragrance compound, γ-decalactone, through β-oxidation.[20] In another instance, an oleate hydratase can use water to convert the oleic acid into 10-hydroxy stearic acid (10-HSA). Additionally, cytochrome P450s catalyze the ω, ω-(1,2, or 3), and α oxidation of fatty acids. Natural fatty acids containing both hydroxyl and ketone functional groups, such as 14-hydroxy-15-ketohexadecanoic acid, are produced by a cytochrome P450 in *Bacillus megaterium*.[21] A final form of oxygenation is through the formation of epoxide rings. One such epoxy fatty acid, vernolic acid, is found as a high component of lipids in the seeds of the *Stokesia*, *Vernonia*, and *Crepis* genera and is produced from linoleic acid through the activity of a P450 monooxygenase. Although native producers are able to accumulate large quantities of vernolic acid in their seeds, transgenic plants engineered with this enzyme exhibited severely impaired production.[22] This difficulty demonstrates the fundamental differences in the lipid metabolisms across species that must be addressed in order to aid efficient metabolic engineering efforts.

A wide variety of functional groups can be introduced into fatty acids through the activity of a class of enzymes known as lipoxygenases (LOXs). These enzymes catalyze the dioxygenation of PUFAs, introducing hydrogen peroxides into the acyl chains. Although the resulting hydroperoxy fatty acids can serve as signaling molecules, they often serve as intermediates in pathways for synthesizing other fatty acids or oxylipins.[23] Once formed, hydroperoxy fatty acids can be cleaved by hydroperoxide lyases (HPLs) into C_6 alcohols and ω-oxo-fatty acids that can be further modified under certain conditions.[24,25] Additionally, a LOX isolated from soybeans has been able to catalyze the formation of pentane rather than C_6 alcohols.[26] This is particularly appealing since pentane has the potential to serve as an alternative biofuel. Additional LOX-derived productions include jasmonates,[24,27] divinyl ester fatty acids,[28] epoxyalcohols,[29] and trihydroxy fatty acids (TriHFAs)[29] that result from the activities of allene oxide synthases (AOSs), divinyl ester synthases (DESs), hydroperoxide isomerases (HPIs), and epoxide hydrolases (EHs), respectively.[24,28,29]

In addition to oxygenation, methylation of fatty acids may be seen. One well-known and well-characterized methylated fatty acid is tuberculostearic acid (TSA). This molecule is formed by the activity of a fatty acid methyltransferase that transfers a methyl group from

S-adenosylmethionine (SAM) to the double bond in oleic acids linked to PLs.[30] Some organisms possess a FAS that produces multimethyl-branched fatty acids by utilizing methylmalonyl-CoA rather than malonyl-CoA as the extender unit.[31] Furthermore, methylations of fatty acids can also lead to the formation of cyclopropane rings within the fatty acids through the activity of cyclopropane synthases.[32] The presence of this ring within the fatty acid chains imparts chemical properties resembling both saturated and unsaturated fatty acids, enabling their use in many commercial applications.[33] Furthermore, the resulting cyclopropane rings in fatty acids are also believed to be an intermediate in the production of the cyclopropene containing sterculic acid found in *Sterculia foetida*.[34]

Although the previously discussed modifications to fatty acids have all involved the addition of carbon, hydrogen, and/or oxygen, sulfur containing groups can be introduced into acyl chains as well. As an example, lipoic acid (6,8-thioctic acid) is an interesting antioxidant molecule and contains a disulfide bond. Lipoic acid is produced in the mitochondria by lipoic acid synthase, which adds two thiol functional groups to caprylic acid.[35]

As demonstrated in this section, nascent fatty acids may be modified with alternative chemistry to include desaturation, hydroxylation, methylation, epoxidation, acetylenation, cyclopropanation, and hydroperoxidation. These modifications are important in the production of downstream natural products important in varied industries with a variety of applications such as lubricants, plasticizers, adhesives, and coatings.[36] Furthermore, the presence of these functional groups enables the use of organic chemistry to modify fatty acids *in vitro* to new products. For instance, crepenynic acid can be converted into chemicals useful in coatings and cold-weather lubricants through alkali isomerization.[15] As a result, modified fatty acids, if produced in meaningful quantities, have the potential to fill diverse chemical applications.

3.4 Conversion of Lipids and Fatty Acids into Biodiesel

In addition to the introduction of functional groups, fatty acids can also be converted into other oleochemicals through enzymatic or chemical modification such as transesterification to fatty acid esters, reduction to alcohols and aldehydes, reduction and deformylation to alkanes/alkenes, decarbonylation to α-olefins, the formation of β-ketones from β-keto acids, and the polymerization of hydroxylated acyl-CoAs to form polyhydroxyalkanoates (PHAs).[37] Furthermore, the basic building blocks of fatty acids such as G3P and acetyl-CoA can be used to synthesize isoprenoids that can serve as a form/substitute of biodiesel.[38] Each of these chemical species serves a number of roles in chemical industries and thus provides attractive targets for renewable production in microbial organisms. This section will focus on bioengineering organisms for the production of biodiesel as an alternative fuel.

Diesel is a petroleum-derived fuel source that, like gasoline, contains a mixture of alkanes and aromatics. One difference between diesel and gasoline fuels is the size of the hydrocarbons contained within them; gasoline contains primarily C_4-C_{12} hydrocarbons whereas the average hydrocarbon in diesel contains 16 carbons.[38] Due to the use of diesel in compression combustion engines, biodiesels must match diesel in several properties including cetane number, viscosity, cloud point, and water miscibility.[39,40] For these reasons, molecules such as fatty acid esters and some isoprenoids have been evaluated for their potential use as biodiesel. Engineering efforts are required to produce these molecules at scale since they only occur in very low levels in native organisms.

Fatty acid esters are of particular interest as fatty acids that are esterified with methanol (FAMEs) and ethanol (FAEEs) can be used as biodiesel. Although current biodiesel is produced by chemical reactions of the alcohols with a variety of plant and animal fats/oils (e.g., soybean oil), there are a variety of enzymes in nature that can catalyze this same reaction. In one example, a fatty acid 0-methyltransferase (FAMT) from *Mycobacterium marinum* is able to methylate FFAs using SAM as the methyl donor. Unfortunately, researchers found that that this enzyme was most active on medium chain fatty acids (C_8-C_{12}) that were both saturated and 3-hydroxylated.[37] As such, the resulting biodiesel would be highly different than conventional soybean biodiesel, which is composed of predominantly long-chain fatty acids (C_{16}-C_{18}) and more than 75% unsaturated fatty acids.[10] Additionally, the process faced limitations due to the thioesterase's poor ability to release the preferred 3-hydroxydecanoic acid from the FAS complex.[41] As such, the use of this pathway for biodiesel production would require protein engineering of the various pathway enzymes, as well as further metabolic engineering to ensure that all of the needed metabolites are produced (e.g., SAM) to be economical.

As an alternate means for biodiesel production, wax ester synthase/acyl-CoA:diacylglycerol acyltransferase (WS/DGAT) enzymes from various organisms have been used to produce FAEEs. These enzymes function by esterifying an alcohol to acyl-CoA, resulting in FAEEs. These enzymes are particularly interesting due to their promiscuity, transferring acyl chains to alcohols ranging in sizes from ethanol to octadecanol when palmitoyl-CoA was used as the acyl donor. Additionally, when FAEEs were produced *in vivo*, the researchers found that the yeast was able to produce FAEEs ranging from C_{12} to C_{18} in chain length, indicating that this promiscuity applies to the fatty acid chains as well. Although the combination of these promiscuous activities was not tested, it seems likely that these enzymes would be able to produce diverse fatty acid esters by combining different fatty acids and alcohols.[42] A demonstration of the promiscuity of these enzymes was performed in *E. coli* using the alcohol *O*-acyltransferase (ATF) from *Saccharomyces cerevisiae* to produce a variety of volatile esters using different acyl-CoAs and alcohols for their formation.[43] Furthermore, an additional study enabled the production of diverse butyrate esters using a strawberry alcohol acyltransferase.[44]

In recent years, there have been numerous studies evaluating the production of heterologous lipid species in *S. cerevisiae*. One such target for production has been FAEEs, which was achieved through the overexpression of heterologous enzymes such as those described above. Unfortunately, even after extensive genetic engineering, FAEE production in *S. cerevisiae* has been limited to 25.4 mg/L of FAEEs.[45] In a different study, the acyl-CoA reductase required for fatty alcohol production was introduced into a strain engineered for fatty acid production to produce fatty alcohols at around 90 mg/L.[46]

Despite the fermentation advantages yeast possess compared to bacteria, the production of these chemicals in *E. coli* has had considerably more success, producing nearly 700 mg/L of FAEEs[47] and more than 350 mg/L of fatty alcohols.[48] This was improved upon by incorporating a sensor-regulator system capable of responding to the production of intermediate metabolites, enabling FAEE production in *E. coli* at 1.5 g/L and 28% of the theoretical yield.[49] Nevertheless, with the current cost of sugar, these chemicals would need to be produced at nearly 95% of the theoretical stoichiometric yield to be competitive with the price conventional fuels.[37] For a comparison, one of the highest production titers of FAEEs via microbial conversion only achieved roughly 9.4% of the maximum theoretical yield, far below the efficiency needed.[47] In order to close this gap, it is likely desirable to not only find the optimal genetic modifications but also to attempt production in an ideal host organism.

3.5 Conversion of Fatty Acids into Additional Oleochemicals

Beyond biodiesel, metabolic engineering has the potential to convert fatty acids and lipids into additional drop-in fuel alternatives through the direct synthesis of alkanes and alkenes. A pathway for alkane biosynthesis was identified by assembling an acyl-ACP reductase from *Synechococcus elongatus* and an aldehyde deformylase from *Nostoc punctiforme* into *E. coli*.[50] In this pathway, an acyl chain linked to the ACP is reduced to its corresponding fatty aldehyde. The released aldehyde is then deformylated to produce an alkane with a reduced carbon length as a result of formate production. The result of this pathway is the production of C_{n-1} alkanes from C_n acyl chains and the predominant lengths in *E. coli* were greater than 15 carbons in length.

As an alternative, another study evaluated a different version of the first enzyme, which enables the production of fatty aldehydes that uses acyl-CoA instead of acyl-ACP.[48,51] This pathway has the potential to produce a more diverse array of chemicals since acyl-CoAs can be subjected to a greater degree of modification than acyl chains linked to ACP. For instance, the pathway for alkane biosynthesis was coupled with the expression of a mutant TesA thioesterase resulting in the production of measurable quantities of C_8-C_{14} alkanes/alkenes.[51] This resulting composition could establish a better drop-in fuel alternative since most of the paraffins and olefins found in gasoline are closer to this range of chain length.

The intersection of β-oxidation and fatty acid synthesis can produce a number of interesting intermediates that can be converted into chemicals. Some of these chemicals include β-keto acids, β-keto alcohols, 1,3-diols, β-hydroxyacids, *trans*-Δ^2-fatty acids, and *trans*-Δ^2-fatty alcohols. By engineering a process for reverse β-oxidation, researchers were able to utilize the intermediate acyl, β-ketoacyl, β-hydroxyacyl, and *trans*-Δ^2-enoyl CoA derivatives for the production of other chemicals.[52] Through strain engineering, researchers were able to convert a butyryl derivative of these products into butanol at a titer of 14 g/L in a bioreactor. In addition, incorporation of a 3-ketoacyl-CoA thiolase capable of accepting a broader range of substrates enabled the production of *n*-alcohols up to 10 carbons in length.[52] Furthermore, by feeding a precursor of propionyl-CoA, the production of odd-chain *n*-alcohols was enabled.[52]

As demonstrated in this section, several pathways for hydrocarbons can be introduced into microorganisms. These resulting chemicals are attractive for a number of fuel and chemical applications and industrial-scale production has the potential to alleviate dependency on petrochemicals. Currently, the bio-production of these chemicals is far too inefficient to be cost-competitive with traditional petroleum derived chemicals and fuels. One particular reason for this inefficiency is host organism engineering, a topic we turn to in this next section.

3.6 Host Organism

The choice of host organism greatly impacts the production level of fatty acids and their derivatives. Although organisms such as *S. cerevisiae* and *E. coli* are appealing due to genetic tractability, they natively produce very low levels of lipids compared to other organisms. For comparison, lipids typically comprise roughly 9.7% of *E. coli*'s mass[47] and less than 15% of *S. cerevisiae*'s biomass;[9] however, there are several organisms such as *Cryptococcus curvatus*, *Lipomyces starkeyi*, and *Rhodotorula glutinis* that can accumulate lipids at more than 20% of their biomass and are referred to as "oleaginous".[9,53–55] Additionally, plant engineering has been explored due to differences in lipid metabolism between plants and microorganisms such as yeast.[56] Finally, there is an interest in enabling lipid and biofuel production in various microalgae and cyanobacteria due to their carbon fixation capacity.[57,58] A summary of various engineering efforts for the production of oleochemicals is provided in Table 3.1. We highlight efforts in each of these hosts below.

E. coli has been a widely utilized organism for metabolic engineering and not surprisingly for fatty acid production. In addition to fatty acids themselves, research has explored the production of short and medium chain alcohols, fatty acid esters, and alkanes/alkenes.[50,59] Despite the advantages associated with *E. coli*, oleochemical production titers have only been in the milligram to low gram per liter range.[59] Additional process considerations such as possible phage contaminations and higher temperatures/pH values have limited the attractiveness of this organism for these chemicals.

One of the most appealing yeast organisms for metabolic engineering is *S. cerevisiae*. Although less modified strains of *S. cerevisiae* typically produce low quantities of TAGs, it

Table 3.1: Applications of metabolic engineering for oleochemical production: Summary of metabolic engineering studies of various organisms for the production of oleochemicals including the organism, modifications, and quantity produced

Organism	Type of molecule	Modifications	Production vessel	Production level	Reference
		Bacteria			
E. coli	*n*-Butanol	Reverse β-oxidation	Bioreactor	14 g/L	Dellomonaco[52]
	Fatty alcohols	Reverse β-oxidation	Flask	0.33 g/L	Dellomonaco[52]
	Fatty acids	Reverse β-oxidation	Bioreactor	7 g/L	Dellomonaco[52]
	β-Ketobutyric acid	Reverse β-oxidation	Flask	500 mg/L	Dellomonaco[52]
	β-Hydroxybutyric acid	Reverse β-oxidation	Flask	150 mg/L	Dellomonaco[52]
	Trans-2-butenoic acid	Reverse β-oxidation	Flask	200 mg/L	Dellomonaco[52]
	FAEEs	Wax-ester synthase	Flask	674 mg/L	Steen[47]
	Short-chain alkanes	Thioesterase + alkane pathway	Bioreactor	580.8 mg/L	Choi[51]
		Yeast			
S. cerevisiae	Fatty acids	Truncated DGA1 overexpression	Flask	1.14 g/L	Kamisaka[60]
	FAEEs	Acyltransferase expression and media optimization	Flask	25 mg/L	Thompson[45]
	FAEEs	WS-DGAT expression	Flask	5 mg/L	Runguphan[46]
	Fatty alcohols	Fatty acyl-CoA reductase expression	Flask	90 mg/L	Runguphan[46]
Y. lipolytica	Fatty acids	DGA1 overexpression and impairing β-oxidation	Bioreactor	25 g/L	Blazeck[10]
	Pentane	Lipoxygenase pathway	Bottle	4.98 mg/L	Blazeck[69]
	EPA	Heterologous desaturases	Flasks	15% dcw	Xue[68]
		Plants			
G. max	Vernolic acid	Heterologous epoxygenase and DGAT2 expression	Plant seeds	25.8% total fatty acids	Li[73]
A. thaliana	Hydroxy fatty acids	Heterologous hydroxylase and DGAT2 expression	Plant seeds	20–25% total fatty acids	Burgal[75]

Organism	Type of molecule	Modifications	Production vessel	Production level	Reference
		Microalgae			
Phaeodactylum tricornutum	EPA	Δ^5 desaturase overexpression	Flask	38.9 mg/g dcw	Peng[81]
		Cyanobacteria			
Synechococcus sp. PCC 7002	Free fatty acids	Thioesterase expression	Bottle with CO_2 feed	131 mg/L	Ruffing[85]
Synechocystis sp. PCC6803	ω-3 fatty acids	Heterologous Δ^6 and Δ^{15} desaturase expression	Flask	8.9 mg/L ALA	Chen[86]
	ω-3 fatty acids	Heterologous Δ^6 and Δ^{15} desaturase expression	Flask	4.1 mg/L SDA	Chen[86]
	Fatty alcohols	Acyl-CoA reductase expression, alkane biosynthesis knockouts	Flask	2.87 mg/g dcw	Yao[87]

was shown that overexpression of DGA1 could increase the organism's lipid content to over 30% of dry cell weight (dcw). By fermenting the cells in a higher concentration of glucose (increased from 2% to 10%), the lipid content was further improved to near 45% dcw with TAGs representing nearly 90% of the cell's lipids. However, despite these improvements, the fermentation only managed to produce 1.14 g/L of lipids using 100 g/L of glucose as raw material (roughly 5% of theoretical yield) after 7 days.[60] While significantly more work is required to transform *S. cerevisiae* into an industrial host for oleochemicals production, it has served as a useful model organism to study enzymes that modify fatty acids. As examples, *S. cerevisiae* has been used to evaluate enzymes for the production of unusual plant fatty acids such as ricinoleic acid,[19] vernolic acid,[15] calendic acid,[12] and crepenyic acid.[15] The low production in this host also highlights fundamental differences between yeast and plant lipid metabolism including lipid mobility and storage—critical targets for future metabolic engineering endeavors.

Beyond *S. cerevisiae*, researchers have begun to focus on oleaginous yeast due to their innate capacity for high lipogenesis. Studies of oleaginous species, similar to work in cyanobacteria, have established that lipogenesis is triggered by the availability of nutrients in the media, especially the carbon to nitrogen ratio (C:N ratio).[61] Under conditions in which this ratio is very high, also referred to as "nitrogen starvation," these organisms seek to produce more ammonium by catabolizing AMP. This action results in an inhibition of isocitrate dehydrogenase, leading to a shutdown of the TCA cycle and subsequent accumulation of citrate. Due to the presence of a unique enzyme in oleaginous yeasts,

ATP-citrate lyase (ACL), accumulated citrate as well as CoA is converted into oxaloacetate and acetyl-CoA. The acetyl-CoA can then be converted into malonyl-CoA, the building block of fatty acids, by ACC.[61] This alternative cycle and the activity of ACL are found to highly control lipogenesis.

One oleaginous organism, in particular, *Y. lipolytica*, is quickly becoming a model organism for oleaginous species. Although *Y. lipolytica* is only able to produce a modest amount of lipids through *de novo* biosynthesis (10-15% content),[10] the organism's genome sequence is published[62] and there are powerful tools for strong gene overexpression available enabling metabolic engineering efforts.[63,64] Several studies have been conducted to elucidate the mechanisms of lipid accumulation in this organism. One important study involved the transcriptomic analysis of *Y. lipolytica* during lipid accumulation to determine how gene expression changes during lipogenesis. A key finding was that lipogenesis is complex and that key enzymes such as AMPD, ACL, ACC, and malic enzyme (MAE) were not actually differentially expressed. [65] These results suggest that lipogenesis, even in oleaginous yeast, is highly complex. This study however did identify DGA2 (diacylglycerol acyltransferase) as an enzyme with the highest fold increase among lipid metabolism enzymes.[65]

Additional studies were also conducted that identified key lipogenesis enzymes through genetic engineering. In one such study, researchers performed gene deletions of the DGA1 and phospholipid:diacylglycerol acyltransferase (LRO1) enzymes, both of which catalyze the formation of TAGs, albeit through different mechanisms. Although the study showed that individual deletions of these enzymes had little impact, the double mutant was severely impaired in its ability to produce TAGs in glucose-containing media.[66] The importance of DGA1 activity for improving lipid accumulation in *Y. lipolytica* was later confirmed in separate studies that both substantially improved lipid synthesis by DGA1 overexpression.[10,67] In addition, one of these studies evaluated several gene knockouts which impaired β-oxidation and further improved lipid accumulation.[10] This has been harnessed to enable substantial titers of fatty acids,[10] EPA,[68] and to enable the first microbial production of pentane.[69]

A variety of photosynthetic organisms have also been explored as a potential source of fatty acids and lipids. The most well-known photosynthetic organisms belong to the plant kingdom of biology. Not only are plants able to harness vast amounts of energy from the sun to produce the biomass, they also serve as native producers of oleochemicals and have been the source of oils for current biodiesel processes. Even though plants and seeds are native producers of key oils, several intrinsic problems limit full-scale industrial production. For instance, some of these plants strongly compete with conventional agriculture and often the desired fatty acids are only produced in seed oil rather than throughout the entire plant.[56] Furthermore, for the case of specialty oleochemicals like ricinoleic acid, the plant also contains a potent toxin called ricin, which must be removed from castor oil before use.[56] For these reasons, there have been efforts to enable the production of these useful fatty acids in more readily cultivated plants such as *Arabidopsis thaliana*,[19,70] *Nicotiana tabacum*,[71] and *Gossypium hirsutum L.*[72] Even with these hosts, plants often possess a disadvantage in that oleic acid tends to build up in plants expressing the heterologous enzymes.[56,73,74] This build-up is believed to be caused by the nonnative host's inability to relocate the heterologous products from the PLs and into TAGs for storage, which was validated by a number of studies that improved the production of chemicals such as ricinoleic acid[75] and vernolic acid[73] in transgenic plants by introducing acyltransferases from the native producers. These endeavors once again highlight the importance of interfacing lipid metabolism properly to create new product profiles.

Microalgae and algae have also gained attention as a photosynthetic host for oleochemical production due to high productivity potential. One study developed a model estimating that if the marginal land in China were used to grow microalgae for biofuel production, the amount of biofuels produced could meet the country's energy needs.[76] In addition, some of these organisms are able to produce substantial qualities of important ω-3 fatty acids such as EPA,[77] and DHA.[78] These traits make designing processes for engineered microalgae to produce desirable biofuels. However, the genetic toolbox is still in the process of being developed. Progress in developing genetic engineering tools for microalgae has included publication of genome sequences[57] as well as developing expression systems, selection markers, transformation procedures, and tools such as RNA interference.[79] These techniques have been applied to enable overexpression of ACC[80] as well as the overexpression of desaturases and elongases to improve the production of EPA[81] and DHA.[82] Likewise, environmental conditions have been studied to determine optimal environments for lipid production. Such studies have determined that important factors for lipid production are light intensity, temperature, salinity, and the availability of nutrients such as nitrogen and iron. However, with the exception of light intensity, most of these conditions tend to impact lipid production differently across species.[83] This suggests that customized conditions are required for each host species.

Cyanobacteria are prokaryotic organisms with a slightly more developed set of genetic tools.[84] The *Synechococcus* genus of cyanobacteria in particular has been studied for oleochemical synthesis with engineering efforts to produce FFAs,[85] ALA,[86] SDA,[86] fatty alcohols,[87] wax esters,[88] and alkanes.[88] Despite the advantages of working with quickly growing photosynthetic organisms, oleochemical production in cyanobacteria remains relatively low with titers in the mg/L range. Consequently, a viable process using cyanobacteria to produce oleochemicals would require considerable improvements to production levels before becoming commercially viable.

A final option for unique oleochemicals has been to engineer native producers. For instance, the *Pseudomonas aeruginosa* PR3 and *Bacillus megaterium* ALA2 strains both natively produce dihydroxy fatty acids and TriHFAs.[89] Additionally, the TriHFAs of *B. megaterium* ALA2 can be cyclized and further hydroxylated into a variety of tetrahydrofuranyl fatty acids (THFAs). This strain is of particular interest due to its ability to produce THFAs from a variety of substrates, including linoleic acid, EPA, and DHA.[89] Although the biochemical pathway for THFAs has been established, the enzyme(s) responsible for THFA and polyhydroxy fatty acids in this organism have not been isolated. Additional strains including an ammonium oxidizing bacteria are able to produce a unique class of lipids, ladderenes that contain multiple fused cyclobutyl rings believed to be formed from the cyclization of methylated fatty acids.[90] These hosts may provide promise either as a source for bioprospecting or as a production host.

3.7 Conclusion

This chapter has discussed the fundamental aspects of lipid metabolism: how lipids are synthesized, stored, utilized, and degraded within organisms. In doing so, we demonstrate the numerous ways in which common fatty acids can be converted into diverse oleochemicals including, but not limited to, alkanes, biodiesel, fatty alcohols, and fatty acids that contain uncommon functional groups. In particular, this chapter emphasized the specificity at which particular fatty acid(s) and form of lipid are recognized as the substrates for these modifying enzymes. Acknowledging these substrate dependencies is critical for any metabolic engineering endeavor to produce oleochemicals. Thus, proper engineering for the bioproduction of oleochemicals will require not only providing the substrates that these enzymes need but also insuring that the resulting products can be either converted into or stored as a stable-form product.

As revealed in this chapter, the host organism chosen for use in oleochemical production is a critical component of any metabolic engineering endeavor. Even though *E. coli* and *S. cerevisiae* are two of the most well understood microbial species, their ability to produce lipids can be easily surpassed by oleaginous organisms. As an alternative to working with *E. coli* or *S. cerevisiae*, it may be more desirable to develop better tools for organisms that are already oleaginous, rather than attempting to introduce this complex phenotype into other species. In addition, a basic understanding of lipid metabolism is required to identify proper hosts to serve as cells for oleochemical production. In addition, it is important to recognize that this chapter has discussed only several biomodified fatty acids; there are many thousands of different species that have been identified, which may be of commercial interest.[16] Investigating the biosynthesis pathways of unusual fatty acids will be necessary to ultimately lead to efficient production of value-added oleochemicals.

Finally, any industrial-scale biotechnological method for producing fuels and chemicals derived from fatty acids must be considered as an entire entity, i.e., a process. This chapter endeavors to break down the metabolic engineering for oleochemical production into several subsystems that include the basic components of lipid metabolism, how organisms can modify fatty acid chains, how fatty acids can be converted into other chemical species, and how hosts have been engineered for oleochemical production as well as some of their advantages and disadvantages. Ultimately, these forces must all be balanced together with an organism that can be well suited for industrial fermentations. In the end, metabolic engineering can enable the efficient production of an array of fuels and chemicals from fatty acid and lipid metabolism.

References

1. Dunn DR. *January 2015 monthly energy review*. Washington, DC: US Department of Energy; 2015.
2. Koch B, Schmidt C, Daum G. Storage lipids of yeasts: a survey of nonpolar lipid metabolism in *Saccharomyces cerevisiae*, *Pichia pastoris*, and *Yarrowia lipolytica*. *FEMS Microbiol Rev* 2014;**38**(5):892–915.
3. Klug L, Daum G. Yeast lipid metabolism at a glance. *FEMS Yeast Res* 2014;**14**:369–88.
4. Duan X, Xiang X, Xie J. Crucial components of mycobacterium type II fatty acid biosynthesis (Fas-II) and their inhibitors. *FEMS Microbiol Lett* 2014;**360**:87–99.
5. Qureshi N, Sathyamoorthy N, Takayama K. Biosynthesis of C30 to C56 fatty acids by an extract of *Mycobacterium tuberculosis* H37Ra. *J Bacteriol* 1984;**157**:46–52.
6. Millar AA, Kunst L. Very-long-chain fatty acid biosynthesis is controlled through the expression and specificity of the condensing enzyme. *Plant J* 1997;**12**:121–31.
7. Oh CS, Toke DA, Mandala S, Martin CE. ELO2 and ELO3, homologues of the *Saccharomyces cerevisiae* ELO1 Gene, function in fatty acid elongation and are required for sphingolipid formation. *J Biol Chem* 1997;**272**:17376–84.
8. Davis MS, Cronan JE. Inhibition of *Escherichia coli* acetyl coenzyme a carboxylase by acyl-acyl carrier protein. *J Bacteriol* 2001;**183**:1499–503.
9. Beopoulos A, Cescut J, Haddouche R, Uribelarrea J, Molina-Jouve C, Nicaud JM. *Yarrowia lipolytica* as a model for bio-oil production. *Prog Lipid Res* 2009;**48**:375–87.
10. Blazeck J, Hill A, Liu L, Knight R, Miller J, Pan A, Otoupal P, Alper HS. Harnessing *Yarrowia lipolytica* lipogenesis to create a platform for lipid and biofuel production. *Nat Commun* 2014;**5**:3131.
11. Yuan GF, Chen XE, Li D. Conjugated linolenic acids and their bioactivities: a review. *Food Funct* 2014;**5**:1360–8.
12. Qiu X, Reed DW, Hong H, MacKenzie SL, Covello PS. Identification and analysis of a gene from *Calendula officinalis* encoding a fatty acid conjugase. *Plant Physiol* 2001;**125**:847–55.
13. Liu L, Hammond EG, Nikolau BJ. In vivo studies of the biosynthesis of [alpha]-eleostearic acid in the seed of *Momordica charantia* L. *Plant Physiol* 1997;**113**:1343–9.
14. Sperling P, Lee M, Girke T, Zähringer U, Stymne S, Heinz E. A bifunctional Δ6-fatty acyl acetylenase/desaturase from the moss *Ceratodon purpureus*. *Eur J Biochem* 2000;**267**:3801–11.
15. Lee M, Lenman M, Banaś A, Bafor M, Singh S, Schweizer M, Nilsson R, Liljenberg C, Dahlqvist A, Gummeson PO, Sjödahl S, Green A, Stymne A. Identification of non-heme diiron proteins that catalyze triple bond and epoxy group formation. *Science* 1998;**280**:915–8.
16. Minto RE, Blacklock BJ. Biosynthesis and function of polyacetylenes and allied natural products. *Prog Lipid Res* 2008;**47**:233–306.
17. Kuklev DV, Domb AJ, Dembitsky VM. Bioactive acetylenic metabolites. *Phytomedicine* 2013;**20**:1145–59.
18. Mutlu H, Meier MAR. Castor oil as a renewable resource for the chemical industry. *Eur J Lipid Sci Technol* 2010;**112**:10–30.
19. Smith M, Moon H, Chowrira G, Kunst L. Heterologous expression of a fatty acid hydroxylase gene in developing seeds of *Arabidopsis thaliana*. *Planta* 2003;**217**:507–16.

20. Waché Y, Aguedo M, Choquet A, Gatfield IL, Nicaud JM, Belin JM. Role of β-oxidation enzymes in γ-decalactone production by the yeast *Yarrowia lipolytica*. *Appl Environ Microbiol* 2001;**67**:5700–4.
21. Hou CT. Biotechnology for fats and oils: new oxygenated fatty acids. *New Biotechnol* 2009;**26**:2–10.
22. Cahoon EB, Ripp KG, Hall SE, McGonigle B. Transgenic production of epoxy fatty acids by expression of a cytochrome P450 enzyme from *Euphorbia lagascae* seed. *Plant Physiol* 2002;**128**:615–24.
23. Brash AR. Lipoxygenases: occurrence, functions, catalysis, and acquisition of substrate. *J Biol Chem* 1999;**274**:23679–82.
24. Mosblech A, Feussner I, Heilmann I. Oxylipins: structurally diverse metabolites from fatty acid oxidation. *Plant Physiol Biochem* 2009;**47**:511–7.
25. Croft K, Juttner F, Slusarenko AJ. Volatile products of the lipoxygenase pathway evolved from *Phaseolus vulgaris* (L.) leaves inoculated with *Pseudomonas syringae* pv. *phaseolicola*. *Plant Physiol* 1993;**101**:13–24.
26. Sanders TH, Pattee HE, Singleton JA. Aerobic pentane production by soybean lipoxygenase isozymes. *Lipids* 1975;**10**:568–70.
27. Worrall D, Holroyd GH, Moore JP, Glowacz M, Croft P, Taylor JE, Paul ND, Roberts MR. Treating seeds with activators of plant defence generates long-lasting priming of resistance to pests and pathogens. *New Phytol* 2012;**193**:770–8.
28. Itoh A, Howe GA. Molecular cloning of a divinyl ether synthase: identification as a CYP74 cytochrome p-450. *J Biol Chem* 2001;**276**:3620–7.
29. Thomas CP, Boeglin WE, Garcia-Diaz Y, O'Donnell VB, Brash AR. Steric analysis of epoxyalcohol and trihydroxy derivatives of 9-hydroperoxy-linoleic acid from hematin and enzymatic synthesis. *Chem Phys Lipids* 2013;**167–168**:21–32.
30. Meena LS, Kolattukudy PE. Expression and characterization of Rv0447c product, potentially the methyltransferase involved in tuberculostearic acid biosynthesis in *Mycobacterium tuberculosis*. *Biotechnol Appl Biochem* 2013;**60**:412–6.
31. Mathur M, Kolattukudy PE. Molecular cloning and sequencing of the gene for mycocerosic acid synthase, a novel fatty acid elongating multifunctional enzyme, from *Mycobacterium tuberculosis* var. *bovis Bacillus Calmette-Guerin*. *J Biol Chem* 1992;**267**:19388–95.
32. Iwig DF, Grippe AT, McIntyre TA, Booker SJ. Isotope and elemental effects indicate a rate-limiting methyl transfer as the initial step in the reaction catalyzed by *Escherichia coli* cyclopropane fatty acid synthase. *Biochemistry* 2004;**43**:13510–24.
33. Yu XH, Rawat R, Shanklin J. Characterization and analysis of the cotton cyclopropane fatty acid synthase family and their contribution to cyclopropane fatty acid synthesis. *BMC Plant Biol* 2011;**11**:97.
34. Bao X, Katz S, Pollard M, Ohlrogge J. Carbocyclic fatty acids in plants: biochemical and molecular genetic characterization of cyclopropane fatty acid synthesis of *Sterculia foetida*. *Proc Natl Acad Sci* 2002;**99**:7172–7.
35. Gomes M, Negrato C. Alpha-lipoic acid as a pleiotropic compound with potential therapeutic use in diabetes and other chronic diseases. *Diabetol Metab Syndr* 2014;**6**:80.
36. Jaworski J, Cahoon EB. Industrial oils from transgenic plants. *Curr Opin Plant Biol* 2003;**6**:178–84.
37. Lennen RM, Pfleger BF. Microbial production of fatty acid-derived fuels and chemicals. *Curr Opin Biotechnol* 2013;**24**:1044–53.
38. Lee SK, Chou H, Ham TS, Lee TS, Keasling JD. Metabolic engineering of microorganisms for biofuels production: from bugs to synthetic biology to fuels. *Curr Opin Biotechnol* 2008;**19**:556–63.
39. Peralta-Yahya PP, Zhang F, del Cardayre SB, Keasling JD. Microbial engineering for the production of advanced biofuels. *Nature* 2012;**488**:320–8.
40. Peralta-Yahya PP, Ouellet M, Chan R, Mukhopadhyay A, Keasling JD, Lee TS. Identification and microbial production of a terpene-based advanced biofuel. *Nat Commun* 2011;**2**:483.
41. Nawabi P, Bauer S, Kyrpides N, Lykidis A. Engineering *Escherichia coli* for biodiesel production utilizing a bacterial fatty acid methyltransferase. *Appl Environ Microbiol* 2011;**77**:8052–61.
42. Shi S, Valle-Rodríguez JO, Khoomrung S, Siewers V, Nielsen J. Functional expression and characterization of five wax ester synthases in *Saccharomyces cerevisiae* and their utility for biodiesel production. *Biotechnol Biofuels* 2012;**5**:7.

43. Rodriguez GM, Tashiro Y, Atsumi S. Expanding ester biosynthesis in *Escherichia coli. Nat Chem Biol* 2014;**10**:259–65.
44. Layton DS, Trinh CT. Engineering modular ester fermentative pathways in *Escherichia coli. Metab Eng* 2014;**26**:77–88.
45. Thompson RA, Trinh CT. Enhancing fatty acid ethyl ester production in *Saccharomyces cerevisiae* through metabolic engineering and medium optimization. *Biotechnol Bioeng* 2014;**111**:2200–8.
46. Runguphan W, Keasling JD. Metabolic engineering of *Saccharomyces cerevisiae* for production of fatty acid-derived biofuels and chemicals. *Metab Eng* 2014;**21**:103–13.
47. Steen EJ, Kang Y, Bokinsky G, Hu Z, Schirmer A, McClure A, Del Cardayre SB, Keasling JD. Microbial production of fatty-acid-derived fuels and chemicals from plant biomass. *Nature* 2010;**463**:559–62.
48. Akhtar MK, Turner NJ, Jones PR. Carboxylic acid reductase is a versatile enzyme for the conversion of fatty acids into fuels and chemical commodities. *Proc Natl Acad Sci* 2013;**110**:87–92.
49. Zhang F, Carothers JM, Keasling JD. Design of a dynamic sensor-regulator system for production of chemicals and fuels derived from fatty acids. *Nat Biotechnol* 2012;**30**:354–9.
50. Schirmer A, Rude MA, Li XZ, Popova E, del Cardayre SB. Microbial biosynthesis of alkanes. *Science* 2010;**329**:559–62.
51. Choi YJ, Lee SY. Microbial production of short-chain alkanes. *Nature* 2013;**502**:571–4.
52. Dellomonaco C, Clomburg JM, Miller EN, Gonzalez R. Engineered reversal of the beta-oxidation cycle for the synthesis of fuels and chemicals. *Nature* 2011;**476**:355–9.
53. Thiru M, Sankh S, Rangaswamy V. Process for biodiesel production from *Cryptococcus curvatus. Bioresour Technol* 2011;**102**:10436–40.
54. Tsakona S, Kopsahelis N, Chatzifragkou A, Papanikolaou S, Kookos IK, Koutinas AA. Formulation of fermentation media from flour-rich waste streams for microbial lipid production by *Lipomyces starkeyi. J Biotechnol* 2014;**189**:36–45.
55. Vieira JPF, Lenczak JL, Rossell CEV, Pradella JGC, Franco TT. Microbial lipid production: screening with yeasts grown on Brazilian molasses. *Biotechnol Lett* 2014;**36**:2433–42.
56. Napier JA. The production of unusual fatty acids in transgenic plants. *Annu Rev Plant Biol* 2007;**58**:295–319.
57. Lu J, Sheahan C, Fu P. Metabolic engineering of algae for fourth generation biofuels production. *Energy Environ Sci* 2011;**4**:2451–66.
58. Nozzi NE, Oliver JWK, Atsumi S. Cyanobacteria as a platform for biofuel production. *Front Bioeng Biotechnol* 2013;**1**:7.
59. Huffer S, Roche CM, Blanch HW, Clark DS. *Escherichia coli* for biofuel production: bridging the gap from promise to practice. *Trends Biotechnol* 2012;**30**:538–45.
60. Kamisaka Y, Kimura K, Uemura H, Yamaoka M. Overexpression of the active diacylglycerol acyltransferase variant transforms *Saccharomyces cerevisiae* into an oleaginous yeast. *Appl Microbiol Biotechnol* 2013;**97**:7345–55.
61. Beopoulos A, Mrozova Z, Thevenieau F, Le Dall MT, Hapala I, Papanikolaou S, Chardot T, Nicaud JM. Control of lipid accumulation in the yeast *Yarrowia lipolytica. Appl Environ Microbiol* 2008;**74**:7779–89.
62. Liu L, Alper HS. Draft genome sequence of the oleaginous yeast *Yarrowia lipolytica* PO1f, a commonly used metabolic engineering host. *Genome Announc* 2014;**2**: e00652-14.
63. Blazeck J, Liu L, Redden H, Alper H. Tuning gene expression in *Yarrowia lipolytica* by a hybrid promoter approach. *Appl Environ Microbiol* 2011;**77**:7905–14.
64. Blazeck J, Reed B, Garg R, Gerstner R, Pan A, Agarwala V, Alper HS. Generalizing a hybrid synthetic promoter approach in *Yarrowia lipolytica. Appl Microbiol Biotechnol* 2013;**97**:3037–52.
65. Morin N, Cescut J, Beopoulos A, Lelandais G, Le Berre V, Uribelarrea JL, Molina-Jouve C, Nicaud JM. Transcriptomic analyses during the transition from biomass production to lipid accumulation in the oleaginous yeast *Yarrowia lipolytica. PLoS One* 2011;**6**: e27966.
66. Athenstaedt K. YALI0E32769g (DGA1) and YALI0E16797g (LRO1) encode major triacylglycerol synthases of the oleaginous yeast *Yarrowia lipolytica. Biochim Biophys Acta Mol Cell Biol Lipids* 2011;**1811**:587–96.

67. Tai M, Stephanopoulos G. Engineering the push and pull of lipid biosynthesis in oleaginous yeast *Yarrowia lipolytica* for biofuel production. *Metab Eng* 2013;**15**:1–9.
68. Xue Z, Sharpe PL, Hong SP, Yadav NS, Xie D, Short DR, Damude HG, Rupert RA, Seip JE, Wang J, Pollak DW, Bostick MW, Bosak MD, Macool DJ, Hollerbach DH, Zhang H, Arcilla DM, Bledsoe SA, Croker K, McCord EF, Tyreus BD, Jackson EN, Zhu Q. Production of omega-3 eicosapentaenoic acid by metabolic engineering of *Yarrowia lipolytica. Nat Biotechnol* 2013;**31**:734–40.
69. Blazeck J, Liu L, Knight R, Alper HS. Heterologous production of pentane in the oleaginous yeast *Yarrowia lipolytica. J Biotechnol* 2013;**165**:184–94.
70. Hatanaka T, Shimizu R, Hildebrand D. Expression of a *Stokesia laevis* epoxygenase gene. *Phytochemistry* 2004;**65**:2189–96.
71. van de Loo FJ, Broun P, Turner S, Somerville C. An oleate 12-hydroxylase from *Ricinus communis* L. is a fatty acyl desaturase homolog. *Proc Natl Acad Sci* 1995;**92**:6743–7.
72. Zhou X, Singh S, Liu Q, Green A. Combined transgenic expression of Δ12-desaturase and Δ12-epoxygenase in high linoleic acid seeds leads to increased accumulation of vernolic acid. *Funct Plant Biol* 2006;**33**:585–92.
73. Li R, Yu K, Hatanaka T, Hildebrand DF. Vernonia DGATs increase accumulation of epoxy fatty acids in oil. *Plant Biotechnol J* 2010;**8**:184–95.
74. Thomæus S, Carlsson AS, Stymne S. Distribution of fatty acids in polar and neutral lipids during seed development in *Arabidopsis thaliana* genetically engineered to produce acetylenic, epoxy and hydroxy fatty acids. *Plant Sci* 2001;**161**:997–1003.
75. Burgal J, Shockey J, Lu C, Dyer J, Larson T, Graham I, Browse J. Metabolic engineering of hydroxy fatty acid production in plants: RcDGAT2 drives dramatic increases in ricinoleate levels in seed oil. *Plant Biotechnol J* 2008;**6**:819–31.
76. Zhang Q, Ma J, Qiu G, Li L, Geng S, Hasi E, Li C, Wang G, Li X. Potential energy production from algae on marginal land in China. *Bioresour Technol* 2012;**109**:252–60.
77. Cardozo KHM, Guaratini T, Barros MP, Falcão VR, Tonon AP, Lopes NP, Campos S, Torres MA, Souza AO, Colepicolo P, Pinto E. Metabolites from algae with economical impact. *Comp Biochem Physiol Part C Toxicol Pharmacol* 2007;**146**:60–78.
78. Valencia I, Ansorena D, Astiasarán I. Development of dry fermented sausages rich in docosahexaenoic acid with oil from the microalgae *Schizochytrium* sp.: influence on nutritional properties, sensorial quality and oxidation stability. *Food Chem* 2007;**104**:1087–96.
79. Radakovits R, Jinkerson RE, Darzins A, Posewitz MC. Genetic engineering of algae for enhanced biofuel production. *Eukaryotic Cell* 2010;**9**:486–501.
80. Dunahay TG, Jarvis EE, Roessler PG. Genetic transformation of the diatoms *Cyclotella cryptica* and *Navicula saprophila. J Phycol* 1995;**31**:1004–12.
81. Peng KT, Zheng CN, Xue J, Chen XY, Yang WD, Liu JS, Bai W, Li HY. Delta 5 fatty acid desaturase upregulates the synthesis of polyunsaturated fatty acids in the marine diatom *Phaeodactylum tricornutum. J Agric Food Chem* 2014;**62**:8773–6.
82. Hamilton ML, Haslam RP, Napier JA, Sayanova O. Metabolic engineering of *Phaeodactylum tricornutum* for the enhanced accumulation of omega-3 long chain polyunsaturated fatty acids. *Metab Eng* 2014;**22**:3–9.
83. Han SF, Jin WB, Tu RJ, Wu WM. Biofuel production from microalgae as feedstock: current status and potential. *Crit Rev Biotechnol* 2014;1–14.
84. Oliver JK, Atsumi S. Metabolic design for cyanobacterial chemical synthesis. *Photosynth Res* 2014;**120**:249–61.
85. Ruffing AM. Improved free fatty acid production in cyanobacteria with *synechococcus* sp. PCC 7002 as host. *Front Bioeng Biotechnol* 2014;**2**:17.
86. Chen G, Qu S, Wang Q, Bian F, Peng Z, Zhang Y, Ge H, Yu J, Xuan N, Bi Y, He Q. Transgenic expression of delta-6 and delta-15 fatty acid desaturases enhances omega-3 polyunsaturated fatty acid accumulation in *Synechocystis* sp. PCC6803. *Biotechnol Biofuels* 2014;**7**:32.
87. Yao L, Qi F, Tan X, Lu X. Improved production of fatty alcohols in cyanobacteria by metabolic engineering. *Biotechnology Biofuels* 2014;**7**:94.

88. Kaiser BK, Carleton M, Hickman JW, Miller C, Lawson D, Budde M, Warrener P, Paredes A, Mullapudi S, Navarro P, Cross F, Roberts JM. Fatty aldehydes in cyanobacteria are a metabolically flexible precursor for a diversity of biofuel products. *PLoS One* 2013;**8**: e58307.
89. Hou CT. Biotransformation of aliphatic hydrocarbons and fatty acids. In: Hou Ching T, editor. *Biocatalysis and biotechnology for functional foods and industrial products*. Boca Raton: CRC Press; 2006. p. 227–65.
90. Strous M, Pelletier E, Mangenot S, Rattei T, Lehner A, Taylor MW, Horn M, Daims H, Bartol-Mavel D, et al. Deciphering the evolution and metabolism of an anammox bacterium from a community genome. *Nature* 2006;**440**:790–4.

CHAPTER 4

Pathway and Strain Design for Biofuels Production

Omar Khalidi, Sanjay Guleria, Mattheos A.G. Koffas

Centre for Biotechnology and Interdisciplinary Studies, Department of Chemical and Biological Engineering, Rensselaer Polytechnic Institute, Troy, New York, USA

4.1 Introduction

Living systems offer a unique solution to targeted chemical synthesis, providing a diverse array of transformations to build inexpensive and sustainable routes for the production of fine and commodity chemicals from a single renewable carbon source like glucose.[1] However, engineering highly efficient enzymatic pathways for industrial production of fuels and chemicals remains an overwhelming challenge in metabolic engineering and synthetic biology.[2] Unbalanced protein expression, low availability of precursors and cofactors, toxic intermediates and end products, as well as overall end product burden may result in poor performance of engineered pathways.[3] Building a synthetic biological system requires optimization of a number of components, including genetic parts, devices and systems, using design tools capable of regulating these components in a predictable and quantitatively controllable manner.[4]

Recent research trends in this field have focused on the development of novel tools that can be utilized to design and optimize biological systems. Balanced expression of enzymes that comprise a metabolic pathway is essential for obtaining optimal performance of the designed biological system and controlling gene expression in a predictable and qualitatively controllable manner is one of the most important design principles.[5]

This chapter will discuss recent advancements in synthetic biology and metabolic engineering that can be implemented for microorganisms to reach theoretical yields in biofuel production. It comprises of two sections. The first section covers pathway design, construction, and optimization, while the second section discusses strain design in order to achieve optimal biofuel production.

Pathway design begins with the initial pathway construction and validation. It covers such topics as determining what biofuel compound to be produced based on the host organism, and constructing the actual biochemical pathway. Next pathway optimization is discussed.

Biotechnology for Biofuel Production and Optimization. http://dx.doi.org/10.1016/B978-0-444-63475-7.00004-2

Pathway optimization is achieved at various steps in the biofuel production process, from mRNA transcription to the actual activity of the enzymes of the biochemical pathway. Protein engineering can be implemented in order to improve enzyme activity, as well as solubility and stability. In addition, the use of protein complexes to increase the product titers in biofuel production and the design tools that control the gene expression at the transcription and translational levels are also discussed in Section 4.2 of this chapter. The Section 4.3 of this chapter discusses strategies to optimize the host organism to produce biofuel compounds at commercially feasible levels. Strategies discussed include knocking out competing pathways that use the same precursors as the biofuel production pathway, pathway upregulation to increase precursor concentration, *in silico* modeling to determine gene deletion and overexpression candidates, as well as strategies to improve biofuel tolerance.

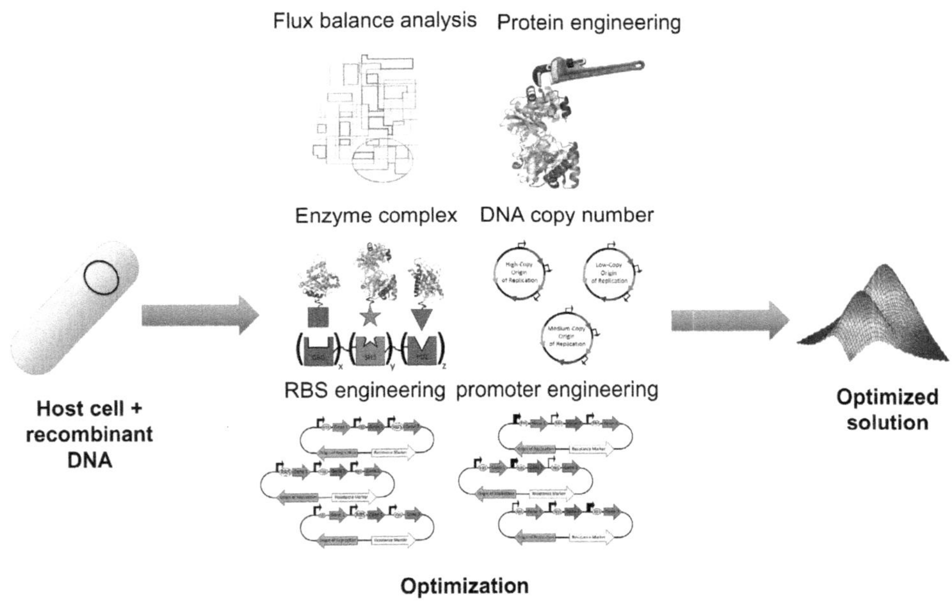

4.2 Pathway Design

4.2.1 Initial Pathway Design

Before designing the pathway, it is important to decide on the target biofuel compound. This is determined based on the metabolic precursors and preexisting metabolic networks in the host organism.[6,7] For a viable biofuel production process, the ideal fuel-producing organism must be able to convert a variety of sugars to fuels anaerobically at near theoretical yields, resist inhibitors generated by biomass pretreatment, and exhibit low product toxicity.[8] Metabolic engineering of model organisms has resulted in high titer of advanced fuels, namely alcohols, fatty acid derivatives, and isoprenoids.[9]

After selecting the biofuel compound and host organism, the next step is pathway assembly to achieve desired biofuel production. The genes to be used for coding the enzymes of the target pathway can be obtained from an array of publicly available databases. The genes used can either be from the same source or from various sources in an attempt to improve product titer. The extensive optimization of both enzyme sources and relative gene expression levels has led to high product titers.[10]

The genes encoding for enzymes that make up the individual pathway components are usually expressed episomally and are assembled onto one or more plasmids, depending on the number of genes necessary to construct the biochemical pathway. These plasmids are transformed into the host organism of choice, which is usually a well-characterized and genetically tractable organism, such as *Escherichia coli* or *Saccharomyces cerevisae.*[11] An alternative organism can be used that is better suited for the desired biofuel compound. However, difficulties will arise due to fewer techniques for genetic manipulation.[12]

Production of the biofuel compound is validated using one of several analytical techniques, such as chromatography, mass spectrometry, or nuclear magnetic resonance. It is likely that low titers will be observed, if at all present. This is to be expected as it is only the initial design, and producing a foreign compound is always inefficient initially. The main purpose is for pathway and product validation. Once production of the biofuel molecule has been verified, pathway optimization must be implemented in order to increase the titer orders of magnitude higher for commercially viable biofuel production.

4.2.2 Pathway Optimization

The initial step in pathway optimization is the construction of the pathway with biochemically characterized enzymes capable of providing high production rates. The next step is pathway optimization, which aims to maximize efficiency of the metabolic networks involved in biofuel production. Efficiency of these pathways is determined by precursor, intermediate, cofactor, and enzyme concentrations. Another important factor is the catalytic activity of the enzymes, which is subjected to regulation at multiple levels, from gene expression to enzyme activity.[13]

However, this process is problematic due to the inability to know the amount of each enzyme needed for pathway optimization a priori. Due to this lack of precision, fine-tuning of the biofuel production pathway becomes very tedious. *In vitro* assays can be used for quantitative analysis,[14] but these results should be treated with caution, as enzymes will always behave differently than *in vivo*.

Another issue that arises is the lack of biochemical characterization of many of the enzymes selected. Genes can easily be found in sequence databases and cloned. However, there are difficulties in protein overexpression and biochemical measurements, which is unfortunate as knowledge of these enzymes' properties would aid in pathway optimization. For example, attempting to increase flux by increasing the copy number of a pathway component would prove to be futile if dealing with an insoluble enzyme or an enzyme that is subjected to strong feedback regulation.

One goal of synthetic biology and metabolic engineering in biofuel production is the creation of these biological systems that robustly produce enzymes at a predetermined level and precise times in order to increase product yield. Enzyme concentrations that are continuously high might lead to the formation of inclusion bodies as well as high concentrations of intermediates that can be toxic to the host organism. Therefore, enzyme levels must be carefully balanced to not only increase flux, but also to maintain intermediate concentrations near their physiological levels to avoid cell toxicity.[15]

Protein engineering

When choosing enzymes to construct a biofuel chemical pathway it is important to consider the intracellular chemical environment of the host organism. An enzyme is more likely to function in the production host if the organism from which it was obtained from has similar intracellular properties such as pH and growth temperature. Genes transplanted between kingdoms are especially prone to compatibility problems.[16] However, this issue can be resolved by engineering the enzyme itself in order to improve its *in vivo* properties such as solubility, stability, selectivity, and enzymatic activity.[17]

If an enzyme in the biochemical pathway is the bottleneck, then screening a library of homologs to find a replacement is a possible solution.[18] There are numerous homologous genes found within genetic databases, each of which codes for enzymes with varying stabilities and catalytic rates. However, homologs may not always be available or may only be a few present in number; and these homologs may not even be high-yielding enzymes. Therefore, it might be better to optimize production by adjusting the copy number, regulation, and enzymatic properties of the components of a pathway, rather than screening libraries of enzymes.[19]

In designing efficient proteins, one may choose to engineer activity, substrate specificity/selectivity, solubility, and stability. Protein engineering was proven useful in the production of specialty chemical triacetic acid lactone.[20] Nair and Zhao[21] engineered a xylose reductase with increased selectivity for D-xylose over L-arabinose. In another study, Zhang et al., 2008[22] engineered 2-keto-isovalerate decarboxylase for preferred substrate specificity toward a nonnatural substrate.

The catalytic activity of the pathway enzymes may also prove to be a limiting factor in product expression. This is due to the fact that not all enzymes exhibit a fast turnover rate, especially if the enzymes are members of a pathway that does not produce large amounts of its product, or generates multiple products.[23] In a recent study, two enzymes from the diterpenoid biosynthetic pathway, geranylgeranyl disphosphate synthase and levopimaradiene synthase (LPS) for levopimaradiene production were engineered for increased protein activity. A homology model structure was used to select 15 residues constituting the binding pocket in order to engineer LPS. This was followed by mutation of these residues to specific residues based on phylogenetic analysis of functionally different LPS enzymes. The mutations which yielded increased levopimaradiene production were combined together, and the resulting mutation led to the production of the most levopimaradiene.[24]

Higher enzyme activity can also be achieved by improving the catalytic efficiency, protein solubility, folding, and stability. Fusion of lipases with a polycationic amino acid tag has been reported to increase the solubility of expressed proteins in *E. coli*.[25] In addition, the formation of specific disulfide bonds has also been shown to facilitate the folding of functional proteins. This has been addressed by using a specialized *E. coli* Origami (DE3) strain or co-expression of the Dsb-family protein (e.g., DsbA) where disulfide bond formation was involved.[26,27]

Enzymes must fold into their native three-dimensional structure in order to function, and proteins often misfold when expressed in a foreign host.[28] An improperly folding pathway enzyme can be replaced or modified in order to improve its folding.

Additionally, for enzymes derived from higher eukaryotic species, posttranslational modifications such as glycosylations and enzyme localization to subcellular compartments (Golgi, endoplasmic reticulum) need to occur for the enzyme to function properly if at all. Sometimes, the proteins responsible for catalyzing these posttranslational modifications can be transplanted into the host organism. Other accessory proteins whose activities are absent in the host cell may also need to be present, with their activity levels optimized.

Further, the pathway can be optimized by increasing the efficiency of the transport system, either through improved transfer of extracellular substrates into the cell or increased export of the products. Transport proteins can be a limiting factor in biofuel production. Recent work has been done involving the directed evolution of xylose transporters with increased affinity toward xylose. This resulted in an improved growth rate of 70% on xylose as the sole carbon source. The caveat here though is that the maximum rate of reaction (V_{max}) decreases.[29] It should also be noted that the trade-off between the Michaelis constant (K_m) and the V_{max} in engineered xylose transporters was also found in cellodextrin transporter engineering.[30] Although it is a viable option to improve biofuel production, strategies aimed at improving transport are underutilized. This is due to the lack of three-dimensional structures of these transport proteins, which prevents the implementation of directed evolution.

Flux through a pathway may also be modulated by feedback regulation of the enzymes by metabolites or other small molecules. Feedback regulation can strongly affect enzymatic activity through product or allosteric inhibition, maintaining concentrations of intermediates within a narrow range in order to prevent toxic levels of small molecules. Negative feedback regulation of an enzyme decreases its activity. When production pathways contain allosterically regulated enzymes, these enzymes can become the bottleneck if the regulating metabolite reaches inhibitory levels. Therefore, engineering feedback-resistant enzymes often enables higher production.[31] In one study, feedback-resistant aspartate kinase/homothreonine dehydrogenase resulted in a fourfold increase in the titer of both 1-propanol and 1-butanol.[32] Furthermore, feedback inhibition can be eliminated by mutating the inhibitory binding pockets, by simply removing the regulatory domain, or by relocating the enzyme to a different compartment of the cell.[33] Use of a truncated version of 3-hydroxy-3-methyl-glutaryl CoA reductase has improved the efficiency of the mevalonate pathway.[34]

Substrate channeling and enzyme complexes

Another way to increase the efficiency of the multienzyme pathway is to localize pathway enzymes in close proximity by means of protein complexes containing sequential biochemical reactions. It has been reported that protein colocalization leads to increased pathway flux by increasing the local metabolite concentration.[35]

This approach is not new as in nature multiple enzyme activities are colocalized in one complex or organelle of its native organism. Direct transfer of the product of an enzyme to another proximate enzyme or cell as its substrate without equilibration with the bulk phase is referred to as substrate channeling.[36] In natural multifunctional enzyme complexes, substrate channeling has been reported in numerous pathways such as the glycolysis pathway,[37] the Calvin cycle,[38] and the tricarboxylcylic acid cycle.[39] Substrate channeling has been proposed to occur by two general mechanisms. One mechanism involves the direct transfer of the intermediate of the first enzyme to the second enzyme, whereas in the second mechanism, the receptive enzyme or cell is in close proximity to the first enzyme.[40] Since the concentration of the product of an enzyme will be higher near its active site, having the next enzyme in the desired pathway in proximity to react might help to increase flux. Colocalization of pathway enzymes could be a useful tool if the pathway enzymes cannot be further optimized, an intermediate is toxic to the host organism, or if the intermediate is a substrate of a competing enzyme.

One way to assemble the protein complex is to fuse the metabolic enzymes that catalyze the successive reactions. A fusion protein of beta-galactosidase and galactose dehydrogenase has been shown to exhibit enhanced reaction rates of 1.5-2.4-fold over free enzyme mixtures.[41] Wang et al.[42] reported that fusion of farnesyl pyrophosphate and farnesene synthase resulted in increased farnesene production. However, there are a limited number of proteins that can be fused without altering the enzyme activity.

Protein scaffolds can be implemented to localize enzymes that catalyze sequential reactions. By designing synthetic protein scaffolds that spatially recruit three mevalonate biosynthetic enzymes, a 77-fold improvement in product titer was observed with low enzyme expression and reduced metabolic load.[43] Synthetic scaffolds thus offer a modular tool for rationally clustering enzymes with stoichiometric control. Scaffolding appears to improve product titers in multiple cases through the formation of intermediate microdomains, where enzyme-scaffold complexes reduce intermediate buildup by limiting the amount of intermediate that can escape the complex.[44]

mRNA transcription optimization

Although intuitively, protein concentration would increase with respect to the mRNA transcripts that encode the protein, this is not always the case. The correlation between mRNA and protein abundances in the cell has been reported to be particularly poor. Furthermore, recent technological advances in the quantitative analysis of mRNA and protein species in complex samples allow the detailed analysis of this pathway at the center of biological systems.[45] However, factors such as DNA copy number and promoter strength have greater bearing on protein expression; therefore, its optimization can be achieved through genetic manipulation.

Another important fact that needs to be considered for heterologous gene expression is whether to go for a plasmid-based gene expression or integrate the pathway into the host genome. Plasmid copy number can be regulated by using plasmids with various origins of replication, which determines a low- or high copy number plasmid. However, plasmid-based gene expression of biofuel pathway may not be efficient as it causes extra metabolic burden on the host cell and, therefore, may lead to low product titers. In addition, plasmids are not always evenly distributed during cell division.[46] Therefore, cells with fewer copy numbers will become more prevalent, leading to a cell population with low overall productivity. On the contrary, in genomic integration, the chromosome is present in very low copy numbers, which may place a very low limit on the pathway copy numbers.[47]

Generally genes involved in the biofuel production pathway are expressed using promoters that function under the control of a synthetic inducer chemical and the gene expression depends upon the concentration of the inducing chemical. This poses a limitation to the use of this strategy when a biofuel producing process is upgraded at the industrial scale and the cost of the inducer becomes an important economic factor in the viability of the biofuel production. This issue can be addressed using constitutive promoters as the constitutive active promoters may provide modular control on the whole biofuel pathway, or individual enzymes that are controlled by a specific constitutive promoter.[48,49] However, constitutive expression of pathway enzymes may present a metabolic burden on the cell which may evolve escape mutants contributing to pathway instability.

Another approach in promoter engineering is the creation of a library of synthetic promoters with a broad range of transcription efficiencies.[50–52] Recently, synthetic hybrid promoter approaches have been utilized to combine core promoters with enhancer elements comprising of the combinations of tandem repeats of upstream activating sequences in yeasts.[53] As per one study, metabolic fluxes have been balanced by controlling transcription efficiencies for optimal production of target molecule. Isoprenoid pathway for the production of taxadiene, has been successfully optimized by using multivariate modular systems harboring different promoter strengths and copy numbers.[54] Zhang et al.[55,56] have reported the development of a dynamic sensor regulator system (DSRS) that uses a naturally occurring transcription factor/synthetic hybrid promoter pair. Key metabolites in fatty acid production pathway could be sensed by DSRS which then dynamically regulates the expression of genes in fatty acid ethyl ester (FAEE) production. Development of strains for fatty acid overproduction using Metabolic Flux Analysis has also been reported.[57]

Another important strategy is synthetic metabolic control in pathway engineering. It can be achieved by engineering novel metabolite sensors with desired input-output relationships.[58] For instance, Liao et al. have designed and applied a regulatory circuit that can sense the glycolytic pathway hallmark metabolite acetyl-phosphate to control the lycopene biosynthetic pathway[59] and generate oscillatory gene expression[60] as well as achieve artificial cell-cell communication.[61] Stress response promoters to improve farnesyl pyrophosphate production, and native quorum-sensing regulon have been rewired in *E. Coli* for autonomous induction of recombinant proteins.[62,63]

Translation optimization

Translation is a well-understood process. However, predicting protein concentration produced from its mRNA transcript is very difficult. Structural elements around the translation initiation region of mRNA are key in determining the amount of protein produced from a particular mRNA sequence.[64–66] According to one strategy, by balancing the expression levels of the operon, a sevenfold increase in titer of mevalonate was observed in the engineered strain as compared to the wild type. In addition, several studies have reported methods for designing synthetic ribosome binding sites (RBSs) and 5′-untranslated regions which can be utilized in controlling posttranscriptional gene expression levels.[67–69] Qi et al., have developed a synthetic RNA processing platform that physically separated linked generic elements using bacterial clustered, regularly interspaced short palindromic repeat (CRISPR)-mediated cleavage to assemble various regulatory elements that might interact with each other in a predictable manner. Using this approach, they successfully designed multigene operons that yielded stable gene expression comparable to that obtained from monocistronic systems or complex regulatory systems.[70]

Recently, the possibility of using RNA-based regulatory tools for modulating gene expression at the posttranscriptional level to design microbes to accomplish specific purposes has been demonstrated. Callura et al., 2012 constructed a genetic switchboard that switched between genetic circuits by independently modulating the expression of multiple genes under the control of orthogonal riboregulators. Using this approach they successfully redirected carbon fluxes in central carbon metabolism to suit their purpose.[71]

The bottleneck in protein production from a transcript is translation initiation. Translation initiation is largely determined by the binding of the ribosomal RNA (rRNA) to the RBS. Binding is controlled by several factors: presence of secondary structures within the mRNA that shield the RBS from the rRNA, as well as the equilibrium binding strength of the RBS to the ribosomal RNA. Since these factors are highly dependent on the mRNA sequence upstream and downstream of the RBS, a single RBS will likely result in widely varying translation initiation rates for different surrounding sequences, which may complicate efforts to precisely tune protein levels across a biofuel production pathway.

The rate of protein elongation can also be engineered. The most common way to do so is via codon optimization. Introducing a foreign metabolic pathway into a heterologous host for the production of various chemicals and fuels often requires re-optimizing the codons of the target genes based on the usage patterns of the host systems.[69] Angov et al., 2011 have developed an algorithm termed codon harmonization, which is just codon usage in the target gene in the heterologous system based on patterns approximated in the native host. By recoding target gene sequences, they were able to express three *Plasmodium falciparum* proteins in *E. coli* at 4- to 1000-fold higher levels in soluble fractions.[72] However, it is important to note that individual genes transcribed within a common operon will have interactions due to the lack of proper understanding of mRNA transcript folding and degradation. Efforts to optimize the translation level of an individual gene could have implications on the expression of other pathway genes in ways that are difficult or impossible to predict.

4.3 Strain Design

The host organism chosen to express the pathway will play a vital role in the biofuel production. The host organism supplies the metabolic precursors and enzyme cofactors. However, there is currently no unmodified host organism that produces chemicals on a commercial scale. Nature did not intend for these microbes to be used as cell factories optimized to produce desired chemicals at the highest possible titers. Therefore, host physiology and biochemistry must be taken into account if high titers are to be achieved. This section of the chapter will focus on host engineering that will lead to improved titers of final products.

The host organism should grow on cheap, renewable feedstock as well as produce high titers of the biofuel compound. One approach is to engineer model organisms that meet these criteria. The advantage of model organisms is their genetic tractability. There are well-established genetic tools as well as a significant amount to knowledge of their metabolic networks. In addition, higher tolerance to biofuels can be engineered. However, the challenges that face model organisms as the optimal host for biofuel production is that their feedstock is limited to simple sugars, as well as the implementation of two pathways, which puts a metabolic burden on the organism.[9]

4.3.1 Deletion of Competing Pathways

The precursors used for biofuel production are usually used for building blocks that will ultimately become components of the cell. Therefore, there is a competition between the engineered biofuel pathway and the native pathways for these resources. An option could be to remove or decrease the activity of competing pathways. For example, the deletion of key enzymes required for alternate pathways relieves production of the biofuel pathway, thus increasing the yield of the final product.

Deletion of enzymes that have no direct effect on precursor pools may also improve titer by eliminating the generation of substrates that may act as competitive inhibitors of an enzyme along the biofuel production pathway. The product yield can also be increased by eliminating pathways that catabolize biofuel precursors.[73]

However, it is worth mentioning here that deletion of the pathways that compete with the engineered pathway may instead decrease the product titers rather than increasing them. This is possible in view of the fact that competing pathways may be important in terms of generating the energy and cofactors, which are indeed essential for overall growth of the host cell. Therefore, the viable option appears to be lowering the expression of the competing pathways.

To increase the yield and productivity, cofactor manipulation has been applied as an essential and powerful tool for metabolic engineering. According to one report, overexpression of the NAD+-dependent formate dehydrogenase (FDH) gene doubled the maximum yield of NADH from 2 to 4 mol NADH/mol glucose consumed, resulting in increased cell density and ethanol production.[74]

In one example, metabolic engineering was performed to increase NADH level in *E. coli* by disrupting the lactate dehydrogenase (LDH), alcohol dehydrogenase (ADH), and fumarate reductase (FRD) genes, while the *Candida boidinii* FDH gene was overexpressed. This metabolic engineering strategy resulted in the construction of *E. coli* strain having capability to produce 15 g/L of butanol.[73]

By eliminating nonessential fatty acid utilization pathway genes (*DGA1*, *LRO1*, *ARE1*, *ARE2*, and *POX1*), a metabolic engineering strategy has been developed resulting in a *Saccharomyces cerevisiae* strain that can overproduce fatty acids and in turn use these for producing FAEEs (biodiesel). Compared to the wild type, there is about a threefold increase in free fatty acid content in a strain devoid of both triacylglycerol and steryl ester formation, a fourfold increase in free fatty acid content in a strain that is incapable of β-oxidation, and a fivefold increase of free fatty acids in a strain lacking all of these nonessential fatty acid utilization pathways.[75]

4.3.2 Pathway Upregulation to Increase Precursor Concentration

Availability of precursors can be increased by the upregulation of the native pathways that produce them. This can ultimately lead to increased product yield. In a recent study, increased isobutanol production in *S. cerevisiae* by the overexpression of genes in valine metabolism has been observed. The yield of isobutanol was increased from 0.16 to 0.97 mg/g glucose by simultaneous overexpression of the valine metabolism genes *ILV2*, *ILV3*, and *ILV5*. Isobutanol yield was further improved twofold by the additional overexpression of the gene *BAT2*, responsible for encoding the cytoplasmic branched-chain amino acid aminotransferase.[76] However, a limitation to the overexpression of these native enzymes is the allosteric regulation, which can inhibit metabolic flux.

4.3.3 In Silico *Modeling*

Due to the difficulty in predicting the effect of gene deletions on cellular metabolism, *in silico* modeling of cellular metabolism is a way to pick unintuitive candidates for gene deletions that may improve biofuel titers. *In silico* methods can also be used not just to pick knockout candidates, but also native genes for upregulation.

OptKnock is a bi-level programming framework for identifying gene knockout strategies for microbial strain optimization for the production of chemicals or biochemicals in *E. coli*. This is accomplished by ensuring that a drain toward growth resources (i.e., carbon, redox potential, and energy) must be accompanied, due to stoichiometry, by the production of a desired product. Based on the OptKnock framework, the most promising gene knockout strategies were identified and their corresponding allowable envelopes of chemical versus biomass production in the context of succinate, lactate, and 1,3-propanediol production in *E. coli*.[77]

A method known as Cipher of evolutionary design (CiED) has been developed to identify genetic perturbations, such as gene deletions and other network modifications that result in optimal phenotypes for the production of end products, such as recombinant natural products. This method further demonstrates the utility of a purely stoichiometric network to predict improved *E. coli* genotypes that more effectively channel carbon flux toward malonyl coenzyme A (CoA) and other cofactors in an effort to generate recombinant strains with enhanced flavonoid production capacity.[78] In this strategy, the engineered *E. coli* strains are constructed first by targeted deletion of native genes predicted by CiED and then second by incorporating selected overexpressions, including those of genes required for the co-expression of the plant-derived flavanones, acetate assimilation, acetyl-CoA carboxylase, and the biosynthesis of coenzyme A[78].

Computational methods have proven invaluable in studying and predicting the deleterious effects of gene deletions. Expression-Dependent Gene Effects is an *in silico* method that can predict deleterious effects resulting from overexpression of either native or foreign metabolic genes.[79] k-OptForce, an optimization-based strain design frame work incorporating the mechanistic details afforded by kinetic models into a genome-scale stoichiometric based modeling formalism has been developed. The study paves the way for the integrated analysis of kinetic and stoichiometric models and enables elucidating system-wide metabolic interventions while capturing regulatory and kinetic effects.[80] OptForce framework for succinate production in *E. coli* using *in silico E. coli* model, iAF1260 has been developed. The method not only recapitulates existing engineering strategies, but also reveals nonintuitive ones that boost succinate production by performing coordinated changes on pathways distant from the last steps of succinate synthesis.[81]

Optgene[82] and RobustKnock[83] are also frameworks to identify deletion candidates to achieve growth-associated production of target compounds. OptReg[84] and EMILiO[85] are frameworks to search for target genes not only for deletions, but also for up- and downregulations. Based on these computational tools, not only bacterial engineering such as lactic acid production in *E. coli*[86] but also vanillin production in yeast[87] and sesquiterpene production in *S. cerevisiae*[88] have been reported.

Although most *in silico* models target gene knockouts, other metabolic engineering strategies involve gene overexpression and partial gene knockdowns. Furthermore, the flux change of a reaction may be attributed to (1) directly engineering genes of the catalyzing enzymes, (2) engineering the availability of reaction precursors and substrates upstream, or (3) eliminating bottlenecks downstream. Another recent approach known as Flux Balance Analysis with Flux Rations (FBrAtio) involves gene overexpression, knockout, and partial knockdown for framing metabolic engineering strategies. This strategy assesses competing enzymes for the same substrate, and allows the distribution of this substrate to be modified and included as a flux ratio constraint in a genome-scale metabolic flux model.[89,90]

Recently, in order to optimize the product titer and productivity, Dynamic Strain Scanning Optimization (DySScO) strategy has been developed, which integrates the Dynamic Flux Balance Analysis method with existing strain algorithms.[91] DySScO strategy has been applied for designing *E. coli* strains for succinate and 1,4-butanediol productions, respectively. Further, DySScO strategy produced strains that balanced the product yield, titer, and productivity, and therefore, is a useful computational tool that finds potential applications in evaluating the economic performance of the designed strains.[91]

4.3.4 Strain Engineering for Improved Biofuel Tolerance

Many types of biofuel products are toxic to microbial strains used in industrial processes, damaging microbial cell structure and function.[92] Biofuels such as ethanol, butanol, and hexane in the cell membrane can cause changes to membrane integrity, structure, and physiological functions.[93–95] It can even cause a change in internal pH levels, affecting cellular metabolism, eventually leading to growth inhibition and cell death.[96] Therefore, the tolerance to the biofuel compound and its biosynthetic intermediates must be taken into consideration when deciding on a potential host organism. If high concentrations of the biofuel product are toxic to the host organism, then it is impossible to produce high titers.

Microbes with low tolerances to biofuels can be engineered in order to resist toxicity. There are several mechanisms and techniques that can be used to do so, such as mutant screening and genetic engineering.[97,98] However, it should be pointed out that increasing biofuel tolerance does not necessarily guarantee improved titer, and could possibly lower it.

Microbes possess certain membrane proteins known as efflux pumps that can excrete hydrophobic compounds from the cell if the intracellular concentration reaches levels that are toxic to the cell. The most studied solvent-tolerant pumps are those from the resistance-nodulation-division family in gram-negative bacteria.[99] These efflux pumps consist of three protein subunits across the inner and outer cell membrane. The inner membrane subunit is responsible for substrate recognition and proton antiport, with the periplasmic subunit acting as a linker between the two subunits, and the outer membrane subunit being responsible for excretion. Also, genes that code for more effective efflux pumps can be incorporated into the host membrane for improved tolerance.[100] However, heterologous pumps may need to be fine-tuned in order to increase compatibility with the host organism.

Another promising method is the use of heat shock proteins. Heat shock proteins play a role in the synthesis, transport, folding, and degradation of proteins. Under stress, they prevent protein aggregation and assist in refolding. It has been discovered that short-chain alcohols and heat shock induce similar stress responses on cells.[100] Therefore, heat shock proteins could be good engineering targets for improving biofuel tolerance, as studies in the past have shown that the overexpression of heat shock proteins improved not only biofuel tolerance, but also yield.[101]

Organisms with solvent-tolerant membranes can actively change membrane composition to block entry of solvents. However, nontolerant strains are unable to do so, allowing solvents to disrupt the cell membrane, eventually leading to cell death. Therefore, host strains can be engineered with these membrane modifications to increase their tolerance to biofuels. Some tolerant strains can shift the composition of the fatty acids in their membrane to block the entry of solvents. A well-studied example is the shift from cis to trans unsaturated fatty acids, which can be catalyzed by the cis-trans isomerase.[102] In addition, modifications to phospholipid head groups or phospholipid chain length have been shown to increase solvent tolerance.[92]

An alternative strategy for tolerance improvement could involve noncoding small regulatory RNA (sRNA) engineering. Although never translated, sRNA's are responsible for gene expression. In a recent study, the sRNA's, DsrA, RprA, and ArcZ, which are activators for RpoS translation, were overexpressed in *E. coli*. RpoS is a regulator of the general stress response, allowing the cell to survive environmental changes as well as cross protection. This overexpression led to a 8500-fold increase in tolerance against carboxylic acid and oxidative stress during active cell growth.[103] An advantage to the overexpression of sRNA's is that it does not impose the metabolic burden onto the cell that overexpression of proteins would have. However, there has been no study published to date about applying this sRNA strategy in order to improve tolerance to biofuels, but early studies have demonstrated that this strategy is generalizable to various bacteria and applicable in developing various physiological traits.[104]

Although increased biofuel tolerance appears to be necessary for biofuel production, it does not seem to be sufficient. Studies with *Clostridium acetobutylicum* have shown that modifications of the fatty acid composition of the membrane have led to increased butanol tolerance. However, this also led to inhibited solvent production.[105] It is also worth pointing out that production yields can exceed tolerance levels. Although cell growth is inhibited at a certain biofuel concentration, production will continue. However, biofuel toxicity still limits production and in order to increase yields to theoretical values, it will be necessary to improve tolerance. Another approach to increase the product tolerance is directed or adaptive evolution. The *E. coli* EMFR9 strain could be evolved to be able to grow in the presence of furfural at high concentration.[106] Similarly, the *S. cerevisiae* M25 strain evolved with transcription factors that could become more tolerant to ethanol.[107] More recently, butanol tolerance of *E. coli* could be increased by using artificial transcription factor libraries that consist of zinc finger DNA-binding proteins and *E. coli* cyclic adenosine monophosphate receptor protein.[108] The selected butanol-tolerant *E. coli* strain could grow in the presence of up to 1.5% (v/v) of butanol.[108] The cellular stress to biofuels can also be reduced by amplification of tolerance-related proteins or addition of protectant in the culture media, which often increases the final titer of the target product.[109]

Recently a novel method called "genome replication engineering assisted continuous evolution" (GREACE), using "mutagenesis coupled with selection" as core principle has been developed to improve microbial tolerance for biofuels production. GREACE has been successfully applied to engineer *n*-butanol and acetate tolerances, two important physiological characteristics for biofuels production.[110]

In conclusion, methods that support the exploration of synergistic approaches to tolerance will be especially valuable. Tolerance engineering will be essential as production of next-generation biofuels continues to improve.

4.4 Future Perspectives

Biofuels are currently attracting increasing attention to provide a solution to the problems of climate change and limited fossil fuels. The major goals of metabolic pathway redesigning for biofuel production include improved product yield, higher product concentration and productivity, and product tolerance. Also, the production strain should be designed in a way that the whole process becomes operationally inexpensive by system-wide optimization of midstream and downstream processes.[109] Advances in metabolic engineering and synthetic biology have developed superior microorganisms for the production of these biofuels. However, due to economic feasibility, biofuel yields must be extremely close to theoretical yields. High product titers of various biofuels have been achieved, though these results are only on the laboratory scale, and many more factors need to be taken into consideration when scaling up to the industrial level. Although this task of fine-tuning microorganisms to produce biofuels at commercial levels is a daunting one, the environmental and economic advantages of biofuels are worth the considerable investment in pathway and strain development.

Acknowledgments

This work was partially supported by grant to Sanjay Guleria from Department of Biotechnology, Government of India under CREST Award (grant no. BT/IN/DBT-CREST Awards/38/SG/2012-13).

References

1. Weeks AM, Chang MCY. Constructing de novo biosynthetic pathways for chemical synthesis inside living cells. *Biochemistry* 2011;**50**:5404–18.
2. Keasling JD. Manufacturing molecules through metabolic engineering. *Science* 2010;**330**:1355–8.
3. Du J, Shao Z, Zhao H. Engineering microbial factories for synthesis of value-added products. *J Ind Microbiol Biotechnol* 2011;**38**:873–90.
4. Zhu L, Zhu Y, Zhang Y, Li Y. Engineering the robustness of industrial microbes through synthetic biology. *Trends Microbiol* 2012;**20**:94–101.
5. Seo SW, Yang J, Min BE, Jang S, Lim JH, Lim HG, et al. Synthetic biology: tools to design microbes for the production of chemicals and fuels. *Biotechnol Adv* 2013;**31**:811–7.
6. McEwen JT, Atsumi S. Alternative biofuel production in non-natural hosts. *Curr Opin Biotechnol* 2012;**23**:744–50.
7. Peralta-Yahya PP, Zhang F, del Cardayre SB, Keasling JD. Microbial engineering for the production of advanced biofuels. *Nature* 2012;**488**:320–8.
8. Huffer S, Roche CM, Blanch HW, Clark DS. *Escherichia coli* for biofuel production: bridging the gap from promise to practice. *Trends Biotechnol* 2012;**30**:538–45.
9. Gronenberg LS, Marcheschi RJ, Liao JC. Next generation biofuel engineering in prokaryotes. *Curr Opin Chem Biol* 2013;**17**:462–71.
10. Santos CNS, Koffas M, Stephanopoulos G. Optimization of a heterologous pathway for the production of flavonoids from glucose. *Metab Eng* 2011;**13**:392–400.
11. Fischer CR, Klein-Marcuschamer D, Stephanopoulos G. Selection and optimization of microbial hosts for biofuels production. *Metab Eng* 2008;**10**:295–304.
12. Alper H, Stephanopoulos G. Engineering for biofuels: exploiting innate microbial capacity or importing biosynthetic potential? *Nat Rev Microbiol* 2009;**7**:715–23.

13. Biggs BW, De Paepe B, Santos CNS, De Mey M, Ajikumar PK. Multivariate modular metabolic engineering for pathway and strain optimization. *Curr Opin Biotechnol* 2014;**29**:156–62.
14. Liu T, Vora H, Khosla C. Quantitative analysis and engineering of fatty acid biosynthesis in *E. coli*. *Metab Eng* 2010;**12**:378–86.
15. Cornishbowden A, Hofmeyr JHS, Cardenas ML. Strategies for manipulating metabolic fluxes in biotechnology. *Bioorg Chem* 1995;**23**:439–49.
16. Bhan N, Xu P, Koffas MAG. Pathway and protein engineering approaches to produce novel and commodity small molecules. *Curr Opin Biotechnol* 2013;**24**:1137–43.
17. Yoshikuni Y, Dietrich JA, Nowroozi FF, Babbitt PC, Keasling JD. Redesigning enzymes based on adaptive evolution for optimal function in synthetic metabolic pathways. *Chem Biol* 2008;**15**:607–18.
18. Peralta-yahya PP, Ouellet M, Chan R, Mukhopadhyay A, Keasling JD, Lee TS. Identification and microbial production of a terpene-based advanced biofuel. *Nat Commun* 2011;**2**:483–8.
19. Yadav VG, Stephanopoulos G. Reevaluating synthesis by biology. *Curr Opin Microbiol* 2010;**13**:371–6.
20. Zha W, Shao Z, Frost JW, Zhao H. Rational pathway engineering of type I fatty acid synthase allows the biosynthesis of triacetic acid lactone from D-glucose *in vivo*. *J Am Chem Soc* 2004;**126**:4534–5.
21. Nair NU, Zhao H. Evolution in reverse: engineering a D-xylose-specific xylose reductase. *ChemBioChem* 2008;**9**:1213–5.
22. Zhang K, Sawaya MR, Eisenberg DS, Liao JC. Expanding metabolism for biosynthesis of nonnatural alcohols. *Proc Natl Acad Sci U S A* 2008;**105**:20653–8.
23. Nobeli I, Favia AD, Thornton JM. Protein promiscuity and its implications for biotechnology. *Nat Biotechnol* 2009;**27**:157–67.
24. Leonard E, Ajikumar PK, Thayer K, Xiao W-H, Mo JD, Tidor B, et al. Combining metabolic and protein engineering of a terpenoid biosynthetic pathway for overproduction and selectivity control. *Proc Natl Acad Sci U S A* 2010;**107**:13654–9.
25. Blank K, Morfill J, Gumpp H, Gaub HE. Functional expression of *Candida antarctica* lipase B in *Escherichia coli*. *J Biotechnol* 2006;**125**:474–83.
26. Di Lorenzo M, Hidalgo A, Haas M, Bornscheuer UT. Heterologous production of functional forms of *Rhizopus oryzae* lipase in *Escherichia coli*. *Appl Environ Microbiol* 2005;**71**:8974–7.
27. Xu Y, Yasin A, Tang R, Scharer JM, Moo-Young M, Chou CP. Heterologous expression of lipase in *Escherichia coli* is limited by folding and disulfide bond formation. *Appl Microbiol Biotechnol* 2008;**81**:79–87.
28. Georgiou G, Valax P. Expression of correctly folded proteins in *Escherichia coli*. *Curr Opin Biotechnol* 1996;**7**:190–7.
29. Young EM, Comer AD, Huang H, Alper HS. A molecular transporter engineering approach to improving xylose catabolism in *Saccharomyces cerevisiae*. *Metab Eng* 2012;**14**:401–11.
30. Ha S-J, Galazka JM, Joong Oh E, Kordić V, Kim H, Jin Y-S, et al. Energetic benefits and rapid cellobiose fermentation by *Saccharomyces cerevisiae* expressing cellobiose phosphorylase and mutant cellodextrin transporters. *Metab Eng* 2013;**15**:134–43.
31. Ishihara A, Asada Y, Takahashi Y, Yabe N, Komeda Y, Nishioka T, et al. Metabolic changes in *Arabidopsis thaliana* expressing the feedback-resistant anthranilate synthase alpha subunit gene OASA1D. *Phytochemistry* 2006;**67**:2349–62.
32. Shen CR, Liao JC. Metabolic engineering of *Escherichia coli* for 1-butanol and 1-propanol production via the keto-acid pathways. *Metab Eng* 2008;**10**:312–20.
33. Báez-Viveros JL, Osuna J, Hernández-Chávez G, Soberón X, Bolívar F, Gosset G. Metabolic engineering and protein directed evolution increase the yield of L-phenylalanine synthesized from glucose in *Escherichia coli*. *Biotechnol Bioeng* 2004;**87**:516–24.
34. Donald K, Hampton R, Fritz I. Effects of overproduction of the catalytic domain of 3-hydroxy-3-methylglutaryl coenzyme A reductase on squalene synthesis in *Saccharomyces cerevisiae*. *Appl Environ Microbiol* 1997;**63**:3341–4.
35. Eriksen DT, Lian J, Zhao H. Protein design for pathway engineering. *J Struct Biol* 2014;**185**:234–42.

36. Geck MK, Kirsch JF. A novel, definitive test for substrate channeling illustrated with the aspartate aminotransferase/malate dehydrogenase system. *Biochemistry* 1999;**38**:8032–7. http://dx.doi.org/10.1021/bi983029c.
37. Graham JWA, Williams TCR, Morgan M, Fernie AR, Ratcliffe RG, Sweetlove LJ. Glycolytic enzymes associate dynamically with mitochondria in response to respiratory demand and support substrate channeling. *The Plant Cell* 2007;**19**:3723–38.
38. Suss KH, Arkona C, Manteuffel R, Adler K. Calvin cycle multienzyme complexes are bound to chloroplast thylakoid membranes of higher plants in situ. *Proc Natl Acad Sci U S A* 1993;**90**:5514–8.
39. Haggie PM, Verkman AS. Diffusion of tricarboxylic acid cycle enzymes in the mitochondrial matrix *in vivo*. Evidence for restricted mobility of a multienzyme complex. *J Biol Chem* 2002;**277**:40782–8.
40. Spivey HO, Merz JM. Metabolic compartmentation. *BioEssays News Rev Mol Cell Dev Biol* 1989;**10**:127–30.
41. Ljungcrantz P, Carlsson H, Mansson MO, Buckel P, Mosbach K, Buelow L. Construction of an artificial bifunctional enzyme, beta-galactosidase/galactose dehydrogenase, exhibiting efficient galactose channeling. *Biochemistry* 1989;**28**:8786–92.
42. Wang C, Yoon S-H, Jang H-J, Chung Y-R, Kim J-Y, Choi E-S, et al. Metabolic engineering of *Escherichia coli* for α-farnesene production. *Metab Eng* 2011;**13**:648–55.
43. Dueber JE, Wu GC, Malmirchegini GR, Moon TS, Petzold CJ, Ullal AV, et al. Synthetic protein scaffolds provide modular control over metabolic flux. *Nat Biotechnol* 2009;**27**:753–9.
44. Lee H, DeLoache WC, Dueber JE. Spatial organization of enzymes for metabolic engineering. *Metab Eng* 2012;**14**:242–51.
45. Maier T, Güell M, Serrano L. Correlation of mRNA and protein in complex biological samples. *FEBS Lett* 2009;**583**:3966–73.
46. Patnaik P. Optimizing initial plasmid copy number distribution for improve*d protein activit*y in a recombinant fermentation. *Biochem Eng J* 2000;**5**:101–7.
47. Huang M, Chen Y, Liu J. Chromosomal engineering of *Escherichia coli* for efficient production of coenzyme Q10. *Chin J Chem Eng* 2014;**22**:559–69.
48. Jensen PR, Hammer K. Artificial promoters for metabolic optimization. *Biotechnol Bioeng* 1998;**58**:191–5.
49. Jensen PR, Hammer K. The sequence of spacers between the consensus sequences modulates the strength of prokaryotic promoters. *Appl Environ Microbiol* 1998;**64**:82–7.
50. Alper H, Fischer C, Nevoigt E, Stephanopoulos G. Tuning genetic control through promoter engineering. *Proc Natl Acad Sci U S A* 2005;**102**:12678–83.
51. Braatsch S, Helmark S, Kranz H, Koebmann B, Jensen PR. *Escherichia coli* strains with promoter libraries constructed by Red/ET recombination pave the way for transcriptional fine-tuning. *BioTechniques* 2008;**45**:335–7.
52. Miksch G, Bettenworth F, Friehs K, Flaschel E, Saalbach A, Twellmann T, et al. Libraries of synthetic stationary-phase and stress promoters as a tool for fine-tuning of expression of recombinant proteins in *Escherichia coli*. *J Biotechnol* 2005;**120**:25–7.
53. Blazeck J, Liu L, Redden H, Alper H. Tuning gene expression in *Yarrowia lipolytica* by a hybrid promoter approach. *Appl Environ Microbiol* 2011;**77**:7905–14.
54. Ajikumar PK, Xiao W-H, Tyo KEJ, Wang Y, Simeon F, Leonard E, et al. Isoprenoid pathway optimization for taxol precursor overproduction in *Escherichia coli*. *Science* 2010;**330**:70–4.
55. Zhang F, Carothers JM, Keasling JD. Design of a dynamic sensor-regulator system for production of chemicals and fuels derived from fatty acids. *Nat Biotechnol* 2012;**30**:354–9.
56. Zhang Z, Donaldson AA, Ma X. Advancements and future directions in enzyme technology for biomass conversion. *Biotechnol Adv* 2012;**30**:913–9.
57. Xu P, Gu Q, Wang W, Wong L, Bower AGW, Collins CH, et al. Modular optimization of multi-gene pathways for fatty acids production in *E. coli*. *Nat Commun* 2013;**4**:1409.
58. Xu P, Wang W, Li L, Bhan N, Zhang F, Koffas MAG. Design and kinetic analysis of a hybrid promoter-regulator system for malonyl-CoA sensing in *Escherichia coli*. *ACS Chem Biol* 2014;**9**:451–8.

59. Farmer WR, Liao JC. Improving lycopene production in *Escherichia coli* by engineering metabolic control. *Nat Biotechnol* 2000;**18**:533–7.
60. Fung E, Wong WW, Suen JK, Bulter T, Lee S, Liao JC. A synthetic gene-metabolic oscillator. *Nature* 2005;**435**:118–22.
61. Bulter T, Lee S-G, Wong WW, Fung E, Connor MR, Liao JC. Design of artificial cell-cell communication using gene and metabolic networks. *Proc Natl Acad Sci* 2004;**101**:2299–304.
62. Dahl RH, Zhang F, Alonso-Gutierrez J, Baidoo E, Batth TS, Redding-Johanson AM, et al. Engineering dynamic pathway regulation using stress-response promoters. *Nat Biotechnol* 2013;**31**:1039–46.
63. Tsao C-Y, Hooshangi S, Wu H-C, Valdes JJ, Bentley WE. Autonomous induction of recombinant proteins by minimally rewiring native quorum sensing regulon of *E. coli*. *Metab Eng* 2010;**12**:291–7.
64. Kudla G, Murray AW, Tollervey D, Plotkin JB. Coding-sequence determinants of gene expression in *Escherichia coli*. *Science* 2009;**324**:255–8.
65. Park YS, Seo SW, Hwang S, Chu HS, Ahn J-H, Kim T-W, et al. Design of 5′-untranslated region variants for tunable expression in *Escherichia coli*. *Biochem Biophys Res Commun* 2007;**356**:136–41.
66. Seo SW, Yang J, Jung GY. Quantitative correlation between mRNA secondary structure around the region downstream of the initiation codon and translational efficiency in *Escherichia coli*. *Biotechnol Bioeng* 2009;**104**:611–6.
67. Na D, Lee S, Lee D. Mathematical modeling of translation initiation for the estimation of its efficiency to computationally design mRNA sequences with desired expression levels in prokaryotes. *BMC Syst Biol* 2010;**4**:71.
68. Salis HM, Mirsky EA, Voigt CA. Automated design of synthetic ribosome binding sites to control protein expression. *Nat Biotechnol* 2009;**27**:946–50.
69. Seo SW, Yang J-S, Kim I, Yang J, Min BE, Kim S, et al. Predictive design of mRNA translation initiation region to control prokaryotic translation efficiency. *Metab Eng* 2013;**15**:67–74.
70. Qi L, Haurwitz RE, Shao W, Doudna JA, Arkin AP. RNA processing enables predictable programming of gene expression. *Nat Biotechnol* 2012;**30**:1002–6.
71. Callura JM, Cantor CR, Collins JJ. Genetic switchboard for synthetic biology applications. *Proc Natl Acad Sci U S A* 2012;**109**:5850–5.
72. Angov E, Legler PM, Mease RM. Adjustment of codon usage frequencies by codon harmonization improves protein expression and folding. *Methods Mol Biol* 2011;**705**:1–13.
73. Shen CR, Lan EI, Dekishima Y, Baez A, Cho KM, Liao JC. Driving forces enable high-titer anaerobic 1-butanol synthesis in *Escherichia coli*. *Appl Environ Microbiol* 2011;**77**:2905–15.
74. Berríos-Rivera SJ, Bennett GN, San K-Y. The effect of increasing NADH availability on the redistribution of metabolic fluxes in *Escherichia coli* chemostat cultures. *Metab Eng* 2002;**4**:230–7.
75. Valle-Rodríguez JO, Shi S, Siewers V, Nielsen J. Metabolic engineering of *Saccharomyces cerevisiae* for production of fatty acid ethyl esters, an advanced biofuel, by eliminating non-essential fatty acid utilization pathways. *Appl Energy* 2014;**115**:226–32.
76. Chen X, Nielsen KF, Borodina I, Kielland-Brandt MC, Karhumaa K. Increased isobutanol production in *Saccharomyces cerevisiae* by overexpression of genes in valine metabolism. *Biotechnol Biofuels* 2011;**4**:21.
77. Burgard AP, Pharkya P, Maranas CD. Optknock: a bilevel programming framework for identifying gene knockout strategies for microbial strain optimization. *Biotechnol Bioeng* 2003;**84**:647–57.
78. Fowler ZL, Gikandi WW, Koffas MAG. Increased malonyl coenzyme A biosynthesis by tuning the *Escherichia coli* metabolic network and its application to flavanone production. *Appl Environ Microbiol* 2009;**75**:5831–9.
79. Wagner A, Zarecki R, Reshef L, Gochev C, Sorek R, Gophna U, et al. Computational evaluation of cellular metabolic costs successfully predicts genes whose expression is deleterious. *Proc Natl Acad Sci U S A* 2013;**110**:19166–71.
80. Chowdhury A, Zomorrodi AR, Maranas CD. k-OptForce: integrating kinetics with flux balance analysis for strain design. *PLoS Comput Biol* 2014;**10**:e1003487.

81. Ranganathan S, Suthers PF, Maranas CD. OptForce: an optimization procedure for identifying all genetic manipulations leading to targeted overproductions. *PLoS Comput Biol* 2010;**6**, e1000744.
82. Patil KR, Rocha I, Förster J, Nielsen J. Evolutionary programming as a platform for *in silico* metabolic engineering. *BMC Bioinformatics* 2005;**6**:308.
83. Tepper N, Shlomi T. Predicting metabolic engineering knockout strategies for chemical production: accounting for competing pathways. *Bioinformatics* 2010;**26**:536–43.
84. Pharkya P, Maranas CD. An optimization framework for identifying reaction activation/inhibition or elimination candidates for overproduction in microbial systems. *Metab Eng* 2006;**8**:1–13.
85. Yang L, Cluett WR, Mahadevan R. EMILiO: a fast algorithm for genome-scale strain design. *Metab Eng* 2011;**13**:272–81.
86. Fong SS, Burgard AP, Herring CD, Knight EM, Blattner FR, Maranas CD, et al. *In silico* design and adaptive evolution of *Escherichia coli* for production of lactic acid. *Biotechnol Bioeng* 2005;**91**:643–8.
87. Brochado AR, Matos C, Møller BL, Hansen J, Mortensen UH, Patil KR. Improved vanillin production in baker's yeast through *in silico* design. *Microb Cell Factories* 2010;**9**:84.
88. Asadollahi MA, Maury J, Patil KR, Schalk M, Clark A, Nielsen J. Enhancing sesquiterpene production in *Saccharomyces cerevisiae through in silico* driven metabolic engineering. *Metab Eng* 2009;**11**:328–34.
89. McAnulty MJ, Yen JY, Freedman BG, Senger RS. Genome-scale modeling using flux ratio constraints to enable metabolic engineering of clostridial metabolism *in silico*. *BMC Syst Biol* 2012;**6**:42.
90. Yen JY, Nazem-Bokaee H, Freedman BG, Athamneh AIM, Senger RS. Deriving metabolic engineering strategies from genome-scale modeling with flux ratio constraints. *Biotechnol J* 2013;**8**:581–94.
91. Zhuang K, Yang L, Cluett WR, Mahadevan R. Dynamic strain scanning optimization: an efficient strain design strategy for balanced yield, titer, and productivity. DySScO strategy for strain design. *BMC Biotechnol* 2013;**13**:8.
92. Ramos JL, Duque E, Gallegos M-T, Godoy P, Ramos-Gonzalez MI, Rojas A, et al. Mechanisms of solvent tolerance in Gram-negative bacteria. *Annu Rev Microbiol* 2002;**56**:743–68.
93. Bowles LK, Ellefson WL. Effects of butanol on *Clostridium acetobutylicum*. *Appl Environ Microbiol* 1985;**50**:1165–70.
94. Ding J, Huang X, Zhang L, Zhao N, Yang D, Zhang K. Tolerance and stress response to ethanol in the yeast *Saccharomyces cerevisiae*. *Appl Microbiol Biotechnol* 2009;**85**:253–63.
95. Sikkema J, de Bont JA, Poolman B. Mechanisms of membrane toxicity of hydrocarbons. *Microbiol Rev* 1995;**59**:201–22.
96. Nicolaou SA, Gaida SM, Papoutsakis ET. A comparative view of metabolite and substrate stress and tolerance in microbial bioprocessing: from biofuels and chemicals, to biocatalysis and bioremediation. *Metab Eng* 2010;**12**:307–31.
97. Dunlop MJ. Engineering microbes for tolerance to next-generation biofuels. *Biotechnol Biofuels* 2011;**4**:32.
98. Lo T-M, Teo WS, Ling H, Chen B, Kang A, Chang MW. Microbial engineering strategies to improve cell viability for biochemical production. *Biotechnol Adv* 2013;**31**:903–14.
99. Putman M, van Veen HW, Konings WN. Molecular properties of bacterial multidrug transporters. *Microbiol Mol Biol Rev* 2000;**64**:672–93.
100. Dunlop MJ, Dossani ZY, Szmidt HL, Chu HC, Lee TS, Keasling JD, et al. Engineering microbial biofuel tolerance and export using efflux pumps. *Mol Syst Biol* 2011;**7**:487.
101. Tomas CA, Welker NE, Papoutsakis ET. Overexpression of groESL in *Clostridium acetobutylicum* results in increased solvent production and tolerance, prolonged metabolism, and changes in the cell's transcriptional program. *Appl Environ Microbiol* 2003;**69**:4951–65.
102. Junker F, Ramos JL. Involvement of the cis/trans isomerase Cti in solvent resistance of *Pseudomonas putida* DOT-T1E. *J Bacteriol* 1999;**181**:5693–700.
103. Gaida SM, Al-Hinai MA, Indurthi DC, Nicolaou SA, Papoutsakis ET. Synthetic tolerance: three noncoding small RNAs, DsrA, ArcZ and RprA, acting supra-additively against acid stress. *Nucleic Acids Res* 2013;**41**:8726–37.

104. Na D, Yoo SM, Chung H, Park H, Park JH, Lee SY. Metabolic engineering of *Escherichia coli* using synthetic small regulatory RNAs. *Nat Biotechnol* 2013;**31**:170–4.
105. Zhao Y, Hindorff LA, Chuang A, Monroe-Augustus M, Lyristis M, Harrison ML, et al. Expression of a cloned cyclopropane fatty acid synthase gene reduces solvent formation in *Clostridium acetobutylicum* ATCC 824. *Appl Environ Microbiol* 2003;**69**:2831–41.
106. Miller EN, Jarboe LR, Yomano LP, York SW, Shanmugam KT, Ingram LO. Silencing of NADPH-dependent oxidoreductase genes (yqhD and dkgA) in furfural-resistant ethanologenic *Escherichia coli*. *Appl Environ Microbiol* 2009;**75**:4315–23.
107. Zhao X, Jiang R, Li N, Yang Q, Bai F. Improving ethanol tolerance of *Saccharomyces cerevisiae* industrial strain by directed evolution of SPT3. *Chin J Biotechnol* 2010;**26**:159–64.
108. Lee JY, Yang KS, Jang SA, Sung BH, Kim SC. Engineering butanol-tolerance in *Escherichia coli* with artificial transcription factor libraries. *Biotechnol Bioeng* 2011;**108**:742–9.
109. Jang Y-S, Park JM, Choi S, Choi YJ, Seung DY, Cho JH, et al. Engineering of microorganisms for the production of biofuels and perspectives based on systems metabolic engineering approaches. *Biotechnol Adv* 2012;**30**:989–1000.
110. Luan G, Cai Z, Li Y, Ma Y. Genome replication engineering assisted continuous evolution (GREACE) to improve microbial tolerance for biofuels production. *Biotechnol Biofuels* 2013;**6**:137.

CHAPTER 5

RNA-Based Molecular Sensors for Biosynthetic Pathway Design, Evolution, and Optimization

Cassandra R. Burke, James M. Carothers
Departments of Chemical Engineering and Bioengineering, Molecular Engineering & Sciences Institute, Center for Synthetic Biology, University of Washington, Seattle, Washington, USA

5.1 Introduction to Biofuel Pathways

Advances in microbial engineering over the past decade have offered an attractive alternative to traditional chemical synthesis of fuels and fuel additives.[1] Isoprenoid, fatty acid, and long-chain alcohol-based biofuels display physical properties similar to petroleum-derived fuels and have been the focus of microbial engineering.[2–4] These chemicals provide a high-energy content, and are compatible with the current transportation and storage infrastructure used for petroleum-derived fuels. These characteristics allow for integration into the existing fuel market, further increasing their commercial viability.[5]

The material discussed in this chapter, although relevant to biofuel pathway design, evolution, and optimization, is applicable to *all* systems where predictable and customizable small molecule sensing is desired.

5.1.1 Current Limitations to Pathway Design and Screening

Despite tremendous improvements in microbial pathway engineering through the design of complex genetic control circuits, industrially relevant titers are still difficult to obtain.

The most successful chemical production has come from heterologous pathway construction in nonnative hosts with no innate regulation.[6] Unregulated expression, however, can lead to poor cellular fitness.[7] High levels of expression of pathway enzymes create a metabolic imbalance and consume vital cellular resources. On the other hand, low levels of expression generate bottlenecks in pathway production and insufficient chemical titers. In addition, heterologous enzymes and pathway intermediates are often toxic to the host and reduce productivity. The focus of biosynthetic pathway engineering

Biotechnology for Biofuel Production and Optimization. http://dx.doi.org/10.1016/B978-0-444-63475-7.00005-4

has, therefore, been shifted to the design of regulatory control systems where the natural survival of the host is balanced with production of the desired chemical compound. The ability to detect pathway metabolites and regulate the expression of heterologous genes would aid in the design, evolution, and optimization of these complex chemical networks (Figure 5.1).

Natural selection is nature's way of enriching a population for a desired phenotype, and often takes many generations. Directed evolution mimics natural selection in the laboratory by screening a large number of genotypes under an induced selective pressure, and identifying those that display a preferred phenotype. The success of this strategy relies on the ability to identify and isolate desired phenotypes from the general population. In terms of metabolic engineering, the goal is to maximize production titers while maintaining overall fitness. Developments in directed evolution phenotype screening have been heavily focused on cellular fitness, as host resistance to toxic pathway intermediates and products is directly correlated to cellular growth and survival rates.[8] Screening for increased production titers is more difficult because only a small fraction of engineered pathway products, specifically those that happen to be chromophores or fluorophores, can be screened using traditional high-throughput methods. Quantification and detection of the majority of small molecules has conventionally required mass spectrometry, a technique that is labor intensive and low throughput by nature.[9] This is a fundamental limitation in directed evolution experiments for biofuel production pathways and, more importantly, metabolic engineering as a whole. This offers yet another capacity where detection and quantification of small molecule metabolites would improve overall pathway productivity.

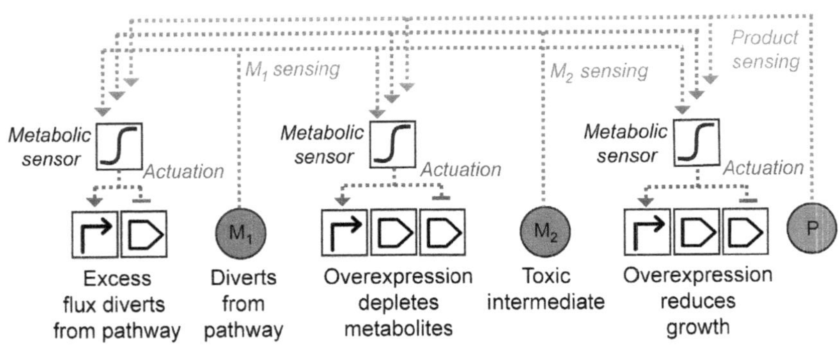

Figure 5.1

Metabolic sensors for the dynamic control of pathway flux. The heterologous gene expression necessary for many synthetically engineered pathways often results in poor host cellular fitness. Over expression of necessary genes depletes cellular metabolites, and many intermediates produced are toxic to the native host. Metabolic sensors specific to pathway intermediates (M_1 and M_2) and products (P) provide the capacity to dynamically control the expression of pathway genes, reducing cellular burden and increasing titers.

5.1.2 Engineered Biosensors for Biofuel Pathway Optimization

Small molecule metabolites have been shown to be extensively involved in cellular signaling and pathway regulation.[10] Naturally occurring metabolite-binding proteins and transcription factors are two of nature's solutions to regulate pathway flux in native hosts by detecting metabolic intermediates. Many of these protein-based sensors have been reengineered to produce microbes that act as environmental reporters.[11] For biosynthetic pathway engineering, sensors that detect metabolic intermediates, rather than environmental chemicals, would provide the capability to dynamically control pathway flux while preserving overall host fitness.

Fatty acids are ideal biodiesel replacements and are suitable for production in microbial hosts because of low overall toxicity and high-energy content. Zhang et al.[12] engineered a fatty acid/acyl-coenzyme A (CoA) sensor from the natural fatty acid-sensing protein FadR; a ligand-responsive transcription factor that regulates the native fatty acid biosynthesis pathway in bacteria by blocking transcription through DNA binding.[13] The DNA-binding activity is directly inhibited by acyl-CoA, an activated form of fatty acid. FadR was put under the control of a synthetic promoter to increase the dynamic range of the natural sensor. When fatty acid is present, it is converted to acyl-CoA, which binds FadR. This cascade results in liberation of the synthetic promoter and induced transcription. This sensor was shown to be responsive to both exogenous and endogenously produced fatty acids in *Escherichia coli*. In addition, the FadR-inspired sensor was engineered into a dynamic sensor-regulated system and increased fatty acid titers nearly threefold to approximately 28% of the projected theoretical maximum in *E. coli*.

Some of the most successfully engineered biosynthetic pathways involve the expression of nonnative genes. Dietrich et al.[14] constructed an alcohol metabolic sensor, a biofuel-relevant chemical, through a nonnative transcription factor-promoter pair. A putative transcription factor and alcohol-controlled promoter from *Thauera butanovora* were engineered into a 1-butanol biosensor in *E. coli*. The sensor was subjected to directed evolution in an engineered microbial strain that contained a heterologous pathway for 1-butanol production. A 35% increase in titer was observed from the best performing variant. In addition, the engineered 1-butanol molecular sensor was shown to effectively aid in screening and selecting desired host phenotypes, specifically increased 1-butanol production coupled with cellular fitness.

Naturally occurring ligand-responsive transcription factors have also been engineered to bind nonnative target metabolites, such as those that are intermediates in biofuel chemical production pathways. Tang and Cirino[15] performed directed evolution experiments that mutated the binding pocket of the *E. coli* AraC protein, a native L-arabinose-binding transcription factor, to bind mevolonate (MEV). MEV is an intermediate in the heterologous isoprenoid pathway, and has been shown to be a limiting step in pathway productivity.[16] The mutated AraC protein was selective toward MEV, and offered a molecular sensor that could directly monitor an intermediate in a biofuel chemical producing pathway.

The molecular sensors discussed in this section show tremendous advances in biosynthetic pathway engineering. The fatty acid/acyl-CoA sensor demonstrates the optimization of pathway production titers through the use of relevant small molecule sensors and dynamic control of pathway flux. The 1-butanol and MEV-engineered sensors offer a unique solution for the quantification of pathway intermediates and products that is necessary for directed evolution phenotype screening. The above sensors were evolved from naturally occurring metabolite-binding proteins and transcription factors. This approach is ideal if the small molecule target of interest has a natural sensor or one that could *easily* be evolved. However, many biosynthetic-pathway-relevant metabolites do not have a natural sensor. An alternative approach to protein metabolic sensors is RNA-based sensors. RNA regulatory elements exist in nature, can be readily engineered and selected, are generally shorter coding sequences, do not need to be translated, and offer the possibility of a high-throughput computational design framework for rational and customizable sensors.

5.2 RNA-Based Regulatory Elements

5.2.1 Natural RNA-Based Metabolite Sensors

Nature has evolved RNA sequences that fold into functional secondary and tertiary structures capable of regulating transcription, translation, splicing, and protein function.[17] A diverse range of small-molecule-responsive RNAs (riboswitches) have been identified in bacteria, and certain plants and fungi. Riboswitches are composed of a metabolite-sensing domain (aptamer) and an expression platform that controls gene expression. RNA aptamers fold into a functional secondary structure that binds a target ligand with high specificity and affinity. Naturally occurring riboswitches that regulate bacterial biosynthetic pathways such as those involved in purine, vitamin, and amino acid metabolism have been identfied.[18]

The metabolite-binding activity of an RNA aptamer is highly dependent on the secondary structure and fold of the nucleic acid. When the target small molecule is bound to the aptamer domain, it has a direct influence on the secondary structure of the expression platform. The family of riboswitches that bind the essential cofactor *S*-adenosyl-methionine (SAM) contains an expression platform that consists of transcription terminator and antiterminator sequences. The secondary structure of the expression platform displays the terminator structure when the SAM ligand is bound, suppressing transcription. However, in the absence of SAM, the terminator is sequestered by the antiterminator sequence resulting in transcriptional read through. Across all the classes of riboswitches, the mechanism of the expression platform has been shown to span transcriptional regulation, as in the case of the SAM riboswitch, translational regulation by sequestration of the ribosomal-binding site

(RBS) shown by the thiamine pyrophosphate-dependent riboswitch, regulation of mRNA degradation rates shown by the gluclosamine-6-phosphate (*glmS*) riboswitch, and alternative splicing control shown by the cyclic di-guanosyl-5′-monophosphate-responsive riboswitch. For a review of the above examples and additional naturally occurring riboswitches, please see Serganov and Nudler.[19]

5.2.2 Synthetic RNA-Based Metabolite Sensors

Engineered RNA regulatory elements and metabolite-responsive sensors have been constructed from known RNA aptamer domains, and have been shown to regulate gene expression by translational, transcriptional, and degradation-dependent control (Figure 5.2).[21]

Synthetic riboswitches have been engineered from *in vitro* selected RNA aptamer domains. Like naturally occurring riboswitches, these synthetic sensors bind a small molecule target and control gene expression through the secondary structure of an expression

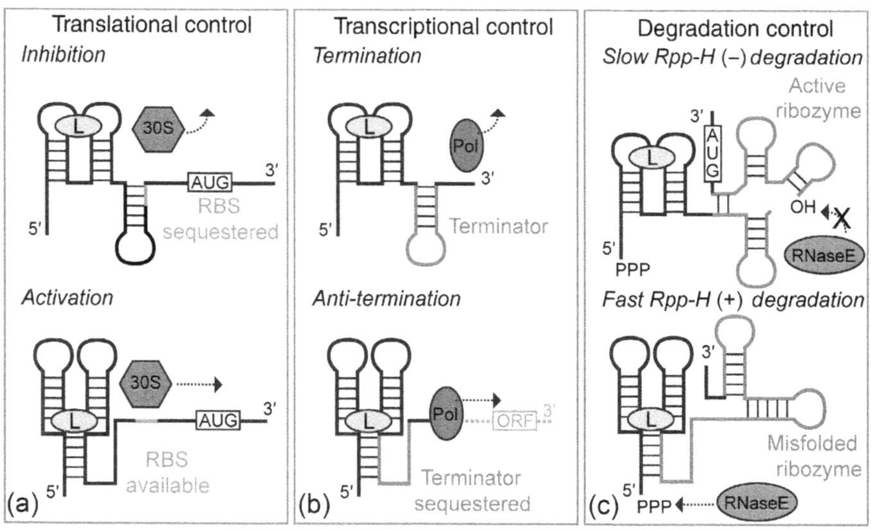

Figure 5.2
Synthetic functional RNAs for translational, transcriptional, and degradation control. Naturally occurring and engineered RNAs that control gene expression consist of a ligand (yellow)-binding aptamer and an expression platform. The composition of the expression platform dictates the actuation method of the sensor. (a) RNA sensors for translational control contain an expression platform where the ribosomal-binding site (RBS) can be sequestered or available to dictate translation. (b) RNAs that control transcriptional regulation consist of an expression platform with terminator and antiterminator sequences. Transcription is inhibited when a terminator secondary structure is formed, and activated when the antiterminator sequence disrupts terminator formation. (c) Carothers et al.[20] controlled mRNA degradation in *E. coli* with an engineered aptazyme sensor. The self-cleavage reaction generates a 5′ hydroxyl, which is degraded slower than the canonical 5′ triphosphate under RNA pyrophosphohydrolase (Rpp-H) degradation.

platform. An RNA aptamer that binds the small molecule ligand theophylline was the first, and currently the most widely used aptamer for riboswitch engineering.[22] Desai and Gallivan[23] designed a riboswitch that provided translational control by the small molecule theophylline. The theophylline aptamer was inserted five base pairs upstream of the RBS of a β-galactosidase reporter gene. The synthetic riboswitch was hypothesized to decrease reporter translation in the presence of the target ligand due to increased secondary structure near the RBS. However, translation activation was increased in the presence of theophylline, and suggested that the secondary structure of the ligand-bound aptamer provided access to the RBS of the reporter gene rather than sequestering it. In addition, cells that contained the synthetic riboswitch could be selectively amplified, demonstrating the potential of RNA sensors as a means for screening phenotypes.

Wachsmuth et al.[24] designed a synthetic riboswitch that regulated transcriptional read through in *E. coli*; the first successful RNA riboswitch-based transcriptional regulator. The theophylline aptamer was used as the sensor, while the expression platform had the capacity to fold into a terminator hairpin structure. Rather than begin with *in vitro* directed evolution experiments, RNA folding algorithms were used to screen the minimum free energy (MFE) structure of the expression platform. In addition, folding simulations depicted by a series of MFE structures calculated for subsequences beginning at the 5′ end, were also implemented. This approach demonstrated the power of *in silico* computational design and screening, and resulted in the *in vitro* testing of far fewer sequences than an entire library. In the absence of theophylline, the secondary structure displays a terminator hairpin and inhibits transcription. However, when theophylline is bound to the riboswitch, the formation of the terminator is hindered providing transcriptional read through. The best sensor was shown to activate gene expression in *E. coli* by approximately 6.5-fold in the presence of the theophylline.

RNA-based molecular sensors called aptazymes have been engineered as well, by combining small-molecule-binding aptamers and catalytic RNAs (ribozymes). Nucleolytic ribozymes that are capable of a self-cleaving reaction have been identified in a wide range of organisms, and are the most common class of ribozyme to be engineered into aptazymes.[25] An ATP-binding aptamer was engineered alongside a hammerhead self-cleaving ribozyme producing a sensor where self-cleavage of the RNA was under the control of ATP binding.[26] Rational incorporation of the ribozyme to the aptamer-generated aptazymes that had catalytic rates that both increased and decreased by the addition of ATP, demonstrating the flexibility of generating RNA sensors that can be positively or negatively regulated by small molecules.

These types of aptazyme sensors have been used to control gene expression in complex engineered biosynthetic pathways by regulating RNA half-life. Carothers et al.[20] engineered an aptazyme that was under the control of *p*-amino-phenylalanine (*p*-AF), a small molecule that has potential applications in polymers such as photoresists, photovoltaic materials, and tissue engineering.[27] The self-cleavage reaction of the aptazyme generates a 5′ hydroxyl, which is degraded slower than the 5′ triphosphate present on the un-cleaved precursor in *E. coli*. The *p*-AF aptazyme was integrated into a small molecule-controlled aptazyme-regulated expression devices (aREDs) and model-driven engineering was shown to quantitatively predict reporter expression in *E. coli*.

In addition to small molecule-controlled RNA-based sensors, such as riboswitches and aptazymes, additional naturally occurring and engineered ligand-independent RNA regulatory elements have been identified. Small RNAs (sRNAs) in prokaryotes regulate gene expression in trans on the posttranscriptional level, and synthetic sRNAs have been shown to knockdown gene expression *in vivo*.[28] In addition, nucleic acid circuits such as strand displacement logic gates have been shown to effectively recognize micro-RNA profiles in living mammalian cells, and perform complex signal processing based on chemical reaction networks.[29,30] These circuits function through Watson-Crick base pairing, and although the cutting-edge work mentioned above was demonstrated with DNA gates, RNA-based gates would function under the same mechanism.[31] The wide diversity of RNA-based regulatory elements, both naturally occurring and synthetically designed, offers an extensive toolbox for engineering metabolite sensors for regulation and optimization of biosynthetic pathways.

5.3 RNA Aptamers for Pathway Intermediates and Products

RNA aptamers can be evolved to bind a target ligand with high specificity and affinity, and thus offer an ideal platform for engineering metabolic sensors.[32] Aside from the functional diversity of RNA regulatory elements discussed in the previous section, RNA offers distinct advantages over protein-based sensors. First, protein sensors require a long genetic program where every amino acid is coded by three nucleotides. Many biosynthetic pathways for biofuel production require numerous exogenous genes, resulting in already very large programs. For example, the fatty acid ethyl ether producing *E. coli* strain that was used in the development of the fatty acid/acyl-CoA sensor discussed earlier, included at least six genes on three plasmids.[33] RNA-based genetic programs are potentially much shorter in length and can be directly coded. In addition, the expression of protein-based sensors entails translation of the genetic program, which consumes valuable cellular resources. In terms of biosynthetic pathway engineering, the fitness of the host is already suboptimal and requiring the expression of an additional protein may be burdensome. RNA-based genetic programs do not need to be translated, conserving cellular resources. Isolation of *de novo* aptamers that bind small molecules of interest is straightforward and does not require an initial candidate to be evolved. Moreover, since the activity of functional RNA is highly dependent on fold and structure, generation of RNA sensors from aptamers could be advanced through the use of computational pipelines that include free energy calculations and co-transcriptional folding simulations.

5.3.1 In Vitro *Selection*

The successful isolation of *de novo* RNA aptamers through *in vitro* selection and evolution experiments has produced aptamers that bind fluorescent dyes, small molecules such as ATP, and protein-small molecule complexes to name a few that are relevant to pathway design and optimization.[34–36] *In vitro* selection or systematic evolution of ligands by exponential enrichment (SELEX) is a method of directed evolution, and begins with a large synthetic or genomic DNA library.[37,38] Current synthesis technologies can typically generate up to a 200-nucleotide long DNA library with a diversity on the order of 10^{12}-10^{15} different sequences. In the most common versions, the nucleotide composition of the synthetic library requires a stretch of random nucleotides that is flanked by two known terminal primer regions for reverse transcription and polymerase chain reaction (PCR). The random region can be supplemented with a stem-tetraloop, which may stabilize active folds and increase the likelihood of obtaining functional molecules.[39] For RNA aptamer selections, the DNA library is transcribed to RNA, most commonly by T7 RNA polymerase, and the RNA pool is exposed to the target ligand that is immobilized on a solid support. Sequences that bind to the ligand are reverse transcribed and amplified for the next round of selection (Figure 5.3). The increased stringency of each round of selection creates an enriched population of RNAs that bind the target ligand.

Since what many people call the first law of directed evolution states, "you get what you screen for" there are certain considerations that must be taken to improve the fidelity of selected sequences.[40] These considerations are particularly important when selecting RNA

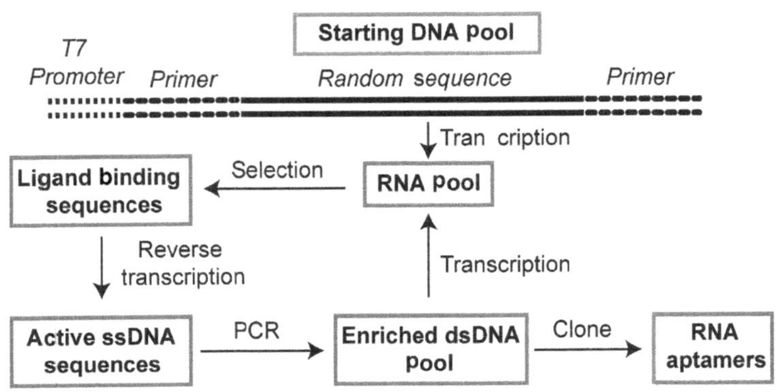

Figure 5.3

In vitro selection for aptamers that bind pathway-relevant metabolites. *In vitro* selection for RNA aptamers (orange) begins with a double-stranded DNA library (blue) that contains a T7 RNA polymerase promoter sequence and two known primer regions that flank a random nucleotide sequence. The library is transcribed to RNA and introduced to target ligand. Those sequences that bind the target are reverse transcribed and amplified. Each round of selection (green) enriches the population for RNAs that bind the target molecule with high specificity and affinity.

aptamers for biosynthetic pathway design, as predictable and well-behaved small molecule binding is essential for precise control of reaction pathways. Mimicking the conditions in which the aptamer is expected to function greatly enhances the quality of the enriched population. Although the aptamers that are relevant to biosynthetic pathway design are expected to function in a host cell, it is impractical to select for high-affinity-binding RNAs *in vivo*. The probability of selecting a high-affinity-binding aptamer is approximately 1 in 10^{20} random sequences and because transformation efficiency in *E. coli* is around 10^9, it is unreasonable to expect to select a high-affinity binder with an *in vivo* selection.[39] Performing an *in vitro* selection in conditions that mimic the intracellular environment, such as temperature and concentrations (particularly salts and magnesium) will help to reduce the optimization needed to transfer the aptamer *in vivo*.[41]

Stringent and diligent counterselections ensure that the aptamer is specific to one target molecule, and not a closely related intermediate. The theophylline aptamer does not bind the small molecule caffeine, which differs by a single methyl group, and this is one of the many reasons it is the most utilized aptamer in riboswitch egnineering.[22] In biofuel-relevant pathways, such as the isopropanol and *n*-butanol production pathways, there are many intermediates that are structurally very similar. In order to generate the most efficient and predictable RNA sensor, the RNA aptamer must be able to discriminate between these closely related molecules. Introducing a structurally similar target ligand as a counterselection will help screen for the specificity of the desired RNA-target interaction. Discarding those RNA sequences that bind this off-target ligand will reduce background-binding effects when the sensor is placed in the context of the desired engineered pathway.

In vitro selections always result in the simplest solution to a given problem.[39] After an RNA aptamer variant has successfully been identified, it is beneficial to reselect this aptamer to improve the binding affinity. Random mutagenesis of a selected aptamer probes the sequence space near the active variant. A reselection of a variant of the *p*-AF aptamer that was engineered into an aptazyme that controlled gene expression in *E. coli* (discussed in Section 5.2.2), was shown to increase binding affinity by approximately 20-fold.[41] Site-specific mutagenesis is also a valuable tool in determining which portions of the RNA secondary structure and sequence are necessary for activity, and which are malleable. This information is useful when engineering RNA parts together, such as aptamers and ribozymes, and integrating these parts into a biologically relevant context, such as a biosynthetic pathway.

5.3.2 Informational Complexity

The secondary structure of functional RNAs is responsible for the activity; however, the relationship between complexity and binding affinity of RNA aptamer-ligand pairs is difficult to quantify. Carothers et al.[39] investigated a set of RNA aptamers that bound the small molecule target guanosine triphosphate (GTP) to assess the relationship between structural complexity and ligand-binding affinity. High-affinity GTP-binding aptamers are structurally more complex, and more rare in a random RNA library than low-affinity aptamers. It was demonstrated that in order to increase the binding affinity of a particular aptamer 10-fold, required 10 additional bits of information (five specified nucleotide positions), and decreased the overall abundance in a random RNA library by 1000-fold. Furthermore, selection of high-affinity aptamers is dependent on the information that is inherent to the chemical nature of the target ligand. Interactions such as hydrogen bonding, planer stacking, and intercalation have been shown to be the primary means of contact between RNA aptamers and target ligands. Small, flexible, and hydrophobic target ligands are unable to form these types of interactions, and it is reasonable to suggest high-affinity RNA aptamers for these targets will be less abundant in a random pool. Large, ridged, and hydrogen bond capable molecules facilitate these interactions well, and are more likely to have high-affinity aptamers in a random library.

Carothers et al.[39,41] formalized a metric for calculating the informational complexity (IC) of RNA structures. This metric can be used to evaluate the probability that a sequence with a specific structural complexity exists in a random RNA library. The IC-binding affinity relationship can be estimated for small molecule ligands from the molecular weight, number of rotatable bonds, and the composition of available functional groups. For example, an RNA-based sensor for a toxic intermediate would require a binding affinity within the toxic concentration range. An estimation of the IC needed to select an aptamer with a binding affinity in this range might show that an RNA aptamer cannot be selected that fits these criteria, or that another pathway intermediate might be more appropriate as a target.

Selecting RNA aptamers for targets in biofuel-relevant pathways provides some additional challenges. Biofuels such as long-chain alcohols, isoprenoids, and fatty acids are small, hydrophobic molecules with limited chemical functionality (Figure 5.4a). High-affinity RNA aptamers for these biofuel molecules are predicted to be rare in a random RNA library. However, many of the intermediates in these pathways are better suited as aptamers targets because their chemical nature is more conducive to RNA interactions. For example, the isopropanol and *n*-butanol production pathways in *E. coli*, contain CoA derivatives, which offer a handle for RNA binding (Figure 5.4b). Extra care must be taken in the counterselection step; however, to ensure that the aptamer is specific to only one CoA derivative and not similar pathway intermediates, or other cellular components. The fatty acid biosynthesis pathway contains CoA derivatives, as well as acyl carrier protein species. It has been shown, that RNA aptamers can be selected for protein-small molecule complexes where the protein provides an additional surface for RNA interactions.[36] This strategy might offer an alternative approach for selecting aptamers for intermediates in the fatty acid biosynthesis pathway, under the stipulation that the protein-small molecule complexes can be purified for an *in vitro* selection.

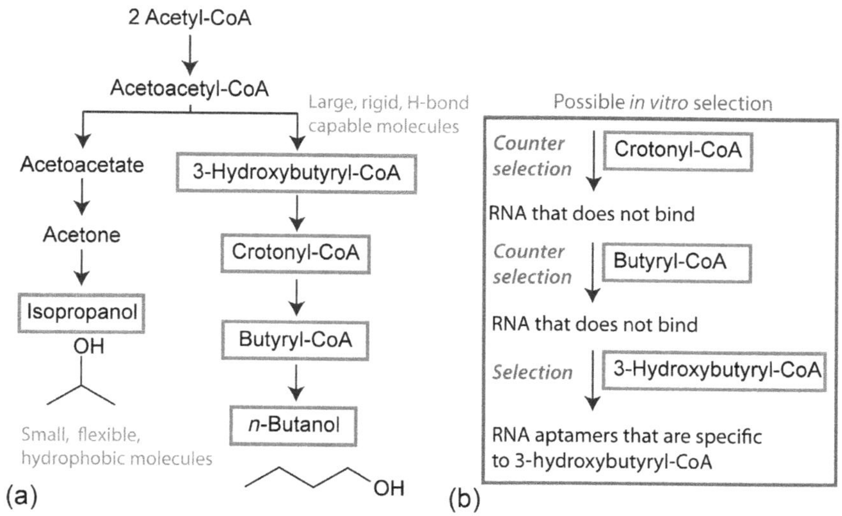

Figure 5.4
Engineered isopropanol and *n*-butanol production pathways in *E. coli*. (a) Random RNA libraries are not anticipated to contain high-affinity aptamers for small, flexible hydrophobic molecules (orange) such as isopropanol and *n*-butanol. Intermediates along the pathway (green) are large, rigid, H-bond capable molecules that are anticipated to have high-affinity RNA aptamers in a random library. (b) In order to generate an RNA aptamer specific to 3-hydroxybutyryl-CoA, a stringent and diligent counterselection against other pathway intermediates is necessary to remove RNA aptamers that bind CoA indiscriminately.

Tools such as IC allow for estimations of the potential binding affinity for a particular ligand target, and can aid in the RNA library and *in vitro* selection design for isolating aptamers with desired characteristics for RNA-based sensors.

5.4 Transforming RNA Aptamers into Sensors

Engineering RNA aptamers into sensors, such as those described in Section 5.2.2, requires an understanding of the relationship between sequence, structure, and function. A complete understanding of which nucleotides and structural elements are necessary for function, and which can be altered, aids in the rational design of RNA small molecule sensors.

5.4.1 In Vitro *Rational Design*

An RNA-based small molecule biosensor requires the integration of an RNA aptamer sensor and an actuator element, such that small molecule ligand binding results in a structural confirmation that generates some measurable output. The two examples discussed in the above sections include riboswitches, where the fold of the expression platform results in genetic control, and aptazymes, where the fold of a ribozyme dictates self-cleavage rates and gene expression. In both instances, the fold is dictated by the presence of a target ligand that stabilizes the aptamer domain and controls the downstream folding trajectory of the actuator. Detailed information about the sequence and structure requirements for both aptamer sensor and actuator function is necessary for engineering robust metabolic sensors with predictable function.

The first aptazymes were generated by combining an ATP-binding aptamer with a hammerhead self-cleaving ribozyme.[26] It was hypothesized that since the ATP aptamer was shown to undergo a significant conformational change upon binding of the small molecule ligand, that this could be exploited to control the confirmation of the ribozyme and therefore control the self-cleavage activity. The stems of the hammerhead ribozyme were known to provide the necessary structural stability to facilitate catalytic self-cleavage; however, their sequence composition could be varied. This effect has been observed in many classes of self-cleaving ribozymes, were the overall secondary structure is conserved for functional stability, but there is no conservation in sequence.[42] The ATP aptazyme was engineered by altering the composition of stem II in the hammerhead ribozyme with the addition an ATP aptamer. The composition of the stem sequence was tuned to generate aptazymes with self-cleavage rates that both increased and decreased with the addition of ATP. This rational design strategy has been shown effective in engineering RNA aptazymes that function in a predictable manner *in vitro*; however, this strategy often requires a significant trail-and-error approach, and integrating these devices into a genetic program to control gene expression *in vivo* has been difficult.

5.4.2 In Vitro *Versus* In Vivo *RNA Folding*

The relationship between RNA structure and function is most apparent when comparing the folding pathways of the same RNA *in vitro* and *in vivo*. Functional RNA selections, such as the ATP-dependent aptazyme, are almost always performed *in vitro*, as the needed library diversity is too large to effectively perform the selection *in vivo*. *In vitro*, the RNA tends to be transcribed, purified, and folded before binding activity is assayed. At the folding step, the RNA is full length and can be renatured such that the structure that is formed in the MFE structure; the most stable structure. It is well known that the MFE structure is rarely achieved when the same RNA is placed in the context of a long transcript *in vivo*.[43] RNA structure formation begins as the RNA is being transcribed, and the addition of subsequent nucleotides affects the local structure. As a result, kinetically trapped RNA structures are displayed more often than an overall MFE structure. For example, the MFE secondary structure of the Avocado Sunblotch Viroid (ASBV) hammerhead ribozyme accurately predicts the active secondary structure confirmation (Figure 5.5a). The context in which the RNA is placed, that is, the surrounding sequence, also contributes to the folding trajectory of the RNA sensor. When the ASBV ribozyme is placed directly in the context of a host genetic program, it misfolds and renders the ribozyme inactive (Figure 5.5b). However, the secondary structure

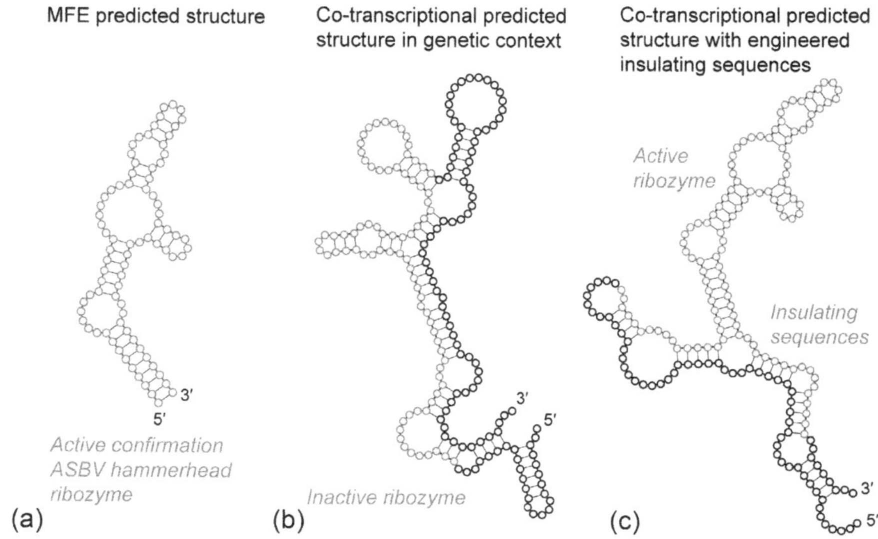

Figure 5.5
RNA co-transcriptional folding is a function of the surrounding sequence space. (a) The minimum free energy (MFE) calculation of the ASBV hammerhead ribozyme (green) displays a correctly folded secondary structure and results in a functional ribozyme. (b) When the same ribozyme is inserted into a genetic program, as it would be in an engineered pathway, the co-transcriptional-predicted structure displays a misfolded and inactive ribozyme. (c) Engineered insulating sequences (orange) specific to the ribozyme and surround sequence can be designed to protect ribozyme folding in the context of a long transcript *in vivo*.

can be protected from surrounding sequence interactions, by engineered insulating sequences (unpublished work W.E. Voje) (Figure 5.5c). In addition, a family of synthetic riboswitches was shown to adopt a kinetic trapping mechanism that resulted in function when the small molecule ligand was introduced co-transcriptionally, as the RNA was being synthesized in cellular extracts, rather than after synthesis was completed.[44] These data suggest that modeling the co-transcriptional folding pathways would provide additional insight into how RNA-based sensors fold *in vivo*. Since the fold is what determines RNA function, engineering devices around co-transcriptional folding trajectories could be an efficient tool for the design, optimization, and integration of devices into engineered biosynthetic pathways.

5.4.3 In Silico *Computational Framework*

The ability to engineer biological pathways in an efficient, robust, and customizable manner would advance the field of synthetic biology. This is necessary not only for the production of biofuels, but pharmaceuticals, novel chemicals, and pathway engineering as a whole. The majority of the advancements in RNA-based sensor technology discussed in this chapter thus far have been experimental *in vitro* and *in vivo* work. Although this has been successful overall, these types of directed evolution and selection experiments are time consuming, low throughput, expensive, and many often need to be performed to generate a single RNA sensor and incorporate it into a single location in host's genetic program. The rate at which these sensor devices are needed for phenotype screening and pathway optimization far exceeds the rate at which they are being engineered, resulting in a bottleneck in pathway production.

Recently, new computational theory has generated helix-based stochastic co-transcriptional folding simulations, which can be used to probe RNA structure formation in a kinetic manner as the RNA is being synthesized. By creating a computational framework where devices are designed, screened, and tested *in silico*, the number of directed evolution and selection cycles would be greatly reduced (Figure 5.6). Stochastic co-transcriptional folding simulations, such as KineFold, model the folding pathway of an elongating RNA as a function of time.[45] These simulations are computationally expensive, as many simulations need to be evaluated in order to generate statically relevant data. As a result, custom software to prepare, submit, and process the large amounts of data on a computational cluster is necessary to formalize an *in silico* pipeline that utilizes these simulations.[46] The overarching goal is a computational framework where any RNA aptamer-expression domain (ribozyme, RBS, nucleic acid circuit, etc.) pair could be specified to generate a virtual RNA library. This library would then be subjected to an *in silico* selection by computationally inexpensive MFE calculations, and co-transcriptional folding simulations, where sequences that fold into the desired functional

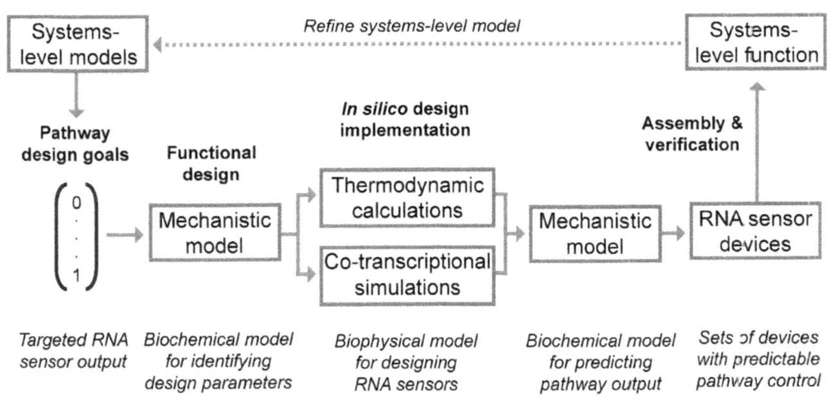

Figure 5.6

Computational framework for the design of RNA-based metabolite sensors specific for integration into biosynthetic pathways. A computational framework for the design of novel RNA-based metabolic sensors is now possible with recent advancements in computational theory. Mechanistic modeling of a proposed biosynthetic pathway provides design parameters for the building and screening of potential RNA devices. Specific devices can be modeled in a desired pathway to predict product output, resulting in sets of devices with reliable control. This systems-level model to function framework, provides a time and cost-efficient method to design devices with predictable *in vivo* pathway function.

target structure would be identified and optimized. These simulations have been shown to be critical for designing programmable RNA-based devices that function in a predictable manner in *E. coli*.[20,46,47] Each of these cases involved a specific RNA-based device in a specific context, but an inclusive and customizable framework has yet to be developed.

Aside from device design, RNA-based sensor integration into a variety of host genetic programs and in various sequence contexts, could be streamlined with a robust computational framework. An issue that plagues RNA sensor design is incorporation of a fully functional device *in vitro* into a genetic program *in vivo*. The folding of the RNA sensor is greatly affected by the surrounding sequence, and often times insertion of a sensor without co-transcriptional folding considerations results in a misfolded and inactive sensor device. The design of insulating sequences to protect the desired RNA secondary structure would eliminate the need for additional directed evolution selections, decreasing the time and cost of device implementation. In addition, devices could easily be transitioned from one genetic context to another without the issues currently associated with repositioning RNA sensors.

All in all, a computational pipeline where systems-level models can be generated specific to pathway-designated goals could aid in the customizability of pathway design. RNA devices with targeted outputs could be generated and screened *in silico*, and their function once incorporated into a pathway could be modeled. This pipeline might provide a necessary tool for the designing and screening of customizable RNA-based sensors for biosynthetic pathway optimization and phenotype screening.

5.5 Modeling Biochemical Pathways

The innate complexity of biological systems has made biosynthetic pathway engineering challenging and often times frustrating. A fundamental goal in synthetic biology is to have the ability to determine which parameters are the most influential to the system, which can be tuned to reach optimal output, and which can be measured for functional validation. RNA-based small molecule sensors have been shown to dynamically regulate engineered biosynthetic pathway flux to promote yields and support host fitness.[20] Mechanistic modeling can investigate the effects of placing an RNA-based metabolic sensor in an engineered pathway, without physically building the system. This approach is advantageous because the effect of having a sensor for all individual intermediates and combinations of intermediates can be studied computationally. In addition, this technique can be tailored to focus on the parameters that are designable and tunable in order to ensure engineering tractability. Mechanistic modeling can provide the proof-of-concept data to demonstrate the biochemical feasibility of an RNA-based metabolic sensor for pathway regulation or phenotype screening. Moreover, the information learned from the modeling can be used as screening considerations when designing and testing RNA-based sensors *in silico* (Figure 5.6).

Coarse-grained mechanistic modeling breaks a complex pathway down into a set of simple chemical reactions with defined kinetic parameters, such as transcription elongation, RNA folding, degradation, and translation rate constants. This framework was implemented to predict reporter gene output in *E. coli* that was under static ribozyme- and dynamic aptazyme-controlled RNA-based sensors.[20] The mechanistic model built to study this system consisted of a set of 25 chemical reactions with 11 effective rate constants, and spanned transcription, translation, RNA-sensor processing, and degradation. The model was evaluated by Monte Carlo sampling of the rate constants within biochemically relevant ranges, and probed the potential design space without individually sampling each possible combination. Sensitivity analysis, including partial correlation coefficients identified parameters and sets of parameters that had the largest influence on pathway output. In this case, it was determined that RNA folding, RNA-sensor processing, and RNA half-life had the largest impact on device function, parameters that can be tuned by altering folding rates and inherent sensor characteristics such as self-cleavage rate. The data derived from the mechanistic model was shown to effectively predict reporter gene output that was under the control of various RNA-sensor devices. In addition, this method was also shown to successfully predict fatty acid ethyl ether output in an *E. coli* strain that was under the control of the dynamic FadR sensor discussed in Section 5.1.2.[12]

The potential of dynamic RNA-based control over static control, such as the control programmed with protein transcription factors, for increased production titers in engineered pathways was demonstrated using a mechanistic modeling approach.[48] A pathway for the production of *p*-aminostyrene (*p*-AS), a vinyl aromatic compound with applications in photolithography and biomedicine, was studied under the control of aREDs (discussed in Section 5.2.2). This pathway is an ideal test bed for solving common pathway engineering control problems, such as intermediate toxicity and decreased cellular fitness due to unregulated expression of pathway genes. Mechanistic models for 728 unique control architectures were generated by inducing control to three nodes along the pathway with three different dynamic aptazyme control modules (aRED, aRED-sRNA, and aRED-transcriptional repressors). It was shown that for aptazymes where there was only a fivefold difference in the ligand activation ratio, the difference between the cleavage rates with and without the target ligand, there was almost a sixfold increase in predicted *p*-AS production titers when compared to static control alone. Mechanistic modeling verifies the potential of RNA-based dynamic sensors and informs as to which control topologies would be the most suited to build and validate.

Mechanistic modeling is a versatile tool that can be implemented to study the effects of an RNA metabolic sensor on the dynamic range and output of a system in a time and cost-efficient manner. These types of calculations can be performed on a standard laptop with the Python package PySCeS and the sensitivity package of R and as a result, provide insight to an engineered pathway without the effort of physically building the system.[49] This method could be implemented to determine if there are any potential bottlenecks in a pathway, or if any unforeseen parameters such as degradation of an intermediate might play a role in pathway output. More importantly, the understanding gained from the model can be used to optimize the *in silico* computational framework implemented to build, screen, and test potential RNA-based metabolic sensors.

5.6 Future Directions for RNA Sensor Design

The computational tools discussed in the previous sections such as co-transcriptional folding simulations and mechanistic modeling offer an opportunity to engineer and study functional RNA in a kinetic manner. This allows for the prospect of designing *de novo* functional *in vivo* RNA-based sensors, rather than optimizing a device that functions *in vitro*, but achieves suboptimal effects in the desired pathway *in vivo*.

5.6.1 Kinetic Proofreading for Increased Fidelity

In order for RNA metabolic sensors to be effective in a biosynthetic pathway, the activity must be specific and sensitive for precise control. Nature has evolved a mechanism called kinetic proofreading to increase the fidelity of small molecule-regulated pathways beyond what would be predicted by thermodynamics alone.[50] The most well-studied example of kinetic proofreading is T-cell recognition[51] (Figure 5.7a). T-cells are responsible for generating an immune response propagated by the binding of foreign antigens to the cell receptors. When the correct antigen binds a given receptor, there is a time delay between binding and signaling. This delay is initiated by a cascade of chemical reactions that render the receptor signal competent. After a certain amount of time, an immune response is issued. However, there are many antigens in a given organism, and they all have similar binding affinities. In the case that an antigen binds a receptor not specific to it, the same set of chemical reactions begins to propagate. However, since the binding affinity is slightly less than the natural antigen, the incorrect antigen dissociates on a time scale that is shorter than what is required to make the receptor single competent. This dissociation results in the rapid conversion of the receptor back to the initial state ensuring that another binding event would reinitiate the time delay. This kinetic proofreading mechanism has been shown to increase the fidelity of the T-cell recognition by almost five orders of magnitude greater than what is predicted by the thermodynamics of antigen binding, resulting in an incredibly efficient and sensitive system.

In the context of RNA-based small molecule sensors, it may be possible to use the same overall design principles to engineer robust and selective sensors (Figure 5.7b). Here, the metabolite-binding affinity is dictated by the secondary structure of the aptamer domain, and for a kinetic proofreading mechanism there must be a kinetic delay before signal propagation. A co-transcriptional multistate folding design strategy could mimic the time delay of the

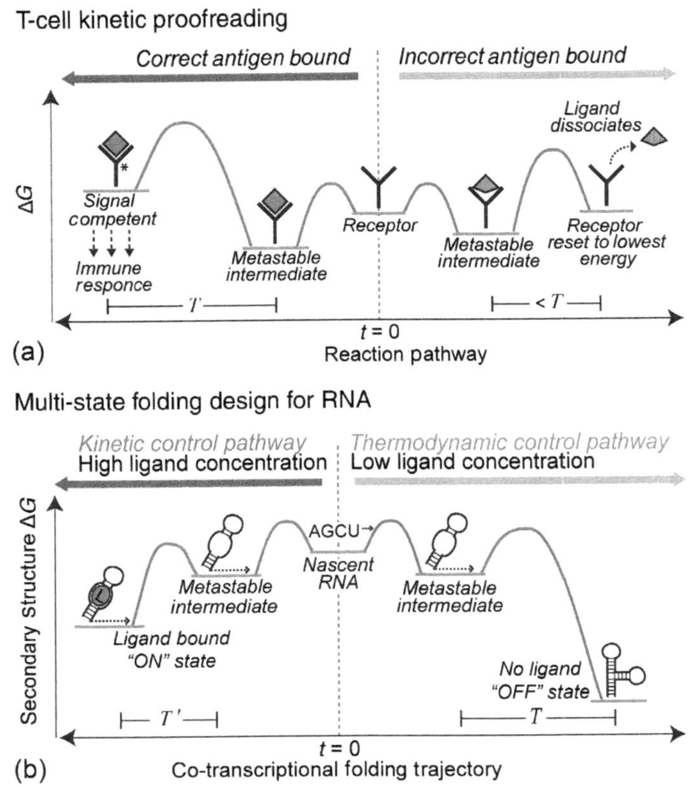

Figure 5.7

Kinetic proofreading and multistate RNA folding design. (a) A kinetic proofreading mechanism accounts for the increased fidelity of T-cell recognition beyond what is calculated from thermodynamics alone. A time delay (T) between antigen binding and an immune response is initiated by the generation of a high energy signal competent state (*). This allows for an incorrect ligand to bind and dissociate on a time scale faster than an immune response (<T) therefore preventing the formation of the signal competent state. (b) A similar kinetic proofreading mechanism could be engineered into RNA-based sensors through multistate RNA folding design. Co-transcriptional RNA folding simulations allow for the ability to engineer metastable intermediate states along the folding trajectory that persist for a determined length of time. By engineering the metastable state in the thermodynamic control pathway, access to a competing kinetic control pathway is granted. For small molecule RNA-based sensors, the presence or absence of the ligand would dictate the folding trajectory and, therefore, the structure and function of the device.

T-cell system, by forcing the RNA to fold into a metastable state that is able to bind the metabolite. In the case where the small molecule is available, it would bind and stabilize the elongating transcript while simultaneously providing access to the expression platform. In the instance where the small molecule target is not available or in a low concentration, the time the metastable state persists would be shorter than the ligand-binding rate. The unstabilized aptamer structure would collapse as it is being transcribed such that the expression platform is not accessible. The competing folding pathways are named based on the free energy (ΔG) of the final secondary structure, where the thermodynamic control pathway has a lower energy final state than the kinetic control pathway.

Many naturally occurring riboswitches, such as the flavin mononucleotide and adenosine riboswitches, function under competing thermodynamic and kinetic folding pathways.[52,53] These competing folding pathways offer an additional level of control upon the system, where time-dependent folding dictates specificity and fidelity. We suggest that the kinetic proofreading mechanism could be implemented by engineering the folding trajectory of the thermodynamic control pathway, and the presence of the metastable ligand-competent state would provide access to the kinetic control pathway. In addition, kinetic parameters such as rate constants for transcription elongation and ligand binding, transcriptional pausing, competition between alternative stable structures and the activation barriers between them have all been shown to play a role in dictating the secondary structure and, therefore, the function.[17,43] Taking into account these kinetic parameters is now made possible by the existence of co-transcriptional folding simulations. Approaches, such as employing kinetic proofreading, could provide the opportunity to compensate for less than ideal ligand-binding rates, by modulating the kinetics of folding with time-dependent metastable states. This would provide alternative tuning knobs for RNA-based sensor design that are not readily available with traditional *in vitro* selection and directed evolution experiments.

5.7 Concluding Remarks

One of the greatest challenges in biosynthetic pathway engineering is mediating the balance between host fitness, and target chemical production. Great efforts have been placed in generating small molecule sensors to aid in the dynamic control of pathway flux, and for screening desired host phenotypes. Naturally occurring protein-based small molecule sensors have been successfully reengineered to increase production titers of microbial pathways. However, many pathway products and intermediates do not have a natural protein sensor, and the demand of small molecule sensors far exceeds the rate at which they are being engineered. RNA-based sensors offer an attractive alternative to protein sensors and their widespread occurrence in nature demonstrates their versatility. The development of engineering frameworks that couple kinetic RNA folding simulations with high-throughput analysis should allow for the rapid construction of customizable RNA-based small molecule sensors.

A high throughput and customizable framework for RNA-based metabolic sensors has the potential to not only improve engineered pathway design and phenotype screening for optimized production titers, but also aid in the design of any system where small molecule sensing is necessary. These types of RNA devices have conceivable applications in medical diagnostics, cellular imaging, and drug delivery in mammalian cells where metabolite sensing is essential for robust and precise activity. A computational pipeline for the design and screening, as well as the ability to predict functional outputs computationally, could lead faster optimized engineered pathways for biofuel production and beyond.

References

1. Hollinshead W, He L, Tang YJ. Biofuel production: an odyssey from metabolic engineering to fermentation scale-up. *Front Microbiol* 2014;**5**:344.
2. Chandran SS, Kealey JT, Reeves CD. Microbial production of isoprenoids. *Process Biochem* 2011;**46**:1703–10.
3. Janssen HJ, Steinbuchel A. Fatty acid synthesis in Escherichia coli and its applications towards the production of fatty acid based biofuels. *Biotechnol Biofuels* 2014;**7**:7.
4. Nozzi NE, Desai SH, Case AE, Atsumi S. Metabolic engineering for higher alcohol production. *Metab Eng* 2014;**25**:174–82.
5. Lee SK, Chou H, Ham TS, Lee TS, Keasling JD. Metabolic engineering of microorganisms for biofuels production: from bugs to synthetic biology to fuels. *Curr Opin Biotechnol* 2008;**19**:556–63.
6. Martin VJ, Pitera DJ, Withers ST, Newman JD, Keasling JD. Engineering a mevalonate pathway in Escherichia coli for production of terpenoids. *Nat Biotechnol* 2003;**21**:796–802.
7. Keasling JD. Manufacturing molecules through metabolic engineering. *Science* 2010;**330**:1355–8.
8. Warnecke TE, Lynch MD, Karimpour-Fard A, Sandoval N, Gill RT. A genomics approach to improve the analysis and design of strain selections. *Metab Eng* 2008;**10**:154–65.
9. Wittmann C. Metabolic flux analysis using mass spectrometry. In: Schuegerl K, Zeng AP, editors. *Advances in biochemical engineering biotechnology. Tools and applications of biochemical engineering science*; 2002, **Vol. 74**, p. 39–64.
10. Li F, Xu W, Zhao S. Regulatory roles of metabolites in cell signaling networks. *J Genet Genomics* 2013;**40**:367–74.
11. van der Meer JR, Belkin S. Where microbiology meets microengineering: design and applications of reporter bacteria. *Nat Rev Microbiol* 2010;**8**:511–22.
12. Zhang F, Carothers JM, Keasling JD. Design of a dynamic sensor-regulator system for production of chemicals and fuels derived from fatty acids. *Nat Biotechnol* 2012;**30**:354–9.
13. Fujita Y, Matsuoka H, Hirooka K. Regulation of fatty acid metabolism in bacteria. *Mol Microbiol* 2007;**66**:829–39.
14. Dietrich JA, Shis DL, Alikhani A, Keasling JD. Transcription factor-based screens and synthetic selections for microbial small-molecule biosynthesis. *ACS Synth Biol* 2013;**2**:47–58.
15. Tang SY, Cirino PC. Design and application of a mevalonate-responsive regulatory protein. *Angew Chem Int Ed Engl* 2011;**50**:1084–6.
16. Pfleger BF, Pitera DJ, Smolke CD, Keasling JD. Combinatorial engineering of intergenic regions in operons tunes expression of multiple genes. *Nat Biotechnol* 2006;**24**:1027–32.
17. Zhang J, Lau MW, Ferre-D'Amare AR. Ribozymes and riboswitches: modulation of RNA function by small molecules. *Biochemistry* 2010;**49**:9123–31.
18. Kang Z, Zhang C, Zhang J, Jin P, Zhang J, Du G, et al. Small RNA regulators in bacteria: powerful tools for metabolic engineering and synthetic biology. *Appl Microbiol Biotechnol* 2014;**98**:3413–24.
19. Serganov A, Nudler E. A decade of riboswitches. *Cell* 2013;**152**:17–24.
20. Carothers JM, Goler JA, Juminaga D, Keasling JD. Model-driven engineering of RNA devices to quantitatively program gene expression. *Science* 2011;**334**:1716–9.
21. Groher F, Suess B. Synthetic riboswitches—a tool comes of age. *Biochim Biophys Acta* 2014;**1839**:964–73.
22. Jenison RD, Gill SC, Pardi A, Polisky B. High-resolution molecular discrimination by RNA. *Science* 1994;**263**:1425–9.
23. Desai SK, Gallivan JP. Genetic screens and selections for small molecules based on a synthetic riboswitch that activates protein translation. *J Am Chem Soc* 2004;**126**:13247–54.
24. Wachsmuth M, Findeiß S, Weissheimer N, Stadler PF, Mörl M. De novo design of a synthetic riboswitch that regulates transcription termination. *Nucleic Acids Res* 2013;**41**:2541–51.
25. Burke CR, Lupták A. Catalytic RNA. In eLS, John Wiley & Sons, Ltd. 2001.

26. Tang J, Breaker RR. Rational design of allosteric ribozymes. *Chem Biol* 1997;**4**:453–9.
27. Carothers JM. Design-driven, multi-use research agendas to enable applied synthetic biology for global health. *Syst Synth Biol* 2013;**7**:79–86.
28. Na D, Yoo SM, Chung H, Park H, Park JH, Lee SY. Metabolic engineering of Escherichia coli using synthetic small regulatory RNAs. *Nat Biotechnol* 2013;**31**:170–4.
29. Hemphill J, Deiters A. DNA computation in mammalian cells: microRNA logic operations. *J Am Chem Soc* 2013;**135**:10512–8.
30. Chen Y-J, Dalchau N, Srinivas N, Phillips A, Cardelli L, Soloveichik D, et al. Programmable chemical controllers made from DNA. *Nat Nanotechnol* 2013;**8**:755–62.
31. Seelig G, Soloveichik D, Zhang DY, Winfree E. Enzyme-free nucleic acid logic circuits. *Science* 2006;**314**:1585–8.
32. Carothers JM, Oestreich SC, Szostak JW. Aptamers selected for higher-affinity binding are not more specific for the target ligand. *J Am Chem Soc* 2006;**128**:7929–37.
33. Steen EJ, Kang Y, Bokinsky G, Hu Z, Schirmer A, McClure A, et al. Microbial production of fatty-acid-derived fuels and chemicals from plant biomass. *Nature* 2010;**463**:559–62.
34. Holeman LA, Robinson SL, Szostak JW, Wilson C. Isolation and characterization of fluorophore-binding RNA aptamers. *Fold Des* 1998;**3**:423–31.
35. Sassanfar M, Szostak JW. An RNA motif that binds ATP. *Nature* 1993;**364**:550–3.
36. Plummer KA, Carothers JM, Yoshimura M, Szostak JW, Verdine GL. In vitro selection of RNA aptamers against a composite small molecule-protein surface. *Nucleic Acids Res* 2005;**33**:5602–10.
37. Ellington AD, Szostak JW. In vitro selection of RNA molecules that bind specific ligands. *Nature* 1990;**346**:818–22.
38. Tuerk C, Gold L. Systematic evolution of ligands by exponential enrichment: RNA ligands to bacteriophage T4 DNA polymerase. *Science* 1990;**249**:505–10.
39. Carothers JM, Oestreich SC, Davis JH, Szostak JW. Informational complexity and functional activity of RNA structures. *J Am Chem Soc* 2004;**126**:5130–7.
40. Schmidt-Dannert C, Arnold FH. International business communications second international symposium on directed evolution of industrial enzymes, San Diego, California, USA, September 14–15, 1998. *Trends Biotechnol* 1999;**17**:135–6.
41. Carothers JM, Goler JA, Kapoor Y, Lara L, Keasling JD. Selecting RNA aptamers for synthetic biology: investigating magnesium dependence and predicting binding affinity. *Nucleic Acids Res* 2010;**38**:2736–47.
42. Webb CH, Riccitelli NJ, Ruminski DJ, Luptak A. Widespread occurrence of self-cleaving ribozymes. *Science* 2009;**326**:953.
43. Lai D, Proctor JR, Meyer IM. On the importance of cotranscriptional RNA structure formation. *RNA* 2013;**19**:1461–73.
44. Mishler DM, Gallivan JP. A family of synthetic riboswitches adopts a kinetic trapping mechanism. *Nucleic Acids Res* 2014;**42**:6753–61.
45. Xayaphoummine A, Bucher T, Isambert H. Kinefold web server for RNA/DNA folding path and structure prediction including pseudoknots and knots. *Nucleic Acids Res* 2005;**33**:W605–10.
46. Thimmaiah T, Voje Jr. WE, Carothers JM. Computational design of RNA parts, devices, and transcripts with kinetic folding algorithms implemented on multiprocessor clusters. *Methods Mol Biol* 2015;**1244**:45–61.
47. Cambray G, Guimaraes JC, Mutalik VK, Lam C, Mai Q-A, Thimmaiah T, et al. Measurement and modeling of intrinsic transcription terminators. *Nucleic Acids Res* 2013;**41**:5139–48.
48. Stevens JT, Carothers JM. Designing RNA-based genetic control systems for efficient production from engineered metabolic pathways. *ACS Synth Biol* 2015;**4**:107–15.
49. Olivier BG, Rohwer JM, Hofmeyr JH. Modelling cellular systems with PySCeS. *Bioinformatics* 2005;**21**:560–1.
50. Hopfield JJ. Kinetic proofreading: a new mechanism for reducing errors in biosynthetic processes requiring high specificity. *Proc Natl Acad Sci U S A* 1974;**71**:4135–9.

51. McKeithan TW. Kinetic proofreading in T-cell receptor signal transduction. *Proc Natl Acad Sci U S A* 1995;**92**:5042–6.
52. Wickiser JK, Cheah MT, Breaker RR, Crothers DM. The kinetics of ligand binding by an adenine-sensing riboswitch. *Biochemistry* 2005;**44**:13404–14.
53. Wickiser JK, Winkler WC, Breaker RR, Crothers DM. The speed of RNA transcription and metabolite binding kinetics operate an FMN riboswitch. *Mol Cell* 2005;**18**:49–60.

CHAPTER 6

Pathway Assembly and Optimization

Miguel Suástegui, Meirong Gao, Zengyi Shao
Department of Chemical and Biological Engineering, The Center for Biorenewable Chemicals, Iowa State University, Ames, Iowa, USA

6.1 Introduction

Pathway assembly represents one of the key aspects in the engineering of microbes for the production of valuable chemicals. With the aid of synthetic biology, recombinant DNA technology, and bioinformatics,[1,2] nowadays it is possible to assemble finely tuned pathways capable of producing a wide variety of natural and nonnatural products.[3,4] The concept of pathway assembly has exponentially evolved with the development of tools that range from efficient and cheap DNA synthesis[1,5] to microfluidics. At the present time, what used to be a time-consuming endeavor is now limited by one's imagination.

There is a commonly accepted "algorithm" for designing heterologous biosynthetic pathways (Figure 6.1). When a target molecule is identified, the first step is to check whether there are natural pathways responsible for its biosynthesis. If the enzymes have not been previously characterized, homolog search[6] is usually performed to help discern the enzymes that could best carry the desired reactions. In this step, phylogenesis[7] of the source organisms and the desired production host is taken into consideration. The functional expression of "difficult enzymes" is more likely to happen if two hosts have a closer evolutional relationship. With this, a group of candidate enzymes can be selected to perform the desired set of reactions toward the molecule of interest. In case that a target molecule does not have precedence in nature, a biosynthetic pathway needs to be designed *de novo*. Generally, this process starts with searching for a pathway that synthesizes a product or an intermediate similar to the target compound in structure.[8–10] Next, how to design feasible routes to complete the remaining conversions is usually challenging and mostly important. Enzyme promiscuity sometimes allows for a direct fusion of two independent pathways; if not, protein engineering needs to be applied, rendering novel activities to the existing biocatalysts to complete the desired conversions.[11–13]

The next important step is to select the organism in which the pathway will be implemented. This represents a pivotal point in the design of the pathway. Considerations such as codon optimization, transcriptional and translational regulation will have to be taken into consideration. The battle horses for heterologous pathway expression are usually the

Biotechnology for Biofuel Production and Optimization. http://dx.doi.org/10.1016/B978-0-444-63475-7.00006-6

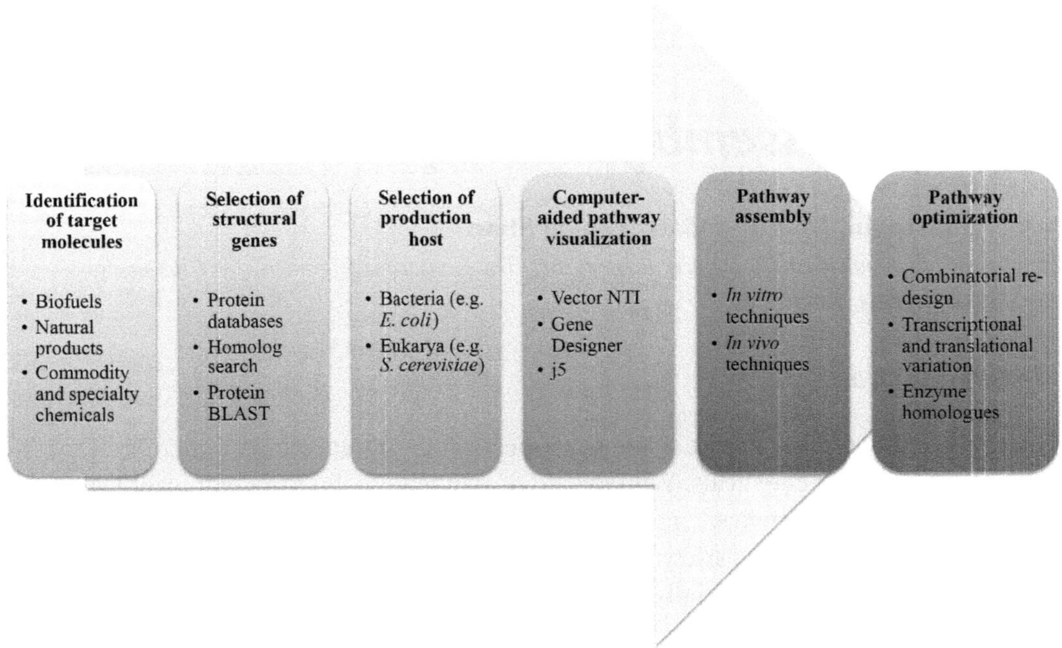

Figure 6.1
Common rationale for pathway design.

bacterium *Escherichia coli* and the yeast *Saccharomyces cerevisiae*, due to their well-understood metabolism and physiology when compared to other organisms. Great efforts have been put into understanding the mechanisms of gene expression and regulation in these organisms, allowing the development of powerful tools that permit quick manipulations in these hosts.[14] Nevertheless, issues like low tolerance to exogenous metabolites have become a drawback in the implementation of metabolic pathways in such model microorganisms. This is mostly common in the heterologous production of biofuels and chemicals.[15] Developing strains with higher tolerance to these metabolites are a strong area of research (see Chapter 9). The advent of technologies like genome sequencing and genome editing have allowed the manipulation of nonconventional microorganisms with outstanding features that make them perfect candidates for the production of toxic compounds.[16]

Once the host organism has been selected, the pathway will be designed first with a computational program such as Vector NTI,[17] j5,[18] and Gene Designer[19] for convenient visualization. The next step consists of putting together the DNA pieces either *in vitro* or *in vivo*. A vast amount of techniques have been developed to carry out this task, all with the same objective: to maximize the assembly efficiency. The assembly efficiency will consist in, among other things, the time required to finalize pathway assembly with minimum errors. Section 6.2 of this chapter is destined to summarize the significance of these techniques.

To finalize the pipeline, the pathway is introduced into the host organism and tested for the desired metabolite using, most commonly, chromatographic tools. After the initial characterization, the pathway may be modified in order to alter, for example, regulatory elements or the initially selected structural genes. Combinatorial design has become a new trend in pathway optimization, allowing the generation of large libraries of pathways with varying features. Elegant techniques have emerged from the need of generating strains with elevated titers in the case of biofuels; however, these techniques can also be applied to engineer the production of nonnatural compounds.[20,21] These combinatorial tools are described and exemplified in Section 6.3 of this chapter.

The last section of this chapter deals with the current challenges and future directions in pathway design. Two of them are predominantly defined here: (1) how to quickly screen for improved strains obtained from combinatorial pathway assembly and (2) how to automate the assembly pipeline in order to maximize the overall efficiency. Emerging state-of-the-art technology that aims to solve these challenges is described, as well as future trends in the dynamic endeavor of pathway assembly.

6.2 Assembly Methods

This section describes the most commonly used DNA assembly methods. Table 6.1 summarizes these techniques depicting the key features and the largest construct that was obtained as reported in the corresponding reference.

6.2.1 *In Vitro* Assembly Based on Homologous Recombination

Unlike the conventional restriction enzyme- and ligase-dependent DNA cloning methods, assembly based on homologous recombination does not require specific restriction sites. This enables the assembly of multiple fragments in a single step. By harnessing the power of homologous recombination, constructs can be assembled *in vitro* from a mixture of DNA fragments with overlapping regions. The assembled plasmids are then introduced into bacteria for enrichment, which finalizes the regeneration of the construct as double stranded, intact molecules. This category of DNA assembly methods provides a multifaceted approach for efficient recombinant DNA construction. Among this category, sequence- and ligation-independent cloning (SLIC), Gibson isothermal assembly, unique nucleotide sequence (UNS), and PaperClip are typical and commonly used approaches.

SLIC

Li et al. developed a SLIC method in 2007.[22] SLIC employs the exonuclease activity of T4 DNA polymerase to chew back the ends of an insert and a vector to reveal 5′ single-stranded overhangs. Two single-stranded stretches will subsequently anneal based on homology to form a circular molecule. Addition of bacterial recombinase, RecA, could facilitate the

Table 6.1: Summary of pathway assembly methods

Methods	Main characteristics	Size of the largest pathway reported (kb)[a]	Uniqueness	Reference
SLIC	*In vitro* homology-based assembly with sticky ends	10.2		22
Gibson Isothermal Assembly		583	Commercial kit available	23
BioBrick/BglBrick		~7	Commercial kit available	24
UNS-guided Assembly		64	Algorithm available; can be used for assembly of sequences with high similarity	25
SLiCE		24	Facilitated by bacterial cell extracts	26
USER		3.2		27
PIPE		7.0		28
LCR		20		29
CPEC		8.4		30
PaperClip		6.0		31
TAR	*In vivo* homology recombination-based assembly	250	Use *S. cerevisiae* as an assembly platform	32
DNA Assembler		52		33
Genome Assembly		(1.08 Mb genome)		34
MAGIC		3.9	Assembly accomplished through bacterial mating	35
Golden Gate	*In vitro* type II restriction enzyme digestion	33		36
MASTER		36.4		37
GoldenBraid Assembly		14.3		38
Pairwise selection assembly	Particularly useful for assembly of highly repeated fragments	91		39
OGAB		13	Use *B. subtilis* as an assembly platform	40
SIRA	*In vitro att-sites* based matching	7.5	Site-specific recombination mediated by serine integrase	41
SSRTA		62.4		42

[a]The size does not indicate the upper limit for each assembly method.

annealing efficiency. Following electroporation to *E. coli*, the gaps in the recombinant molecule can be repaired by the endogenous homology recombination enzymes such as resolvases, nucleases, and polymerases. To stimulate an efficient recombination *in vitro*, 30-40 bp homology overlapping regions and a molar ratio of 2-4:1 between an insert and a vector are recommended.

To prove the efficiency of SLIC, as many as ten fragments with 40-bp overlapping regions were assembled to generate a circular plasmid carrying a kanamycin-resistant gene. Among the fragments, nine were protein-coupled receptor (PCR)-generated DNA molecules with sizes ranging from 275 to 980 bp and the other one was a 3.1-kb linearized plasmid. It was observed that almost 20% of the colonies obtained after transformation had correct restriction digestion pattern.[22]

Gibson Isothermal Assembly

Gibson Isothermal Assembly is one of the most widely used DNA assembly methods due to its ease of use, flexibility, and capacity for constructing large vectors (Figure 6.2). It simplifies the recombination of overlapping DNA molecules in a single step while maintaining high efficiency.[23] Very similar to SLIC, the assembly process begins with an enzyme cocktail (including an exonuclease, a DNA polymerase, and a DNA ligase) mixed with a set of overlapping DNA fragments. The function of the exonuclease is to remove nucleotides from the 5′ ends of double-stranded DNA (dsDNA) molecules, thus exposing complementary single-stranded 3′ overhangs for homologous annealing. The DNA polymerase fills in the remaining gaps after annealing, and the DNA ligase seals the nicks in the assembled DNA construct. After a short incubation period, double-stranded fully sealed DNA molecules are obtained. Gibson Assembly® Master Mix is commercially available, making this technique amenable to high-throughput assembly automation.

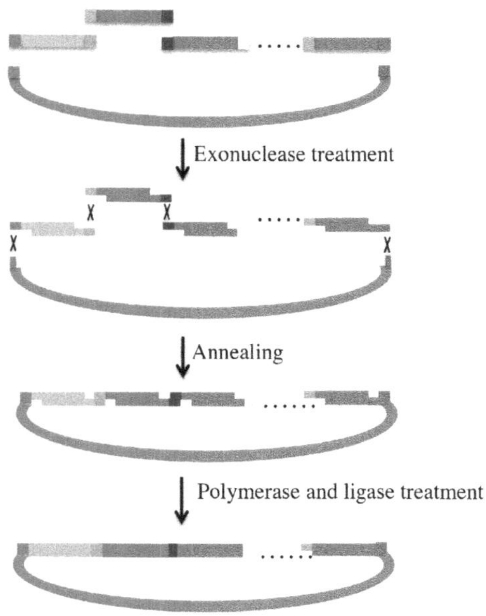

Figure 6.2
Gibson Isothermal Assembly *in vitro*. Exonuclease is employed to generate ssDNA overhangs. Polymerase and ligase are used to repair annealed fragments and fill gaps.

UNS

On the basis of Gibson Assembly, UNS assembly introduces a computational algorithm to design short unique overlaps (40 bp) between adjacent DNA fragments. This allows to drive an ordered assembly with high efficiency, especially when repeated sequences are involved.[43,44] The algorithm is built upon the following premises: UNSs are balanced on GC content, have no secondary hairpin structure, avoid promiscuous hybridization, and evade homology to start codons, promoter-like sequences, and commonly used restriction sites. Although the principle guiding UNS assembly is identical to Gibson Assembly and very similar to SLIC, it uniquely offers standard overlapping parts that can be easily repurposed among synthetic pathways and circuits. Therefore, this technology promotes a philosophy of standardization and insulation as emphasized in synthetic biology. To append the UNSs to the DNA fragments to be assembled, they can be cloned from preassembled vectors, from PCR amplification with primers containing the UNSs, or simply by direct synthesis of the cassettes with the corresponding UNSs.

PaperClip

PaperClip is a DNA assembly method based on bridging oligos attached to the ends of the fragments to be assembled.[31] This flexible technique allows the assembly of DNA fragments without the need of restriction digestion, or specific primers to add end homology. The attached oligos (Clips) drive an orderly assembly process, enabling combinatorial assembly of fragments.

Four single-stranded oligos are required for every part to be assembled. These oligos contain homology with the assembled part as well as 3-bp sticky ends to allow the formation of a full Clip. Pairs of upstream and downstream oligos of each DNA part are first annealed separately to form the upstream and downstream half Clips. Then the downstream half Clip of one DNA molecule and the upstream half Clip of its adjacent downstream molecule, sharing the designed complementary trinucleotide sequence, are ligated to form an intact Clip. The assembly reaction containing a destination backbone, DNA fragments, and Clip can be performed either in cell extract of *E. coli* expressing recombination protein or through two-step PCR[30] to maintain high efficiency.

The key highlight of this method is the direct assembly of existing DNA parts without further cloning and PCR amplification with the help of four single-stranded oligonucleotides. Moreover, the four single-stranded oligonucleotides, the half Clips, and the full Clips can be stored and reused to the user's convenience. This feature is a great advantage in automated DNA assembly processes.[31]

6.2.2 In Vivo *Assembly Based on Homologous Recombination*

A series of assembly methods based on *in vivo* homologous recombination have been developed. These include the transformation-associated recombination (TAR),[45]

DNA Assembler,[46] and mating-assisted genetically integrated cloning (MAGIC),[35] all of which take advantage of the endogenous homologous recombination machinery in biological cells (Table 6.1).

DNA Assembler

Shao et al. developed a powerful assembly tool, named DNA assembler, which allows single-step assembly of biochemical pathways *in vivo*.[46,47] This method relies on the native homologous recombination machinery of *S. cerevisiae* to anneal multiple overlapping DNA sequences. The length of overlaps should be at least 40 bp to maintain high efficiency. In this approach, multiple target DNA fragments and a linearized vector are co-transformed into *S. cerevisiae* through electroporation (Figure 6.3a). The transformants are selected from amino acid dropout media according to the auxotrophic marker carried by the vector, and the correct constructs are confirmed by restriction digestion or sequencing. This method usually takes one more week than *in vitro* assembly because the assembled plasmids in yeast need to be retransformed to *E. coli* for DNA enrichment before being confirmed. However, because the assembly is conducted inside yeast cells, the power of *S. cerevisiae* homology recombination can be fully picked up, making the assembly accuracy and capacity significantly higher than any of the *in vitro* assembly methods. It has been demonstrated to construct a plasmid as large as 52 kb[33] although the upper limit could be far beyond as indicated in genome assembly based on the same principle. Shao et al. also described the versatility of this approach in manipulating large natural product gene clusters, including achieving heterologous expression, single or multiple point mutations, scar-less gene deletion, synthesis of natural product derivatives[47,48] and artificial gene cluster refactoring.[33,49]

In addition, DNA assembler also enables direct chromosome-based assembly if the two outermost fragments are designed to anneal to the selected locus on the yeast chromosome (Figure 6.3b). More of this technique in the context of combinatorial pathway assembly is described in Section 6.3 of this chapter.

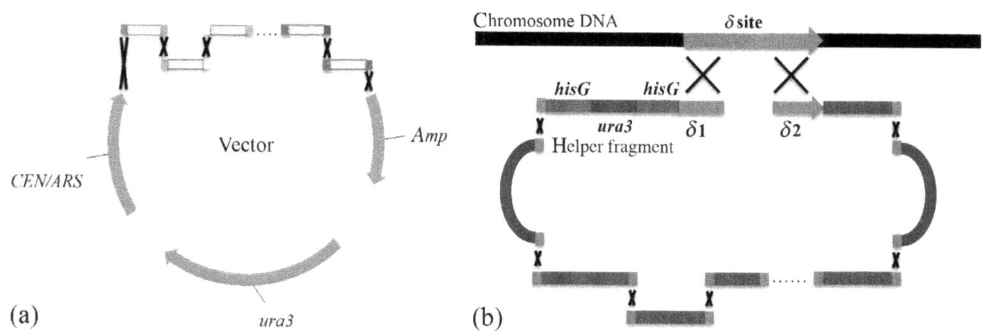

Figure 6.3
DNA assembler based on *in vivo* assembly. (a) Vector-based assembly and (b) chromosome-based assembly.

Genome assembly

A key milestone in DNA assembly history was achieved in 2008 with the invention of the Gibson Assembly. Daniel Gibson reconstituted the complete synthetic 583-kb genome of *Mycoplasma genitalium*.[50] The whole genome was split into 101 fragments with sizes ranging from 5 to 7 kb and overlapping regions ranging from 80 to 360 bp. Four quarter genomes were individually assembled through Gibson Assembly, and then transferred to *S. cerevisiae* to reconstitute the whole genome *in vivo*. It is the yeast *in vivo* homologous recombination that ensures such a superior efficiency for large DNA fragment assembly when compared to *in vitro* assembly.

In 2010, Gibson et al. further assembled the 1.08-Mb *Mycoplasma mycoides* genome and more importantly, transplanted it to the "shell" of *Mycoplasma capricolum* whose genome had been removed.[51] The newly transplanted cell demonstrated the phenotype directed by the synthesized genome rather than the recipient cell and was capable of continuous self-replication. Beyond the numbers game, the significance of this achievement is, for the first time, life was synthesized purely based on chemical building blocks. It can be imagined that, in the future, valuable creatures can be digitally designed and created according to human desires, despite numerous bioethical controversies it will likely initiate.

The most recent groundbreaking progress in genome assembly has been made through synthesizing the first eukaryotic chromosome, *S. cerevisiae* chromosome III, and moreover removing dispensable junk sequences to give a remarkable 14% reduction in chromosome size.[34] Starting off with the efforts of 60 undergraduate students enrolled in the "Build a Genome" project 8 years ago, the project has quickly sparked a global collaboration, named Sc2.0. This initial success in minimizing one yeast chromosome without losing major functions will lead to the reconstruction of the other 15 yeast chromosomes. Once completed, Sc2.0 will push the boundaries of synthetic biology through creating synthetic yeast with a minimized genome that will bolster development of more efficient industry processes for producing biofuels, chemicals, and vaccines.

6.2.3 Non-Homology-Based Assembly

Non-homology-based assembly relies on type II restriction enzymes or serine integrases that recognize and cleave specific DNA molecules at fixed sites. This leads to 5′ or 3′ DNA overhangs with potentially infinite combination of nucleotides. On the basis of this property, a series of pathway assembly tools have been built such as Golden Gate Assembly,[52,53] Ordered Gene Assembly in *Bacillus subtilis* (OGAB),[40] Serine integrase recombinational assembly (SIRA),[41] GoldenBraid,[38] and the pairwise selection assembly method[39] (Table 6.1).

Golden Gate Assembly

The uniqueness of Golden Gate Assembly is the recognition sites of type II restriction enzymes which are removed during digestion, therefore, allowing digestion and ligation to take place in the same pool.[52] As shown in Figure 6.4, the fragments to be assembled are flanked with type II cleavage and the recognition sites. Incubation of these fragments with DNA ligase and a single type II restriction enzyme leads to an irreversible removal of the recognition sites. By designing unique overhangs, Golden Gate Assembly allows an orderly assembly of multiple fragments simultaneously. It has been shown that nine fragments were efficiently assembled into the recipient vector using Golden Gate method.[53] More importantly, unlike homology-based assembly, Golden Gate Assembly has been proven to be particularly useful for building large genes or pathways with repeated sequences. For example, it was used to prepare transcription activator-like effector nucleases (TALENs) for genome editing, whereas homology-based assembly methods were useless because TALENs usually have 12-23 repeat modules.[54]

OGAB

Besides *E. coli* and *S. cerevisiae*, other microorganisms possessing unique characters beneficial to DNA assembly have been subject of intense research.[40,55,56] OGAB is one of these examples.[40] Unlike *E. coli*, *B. subtilis* prefers to uptake linear-form DNA and has

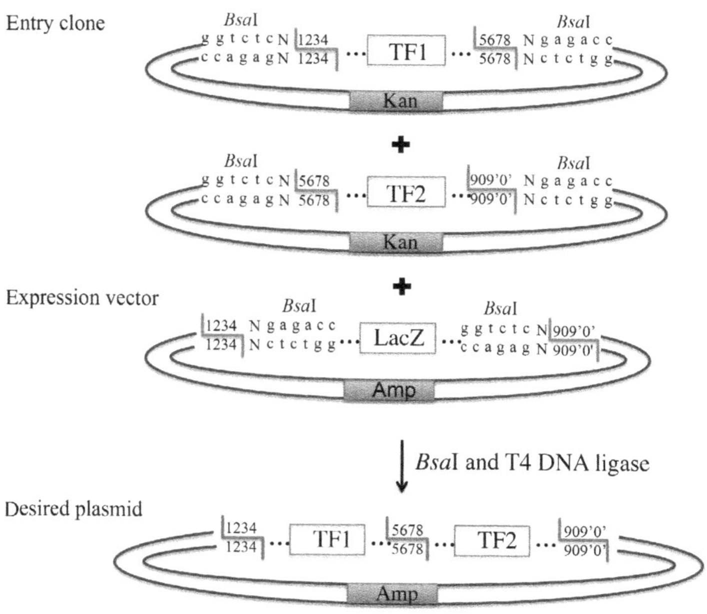

Figure 6.4

Golden Gate Assembly. The final desired plasmid is resistant to *BsaI* digestion. Blue elbow connectors show the *BsaI* restriction digestion sites. TF1, target fragment 1; TF2, target fragment 2. Numbers 0-9 (including 0′ and 9′) denote any nucleotide sequence.

nucleases located on its cell membranes to process the attached dsDNA into single-stranded DNA (ssDNA). It can incorporate as many as 50 ssDNAs per cell simultaneously. Once inside the cell, the single DNA strands will pair with their complementary fragments to form desired circular plasmids. This is highly facilitated by the endogenous recombination machinery in *B. subtilis*. In fact, within the ligation mixture, after type II enzyme and DNA ligase treatment, target fragments will line up according to sticky end matching, forming linear concatemers, with a much higher abundance compared to the circularized form. The ability of *B. subtilis* to uptake these linear concatemers and circularize them inside cells significantly increases assembly capacity. Using OGAB, it was shown that 98 out of 100 examined colonies were correct when five-gene cassettes were ligated with a linearized backbone.

SIRA

SIRA requires the recruitment of a serine integrase and a recombination directionality factor (RDF) to accomplish site-specific recombination.[41] The serine integrase alone is able to cut DNA fragments at *attB* and *attP* sites in synchrony, and re-ligate them to form two new sites, *attL* and *attR*. However, the addition of RDF in assembly mixtures can completely reverse the reaction. To maintain efficient recombination, exact matching of 2-bp overlaps of adjacent fragments is required. Orthogonal recombination sites are necessary for ordered recombination of multiple fragments.

High flexibility is the main advantage of this method. DNA fragments can be deleted from the recombined vectors or substituted by one or more fragments. A two-step replacement strategy was developed to improve the efficiency of fragment substitution and addition. In the first step, a target gene was replaced by an antibiotic resistance cassette through serine integrase- and RDF-dependent *attL*/*attR* recombination. Then, with serine integrase alone, three DNA fragments with the appropriate flanking sites replaced the antibiotic resistance marker. SIRA was also utilized to construct metabolic pathways, alter gene order to balance expression levels, and optimize synthetic genetic pathways.[41]

6.2.4 Tools for Genetic Block Visualization

The development of computational biology and bioinformatic tools has immensely aided the modeling and prediction of a myriad of intricate biological mechanisms. From computational genomics[57] to evolutionary biology,[58] cutting-edge computational tools allow scientists an *in silico* first approach to the design of wet-lab experiments, lowering the possibility of introducing the rather inevitable human errors. This is precisely what has been achieved with the visualization tools for pathway assembly. Nowadays, one can virtually manipulate DNA parts as building blocks with a friendly user interface program, allowing the elaboration of complex gene circuits and metabolic pathways. This also opens the door for the accurate

Table 6.2: Computational tools for visualization and design of DNA constructs

Software	Source	Reference
Vector NTI	Life Technology	17
j5	http://j5.jbei.org	18
GeneDesigner	https://www.dna20.com/resources/genedesigner	19
GeneDesign	http://54.235.254.95/gd/index.html	59
Gene Composer	http://www.genecomposer.net/	60

design of intricate combinatorial pathway constructs, permitting to precisely keep track of restriction sites, homology regions, as well as to seamlessly design primers. Table 6.2 contains the current most popular computational tools for design and visualization of DNA circuits. Here we will pay special attention to j5 since it is a free, open-source web software that anyone can access from a computer with an Internet connection.

j5 DNA assembly software was first introduced in 2012 by Hillson et al. with the goal of automating the design of scar-less multipart constructs.[18] This software was designed with the capability of adopting the requirements of common assembly techniques like isothermal Gibson[50] and Golden Gate.[52] A great feature enabled in this software is the cost-analysis algorithm that compares the cost of synthesizing a DNA fragment to laborious PCR assembly of the same fragment. This software also contains algorithms to create precise oligonucleotides to provide homology regions for the case of SLIC,[22] Gibson,[23] and circular polymerase extension cloning (CPEC)[30] (Table 6.1). One of the criteria for this design relies on avoiding the creation of fragments that share high similarity in order to prevent unspecific assemblies. Moreover, for Golden Gate Assembly, an algorithm built in the software allows the selection of appropriate overhang combinations to circumvent potential self-cohesive or off-target annealing events.

j5 and other software like GeneDesig[59] and Vector NTI[17] do not focus on the biological activity of the DNA fragments to be assembled. As a visualization tool, these software solely aid in the assembly design of the genetic constructs. Nevertheless, tools like the ribosome binding site (RBS) calculator, greatly complement these software since their main purpose is to generate DNA variants of regulatory fragments with tailored biological functionalities. The junction of software like these, together with robotics and microfluidics, open the door for the implementation of combinatorial approaches for assembly of pathways to exploit biological diversity. The concept of combinatorial design is introduced in the following section.

6.3 Combinatorial Assembly

Pathway assembly represents an important step in the engineering of microbes toward the production of value-added compounds. However, after introducing the heterologous genes into the host, an ample room for optimization remains available. Combinatorial pathway assembly is a concept that has been recently coined as the product of the desire to exploit variability in the components of a pathway. For example, this variability, commonly introduced by some types of mutagenesis, can be targeted to create libraries of promoters with different strengths. As a result, a more efficient pathway may be generated due to an optimal balance in gene expression.

As it will be depicted, all of the combinatorial assembly techniques make use of the assembly methods mentioned in the previous section. It is in this case that the rapidness of assembly plays a very important role. Only the most efficient techniques, generally the ones that allow assembly with only a few steps, will permit the wide application of combinatorial design. Techniques limited by restriction digestion and multiple cloning steps may not be considered as an option when a combinatorial approach will be put into play. Table 6.3 displays the predominant combinatorial techniques described in this section.

This section will describe several approaches to assemble pathways for the production of biofuels and related compounds using combinatorial techniques. Also, an extension to the combinatorial approach toward production of natural products will be included. The intrinsic

Table 6.3: Techniques for combinatorial pathway design

Combinatorial technique	Assembly foundation	Control level	Applications
δ-Cocktail integration	Integration of genes into δ-sequence in yeast genome	- Transcriptional (*variability in integration events*)	- Integration of cellulolytic enzymes in *S. cerevisiae*[61] - Introduction of five-gene isobutanol pathway and eight-gene amorpha-4, 11-diene pathway in *S. cerevisiae*[62]
COMPACTER	DNA Assembler	- Transcriptional (*promoter variants*)	- Construction of an enhanced three-gene xylose and a two-gene cellobiose-utilization pathway in *S. cerevisiae* controlled by a promoter library[20]
Single-step linker-based assembly	Gibson Assembly	- Transcriptional (*promoter variants*) - Enzyme variability	- Restoring the acetate utilization pathway in *E. coli*[63]
Single-strand assembly		- Transcriptional (*promoter library*) - Translational (*RBS library*) - Enzyme variability	- Combinatorial assembly of variant promoters, RBS sequences, and reporter genes as proof of concept[64]
MoClo	Golden Gate	- Transcriptional	- Creation of an 11-transcription unit construct conformed by 44 individual basic units[36]
GB2.0		- Transcriptional - Translational	- Standardized modular assembly method for gene expression in plants[65]
ePathBrick	BioBrick	- Transcriptional - Translational	- Flavonoid production pathway in *E. coli*[66] - Engineering of the fatty acid central metabolism in *E. coli* leading to the highest reported productivity[67]

modularity of the pathways for natural product production allows combinatorial design to be exploited at its fullest.

6.3.1 *In Vivo* Combinatorial Assembly

Cocktail assembly and chromosomal integration of metabolic pathways

An interesting feature in the genomic DNA of *S. cerevisiae* is the presence of retrotransposon sequence denominated δ-site. This sequence has more than 100 copies spread out on the genome and has been used as recombination sites for the integration of heterologous genes in the yeast since 1990,[68] and continues to be exploited for metabolic engineering studies.[46,61,69]

In 2010, a technique denominated cocktail δ-integration was developed to introduce cellulose-degrading genes from the fungus *Trichoderma reesei* into the δ-sites of *S. cerevisiae*. *T. reesei* is a fungus known for possessing various enzymes capable of degrading cellulose.[61] Depending on the environment, this fungus can balance the expression of such genes to efficiently degrade cellulose.[70] The dynamic nature of this process has not permitted to discern the appropriate expression levels of the genes required in this process. The goal of cocktail δ-integration was to develop a *S. cerevisiae* strain capable of degrading cellulose more efficiently without knowledge *a priori* of the required balanced expression levels of the genes. They incorporated three genes, namely, endoglucanase (EG), cellobiohydrolase (CBH), and β-glucosidase (BGL) with different auxotrophic markers into multiple δ-sites of the yeast. After only two rounds of integration a strain was obtained with a higher activity of degrading phosphoric acid swollen cellulose. This increase was merely due to a balanced expression of the three genes, as opposed to a strain that simply contained the three genes overexpressed which had 13% less activity than the resulting variant with a balanced expression.

Although this technique does not necessarily go through an assembly line to create a continuous pathway, the fact that an enhancement was observed in the degradation of cellulose correlates with one of the fundamental drives of combinatorial pathway assembly. This is, to optimize a pathway, the expression of the genes involved has to be finely balanced as opposed to the common thought of maximizing the expression of all genes simultaneously.

Recently, Yuan and Ching[62] took the foundation of cocktail δ-integration to engineer *S. cerevisiae* for the production of two metabolites of interest in the biorenewables field, namely, isobutanol (a potential biofuel) and amorpha-4,11-diene (a precursor for artemisinin, the potent antimalarial medicine). They modified the δ-integration technique by introducing two coding sequences (CDS) per integration cassette driven by a bidirectional promoter. Also, each cassette contained a gene for phleomycin resistance, allowing them to control the integration frequency by varying the antibiotic concentration.

Proceeding with this idea they assembled five genes involved in isobutanol production into the yeast genomic DNA. After a single round of integration, they obtained a strain capable of producing 600 mg/L of isobutanol, an impressive increase, since the reference strain, containing the pathway in a multiple copy plasmid, was only able to produce around 6 mg/L. Following the same rationale, the mevalonate pathway consisting of eight genes was integrated into the δ-sites with 100% efficiency (all randomly selected mutants contained the full pathway). Although their final strain did not reach to the maximal level as reported in literature with a different yeast strain as the host,[71,72] they did consistently observe that the strain with higher amorpha-4,11-diene accumulation gave a low transcription abundance of the gene susceptible to cause feedback inhibition specific for this pathway. This allowed concluding that, indeed, their technique can go beyond what was originally shown in the DNA Assembler method for chromosome-based pathway assembly,[46] and be utilized as a combinatorial way to fine tune the expression levels of individual genes to obtain more efficient phenotypes.

The combinatorial rationale in the aforementioned technique pertains to balancing the expression of the genes constituting the pathway. Work like this can help decipher the ways in which the hosts naturally able to produce the metabolite of interest balance the expression of their genes. Nevertheless, it is important to keep in mind that, in most cases, metabolic pathways have to be tailored to a specific host. That is, it is common that a pathway resulting from a combinatorial approach does not yield the same results when transferred between strains even if they are from the same species. Du et al.[20] demonstrated this when they engineered an efficient xylose- and cellobiose-utilization pathways in strains of *S. cerevisiae* with different genetic backgrounds. Similarly to the cocktail integration technique, this group targeted combinatorial expression levels in their constructed pathways. However, unlike having multiple integrations into the yeast chromosomes, they created libraries of mutant promoters with confirmed varying strengths obtained by nucleotide analog mutagenesis.[73]

The name of this combinatorial approach is COMPACTER, which stands for "customized optimization of metabolic pathways by combinatorial transcriptional engineering" and it is based on the DNA Assembler[46] technique described in the previous section. To analyze the colony variants, they developed a rigorous screening method that allowed them to select strains with higher pathway activity based on colony sizes. Subsequential refinement of the selected mutants helped rule out the possibility of background mutations and host adaptation to xylose or cellobiose. For each strain studied, they were able to obtain a mutant pathway that enabled optimal utilization of either xylose or cellobiose as sole carbon source.

One of the most interesting discoveries in this work is that, when the optimized pathways were swapped between strains, the performance in carbon consumption was different. This means that the resulting pathways from their combinatorial assembly were tailored specifically for the original strain. For example, in the case of xylose utilization in the laboratory strain, a higher expression of xylitol dehydrogenase (XDH) gene versus that of xylose reductase (XR) allowed a better flux of carbon through the pathway, which differed drastically from the industrial strain. It is then valid to establish that when optimizing a pathway, not only the promoter strength has to be balanced among the genes, but also the genetic background of the strain in use can play an important role in the success of such a pathway.

With the previous examples, combinatorial pathway assembly has been exemplified at the level of transcriptional regulation by either generating promoter libraries or by integrating multiple copies of the genes into the host chromosomes. Both cases have shown that rather than overexpressing all the genes, a balanced ratio on the level of heterologous transcripts results in better multigene pathway performance. Nevertheless, in pathway optimization, there are other levels of control that can be modulated following a combinatorial approach. Translational optimization in combinatorial pathway design adds a second level of control to enhance heterologous production of biofuels and chemicals. This approach includes, but is not limited to, modulating the RBS sequences,[74,75] combining enzyme homologs,[76] or implementing libraries of mutant enzymes. The following example demonstrates how an efficient xylose-utilization pathway was constructed by exploring thousands of combinations of enzyme homologs with various features. This approach, again, took advantage of the DNA Assembler[46] method discussed in Section 6.2.2 of this chapter.

There are multiple microorganisms that are capable of utilizing xylose as a carbon source. This invaluable feature is enabled by the presence of three key enzymes, XR, XDH, and xylulose kinase (XK). Although this set of enzymes share a certain level of homology among xylose-utilizing microbes, they all display differences in cofactor selectivity. These differences were explored by Kim et al.[76] through assembling around 8000 xylose-utilization pathway variants as a result of the combination of multiple enzyme homologs of the three enzymes originated from *Scheffersomyces stipitis*, a yeast well known for its exceptional assimilation of xylose as a carbon source.[77–79]. Their pathway library consisted of 20 XR, 22 XDH, and 19 XK homologs from various hosts, yielding a total of 8360 possible combinations. Also they tested the performance of the pathways in different *S. cerevisiae* strains, such as laboratory (InvSc1) and industrial (ATCC 4124) strains. Coupled with a screening method based on colony size, xylose consumption, and ethanol production, they were able to obtain combinations that allowed efficient xylose utilization compared to other highly efficient xylose-utilizing strains. It was observed that activities among the enzymes involved were, XKS>XDH>XR, represented by specific activities, which at least set a starting point for further understanding the optimal balance in expression level for this three-gene pathway. They also proved that the performance of the pathway was strain dependent as well as carbon source dependent. Once again, it was proven that a combinatorial approach allows a fast way to exploit random combinations of pathway elements that yield better performances.

6.3.2 In Vitro *Combinatorial Assembly*

The following examples show combinatorial pathway design based on *in vitro* assembly techniques. So far, we have shown how two levels of pathway assembly variability can be independently targeted by combinatorial techniques. Could it be possible to merge both into one to create an even larger pool of pathway variants?

In 2010 Ramon and Smith[63] developed a technique founded upon isothermal assembly (Gibson),[23] which targeted the combinatorial assembly of promoters and gene cassettes. The fragments involved in the construct were flanked with 40-bp unique linker sequences to drive the assembly in an orderly manner. This opened the combination at two control levels: one referring to the promoter strength and the other to the variable features displayed by homolog genes. The combinatorial assembly was done *in vitro* and then transferred to the host organism, unlike the previous examples, in which the library of pathways was directly assembled and screened in the same organism. As an example to prove the functionality of their method, they restored the acetate assimilation pathway, consisting of the genes acetate kinase (*ackA*) and phosphotransacetylase (*pta*), into a mutant *E. coli* strain incapable of growing on acetate as sole carbon source. In total, they selected three different promoters and four homologs for each enzyme to yield a total of 144 combinations. They were able to select positive transformants from the plates containing acetate as sole carbon source. However, the distribution analysis of their mutants showed misrepresentation of some combinations that were later attributed to uneven concentrations of individual cassettes before the assembly. Predominance of some combinations could also be originated from better functioning pathways; showing again that combinatorial assembly can function as a tool to efficiently find promoters and gene homologs that display better performance in a multigenic pathway.

Two levels of combinatorial control have been studied in the previous examples; expectedly, we could add more variations onto the pathway assembly to enlarge the possibilities of obtaining strains with new and better phenotypes. Recently, Coussement et al.[64] showed that targeting all three levels of pathway control in *E. coli* was feasible using fluorescent reporter gene as a proof of concept. The library of assembled modules contained variability in promoters, RBS sequences, and CDS. To enable this combinatorial approach, they relied on Gibson Assembly[23] of oligonucleotides encoding different regions of the construct. They named this approach as the single-stranded assembly technique. Since they targeted variability in all three elements of their target, they devised three ways in which their oligonucleotides could be assembled, and tested for the one yielding the highest efficiency. In their first case, a single-very long oligonucleotide contained all three sections namely, promoter, RBS, and CDS. Their second scheme partitioned the sequence in three smaller oligos, while their last scheme partitioned the sequence into two oligos of similar sizes. Their assembly analysis showed that the third scheme had the highest efficiency followed by the second scheme with almost 100% efficiency, while the first scheme had only around 25% efficiency.

When implementing the third scheme into the combinatorial assembly with varying promoter and RBS, they were able to observe that the levels of fluorescence resulting from these combinations exceeded the values obtained when just the promoter or the RBS were individually subjected to variability. Finally, when incorporating two reporter genes in control of a promoter library, they observed good orthogonality in the expression of the genes. They were able to conclude that the method allows to independently and in parallel tune the transcript levels of the genes. However, even though their best scheme showed excellent assembly efficiencies, longer oligonucleotides may be required to assemble longer pathways, thus limiting the potential of this technique to small pathways, or to multiple steps of assembly, a common drawback of *in vitro* assembly techniques.

Other combinatorial techniques like ePathBrick[66] have shown potential in enabling combinatorial pathway design. Based on the BioBrick[80] assembly method, this technique expands the common isocaudomer pair, *Xba*I and *Spe*I, to two more, namely, *Avr*II and *Nhe*I. This combination of restriction sites, strategically introduced in the ePathBrick vector, allows for modular assembly and further pathway optimization. Although this technique allows a precise design of the final construct, the fact that only one gene can be added per assembly step conveys a very time-consuming practice, compared to homologous recombination strategies, where a single reaction may allow the assembly of multiple parts.

There are two main examples in which this method has been implemented for the production of metabolites. The first one involved the combinatorial assembly of pathways leading to flavonoid production.[66] In this case, the three genes constituting the pathway were arranged in operon, semioperon, or monocistronic cassettes allowing for nine possible combinations with a further diversification to 54 pathway variants by swapping the order of the genes. In the second study, this technique was applied for better understanding of the dynamic regulation of fatty acid production in *E. coli*. By modulating the expression levels of the enzymes involved in the central metabolism of fatty acids in a combinatorial way as well as varying RBS, they obtained the highest productivity of fatty acids in *E. coli* in fed-batch fermentation.[67]

Further, *in vitro* combinatorial techniques have originated from well-established assembly techniques. For example, the Golden Gate Assembly technique has allowed the origination of two combinatorial assembly techniques, namely MoClo[36] and GoldenBraid.[38] The main aim of these techniques is to standardize the way combinatorial pathways can be assembled. In MoClo and GoldenBraid, they subdivided all the genetic elements that will confer tiers or levels on their DNA composite, starting from level zero, which consists of individual genetic elements that can include, but are not limited to RBS, CDS, terminators, and signal peptides. In a series of one-pot assembly reactions, they combine the level-zero parts to form a level-one composite. Successively, the level-one constructs can be combined to form level-two assemblies. This is enabled by flanking each module with fusion sites and type II restriction sites (*Bsa*I and *Bpi*I); destination vectors allowed for different selection methods of positive transformants.

In 2013, an update to GoldenBraid was published and was denominated GoldenBraid 2.0 (GB2.0).[65] Similarly to the two previous examples, the inventors of GB2.0 made an analogy of their technique to the foundation of English grammar. Their level-zero parts function just like morphemes and grammemes as units in linguistics that allow the construction words to complete phrases. Mainly, GB2.0 was developed as a standardized procedure to assemble pathways for plant metabolic engineering with combinatorial applicability. As an updated version of GoldenBraid, GB2.0 incorporates several new tools that simplify the overall assembly process. First, they introduced a *Btg*ZI site into the entry vector. The ability of this enzyme to cut 10 nucleotides from its recognition site enables the capability of a "dual-release" option for each GBpart, speeding up the construction of multigenic assemblies. Also they have arranged a collection of standard parts compatible with their technique available as an open-source, as well as web applications that allow a friendly interface for *in silico* pathway design. The authors envisioned their technique as a revolutionary method that will allow fine automation of the assembly process. They claimed that the use of their technique will potentialize the advancement in, not only plant synthetic biology, but also for microorganisms, since their method facilitates exchangeability of genetic parts.

The examples described so far have shown the potential of combinatorial assembly for the generation of microbial strains with new and better phenotypic features. From simple bacterial implementations to plant synthetic biology, the assembly of pathways is definitely not limited to a certain organism. Moreover, the power of combinatorial pathway assembly goes beyond the production of biofuels and related compounds. Natural products represent the perfect example for which a combinatorial approach is needed to enable exploration of the vast pool of metabolite variants. The current availability of DNA manipulation techniques and bioinformatic tools open the door for the discovery of new natural and nonnatural products.

The importance of natural products relies on the implementation of such metabolites as new pharmaceutical drugs. In previous decades, high percentages of cancer drugs have been obtained directly from natural products or their derivatives. They can also be utilized as antibacterial and antiviral drugs.[81] Although the discovery of natural products is a topic of great importance, the intricate regulations embedded in gene clusters involved in the production of such secondary metabolites hamper quick advancement in this area. New DNA technologies have to be implemented to ease drug discovery and one of them is fast combinatorial ways to assemble gene clusters or independent enzymes involved in secondary metabolism.

A classic example of combinatorial assembly for natural products is the scrambling of enzymes belonging to the family of type I polyketide synthase (PKS). The enzymes involved in this family resemble an industrial assembly line in which each module carries out a specific reaction. This feature allows to easily apply the concept of combinatorial design, in which each module can be removed, mutated, or substituted by putative enzymes to finally yield the desired product or even create unnatural ones. There are various examples in which these approaches have been followed. Menzella et al.[82] developed a combinatorial approach to assemble codon-optimized PKS modules to produce unnatural triketide lactones in *E. coli*. They designed 11 donor and 14 acceptor modules from which 72 out of the 154 possible combinations were able to produce a potential unnatural triketide lactone. In addition, in 2007 Katsuyama et al.[83] targeted type III PKS and produced a series of metabolites from the stilbene and flavonoid family by combining enzymes from different nonmicrobial sources. By mixing enzymes involved in precursor formation and post-PKS tailoring, the authors enabled the diversification of the originally targeted products. From the 87 resulting metabolites, 36 had no trace of being already reported in literature.

Most of the research in this field has been done with the aid of classic DNA cloning methodologies, in which time-consuming steps limited the research to case-by-case scenarios. The development of new assembly techniques and synthetic biology tools, hence, represents a great potential for new natural product discovery. Even though there are only a few examples of applying such techniques in this area, it is not hard to foresee the great potential that they possess. For example, a recently developed synthetic biology strategy based on a plug-and-play scaffold allowed for the heterologous production of spectinabilin in *Streptomyces lividans*.[33] This technique allowed the bypass of endogenous regulation by setting the expression of the genes in the cluster to constitutive promoters. Founded upon the DNA Assembler,[46] this technique has also allowed the production of polycyclic tetramate macrolactams and the discovery of new derivatives from the same family.[49]

It is evident that there are advantages and disadvantages associated with the selected technology for pathway assembly. Nevertheless, the methods exemplified in this text have demonstrated vast improvements in terms of standardization, efficacy, and versatility. Nowadays, the limitations do not precisely rely on the assembly but on developing fast screening techniques that permit to easily explore the multiple phenotypes generated along the assembly process. This means that, to leverage rapid strain engineering enabled by combinatorial pathway assembly, high-throughput screening techniques must be readily available. As it will be mentioned in the following section, the development of high-throughput screening techniques goes hand in hand with the creation of combinatorial pathways. Lacking such rapid screening methods might become the limiting factor in pathway optimization.

6.4 Challenges and Future Direction

The classic and most common techniques for metabolite detection are based on chromatographic methods. However, when we think of combinatorial assembly and creation of pathway libraries, excessive dependence on chromatographic separations causes elevated time consumption and monetary expenses. That is why a combinatorial assembly approach is usually coupled with a fast or high-throughput screening method. A good example could be a screening method based on cell growth. As studied in the previous section, in the development of COMPACTER,[20] the initial screening method entailed selecting colonies with larger sizes in solid media containing xylose as the sole carbon source. Narrowing down the collection of the potential positive mutants with this approach allowed a more feasible screening with chromatographic methods. Like this, the need for high-throughput screening methods has led to the development of technologies, including but not limited to, molecular biosensors, microfluidics, and liquid-handling robotics coupled with user-friendly programming languages. This section briefly introduces these topics with its most recent applications toward the development of engineered microbes (Figure 6.5).

The development of molecular biosensors is a topic of high interest due to the ability of creating molecular machineries capable of detecting specific metabolites. In yeast and bacterium, molecular sensing has been implemented with transcriptional elements that

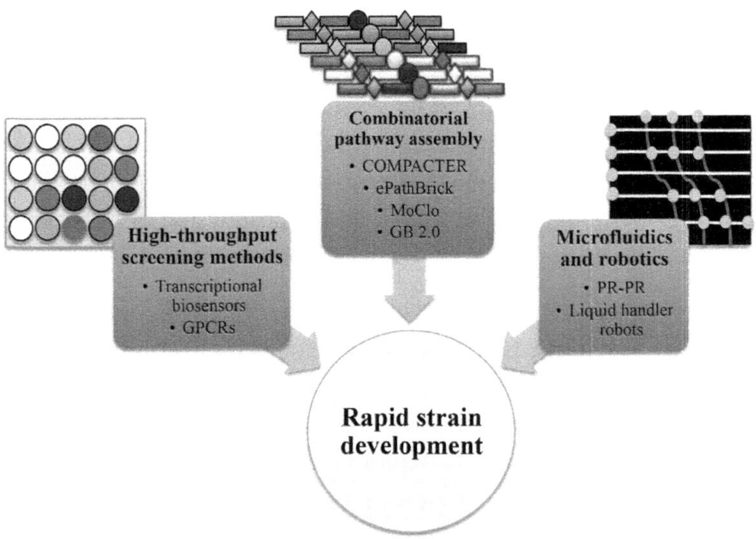

Figure 6.5

Combination of technologies required for advanced strain development for the production of biofuels. The convergence of combinatorial pathway assembly, high-throughput screening techniques, and microfluidics and robotics will lead to a fast advancement in the development of strains capable of producing biofuels and other high-value compounds.

activate the expression of a reporter gene or repress the expression of a toxic gene in the presence of a target compound. The reporters can generate different signals like optical (fluorescence, luminescence, and colorimetric),[84,85] or electrochemical.[86,87] Another classification of molecular biosensors is the transcriptionally independent activators. A good example is the well-known G-protein-coupled receptors (GPCRs). These receptors, localized on the cell membranes, are capable of triggering a cascade response when they bind to the molecule of interest, leading to transcription of reporter genes. The modularity of GPCRs makes them a versatile tool subject to protein engineering for the detection of a myriad of very specific metabolites.[88–90] For a comprehensive review on yeast biosensors, refer to "Yeast-Based Biosensors: Design and Applications" by Adeniran et al.[91]

Other types of biosensors that do not rely on transcriptional operators have been implemented in yeast to analyze the levels of internal metabolites. This can be applied to detect potential metabolite imbalances caused by the introduction of heterologous pathways. Common metabolic targets in these processes are NADH and NADPH, which are cofactors for oxido-reducing enzymes. Maintaining the right balance of these cofactors can ultimately ensure the correct performance of a heterologous pathway.[92–94] Electrical devices can readily measure the variability of these compounds by pumping electrons to electrodes by means of special mediators,[87] thereby allowing a fast detection of the abundance of such cofactors.

The creation of extensive mutant pathway libraries and the screening of such mutants have been immensely aided by the development of liquid-handling robotics and microfluidic technologies. The utilization of these techniques can reduce the introduction of human errors, operational time, and working volumes in great amounts. For example, recent advances in microfluidics have allowed the development of an ultrahigh-throughput method for directed evolution applications.[95] In this case, a library of 10^8 yeast cells carrying mutated horseradish peroxidases was screened on a cell-by-cell basis in about 10 h, limiting the consumption of reagents to 150 μl. Although this technology has not been applied to enhance the production of biofuels in microbes, it is not hard to envision its applicability to engineering proteins and further assemble them in a combinatorial pathway.

In the realm of pathway assembly, robotics and microfluidics have had a great impact as well. Recently, a programming language denominated PR-PR, developed by Linshiz et al.,[96] enabled cross talk between liquid handler robots, microfluidic systems, and manual operations. As a proof of their technique, they assembled a combinatorial DNA library composed of a common vector backbone, four different promoters, and bicistronic variants coupled with the green fluorescence protein reporter gene. To design their combinatorial blocks for assembly, they used the j5 software.[18] They based their assembly rationale upon two techniques described in the previous sections: Golden Gate

DNA Assembly and Gibson DNA Assembly. The robotic and microfluidics array allowed performing all the PCR reactions needed by implementation of the PR-PR program to confirm correct assemblies.

The technology described in this chapter encompasses trending methods for pathway engineering and optimization. Although efficient biofuel production in microbes does not rely solely on the introduction of heterologous pathways, it represents a big part of the engineering foundation to obtain industrial-level microbial factories. Nonetheless, to increase the pace of these advancements, several things have to be enabled. One of them is to develop truly combinatorial assembly methods, for example, methods that will allow mixing and matching genetic parts in a few steps and relying on inexpensive chemicals. Developing fast screening techniques is also indispensable for the efficient selection of mutants with desired phenotypes. And finally, it is important to develop assembly techniques capable of being adopted by a programing language for implementation in liquid handlers or microfluidic devices. By transferring as most wet-lab work to a machine as possible, we are reducing potential errors, time, and money expenditures. But most importantly, we are ensuring that other researchers can easily adopt such technologies, which opens the door for fast advancements in the already palpable aims to substitute petroleum-derived fuels to its biological counterparts.

References

1. Kahl LJ, Endy D. A survey of enabling technologies in synthetic biology. *J Biol Eng* 2013;**7**:13.
2. Priscilla EMP, Ron W. The second wave of synthetic biology: from modules to systems. *Nat Rev Mol Cell Biol* 2009;**10**:410.
3. Rabinovitch-Deere CA, Rodriguez GM, Oliver JWK, Atsumi S. Synthetic biology and metabolic engineering approaches to produce biofuels. *Chem Rev* 2013;**113**:4611–32.
4. Lee SK, Chou H, Ham TS, Lee TS, Keasling JD. Metabolic engineering of microorganisms for biofuels production: from bugs to synthetic biology to fuels. *Curr Opin Biotechnol* 2008;**19**:556–63.
5. Robert C. The changing economics of DNA synthesis. *Nat Biotechnol* 2009;**27**:1091.
6. Boratyn GM, Schäffer AA, Agarwala R, Altschul SF, Lipman DJ, Madden TL. Domain enhanced lookup time accelerated BLAST. *Biol Direct* 2012;**7**:12.
7. Jing F, Cantu D, Tvaruzkova J, Chipman J, Nikolau BJ, Yandeau-Nelson MD. Phylogenetic and experimental characterization of an acyl-ACP thioesterase family reveals significant diversity in enzymatic specificity and activity. *BMC Biochem* 2011;**12**:44.
8. Dellomonaco C, Clomburg JM, Miller EN, Gonzalez R. Engineered reversal of the β-oxidation cycle for the synthesis of fuels and chemicals. *Nature* 2011;**476**:355.
9. Ren Y, Perepelov AV, Wang H, Zhang H, Knirel YA, Wang L, et al. Biochemical characterization of GDP-l-fucose de novo synthesis pathway in fungus *Mortierella alpina*. *Biochem Biophys Res Commun* 2010;**391**:1663–9.
10. Bloch SE, Schmidt-Dannert C. Construction of a chimeric biosynthetic pathway for the *de novo* biosynthesis of rosmarinic acid in *Escherichia coli*. *ChemBioChem* 2014;**15**(16):2393–401.
11. Lane MD, Seelig B. Advances in the directed evolution of proteins. *Curr Opin Chem Biol* 2014;**22**:129–36.
12. Li Y, Cirino PC. Recent advances in engineering proteins for biocatalysis. *Biotechnol Bioeng* 2014;**111**:1273–87.
13. Pandya C, Farelli JD, Dunaway-Mariano D, Allen KN. Enzyme promiscuity: engine of evolutionary innovation. *J Biol Chem* 2014;**289**(44):30229–36.

14. Redden H, Morse N, Alper HS. The synthetic biology toolbox for tuning gene expression in yeast. *FEMS Yeast Res* 2014;**1**. http://dx.doi.org/10.1111/1567-1364.12188.
15. Royce LA, Liu P, Stebbins MJ, Hanson BC, Jarboe LR. The damaging effects of short chain fatty acids on *Escherichia coli* membranes. *Appl Microbiol Biotechnol* 2013;**97**:8317.
16. Xiao H, Shao ZY, Jiang Y, Dole S, Zhao H. Exploiting *Issatchenkia orientalis* SD108 for succinic acid production. *Microb Cell Fact* 2014;**13**:121.
17. Lu G, Moriyama EN. Vector NTI, a balanced all-in-one sequence analysis suite. *Brief Bioinform* 2004;**5**:378.
18. Hillson NJ, Rosengarten RD, Keasling JD. j5 DNA assembly design automation software. *ACS Synth Biol* 2012;**1**:14.
19. Villalobos A, Ness J, Gustafsson C, Minshull J, Govindarajan S. Gene designer: a synthetic biology tool for constructing artificial DNA segments. *BMC Bioinform* 2006;**7**:285.
20. Du J, Yuan Y, Si T, Lian J, Zhao H. Customized optimization of metabolic pathways by combinatorial transcriptional engineering. *Nucleic Acids Res* 2012;**40**:142.
21. Cobb RE, Ning JC, Zhao H. DNA assembly techniques for next-generation combinatorial biosynthesis of natural products. *J Ind Microbiol Biotechnol* 2014;**41**:469–77.
22. Li MZ, Elledge SJ. Harnessing homologous recombination in vitro to generate recombinant DNA via SLIC. *Nat Methods* 2007;**4**:251.
23. Gibson DG, Young L, Chuang R, Venter JC, Hutchison CA, Smith HO. Enzymatic assembly of DNA molecules up to several hundred kilobases. *Nat Methods* 2009;**6**:343.
24. Anderson JC, Dueber JE, Leguia M, Wu GC, Goler JA, Arkin AP, et al. BglBricks: a flexible standard for biological part assembly. *J Biol Eng* 2010;**4**:1.
25. Guye P, Li Y, Wroblewska L, Duportet X, Weiss R. Rapid, modular and reliable construction of complex mammalian gene circuits. *Nucleic Acids Res* 2013;**41**:156.
26. Zhang Y, Werling U, Edelmann W. SLiCE: a novel bacterial cell extract-based DNA cloning method. *Nucleic Acids Res* 2012;**40**:e55.
27. Bitinaite J, Rubino M, Varma KH, Schildkraut I, Vaisvila R, Vaiskunaite R. USER™ friendly DNA engineering and cloning method by uracil excision. *Nucleic Acids Res* 2007;**35**:1992–2002.
28. Stevenson J, Krycer JR, Phan L, Brown AJ. A practical comparison of ligation-independent cloning techniques. *PLoS ONE* 2013;**8**, e83888.
29. De Kok S, Stanton LH, Slaby T, Durot M, Holmes VF, Patel KG, et al. Rapid and reliable DNA assembly via ligase cycling reaction. *ACS Synth Biol* 2014;**3**:97.
30. Quan J, Tian J. Circular polymerase extension cloning of complex gene libraries and pathways. *PLoS ONE* 2009;**4**, e6441.
31. Trubitsyna M, Michlewski G, Cai Y, Elfick A, French CE. PaperClip: rapid multi-part DNA assembly from existing libraries. *Nucleic Acids Res* 2014;**42**, e154.
32. Natalay K, Vladimir L. TAR cloning: insights into gene function, long-range haplotypes and genome structure and evolution. *Nat Rev Genet* 2006;**7**:805.
33. Shao Z, Rao G, Li C, Abil Z, Luo Y, Zhao H. Refactoring the silent spectinabilin gene cluster using a plug-and-play scaffold. *ACS Synth Biol* 2013;**2**:662.
34. Annaluru N, Muller H, Mitchell LA, Ramalingam S, Stracquadanio G, Richardson SM, et al. Total synthesis of a functional designer eukaryotic chromosome. *Science* 2014;**344**:55.
35. Li MZ, Elledge SJ. MAGIC, an in vivo genetic method for the rapid construction of recombinant DNA molecules. *Nat Genet* 2005;**37**:311.
36. Weber E, Engler C, Gruetzner R, Werner S, Marillonnet S. A modular cloning system for standardized assembly of multigene constructs. *PLoS ONE* 2011;**6**.
37. Chen WH, Qin ZJ, Wang J, Zhao GP. The MASTER (methylation-assisted tailorable ends rational) ligation method for seamless DNA assembly. *Nucleic Acids Res* 2013;**41**, e93.
38. Sarrion-Perdigones A, Falconi EEl, Zandalinas SI, Juárez P, Fernández-del-carmen A, Granell A, et al. GoldenBraid: an iterative cloning system for standardized assembly of reusable genetic modules. *PLoS ONE* 2011;**6**. http://dx.doi.org/10.1371/journal.pone.0021622.

39. Blake WJ, Chapman BA, Zindal A, Lee ME, Lippow SM, Baynes BM. Pairwise selection assembly for sequence-independent construction of long-length DNA. *Nucleic Acids Res* 2010;**38**:2594–602.
40. Tsuge K, Matsui K, Itaya M. One step assembly of multiple DNA fragments with a designed order and orientation in *Bacillus subtilis* plasmid. *Nucleic Acids Res* 2003;**31**, e133.
41. Colloms SD, Merrick CA, Olorunniji FJ, Stark WM, Smith MCM, Osbourn A. Rapid metabolic pathway assembly and modification using serine integrase site-specific recombination. *Nucleic Acids Res* 2014;**42**, e23.
42. Zhang L, Zhao G, Ding X. Tandem assembly of the epothilone biosynthetic gene cluster by in vitro site-specific recombination. *Sci Rep* 2011;**1**:141.
43. Torella JP, Boehm CR, Lienert F, Chen J-H, Way JC, Silver PA. Rapid construction of insulated genetic circuits via synthetic sequence-guided isothermal assembly. *Nucleic Acids Res* 2014;**42**:681.
44. Torella JP, Lienert F, Boehm CR, Chen J, Way JC, Silver PA. Unique nucleotide sequence-guided assembly of repetitive DNA parts for synthetic biology applications. *Nat Protoc* 2014;**9**:2075.
45. Ma H, Kunes S, Schatz PJ, Botstein D. Plasmid construction by homologous recombination in yeast. *Gene* 1987;**58**:201–16.
46. Shao Z, Zhao H, Zhao H. DNA assembler, an *in vivo* genetic method for rapid construction of biochemical pathways. *Nucleic Acids Res* 2009;**37**, e16.
47. Shao Z, Luo Y, Zhao H. DNA assembler method for construction of zeaxanthin-producing strains of *Saccharomyces cerevisiae*. *Methods Mol Biol* 2012;**898**:251–62.
48. Shao Z, Luo Y, Zhao H. Rapid characterization and engineering of natural product biosynthetic pathways via DNA assembler. *Mol Biosyst* 2011;**7**:1056–9.
49. Luo Y, Huang H, Liang J, Wang M, Lu L, Shao Z, et al. Activation and characterization of a cryptic polycyclic tetramate macrolactam biosynthetic gene cluster. *Nat Commun* 2013;**4**:2894.
50. Gibson DG, Benders GA, Andrews-Pfannkoch C, Denisova EA, Baden-Tillson H, Zaveri J, et al. Complete chemical synthesis, assembly, and cloning of a *Mycoplasma genitalium* genome. *Science* 2008;**319**:1215.
51. Gibson DG, Glass JI, Lartigue C, Noskov VN, Chuang R, Algire MA, et al. Creation of a bacterial cell controlled by chemically synthesized genome. *Science* 2010;**329**:52.
52. Engler C, Kandzia R, Marillonnet SA. One pot, one step, precision cloning method with high throughput capability. *PLoS ONE* 2008;**3**, e3647.
53. Engler C, Gruetzner R, Kandzia R, Marillonnet S. Golden gate shuffling: a one-pot DNA shuffling method based on type IIs restriction enzymes. *PLoS ONE* 2009;**4**, e5553.
54. Weber E, Gruetzner R, Werner S, Engler C, Marillonnet S. Assembly of designer TAL effectors by Golden Gate cloning. *PLoS ONE* 2011;**6**, e19722.
55. Gao S, Han L, Zhu L, Ge M, Yang S, Jiang Y, et al. One-step integration of multiple genes into the oleaginous yeast *Yarrowia lipolytica*. *Biotechnol Lett* 2014;**36**(12):2523–8.
56. Cao P, Wang L, Zhou G, Wang Y, Chen Y. Rapid assembly of multiple DNA fragments through direct transformation of PCR products into *E. coli* and *Lactobacillus*. *Plasmid* 2014;**76**:40–6.
57. Koonin EV. Computational genomics. *Curr Biol* 2001;**11**(5):R155–8.
58. Torday JS. Evolutionary biology. US: John Wiley and Sons Inc.; 2012.
59. Richardson SM, Wheelan SJ, Yarrington RM, Boeke JD. GeneDesign: rapid, automated design of multikilobase synthetic genes. *Genome Res* 2006;**16**:550.
60. Lorimer D, Raymond A, Walchli J, Mixon M, Barrow A, Wallace E, et al. Gene Composer: database software for protein construct design, codon engineering, and gene synthesis. *BMC Biotechnol* 2009;**9**:36.
61. Yamada R, Taniguchi N, Tanaka T, Ogino C, Fukuda H, Kondo A. Cocktail delta-integration: a novel method to construct cellulolytic enzyme expression ratio-optimized yeast strains. *Microb Cell Fact* 2010;**9**:32.
62. Yuan J, Ching CB. Combinatorial assembly of large biochemical pathways into yeast chromosomes for improved production of value-added compounds. *ACS Synth Biol* 2015;**4**(1):23–31.
63. Ramon A, Smith H. Single-step linker-based combinatorial assembly of promoter and gene cassettes for pathway engineering. *Biotechnol Lett* 2011;**33**:549–55.

64. Coussement P, Maertens J, Beauprez J, Van Bellegem W, De Mey M. One step DNA assembly for combinatorial metabolic engineering. *Metab Eng* 2014;**23**:70–7.
65. Sarrion-Perdigones A, Vazquez-Vilar M, Palaci J, Castelijns B, Forment J, Ziarsolo P, et al. GoldenBraid 2.0: a comprehensive DNA assembly framework for plant synthetic biology. *Plant Physiol* 2013;**162**:1618.
66. Xu P, Vansiri A, Bhan N, Koffas MAG. ePathBrick: a synthetic biology platform for engineering metabolic pathways in *E. coli*. *ACS Synth Biol* 2012;**1**:256.
67. Peng X, Gu Q, Wang W, Wong L, Bower A, Collins C, et al. Modular optimization of multi-gene pathways for fatty acids production in *E. coli*. *Nat Commun* 2013;**4**:1409.
68. Sakai A, Shimizu Y, Hishinuma F. Integration of heterologous genes into the chromosome of *Saccharomyces cerevisiae* using a delta sequence of yeast retrotransposon Ty. *Appl Microbiol Biotechnol* 1990;**33**:302–6.
69. Lee FW, Da Silva NA. Sequential delta-integration for the regulated insertion of cloned genes in *Saccharomyces cerevisiae*. *Biotechnol Prog* 1997;**13**:368–73.
70. Stricker AR, Mach RL, de Graaff LH. Regulation of transcription of cellulases- and hemicellulases-encoding genes in *Aspergillus niger* and *Hypocrea jecorina* (*Trichoderma reesei*). *Appl Microbiol Biotechnol* 2008;**78**:211–20.
71. Newman JD, Marshall J, Chang M, Nowroozi F, Paradise E, Pitera D, et al. High-level production of amorpha-4,11-diene in a two-phase partitioning bioreactor of metabolically engineered *Escherichia coli*. *Biotechnol Bioeng* 2006;**95**:684–91.
72. Tsuruta H, Paddon CJ, Eng D, Lenihan JR, Horning T, Anthony LC, et al. High-level production of amorpha-4,11-diene, a precursor of the antimalarial agent artemisinin, in *Escherichia coli*. *PLoS ONE* 2009;**4**, e4489.
73. Ueda H, Nakanishi M, Suzuki E, Nagamune T. Enhancement of mutation frequency with nucleotide triphosphate analogs in PCR random mutagenesis. *J Ferment Bioeng* 1995;**79**:303–5.
74. Salis HM. The ribosome binding site calculator. *Methods Enzymol* 2011;**498**:19–42.
75. Levin-Karp A, Barenholz U, Bareia T, Dayagi M, Zelcbuch L, Antonovsky N, et al. Quantifying translational coupling in *Escherichia coli* synthetic operons using RBS modulation and fluorescent reporters. *ACS Synth Biol* 2013;**2**:327.
76. Kim B, Du J, Eriksen DT, Zhao H. Combinatorial design of a highly efficient xylose-utilizing pathway in *Saccharomyces cerevisiae* for the production of cellulosic biofuels. *Appl Environ Microbiol* 2013;**79**:931–41.
77. Jeffries TW, Jin YS. Metabolic engineering for improved fermentation of pentoses by yeasts. *Appl Microbiol Biotechnol* 2004;**63**:495–509.
78. Jeffries TW. Engineering yeasts for xylose metabolism. *Curr Opin Biotechnol* 2006;**17**:320–6.
79. Jeffries TW, Grigoriev IV, Grimwood J, Laplaza JM, Aerts A, Salamov A, et al. Genome sequence of the lignocellulose-bioconverting and xylose-fermenting yeast *Pichia stipitis*. *Nat Biotechnol* 2007;**25**:319–26.
80. Brown C. BIOBRICKS: new design foundation-synthesized DNA to assemble as circuits from library of cells. *Electron Eng Times* 2004;**1**.
81. Saleem M, Nazir M, Ali MS, Hussain H, Lee YS, Riaz N, et al. Antimicrobial natural products: an update on future antibiotic drug candidates. *Nat Prod Rep* 2010;**27**:238–54.
82. Menzella HG, Reid R, Carney J, Chandran SS, Reisinger SJ, Patel KG, et al. Combinatorial polyketide biosynthesis by *de novo* design and rearrangement of modular polyketide synthase genes. *Nat Biotechnol* 2005;**23**:1171–6.
83. Katsuyama Y, Funa N, Miyahisa I, Horinouchi S. Synthesis of unnatural flavonoids and stilbenes by exploiting the plant biosynthetic pathway in *Escherichia coli*. *Chem Biol* 2007;**14**:613–21.
84. Bakhrat A, Eltzov E, Finkelstein Y, Marks RS, Raveh D. UV and arsenate toxicity: a specific and sensitive yeast bioluminescence assay. *Cell Biol Toxicol* 2011;**27**:227–36.
85. Bermejo C, Haerizadeh F, Takanaga H, Chermak D, Frommer WB. Optical sensors for measuring dynamic changes of cytosolic metabolite levels in yeast. *Nat Protoc* 2011;**6**:1806–17.

86. Baronian KH, Downard AJ, Lowen RK, Pasco N. Detection of two distinct substrate-dependent catabolic responses in yeast cells using a mediated electrochemical method. *Appl Microbiol Biotechnol* 2002;**60**:108–13.
87. Heiskanen A, Coman V, Kostesha Na, Sabourin D, Haslett N, Baronian K, et al. Bioelectrochemical probing of intracellular redox processes in living yeast cells-application of redox polymer wiring in a microfluidic environment. *Anal Bioanal Chem* 2013;**405**:3847–58.
88. Chang S, Ross EM. Activation biosensor for G protein-coupled receptors: a FRET-based m1 muscarinic activation sensor that regulates Gq. *PLoS ONE* 2012;**7**, e45651.
89. Katritch V, Cherezov V, Stevens RC. Structure-function of the G protein coupled receptor superfamily. *Annu Rev Pharmacol Toxicol* 2013;**53**:531–56.
90. Ault AD, Broach JR. Creation of GPCR-based chemical sensors by directed evolution in yeast. *Protein Eng Des Sel* 2006;**19**:1–8.
91. Adeniran A, Sherer M, Tyo KE. Yeast-based biosensors: design and applications. *FEMS Yeast Res* 2014;http://dx.doi.org/10.1111/1567-1364.12203.
92. Hou J, Hou J, Suo F, Wang C, Li X, Shen Yu, et al. Fine-tuning of NADH oxidase decreases byproduct accumulation in respiration deficient xylose metabolic *Saccharomyces cerevisiae*. *BMC Biotechnol* 2014;**14**:13.
93. Matsuda F, Ishii J, Kondo T, Ida K, Tezuka H, Kondo A. Increased isobutanol production in *Saccharomyces cerevisiae* by eliminating competing pathways and resolving cofactor imbalance. *Microb Cell Fact* 2013;**12**:119.
94. Shi A, Zhu X, Lu J, Zhang X, Ma Y. Activating transhydrogenase and NAD kinase in combination for improving isobutanol production. *Metab Eng* 2013;**16**:1–10.
95. Agresti JJ, Antipov E, Abate A, Ahn K, Rowat A, Baret J, et al. Ultrahigh-throughput screening in drop-based microfluidics for directed evolution. *Proc Natl Acad Sci U S A* 2010;**107**:4004.
96. Linshiz G, Stawski N, Goyal G, Bi C, Poust S, Sharma M, et al. PR-PR: cross-platform laboratory automation system. *ACS Synth Biol* 2014;**3**:515.

CHAPTER 7

Design of Dynamic Pathways

D. Liu*, G.J. Bentley†, K. Chu*, F. Zhang*,†
**Department of Energy, Environmental and Chemical Engineering, Washington University in St. Louis, Saint Louis, Missouri, USA †Division of Biological & Biomedical Sciences, Washington University in St. Louis, Saint Louis, Missouri, USA*

Summary

Advances in metabolic engineering and synthetic biology have enabled microbial production of many useful chemicals. However, pathway productivities and yields are often limited by metabolic imbalances, which inhibit cell growth and chemical production. Common strategies to balance metabolic pathways involve tuning promoter and RBS strength, which allow static control over enzyme expression. These static controllers are generally only optimized for a particular condition and cannot adjust themselves to environmental variations, which could lead to suboptimal production. On the other hand, environmental variations are common during large-scale production, where fluctuations in nutrient concentrations, oxygen levels or pH can deviate unregulated pathways from their optimal productivities.

By contrast, natural metabolic pathways are dynamically regulated according to cell metabolic status. These regulatory systems allow cells to adjust gene expression and metabolite concentrations to desirable levels, balancing the pathway when environment changes and improving cell fitness. This has inspired synthetic biologists to engineer synthetic regulatory systems to dynamically regulate engineered metabolic pathways. Several successful examples have been demonstrated that dynamic regulation could better balance a pathway, improving titers, yields, and productivities.

In this chapter, we will discuss current technologies to design and build dynamic regulatory systems. Dynamic regulation at various levels (transcriptional, translational, and posttranslational) using different design principles will be discussed. We will also discuss available tools for fine-tuning and provide a general guide to design dynamic pathways based on current knowledge. At last, future directions in the field of dynamic regulation will be discussed. With the development of more sophisticated synthetic biology tools and deeper understanding of metabolic dynamics, we expect dynamic regulated pathways will become mainstream in future metabolic engineering, improving productivities, yields, and robustness of engineered microbial cell factories.

Biotechnology for Biofuel Production and Optimization. http://dx.doi.org/10.1016/B978-0-444-63475-7.00007-8

7.1 Dynamic Regulation Basics

7.1.1 Importance of Dynamic Regulation

Current advances in metabolic engineering have enabled microbial organisms to produce valuable chemicals[1–5]. Microbes allow for fast replication, facile genetic manipulation, and most importantly, cost-effective fermentation for chemical production. Furthermore, microbial production of chemicals avoids the high temperature/pressure conditions, expensive catalysts, and toxic byproducts, which are often inevitable in industrial chemical synthesis. In order to maximize economic viability, productivities and yields must surpass a threshold at which microbial production of chemicals is justified. For valuable and difficult-to-obtain compounds, such as the antimalarial drug artemisinin, relatively low productivities and yields may justify microbial production over the expensive, labor-intensive method of extraction from the plant *Artemisia annua*[2]. The threshold at which microbial production of low-value bulk chemicals and biofuels is economically justified is much higher than high-value chemicals like artemisinin. Despite the higher threshold for economical industrial production, there is a great need for alternative fuel sources. Particularly, the declining availability of easily accessible crude oil and the increasing use of extreme extraction methods, such as hydraulic fracturing, have increased the demand for renewable alternatives, including microbial production of biofuels.

Although encouraging, microbial production of biofuels and chemicals currently suffers from two fatal problems. First, the manipulation of metabolic pathways to produce a desired compound inherently creates imbalances in the natural metabolic network; a network which was evolved for cell growth and replication, not chemical production. Alteration of the natural metabolism by metabolic engineering may affect cell growth, causing buildup of toxic intermediates, and ultimately, a decrease in production of the desired compounds. Second, microbial strains are often engineered and tested under well-defined and well-controlled conditions, and cannot withstand the fluctuating pH, temperature, oxygen, and/or osmolarity that are often encountered in large bioreactors. Thus, strains optimized in the laboratory may not behave as expected once scaled-up for production.

Researchers have attempted to optimize engineered systems by modifying classical gene regulatory units, such as the gene copy number, the promoter strength, and ribosome-binding sites (RBS)[6,7]. However, the engineered strains assume static intracellular conditions, and cannot adapt to changing conditions or imbalanced intermediate levels. An engineered metabolic pathway should ideally be dynamically regulated in response to metabolic flux or environmental changes. The ubiquity of dynamic regulation in nature substantiates the untapped potential for dynamic regulation in metabolic engineering. This chapter will introduce the utility of incorporating dynamic regulatory elements into engineered metabolic pathways to improve the productivity of the strain as well as its stability.

7.1.2 Dynamic Regulation in Natural Systems

In nature, organisms spanning from simple microbes to complex, multicellular animals rely on various levels of dynamic regulation to maintain metabolic homeostasis, basic cellular functions, and to adapt to ever-changing intracellular and extracellular environments. As a simple example, *Escherichia coli* utilizes a plethora of transcriptional regulatory networks: 34 coherent feedforward loops, 6 incoherent feedforward loops, 68 operons under single-input module control, and 203 pairs of operons regulated by the same two transcription factors (TFs)[8]. These numbers, limited by current investigational techniques, are likely a vast underestimate of the actual abundance of transcription-level natural dynamic regulation in *E. coli.* The number of documented dynamic regulation suggests that this is an effective regulatory mechanism for maintaining cell homeostasis in response to naturally fluctuating environments. Many microorganisms other than *E. coli* use similar regulatory networks to regulate a variety of pathways. For example, seven out of eight genes in the *S. cerevisiae* leucine biosynthesis pathway are positively regulated by the TF Leu3, which is in turn activated by a pathway intermediate—α-isopropyl-malate (αIPM). As a result, leucine biosynthesis is positively controlled by the intermediate branching from the common precursor of branched-chain amino acid biosynthesis[9]. Alternatively, over 100 diverse bacterial genomes contain a prototypical negative feedback loop in which the product, tryptophan, represses transcription of the *trp* operon[10].

Figure 7.1 demonstrates some common naturally occurring regulatory topologies based on a simplified metabolic pathway, in which substrate (S) is converted to product (P) via the intermediate (I). The biosynthesis of I and the conversion of I to P can be either upregulated

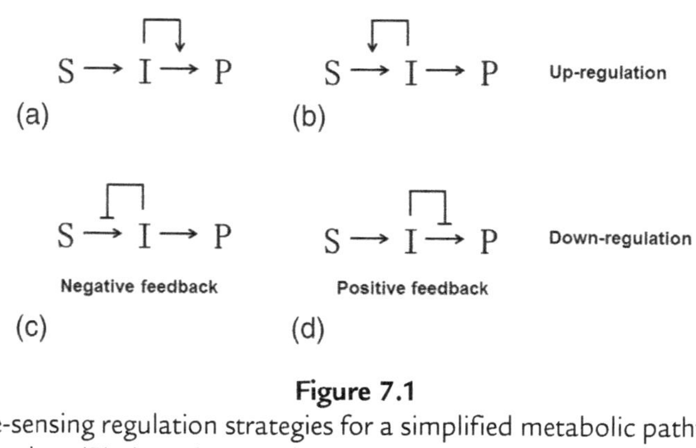

Figure 7.1
The intermediate-sensing regulation strategies for a simplified metabolic pathway. Substrate (S) is converted to product (P) via an intermediate (I), and the upstream or downstream gene of the intermediate can be either upregulated or downregulated. (a) and (b) show the upregulation strategies that create either a negative feedback (a) or a positive feedback (b) circuit. (c) and (d) show the downregulation strategies that create either a negative feedback (c) or a positive feedback (d) circuit.

(Figure 7.1a and b) or downregulated (Figure 7.1c and d) by I. From the point of view of the intermediate metabolite, both upregulation of a downstream pathway (Figure 7.1a) and downregulation of an upstream pathway (Figure 7.1c) create negative feedback loops, since they both act to decrease the level of I. On the other hand, upregulation of an upstream pathway (Figure 7.1b) and downregulation of a downstream pathway (Figure 7.1d) lead to positive feedback loops, since they both act to increase the level of I.

In practice, each type of regulatory topology has its specific benefits. **Negative autoregulation,** in which a TF represses its own transcription, is a key regulatory mechanism used by almost half of all TFs in *E. coli*. In addition to limiting wasteful gene and TF expression, negative autoregulation has been shown to increase the dynamic range of the input required to activate the target gene[11]. This feature is desirable when building synthetic dynamic regulation systems, because it minimizes the need to tune the input dynamic range. Another important characteristic of negative autoregulation is that it can decrease the response time of a genetic circuit[12]. When a strong promoter is used, a fast initial rise in protein production is generated, and as the TF level increases, it reaches the repression threshold for its own promoter, thus decreasing its own production rate. In this case, the negatively autoregulated TF shuts off its transcription after the initial rise, quickly reaching a steady state[12]. Furthermore, negative autoregulation can also reduce variations in protein levels across a cell population, another desired feature in an engineered dynamic regulatory system[13,14].

Negative feedback loops, as shown in Figure 7.1a and c may also be of use in designing a dynamic circuit. In general, negative feedback can prevent cells from synthesizing unneeded proteins, thus conserving carbon and energy. Moreover, under some conditions, negative feedback loops may reduce noise in metabolic systems. This reduction in noise helps to avoid stochastic deviation from the expected metabolic state[15]. In a metabolically engineered system, reduction of noise and improved stability of metabolic states helps achieve maximal product production[16].

Another common regulation strategy in nature is a **Feed forward loop**. Feed forward loops contain two TFs; the first TF regulates the second TF, and both the first and the second TF regulate the same target gene. The activity of the first TF directly on the target gene is the direct loop of activation, while the activity of the second transcription on the target gene is the indirect loop of activation. Coherent feed forward loops, i.e., loops acting in concert, lead to product amplification due to the cooperative direction of the indirect and direct loops. Alternatively, incoherent feedforward loops, i.e., loops that are not acting in concert, contain indirect and direct loops oriented in opposite directions, leading to repression[17].

Dynamic regulation is not only beneficial to natural systems; it can also be very powerful for engineered pathways and synthetic devices. One such example is a metabolic negative feedback loop based on a malonyl-CoA sensor-actuator in *E. coli*. When this dynamic regulatory circuit was used for the production of free fatty acids, there was a 34%

improvement in free fatty acid titers obtained, compared to a strain lacking the dynamic control loop[18]. Similarly, a dynamic regulatory circuit was utilized to improve fatty acid ethyl ester biodiesel formation threefold[1]. These will be discussed in detail in the following sections. The concept of harnessing TFs to dynamically regulate a metabolic pathway can be extended to a plethora of other pathways.

7.1.3 Fermenter Conditions and Dynamic Regulation

Pathway productivities and yields are not only affected by imbalanced gene expression and the accumulation of toxic intermediates or proteins, but also by constant micro-environmental changes during fermentation. In nature, organisms have evolved a variety of mechanisms to respond to microenvironmental changes. For example, many sigma factors respond to a variety of stresses, such as pH and osmolarity changes, in addition to the more subtle changes such as polarity, division, adjustment of motility, quorum sensing, etc. An engineered system without dynamic regulation effectively locks metabolic processes in a static state, even if this is detrimental to the cell. As such, these statically engineered pathways may not effectively produce the desired compound(s) when reactions are scaled up to large reactors or they may undergo mutation to remove the metabolic lock. For example, if glucose and/or oxygen levels dip while producing a target compound, a dynamically engineered strain would fare better than a comparable strain with constitutive pathway expression as cellular metabolism would not be taxed by excessive pathway expression when nutrient sources decline. Figure 7.2 demonstrates how pH, oxygen availability, and sugar sources change over time in a bioreactor[19]. Although work has been done in attempts to improve these conditions, and despite sophisticated microreactors, conditions still wildly fluctuate over time. Dynamic pathway engineering represents a promising answer to maximize product formation in a fluctuating fermenter environment.

Beyond dynamic response to pathway intermediates, strains can be engineered to dynamically respond directly to reactor conditions to maximize production and to limit cellular stress and potential mutation. For example, if a particular engineered process was closely tied to glycolysis (associated with the availability of oxygen), the pathway could anticipate oxygen fluctuations in the reactor using a sensor-actuator system. An oxygen-sensing heme protein can be expressed in the host cell, which regulates key pathway genes in response to oxygen levels, therefore allowing dynamic regulation in response to oxygen[20]. With today's availability of molecular biological tools, the possibilities are endless. With some creativity, researchers can develop ways for existing natural regulatory mechanisms to work in harmony with dynamically regulated synthetic pathways, conferring a multitude of benefits derived from dynamic regulation.

The remainder of this chapter will discuss synthetic dynamic systems at the molecular level as well as provide some techniques, current resources, and some ideas for future dynamically regulated engineered microbes.

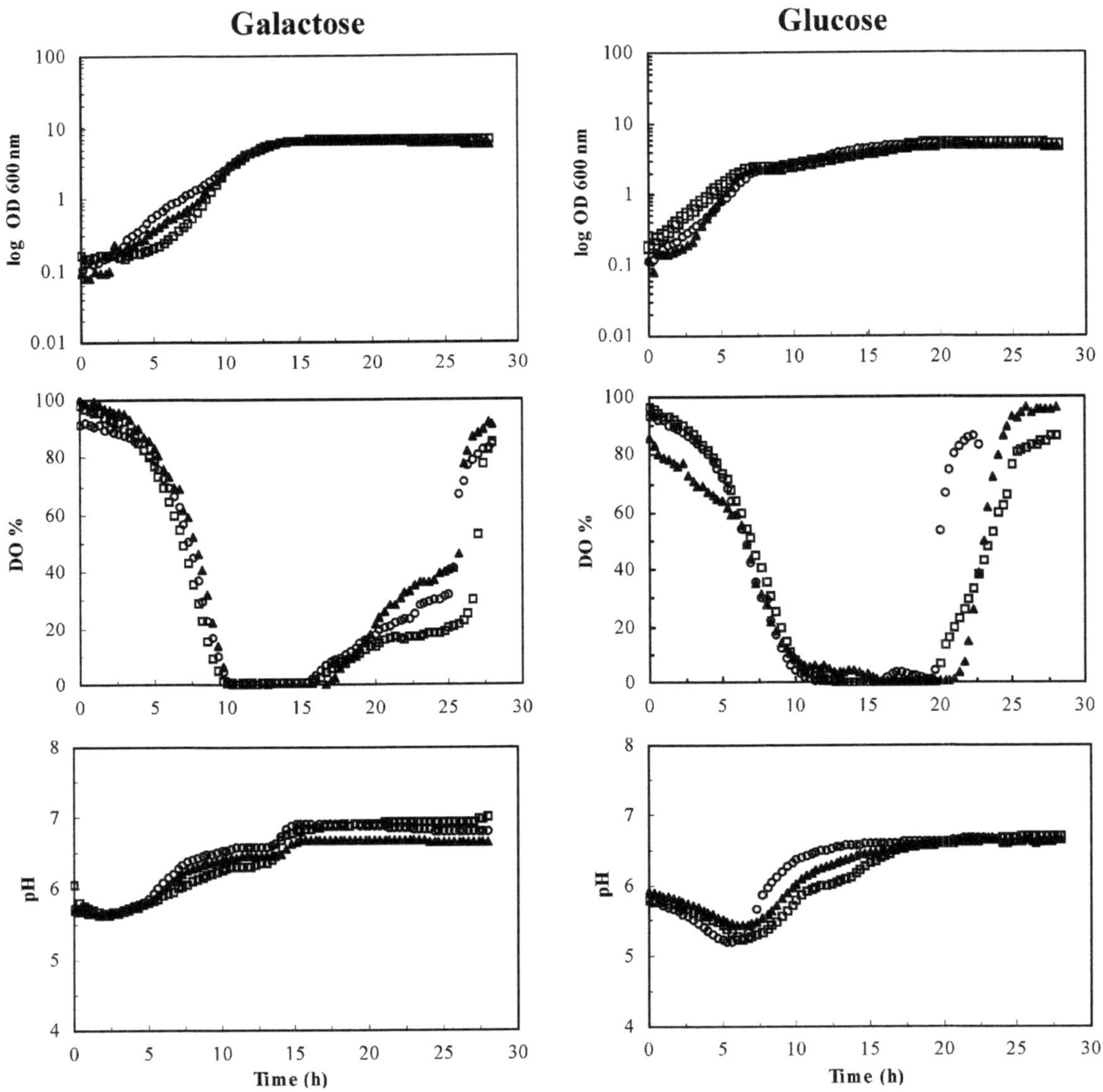

Figure 7.2

Variation of bioreactor conditions in a controlled microbioreactor. *S. cerivisae* was cultivated in a controlled bioreactor with either galactose or glucose as the carbon source. OD, dissolved oxygen (DO%), and pH were measured *in situ* over time. *Figure reprinted with generous permission granted by the author, A.J. Sinskey.*

7.2 Current Progress on Dynamic Pathway Designs

Advances in synthetic biology have enabled the construction of metabolic control circuits to dynamically regulate enzyme expression levels according to cellular metabolic status. The advantages of dynamic regulation to both balance an engineered pathway and to improve chemical production have been demonstrated through experimental and computational approaches. In this section, we will focus on introducing current technologies to build and to simulate dynamically regulated pathways.

7.2.1 Experimental Approach

To dynamically regulate a pathway, sensor-actuators that can modulate gene expression in response to metabolite levels need to be engineered. A sensor can be any kind of biomolecule such as a protein or RNA that interacts with the target molecule. Once a sensor is available, an actuator is then needed to transmit the signal from the sensor to control gene expression at the (1) transcriptional level by promoter engineering based on metabolite-responsive TFs, (2) transcriptional and translational level by metabolite-controlled riboswitches, and (3) posttranslational level by engineering metabolite-regulated proteins/enzymes. To regulate a particular pathway of interest, an engineered sensor-actuator needs to respond to a pathway-relevant metabolite. In the following subsections we will discuss various strategies used to engineer sensor-actuators based on the abovementioned types of molecular interactions. We will also discuss how they are applied to dynamically regulate metabolic pathways.

Transcriptional Regulation Based on TF-Metabolite Interaction

While a sensor-actuator can be constructed at different levels, dynamic regulation at the transcriptional level based on TFs is most common. TFs are proteins that bind to specific DNA sequences to control the rate of transcription. A TF can act as either an activator or a repressor depending on its interaction with the RNA polymerase. Furthermore, a TF can be activated (or deactivated) by small molecules using different mechanisms including ligand binding, phosphorylation, or interaction with other TFs. In this subsection, we will focus on ligand-regulated TFs because they have direct links to metabolite status.

Figure 7.3 shows how TFs control gene expression in response to ligands. The TF can be either activated or deactivated upon ligand binding to enhance or repress transcription (Figure 7.3a). As an example, Figure 7.3b and c shows the detailed mechanisms of two metabolite-regulated TFs. In Figure 7.3b, the absence of metabolite (M) allows the TF R to bind to its cognate operator site on the promoter region, creating steric hindrance which blocks the binding of RNA polymerase, thereby inhibiting transcription. In the presence of M, M binds to R, triggering a conformation change in R, which in turn releases R from the promoter and turns on transcription. In another case as shown in Figure 7.3c, transcriptional activator A promotes the binding of RNA polymerase to the promoter, increasing gene expression level. The presence of the metabolite antagonizes the RNA polymerase recruitment activity of A, leading to decreased transcription.

TF-based sensor-regulators are commonly used to engineer inducible promoters. In this case, an inducer-regulated TF is often constitutively expressed to a proper level, and its

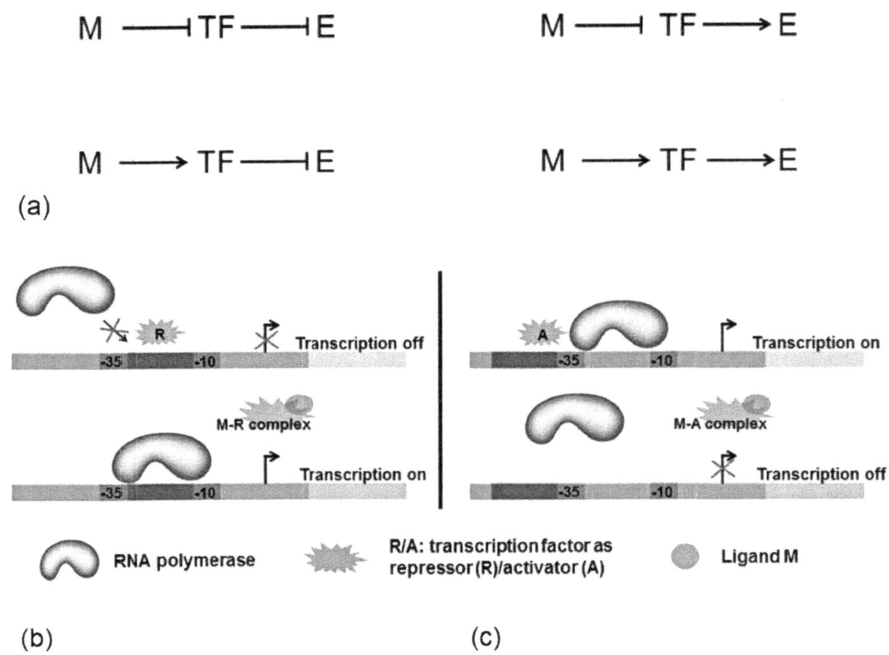

Figure 7.3
Transcriptional regulation strategies and mechanisms based on ligand-regulated transcription factors. (a) M represents ligand, TF represents transcription factor, and *E* represents enzyme. Ligand binding can either activate (represented by arrows) or deactivate (represented by blunt-end arrows) the transcription factor to enhance or repress gene expression. (b) Ligand binding deactivates a repressor and enhances transcription. (c) Ligand binding deactivates an activator and represses transcription.

corresponding operator sequence is inserted into a hybrid promoter. This promoter then responds to the level of the TF to control expression of a gene of interest. The addition of an inducer to the culture medium either activates or deactivates the TF activity, affecting the transcription initiation rate. The design principles of promoter engineering have been extensively studied, enabling efficient design of promoters with desired responses[21]. For example, some commonly used inducible promoters such as P_{LacUV5} and P_{T7} were created by inserting the *lacO* site in the promoter region. Without inducer, the constitutively expressed repressor LacI represses these promoters. The expression levels of these promoters can be variably controlled at different levels by titrating the inducer, and serve as robust methods to control gene expression.

Based on the same concept, metabolite-responsive TFs have been engineered to create sensor-actuators that modulate gene expression dynamically in response to metabolite levels. In one of the first examples, an acyl-CoA sensor-actuator was engineered based on an *E. coli* acyl-CoA-responsive TF, FadR[1]. Several synthetic promoters were designed by inserting the cognate FadR binding sequence near the −10 and −35 region of a phage promoter. Binding of FadR creates steric hindrance, blocking the binding of RNA polymerase to the synthetic promoter. The presence of acyl-CoA antagonizes the binding of FadR to the promoter, thus initiates transcription. This acyl-CoA-based sensor-actuator was used to upregulate acyl-CoA biosynthesis, ethanol production, and the expression of a downstream enzyme that condenses ethanol and acyl-CoAs to fatty acid ethyl esters (FAEEs), a diesel fuel replacement (Figure 7.4a). The acyl-CoA-based dynamic regulation allowed the downstream pathway to be activated only when there is sufficient acyl-CoA, preventing the production of unnecessary proteins and ethanol at the early stage of fermentation. As a result, FAEE titer was increased 3-fold, reaching 1.5 g/L.

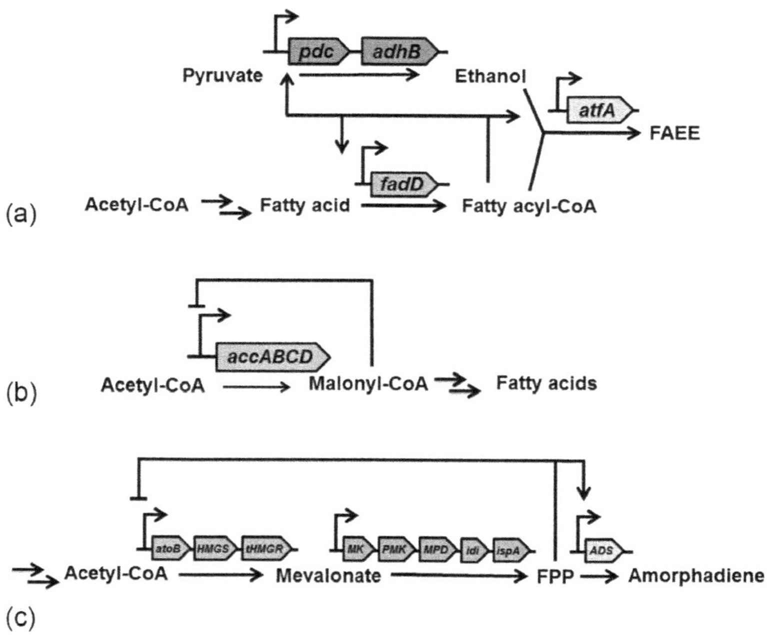

Figure 7.4
Examples of dynamically regulated synthetic pathways. (a) A dynamic upregulation system for FAEE production. Synthesis of FAEE and its precursors, ethanol and fatty acyl-CoA, are simultaneously activated by a key intermediate, acyl-CoA. This regulation system dynamically turns on downstream enzyme synthesis only when the upstream pathway accumulates enough key intermediates. (b) A negative feedback circuit to regulate fatty acid production. In this circuit, a rate limiting but toxic enzyme (acetyl-CoA carboxylase, encoded by *accABCD*) is negatively regulated by its product, malonyl-CoA. This circuit downregulates *accABCD* expression under high malonyl-CoA levels, alleviating toxicity while providing enough building blocks for fatty acids. (c) A dynamic regulation circuit for amorphadiene production. This circuit turns down FPP synthesis and activates the conversion of FPP to the product when there is sufficient FPP, thus avoiding the accumulation of the toxic intermediate.

While this study designed a system to upregulate enzyme expression, under some conditions it is desirable to downregulate certain enzyme levels. As is the case with natural pathway regulation, downregulation of enzyme expression at high intermediate/product levels saves carbon and energy. While high carbon usage efficiency is desired in an engineered metabolic pathway, this may require very different control strategies from the natural regulation pathways to maximize production. For example, in natural systems an end product often inhibits the first enzyme of the pathway when it reaches the threshold for its cellular function to save carbon. However, this commonly used bare minimum strategy from nature would hardly benefit production in an engineered pathway. This difference between natural and synthetic pathway regulation is caused by their different purposes. While natural regulation generally improves cell fitness to maximize the population, the purpose of synthetic pathway regulation is to maximize product synthesis. Therefore, completely different regulation architectures are needed for metabolic engineering purposes. For metabolic engineering purposes, it may be necessary to overexpress an enzyme that is toxic to the cell, but essential to synthesize the product or maximize production. This further complicates the disparity between natural and engineered regulatory architectures and supports the development of logically designed, dynamically regulated engineered pathways. Intuitively, when an enzyme or intermediate is toxic, downregulation is desired as it not only benefits cell growth, but it also saves carbon and energy. This idea was demonstrated in an engineered *E. coli* fatty acid biosynthetic pathway. In this pathway, fatty acid production can be enhanced by overexpressing acetyl-CoA carboxylase (encoded by *acc*), since it improves the concentration of the fatty acid building block, malonyl-CoA. However, overexpression of *acc* inhibits cell growth. In this study, a malonyl-CoA sensor-actuator was engineered to create a negative feedback loop to dynamically control the *acc* transcription rate (Figure 7.4b), which allowed the acetyl-CoA carboxylase level to be dynamically downregulated when cells accumulated high malonyl-CoA levels[18]. This circuit was able to recover cell growth and enhance fatty acid production by 34% more than without dynamic regulation.

While these systems are based on sensors that can be readily engineered from existing TFs, sometimes TFs that respond to the target metabolite may not be available. Under these conditions, transcriptional level sensors can be identified by transcriptomics. In one example, when the heterologous mevalonate-based isoprenoid biosynthetic pathway is expressed in *E. coli*, it has been found that five out of eight intermediates (including 3-hydroxy-3-methylglutaryl-CoA (HMG-CoA), mevalonate pyrophosphate (Mev-PP), isopentenyl pyrophosphate (IPP), dimethylallyl pyrophosphate (DMAPP), and farnesyl pyrophosphate (FPP)) are toxic. However, no natural sensor is available for these compounds. To identify a sensor for FPP, whole-genome transcriptional analysis was performed to analyze the cell's transcriptional response to the accumulation of FPP. Promoters that respond to FPP accumulation were identified and used to regulate amorphadiene synthesis. Specifically, FPP-repressive promoters were used to downregulate genes that convert acetyl-CoA to

FPP, and FPP-activated promoters were used to upregulate the enzyme that convert FPP to amorphadiene (Figure 7.4c). As a result, the production of amorphadiene was increased by twofold[22].

The experimental designs of dynamic pathway regulation mentioned above are all built upon transcriptional regulation, indicating that this strategy has a lot of potential in engineering control circuits to regulate metabolic pathways. Indeed, from several aspects, transcription-level regulation is advantageous compared to other regulatory strategies. First, nature has a large pool of natural TFs that interact with a variety of metabolites, offering a wide range of choices to engineer sensor-actuators for the target metabolites. Second, transcriptional regulation is versatile and can be engineered to both upregulate and downregulate gene expression. Third, the input-output range of transcriptional regulation can be readily tuned (see Section 7.3), enabling various regulatory dynamics. Finally, from an energy standpoint, transcriptional regulation is the most energy efficient because it not only optimizes the concentration of metabolites, but also optimizes the concentration of proteins and mRNAs.

Transcriptional and Translational Control Based on RNA-metabolite Interaction

Besides controlling transcription initiation, gene expression can also be regulated by metabolites at both transcriptional and translational levels based on RNA-metabolite interactions.

Riboswitches are RNA-based control elements that are commonly found in natural systems. In bacteria, riboswitches are found exclusively in the 5'-untranslated regions (UTRs) of target genes[23]. Moreover, riboswitches that can respond to a variety of metabolites have also been engineered [24–26]. A riboswitch usually consists of two domains: an aptamer domain that interacts with a signaling molecule (such as a metabolite), and an actuator domain that transduces a signal into a change in gene expression level. These conformation changes can control gene expression through a few mechanisms. At the transcriptional level, riboswitches can control transcription elongation. Ligand binding to the aptamer either facilitates or disrupts the formation of a terminator structure in the 5' UTR to prevent the synthesis of long mRNAs, creating transcriptional repression or activation (Figure 7.5) [23]. For example, it was found that flavin mononucleotide (FMN) biosynthesis and its transport in *Bacillus subtilis* were regulated by riboswitches. The binding of FMN causes premature transcription termination of the *ribDEAHT* operon and thus decreases gene expression levels in the presence of FMN[27]. At the translational level, riboswitches can control gene expression by modulating the secondary structure of mRNAs by either presenting or sequestering the RBS sequence from the 30S ribosomal subunit (Figure 7.5)[23,28–30]. The first examples of riboswitches identified were *E. coli* mRNAs encoding thiamine biosynthesis enzymes which can interact with thiamine or its pyrophosphate derivative[31]. The metabolite-mRNA complex forms a structure that sequesters the ribosome-binding site, thus decreasing gene expression. Moreover, riboswitches can also regulate gene expression by controlling mRNA stability. For example, an RNA device that

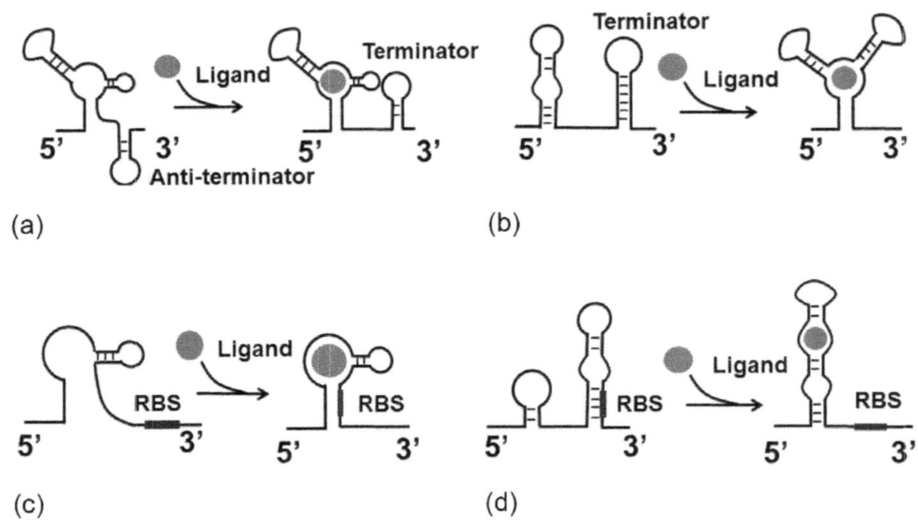

Figure 7.5

Riboswitch-mediated gene regulation mechanisms. Transcription termination (a, b) and translation initiation (c, d) can be controlled by riboswitches. At the transcriptional level, binding of a ligand can either facilitate (a) or disrupt (b) the formation of a terminator structure, leading to transcription termination (a) or transcription activation (b). At the translational level, binding of a ligand can either sequester (c) or present (d) RBS from the 30S ribosomal subunit, which inhibits (c) or initiates (d) translation.

couples RNA aptamers and Rnt1p hairpins has been engineered based on an RNase III enzyme [32]. The self-cleavage activity was inhibited by a conformation change triggered by binding to the ligand (theophylline), thus stabilizing the transcript.

All the mechanisms mentioned above can be used to engineer riboswitches that respond to a variety of small molecules, providing alternative tools and broadening the variety of metabolites that can be targeted to regulate metabolic pathways. However, because RNA-metabolite interactions rely largely on the mRNA secondary structure, swapping riboswitches to control different genes must be done with extreme care. Different genes may cause mRNAs to fold differently, affecting the proper regulatory function of riboswitches. Currently, it is still challenging to accurately predict the secondary structure of RNA. Furthermore, RNA folding kinetics is another important concern. In order for translational regulation to work properly, a regulatory riboswitch has to fold before translation starts and before RNA degradation occurs. In addition, design of RNA-based sensor-regulators is limited by the ability to engineer aptamers for target metabolites. Due to the limited chemical functional groups of RNA, RNA aptamers only recognize certain categories of molecules. Expending their ligand diversity and tuning the binding affinity with their ligands is very challenging. Due to these limitations, engineering of RNA-based dynamic regulatory systems has been scarcely, if ever, reported.

Posttranslational Control Based on Protein-Metabolite Interaction

Finally, gene expression can be controlled at the posttranslational level by modulating enzyme activities based on protein-metabolite interaction.

Posttranslational regulation is very common in nature. Activities of many enzymes are allosterically regulated by small molecules. Oftentimes in *E. coli*, the first enzymes of a biosynthesis pathway are inhibited by the end product of that very same pathway. Such negative regulation provides a rapid feedback of pathway flux, avoiding the accumulation of metabolites. Compared with regulation at other levels, posttranslational-level control directly regulates the protein activity, thus minimizing the delay time in metabolic response. Therefore, posttranslational regulation is believed to be a quick way to adjust pathway activity. However, since it does not avoid the synthesis of extra mRNA or protein molecules, this type of regulation is the least efficient in terms of carbon and energy usage.

In general, posttranslational control requires a protein that has a ligand binding domain to interact with a small molecule and an output domain to control the enzyme activity based on the protein-ligand interaction. Synthetic posttranslational regulation can be engineered by the functional fusion of a ligand binding domain to an enzyme catalytic domain. This strategy has been demonstrated in the case of β-lactamase (BLA) [33–35]. For this sensor-actuator, the BLA protein was fused with a maltose binding protein (MBP). Maltose binding to MBP triggers a conformational change of the chimeric protein and allosterically activates BLA. The key in engineering posttranslational control is the connection between the ligand-sensing domain and the catalytic domain, which requires not only both of the domains to be functional after fusion, but also the catalytic activity to be controlled by ligand binding, a substantial challenge in protein engineering. Hopefully, with the development of protein computational design, this posttranslational regulation will become a more accessible approach.

Summary of Experimental Approaches

In this section, we have discussed several strategies to experimentally construct and engineer sensor-actuators. These strategies provide multiple levels of interactions to engineer sensor-actuators that could respond to a wide variety of pathway metabolites, which can be used to regulate different metabolic pathways. Different types of regulations could lead to different behaviors of the regulated pathways according to the inherent molecular interaction and regulatory mechanism. These features are important in choosing the desired regulation strategy (Table 7.1).

To build dynamic regulation systems, the desired regulation strategy needs to be determined, including which metabolites to sense, which enzymes to regulate, how to regulate pathway genes (downregulation or upregulation), and which regulatory parameters (regulation strength and dynamics) to choose. To determine a regulation strategy, a thorough understanding of the pathway and the dynamic behavior of regulatory systems is required. Once the desired

Table 7.1: Comparison of sensor-actuators at different gene expression stages

	Response speed	Energy/carbon efficient
Transcriptional regulation	+	+++
Translational regulation	++	++
Posttranslational regulation	+++	+

regulation strategy is chosen, sensor-actuators specific to the desired pathway metabolites can be engineered to control gene expression. These controllers may need to be tuned in order to respond to the target metabolite at biologically relevant levels. Furthermore, the strength of regulation at different target concentrations also needs to be considered. Regulation with undesired strength could lead to insensitive or ultra-sensitive responses. Moreover, the dynamic properties associated with different types of controllers are also important as they directly affect metabolite dynamics, which determines pathway productivities. With all these aspects taken into account, dynamic pathway regulation can be engineered to improve pathway productivities.

7.2.2 Computational Approach

While the power of dynamic regulation to improve pathway productivity has been demonstrated experimentally, it often requires extensive testing and tuning in order to obtain the desired (or optimal) pathway flux. Alternatively, computational models could be useful to simulate the behaviors of such complex systems, helping to screen for the optimal control strategies and biochemical parameters, therefore, providing a powerful tool to study dynamic regulatory systems.

Computational methods have been widely used to study different natural regulation architectures. Since natural regulatory networks are extremely complicated, the basic patterns that occur frequently in these networks have been identified and extensively studied. Particularly, their effects on gene expression dynamics[12], gene expression noise[14,36], and the input-output response curves[11,37] have been explored. To simulate different regulation patterns, typically a series of kinetic equations are written that describe the intracellular concentrations of the TF, the inducer, and the reporter. For example, in the case of the negative autoregulation architecture, computational models have been established in a few studies that demonstrate that negative autoregulation can shorten response time, stabilize gene expression against noise, linearize the input-output response curve, and increase the input dynamic range[11,12,14,37]. This allows for a better understanding of the functions of different regulation patterns and provides insights on the design of synthetic regulatory circuits.

While the models in the studies mentioned above focus on exploring the features of one individual regulation pattern, the effects of regulation architectures on a metabolic pathway have also been studied using the computational approach. In one study, it was found in a number of yeast amino acid pathways controlled by the intermediate metabolite activation

(IMA) architecture, the enzyme immediately downstream of the metabolite responds strongest to the metabolite, while the enzymes upstream of the intermediate respond relatively weakly. To understand this pattern, kinetic models that describe the concentrations of enzymes, metabolites, the TF, and the product were used to simulate the pathway. The optimal gene expression patterns were obtained by minimizing the cost function, which represented an estimate of growth penalty. It was found under IMA it is highly favorable to have strong control on the enzymes downstream of the intermediate[38]. The knowledge obtained by simulations of the natural regulatory pathways will contribute to the design principles of synthetic dynamic pathways, and at the same time the models can also be adjusted to simulate synthetic dynamic pathways.

Similarly, to simulate synthetic dynamic pathways, a series of equations that describe cell growth, the biosynthetic rates of pathway metabolites, enzymes, and products need to be assembled. However, while the goal of a natural pathway is often to optimize the cost function[38,39], for synthetic dynamic pathways the goal is to optimize production of the target compound. In one study, a model was constructed to simulate the effects of a closed loop to control efflux pump expression. The expression of efflux pumps exports biofuel molecules out of the cell to lower toxicity [40]. However, there is a trade-off due to the toxicity from their own expression. This necessitates their expression to be well controlled to achieve the maximal production. Four control strategies were considered and each controller performance, including biofuel production, cellular toxicity, and noise control ability, was compared. This allows for a prediction of how different control strategies affect pump overexpression, and gives a guide to the experimental design.

Besides simulating different control topologies, the effects of control parameters can also be investigated by a sensitivity analysis. In the case of a negatively controlled efflux pump, cell viability is found to be insensitive to many system parameters, including the repressor degradation rate and the repressor activation constant, but is only affected by a few factors such as biofuel export rate and biofuel production rate[41]. This allows researchers to focus on tuning the biofuel export rate and expression levels of genes encoding biofuel production to optimize the regulatory system for biofuel production. Overall, these computational simulation results provide useful insights in designing and optimizing different regulatory architectures to improve pathway productivity.

In general, computational models may provide a general view of which regulatory topology is more effective and which parameters are more sensitive. On the other hand, efficient construction and fine-tuning of a regulatory system under the guidance of models requires experimental tools. This necessitates the advance in new experimental tools, including a general method to engineer biosensors that could detect any metabolite of interest, actuators that could transmit the signal to regulate pathway activity, and efficient tuning strategies that could modulate the input-output range of sensor-actuators to the desired levels.

7.3 Tools to Engineer Dynamic Pathways

With more successful experimental and computational examples that demonstrate the benefits of dynamic pathway regulation, this field is now gaining wide research attention and quickly developing. These advances expand the variety of tools that can be used to better engineer dynamic pathways. These tools will be discussed in this section. Specifically, we will focus on experimental methods to build metabolite-responsive sensor-actuators, the process of dynamic pathway assembly, and computational tools that facilitate the understanding of dynamic processes.

7.3.1 Tools to Engineer Sensor-Actuators

A key element to engineering a dynamic pathway is a sensor-actuator that can regulate pathway activity in response to the metabolite of interest. In the previous section, different mechanisms and examples of sensor-actuators were discussed. In this section the tools that are available to design and construct sensor-actuators that can respond to a variety of metabolites will be examined.

Tools to Engineer Sensor-Actuators Based on TFs

At the transcriptional level, a large toolbox is available in which TFs for a large variety of metabolites have already naturally evolved in a broad spectrum of microorganisms. Some naturally occurring TFs and their cognate ligands are summarized in Table 7.2.[42–80]

With these TFs, sensor-actuators that respond to the corresponding ligands can be engineered through promoter engineering (refer to Section "Transcriptional Regulation Based on TF-Metabolite Interaction" for the design principles).

Besides the TFs identified from the nature toolbox, TFs responsive to a target ligand can also be created through protein engineering. Multiple-site saturation mutagenesis and fluorescence-activated cell sorting (FACS) can be applied to engineer and select for the TFs that respond to a particular ligand. In one example, five codons encoding the binding pocket of the *E. coli* AraC protein were subjected to simultaneous saturation mutagenesis. To select for the AraC variant that responds to mevalonate, the gene encoding a green fluorescent protein (GFPuv) was placed under the control of a P_{BAD} promoter. FACS was then used to isolate clones that had significantly higher GFP expression in the presence of mevalonate[81]. This general approach should also be useful to develop sensors that respond to a variety of compounds[82,83].

Tools to Engineer Sensor-Actuators Based on Riboswitches

To engineer riboswitches that respond to various metabolites, various tools have been developed to engineer aptamers and actuators.

Table 7.2: List of ligand-responsive transcription factors (LTFs) and their cognate ligands

Category	Ligand	LTFs	Pathway	Activator (A) or Repressor (R)	References
Lipids	Acyl-CoA	FadR	Fatty acid	A/R	42
	Unsaturated acyl-ACP, acyl-CoA	FabR	Fatty Acid	R	43
Amino acids	Arginine	ArgR	Arginine biosynthesis	R	44
	N-acetylserine	CysB	Cysteine biosynthesis, sulfur utilization	A/R	45
	Glycine	GcvR	Glycine cleavage	R	46
	Glycine	GcvA	Glycine cleavage	A/R	47
	Cysteine	MetR	Methionine biosynthetic pathway	R	48
	Acetoacetate, acetohydrobutyrate	IlvY	Isoleucine and valine biosynthesis	A	49
	Leucine	Lrp	Leucine response, amino acid limitation, global regulator	A/R	51
	Diaminopimelate	LysR	Lysine biosynthesis	R	52
	D-serine	DsdC	D-serine deaminase	A/R	53
	Tyrosine, phenylalanine	TyrR	Tyrosine biosynthesis	R	50
	Tryptophan	TrpR	Tryptophan biosynthesis	R	54
Nucleic acid-related compounds	*O*-methylgaunine	Ada	Alkylated DNA reparation	A	55
	Deoxyribonucleosides	DeoR	Deoxyribonucleoside catabolic pathway	R	56
	Hypoxanthine, guanine	PurR	Purine metabolism	R	57
	Uric acid	HucR	Purine degradation pathway	R	58
Carbohydrates	Arabinose	AraC	Arabinose catabolism and transport	A/R	59
	Fructose 1 (6)-phosphate	FruR	Fructose metabolism	A/R	60
	Galactose	GalR	Galactose catabolism	A/R	61
	Allolactose, IPTG	LacI	Lactose transport and catabolism	R	62
	Maltodextrin	MalT	Maltose catabolism and transport	A	63
	Melibiose	MelR	Melibiose degradation	R	64
	Rhamnose	RhaS	Rhamnose degradation and transport	A	65
	Ribose	RbsR	Ribose catabolism and transport	R	66
	D-allose	RpiR	Ribose catabolism and transport	R	67

(Continued)

Table 7.2: List of ligand-responsive transcription factors (LTFs) and their cognate ligands—cont'd

Category	Ligand	LTFs	Pathway	Activator (A) or Repressor (R)	References
Central metabolites and environment-related compounds	Pyruvate	PhdR	Pyruvate dehydrogenase	A/R	68
	Formate	FhlA	Formate metabolism	A	69
	Oxygen	Fnr	Aerobic, anaerobic respiration, osmotic balance	A	70
	NAD/ATP	NadR	NAD biosynthesis	R	71
	H_2O_2	OxyR	Oxidative stress	A/R	72
	Glc-Nac-6-phosphate	NagC	Regulating phosphotransferase system	R	73
	cAMP	Crp	Catabolite repression, global regulation	A/R	74
	Copper ion	CueR	Copper resistance	A/R	75
	Iron	FUR	Iron homeostasis	A/R	76
	Glycerol, glycerol-3-phosphate	GlpR	Glycerol-3-phosphate pathway	R	77
	Acetic acid	CidR	Murein hydrolase regulation	R	78
	Nickel	NikR	Nickel homeostasis	R	79
	Superoxide/NO	SoxR	Superoxide stress	R	80
Lipids	Acyl-CoA	FadR	Fatty acid	A/R	42
	Unsaturated acyl-ACP, acyl-CoA	FabR	Fatty Acid	R	43
Amino acids	Arginine	ArgR	Arginine biosynthesis	R	44
	N-acetylserine	CysB	Cysteine biosynthesis, sulfur utilization	A/R	45
	Glycine	GcvR	Glycine cleavage	R	46
	Glycine	GcvA	Glycine cleavage	A/R	47
	Cysteine	MetR	Methionine biosynthetic pathway	R	48
	Acetoacetate, acetohydrobutyrate	IlvY	Isoleucine and valine biosynthesis	A	49
	Leucine	Lrp	Leucine response, amino acid limitation, global regulator	A/R	51
	Diaminopimelate	LysR	Lysine biosynthesis	R	52
	D-serine	DsdC	D-serine deaminase	A/R	53
	Tyrosine, phenylalanine	TyrR	Tyrosine biosynthesis	R	50
	Tryptophan	TrpR	Tryptophan biosynthesis	R	54

Nucleic acid-related compounds	*O*-methylgaunine	Ada	Alkylated DNA reparation	A	55
	deoxyribonucleosides	DeoR	Deoxyribonucleoside catabolic pathway	R	56
	Hypoxanthine, guanine	PurR	Purine metabolism	R	57
	Uric acid	HucR	Purine degradation pathway	R	58
Carbohydrates	Arabinose	AraC	Arabinose catabolism and transport	A/R	59
	Fructose 1 (6)-phosphate	FruR	Fructose metabolism	A/R	60
	Galactose	GalR	Galactose catabolism	A/R	61
	Allolactose, IPTG	LacI	Lactose transport and catabolism	R	62
	Maltodextrin	MalT	Maltose catabolism and transport	A	63
	Melibiose	MelR	Melibiose degradation	R	64
	Rhamnose	RhaS	Rhamnose degradation and transport	A	65
	Ribose	RbsR	Ribose catabolism and transport	R	66
	D-allose	RpiR	Ribose catabolism and transport	R	67
Central metabolites and environment-related compounds	Pyruvate	PhdR	Pyruvate dehydrogenase	A/R	68
	Formate	FhlA	Formate metabolism	A	69
	Oxygen	Fnr	Aerobic, anaerobic respiration, osmotic balance	A	70
	NAD/ATP	NadR	NAD biosynthesis	R	71
	H_2O_2	OxyR	Oxidative stress	A/R	72
	Glc-Nac-6-phosphate	NagC	Regulating phosphotransferase system	R	73
	cAMP	Crp	Catabolite repression, global regulation	A/R	74
	Copper ion	CueR	Copper resistance	A/R	75
	Iron	FUR	Iron homeostasis	A/R	76
	Glycerol, glycerol-3-phosphate	GlpR	Glycerol-3-phosphate pathway	R	77
	Acetic acid	CidR	Murein hydrolase regulation	A/R	78
	Nickel	NikR	Nickel homeostasis	R	79
	Superoxide/NO	SoxR	Superoxide stress	R	80

RNA can form versatile structures, which can be tailored and selected to detect a wide array of metabolites. A well-established strategy to select RNA structures with high binding affinities to a given ligand is through *in vitro* selection experiments[84]. This process begins by generating large random RNA sequence libraries and immobilizing the target molecule on a solid support. To select for a sequence with high affinity to a target molecule, a library is incubated with the immobilized molecules and the bound sequences are isolated based on an affinity chromatography. These sequences are then subjected to RT-PCR and the products are used as the input library for the next round of selection. By iterating this cycle with increasing selection stringencies, sequences with high affinities can be systematically evolved and enriched. This strategy has been used to generate RNA aptamers for many metabolites, including aromatic ligands, amino acids, and oligosaccharides[84,85].

However, there are a few challenges that limit the application and effectiveness of this strategy. First, the target molecule needs to have chemical functional groups that can be linked to a solid support. Under conditions when such functional groups are not available, the molecule can be chemically modified to add the functional group to the ligand. However, this modification could change the interaction environment of the ligand with the RNA, which might affect selection[86]. Second, the RNA library design method needs to be optimized to improve the efficiency of selection[87]. Third, RNA structures identified by the *in vitro* selection process may not work well *in vivo*. This necessitates the coupling of *in vivo* screening method to make sure it works in the specific cellular environment[88].

Besides selection-based methods, RNA aptamers can also be engineered through rational design. RNA secondary structure folding can be predicted based on the thermodynamic parameters of nucleotide base-pairing and other factors, including steric interaction, quaternary structure, and slip-structure interconversion[89]. Although it is still very difficult to engineer aptamers from scratch, there are tools available to rationally tune the aptamer response. In one example, riboregulators that respond to theophylline were rationally tuned to control gene expression in eukaryotic cells[90]. Prediction of RNA structures facilitated the design of riboregulators that respond to theophylline in different manners (positive and negative response). Furthermore, the theophylline response range was tuned rationally by designing nucleotide sequences with different Gibbs free energy changes. This rational design strategy has allowed tunable aptamer responses[90].

RNA-based sensor-actuators can be engineered using different design principles as discussed in Section "Transcriptional and Translational Control Based on RNA-metabolite Interaction". To make the design of RNA devices more extensible, researchers have developed a framework that allows the modular assembly of aptamers, linkers, and gene regulatory components. In one example, a ribozyme switch system was constructed in *S. cerevisiae*[91]. This system standardizes the signal transmission between the aptamer and the actuator domain, which facilitates the design of other ligand-responsive RNA devices by aptamer swapping, and thus avoids redesign of the whole device.

Tools to Engineer Sensor-Actuators Based on Protein Activity

A third option to engineer sensor-actuators involves direct modulation of protein activity upon ligand binding, which requires tools to engineer proteins that have a ligand binding domain and a protein activity domain.

As mentioned in Section "PostTranslational Control Based on Protein-Metabolite Interaction", protein activity-based sensor-actuators can be engineered by combining a ligand binding domain and a protein activity domain. The ligand binding domain can be obtained from naturally existing proteins that interact with the target ligand, or it can be engineered by modifying the ligand binding pocket of a protein that interacts with a structurally similar ligand. For screening purposes, the protein activity domain can be either a reporter protein that gives a measurable readout, or a protein that affects a selectable phenotype. For dynamic pathway regulation purposes, protein activity-based sensor-actuators can also be created by choosing proteins whose activities determine pathway flux as the output domain. Once the ligand binding and the protein activity domain are chosen, they can be linked through domain insertion. This can be done through creating a random domain insertion library or rational design of the insertion sites[33,92–94]. One of the earliest examples of creating a domain insertion library was based on DNaseI, which introduces random breaks to the target DNA to insert the output sequence[33]. However, the non-specificity of DNaseI decreases the library quality by cutting multiple sites that cause deletions on the target DNA. As an alternative, a transposon-based directed evolution method for coupling protein functions was later developed, which was able to avoid mutations on the target DNA[95].

Another approach to create protein-based sensor-actuators is to use inteins, which are short peptides in a protein that can cut themselves and ligate the remaining parts together covalently. To engineer an intein-based sensor-actuator, a ligand binding domain can be inserted between a pair of split inteins, whose termini are fused to a split reporter (such as split β-lactamase) or a split pathway enzyme. Binding of a ligand triggers a conformation change of the peptide, bringing the two split intein fragments together, which in turn constructs an active enzyme. The level of the active spliced protein can be used as an indicator of the ligand concentration or a controller that regulates enzyme expression in response to the ligand[86].

7.3.2 Tools to Assemble Dynamic Regulatory Pathways

With the above-mentioned tools in mind, it is vital to employ a logical process to construct a dynamic regulatory pathway using available approaches. For this section, TF-based sensor-actuators that apply dynamic regulation to a system will be the focus due to the relative ease and the low-energy cost to the engineered pathway. The basic overall logical processes can be applied to any of the above-discussed regulatory techniques.

Logical Process to Assemble a Dynamic Regulatory Pathway

Figure 7.6 contains a logical flowchart to determine whether applying dynamic regulation to an engineered pathway may work to increase productivity of a given pathway. To determine which step may be the best candidate for dynamic regulation in the given system, follow the questions in Figure 7.6. A bottleneck step, restricted by low enzyme activity or low precursor concentration, may be upregulated dynamically or a toxic enzyme may be downregulated in response to dynamic cellular flux (such as *acc* in the fatty acid biosynthesis pathway)[1,18,22]. In either case, the TF regulating that step can be used to confer either upregulation or downregulation depending on how the dynamic regulatory circuit is constructed. Downregulation or upregulation using a naturally repressive or activating TF, respectively, is relatively straightforward in the initial design. A repressing TF can downregulate a toxic enzyme, while an activating TF can upregulate a bottleneck step so that flux through the downstream steps is improved. In both cases, overproduction of unnecessary proteins and metabolites will be avoided, allowing for improved production.

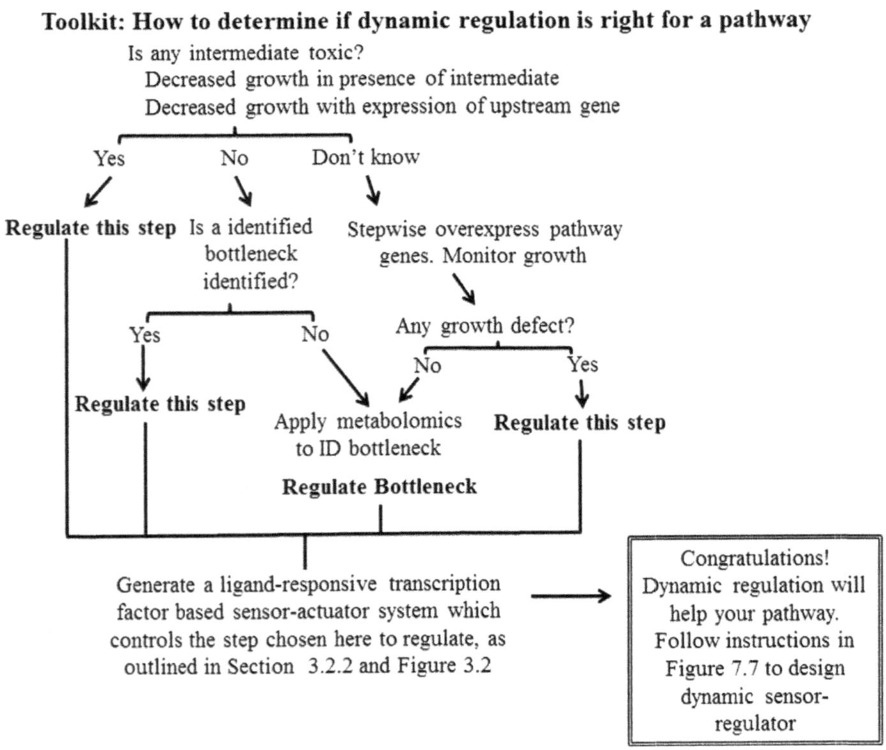

Figure 7.6

Flow chart to identify the point at which dynamic regulation can be applied to a pathway. Start at the top of the figure, and answer the questions to determine the appropriate step to regulate.

After identifying the point for regulation (Figure 7.6), one may next assemble the dynamic metabolic pathway using the resources included in this chapter.

Basic Assembly of a TF-based Sensor-Actuator

1. First, identify a candidate TF based on the table described in Section "Tools to Engineer Sensor-Actuators Based on TFs" and the resources listed in Figure 7.7. The TF chosen may be endogenous to the organism on which the dynamic pathway will be built, or heterologous. Both heterologous and endogenous TFs each have their own pros and cons. A heterologous TF can be introduced without significant concern over potential reactivity with native genes or promoter regions, but they need to fold and function properly in the heterologous host. A native TF, on the other hand, can be directly used without concern for codon optimization or potential functionality differences in a heterologous organism, but potential interference with off-target native pathways needs to be considered.

   If the identified TF does not work in the desired direction (e.g., TF is an activator when a repressor is needed), a secondary layer of regulation can be constructed. To allow an activating TF to, for example, repress a toxic gene, one can place a heterologous repressor under the control of the activator so that when the activator senses the toxic by-product, the repressor is activated. This repressor can then be used to repress the expression of a target gene, creating a negative regulation using an activating TF. Due to the added complexity, for the sake of this section, the focus will be limited to direct activators or repressors performing their respective order of regulation.

2. Once a metabolite-sensing TF has been identified, the promoter binding sequence of the metabolite-sensing TF can be incorporated into a synthetic hybrid promoter and cloned 5' of the target gene that needs to be regulated (either positive or negative). The strategy to design a hybrid promoter is demonstrated in Figure 7.7.
3. With a hybrid promoter designed, the responsiveness of the TF to the metabolite and its signal transmission to the hybrid promoter can be tested by placing a fluorescent reporter under the control of the hybrid promoter. With increasing metabolite levels (through exogenous addition or endogenous accumulation), the signal should either increase or decrease depending on whether the TF is an activator or repressor, respectively.
4. If metabolite titration does not generate a corresponding fluorescent dose-response, the system may need to be tuned. If no fluorescence is observed, the promoter strength of the repressor or activator should be reduced or increased, respectively, to allow for fluorescence readout from the sensor-actuator. If the response is saturated immediately, the opposite approach can be taken. These potential issues will be addressed in Section "Tools to Tune the Sensor-Actuator".
5. Finally, once the system has been validated to successfully control the expression of a reporter in response to the ligand, the dynamic range of the system can be measured and tuned. The next subsection will discuss techniques to tune sensor-actuators in a biological setting.

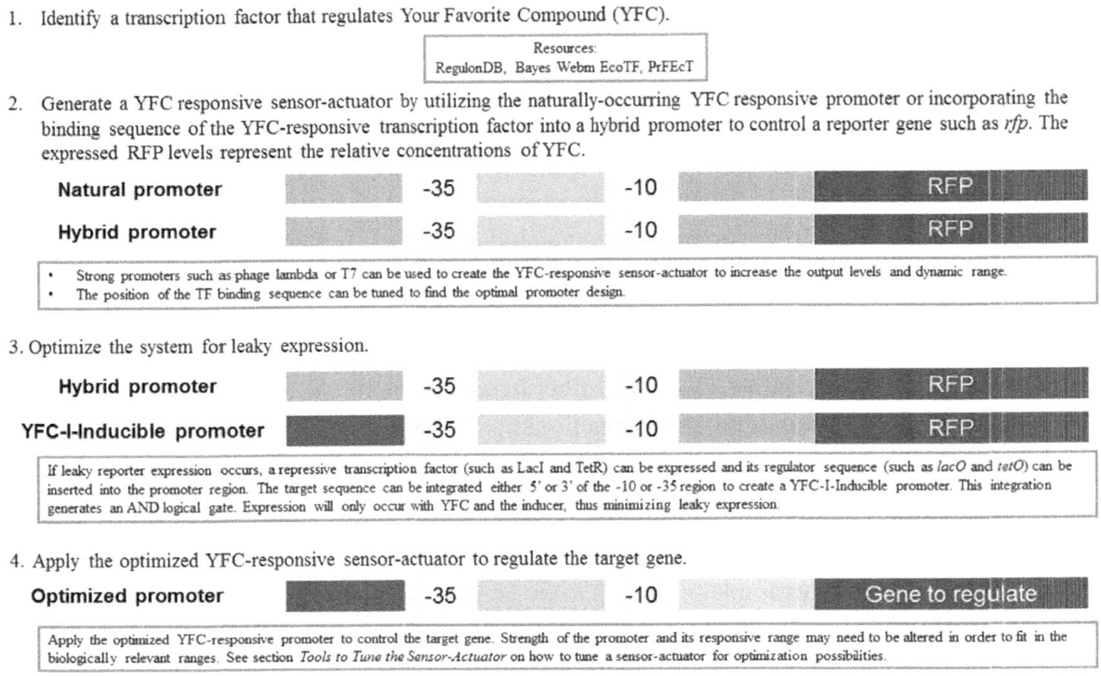

Figure 7.7

Strategy to generate and optimize a metabolite-responsive sensor-actuator. Step 1 includes the identification of the YFC-responsive transcription factor. Step 2 outlines options for generating a hybrid promoter containing the binding sequence of YFC-responsive transcription factor integrated into a phage promoter sequence. Responsiveness of the hybrid promoter can be tested by titrating YFC either endogenously or exogenously and measuring RFP output. To decrease the leaky expression and tune the dynamic range, a second regulator sequence can be incorporated to generate a logical AND gate. With the AND gate, expression of the downstream gene can only occur in the presence of the inducer and YFC. The engineered sensor-actuator can then be used to regulate the gene of interest. Further tuning of the circuit may be required and can be performed using the guidelines outlined in Section "Tools to Tune the Sensor-Actuator".

Tools to Tune the Sensor-Actuator

Once a sensor has been designed using the criteria above, it may be necessary to tune the response to reflect the regulation required by the system. The amount of input required should be realistic for the given system. For example, if the concentration of a ligand required to activate its TF is higher than the ligand's possible intracellular concentration, the sensitivity of the sensor-regulator has to be tuned to match the relevant cellular concentration of the metabolite. On the other hand, if a sensor-regulator is hyper-reactive (a small amount of ligand is sufficient to saturate the sensor-regulator), the regulatory system will not work dynamically. In either case, the sensor-actuator needs to be tuned based on the actual or estimated intracellular ligand concentration and the K_d of molecular interactions. A single or combination of the tuning approaches discussed below can be used to logically tune the system.

A variety of approaches may be taken to tune the sensor-actuator system. The levels at which the system may be tuned are at the same levels of regulation discussed in Section 7.2. Namely, tuning may be done on the level of plasmid copy number, the promoter strength, the RBS strength, or as a posttranslational modification. In tuning a system, the output can be readily measured by placing a fluorescence protein under the control of the promoter. However, when using a fluorescent marker as a proxy for a toxic enzyme, caution should be used as the range optimized with the marker may not reflect the optimal conditions for the enzyme in the complete system. For example, the fluorescent protein does not enzymatically contribute to the engineered pathway, while the target gene inherently contributes to the designed pathway. When the sensor-actuator is applied to the target gene, and not simply a fluorescent marker, there may be unanticipated effects not possibly identified through a fluorescent marker alone. In this case, a fluorescent marker is limited in utility to serve as a crude output for sensor-actuator sensitivity and basic dynamic range. The intact system may require further tuning to identify the dynamic range relevant to enzyme and metabolite levels in the intact system.

In planning for sensor-actuator tuning, it is helpful to determine which level of regulation will work best for the system at hand. Modification of the sensor TF's cognate promoter can tune the sensitivity of the response on the level of transcriptional regulation. Furthermore, the dynamic range of the sensor-actuator can be tuned by modification of the RNAP binding site sequences and TF binding sequences due to changes in their binding affinities.

Despite these advantages of tuning TF binding sites, altering RBS sequences may represent a simpler method to tune a system. In a system with a wide dynamic range and a tight promoter, RBS tuning can serve to further fine-tune the system[6]. Previously characterized RBS sequences and the RBS calculator offer a quantitative alternative to promoter optimization, despite the increased energy burden[96].

In addition, posttranslational tuning can be performed by engineering a degradable protein system. However, this system may be the least desirable energetically, since not only is the protein synthesized then degraded, which siphons off amino acids and ATP, but this also requires inclusion of degradation machinery, which may interfere with natural protein degradation systems[97].

7.3.3 Computational Tools to Simulate the Sensor-Actuator Response and the Metabolic Control Circuit Behavior

This subsection will discuss various computational tools that have been developed to facilitate the understanding and fine-tuning of both sensor-actuators and dynamic pathways.

Computational Tools to Simulate the Sensor-Actuator Response

The response of sensor-actuators can be described by transfer functions, which correlate input signal levels to their corresponding outputs. Since sensor-actuators are based on simple molecular interactions, the transfer functions can be derived based on mathematical equations that describe these processes.

A thermodynamic model that has been widely used in synthetic biology to describe the response of a TF-based sensor-actuator will be used as an example[18,98]. Consider the scenario where a sensor-actuator was engineered by inserting two regulator sequences in the promoter region, and a repressor TF binds to the regulator sequence and represses gene expression. This sensor-actuator is induced by accumulating the ligand M. M binds to the TF and thus releases its binding from the promoter, initiating transcription. Therefore, at equilibrium, the promoter has five possible states (Figure 7.8): (a) the promoter is free with no TF or RNA polymerase binding; (b) the first regulator sequence is occupied by the TF; (c) the second regulator sequence is occupied by the TF; (d) both regulation sequences are occupied by the TF; (e) the promoter is bound with the RNA polymerase.

The equilibrium of one TF binding to the promoter (state b and c) can be described as:

$$K_1 = \frac{\mathrm{TF-Pt}}{\mathrm{TF_f} \cdot \mathrm{Pt_f}} \tag{7.1}$$

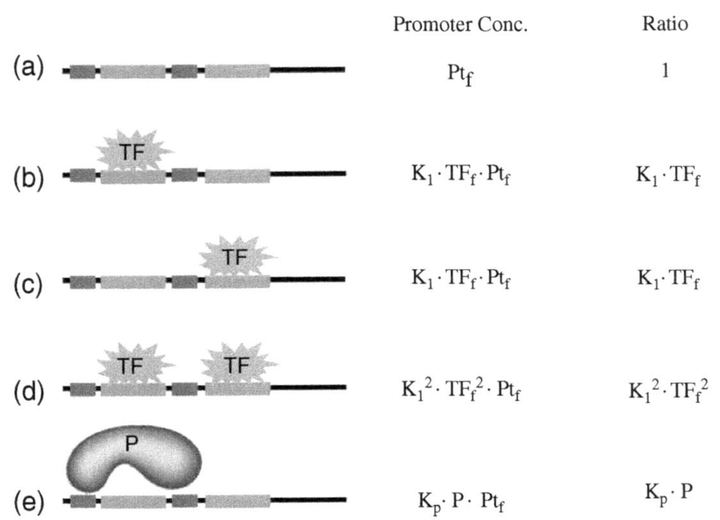

Figure 7.8

Schematic of possible states of a promoter sensor-actuator. The transcription factor (TF) competes with the RNA polymerase (*P*) binding to DNA. Two regulator sequences (represented by long bars) are cloned into a synthetic promoter. The two small bars represent the −35 and −10 region of a prokaryotic promoter. The promoter could exist in one of the five states, and the concentration of each state and the relative ratios are listed. Transcription only happens in state e and serves as the ON state.

where TF-Pt, TF_f, Pt_f represent the concentrations of the TF-promoter complex, free TF, and free promoter, respectively; K_1 is the association constant of the TF to the promoter.

When both regulator sequences are occupied by the TF (state d), the equilibrium can be described as:

$$K_1^2 = \frac{\mathrm{TF-TF-Pt}}{\mathrm{TF_f}^2 \cdot \mathrm{Pt_f}} \tag{7.2}$$

where TF-TF-Pt represents the concentration of promoters that are bound by two TFs.

The equilibrium between the RNA polymerase and the promoter (state e) can be described as:

$$K_\mathrm{p} = \frac{P - \mathrm{Pt}}{P \cdot \mathrm{Pt_f}} \tag{7.3}$$

where P and P-Pt represent the concentration of RNA polymerase and promoter-RNA polymerase complex; K_p is the association constant of the RNA polymerase to the promoter.

Therefore, if we set the maximum expression level of the sensor-actuator to be P_{max}, the expression level of the sensor-actuator (P_M) can be determined by the ratio of the ON state:

$$P_\mathrm{M} = P_\mathrm{max} \frac{K_\mathrm{p} P}{1 + K_\mathrm{p} P + 2K_1 \mathrm{TF_f} + K_1^2 \mathrm{TF_f}^2} \tag{7.4}$$

The TF_f level can be obtained by the equilibrium between M and TF:

$$\mathrm{TF_f} = \frac{K_\mathrm{d}^n}{M^n + K_\mathrm{d}^n} \mathrm{TF_t} \tag{7.5}$$

where M and TF_t represent the concentration of the free ligand and the total intracellular concentration of TF; n is the cooperativity constant; K_d is the dissociation constant of M binding to TF.

The transfer function that directly correlates the output levels to ligand concentrations can be obtained by combining Equations (7.4) and (7.5):

$$P_\mathrm{M} = P_\mathrm{max} \frac{K_\mathrm{p} P}{1 + K_\mathrm{p} P + 2K_1 \dfrac{K_\mathrm{d}^n}{M^n + K_\mathrm{d}^n} \mathrm{TF_t} + K_1^2 \left(\dfrac{K_\mathrm{d}^n}{M^n + K_\mathrm{d}^n} \mathrm{TF_t} \right)^2} \tag{7.6}$$

Using similar derivations, transfer functions that describe different regulation scenarios can also be obtained. Experimentally, the transfer functions can be determined by measuring the readout (such as fluorescent protein levels) of each sensor-actuator at different ligand concentrations. The obtained response curve can be fitted to Equation (7.6), and allows a quantitative prediction on how to tune the biochemical parameters to achieve the desired sensor-actuator response curve.

Computational Tools to Simulate the Behavior of the Metabolic Control Circuits

The effects of different control strategies on pathway productivities can be simulated by a series of ordinary differential equations (ODEs) that describe cell growth, enzyme induction, intermediate accumulation, product formation, etc. As an example, we consider a simple linear pathway with three enzymatic steps, where TF-S_3 complex represses the expression of E_1 and E_2 (Figure 7.9). The model can be formulated as follows (adapted from previous studies)[38,40]:

$$\frac{dE_i}{dt} = c_i + b_i \frac{1}{a_i + \mathrm{TF}} - kE_i \tag{7.7}$$

$$\mathrm{TF} = \frac{S_3}{K_f + S_3} \tag{7.8}$$

$$\frac{dS_i}{dt} = E_{i-1} v_{i-1} \frac{S_{i-1}}{K_{i-1} + S_{i-1}} - E_i v_i \frac{S_i}{K_i + S_i} - kS_i \tag{7.9}$$

$$\frac{dS_1}{dt} = c_s - E_1 v_1 \frac{S_1}{K_1 + S_1} - kS_1 \tag{7.10}$$

$$\frac{dX}{dt} = \mu X \left(1 - \frac{X}{X_{\max}}\right) - \sum \delta_{E_i} \frac{E_i}{t_{E_i} + E_i} - \sum \delta_{S_i} \frac{S_i}{t_{S_i} + S_i} \tag{7.11}$$

$$\frac{dP}{dt} = E_3 v_3 \frac{S_3}{k_3 + S_3} - kP \tag{7.12}$$

$$P(\mathrm{total}) = \int_0^T (P \times X)\,dt \tag{7.13}$$

Parameters used in this model are summarized as follows:

E_i (i = 1, 2, 3): The concentrations of the enzymes of the pathway.

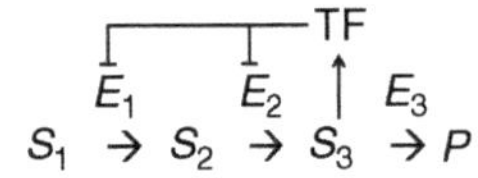

Figure 7.9
Schematic of a dynamically regulated three-enzyme pathway. S_3-TF complex represses the transcription of E_1 and E_2. E_1, E_2, and E_3 represent enzymes; S_1, S_2, and S_3 represent metabolites; P represents the product.

TF: The fraction of the active TF (bound by S_3).

S_i (i=2, 3): The concentrations of each metabolite in the pathway.

P: The concentration of the final product.

c_i (i=1,2,3): The constitutive basal rate of expression of each enzyme.

b_i (i=1,2): The maximum rate of expression induced by TF.

a_i (i=1,2): The fraction of TF needed for half-maximum activation.

k: The sum of the degradation and dilution rates of the metabolites and enzymes. In the exponentially growing cells, dilution due to cell division is regarded as the major component.

K_f: The concentration of S_3 necessary for half-maximal TF activity.

v_i: The catalytic constants of each enzyme.

K_i: The Michaelis-Menten constant of each enzyme.

c_s: The synthesis rate of S_1.

X: Cell density.

X_{max}: The maximum biomass supported by the nutrients supplied in the medium, assuming no toxicity.

μ: The specific growth rate of the cells.

δ_{E_i} : The toxicity coefficient of enzyme E_i.

t_{E_i} : The concentration of E_i that causes its half-maximum toxicity to cell growth.

δ_{s_i} : The toxicity coefficient of metabolite S_i.

t_{s_i} : The concentration of S_i that causes its half-maximum toxicity to cell growth.

Parameters that represent the enzyme or metabolite synthesis rate (c_i, c_s), enzyme kinetics constants (v_i, K_i), binding constants of TF to metabolites (K_f), and cell dilution rate (k) can be obtained from an enzyme database or direct experimental measurements. δ_{E_i}, t_{E_i}, δ_{s_i}, t_{s_i}, μ, and X_{max} can be determined by fitting the growth Equation (7.11) to an experimentally measured cell growth curve. Two critical parameters that determine the effects of the regulation on enzyme expression, b_i and a_i, can be tuned experimentally.

With these parameters, this set of equations can be mathematically solved to obtain the production profile under this regulation strategy. Similarly, models that describe other regulation scenarios can also be assembled using equations similar to the ones shown above. By solving each set of equations with different experimentally tunable parameters, the desired parameters of each control strategy that generates the optimal productivity can be identified. Moreover, by comparing the optimal productivity of different regulation strategies, the optimal regulation architecture can be readily identified. These equations will also provide the dynamic information of enzymes and metabolites, which also helps to understand the dynamic effects of synthetic regulatory circuits. Overall, the results obtained from the model can guide the design and tuning of dynamic regulation to achieve optimal productivity, as well as to illustrate the effects of different regulatory topologies on metabolite/enzyme dynamics.

7.4 Discussion

So far we have discussed the importance of dynamic regulation and summarized some available tools to apply this method to regulate metabolic pathways. Dynamic regulatory systems detect the level of critical metabolites and modulate pathway flux in each cell according to its own metabolic status. As cell metabolism changes over time and across space, this method could be particularly useful to improve production by optimizing pathway activity in each cell under changing conditions.

However, metabolic changes can occur at different time scales. In a large-scale fermenter, cell metabolism may change with microenvironmental changes, which depends on how fast a cell moves from one environment to another (e.g., from a low oxygen region to a high oxygen region). Several factors may affect the frequency of microenvironmental changes, including cell density, mixing speed, fermenter size and shape, etc. To design dynamic regulatory systems to respond to these changes, one has to consider the response time needed for the regulatory system (defined as the time from the moment of metabolite sensing to the moment of adjusting the pathway flux to a desired level). In general, a quick response is preferred from the control standpoint. Of all the regulation stages, posttranslational regulation provides the quickest response to adjust pathway flux (discussed in Section "Summary of Experimental Approaches"). However, posttranslational regulation does not prevent the synthesis of extra proteins and thus may be not ideal if the burden of protein overexpression is more deleterious than overproduction of metabolites. By contrast, transcriptional regulation takes time to turn on/off protein synthesis (estimated to be a few minutes at the single molecule level), which may be sufficient under various conditions.

Contrarily, from the control theory point of view, strong regulation could lead to instability, which is unfavorable for many biological systems. A common observation for strong regulation is the overshoot of metabolite and/or enzyme levels before reaching a steady state. Over-regulation could also lead to dampened/sustained oscillations depending on system response time. This could lead to a trade-off between regulation response time and instability. Therefore, circuit design strategies that enable fast and stable kinetics are needed for optimal regulation dynamics.

7.5 Future Trends

While in the past it has often been suggested to delete or overexpress enzymes to overwhelm the native regulation to enhance production, under some conditions natural regulations can be utilized [99] and synthetic regulatory circuits can be introduced to improve production [1,4,18]. With advances in the field of metabolic engineering and synthetic biology, dynamic regulation will find increasingly more applications.

With such promise in the field of dynamic pathway regulation, more tools are being developed to apply dynamic regulation to improve the production of a broad range of products. More experimental strategies have been developed to engineer novel sensor-actuators that respond to a variety of metabolites in different metabolic pathways. Additionally, computational simulations have been used to analyze different regulation systems, providing predictions on the behaviors of different regulation strategies[100,101]. These tools will allow for the application of dynamic regulation in many other metabolic pathways.

A key factor in extending dynamic regulation to a variety of metabolic pathways is to understand the metabolic dynamics (defined as metabolite concentration change over time), since it directly determines pathway productivity. However, while the ability of dynamic regulation to improve productivity has been demonstrated in several pathways, very little is known about how synthetic regulation affects metabolic dynamics and how metabolic dynamics improves pathway productivity. Although metabolic dynamics in natural pathways has been extensively investigated, most studies focus on its effects on cell fitness and adaption of cells to different environments. However, the goal of synthetic regulation in metabolic engineering is to maximize pathway productivities and yields, which requires different metabolic dynamics from what is studied in natural systems. As such, the effects of synthetic regulation on metabolic dynamics have rarely been studied. Characterizing metabolic dynamics under synthetic regulation will allow us to better understand how different regulatory topologies affect global metabolic dynamics, making dynamic pathways more robust and productive.

In summary, this chapter introduced dynamic pathway regulation as a recently developed synthetic biology tool to optimize engineered metabolic pathways to improve production. By modulating gene expression levels in response to metabolite concentrations, metabolic pathways can be dynamically balanced according to the metabolic status of a cell. It has been demonstrated through both experimental and computational approaches that dynamic pathway regulations improve cell growth and production. Some available tools to construct a dynamically regulated pathway and methods to fine-tune these systems were also discussed. With further advances in metabolic engineering and synthetic biology, dynamic pathway regulation will serve as a powerful way to make the microbial production of biofuels, pharmaceuticals, and materials economically viable.

Acknowledgments

We thank all of our group members for their discussions. We also thank the Sinskey Lab for giving us permission to use their published figure in this chapter. This work was supported by the National Science Foundation (MCB1453147, MCB1331194), the Defense Advanced Research Projects Agency (D13AP00038), and the International Center for Advanced Renewable Energy and Sustainability (I-CARES).

References

1. Zhang FZ, Carothers JM, Keasling JD. Design of a dynamic sensor-regulator system for production of chemicals and fuels derived from fatty acids. *Nat Biotechnol* 2012;**30**:354–9.
2. Ro DK, Paradise EM, Ouellet M, Fisher KJ, Newman KL, Ndungu JM, et al. Production of the antimalarial drug precursor artemisinic acid in engineered yeast. *Nature* 2006;**440**:940–3.
3. Li H, Opgenorth PH, Wernick DG, Rogers S, Wu TY, Higashide W, et al. Integrated electromicrobial conversion of CO2 to higher alcohols. *Science* 2012;**335**:1596.
4. Farmer WR, Liao JC. Improving lycopene production in Escherichia coli by engineering metabolic control. *Nat Biotechnol* 2000;**18**:533–7.
5. Handke P, Lynch SA, Gill RT. Application and engineering of fatty acid biosynthesis in Escherichia coli for advanced fuels and chemicals. *Metab Eng* 2011;**13**:28–37.
6. Oyarzun DA, Stan GBV. Synthetic gene circuits for metabolic control: design trade-offs and constraints. *J R Soc Interface* 2013;**10**. Article Number: 20120671.
7. Karim AS, Curran KA, Alper HS. Characterization of plasmid burden and copy number in Saccharomyces cerevisiae for optimization of metabolic engineering applications. *FEMS Yeast Res* 2013;**13**:107–16.
8. Shen-Orr SS, Milo R, Mangan S, Alon U. Network motifs in the transcriptional regulation network of Escherichia coli. *Nat Genet* 2002;**31**:64–8.
9. Kohlhaw GB. Leucine biosynthesis in fungi: entering metabolism through the back door. *Microbiol Mol Biol Rev* 2003;**67**:1–15.
10. Xie G, Keyhani NO, Bonner CA, Jensen RA. Ancient origin of the tryptophan operon and the dynamics of evolutionary change. *Microbiol Mol Biol Rev* 2003;**67**:303–42.
11. Madar D, Dekel E, Bren A, Alon U. Negative auto-regulation increases the input dynamic-range of the arabinose system of Escherichia coli. *BMC Syst Biol* 2011;**5**:111.
12. Rosenfeld N, Elowitz MB, Alon U. Negative autoregulation speeds the response times of transcription networks. *J Mol Biol* 2002;**323**:785–93.
13. Alon U. Network motifs: theory and experimental approaches. *Nat Rev Genet* 2007;**8**:450–61.
14. Becskei A, Serrano L. Engineering stability in gene networks by autoregulation. *Nature* 2000;**405**:590–3.
15. Levine E, Hwa T. Stochastic fluctuations in metabolic pathways. *Proc Natl Acad Sci U S A* 2007;**104**:9224–9.
16. Oyarzun DA, Lugagne JB, Stan GB. Noise propagation in synthetic gene circuits for metabolic control. *ACS Synth Biol* 2015;**4**:116–25.
17. Mangan S, Alon U. Structure and function of the feed-forward loop network motif. *Proc Natl Acad Sci U S A* 2003;**100**:11980–5.
18. Liu D, Xiao Y, Evans B, Zhang F. Negative feedback regulation of fatty acid production based on a malonyl-CoA sensor-actuator. *ACS Synth Biol* 2013;**4**:132.
19. Boccazzi P, Zhang Z, Kurosawa K, Szita N, Bhattacharya S, Jensen KF, et al. Differential gene expression profiles and real-time measurements of growth parameters in Saccharomyces cerevisiae grown in microliter-scale bioreactors equipped with internal stirring. *Biotechnol Prog* 2006;**22**:710–7.
20. Yang JH, Kim KD, Lucas A, Drahos KE, Santos CS, Mury SP, et al. A novel heme-regulatory motif mediates heme-dependent degradation of the circadian factor period 2. *Mol Cell Biol* 2008;**28**:4697–711.

21. Cox RS, Surette MG, Elowitz MB. Programming gene expression with combinatorial promoters. *Mol Syst Biol* 2007;**3**:145.
22. Dahl RH, Zhang F, Alonso-Gutierrez J, Baidoo E, Batth TS, Redding-Johanson AM, et al. Engineering dynamic pathway regulation using stress-response promoters. *Nat Biotechnol* 2013;**31**:1039–46.
23. Montange RK, Batey RT. Riboswitches: emerging themes in RNA structure and function. *Annu Rev Biophys* 2008;**37**:117–33.
24. Desai SK, Gallivan JP. Genetic screens and selections for small molecules based on a synthetic riboswitch that activates protein translation. *J Am Chem Soc* 2004;**126**:13247–54.
25. Ogawa A, Maeda M. An artificial aptazyme-based riboswitch and its cascading system in E. coli. *Chembiochem* 2008;**9**:206–9.
26. Suess B, Fink B, Berens C, Stentz R, Hillen W. A theophylline responsive riboswitch based on helix slipping controls gene expression in vivo. *Nucleic Acids Res* 2004;**32**:1610–4.
27. Winkler WC, Cohen-Chalamish S, Breaker RR. An mRNA structure that controls gene expression by binding FMN. *Proc Natl Acad Sci U S A* 2002;**99**:15908–13.
28. Nudler E, Mironov AS. The riboswitch control of bacterial metabolism. *Trends Biochem Sci* 2004;**29**:11–7.
29. Winkler WC. Riboswitches and the role of noncoding RNAs in bacterial metabolic control. *Curr Opin Chem Biol* 2005;**9**:594–602.
30. Winkler WC, Breaker RR. Genetic control by metabolite-binding riboswitches. *Chembiochem* 2003;**4**:1024–32.
31. Winkler W, Nahvi A, Breaker RR. Thiamine derivatives bind messenger RNAs directly to regulate bacterial gene expression. *Nature* 2002;**419**:952–6.
32. Babiskin AH, Smolke CD. Synthetic RNA modules for fine-tuning gene expression levels in yeast by modulating RNase III activity. *Nucleic Acids Res* 2011;**39**:8651–64.
33. Guntas G, Ostermeier M. Creation of an allosteric enzyme by domain insertion. *J Mol Biol* 2004;**336**:263–73.
34. Guntas G, Mitchell SF, Ostermeier M. A molecular switch created by in vitro recombination of nonhomologous genes. *Chem Biol* 2004;**11**:1483–7.
35. Guntas G, Mansell TJ, Kim JR, Ostermeier M. Directed evolution of protein switches and their application to the creation of ligand-binding proteins. *Proc Natl Acad Sci U S A* 2005;**102**:11224–9.
36. Stekel DJ, Jenkins DJ. Strong negative self regulation of Prokaryotic transcription factors increases the intrinsic noise of protein expression. *BMC Syst Biol* 2008;**2**:6.
37. Nevozhay D, Adams RM, Murphy KF, Josic K, Balazsi G. Negative autoregulation linearizes the dose–response and suppresses the heterogeneity of gene expression. *Proc Natl Acad Sci U S A* 2009;**106**:5123–8.
38. Chubukov V, Zuleta IA, Li H. Regulatory architecture determines optimal regulation of gene expression in metabolic pathways. *Proc Natl Acad Sci U S A* 2012;**109**:5127–32.
39. Zaslaver A, Mayo AE, Rosenberg R, Bashkin P, Sberro H, Tsalyuk M, et al. Just-in-time transcription program in metabolic pathways. *Nat Genet* 2004;**36**:486–91.
40. Dunlop MJ, Keasling JD, Mukhopadhyay A. A model for improving microbial biofuel production using a synthetic feedback loop. *Syst Synth Biol* 2010;**4**:95–104.
41. Harrison ME, Dunlop MJ. Synthetic feedback loop model for increasing microbial biofuel production using a biosensor. *Front Microbiol* 2012;**3**:360.
42. Fujita Y, Matsuoka H, Hirooka K. Regulation of fatty acid metabolism in bacteria. *Mol Microbiol* 2007;**66**:829–39.
43. Zhang YM, Marrakchi H, Rock CO. The FabR (YijC) transcription factor regulates unsaturated fatty acid biosynthesis in Escherichia coli. *J Biol Chem* 2002;**277**:15558–65.
44. Caldara M, Charlier D, Cunin R. The arginine regulon of Escherichia coli: whole-system transcriptome analysis discovers new genes and provides an integrated view of arginine regulation. *Microbiology* 2006;**152**:3343–54.
45. Lochowska A, Iwanicka-Nowicka R, Plochocka D, Hryniewicz MM. Functional dissection of the LysR-type CysB transcriptional regulator—regions important for DNA binding, inducer response, oligomerization, and positive control. *J Biol Chem* 2001;**276**:2098–107.

46. Heil G, Stauffer LT, Stauffer GV. Glycine binds the transcriptional accessory protein GcvR to disrupt a GcvA/GcvR interaction and allow GcvA-mediated activation of the Escherichia coli gcvTHP operon. *Microbiology* 2002;**148**:2203–14.
47. Ghrist AC, Heil G, Stauffer GV. GcvR interacts with GcvA to inhibit activation of the Escherichia coli glycine cleavage operon. *Microbiology* 2001;**147**:2215–21.
48. Weissbach H, Brot N. Regulation of methionine synthesis in Escherichia-Coli. *Mol Microbiol* 1991;**5**:1593–7.
49. Rhee KY, Senear DF, Hatfield GW. Activation of gene expression by a ligand-induced conformational change of a protein-DNA complex. *J Biol Chem* 1998;**273**:11257–66.
50. Pittard J, Camakaris H, Yang J. The TyrR regulon. *Mol Microbiol* 2005;**55**:16–26.
51. Chen S, Calvo JM. Leucine-induced dissociation of Escherichia coli Lrp hexadecamers to octamers. *J Mol Biol* 2002;**318**:1031–42.
52. Stragier P, Richaud F, Borne F, Patte JC. Regulation of diaminopimelate decarboxylase synthesis in Escherichia-Coli. 1. Identification of a Lysr gene encoding an activator of the Lysa Gene. *J Mol Biol* 1983;**168**:307–20.
53. Mcfall E, Heincz MC. Identification and control of synthesis of the Dsdc activator protein. *J Bacteriol* 1983;**153**:872–7.
54. Gunsalus RP, Yanofsky C. Nucleotide-sequence and expression of Escherichia-Coli Trpr, the structural gene for the Trp aporepressor. *Proc Natl Acad Sci U S A* 1980;**77**:7117–21.
55. Landini P, Volkert MR. Regulatory responses of the adaptive response to alkylation damage: a simple regulon with complex regulatory features. *J Bacteriol* 2000;**182**:6543–9.
56. Zeng G, Ye SZ, Larson TJ. Repressor for sn-glycerol 3-phosphate regulon of Escherichia coli K-12: Primary structure and identification of the DNA-binding domain. *J Bacteriol* 1996;**178**:7080–9.
57. Choi KY, Zalkin H. Structural characterization and corepressor binding of the Escherichia-Coli purine repressor. *J Bacteriol* 1992;**174**:6207–14.
58. Wilkinson SP, Grove A. HucR, a novel uric acid-responsive member of the MarR family of transcriptional regulators from Deinococcus radiodurans. *J Biol Chem* 2004;**279**:51442–50.
59. Miyada CG, Stoltzfus L, Wilcox G. Regulation of the Arac gene of Escherichia-Coli—catabolite repression. Auto-regulation, and effect on arabad expression. *Proc Natl Acad Sci U S A* 1984;**81**:4120–4.
60. Ramseier TM, Negre D, Cortay JC, Scarabel M, Cozzone AJ, Saier MH. In-vitro binding of the pleiotropic transcriptional regulatory protein. Frur, to the Fru, Pps, Ace, Pts and Icd Operons of Escherichia-Coli and Salmonella-typhimurium. *J Mol Biol* 1993;**234**:28–44.
61. Weickert MJ, Adhya S. Control of transcription of Gal repressor and isorepressor genes in Escherichia-Coli. *J Bacteriol* 1993;**175**:251–8.
62. Balaeff A, Mahadevan L, Schulten K. Structural basis for cooperative DNA binding by CAP and lac repressor. *Structure* 2004;**12**:123–32.
63. Schneider E, Freundlieb S, Tapio S, Boos W. Molecular characterization of the malt-dependent periplasmic alpha-amylase of Escherichia-Coli encoded by mals. *J Biol Chem* 1992;**267**:5148–54.
64. Wade JT, Belyaeva TA, Hyde EI, Busby SJW. A simple mechanism for co-dependence on two activators at an Escherichia coli promoter. *EMBO J* 2001;**20**:7160–7.
65. Wickstrum JR, Skredenske JM, Kolin A, Jin DJ, Fang J, Egan SM. Transcription activation by the DNA-binding domain of the AraC family protein RhaS in the absence of its effector-binding domain. *J Bacteriol* 2007;**189**:4984–93.
66. Mauzy CA, Hermodson MA. Structural and functional analyses of the repressor. Rbsr, of the ribose operon of Escherichia-Coli. *Protein Sci* 1992;**1**:831–42.
67. Sorensen KI, HoveJensen B. Ribose catabolism of Escherichia coli: characterization of the rpiB gene encoding ribose phosphate isomerase B and of the rpiR gene, which is involved in regulation of rpiB expression. *J Bacteriol* 1996;**178**:1003–11.
68. Ogasawara H, Ishida Y, Yamada K, Yamamoto K, Ishihama A. PdhR (pyruvate dehydrogenase complex regulator) controls the respiratory electron transport system in Escherichia coli. *J Bacteriol* 2007;**189**:5534–41.

69. Schlensog V, Lutz S, Bock A. Purification and DNA-binding properties of Fhla, the transcriptional activator of the formate hydrogenlyase system from Escherichia-Coli. *J Biol Chem* 1994;**269**:19590–6.
70. Overton T, Reid EGF, Foxall R, Smith H, Busby SJW, Cole JA. Transcription activation at Escherichia coli FNR-dependent promoters by the gonococcal FNR protein: Effects of a novel S18F substitution and comparisons with the corresponding substitution in E-coli FNR. *J Bacteriol* 2003;**185**:4734–47.
71. Raffaelli N, Lorenzi T, Mariani PL, Emanuelli M, Amici A, Ruggieri S, et al. The Escherichia coli NadR regulator is endowed with nicotinamide mononucleotide adenylyltransferase activity. *J Bacteriol* 1999;**181**:5509–11.
72. Zheng M, Aslund F, Storz G. Activation of the OxyR transcription factor by reversible disulfide bond formation. *Science* 1998;**279**:1718–21.
73. Plumbridge J, Pellegrini O. Expression of the chitobiose operon of Escherichia coli is regulated by three transcription factors: NagC, ChbR, and CAP. *Mol Microbiol* 2004;**52**:437–49.
74. Zheng DL, Constantinidou C, Hobman JL, Minchin SD. Identification of the CRP regulon using in vitro and in vivo transcriptional profiling. *Nucleic Acids Res* 2004;**32**:5874–93.
75. Yamamoto K. The hierarchic network of metal-response transcription factors in Escherichia coli. *Biosci Biotechnol Biochem* 2014;**78**:737–47.
76. Lee JW, Helmann JD. Functional specialization within the Fur family of metalloregulators. *Biometals* 2007;**20**:485–99.
77. Zhao NY, Oh W, Trybul D, Thrasher KS, Kingsbury TJ, Larson TJ. Characterization of the interaction of the Glp repressor of Escherichia-Coli K-12 with single and tandem Glp operator variants. *J Bacteriol* 1994;**176**:2393–7.
78. Yang SJ, Rice KC, Brown RJ, Patton TG, Liou LE, Park YH, et al. A LysR-type regulator. CidR, is required for induction of the Staphylococcus aureus cidABC operon. *J Bacteriol* 2005;**187**:5893–900.
79. Phillips CM, Schreiter ER, Guo Y, Wang SC, Zamble DB, Drennan CL. Structural basis of the metal specificity for nickel regulatory protein NikR. *Biochemistry* 2008;**47**:1938–46.
80. Pomposiello PJ, Demple B. Redox-operated genetic switches: the SoxR and OxyR transcription factors. *Trends Biotechnol* 2001;**19**:109–14.
81. Tang SY, Cirino PC. Design and application of a mevalonate-responsive regulatory protein. *Angew Chem Int Ed* 2011;**50**:1084–6.
82. Tang SY, Qian S, Akinterinwa O, Frei CS, Gredell JA, Cirino PC. Screening for Enhanced triacetic acid lactone production by recombinant Escherichia coli expressing a designed triacetic acid lactone reporter. *J Am Chem Soc* 2013;**135**:10099–103.
83. Gredell JA, Frei CS, Cirino PC. Protein and RNA engineering to customize microbial molecular reporting. *Biotechnol J* 2012;**7**:477–99.
84. Hermann T, Patel DJ. Biochemistry—adaptive recognition by nucleic acid aptamers. *Science* 2000;**287**:820–5.
85. Jenison RD, Gill SC, Pardi A, Polisky B. High-resolution molecular discrimination by Rna. *Science* 1994;**263**:1425–9.
86. Michener JK, Thodey K, Liang JC, Smolke CD. Applications of genetically-encoded biosensors for the construction and control of biosynthetic pathways. *Metab Eng* 2012;**14**:212–22.
87. Famulok M. Molecular recognition of amino-acids by Rna-aptamers - an L-citrulline binding Rna motif and its evolution into an L-arginine binder. *J Am Chem Soc* 1994;**116**:1698–706.
88. Sinha J, Reyes SJ, Gallivan JP. Reprogramming bacteria to seek and destroy an herbicide. *Nat Chem Biol* 2010;**6**:464–70.
89. Bang GS, Cho S, Lee N, Lee BR, Kim JH, Kim BG. Rational design of modular allosteric aptamer sensor for label-free protein detection. *Biosens Bioelectron* 2013;**39**:44–50.
90. Bayer TS, Smolke CD. Programmable ligand-controlled riboregulators of eukaryotic gene expression. *Nat Biotechnol* 2005;**23**:337–43.
91. Win MN, Smolke CD. A modular and extensible RNA-based gene-regulatory platform for engineering cellular function. *Proc Natl Acad Sci U S A* 2007;**104**:14283–8.

92. Baird GS, Zacharias DA, Tsien RY. Circular permutation and receptor insertion within green fluorescent proteins. *Proc Natl Acad Sci U S A* 1999;**96**:11241–6.
93. Betton JM, Jacob JP, Hofnung M, Broome-Smith JK. Creating a bifunctional protein by insertion of beta-lactamase into the maltodextrin-binding protein. *Nat Biotechnol* 1997;**15**:1276–9.
94. Collinet B, Herve M, Pecorari F, Minard P, Eder O, Desmadril M. Functionally accepted insertions of proteins within protein domains. *J Biol Chem* 2000;**275**:17428–33.
95. Edwards WR, Busse K, Allemann RK, Jones DD. Linking the functions of unrelated proteins using a novel directed evolution domain insertion method. *Nucleic Acids Res* 2008;**36**:e78.
96. Salis HM, Mirsky EA, Voigt CA. Automated design of synthetic ribosome binding sites to control protein expression. *Nat Biotechnol* 2009;**27**:946–50.
97. Groll M, Bochtler M, Brandstetter H, Clausen T, Huber R. Molecular machines for protein degradation. *Chembiochem* 2005;**6**:222–56.
98. Moon TS, Lou CB, Tamsir A, Stanton BC, Voigt CA. Genetic programs constructed from layered logic gates in single cells. *Nature* 2012;**491**:249–53.
99. Zhang FZ, Ouellet M, Batth TS, Adams PD, Petzold CJ, Mukhopadhyay A, et al. Enhancing fatty acid production by the expression of the regulatory transcription factor FadR. *Metab Eng* 2012;**14**:653–60.
100. Stevens JT, Carothers JM. Designing RNA-based genetic control systems for efficient production from engineered metabolic pathways. *ACS Synth Biol* 2014;**4**:107–15.
101. Franco E, Giordano G, Forsberg PO, Murray RM. Negative autoregulation matches production and demand in synthetic transcriptional networks. *ACS Synth Biol* 2014;**3**:589–99.

CHAPTER 8

Applications of Constraint-Based Models for Biochemical Production

Cameron Cotten, Jennifer L. Reed
Department of Chemical and Biological Engineering, University of Wisconsin-Madison, Madison, Wisconsin, USA; Great Lakes Bioenergy Research Center, University of Wisconsin-Madison, Madison, Wisconsin, USA

8.1 Introduction

Constraint-based metabolic modeling has been an important technique to broaden and deepen our understanding of microbial metabolism and regulation. Genome-scale metabolic models enable global analysis of microbial cells by considering all metabolic reactions simultaneously. The global, integrated view that metabolic models provide has been important as we move from studying individual cellular components to studying systems-level behaviors that arise from interactions between these components. The applications of constraint-based metabolic modeling are broad, and a key application has been the study and design of microbes to produce biofuels and other chemicals. Biofuels are metabolic products, and knowledge of how metabolism operates is critical to understanding and improving biofuel production by microbes.

This chapter reviews how genome-scale metabolic models are built and how constraint-based modeling approaches have been used in metabolic engineering to improve biofuel production. This chapter first introduces genome-scale metabolic reconstructions, which are comprehensive listings of all the reactions, compounds, and genes that are part of cellular metabolism in a particular organism. Constraint-based modeling methods are then presented, which can generate descriptions and predictions about cellular metabolism. Descriptively focused constraint-based methods that integrate experimental data to describe intracellular fluxes are reviewed. Additionally, predictive methods that design strains for chemical and biofuel production are discussed. Finally, case studies are presented highlighting recent successful applications of constraint-based modeling to engineer chemical production strains.

8.2 Genome-Scale Metabolic Reconstructions

8.2.1 Elements of a Genome-Scale Metabolic Reconstruction

Cellular metabolism can be viewed as a large network of reactions. The biological purpose of this reaction network is to convert carbon and energy from substrates into chemical energy,

Biotechnology for Biofuel Production and Optimization. http://dx.doi.org/10.1016/B978-0-444-63475-7.00008-X

reducing power, and building blocks needed for cellular growth (e.g., amino acids, nucleic acids, cell membrane components, enzyme cofactors, and ATP) and survival (e.g., energy storage molecules and antibiotics). A *genome-scale metabolic reconstruction* contains the following information about metabolism in a specific organism[1]:

- A set of the *cellular compartments* present in the organism. Prokaryote models often only include the cytosol and periplasm in their list of compartments, while models for eukaryotes, such as yeast, contain many compartments such as the cytoplasm, mitochrondria, peroxisome, endoplasmic reticulum, Golgi apparatus, and nucleus. All reconstructions also have an artificial compartment that represents the cell's extracellular environment (e.g., medium), referred to as the *extracellular compartment*.
- A set of the *metabolites* present in the organism. In addition to low molecular weight intermediates that make up catabolic and anabolic pathways, there are also a number of higher molecular weight peptides, lipids, cofactors (e.g., NADH and pyridoxal-5-phosphate), and energy carriers (e.g., ATP). Information about a metabolite is also captured, including its molecular formula and the compartment in which the metabolite is found. Separate metabolite listings are made for each compartment in which the metabolite appears.
- A set of metabolic *reactions* (including enzymatic, spontaneous, and transport reactions) that can occur in the organism. As its name implies, a genome-scale metabolic reconstruction should contain all of the metabolic reactions associated with genes in the genome of a species. Nonmetabolic reactions, such as DNA polymerase and DNA repair, should not be included. A reaction listing contains the following information:
  - The metabolites that the reaction produces or consumes, and the proportion in which these metabolites are used. This is called the reaction's *stoichiometry*. Separate reaction listings are required for reactions that occur in multiple compartments.
  - The *reversibility* of the reaction. If the reaction is irreversible under cellular conditions (based on thermodynamics or enzyme characterization) then it should be marked as *irreversible*. In some instances, a reaction occasionally must be marked as irreversible so that the corresponding model matches observed metabolic phenotypes, despite being reversible in principle. This could be due to metabolite concentrations [e.g., ATP, ADP, and NAD(P)H] and/or transcriptional regulation that prevents the reaction from operating in both directions *in vivo*.
  - The gene-protein-reaction (GPR) association, which includes a list of *genes* that encode proteins that catalyze each reaction. In some cases, multiple genes are required to perform a reaction (i.e., *enzyme subunits*), and in other cases, multiple genes can be used interchangeably to perform a reaction (i.e., *isozymes*). This information should be included into the GPRs, which are often written as Boolean statements (using "and" and "or" operators). GPRs will be absent for spontaneous reactions or for reactions whose associated genes are unknown.

  - Some models include *confidence* scores for each reaction. These confidence levels range from high (e.g., protein was purified and biochemically assayed) to low (e.g., reaction is needed to match experimental observations, but no gene responsible for the reaction can be found in the organism's genome).
- A *biomass reaction* that contains the amounts of metabolites that are found in 1 gram dry weight (gDW) of cells. In practice, the biomass reaction is included in the list of reactions. The reactants in the biomass reaction should be the terminal products of anabolic pathways such as amino acids and phospholipids (and not intermediates or precursors). Energy, in the form of ATP, that is used to produce biomass (e.g., for polymerizing amino acids) should also be added to the biomass reaction. Other products (e.g., antibiotics, fermentation by-products, and secreted proteins) that are not constant parts of cells should not be included in the biomass reaction. Typical scaling for the biomass reaction stoichiometry is mmol/gDW.

8.2.2 Construction of a New Genome-Scale Metabolic Reconstruction

If a genome-scale metabolic reconstruction of a species of interest has already been developed, then the reconstruction can be downloaded in a format that can be used by the preferred modeling software. The current standard for the field is the Systems Biology Markup Language (SBML), and tools exist for parsing SBML-compliant files into a number of modeling software packages. Reconstructions exist for a large variety of biofuel-producing organisms including *Escherichia coli*,[2] *Saccharomyces cerevisiae*,[3] *Clostridium acetobutylicum*,[4,5] *Corynebacterium glutamicum*,[6,7], and *Zymomonas mobilis*.[8,9] A list of available reconstructions is maintained by the Systems Biology Research Group[10] and many models are also available from the BioModels database.[11]

The process for generating a new metabolic reconstruction has been established (Figure 8.1). A number of software packages, databases, and protocols already exist to guide the reconstruction process.[1,12–19] The reconstruction process involves mapping genes to metabolic reactions, generating a biomass reaction, and refining the metabolic content based on comparisons between model predictions and experimental observations.

An initial list of candidate reactions and GPRs can be acquired from databases (such as Model SEED[15] and KEGG[16]) or software (such as Pathway Tools,[17] AUTOGRAPH,[18] and RAVEN[19]). The list of candidate reactions and metabolites should be checked to ensure reactions are elementally balanced, have correct reaction reversibility, are not missing metabolites, and use the correct cofactors and energy carriers. In addition, the draft list of metabolites should be refined by adding or correcting molecular formulas, replacing generic metabolites (e.g., electron carrier) with specific metabolites (e.g., NAD), and removing metabolites that are not present in the organism of interest. Finally, some pathways reported in the biochemical literature may be missing from metabolic databases and need to be manually added.

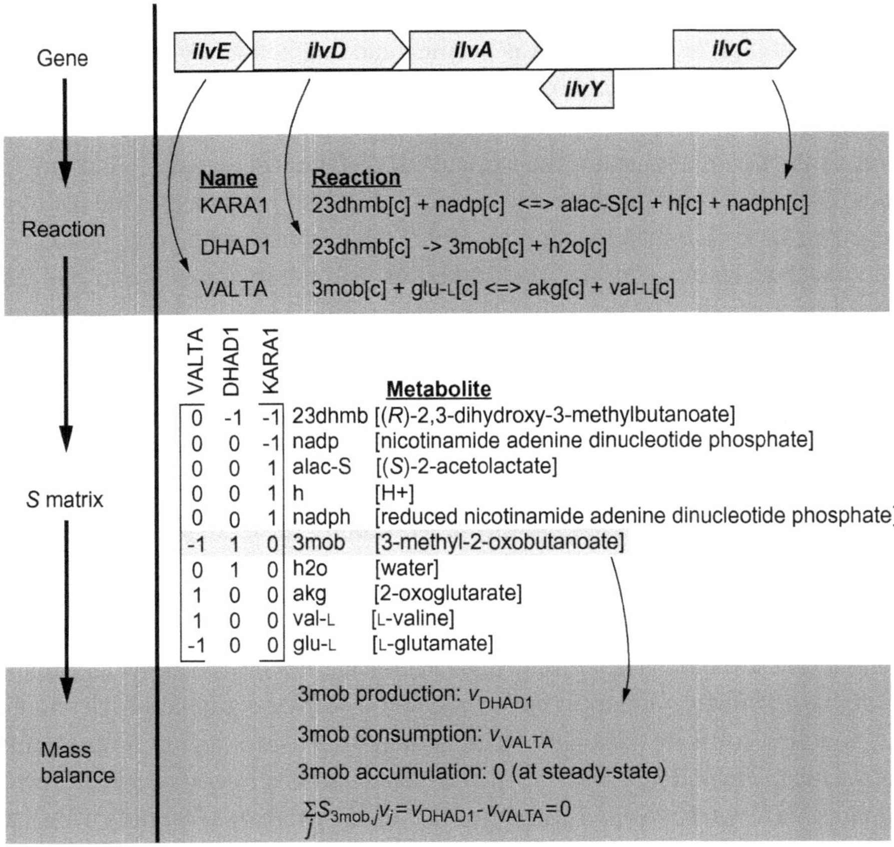

Figure 8.1

Genome-scale metabolic reconstructions. Genome-scale metabolic reconstructions contain information about metabolic genes, reactions, and metabolites in a given organism. Metabolic genes are first identified from an organism's genome annotation. Reactions associated with these metabolic genes can be extracted from online databases or the literature. These reactions are entered into a stoichiometric matrix, where each reaction corresponds to a different column of the matrix, and each metabolite to a different row. Mass balances for each metabolite can be constructed for each metabolite (from the corresponding row in the stoichiometric matrix) to give linear relationships between metabolic fluxes.

The initial reconstruction is used to build a draft model, where a biomass reaction is added (which includes biomass ATP requirements) based on experimental data.[20] This draft model is used to predict growth phenotypes under different nutrient conditions which are compared to experimental data. Growth phenotyping experiments are inexpensive, high throughput, and provide information that can be easily used to refine the models. Matching metabolic model growth predictions to experimental growth phenotypes is a key step in the reconstruction process.[20] Automated methods can be used to identify the smallest set of reactions to add to (or remove from) the draft model to resolve discrepancies between model predictions and experimental results.[21–23] Recent methods have incorporated functional genomic and flux data to increase the accuracy of these gap-filling methods.[24]

8.3 Foundations of Constraint-Based Modeling

8.3.1 Mass Balances

Most metabolites in the metabolic network can be produced and consumed by multiple reactions. It is, therefore, useful to consider the *mass balance* of a metabolite, or the sum of all fluxes through reactions that produce or consume a particular metabolite. The mass balance for a metabolite can be written as:

$$(\text{Total production rate}) - (\text{Total consumption rate}) = (\text{Accumulation}) \tag{8.1}$$

Here the intracellular accumulation rate for a given metabolite is the rate of production of the metabolite minus the rate of consumption of overall reactions in the metabolic network. Constraint-based models assume the cells are at steady state, in which case there is no net accumulation of any metabolite in the cell. As such, constraint-based models can only be used to model conditions under which metabolism is at a steady (or pseudo steady) state, such as during exponential growth or stationary phase in batch cultures, and steady-state growth in chemostats. At steady-state, the mass balance equation simplifies to:

$$(\text{Total production rate}) - (\text{Total consumption rate}) = 0 \tag{8.2}$$

8.3.2 Common Physical and Chemical Constraints

Constraint-based models use the stoichiometry from the reaction list to relate the total production and consumption rates (used in the mass balances) to the *fluxes* (v) through each reaction (j) in the metabolic model. These fluxes are the variables in the model and are typically reported with units of mmol/gDW/h, except for flux through the biomass reaction which has units of 1/h. The stoichiometry of a particular reaction indicates how many of each metabolite are consumed or produced per unit flux. The stoichiometry for all the reactions in the network can be stored in a *stoichiometric matrix* (S), where rows correspond to metabolites and columns correspond to reactions. Reactants have negative coefficients in the S matrix and products have positive coefficients. Since the S matrix captures how metabolites are produced and consumed by different reactions, the steady-state mass balances for all metabolites can be written as a system of linear equations using

$$S \cdot v = 0 \tag{8.3}$$

This system of equations (one for each metabolite) is referred to as *mass balance constraints* (Figure 8.1). Steady-state *flux distributions* can be found by solving Equation (8.3) using linear algebra techniques (e.g., calculating null space basis vectors of S); however, the problem is often under determined since there are normally more reactions (variables) than metabolites (equations). As a result there are many possible steady-state flux distributions through a metabolic network.

Additionally, the fluxes through some reactions can be bounded by thermodynamics or measured capacities. The fluxes for irreversible reactions should be constrained to be nonnegative by

$$v_j \geq 0 \ \forall j \in \mathrm{Irr} \tag{8.4}$$

This has historically been referred to as a *thermodynamic constraint*. While the set of irreversible reactions (Irr) has been historically based on the biochemical characterization of enzymes, recent efforts have included Gibbs free energy of reaction calculations directly into the models to limit reaction directionality (see Section 8.4.5 for details). It is also common to include upper and lower bounds for all reactions (J) using

$$v_j^{\min} \leq v_j \leq v_j^{\max} \ \forall j \in J \tag{8.5}$$

Here the flux limit parameters, $v_j^{\min}$ and $v_j^{\max}$, are set based on measured flux values (e.g., glucose or oxygen uptake rates) or set to values exceeding typical flux values (oftentimes—1000 and 1000 mmol/gDW/h, respectively). The flux limits are often referred to as *enzyme capacity* constraints. Solving Equations (8.3)–(8.5) simultaneously cannot be done using linear algebra techniques, and so other approaches must be used to find flux distributions that satisfy the mass balance, thermodynamic, and enzyme capacity constraints.

8.4 Constraint-Based Strain Evaluation Techniques

A number of approaches can be used to predict intracellular flux distributions that take different types of experimental data (e.g., medium conditions, gene expression, or metabolite concentrations) into account. This section describes elementary modes and how they can be used to describe all steady-state flux distributions. In addition, flux balance analysis (FBA) and other constraint-based approaches are described that use optimization to predict intracellular fluxes by integrating experimental data into the models. These intracellular flux predictions give a wealth of information about an existing strain, including which pathways are being used and potential bottleneck reactions that are preventing a desired behavior.

8.4.1 Elementary Modes

Elementary modes are defined as nondecomposable flux vectors that satisfy steady-state mass balance and thermodynamic constraints (Equations 8.3 and 8.4). These elementary modes (and related extreme pathways[25]) define the metabolic capabilities of a cell, where cells could use flux distributions that correspond to individual elementary modes or linear combinations of elementary modes. The number of elementary modes for a reaction network can be large (e.g., ~43,000 for a reaction network with 110 reactions and four carbon sources[26]) and enumerating all elementary modes is an NP hard problem. As a result, elementary modes are typically generated for smaller reaction networks (e.g., ~100 reactions). For larger genome-scale networks a subset of elementary modes can be found using optimization, where the elementary modes using the fewest reactions are identified first.[27]

8.4.2 Flux Balance Analysis

For some organisms, the flux distribution that enables the fastest growth of the organism is close to the experimentally measured flux distribution. This is reasonable, since natural selection favors faster growing strains. FBA uses optimization to identify a flux distribution that maximizes an objective function and satisfies mass balance (Equation 8.3), thermodynamic (Equation 8.4), and enzyme capacity constraints (Equation 8.5). In most FBA simulations, the growth rate (i.e., flux through the biomass reaction, v_{Bio}) is maximized[28]; however, other objectives have been used.[29] FBA is formulated as the following optimization problem:

$$\max_v \sum_{j \in J} c_j v_j \tag{8.6}$$

such that

$$\sum_{j \in J} S_{ij} v_j = 0 \quad \forall i \in I \tag{8.7}$$

$$v_j \geq 0 \quad \forall j \in \mathrm{Irr} \tag{8.8}$$

$$v_j^{\min} \leq v_j \leq v_j^{\max} \quad \forall j \in J \tag{8.9}$$

Here, I and J are the sets of metabolites and reactions in the metabolic model, v are the flux variables, and S is the stoichiometric matrix. The set Irr refers to the subset of reactions that are irreversible. The objective function can be maximized (as in Equation 8.6) or minimized, and in many cases is set to v_{Bio} by making all c_j values equal to 0 except for c_{Bio}, which is set to 1. However, FBA can also be used to calculate maximum theoretical yields (and the flux distributions needed to achieve them) by instead setting the objective function to be the product secretion flux.

FBA can be solved by existing optimization solvers (e.g., CPLEX) that solve *linear programs* (LP) since the constraints and objective function are all linear equations, and because all the variables (in this case just v) are continuous. A variety of software packages are available to solve FBA problems and other constraint-based methods,[19,30] and online interfaces are also available.[13,15]

Though simple, FBA is an effective tool for a variety of metabolic engineering applications. It was observed that single deletion mutants of *E. coli* achieved growth rates predicted by FBA after adaptive evolution.[31] Instead of maximizing biomass production, it is also possible to maximize production of a biofuel of interest and add a minimum growth rate as a constraint (by making $v_{\mathrm{Bio}}^{\min}$ nonzero). By varying the minimum growth rate, trade-offs between

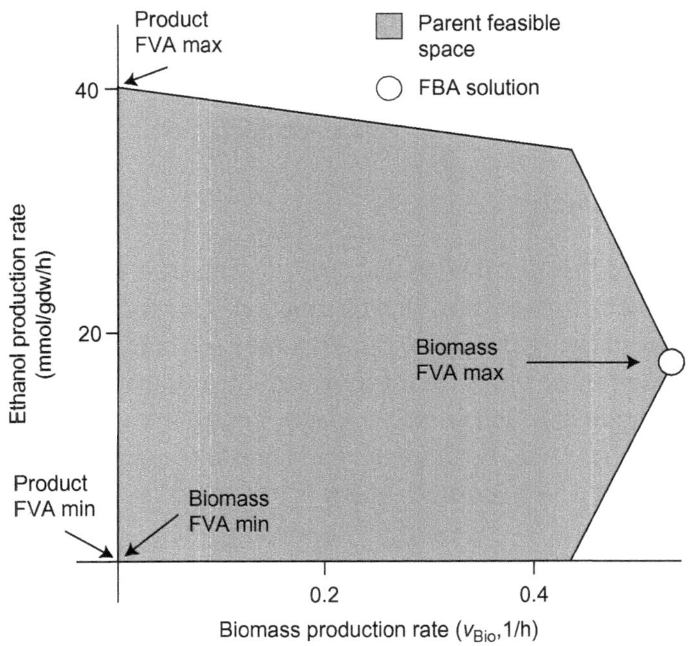

Figure 8.2

Flux balance and flux variability analysis. The figure shows (in gray) the possible combinations of biomass and ethanol production that can be achieved which satisfy mass balance, thermodynamic, and capacity constraints. The feasible space shown here is derived from the iJO1366 *E. coli* model for glucose anaerobic conditions. FBA finds a feasible flux distribution (white circle) that optimizes an objective function, in this example maximizes biomass production (i.e., growth) rate. Flux variability analysis (FVA) finds the maximum and minimum value for each flux, which still satisfies the imposed constraints. The FVA solutions for the biomass and ethanol production rates are indicated with arrows.

biomass and biofuel production can be elucidated using FBA, and pathways that are used for biomass and biofuel production can be determined (Figure 8.2). This information can be useful to identify target growth and production rates to maximize productivity or titers.[32] It is also possible to determine the maximum theoretical yield of any fuel molecule from any feedstock very easily by removing the minimum growth requirement and maximizing biofuel production. This can also be helpful when investigating production of biofuels from different carbon sources or under different conditions (e.g., aerobic vs. anaerobic).

Since there can be alternate FBA solutions (i.e., multiple flux distributions with the same value for the objective function) approaches have been developed to characterize them. Flux variability analysis (FVA) is one such approach that finds the minimum and maximum flux value through each reaction in the network. FVA is essentially an FBA problem, where each flux is minimized or maximized individually by changing the FBA objective (Equation 8.6). FVA gives tighter bounds on each reaction than the global, arbitrarily large bounds on each reaction given in Equation (8.9). This information has been used in a number of studies to characterize the flexibility and range of possible flux distributions in a metabolic network.[33–35] Another approach, called *Parsimonious flux balance analysis* (pFBA), was developed to find a solution that maximizes the FBA objective function and also minimizes the total flux through the network.[36] The method first solves FBA to find the optimal value for the objective function, and then solves another linear program to find the flux distribution that minimizes the total flux through all metabolic reactions while satisfying an additional constraint that the original FBA objective function equals the optimal value. The method has been justified biologically since higher flux levels would have higher enzyme costs needed to achieve them. In practice, pFBA has outperformed FBA and other methods at predicting intracellular flux distributions.[29,36,37] Many of the methods that will be discussed in the rest of this chapter include modifications to the objective function or constraints used in FBA.

8.4.3 Methods Incorporating Expression Data

Methods for incorporating mRNA levels into constraint-based modeling have been extensively developed since transcriptomic data is relatively inexpensive and easy to obtain. These methods use mRNA data to impose additional constraints on reaction fluxes, such as blocking fluxes through reactions when associated transcript levels fall below a given threshold. These expression-based constraints reduce the solution space of feasible flux distributions, as well as identify cases where metabolic regulation blocks desirable cellular behavior. Interested readers should read a more detailed review and analysis of the different methods.[37]

In general, transcriptomic constraint-based methods make changes to the upper and lower bounds of reactions (Equation 8.9). Methods differ based on the types of changes they make to the flux bounds and how these changes are inferred from expression data. One approach, developed by Akesson et al.,[38] defined a threshold for "low" gene expression and removed reactions associated with genes whose expression fell below this threshold from the model (Figure 8.3). Using this approach, the authors were able to improve estimates of uptake rates by yeast grown in batch culture.[38] The GIMME method uses a similar low expression cutoff, but instead minimizes a weighted linear combination of fluxes through reactions associated with lowly expressed genes (where the weights corresponded to how far below the cutoff the gene's expression was)[39]. Another related method, developed by Shlomi et al., used two thresholds for genes with "low" and "high" expression. The method then finds a flux distribution with flux through reactions associated with highly expressed genes and no flux through reactions associated with lowly expressed genes. This method can be used to predict flux distributions when maximization of growth is not a good predictor of cell behavior.[40,41]

Some methods instead compared transcriptomic data from multiple environmental or genetic conditions to modify flux bounds. The MADE algorithm uses gene expression

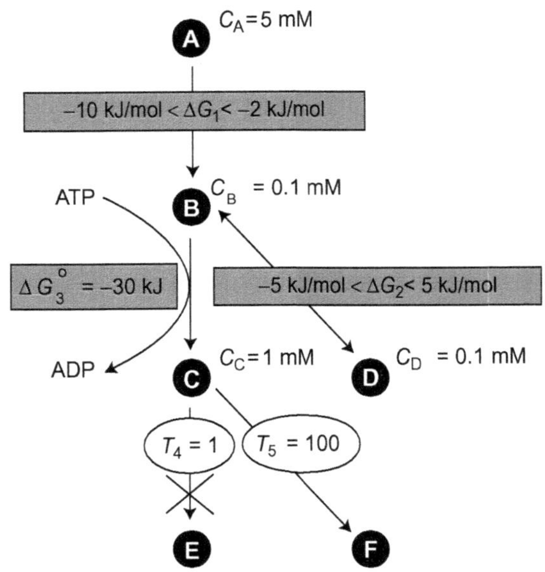

Figure 8.3

Methods incorporating metabolite and mRNA transcript levels. Incorporating additional experimental data can produce more accurate constraint-based model predictions. The reaction from A to B only proceeds in the forward direction because the concentration of A (C_A) is much larger than the concentration of (C_B). The corresponding estimate of ΔG is always negative (boxed values) indicating the reaction proceeds in one direction. In contrast, the reaction between B and D can proceed in either direction because the estimate for ΔG is near zero. The reaction from B to C is powered by the hydrolysis of ATP to ADP and thermodynamic estimates indicate the reaction can only proceed in the forward direction. mRNA transcript data can also be used to prevent reactions from occurring. The expression level of the gene associated with the reaction from C to E (T_4) falls below a user-defined cutoff (in this case 10), and so the flux through this reaction can be constrained to zero (marked with X). However, expression for the gene associated with the reaction from C to F (T_5) is above the cutoff and so the associated flux is not constrained to zero.

data from multiple conditions to predict flux distributions that are consistent with changes in the transcriptome over different conditions (time points, media, etc.).[42] In this method, a minimum flux through the biomass reaction is enforced (i.e., $v_{\mathrm{Bio}} \geq v_{\mathrm{Bio}}^{\min}$) for conditions where growth was observed, and each reaction's ability to carry flux is toggled on (or off) between conditions based on whether there was a statistically significant increase (or decrease) in the corresponding gene transcripts in the respective conditions. The PROM algorithm uses multiple gene expression data sets to calculate conditional probabilities that a gene is expressed when a transcription factor is not expressed. These conditional probabilities are then used to predict fluxes for a transcription factor knockout mutant by scaling the flux bounds by the probability.[43] The method was used to successfully predict the growth phenotypes of a large number of transcription factor knockouts. This approach may be useful for biofuels applications since transcription factors are often knockout targets in engineered targets.[44–47]

Many of the transcriptomic methods introduce *binary variables* (that can only be 0 or 1) into the optimization problem, and as a result require the use of solvers that can solve *mixed integer linear programs* (MILP). Additional software is required beyond the software that is used to solve FBA problems, and even with the appropriate software, MILP models take much more time to solve than LP models of similar size. Some of the methods described in this section are included in the TIGER[48] and COBRA[30] toolboxes for MATLAB.

8.4.4 Methods Incorporating ^{13}C-MFA Data

Metabolic flux analysis (MFA) is the current standard to accurately and precisely estimate fluxes through intracellular reactions.[49] With MFA, ^{13}C labeled substrates are fed to an organism and the distribution of labeled carbon atoms in amino acids and other metabolites can be used to determine fluxes through central metabolic pathways. It is possible to incorporate MFA results into constraint-based models to gain additional insight beyond the original metabolic flux experiment. MFA can determine flux ratios, or fractions of carbon flux that are sent down different pathways at key branch points in the metabolic network. Some studies suggest that these flux ratios are invariant over numerous genetic perturbations in a variety of organisms including *E. coli*, *S. cerevisiae*, and *Bacillus subtilis*.[50–52] These flux ratios can easily be added to an FBA problem by adding inequality constraints to the FBA problem, for example:

$$v_{\mathrm{PPP}} \geq \alpha v_{\mathrm{PTS}} \tag{8.10}$$

where v_{PTS} is the flux of glucose into the cell through the transport system, v_{PPP} is the flux into the pentose phosphate pathway through glucose-6-phosphate dehydrogenase, and α is the MFA-determined flux ratio. It has been recently shown that the addition of flux ratio constraints to FBA can greatly improve model predictions for *C. acetobutylicum* which has many alternative side-product production pathways.[53]

8.4.5 Methods Incorporating Metabolomics Data

Absolute concentrations of metabolites are another common metabolic measurement. Thermodynamic constraints have been the main way metabolomics data have been incorporated into constraint-based models. The possible direction of each reaction can be determined with information about the concentrations of products and reactants and changes in Gibbs free energies of the reaction. These thermodynamic constraints replace the previous reaction directionality constraints (Equation 8.8 and Figure 8.3). Various objective functions have been used in conjunction with these new thermodynamic constraints including maximization of biomass production, minimization of total flux, minimization of total metabolite concentration, and fit of metabolite concentration variables to experimentally measured metabolite levels.[54–56] Gibbs free energy values are needed for the new thermodynamic constraints and these are often estimated using group contribution theory.[57]

Methods incorporating thermodynamics have yielded a variety of information about metabolic networks. Transport of sugars and other compounds across cell membranes is strongly affected by thermodynamics and metabolite concentrations, since concentrations on both sides of the membrane affect transport directionality. Accordingly, a method that incorporated thermodynamic constraints was able to predict the extracellular concentrations of lactate and oxygen that do not allow metabolic flux in human erythrocytes.[54] Modeling of metabolite concentrations using thermodynamic constraints can also recapitulate experimentally measured metabolite concentrations.[55] It also appears possible to replace the irreversibility constraints in FBA (Equation 8.8) with detailed thermodynamic constraints after including concentration measurements for a few key metabolites.[56] Since databases of reactions occasionally give incorrect reaction directions and reversibility, detailed thermodynamic constraints may be a valid alternative that does not rely on manual database curation.

A number of other constraint-based methods use metabolite (and enzyme) concentration measurements to limit flux values based on kinetic expressions. Most methods use enzyme kinetics to constrain transporter uptake rates based on Michaelis-Menten rate laws and extracellular concentrations of metabolites.[58] This improves FBA, since limits on carbon and oxygen uptake in FBA are generally fixed and do not change in response to media concentrations. A number of methods use kinetic expressions for transporters to model batch cultures, where concentrations of substrates change over the course of the culture.[59,60] These methods can recapitulate a number of important aspects of batch growth, including diauxic shifts and the release and subsequent uptake of compounds such as acetate.[58]

Kinetic expressions can also be applied to internal reactions, and methods have generally used approximate expressions for kinetic rate laws. A kinetic model using lin-log kinetics constructs linear relationships between fluxes and metabolite concentrations, and is somewhat similar to the linear relationships that are used in thermodynamic constraints.[61] A lin-log kinetic model applied to a yeast genome-scale metabolic model was able to recapitulate reactions which control flux through central carbon metabolism.[62] Mass-action kinetic expressions are more difficult to integrate into constraint-based models since they introduce nonlinear equations, which are not easy to solve reliably. One method presolved FBA for a desired flux distribution, then solved the nonlinear mass-action kinetic expressions to find metabolite concentration profiles that were consistent with the desired flux distribution.[63] Databases of kinetic information about metabolic enzymes are publicly accessible and can provide data for parameterizing kinetic models or verifying results from these models.[64] More detailed kinetic expressions have been used recently in constraint-based models. These models have captured many of the secondary aspects of enzyme kinetics beyond mass-action kinetics, including active site occupancy (Michaelis-Menten kinetics) and inhibition/activation by metabolites not involved in the reaction directly.[65] Precise estimates from detailed kinetic expressions can replace some of the global upper and lower bounds on fluxes ($v_j^{\min}$ and $v_j^{\max}$) in FBA.

8.5 Constraint-Based Strain Design Techniques

As discussed in the last section there are a number of methods that can be used to predict fluxes in an existing strain. This information is important, but it does not always specify what genetic changes are needed to improve chemical production. There are generally three types of approaches to find such metabolic engineering strategies. The first set of approaches make predictions for a given set of mutations, and by applying these approaches for different sets of mutations the best strategies can be identified. The second set of approaches algorithmically determine the best genetic changes to make without having to manually test all possible strategies. The third type of approach analyses elementary modes to eliminate undesirable metabolic functionalities.

8.5.1 Gene Deletion Prediction Methods

It is possible to predict deletion mutant phenotypes using FBA. In this case, a set of disabled reactions is predetermined based on what genes are deleted, and the reactions' fluxes are set to zero. Because FBA predicts phenotypes for strains optimized for biomass production, FBA can mispredict behaviors of strains with gene deletions that have not undergone adaptive evolution. Instead, a number of methods have also been developed to predict metabolic fluxes immediately after gene deletions. The first of these methods was minimization of metabolic adjustment (MOMA).[66] MOMA assumes that the flux distribution immediately after gene deletions will be as close as possible to the flux distribution before the deletions. Here, the Euclidean distance between the previous flux and the mutant flux distribution is used (Figure 8.4). MOMA uses the same FBA constraints (Equations 8.7–8.9) and a new objective function:

$$\min_v \sum_j \left(v_j - v_j^{\mathrm{PS}}\right)^2 \tag{8.11}$$

Here, v_j^{PS} is the flux through reaction j in the parental strain before gene deletions and is a fixed parameter usually predetermined by FBA. Fluxes associated with deleted genes are also constrained to be zero. Compared to FBA, MOMA was better at predicting fluxes for an *E. coli* pyruvate kinase deletion strain and growth phenotypes for various *E. coli* mutants.[66] MOMA was also used to refine a yeast genome-scale metabolic model by comparing MOMA predictions to growth phenotypes of mutants from a single gene deletion library.[67]

Other metrics can be used to quantify the distance between flux distributions for the parental strain and a deletion mutant. Regulatory on/off minimization (ROOM) minimizes the number of fluxes that must change from the parental strain using a threshold for significant

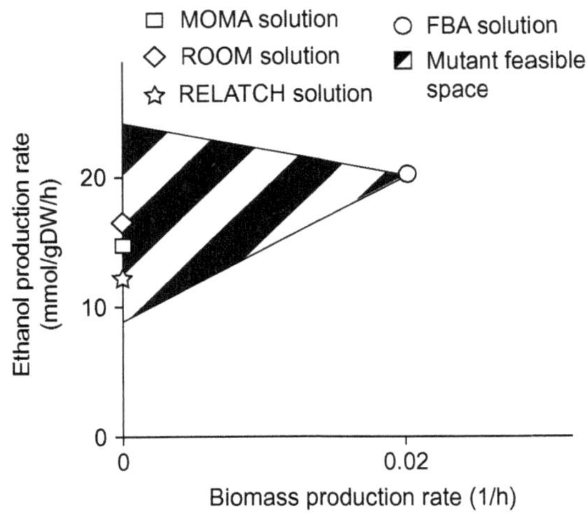

Figure 8.4

Deletion prediction methods. Deleting metabolic genes reduces the number of feasible flux distributions, and different methods can be used to predict which of these possible flux distributions a cell will utilize. Here, results are shown for a deletion mutant of *E. coli* (Δ*nuoN* Δ*eno* Δ*fpr*) under glucose anaerobic conditions predicted by iJO1366. The majority of the parental strain's feasible space (shown in Figure 8.2) is no longer available due to the deletions. Note the change in scale and shape of the feasible space compared to Figure 8.2. Only the flux distributions shown in the hashed region are possible for the deletion mutant. The FBA prediction (maximizing growth rate) for this mutant is shown as a circle. The MOMA predicted flux distribution (rectangle) has the shortest Euclidean distance to the parental strain's flux distribution (circle in Figure 8.2). RELATCH predicts that the flux distribution (star) with the shortest relative distance the parental flux distribution, while ROOM predicts a flux distribution (diamond) with the fewest number of flux changes.

flux changes. ROOM was shown to better predict fluxes in five *E. coli* deletion mutants compared to MOMA.[68] Another method based on relative flux changes (RELATCH) uses the objective:

$$\min_v \sum_j \frac{\left(v_j - v_j^{\text{PS}}\right)^2}{\left(v_j^{\text{PS}}\right)^2} + \alpha \sum_j \sum_{n \in N_{\text{inactive}}} V_{j,n} \tag{8.12}$$

Here $V_{j,n}$ is the enzyme contribution, defined as the flux that an enzyme (n) contributes for a reaction (j). In RELATCH, enzyme contributions that were inactive in the parental strain that become active (i.e., nonzero) in the mutant strain (sometimes referred to as latent pathway activation) are penalized in the objective. Large changes in enzyme contributions that are already active in the parental strain are also limited by additional constraints. This method was shown to more accurately predict fluxes than FBA, MOMA, and ROOM for gene deletion mutants and for parental strains in new growth conditions.[69] MOMA and RELATCH both require a nonlinear programming or quadratic programming solver due to their nonlinear objective functions, while ROOM requires an MILP solver.

8.5.2 Bilevel Optimization Strain Design Methods

Although it is possible to predict fluxes in adaptively evolved and unevolved gene deletion mutants, choosing what gene deletions are needed to increase chemical production can be a difficult problem. As the number of gene deletions increases, the number of possible combinations of gene deletions increases exponentially. Thus, methods that can more efficiently find the deletion mutants with the highest production are used instead. Bilevel optimization methods embed a constraint-based model (with mass balance constraints, thermodynamic constraints, capacity constraints, and biological objective), such as FBA, as an inner problem into another outer optimization problem. The outer problem finds gene or reaction deletions that optimize the desired metabolic engineering goal (e.g., maximize production rate or yield). The outer optimization problems pass deletions to the embedded constraint-based model, which then predicts and returns the corresponding flux distribution to the outer optimization problem. Many of these bilevel problems use binary decision variables to indicate gene or reaction deletions and must be solved using MILP solvers.

Methods for finding gene deletions and additions

The first bilevel optimization strain design method was OptKnock, which uses FBA as an inner problem to find reaction deletion strategies that have the highest chemical production at the mutant's maximal rate.[70] OptKnock strategies are often described as having coupled growth and chemical production, indicating that chemical production should increase as growth rate increases (Figure 8.5). MILP solvers can quickly search through all possible reaction deletion combinations to find the simplest deletion strategies with highest chemical production. OptKnock has been used to find deletion strategies for producing a large variety of chemicals in a number of organisms (see Section 8.6.1 for an example). Several methods were subsequently proposed to find deletion mutants more quickly,[71–74] to use gene deletions instead of reaction deletions,[72,75,109] to consider transcriptional regulatory effects,[109] and to consider multiple design criteria.[76]

All of these methods use FBA to predict mutant metabolic fluxes, and so the resulting deletion mutants may need to be adaptively evolved experimentally so that a high chemical production phenotype is achieved. To design strains that would immediately have high chemical production, without having to evolve the mutants, then the bilevel optimization problem needs to use MOMA, ROOM, or RELATCH as an inner problem (instead of FBA). To date, MOMA has been included into two bilevel strain design problems which have been solved using genetic algorithms (OptGene)[75] or optimization solvers (BiMOMA).[74] BiMOMA proposed *E. coli* deletion mutants with 40-50% of the theoretical maximum yields for glutamate and pyruvate, where MOMA was used to predict fluxes in mutants immediately after gene deletions.

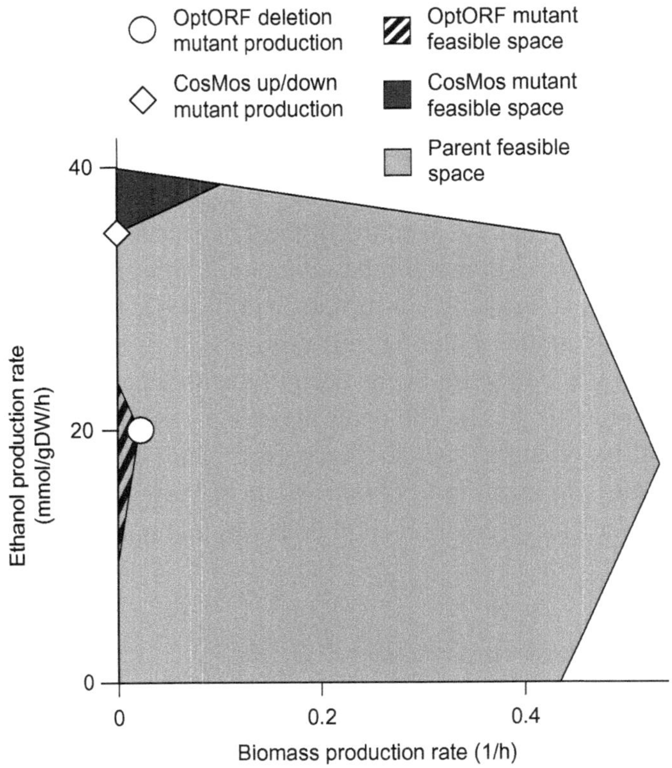

Figure 8.5
Strain design methods. Strain design methods differ in how they predict flux distributions for mutant strains (e.g., using FBA or MOMA) and how they model changes to the organism (e.g., flux deletion or up/downregulation). OptKnock and OptORF find deletions that create a coupling between production of biomass production and the desired chemical. The mutant shown in this figure and also in Figure 8.4 was found by OptORF and has a higher production rate of ethanol (circle) than the parental strain at their corresponding maximum biomass production rates. In contrast, OptForce and CosMos consider reaction up- and downregulations in addition to reaction deletions, and design mutants which have high minimum chemical production rates (for all growth rates). The black area represents flux distributions that are available to a mutant involving reaction deletions, upregulations, and downregulations identified by CosMos. CosMos compares mutants with different modification strategies based on the flux distribution with the lowest chemical production (diamond) in each mutant's feasible space.

In addition to deletions, methods to identify nonnative reactions that could be added to the metabolic network to improve chemical production have also been developed. OptStrain first identifies the minimal number of reactions from a database of known nonnative reactions (e.g., KEGG[16] or BRENDA[64]) needed to achieve the highest yield.[77] These reactions are subsequently added to the model and then OptKnock is used to find any necessary reaction deletions to couple growth and chemical production. Instead, SimOptStrain considers additions and deletions simultaneously and as a result can find strategies with potentially higher production at the corresponding mutant's maximum growth.[74]

Methods for finding genes to overexpress or repress

Strain engineering often involves gene expression changes along with gene deletions. Gene expression can be modified experimentally by changing promoters, ribosome binding sites, and plasmid copy numbers or modulating expression of transcription factors. A number of strain design algorithms have incorporated flux increases or decreases (in addition to deletions) as engineering options. We will refer to these options as reaction upregulations and downregulations due to how they are modeled. However, these reaction upregulations or downregulations must be accomplished experimentally by modulating gene expression.

Reaction upregulations and downregulations are represented in the model by altering the flux bounds of reactions, Equation (8.9). The reactions with bound modifications can be written as:

$$v_j \geq v_j^{\min} + \alpha_j \tag{8.13}$$

$$v_j \leq v_j^{\max} - \beta_j \tag{8.14}$$

Here a_j and β_j are variables (or predetermined parameters) that can be adjusted by the strain design method to alter the flux bounds from the default values ($v_j^{\min}$ and $v_j^{\max}$, which are often 0 or large positive/negative values). An inner FBA objective minimizing desired chemical production is used (since adaptive evolution could undo desired flux changes), while the outer problem maximizes desired chemical production (Figure 8.5). The resulting bilevel algorithms find flux modifications that will eliminate flux distributions with low chemical production irrespective of growth rate. OptForce uses FVA bounds of the parental strain and of an overproducing strain to generate predetermined values for a_j and β_j that can be used to modify the flux bounds.[35] The method was successfully used to engineer *E. coli* to produce fatty acids, a precursor of biodiesel (see Section 8.6.2).[78] OptForce has been recently extended to incorporate enzyme kinetics (k-OptForce), a key consideration when enzymes are up- or downregulated.[79] Alternatively, the CosMos method (for continuous modifications) uses continuous variables for upregulation and downregulation, rather than relying on FVA bounds to preselect potential changes to flux bounds.[80] CosMos was able to find groups of reactions to upregulate and downregulate that were significantly smaller than the groups of reactions that were found by OptForce.

Instead of representing upregulation and downregulation as changes in flux bounds, the Redirector method adds terms to the FBA objective function (Equation 8.6) to find FBA solutions that incorporate increasing or decreasing amounts of those fluxes.[81] The method combines bilevel optimization solutions with iterative searches to identify alternate engineering targets. Redirector generated all the known targets for producing free fatty acids that were known at the time. Other methods have also determined metabolic targets for upregulation and downregulation without using bilevel optimization. The FSEOF method sets production of a target chemical to a significant portion of its maximum theoretical yield, and identifies fluxes that increase with increasing product formation.[82] The method was also improved by considering FVA of overproducing strains and genomic context of altered genes (FVSEOF).[83] These methods aided in the design of strains that overproduced lycopene and putrescine.

8.5.3 Minimal Metabolic Functionality and Minimal Cut Sets

As described earlier, elementary modes are flux vectors that describe the metabolic capabilities of cell. Elementary modes can also be analyzed to identify strategies for enhancing chemical production. Deleting genes or reactions prevents any elementary mode utilizing those reactions from being used. Once elementary modes are calculated they can be analyzed to find reaction deletions that eliminate specified network functions (e.g., by-product formation or low product yields). The minimal set of irreducible, nondecomposable reaction deletions that prevent such specified network functions are referred to as *minimal cut sets* (MCSs).[84] MCSs can be calculated by finding the smallest number of reaction deletions to block all undesirable elementary modes with the specified network function,[84] or by finding the elementary modes to the dual of the network.[85] However, MCSs might eliminate desired network functionalities (e.g., growth) along with the specified network functionality (e.g., by-product formation). As a result, *constrained minimal cut sets* (cMCSs) can be calculated that eliminate a specified network functionality while retaining some other desired network functionality[86]; for example, eliminate low product yield solutions while retaining growth solutions. OptKnock solutions, where product formation and growth are coupled, can also be found using cMCSs.[86] Recently, optimization-based methods for identifying the smallest MCSs and cMCSs in genome-scale networks were developed and strategies found for enhancing ethanol, fumarate, and serine in *E. coli* were proposed.[87] Similar to cMCS analysis, *minimal metabolic functionality* (MMF) finds deletions that eliminate undesirable elementary modes while keeping a subset of desirable elementary modes.[88,89] MMF is an iterative approach, where the impact of each deletion is evaluated at each iteration. The reaction that eliminates the most undesirable modes while still allowing high product yield and high growth modes is selected. Unlike calculating MCSs and cMCSs, MMF requires the full set of elementary modes be determined. As a result, MMF is still utilized for smaller metabolic networks. MMF was successfully used to engineer *E. coli* strains that produce ethanol from glucose and xylose,[88] ethanol from glycerol,[90] or carotenoids.[91]

8.6 Case Studies

8.6.1 Design of a 1,4-Butanediol Producing *E. coli* Strain

OptKnock was the first strain design method in constraint-based modeling. Though many improvements and alternatives to OptKnock have been proposed, OptKnock remains one of the most used methods in the field. The first patent granted based on results from OptKnock was a strain of *E. coli* that produced succinate.[92] The holding company, Genomatica, developed and patented a strain producing 1,4-butanediol (BDO) in 2011.[93] In partnership with DuPont Tate & Lyle, Genomatica built an industrial-scale plant that produced 5 million pounds of BDO in 5 weeks. The BASF chemical company licensed the BDO strain and fermentation process to construct a "world-scale production facility" of BDO in early 2013.[94]

Before Genomatica's BDO strain was developed, BDO was exclusively produced from petrochemical sources. In one route, acetylene reacts with formaldehyde at high pressures and over heavy metal acetylides to produce 1,4-butynediol, which is hydrogenated to BDO. Other routes start with propylene oxide, maleic anhydride, butadiene, allyl acetate, and succinic acid. In contrast to the petrochemical routes, the *E. coli* catalyst in the Genomatica's BDO process uses biomass-derived sugars (including glucose and xylose) as a feedstock. A number of important products are produced directly from BDO, including gamma-butyrolactone and tetrahydrofuran (THF). The worldwide market for BDO is about 2 billion USD per year, and a large portion of the BDO produced worldwide is ultimately used in plastics and elastic fibers after conversion to THF, while another portion is used as an industrial solvent.

A number of constraint-based strain design methods were used to develop the BDO-producing strain. Since BDO is not a natural product in any known organism, it was necessary to design an entirely new pathway using biochemical transformation rules for functional groups extracted from databases of known reactions.[95] Starting from metabolites in the iJR904 model for *E. coli*,[96] biochemical transformations to novel metabolites were predicted by applying chemical transformation rules to the functional groups of native *E. coli* metabolites. Nonnative, but known reactions were also considered to construct novel pathways to BDO. Ten thousand unique pathways were generated that could produce BDO from native *E. coli* metabolites, each using some reactions that were not yet reported in metabolic databases. The pathways were next filtered using criteria such as BDO yield, pathway length, number of nonnative steps, number of novel steps, and thermodynamic feasibility. After a pathway to produce BDO was selected based on these criteria, OptKnock was used to couple BDO production to biomass production. The engineered strain was able to produce BDO, and OptKnock-designed deletion mutants showed increased BDO production. In addition to its patent for strains that produce BDO, Genomatica also holds numerous patents for *E. coli* biocatalysts to produce ethylene glycol, propylene, alkenes, *n*-propanol, and methacrylic acid.[97–100] This illustrates that constraint-based modeling approaches are flexible and can be applied to different organisms, feedstocks, and desired products.

8.6.2 Design of Fatty Acid Producing E. coli *Strain*

The OptForce method has been successfully used to design novel *E. coli* strains to produce succinate,[35] 1-butanol,[101] and malonyl-CoA.[102] OptForce tries to increase flux to the desired product and decrease flux through competing pathways by proposing reaction deletions, upregulations, and downregulations. Using reaction upregulations and downregulations, OptForce can design strains with higher yields (e.g., up to 100% of maximum theoretical yield for succinate) at any growth rate, compared to strategies found with OptKnock.

OptForce was successfully used to develop a fatty acid overproducing strain.[78] Fatty acids can be converted to diesel replacements using a number of organic and inorganic methods and accordingly have become a target for metabolic engineering. An important precursor for fatty acid biosynthesis is malonyl-CoA, which contributes two carbons to the growing alkane chain during each round of the elongation cycle. OptForce was used to determine reaction deletions, upregulations and downregulations to increase malonyl-CoA production in *E. coli*. These modifications were implemented experimentally and shown to increase intracellular malonyl-CoA levels up to fourfold.[102] To overproduce fatty acids, a fatty acid degradation enzyme (FadD) was deleted, an elongation enzyme (FabZ) was upregulated, and a medium chain thioesterase (which cleaves a growing fatty acid chain from its carrier protein) was inserted, which was enough to increase fatty acid production by 3.5-fold.[78]

8.6.3 Evaluation of S. cerevisiae for Ethanol Production

S. cerevisiae naturally produces ethanol at high yields from glucose and exhibits high ethanol tolerance. These properties are why *S. cerevisiae* is used in fermentations to produce ethanol. However, xylose, a predominant sugar in lignocellulosic biomass, is not readily converted into ethanol. Engineering efforts in *S. cerevisiae* have focused on improving sugar utilization and ethanol production. Constraint-based metabolic models of *S. cerevisiae* have been used with transcriptomic data sets and regulatory models to describe fluxes under different conditions and predict responses to genetic and media perturbations.

A variety of methods have been applied in *S. cerevisiae* to account for transcriptional regulatory processes. Many of the methods incorporating transcriptomic data into constraint-based models were developed for yeast models.[38,42] Using gene expression data, Akesson et al. were able to more accurately predict biomass, ethanol, and glycerol yields from yeast grown on glucose, compared to FBA alone.[38] This study highlighted the importance of regulatory limitations in yeast. The MADE method was used to describe the regulatory and metabolic shifts that occur after switching yeast from glucose-rich media to glycerol-rich media.[42] The results of MADE indicate that it may be possible to identify how regulatory processes limit catabolism of different carbon sources, which may be important for biofuel production from lignocellulosic biomass which is made up of complex mixtures of substrates.

Due to the importance of regulatory effects on yeast metabolism, a genome-scale yeast regulatory model has been developed that accounts for interactions between transcription factors and their metabolic gene targets.[103] The model predicted the effects of transcription factor deletions and upregulations on mRNA levels and growth phenotypes. Smaller regulatory and metabolic network models have been combined to simulate the response of yeast to osmotic stress.[104] Regulatory models could be developed to represent additional stresses, caused by plant hydrolysate and biofuels, which are common during biofuel production. Since regulatory network modifications are commonly used in metabolic engineering,[44–47] regulatory models may be useful for suggesting new strategies to improve biofuel production.[109]

Constraint-based models have been used to identify genetic perturbations to improve ethanol production from glucose and xylose. To improve ethanol yields, Bro et al. computationally identified a heterologous enzyme (E.C. 1.2.1.9) that produces an NADPH instead of an NADH as part of glycolysis. Expressing this enzyme-increased ethanol production and decreased glycerol production experimentally.[105] The importance of cofactor (NAD(P)H) balancing in xylose catabolism was also investigated computationally as a strategy to improve ethanol yields and *in silico* results were consistent with experimental observations of increased ethanol yields.[106] Reengineering enzymes to use different cofactors is challenging, and constraint-based modeling can successfully identify what enzymes should be reengineered to increase product yields.[107]

Dynamic models of batch and fed batch cultures have also been used to model ethanol production in yeast, since they can account for the effects of media conditions on metabolic fluxes. These dynamic models use extracellular glucose and oxygen concentrations and kinetics for glucose and oxygen uptake to constrain the respective uptake fluxes in a yeast constraint-based model. Studies found complex relationships between choice of heterologous enzymes, fed batch operating parameters, and media on ethanol productivity and yield.[59,60] Imbedding dynamic models in a bilevel optimization algorithm computationally is challenging, but enables optimal fine tuning of fed batch operating parameters.[108] A dynamic bilevel approach is similar in concept to bilevel strain design methods, but uses fermentation operating parameters as variables instead of genetic modifications. Dynamic FBA will be important for continued efforts to improve yeast strains for biofuel production.

8.7 Concluding Remarks

This chapter reviewed genome-scale metabolic reconstructions, constraint-based modeling, and their applications in metabolic engineering of microbes for biofuel and biochemical production. Constraint-based modeling methods can incorporate various types of experimental data to describe metabolic fluxes in parental and engineered strains. Additional constraint-based methods can predict the effects of genetic perturbations (deletions, additions, and up/downregulations) to identify metabolic interventions that would improve uptake or secretion rates. As discussed, these modeling approaches aided metabolic engineering efforts in a variety of organisms to produce biofuels and other valuable chemicals. Metabolic models provide a unique integrated view of the cell that is critical to understanding and improving biofuel production by microbes.

Acknowledgments

This work was funded by the US Department of Energy Great Lakes Bioenergy Research Center (DOE BER Office of Science DE-FC02-07ER64494) and the Keck Foundation.

References

1. Thiele I, Palsson BØ. A protocol for generating a high-quality genome-scale metabolic reconstruction. *Nat Protoc* 2010;**5**:93–121.
2. Orth JD, Conrad TM, Na J, Lerman JA, Nam H, Feist AM, et al. A comprehensive genome-scale reconstruction of Escherichia coli metabolism—2011. *Mol Syst Biol* 2011;**7**:535.
3. Mo ML, Palsson BO, Herrgard MJ. Connecting extracellular metabolomic measurements to intracellular flux states in yeast. *BMC Syst Biol* 2009;**3**:37.
4. Lee J, Yun H, Feist AM, Palsson BO, Lee SY. Genome-scale reconstruction and in silico analysis of the Clostridium acetobutylicum ATCC 824 metabolic network. *Appl Microbiol Biotechnol* 2008;**80**:849–62.
5. Senger RS, Papoutsakis ET. Genome-scale model for Clostridium acetobutylicum: part I. Metabolic network resolution and analysis. *Biotechnol Bioeng* 2008;**101**:1036–52.
6. Kjeldsen KR, Nielsen J. In silico genome-scale reconstruction and validation of the Corynebacterium glutamicum metabolic network. *Biotechnol Bioeng* 2009;**102**:583–97.
7. Shinfuku Y, Sorpitiporn N, Sono M, Furusawa C, Hirasawa T, Shimizu H. Development and experimental verification of a genome-scale metabolic model for Corynebacterium glutamicum. *Microb Cell Fact* 2009;**8**:43.
8. Lee KY, Park JM, Kim TY, Yun H, Lee SY. The genome-scale metabolic network analysis of Zymomonas mobilis ZM4 explains physiological features and suggests ethanol and succinic acid production strategies. *Microb Cell Fact* 2010;**9**:94.
9. Widiastuti H, Kim JY, Selvarasu S, Karimi IA, Kim H, Seo JS, et al. Genome-scale modeling and in silico analysis of ethanologenic bacteria Zymomonas mobilis. *Biotechnol Bioeng* 2011;**108**:655–65.
10. Systems Biology Research Group. *Available predictive genome-scale metabolic network reconstructions*; 2009. Available from: http://systemsbiology.ucsd.edu/InSilicoOrganisms/OtherOrganisms [7/17/2014].
11. Li C, Donizelli M, Rodriguez N, Dharuri H, Endler L, Chelliah V, et al. BioModels Database: an enhanced, curated and annotated resource for published quantitative kinetic models. *BMC Syst Biol* 2010;**4**:92.
12. Hamilton JJ, Reed JL. Software platforms to facilitate reconstructing genome-scale metabolic networks. *Environ Microbiol* 2014;**16**:49–59.
13. Caspi R, Altman T, Billington R, Dreher K, Foerster H, Fulcher CA, et al. The MetaCyc Database of metabolic pathways and enzymes and the BioCyc collection of Pathway/Genome Databases. *Nucleic Acids Res* 2014;**42**:D459–71.
14. Schellenberger J, Park JO, Conrad TM, Palsson BØ. BiGG: a Biochemical Genetic and Genomic knowledgebase of large scale metabolic reconstructions. *BMC Bioinform* 2010;**11**:213.
15. Henry CS, DeJongh M, Best AA, Frybarger PM, Linsay B, Stevens RL. High-throughput generation, optimization and analysis of genome-scale metabolic models. *Nat Biotechnol* 2010;**28**:977–82.
16. Kanehisa M, Goto S, Sato Y, Kawashima M, Furumichi M, Tanabe M. Data, information, knowledge and principle: back to metabolism in KEGG. *Nucleic Acids Res* 2014;**42**:D199–205.
17. Karp PD, Paley SM, Krummenacker M, Latendresse M, Dale JM, Lee TJ, et al. Pathway tools version 13.0: integrated software for pathway/genome informatics and systems biology. *Brief Bioinform* 2010;**11**:40–79.
18. Notebaart RA, van Enckevort FH, Francke C, Siezen RJ, Teusink B. Accelerating the reconstruction of genome-scale metabolic networks. *BMC Bioinform* 2006;**7**:296.

19. Agren R, Liu L, Shoaie S, Vongsangnak W, Nookaew I, Nielsen J. The RAVEN toolbox and its use for generating a genome-scale metabolic model for Penicillium chrysogenum. *PLoS Comput Biol* 2013;**9**, e1002980.
20. Feist AM, Henry CS, Reed JL, Krummenacker M, Joyce AR, Karp PD, et al. A genome-scale metabolic reconstruction for Escherichia coli K-12 MG1655 that accounts for 1260 ORFs and thermodynamic information. *Mol Syst Biol* 2007;**3**:121.
21. Reed JL, Patel TR, Chen KH, Joyce AR, Applebee MK, Herring CD, et al. Systems approach to refining genome annotation. *Proc Natl Acad Sci U S A* 2006;**103**:17480–4.
22. Satish Kumar V, Dasika MS, Maranas CD. Optimization based automated curation of metabolic reconstructions. *BMC Bioinform* 2007;**8**:212.
23. Thiele I, Vlassis N, Fleming RM. fastGapFill: efficient gap filling in metabolic networks. *Bioinformatics* 2014;**30**(17):2529–31.
24. Vitkin E, Shlomi T. MIRAGE: a functional genomics-based approach for metabolic network model reconstruction and its application to cyanobacteria networks. *Genome Biol* 2012;**13**:R111.
25. Schilling CH, Letscher D, Palsson BO. Theory for the systemic definition of metabolic pathways and their use in interpreting metabolic function from a pathway-oriented perspective. *J Theor Biol* 2000;**203**:229–48.
26. Stelling J, Klamt S, Bettenbrock K, Schuster S, Gilles ED. Metabolic network structure determines key aspects of functionality and regulation. *Nature* 2002;**420**:190–3.
27. de Figueiredo LF, Podhorski A, Rubio A, Kaleta C, Beasley JE, Schuster S, et al. Computing the shortest elementary flux modes in genome-scale metabolic networks. *Bioinformatics* 2009;**25**:3158–65.
28. Orth JD, Thiele I, Palsson BØ. What is flux balance analysis? *Nat Biotechnol* 2010;**28**:245–8.
29. Schuetz R, Kuepfer L, Sauer U. Systematic evaluation of objective functions for predicting intracellular fluxes in Escherichia coli. *Mol Syst Biol* 2007;**3**:119.
30. Schellenberger J, Que R, Fleming RM, Thiele I, Orth JD, Feist AM, et al. Quantitative prediction of cellular metabolism with constraint-based models: the COBRA Toolbox v2.0. *Nat Protoc* 2011;**6**:1290–307.
31. Fong SS, Palsson BO. Metabolic gene-deletion strains of Escherichia coli evolve to computationally predicted growth phenotypes. *Nat Genet* 2004;**36**:1056–8.
32. Zhuang K, Yang L, Cluett WR, Mahadevan R. Dynamic strain scanning optimization: an efficient strain design strategy for balanced yield, titer, and productivity. DySScO strategy for strain design. *BMC Biotechnol* 2013;**13**:8.
33. Vu TT, Stolyar SM, Pinchuk GE, Hill EA, Kucek LA, Brown RN, et al. Genome-scale modeling of light-driven reductant partitioning and carbon fluxes in diazotrophic unicellular cyanobacterium Cyanothece sp. ATCC 51,142. *PLoS Comput Biol* 2012;**8**:e1002460.
34. Mahadevan R, Schilling CH. The effects of alternate optimal solutions in constraint-based genome-scale metabolic models. *Metab Eng* 2003;**5**:264–76.
35. Ranganathan S, Suthers PF, Maranas CD. OptForce: an optimization procedure for identifying all genetic manipulations leading to targeted overproductions. *PLoS Comput Biol* 2010;**6**, e1000744.
36. Lewis NE, Hixson KK, Conrad TM, Lerman JA, Charusanti P, Polpitiya AD, et al. Omic data from evolved E. coli are consistent with computed optimal growth from genome-scale models. *Mol Syst Biol* 2010;**6**:390.
37. Machado D, Herrgärd M. Systematic evaluation of methods for integration of transcriptomic data into constraint-based models of metabolism. *PLoS Comput Biol* 2014;**10**, e1003580.
38. Akesson M, Forster J, Nielsen J. Integration of gene expression data into genome-scale metabolic models. *Metab Eng* 2004;**6**:285–93.
39. Becker SA, Palsson BO. Context-specific metabolic networks are consistent with experiments. *PLoS Comput Biol* 2008;**4**: e1000082.
40. Zur H, Ruppin E, Shlomi T. iMAT: an integrative metabolic analysis tool. *Bioinformatics* 2010;**26**:3140–2.
41. Shlomi T, Cabili MN, Herrgärd MJ, Palsson BØ, Ruppin E. Network-based prediction of human tissue-specific metabolism. *Nat Biotechnol* 2008;**26**:1003–10.
42. Jensen PA, Papin JA. Functional integration of a metabolic network model and expression data without arbitrary thresholding. *Bioinformatics* 2011;**27**:541–7.

43. Chandrasekaran S, Price ND. Probabilistic integrative modeling of genome-scale metabolic and regulatory networks in Escherichia coli and Mycobacterium tuberculosis. *Proc Natl Acad Sci U S A* 2010;**107**:17845–50.
44. Atsumi S, Liao JC. Metabolic engineering for advanced biofuels production from Escherichia coli. *Curr Opin Biotechnol* 2008;**19**:414–9.
45. Alper H, Moxley J, Nevoigt E, Fink GR, Stephanopoulos G. Engineering yeast transcription machinery for improved ethanol tolerance and production. *Science* 2006;**314**:1565–8.
46. Zhang H, Chong H, Ching CB, Song H, Jiang R. Engineering global transcription factor cyclic AMP receptor protein of Escherichia coli for improved 1-butanol tolerance. *Appl Microbiol Biotechnol* 2012;**94**:1107–17.
47. Zhang F, Ouellet M, Batth TS, Adams PD, Petzold CJ, Mukhopadhyay A, et al. Enhancing fatty acid production by the expression of the regulatory transcription factor FadR. *Metab Eng* 2012;**14**:653–60.
48. Jensen PA, Lutz KA, Papin JA. TIGER: toolbox for integrating genome-scale metabolic models, expression data, and transcriptional regulatory networks. *BMC Syst Biol* 2011;**5**:147.
49. Kim HU, Kim TY, Lee SY. Metabolic flux analysis and metabolic engineering of microorganisms. *Mol Biosyst* 2008;**4**:113–20.
50. Sauer U, Lasko DR, Fiaux J, Hochuli M, Glaser R, Szyperski T, et al. Metabolic flux ratio analysis of genetic and environmental modulations of Escherichia coli central carbon metabolism. *J Bacteriol* 1999;**181**:6679–88.
51. Blank LM, Kuepfer L, Sauer U. Large-scale 13C-flux analysis reveals mechanistic principles of metabolic network robustness to null mutations in yeast. *Genome Biol* 2005;**6**:R49.
52. Fischer E, Sauer U. Large-scale in vivo flux analysis shows rigidity and suboptimal performance of Bacillus subtilis metabolism. *Nat Genet* 2005;**37**:636–40.
53. McAnulty MJ, Yen JY, Freedman BG, Senger RS. Genome-scale modeling using flux ratio constraints to enable metabolic engineering of clostridial metabolism in silico. *BMC Syst Biol* 2012;**6**:42.
54. Hoppe A, Hoffmann S, Holzhutter HG. Including metabolite concentrations into flux balance analysis: thermodynamic realizability as a constraint on flux distributions in metabolic networks. *BMC Syst Biol* 2007;**1**:23.
55. Tepper N, Noor E, Amador-Noguez D, Haraldsdottir HS, Milo R, Rabinowitz J, et al. Steady-state metabolite concentrations reflect a balance between maximizing enzyme efficiency and minimizing total metabolite load. *PLoS One* 2013;**8**, e75370.
56. Hamilton JJ, Dwivedi V, Reed JL. Quantitative assessment of thermodynamic constraints on the solution space of genome-scale metabolic models. *Biophys J* 2013;**105**:512–22.
57. Jankowski MD, Henry CS, Broadbelt LJ, Hatzimanikatis V. Group contribution method for thermodynamic analysis of complex metabolic networks. *Biophys J* 2008;**95**:1487–99.
58. Mahadevan R, Edwards JS, Doyle 3rd FJ. Dynamic flux balance analysis of diauxic growth in Escherichia coli. *Biophys J* 2002;**83**:1331–40.
59. Hjersted JL, Henson MA. Steady-state and dynamic flux balance analysis of ethanol production by Saccharomyces cerevisiae. *IET Syst Biol* 2009;**3**:167–79.
60. Hjersted JL, Henson MA, Mahadevan R. Genome-scale analysis of Saccharomyces cerevisiae metabolism and ethanol production in fed-batch culture. *Biotechnol Bioeng* 2007;**97**:1190–204.
61. Smallbone K, Simeonidis E, Broomhead DS, Kell DB. Something from nothing: bridging the gap between constraint-based and kinetic modelling. *FEBS J* 2007;**274**:5576–85.
62. Smallbone K, Simeonidis E, Swainston N, Mendes P. Towards a genome-scale kinetic model of cellular metabolism. *BMC Syst Biol* 2010;**4**:6.
63. Jamshidi N, Palsson BØ. Mass action stoichiometric simulation models: incorporating kinetics and regulation into stoichiometric models. *Biophys J* 2010;**98**:175–85.
64. Schomburg I, Chang A, Placzek S, Sohngen C, Rother M, Lang M, et al. BRENDA in 2013: integrated reactions, kinetic data, enzyme function data, improved disease classification: new options and contents in BRENDA. *Nucleic Acids Res* 2013;**41**:D764–72.

65. Cotten C, Reed JL. Mechanistic analysis of multi-omics datasets to generate kinetic parameters for constraint-based metabolic models. *BMC Bioinform* 2013;**14**:32.
66. Segre D, Vitkup D, Church GM. Analysis of optimality in natural and perturbed metabolic networks. *Proc Natl Acad Sci U S A* 2002;**99**:15112–7.
67. Snitkin ES, Dudley AM, Janse DM, Wong K, Church GM, Segre D. Model-driven analysis of experimentally determined growth phenotypes for 465 yeast gene deletion mutants under 16 different conditions. *Genome Biol* 2008;**9**:R140.
68. Shlomi T, Berkman O, Ruppin E. Regulatory on/off minimization of metabolic flux changes after genetic perturbations. *Proc Natl Acad Sci U S A* 2005;**102**:7695–700.
69. Kim J, Reed JL. RELATCH: relative optimality in metabolic networks explains robust metabolic and regulatory responses to perturbations. *Genome Biol* 2012;**13**:R78.
70. Burgard AP, Pharkya P, Maranas CD. Optknock: a bilevel programming framework for identifying gene knockout strategies for microbial strain optimization. *Biotechnol Bioeng* 2003;**84**:647–57.
71. Ohno S, Shimizu H, Furusawa C. FastPros: screening of reaction knockout strategies for metabolic engineering. *Bioinformatics* 2014;**30**:981–7.
72. Lun DS, Rockwell G, Guido NJ, Baym M, Kelner JA, Berger B, et al. Large-scale identification of genetic design strategies using local search. *Mol Syst Biol* 2009;**5**:296.
73. Egen D, Lun DS. Truncated branch and bound achieves efficient constraint-based genetic design. *Bioinformatics* 2012;**28**:1619–23.
74. Kim J, Reed JL, Maravelias CT. Large-scale bi-level strain design approaches and mixed-integer programming solution techniques. *PLoS One* 2011;**6**: e24162.
75. Patil KR, Rocha I, Forster J, Nielsen J. Evolutionary programming as a platform for in silico metabolic engineering. *BMC Bioinform* 2005;**6**:308.
76. Tervo CJ, Reed JL. Expanding metabolic engineering algorithms using feasible space and shadow price constraint modules. *Metab Eng Commun* 2014;**1**:1–11.
77. Pharkya P, Burgard AP, Maranas CD. OptStrain: a computational framework for redesign of microbial production systems. *Genome Res* 2004;**14**:2367–76.
78. Ranganathan S, Tee TW, Chowdhury A, Zomorrodi AR, Yoon JM, Fu Y, et al. An integrated computational and experimental study for overproducing fatty acids in Escherichia coli. *Metab Eng* 2012;**14**:687–704.
79. Chowdhury A, Zomorrodi AR, Maranas CD. k-OptForce: integrating kinetics with flux balance analysis for strain design. *PLoS Comput Biol* 2014;**10**, e1003487.
80. Cotten C, Reed JL. Constraint-based strain design using continuous modifications (CosMos) of flux bounds finds new strategies for metabolic engineering. *Biotechnol J* 2013;**8**:595–604.
81. Rockwell G, Guido NJ, Church GM. Redirector: designing cell factories by reconstructing the metabolic objective. *PLoS Comput Biol* 2013;**9**, e1002882.
82. Choi HS, Lee SY, Kim TY, Woo HM. In silico identification of gene amplification targets for improvement of lycopene production. *Appl Environ Microbiol* 2010;**76**:3097–105.
83. Park JM, Park HM, Kim WJ, Kim HU, Kim TY, Lee SY. Flux variability scanning based on enforced objective flux for identifying gene amplification targets. *BMC Syst Biol* 2012;**6**:106.
84. Klamt S, Gilles ED. Minimal cut sets in biochemical reaction, networks. *Bioinformatics* 2004;**20**:226–34.
85. Ballerstein K, von Kamp A, Klamt S, Haus UU. Minimal cut sets in a metabolic network are elementary modes in a dual network. *Bioinformatics* 2012;**28**:381–7.
86. Hadicke O, Klamt S. Computing complex metabolic intervention strategies using constrained minimal cut sets. *Metab Eng* 2011;**13**:204–13.
87. von Kamp A, Klamt S. Enumeration of smallest intervention strategies in genome-scale metabolic networks. *PLoS Comput Biol* 2014;**10**, e1003378.
88. Trinh CT, Unrean P, Srienc F. Minimal Escherichia coli cell for the most efficient production of ethanol from hexoses and pentoses. *Appl Environ Microbiol* 2008;**74**:3634–43.
89. Trinh CT, Carlson R, Wlaschin A, Srienc F. Design, construction and performance of the most efficient biomass producing E. coli bacterium. *Metab Eng* 2006;**8**:628–38.

90. Trinh CT, Srienc F. Metabolic engineering of Escherichia coli for efficient conversion of glycerol to ethanol. *Appl Environ Microbiol* 2009;**75**:6696–705.
91. Unrean P, Trinh CT, Srienc F. Rational design and construction of an efficient E. coli for production of diapolycopendioic acid. *Metab Eng* 2010;**12**:112–22.
92. Burgard AP, Van Dien SJ. Methods and organisms for the growth-coupled production of succinate. Patent No. WO/2007/030830 (March 15, 2007).
93. Burk MJ, Van Dien SJ, Burgard AP, Niu W. Compositions and methods for the biosynthesis of 1,4-butanediol and its precursors. U.S. Patent No. 8,889,399 (Nov 18, 2014).
94. Genomatica. *Genomatica and DuPont Tate & Lyle Bio Products successfully produce 1,4-butanediol (BDO) on commercial scale*; 2013. September 8, 2014. Available from: http://www.genomatica.com/news/press-releases/successful-commercial-production-of-5-million-pounds-of-bdo/.
95. Yim H, Haselbeck R, Niu W, Pujol-Baxley C, Burgard A, Boldt J, et al. Metabolic engineering of Escherichia coli for direct production of 1,4-butanediol. *Nat Chem Biol* 2011;**7**:445–52.
96. Reed JL, Vo TD, Schilling CH, Palsson BØ. An expanded genome-scale model of Escherichia coli K-12 (iJR904 GSM/GPR). *Genome Biol* 2003;**4**:R54.
97. Burgard AP, Burk MJ, Osterhout RE, Pharkya P. Microorganisms for the production of methacrylic acid. Google Patents; 2012.
98. Burgard AP, Osterhout RE, Sun J, Pharkya P. Microorganisms for producing ethylene glycol and methods related thereto. Google Patents; 2013.
99. Burgard AP, Osterhout RE, Sun J, Pharkya P. Microorganisms for producing n-propanol 1, 3-propanediol, 1,2-propanediol or glycerol and methods related thereto. Google Patents; 2013.
100. Burk MJ, Burgard AP. Microorganisms and methods for the biosynthesis of propylene. Google Patents; 2013.
101. Ranganathan S, Maranas CD. Microbial 1-butanol production: identification of non-native production routes and in silico engineering interventions. *Biotechnol J* 2010;**5**:716–25.
102. Xu P, Ranganathan S, Fowler ZL, Maranas CD, Koffas MA. Genome-scale metabolic network modeling results in minimal interventions that cooperatively force carbon flux towards malonyl-CoA. *Metab Eng* 2011;**13**:578–87.
103. Herrgård MJ, Lee BS, Portnoy V, Palsson BO. Integrated analysis of regulatory and metabolic networks reveals novel regulatory mechanisms in Saccharomyces cerevisiae. *Genome Res* 2006;**16**:627–35.
104. Lee JM, Gianchandani EP, Eddy JA, Papin JA. Dynamic analysis of integrated signaling, metabolic, and regulatory networks. *PLoS Comput Biol* 2008;**4**, e1000086.
105. Bro C, Regenberg B, Forster J, Nielsen J. In silico aided metabolic engineering of Saccharomyces cerevisiae for improved bioethanol production. *Metab Eng* 2006;**8**:102–11.
106. Ghosh A, Zhao H, Price ND. Genome-scale consequences of cofactor balancing in engineered pentose utilization pathways in Saccharomyces cerevisiae. *PLoS One* 2011;**6**, e27316.
107. King ZA, Feist AM. Optimal cofactor swapping can increase the theoretical yield for chemical production in Escherichia coli and Saccharomyces cerevisiae. *Metab Eng* 2014;**24**:117–28.
108. Hjersted JL, Henson MA. Optimization of fed-batch Saccharomyces cerevisiae fermentation using dynamic flux balance models. *Biotechnol Prog* 2006;**22**:1239–48.
109. Kim J, Reed JL. OptORF: Optimal metabolic and regulatory perturbations for metabolic engineering of microbial strains. *BMC Syst Biol* 2010;**4**:53.

CHAPTER 9

Biotechnological Strategies for Advanced Biofuel Production: Enhancing Tolerance Phenotypes Through Genome-Scale Modifications

A.L. Halweg-Edwards*, M.C. Bassalo†, J.D. Winkler*, R.T. Gill*,‡

**Department of Chemical and Biological Engineering, University of Colorado Boulder, Boulder, Colorado, USA †Department of Molecular, Cellular, and Developmental Biology, University of Colorado Boulder, Boulder, Colorado, USA ‡Renewable and Sustainable Energy Institute, University of Colorado Boulder, Boulder, Colorado, USA*

9.1 Introduction

The United States Energy Independence and Security Act (EISA) of 2007 updated the Renewable Fuel Standard[1] (RFS2) that was initiated in 2005. RFS2 requires that petroleum refiners and importers producing and/or importing gasoline or diesel in all states (except Alaska) steadily replace petroleum-based transportation fuels with biofuels such that nearly 20% (36 billion gallons) of the transportation fuel sold in the United States by 2022 will comprise a defined proportion of renewable fuels, advanced biofuels, biomass-based diesel, and cellulosic biofuel. While congressional enthusiasm for biofuels is encouraging and undoubtedly incentivizes companies to invest in increased production capacity for advanced biofuels, the obligated parties in the transportation fuel sector have had difficulty meeting the annual requirements.

The four major production facilities, operated by DuPont, POET-DSM Advanced Biofuels, Abengoa, and BlueFire Ethanol offer the best near-term prospects for cellulosic biofuel production at a meaningful scale. However, given that the maximum combined production capacity for these facilities is ~120 million gallons/year and annual US consumption of transportation fuels is approximately 130 billion gallons/year,[2] it is unlikely that cellulosic biofuel production will reach much more than 0.1% of consumer demand by 2022. In light of the practical limits of technology development, EISA allows the Environmental Protection Agency (EPA) Administrator to reduce RFS2 requirements each year to account for appropriate infrastructure development at fueling stations and second-generation biofuel production plants

Biotechnology for Biofuel Production and Optimization. http://dx.doi.org/10.1016/B978-0-444-63475-7.00009-1

if the Administrator determines that there is inadequate domestic supply of renewable fuel.[3] Accordingly, the EPA has made striking amendments to cellulosic biofuel, advanced biofuel, and biodiesel requirements for the past 2 years. For example, the EPA retroactively reduced the cellulosic biofuel requirement for 2013 from 1 billion to 6 million gallons and is proposing to reduce the 2014 requirement from 1.75 billion to 17 million gallons.[3,4]

While it is possible that RFS2 will be repealed or significantly amended, lawmakers and the broader consumer base have recognized that nonrenewable sources of transportation fuels that rely heavily on petroleum import are not economically viable for the long term.[5,6] Similarly, the difficulty of extracting and purifying usable material from tar sands and shale oil reserves makes these alternative sources unattractive compared with "light sweet crude oil," which is no longer readily available.[5,7,8] Therefore, legislative pressure to push technology development of renewable routes for infrastructure-compatible advanced biofuels is projected to offer notable long-term economic and security gains. In the event that proposed 2014 standards are kept constant for the next several years, the Congressional Budget Office anticipates that the cost of E85 gasoline, which can be used by "flex-fuel" vehicles, will decrease by as much as 50% by 2022.[4] Similarly, the EPA predicts that carbon emissions associated with ethanol produced through enzymatic treatment and fermentation of corn stover, the material commonly left in fields following corn harvest, would be 129% lower than those associated with gasoline production if technologies expected to be available by 2022 are realized.[4]

While fueling stations and advanced biofuel production facilities continue to upgrade infrastructure over the next decade to accommodate higher ethanol fuel blends and advanced biofuels, it is clear that several grand challenges have emerged for metabolic engineers and the synthetic biology community in order to reduce biofuel production cost. First, microbial strains for industrial-scale production of advanced biofuels need to be built and optimized to improve biofuel process economics. Second, the methodological repertoire of DNA assembly techniques and multiplexed, genome-scale editing and tracking technology should be expanded and generalized to enable the design of the requisite biocatalysts needed. Third, computational tools allowing comprehensive strain design, metabolic modeling, and techno-economic analysis of biotechnological workflows need to be established in order to facilitate recursive engineering processes through an automated design-build-test framework. Among these grand challenges, we will focus this chapter on progress that has been made toward strain development and genome engineering technologies and highlight the role that toxicity tolerance phenotypes play in strain engineering.

9.2 Advanced Biofuel Candidates

First-generation biofuels, which are made directly from sugarcane juice or corn starch, have the unwanted side-effect of competing for valuable resources, including water and fertilizer, which would otherwise be used for food production. As a result, EISA has placed a cap on the amount of biofuels sold in the United States that are derived from corn starch, sugar beet, and sugarcane.[9–11] In contrast, second-generation biofuels, which are produced from lignocellulosic biomass, extend production technology to allow biofuels and other important commodity chemicals to be synthesized directly from recalcitrant plant material while not requiring additional land and resource appropriation. Nevertheless, both technological platforms involve identification and optimization of metabolic pathways leading from sugar to a target molecule. The following section centers on the various types of advanced biofuels and the progress that has been made to date for optimizing production strains.

9.2.1 Alcohols

The use of alcohol as a fuel is not a new concept. Henry Ford's first automobile, the quadricycle, was designed to run on pure ethanol, but the US tax on alcohol made ethanol a cost-prohibitive choice for the Model T and subsequent designs.[12] In contrast, the newly discovered petroleum-derived gasoline was abundant and the clear economic favorite. Compared to gasoline and diesel, alcohols targeted as advanced biofuels have a lower energy density, resulting in less energy released per unit mass of combusted fuel[13–15] (Table 9.1). However, alcohols also have a higher octane rating than gasoline and diesel,[53] meaning that they can withstand higher compression pressure before combusting, burning more slowly and allowing more energy to be extracted per unit volume. Therefore, the superior fuel economy offered by alcohols offsets the lower energy density of these compounds with respect to gasoline, making them useful fuel additives as well as advanced biofuel candidates.[16,54]

Ethanol currently stands as the dominant commercial biofuel partly due to the long history of yeast domestication for food-grade fermentation and partially due to the nearly quantitative yield of ethanol from glucose that can be achieved today. However, the relatively low energy density and high hygroscopy are major drawbacks, presenting major limitations associated with a biofuel pipeline centered on ethanol. Therefore, the community is turning to heavier, branched chain alcohols that can be produced biologically as advanced biofuel candidates. Through this section, we will discuss the microbial fermentation pathways to ethanol as well as the metabolic engineering challenges associated with producing ethanol from cellulosic biomass. Then we will discuss other advanced biofuel candidates and the progress that has been made toward designing and building microbial production strains.

Fermentation pathways to ethanol

Wild-type *Escherichia coli* uses a mixed acid fermentation process to maintain redox balance in the cell, producing a mixture of CO_2, H_2, ethanol, lactate, acetate, formate, and succinate under anoxic conditions[55] (Figure 9.1). Through the Embden-Meyerhof-Parnas (EMP) pathway (commonly "glycolysis"), 1 mol equivalent of glucose-6-phosphate is catabolized, generating 2 equivalents each of ATP, NADH, and pyruvate. Under anaerobic conditions, the citric acid cycle is downregulated leading to an incomplete oxidation of sugars and formation

Table 9.1: Biofuel production statistics. This table summarizes the properties and current production statistics for several important biofuels. Cross represents toxicity relative to the base strain.

	Fuel	Relative octane number (RON)	Energy density (MJ/L)	Titer (g/L)	Tolerance† (% v/v)	Microorganism
Petrofuels	Gasoline	90-100[16,17]	34.7[18]			
	E10 gasoline	-	33.2[19]			
	E85 gasoline	102-105	25.6[19]			
	D2 diesel	-	38.3[18]			
	Biodiesel (FAMEs)	-	33-37[18,19]			
Alcohols (biogasoline)	Ethanol	109	23.6[18]	128[20]	28[21]	*S. cerevisiae*
				60[22]	16[23]	*Z. mobilis*
				50[24,25]	5[26]	*E. coli*
	Propanol	118	27[27]	7[28]*	5[26]	*E. coli*
	Isopropanol	118	23.9[29]	143[30]**	-	*E. coli*
	1-Butanol	96[31]	29.2[32]	0.016[33]	2.0[18]	*S. cerevisiae*
				30[34]**	1.5[18]	*E. coli*
				21[35]	2[36]	*C. acetobutylicum*
	Isobutanol	113[31]	26.4[32]	50[37]**	1.5[38]	*E. coli*
				1.6[39]	-	*S. cerevisiae*
				72.7[40]**	2.5[41]	*C. glutamicum*
	3-Methylbutanol	102[42]	30.5[32]	9.5[43]**	0.1[43]	*E. coli*
Terpenes (biodiesels and jet fuels)	3-Methyl-3-butenol	102[42]		0.055[42]	-	*E. coli*
	3-Methyl-2-butenol	92[42]		0.01[42]	-	*E. coli*
	Farnesene	-	39.5	1.1[44]	1.1	*E. coli*
				0.75[44]	-	*S. cerevisiae*
	Bisabolene	-	-	0.9[45]	>20[45]	*E. coli*
				0.9[45]	>20[12]	*S. cerevisiae*
	Limonene	-	38.9[18]	0.03[46]	0.5[21]	*E. coli*
	Pinene	-	37.8[47]	0.06[21]	0.025[21]	*E. coli*
Fatty acids (biodiesels)	Medium chain (C_8-C_{18}) FAAs	-	30-35[48]	1.73[49,50]	≥0.3[51]***	*E. coli*
	Alkanes (C_{10}-C_{15})	-	47 (Pentadecane)	0.5[52]	-	*E. coli*

†Toxicity to the base strain (prior to engineering enhanced tolerance)
*Co-production of propanol (7 g/L) with ethanol (31 g/L)
**High titers were achieved through continuous product removal during biofuel production.
***Endogenous levels measured during production.

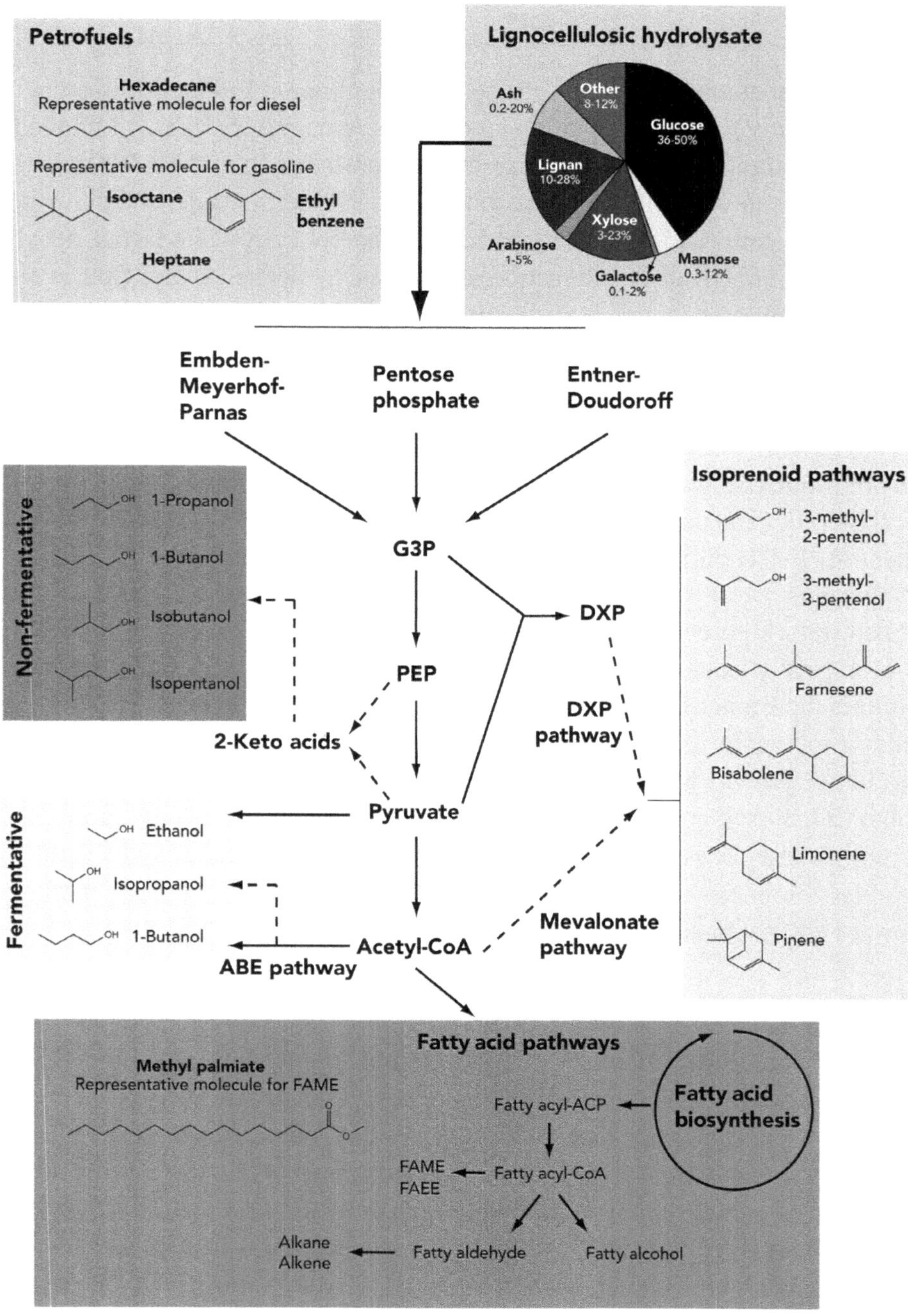

Figure 9.1

Metabolic pathways to advanced biofuels. The chemical structures of representative petroleum-derived fuels are presented in the top left orange box. The composition of lignocellulosic biomass as reported by Chandrakant and Bisaria[56] is shown as a pie chart in the upper right green box. Pentose and hexose sugars are processed by microorganisms through glycolysis (Embden-Meyerhof-Parnas pathway), the pentose phosphate pathway, and/or the Entner-Doudoroff pathway, resulting in a variety of fermentative and nonfermentative products depending on the organism. *E. coli*, *S. cerevisiae*, *Z. mobilis*, *C. acetobutylicum*, *C. glutamicum*, and other microbes have been engineered to produce a variety of advanced biofuels using cellulosic hydrolysate as feedstock.

of mixed-acid fermentation products.[57] Phosphoenolpyruvate (PEP) is processed by PEP-carboxylase (*ppc*), malate dehydrogenase (*mdh*), fumarase (*fumB*), and fumarate reductase (*frd*) resulting in succinate formation and consumption of 2 NADH equivalents. Alternatively, pyruvate kinase (*pyk*) dephosphorylates PEP to pyruvate, and pyruvate formate lyase (*pflB*) shunts pathway flux toward formate and acetyl-CoA while lactate dehydrogenase (*ldhA*) consumes 1 NADH during reduction of pyruvate to lactate. Acetyl-CoA is converted to acetate by phosphotransacetylase (*pta*) and acetate kinase (*ack*). Acetyl-CoA is also used by alcohol dehydrogenase (*adhE*), which consumes 2 equivalents of NADH during production of ethanol.

In contrast to *E. coli*, yeast and some ethanologenic bacteria ferment sugars through the homoethanologenic pathway. Following 1 molecule of pyruvate formation, pyruvate decarboxylase (*pdc*) and an alcohol dehydrogenase isozyme (*adhA* or *adhB*) consumes 1 equivalent of NADH during production of 1 equivalent of ethanol. Brewer's yeast, and other ethanologenic strains including *Saccharomyces cerevisiae*, process glucose through the EMP pathway (Figure 9.1). In contrast, the proteobacterium *Zymomonas mobilis*, which was originally isolated as a contaminant in alcoholic beverages including beer[58] and cider,[23] utilizes the Entner-Doudoroff (ED) pathway to convert 1 molecule of glucose into 2 molecules of pyruvate, and 1 molecule each of ATP and NADH (Figure 9.1). Therefore, organisms like *Z. mobilis* and *S. cerevisiae* perform a nearly quantitative conversion of glucose into CO_2 and ethanol.[59,60] Moreover, because the ED pathway only produces 1 mol ATP per mol glucose, *Z. mobilis* maintains a relatively high flux through this pathway, resulting in less biomass production than observed in yeast or *E. coli* while shunting more carbon toward ethanol biosynthesis.

Ethanol production in S. cerevisiae *and* Z. mobilis

The naturally ethanologenic microbes *S. cerevisiae* and *Z. mobilis* have historically been the preferred workhorses of the commercial alcohol industry. As of 2012, it was estimated that ~6.3 billion gallons of fuel-grade bioethanol were produced from Brazilian manufactures alone[61] through fermentation of sugarcane by selected strains of *S. cerevisiae*. Yeast offer the added advantage of growth at low pH, allowing cells to withstand washes with dilute sulfuric acid (pH 2.0-2.5) every 11-12 h, allowing fermentation tanks to be cleaned and providing a nice selection against contaminating bacteria.[62,63] Similarly, companies like DuPont have partnered with the National Renewable Energy Laboratory (NREL)[64] to engineer pentose sugar utilization pathways in *Z. mobilis* aiming to improve fermentation of cellulosic hydrolysate into ethanol. Wild-type *Z. mobilis* strains can produce up to 12% (w/v) ethanol at a rate nearly 2.5-fold higher than its industrial counterpart *S. cerevisiae*[65–67] and tolerates up to 120 g/L ethanol. While *S. cerevisiae* and *Z. mobilis* have been targeted for more than three decades as platform strains for cellulosic ethanol production, both microbial chassis strains present similar challenges for industrial microbiologists and each offers slightly different advantages.

The brewer's yeast, *S. cerevisiae*, can withstand fairly narrow range of growth temperatures (30-35 °C). In the 1980s, biotechnologists began selecting for yeast strains that produce ethanol while growing at elevated temperatures in order to cut the cost of cooling fermentation tanks.[68] In one example, a strain that converted 15% glucose in the media to 6.4% ethanol at 40 °C was isolated.[68] Unfortunately, these strains were easily overtaken by "wild" yeast that accompanied sugarcane juice into the fermentation tanks.[62,69] Therefore, during the 1990s, Brazilian manufactures selected wild yeast strains that preferentially thrive under industrial fermentation conditions while also maintaining high ethanol production efficiency. The PE-2 strain, which grows well at 42 °C, emerged as the most persistent and strongest ethanol producer, achieving a 92% theoretical yield,[63] and, as of 2009, was cited as being used by ~30% of Brazilian distilleries accounting for 10% of global bioethanol production.[70]

While yeast are extremely efficient at fermenting glucose and various other hexose sugars, these microbes cannot readily process pentose sugars. Currently, sugarcane accounts for ~60% of bioethanol production cost in Brazil.[71] Similarly, the cost of bioethanol in the United States is tightly coupled to the cost of corn, where 36% of US grown corn was sold to distilleries in 2014.[72] In contrast, lignocellulosic biomass is abundant and a common waste-product of industrial agriculture,[73] making it a highly cost-effective alternative to sugarcane and corn starch as a biofuel feedstock. Osborn and Chen[74,75] demonstrated that this material can be readily hydrolyzed by dilute acids and enzymes to produce a mixture of mono- and di-saccharides. Glucose typically makes up 35-50% of lignocellulosic composition, while xylose, a pentose sugar that is not fermentable by brewer's yeast, is generally present at 15-20% (Figure 9.1). Mannose, galactose, and arabinose constitute ~5% of the remaining sugars while the non-carbohydrates, lignan and ash, make up the remaining 15-40% of lignocellulosic biomass.[56] Consequently, one major challenge for engineers over the past two decades has been to develop yeast strains that can metabolize the pentose sugars of cellulosic hydrolysate.

US manufacturers and researchers have applied a range of engineering methodology to improve xylose uptake and metabolism in *S. cerevisiae*.[76] The most common approach has been to implant the xylose transportation and metabolism machinery from microbes that naturally ferment xylose into *S. cerevisiae*.[76] Two heterologous xylose-assimilating pathways are frequently used for xylose uptake and connection to the pentose phosphate pathway: Xylose isomerase[77–81] (*xylA*) or xylose reductase (*XR*) plus xylitol dehydrogenase[82–84] (*XDH*). Both pathways require overexpression of xylulokinase (*XK*) which phosphorylates xylulose to xylulose-5P, a substrate for the pentose phosphate pathway. As with sugarcane fermentation, researchers have observed that industrial strains, such as PE-2, perform better than their laboratory counterparts when expressing the xylose-assimilating pathway XR/XDH/XK, as the industrial strains showed both faster xylose consumption rates and higher ethanol yields.[85]

As with *S. cerevisiae*, application of *Z. mobilis* for industrial ethanol production has been hampered by the relatively low variety of carbon sources that this organism is capable of fermenting, namely, glucose, fructose, and sucrose.[59] Xylose and arabinose transport and metabolic pathways in *Z. mobilis* were developed over the past 3 decades using recombinant DNA cloning techniques, standard mutagenesis reagents,[86] transposon mutagenesis,[87–89] and adaptive laboratory evolution experiments.[90–94] In the mid-1990s, researchers at the NREL engineered variants of *Z. mobilis* capable of utilizing xylose as a carbon source[95] by expressing xylose assimilation and pentose phosphate pathway enzymes from *E. coli*, resulting in a strain that fermented xylose to ethanol at 86% theoretical yield. Later, Zhang et al. engineered an arabinose-utilizing variant by expressing the arabinose metabolic machinery from *E. coli*, resulting in a strain that produced ethanol at 98% theoretical yield.[96] Finally, all seven genes required for pentose sugar utilization were placed on the *Z. mobilis* genome, producing strain AX101, which could utilize both pentose and hexose sugars for ethanol production.[97] The utility of *Z. mobilis* as a production chassis has been demonstrated by DuPont in a pilot-scale biorefinery plant that utilized this ethanologen to produce 680 gallons of cellulosic ethanol per acre of fermented switchgrass.[98] Moreover, *Z. mobilis* has been engineered to produce other valuable compounds including sorbitol, bionic acid, levan, succinic acid, isobutanol, and a variety of isoprenoids.[99]

Ethanol production in E. coli

Though *E. coli* lacks the native ability to produce ethanol as the primary fermentation product, it does have the ability to ferment both hexose and pentose sugars. Moreover, the variety of biotechnological applications stemming from each of *E. coli*'s fermentation products,[75,100,101] the lack of requirement for complex growth factors, and the long history of industrial use of *E. coli* as a production strain for recombinant proteins has resulted in perhaps the largest genetic engineering toolbox that exists for any one organism.[102] Therefore, *E. coli* has been targeted as an industrial chassis strain for over three decades (the reader is referred to extensive reviews[24,103–106] on the topic). The construction of *E. coli* strains that selectively produce ethanol was pioneered by Lonnie Ingram in the 1980s and represents one of the first successful applications of metabolic engineering.[107]

Sugar fermentation in *E. coli* typically results in a 2:1:1 molar ratio of formate-ethanol-acetate, depending on the relative concentration of *ldhA*, *pfl*, and *ppc*.[55,57] Notably, Ingram and coworkers have shown that the fermentation pathways utilized by *E. coli* are less important for cell's survival than the ultimate reestablishment of redox balance. Consequently, the cell's glucose fermentation pathways can almost entirely be collapsed into one major vein leading to ethanol by heterologously expressing the ethanologenic pathway from *Z. mobilis*. In 1987, Ingram et al. constructed an inducible expression vector, pLOI297, comprising *pdc* and *adhB* from *Z. mobilis* under control of the *lac* promoter on the pUC18 backbone.[107] Because *pdc* K_M for pyruvate is 5- to 18-fold lower than *pfl* and *ldhA*, respectively, *pdc* effectively competes for pyruvate consumption.[108] Therefore, even though no genetic modifications were made to the parent *E. coli* strain harboring this expression vector, 43 g/L ethanol was produced, representing >95% of the fermentation products. Because plasmids vary in copy number and are easily selected out of the population, Ingram and coworkers created a strain with greater genetic stability by directly editing the genome of *E. coli* strain W (ATCC 9637).[24,109,110] The chromosomal copy of *pfl* was replaced with the promoterless *Z. mobilis* ethanol biosynthetic cassette using homologous recombination,[111] and the wild-type *fdr* was replaced with an inactive mutant using Tn*10* transposition. The resulting ethanologenic strain, KO11 (ATCC 55124), produced 53 and 42 g/L ethanol from glucose and xylose, respectively, representing 74% and 83% theoretical yield.[109,112] The inherent utility of this *E. coli* strain as a fermentation platform is underscored by the demonstration that KO11 can convert a wide range of cellulosic hydrolysate substrates into ethanol at nearly 100% theoretical yield.[75,100,101]

Advanced biofuels

While ethanol can be used as a liquid fuel, current fueling stations as well as the majority of combustion engines, shipping, and aviation vehicles cannot support fuel blends containing >10% ethanol.[113] Therefore, pathways that produce biofuels with higher energy densities, better combustion efficiencies, lower recovery costs, and are compatible with existing liquid fuel infrastructures need to be optimized for cost-competitive production.[114] These advanced biofuel candidates (Table 9.1) include the heavier branched-chain alcohols: butanol (1-butanol, *n*-butanol), isobutanol (2-methyl-1-propanol), isopentanol (3-methyl-1-butanol), and 2-methyl-1-butanol; the isoprenoid-derived terpenes: isopentanol (3-methyl-3-buten-1-ol; 3-methyl-2-buten-1-ol), farnesene, and bisabolene; and fatty acid ethyl esters (biodiesel). These "drop-in" biofuels provide similar performance as petroleum-based fuels, and are therefore compatible with existing liquid fuel infrastructures. For these reasons, biological production of these advanced biofuel candidates has been attempted for the past decade using fermentation pathways from organisms that naturally produce some of these fuels, such as Clostridia, as well as nonfermentative pathways in *S. cerevisiae*, *E. coli*, and other organisms.

Higher alcohols: Acetone-butanol-ethanol fermentation pathway

Clostrida are obligately anaerobic bacteria that produce acetone, butanol, and ethanol in a 3:6:1 ratio using the eponymous acetone-butanol-ethanol (ABE) fermentation pathway[115] (Figure 9.1). During the first part of the twentieth century, ABE fermentation in Clostrida was exploited for production of acetone,[115] which was used to produce cordite, a low-grade explosive that served as a replacement for gunpowder during World War I (WWI). Today, industrial microbiologists recognize solventogenic Clostridia for their potential to be good producers of butanol,[116–118] which can be used as a "drop-in" biofuel. Compared to ethanol, butanol offers ~25% more energy per liter than ethanol[10] and has superior combustion characteristics, engine performance, and exhaust emissions in both spark- and compression-ignition engines.[31] Moreover, butanol is less hygroscopic than ethanol, providing better tolerance to water contamination in the vehicle fuel systems and pipelines, allowing current combustion engines and fueling stations to handle gasoline blends containing up to 85% butanol. A major advantage to butanol production in Clostrida is the wide range of carbon sources fermentable by these microbes, which include glucose, xylose, arabinose, galactose, and cellobiose.[119–121]

As with *E. coli* and *S. cerevisiae*, Clostridia employ the EMP pathway to break hexose sugars into pyruvate, ATP, and NADH, and like *E. coli*, Clostridia use the pentose phosphate pathway to break pentose sugars into glycerol-3-phosphate, which can be fed back into EMP pathway for pyruvate and ATP generation. Following pyruvate formation, clostridial solvent production occurs through biphasic fermentation comprised of acidogenic and solventogenic phases. Acetate and butyrate (along with H_2 and CO_2) are produced from pyruvate during the acidogenic phase under exponential growth conditions,[122,123] and these precursors are reassimilated into acetone, butanol, and ethanol during the solventogenic phase, which occurs during stationary growth. Although counterintuitive, knocking-out or knocking-down flux through acetone production does not selectively increase flux to 1-butanol, and this strategy results in increased acid production as well as reduced 1-butanol formation.[124–126]

An alternative strategy has been to produce butanol from clostridia that naturally produce high titers of butyrate by overexpressing aldehyde and alcohol dehydrogenases. In one example, the *Clostridium acetobutylicum adhE2* gene was overexpressed in the high butyrate-producing strain, *C. tyrobutyricum*, resulting in 21 g/L of butanol.[35] A similar strategy involves decreasing flux through acidogenic pathways while overexpressing *adhE*.[35] Using this approach, *C. acetobutylicum* ATCC 824 was successfully converted to a strain producing 19 g/L of 1-butanol by decreasing flux toward acetate and butyrate by knocking out phosphotransacetylase (*pta*) and butyrate kinase (*buk*), respectively, and overexpressing an engineered variant of butyraldehyde dehydrogenase (*adhE*), which utilizes NADPH instead of NADH.[127] For Clostridia, ~20 g/L of butanol represents the toxicity limit, and recent efforts have focused on increasing tolerance to acids and solvents in butanol-producing strain.[34]

Although there has been significant progress toward engineering Clostridia for selective 1-butanol production, the difficulty of performing genetic manipulations in these microbes as well as the biphasic nature of solventogenesis has resulted in rather slow strain improvements. In some cases, engineers have turned to importing the ABE pathway from Clostridia to more genetically malleable organisms like *E. coli*, resulting in strains that produce up to 15 g/L (88% theoretical maximum) of 1-butanol in batch fermentation.[128–130]

Nonfermentative routes to branched chain alcohols

An alternative approach to taking advantage of the natural fermentation pathways utilized by microbes is to engineer nonfermentative pathways for the production of short-chain alcohols as has been demonstrated over the past decade in efforts spearheaded by James Liao's group. The logic behind this strategy is centered on the fact that 2-keto acids, which are metabolic intermediates of amino acid metabolism, can be decarboxylated to 1-aldehydes that are then converted to the corresponding alcohols via alcohol dehydrogenase (Figure 9.1). Various branched and linear chain alcohols have been produced through modification of this pathway including 1-propanol,[131,132] 1-butanol,[132] isobutanol,[37,38,133] 2-methyl-1-butanol,[134] and 3-methyl-1-butanol.[43] Moreover, the chain length of 2-keto acids can be extended by shunting intermediates through the leucine biosynthetic pathway, *leuABCD*, which adds one carbon every cycle,[135] and this approach has lead to the production of various 6-carbon alcohols (4-methyl-1-pentanol, 3-methyl-1-pentanol, and 1-hexanol).[135]

A key commonality of these engineered pathways is the utilization of the promiscuous *Lactococcus lactis* 2-ketoacid dehydrogenase (*kivD*) and *S. cerevisiae* alcohol dehydrogenase (*adh2*).[38] Engineering strategies involving these enzymes have been successfully employed in a wide range of organisms including *E. coli*,[38] *S. cerevisiae*,[136] *Bacillus subtilis*,[137,138] *Clostridium cellulolyticum*,[139] *Corynebacterium glutamicum*,[41,140] and *Ralstonia eutropha*,[141] demonstrating the utility of this approach. In the seminal report by Atsumi et al., classical recombinant DNA technology coupled with rational metabolic pathway designs resulted in 9.3 g/L (86% theoretical yield) isobutanol production after 24 h in *E. coli*. This result was later improved to 13.4 g/L isobutanol produced in 24 h (103% theoretical yield) by reversing the cofactor specificity of *adh2* from NADPH to NADH through rational protein engineering,[142] thereby relieving the cofactor imbalance created by producing isobutanol under anaerobic conditions.

Aside from utilization of the leucine/isoleucine biosynthetic pathways, 2-ketobutyrate, a critical 2-ketoacid intermediate for the production of 1-propanol, 1-butanol, 2-methyl-1-butanol, and 3-methyl-1-pentanol, can be produced to a high titer through the threonine degradation pathway (Figure 9.1). Overexpression of threonine dehydratase (*ilvA*) and citramalate synthase (*cimA*) along with deletion of enzymes that compete for pyruvate (*ilvI*, *ilvH*, *ilvB*, and *ilvN*) results in accumulation of 2-ketobutyrate. Under this genetic background, the endogenous acetate kinase/propionate kinase II (*ackA*), acetyl-CoA:acetoacetyl-CoA synthase (*atoDA*), and an aerobically functional mutant alcohol/aldehyde dehydrogenase (*adhE*^{mut}) were overexpressed, resulting in the production of >10 g/L 1-propanol.[131] Because the octane number of a fuel tends to increase with the degree of saturation and chain branching,[128–130] molecules like isobutanol offer better fuel economy than their straight-chain counterparts,[31,143] suggesting that nonfermentative pathways to biofuels likely have more utility than naturally occurring fermentation pathways.

9.2.2 Isoprenoids

Isoprenoids, also referred to as terpenes and/or terpenoids, comprise a large (>40,000 known compounds) and structurally diverse class of molecules that are biosynthesized from assimilation of five-carbon isoprene units (Figure 9.1). Historically, isoprenoids have been valued for their fragrance and pharmaceutical properties. For example, the terpenes produced by hops (myrcene, β-pinene, β-caryophyllene, and α-humulene) are partially responsible for the enticing aroma of some beers,[144] and the antimalarial drug, artemisinin, is an isoprenoid naturally produced by the herb *Artemisia annua* (commonly referred to as sweet wormwood).[145] Because isoprenoids, like artemisinin, are highly valued in much of the world, a considerable effort has been devoted over the past two decades to engineer the common microbial workhorses *E. coli* and *S. cerevisiae* for the production of isoprenoids.[146–149] Today, some isoprenoids have gained attention due to their inherent fuel properties.

While the terms "isoprenoid, terpenoid, and terpene" are often used interchangeably, "isoprenoid" is generally an umbrella term describing all compounds derived from either of the activated forms of isoprene, isopentenyl pyrophosphate (IPP), and its isomer dimethylallyl pyrophosphate (DMAPP), or a combination of the two. "Terpenoid" refers to all isoprenoids made up of between 1 and 5 isoprene units (including those with highly modified backbones), and "terpene" refers to terpenoid olefins. The terpenoids are classified by the number of isoprene units used to make the structure's backbone such that terpenoids made from 1-, 2-, 3-, 4-, and 5-isoprene units are referred to as hemi-, mono-, sesqui-, di-, and triterpenes, respectively.[149] Two different pathways are used to generate the activated isoprene units IPP and DMAPP.[149] Through the mevalonate (MVA) pathway, which is used by all eukaryotes (including *S. cerevisiae*) and archaea, IPP is synthesized from acetyl-CoA through the MVA intermediate. In contrast, plant plastids and eubacteria, including *E. coli*, use the deoxyxylulose-5-phosphate (DXP) pathway.[149] The activated isoprenes IPP and DMAPP can be hydrolyzed or condensed in a head-to-tail orientation to produce terpenes. Both intermediates and the final terpene products have been tapped as possible advanced biofuel candidates if biosynthetic production becomes economically viable.

Microbial engineering for production of several isoprenoid-derived biofuel candidates has been achieved over the past decade (the reader is referred to several excellent reviews[15,150]), and a subset of notable examples are briefly described here. Isopentanol was produced in *E. coli* by dephosphorylating IPP using a heterologously expressed pyrophosphatase from *B. subtilis*, resulting in a strain that produced 1.3 mM isopentanol after 43 h.[151] The sesquiterpene farnesene (Biofene, Amyris Biotechnologies, Inc.) is currently being used as a jet fuel by both commercial and military airlines and can be hydrogenated to the biodiesel farnesane.[44] Both *E. coli* and *S. cerevisiae* have been engineered to produce farnesene from farnesyl pyrophosphate using a farnesene synthase, producing 1 g/L after 120 h and 750 mg/L after 72 h of growth, respectively. The hydrogenation of another sesquiterpene, bisabolene, has been shown to be superior to D2 diesel,[45] the fuel used by conventional diesel-powered vehicles. Both *E. coli* and *S. cerevisiae* have been engineered to produce bisabolene by heterologous expression of the *Abies grandis* (fir tree) bisabolene synthase, each producing 900 mg/L bisabolene after 73 h of growth.[45] The monoterpenes pinene, sabinene, terpinene, and limonene have also been touted as next-generation jet fuels.[152,153] Both pinene and limonene have been produced in *S. cerevisiae*[154] and *E. coli*[155] at <5 mg/g Dry Cell Weight (DCW).

9.2.3 Fatty Acids

Biodiesel, defined as mono-alkyl esters of long-chain fatty acids derived from renewable feedstock, such as vegetable oil, animal fats, algae, etc., is considered as a possible substitute of conventional diesel fuel. Biodiesel usually consists of fatty acid methyl/ethyl esters obtained from triglycerides by transesterification with methanol/ethanol, respectively.[156,157] Notably, biodiesel has many similar properties to petroleum-derived diesel and can blend with conventional diesel in any proportion. For example, the power output from biodiesel is nearly identical to conventional diesel, and the CO, CO_2, and total hydrocarbon emissions from biodiesel are reduced in comparison to conventional diesel because of the oxygen content which leads to more complete combustion.[158,159] Moreover, biodiesel has superior safety properties with respect to D2 diesel fuel because of the comparatively low sulfur content as well as the significantly higher flash point temperatures (~ 130 °C for generic biodiesel compared with ~60 °C for petroleum-derived diesel).

In contrast to cellulosic biofuels, US biodiesel production reached nearly 1.8 billion gallons in 2013, far exceeding RFS2 mandates and allowing some facilities to offset their advanced biofuel requirements with their biodiesel production surplus.[3] To date, biofuels suitable for internal combustion engines such as diesel engines are derived from soybean oil and "yellow grease" (recycled cooking oils).[160] Typically, C_{16}-C_{18} fatty acid methyl esters (FAMEs) are formed from the transesterification of vegetable oil using either chemical (acidic or alkali conditions) or enzymatic treatment with lipases.[161] Although FAMEs are a good surrogate for diesel fuel,[162] drawbacks include a relatively high melting point, predisposition to biofouling,[163] and the corrosive nature of acidic oxidation productions[164] (i.e., formic, acetic, propionic, palmitic, stearic, and oleic acids). Second-generation biodiesel is being produced through a combined hydrogenation and deoxygenation processes, such as the NExBTL process, yielding the corresponding alkanes or paraffins directly from vegetable oils or animal fats.[165] Because of the nature of the starting material, both methods yield a complex and heterogeneous mixture of products that may vary from batch to batch. This product variability can complicate making a fuel with defined specifications or requirements. Moreover, used cooking oils are also widely used for the production of soaps, cosmetics, rubber, detergents, and animal feedstocks, resulting in sharp price inflation for yellow grease. As a result, there are needs for alternative methods for production of diesel fuel components which can be made purely, reliably, and reproducibly.

Phosphorylated triacylglycerols are a major component of cell membranes, constituting ~30-70% of microbial DCW.[166] Therefore, an increasingly common alternative to using waste oil products involves microbial production of fatty acid-derived biofuels. This biosynthetic route offers the advantage of bypassing triacylglycerol formation (thus eliminating downstream chemical or enzymatic processing) by directly converting fatty acids to the desired fuels *in vivo*. Three similar and phylogenetically distinct biosynthetic pathways exist for fatty acid biogenesis. The mammalian fatty acid synthase (mFAS) is an iterative, modular polyketide synthase (PKS) that is structurally and phylogenetically related[167,168] to the noniterative, multimodular PKSs found in soil bacteria that synthesize high-valued compounds used as antibiotics and chemotherapies. Unlike prototypical PKSs, mFAS fully reduces the polymerized acyl chain, producing C_{16}-C_{18} fatty acids. In contrast with mFAS, microbial fatty acid synthases are phylogenetically distinct from PKSs and divided into two types based on sequence and structural characteristics. Type I FASs, which are found in fungi and some bacteria, are extremely large barrel-shaped protein complexes (~2 MDa) comprised of unique α and β subunits that form $\alpha_6\beta_6$ hexameric complexes.[169–171] Type II FASs, like those found in *E. coli*, are comprised of 10 distinct and dissociated enzymes (FabA, FabB, FabD, FabF, FabG, FabH, FabI, FabZ, ACP, and TesA). Both type I and type II FASs systems biosynthesize C_{14}-C_{18} fatty acids through polymerization of acetyl and malonyl units derived from acetyl-CoA and malonyl-CoA, respectively. Malonyl-CoA is produced from acetyl-CoA through the action of acetyl-CoA carboxylase, an ATP-utilizing enzyme complex that is required in both type I and type II FAS systems.[172,173]

Due in part to the structural characteristics of the *E. coli* FAS system, the majority of fatty acid-derived biofuel engineering efforts has focused on this chassis organism. Downregulation of fatty acid degradation pathways along with overexpression of thioesterases, which catalyze acyl chain release from the acyl carrier protein (ACP), causes dysregulation of fatty acid biosynthesis, allowing accumulation of free fatty acids of short and medium chain lengths (C_8-C_{18}).[174] The resulting fatty acids can be converted to the corresponding alcohols by overexpressing fatty acyl-CoA reductase[174,175] along with an acyl-CoA reductase, yielding as much as 1.73 g/L fatty alcohols.[49,50] Although alcohols and FAMEs are good substitutes for petroleum-derived fuels, conventional fuels are comprised of unmodified alkanes and alkenes. Recently, biosynthetic pathways for production of alkanes have been discovered in cyanobacteria.[176] Free fatty acyl-ACPs are reduced, and corresponding alkanes and alkenes are produced following removal of the terminal carbonyl groups from the fatty aldehydes. Expression of the cyanobacterial alkane operon in *E. coli* successfully yields recombinant strains that can produce medium length alkanes (C_{13}-C_{17}) in mixtures >300 mg/L that are dominated by pentadecane and heptadecene.[176] Recently, mixtures of short alkanes (gasoline compounds) up to 580 mg/L have been produced in *E. coli* by overexpressing the *C. acetobutylicum* acyl-CoA reductase, the *Arabidopsis thaliana* fatty aldehyde decarbonylase, and a modified thioesterase domain in an *E. coli* strain with downregulated fatty acid degradation and unsaturated fatty acyl-ACP biosynthetic pathways.[52]

9.3 Microbial Production Limitations

The cost competitiveness of biofuel production processes depends on several factors including biomass cost, pretreatment ($0.30-0.40/gal), production plant maintenance costs (including heating/cooling fermentation chambers), product titer, and product separation

(\$0.17-0.20/gal).[114] Final product titer is a critical cost determinant because it affects both the downstream recovery cost and defines the size and maximum production capacity of the processing plant. As a biological entity, the maximum titer from a microbial production chassis is limited by the toxicity of feedstock components, industrial fermentation conditions, and both the products and intermediates produced from the engineered pathways (i.e., high temperatures, pH fluctuations, osmotic pressure, and high solvent concentration). It is widely recognized that production titer can be increased by cross-engineering tolerance phenotypes to feedstock and biofuel compounds.[177–179]

In contrast to the first-generation biofuels (those derived from sugar or corn starch), the second-generation biofuel production is hampered by the wide range of toxic compounds found in lignocellulosic hydrolysate (Figure 9.2). Raw substrates (grasses, corn stover, wood pulp, etc.) are chemically and enzymatically pretreated to break the fibrous material down into fermentable sugars.[180] While not covered in this chapter, it is worth noting that cellulosic pretreatment is currently the most expensive processing step (next to product recovery by distillation) in cellulosic biofuel production.[181] Generally, pretreatment methods are a combination of physical agitation and chemical treatment with acid (H_2SO_4) or base

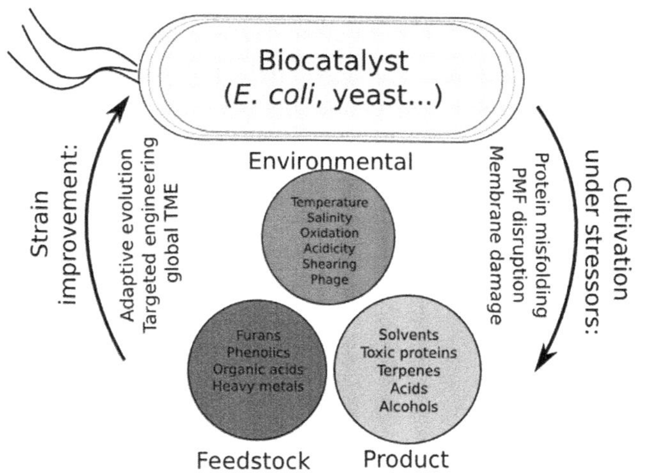

Figure 9.2

Toxicity of feedstock components and biofuels limits productivity. Microorganisms are inherently limited by their sensitivity abiotic to growth conditions (temperature and aeration conditions) as well as their susceptibility to toxic compounds including the targeted biofuel compounds and the acids, aldehydes, furans, and heavy metals found in lignocellulosic hydrolysate. Protein misfolding, PMF disruption, and membrane damage are often observed in cells that have been exposed to toxic compounds and therefore can be diagnostic of the toxicity phenotype. A variety of methods have been demonstrated for engineering enhanced tolerance to toxic growth conditions and compounds including adaptive evolution, targeted genome-engineering, and global transcription machinery engineering. These approaches have been shown to increase biofuel productivity through directly enhancing tolerance.

(ammonia or NaOH). Although, chemical treatments promote hydrolysis and improve recovery of both pentose and hexose sugars,[180] a variety of toxic by-products are also released. These include acids, aldehydes, alcohols, heavy metals, and furan derivatives.[182]

In general, the toxicity of a compound is correlated with its octanol/water partition coefficient, $\log P_{ow}$, which is a measure of hydrophobicity. Organic acids (acetic and formic) can permeate the cell membrane in their undissociated form and disrupt the membrane potential by dissociating to the anionic and proton constituents once they are inside the cell.[183] Acid dissociation in the cytoplasm also disrupts macromolecular synthesis and reduces the intracellular pools of some amino acids.[184] Aldehydes, including those derived from furan derivatives, are known to cross-link proteins and denature polynucleotides.[185] Some hydrolysate constituents such as furfural act synergistically to enhance the toxicity of other compounds and act independently to inhibit central metabolic enzymes like glycolytic dehydrogenases,[186] which are often critical for biofuel production. Finally, phenolic compounds tend to disrupt hydrophobic protein-protein interactions and are usually more toxic than aliphatics containing the same functional group because these compounds are more hydrophobic and tend to have more damaging effects on the membrane.[177]

Aside from the inhibitory constituents presented in the feedstock, toxicity of the biofuel itself can often be a limiting factor for production. As with weak acids and some phenolic compounds, most biofuels are known to cause damage to the cell membrane,[187–189] compromising the cell's structural integrity and changing the internal pH which, in turn, affects a variety of critical metabolic pathways and protein structures. Notably, a high intracellular biofuel concentration can sometimes cause feedback inhibition, resulting in lower production rate. Therefore, even in cases for which the biofuel is not toxic, engineering machinery to efficiently transport the biofuel out of the cell can decrease production cost.

9.3.1 Engineering Tolerance

Historically, engineers enhance chassis strain tolerance by selecting for strains with fortuitous genomic mutations that allow increased growth in the presence of inhibitors. For example, the yeast PE-2, which was naturally selected over years of bioethanol production in Brazilian fermentation tanks was found to have acquired higher copy numbers of vitamin B6 and B1 biosynthetic genes (*SNO* and *SNZ*), resulting in an increased tolerance to the high osmotic pressure caused by the high sugar concentration in the feedstock.[190] Similarly, Ingram and coworkers were able to extend the limits of ethanol production in KO11 by applying an alternating selection scheme to screen for randomly occurring mutants that exhibited both an increased tolerance for up to 50 g/L ethanol in the growth media and production of high ethanol titers. These strains were further improved using a 3-month evolution experiment resulted in strain LY01, which showed decreased sensitivity to toxic aldehydes and produced 61 and 63 g/L ethanol from glucose and xylose, respectively, representing 85% and 89%

theoretical yield.[112] Finally, an extended lab-directed evolution experiment was performed to enhance *Z. mobilis* 8b tolerance to feedstock compounds (furfural and acetate), resulting in a strain that showed >2-fold improvement of xylose utilization (from 40% to 86% xylose consumption) during cofermentation with glucose.[191]

While it is possible to select for naturally occurring variants that survive the given selection pressure, tools for increasing the mutation/adaptation rate are used to accelerate the generation of improved strains. Chemical mutagens and UV irradiation have been recognized for over 30 years as tools for increasing mutation frequency in a population in the context of strain improvement. However, single, randomly dispersed mutations that are generated by chemical or UV treatment have the tendency to be either silent (imparting no change) or undesirably debilitating, making the chances of finding beneficial mutations still relatively rare. In contrast, when multiple mutations are generated simultaneously, synergistic effects allow rapid evolution of tolerance phenotypes. For example, Zhang et al. showed that recursive genomic recombination, promoted through mixed protoplast fusion in a population of chemically mutated variants, of the antibiotic producing *Streptomyces fradiae* resulted in mutants that produced 6- to 8-fold more tylosin than the base strain.[192] Other examples of breeding to improve tolerance abound in the literature; in 2009, three rounds of genome shuffling were applied to a starting pool of *S. cerevisiae* SM-3 (an industrial brewer's yeast strain[193]) generating mutant libraries, which were screened for growth at 35-55 °C on agar plates containing 10-25% (v/v) ethanol. One variant, F34, was selected for further characterization and found to sustain growth in liquid culture at 45-48 °C while consuming 20% (w/v) glucose, producing 99 g/L ethanol, and tolerating 25% (v/v) ethanol in the growth media.[194] Genome shuffling (regardless of the underlying method) has also been successfully applied to improve a range of microbial tolerance phenotypes, including osmotic pressure[195] and lactic acid tolerance.[196] A conceptually related approach relying on DNA transfer via conjugative plasmids has also been explored in search of antibiotic resistance phenotypes.[197,198] These "genome shuffling" techniques have effectively recapitulated many rounds of serial mutation in a much shorter time period, underscoring the power of applied evolution.

Repeated laboratory adaptation experiments such as those described above clearly demonstrate that there is no single gene that confers tolerance to toxic compounds or nonnatural growth conditions; it is common to see a panel of mutations that are accumulated in the final variant. For example, the expression level of >200 Open Reading Frame (ORF) was significantly different in the ethanol-tolerant *E. coli* strain LY01 as compared to KO11, including genes involved in protective osmolyte (glycine and betaine) biosynthesis, antibiotic resistance proteins, and cell envelope components.[199] Adaptive evolution studies examining genetic acquisition of tolerance phenotypes typically find a range of mechanisms that improve cellular growth under stress.[195,200] Because many of the mutations found in the genomes of organisms with tolerance phenotypes cannot be rationally predicted and several mutations have synergistic effects, engineers have begun to apply genome-scale methods in order to increase both the rate and the probability of finding genetic variations that increase tolerance. These tools typically perturb genes that act as global regulators or engineer every gene in the genome simultaneously. The pooled library of variants created from these methods is then exposed to a growth competition screen to identify improved clones for downstream analysis. The logic behind these approaches is centered on the observation that if the phenotype of interest is linked to growth, then strains exhibiting a growth advantage under a given selection condition will dominate the population after only a small number of dilution cycles. Several genome-scale methods are explored below.

Adaptive laboratory evolution, perhaps the original genome-scale approach for strain engineering, is often used to improve poorly characterized microbial phenotypes[94] due to the lack of mechanistic knowledge and genetic tools required. Phenotypes acquired through adaptation often result from complex interactions between mutations that are difficult to unravel, rendering it challenging to reverse-engineer the phenotype of interest. All adaptive evolution experiments rely on the fundamental principle that randomly acquired mutations can improve the fitness of individual cells in the population, which then grow in frequency in the evolving population (presuming they survive genetic drift). With the advent of next-generation sequencing, these studies are generating increasing amounts of genetic information concerning most tolerance phenotypes of interest. However, due to the large genomic space available for mutation, most random mutations have little if any beneficial effect on cell growth; other methods that seek to narrow the potential targets for mutation have been developed.

In 2006, Alper et al. reported the "global transcription machinery engineering" (gTME) technique.[201] Mutations along *SPT15*, the *S. cerevisiae* gene coding for the global transcription factor Spt15p, were generated with error-prone Polymerase Chain Reaction (PCR). This allowed multiple mutations to accumulate on the transcription factor, stacking the deck in favor of selecting unpredictable mutations with synergistic effects. A pooled library of *S. cerevisiae* containing mutant *SPT15* was subjected to growth under elevated ethanol (5-6%) and glucose (10-12%) conditions. The best mutant isolated, spt15-300, showed improved growth in 12.5% and 15% ethanol and was found to produce 40 g/L ethanol (98% theoretical yield) compared with 35 g/L (86% theoretical yield) produced from the base strain. Notably, spt15-300 contained three point mutations, which resulted in differential regulation of hundreds of genes. Although no particular metabolic pathway or network was significantly up- or downregulated with respect to the base strain, enzymes belonging to certain functional groups such as oxidoreductases, cytoplasmic proteins, amino acid metabolism, and electron transport were enriched in the differentially expressed set. To determine whether the upregulated genes acted independently or in concert, individual knockouts were examined. All gene knockouts that were tested in the absence of spt15-300 showed identical tolerance to stressors as that of the base strain. Moreover, each of the single and double mutants that were built based on the three sites identified in spt15-300 were at least fivefold less phenotypically fit in comparison to the triple mutant, underscoring the unpredictable and synergistic effects of multiple, simultaneous mutations.

Because gTME focuses on generating a small number of mutations in a global transcription factor, the library effectively allows examination of differential gene regulation across the genome. Lynch et al. applied a similar logic when searching for *E. coli* mutations that would confer elevated tolerance to 1-naphthol[202] and 3-hydroxypropionic acid[203] (3HP), two compounds widely used in the chemical industry, as well as ethanol.[204] Using Scalar Analysis of Library Enrichments[205] (SCALEs), five plasmid libraries were constructed in which sections of genomic DNA from the *E. coli* K12 strain with defined lengths (0.75-1.5, 1.5-3.0, 3.0-5.0, 5.0-10.0, and 10.0-20.0 kbp) were inserted into a low copy vector backbone. The plasmid libraries were then transformed into a highly competent commercial *E. coli* strain Mach1-T1^R (Invitrogen) and pooled to create a population that constitutively expressed genes and operons from the five representative K12 genomic libraries. The pooled library was then grown to mid-log phase, sampled, and subjected to the growth selection pressure. Fitness of the population was quantitatively tracked during each growth selection using mathematical models relating gene copy number to fitness. Growth of the library in the presence of 1-naphthol pulled out recombinants enriched with genes coding for the natural efflux machinery, acrAB, which is known to have broad substrate specificity and be expressed under solvent exposure[206–208] (*mar*, *sox*, and *rob*), and the phage shock protein (*psp*) operon which has also been shown to increase survival in response to solvent shock.[209] When the SCALEs library was grown in the presence of 15 or 30 g/L ethanol, recombinants enriched with genes involved in amino acid metabolism (*tilS*, *yaeQJ*, *trmA*), importer/exporter pumps (*yhfTU*, *yicE*, *yijD*), transcriptional regulation (*zur*, *rof*), and primary metabolism (*lpcA*, *fadE*, *serA*, *arnBC*) exhibited the fastest growth rates.[204] Similarly, general stress response proteins were isolated when ethanologenic *E. coli* strains were grown with exogenously added ethanol (15 g/L). The *uspC-otsA-otsB* genomic region, which encodes a universal stress protein (upsC) and an osmolyte (trehalose) biosynthetic pathway (otsA-otsB), was found to confer up to 75% improvement in ethanol tolerance and productivity.[210] In contrast, growth against 3HP toxicity enriched for genes from the chorismate and threonine super-pathways, suggesting that 3HP toxicity is likely a result of chorismate and threonine pathway inhibition rather than the global, nonspecific effects elicited by alcohols and other biofuels. Accordingly, when the base strain was grown in media supplemented with chorismate and threonine pathway intermediates, cells displayed an increased tolerance to 3HP toxicity.

To expand on the SCALEs approach, the Gill lab developed trackable multiplex recombineering (TRMR), a method for simultaneously mapping genetic modifications that affect the acquisition of complex phenotypes such as tolerance to toxic compounds.[211] Two panels of dsDNA sequences were designed such that (i) each sequence would replace the promoter and ribosome-binding site (RBS) regions of a single gene along the *E. coli* chromosome (4077 genes), (ii) each sequence contained a unique 20 nucleotide barcode,[212] (iii) each contained either a strong and repressible expression promoter (called "up" cassettes) or replaced the native RBS with an inert sequence (called "down" cassettes), and (iv) a selectable marker, the blasticidin-S resistance gene. Thus, when the "up" and "down" cassettes were introduced into the genome of *E. coli* MG1655, a pool of >8000 possible variants were created. This "TRMR" library was then subjected to growth against a panel of inhibitory compounds, including lignocellulosic hydrolysate,[211,213] acetate,[213] and ethanol.[204] Growth on high concentrations of lignocellulosic hydrolysate (20-40%) was selected for differentially expressed alleles involved in secondary metabolism (*ispF* up, *dxs* up), vitamin metabolic processes (*nadD* up, *menD* up, *apbE* up, *pabC* up, *dxs* up and *ribB* up), and antioxidant activity (*tpx* up, *bcp* up, and *ahpC* up).[211] Notably, *ahpC* had previously been shown to impart resistance to solvents[214] and various antioxidants.[215,216] When the TRMR library was subjected to growth in the presence of 10 g/L acetate, alleles involve with chemotaxis (*tap* up), secondary metabolism (*tqsA* down), ion and peptide importer/exporter pumps (*clcB* down, *znuC* up, and *ddpA* up), and pyrimidine recycling (*deoA* down)[213] were enriched.

Although genotype-phenotype relationships eliciting tolerance are generally not predictable *a priori*, engineers have begun to take a more rational approach to engineering tolerance and started to explore the range of functionality offered by efflux pumps, natural cellular export systems that directly excrete compounds from the cell. In a recent report,[21] a prioritized library of 43 efflux pumps exhibiting homology to the solvent-resistance pump TtgB from *Pseudomonas putida* were heterologously expressed in a strain of *E. coli* lacking the native multidrug-efflux pump acrAB. Using a competitive growth assay, a pool of variants containing an even distribution of cells expressing each efflux pump was subjected to growth against a panel of 7 biofuels, including both alcohols and terpenes. Although no pumps were found to increase tolerance to the tested alcohols (butanol and isopentanol), three pumps were found to be active against the tested terpenes, including *E. coli*'s acrAB. When the previously uncharacterized pump from *Alcanivorax borkumensis*, an alkane degrading marine bacterium, was overexpressed in a limonene-producing *E. coli* strain, a twofold improvement in limonene production was observed (60 mg/L). Similarly, a recent report demonstrated that bioethanol production in *S. cerevisiae* can be significantly improved by increasing K^+ concentration and the reducing acidity of the fermentation media.[20] The team, led by Gregory Stephanopoulos, genetically enhanced K^+ and H^+ regulation by overexpressing the K^+ importer and H^+ exporter pumps, *TRK1* and *PMA1*, resulting in a *S. cerevisiae* laboratory strain that outperforms the Brazilian production strain PE-2 by ~5%. These targeted approaches are most useful if there is evidence to support particular toxicity mechanisms that can be addressed through the manipulation or importation of a small number of genes.

9.4 The Genome Engineering Toolbox

The advancements in metabolic engineering for biofuel production as described in the previous sections have been possible due to concomitant advancements in the emerging field of synthetic biology, a discipline that aims to couple engineering principles with the design of biological parts and genetic circuits using synthetic DNA.[217,218] Over the past three decades, many tools have been developed that allow engineers to chemically synthesize and assemble DNA molecules as well as directly edit microbial genomes. Methods for constructing extrachromosomal plasmids have been extensively reviewed[218,219] and are not covered in this chapter. Notably, the ability to easily assemble plasmids to rapidly test a novel pathway has been a cornerstone of synthetic biology (as exemplified by Ingram's first ethanologenic *E. coli* strains), and production strains that rely on extrachromosomal expression vectors are often quite suitable for commercial production of high-value products. This is because multicopy plasmids often produce much higher levels of the desired pathway components in comparison to single gene copies integrated into the chromosome. However, the utility of plasmids has several important limitations including (i) the burden of plasmid propagation and maintenance, (ii) copy number variability (resulting in a nonhomogeneous cell population), and (iii) the requirement for expensive feedstock components (such as antibiotics) to maintain recombinants. Therefore, constructing a reliable strain for industrial biofuel production generally requires integration of foreign genes into the base strain's chromosome. Moreover, optimization of flux through an engineered pathway usually requires inhibition of competing pathways, which also requires genomic editing technology. The final section of this chapter will be devoted to tools used for editing the base strain genome.

The first attempts to integrate DNA fragments into the *E. coli* genome relied on phage-based approaches, exploring transduction as a means to transfer genetic characters between strains.[220,221] Later, plasmids containing phage integration sites were developed to allow high efficient integration of large size DNA fragments into the *E. coli* genome.[222] These approaches rely on the recombination between phage-specific attachment sites (*attP*) and a chromosomal *attB* complementary site. Due to this requirement, such approaches offer little flexibility over chromosomal position and/or require preparatory steps to integrate the *attB* site into the desired position. Another major drawback is that there is little control over the sequence that will be integrated; for example, the whole vector, including the origin of replication, is often cointegrated. Although there are strategies that overcome this limitation by removing undesired integrated sequences,[223,224] they add extra steps to the method making them labor-intensive.

Relying on *E. coli* recombination systems for engineering the genome has been an extremely useful engineering tool but is challenged by the activity of endogenous exonucleases, which limits the efficiency of *E. coli* transformation with linear pieces of DNA.[225] Therefore, recombination-proficient strains were developed to allow recombination from linear pieces of DNA through inactivation of key exonucleases, like recBCD.[226,227] Despite its role as an exonuclease, recBCD is also involved in recombination so that a sbcBC mutation is required to restore recombination proficiency in recBC mutants.[227] recBCD works as part of the recombination pathway when it recognizes the specific sequence 5′-GCTGGTGG-3′ (called Chi site).[228] With Chi site recognition, recBCD recruits recA,[229] an ssDNA binding protein that promotes strand invasion to complementary DNA in the cell (Figure 9.3a). Therefore, recBCD

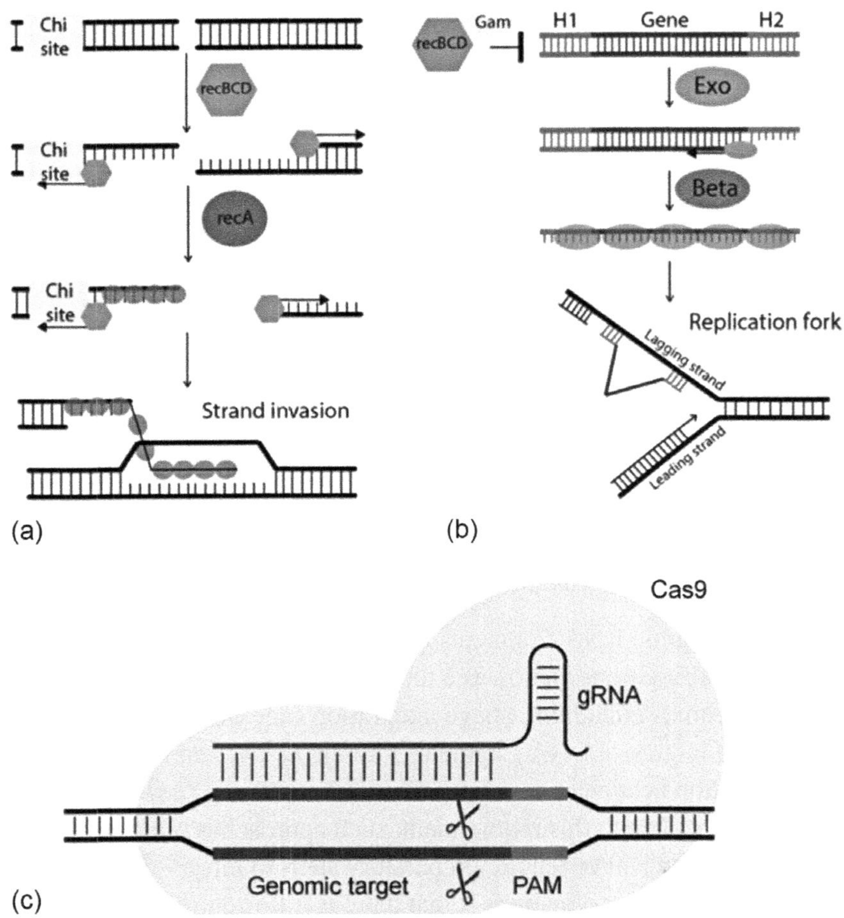

Figure 9.3

Genome engineering strategies. (a) Native recBCD-recA recombination pathway in *E. coli*. Upon recognition of Chi-site, recBCD stops degradation and recruits recA, which promotes strand invasion to a complementary DNA to repair the lesion. (b) Lambda red mechanism. Exo inhibits recBCD from degrading the dsDNA template. Exo is a 5′-3′ exonuclease that generates an ssDNA intermediate. Beta is an ssDNA binding protein that promotes strand annealing to complementary DNA in the open replication fork. (c) CRISPR/Cas system for genome editing. The Cas9 endonuclease is targeted by a guide RNA (gRNA) to the target site and introduces a double-strand break. The consensus PAM sequence (5′-NGG-3′) is required for Cas9 binding and cleavage.

works as both defense against linear dsDNA in the cell and as a repair for double-strand break lesions in the *E. coli* chromosome. The evolutionary conservation of recA among bacteria makes it an attractive mechanism for genome engineering methods, since it could function in other bacterial species.[229] However, the recBC requirement of the chi site to recruit recA limits this pathway to specific sequences.

An important breakthrough in the field came with the introduction of exogenous recombination systems in *E. coli*, particularly the lambda red and recET systems. Both systems come from bacteriophages and are capable of integrating DNA fragments into the *E. coli* genome when the involved proteins are expressed in the cell. The bacteriophage lambda has a recombination system (the Lambda Red system) containing three proteins[230]: gam, exo, and beta. Gam is an inhibitor of recBCD, thereby preventing this complex from degrading the dsDNA fragment that will be integrated. Exo is a 5′-3′ exonuclease that binds to the dsDNA and generates an ssDNA intermediate. Beta binds to the ssDNA and promotes strand annealing to a complementary sequence. Since beta can only promote strand annealing, the current proposed mechanisms suggest that the ssDNA intermediate anneals to exposed complementary ssDNA in the lagging strand of the open replication fork[231](Figure 9.3b). Previous studies demonstrated that recA is not required for lambda red integration, but its presence improves integration efficiency.[230] The lambda red system is a hallmark of genetic engineering in *E. coli* (commonly referred to as "recombineering") and has allowed researchers to easily insert and delete genes from the chromosome using PCR products.[232] However, the method requires selectable markers due to the low recombination efficiency ($\sim 5 \times 10^{-4}$ recombinants per viable cell for a 1 kb gene[233]) and is limited to a maximum of up to 2.5 kb DNA fragments.[234]

recET is a recombination system that has similar functions to lambda red but appears to work in a mechanistically distinct manner. recE is an exonuclease and recT is an ssDNA-binding protein, analogous to lambda red exo and beta functions, respectively.[235] Since the system cannot inhibit recBCD exonuclease activity by itself, the lambda gam function is often employed in engineering efforts that utilize the recET system.[235,236] Despite the similarity in functions, recent studies demonstrated that the recET system performs better for linear-linear recombination, opposed to lambda red (better for linear-circular recombination). Due to the efficient linear-linear recombination, recET is used for direct cloning of DNA fragments straight from a genome, and this approach proved to be a useful strategy for bioprospecting.[236,237]

Taking advantage of the ease to use phage recombination functions, the Church group leveraged the lambda red system to develop a multiplex genome-engineering approach called MAGE.[238] MAGE (multiplex automated genome engineering) relies on recursive rounds of transformation with designed oligos that target multiple genomic sites, creating diversity in a combinatorial fashion. The depth of the diversity generated depends on (1) the degree of variation desired at each locus, (2) the number of sites targeted in the genome, and (3) the number of MAGE rounds performed. The approach automated all the steps, thus rendering the total number of MAGE cycles to be performed not a significant issue. Using MAGE, the group was able to simultaneously target 20 endogenous genes to introduce degenerate RBS sequences and improve lycopene production by fivefold within 3 days.

MAGE was proven useful for multiple studies (e.g., genome-wide codon replacement[239] and promoter-size replacements[240]), but its genome engineering capability is limited to a few base-pairs. This limitation is inherent to the lambda red system, as the recombination efficiency decreases significantly with the amount of nonhomologous sequence (insertions or mismatches) introduced. Therefore, gene-size insertions require selectable markers, as exemplified by TRMR. Another drawback of MAGE regards the requirement for strains with deficient mismatch-repair machinery (ΔmutS), resulting in a high background of mutations.

Other recombination mechanisms can be employed for genome engineering efforts, and several studies took advantage of the Cre-lox system. This system relies on a site-specific recombinase (Cre recombinase) that recognizes a specific 34 bp site (loxP). The loxP sites comprise an 8 bp non-palindromic core region flanked by 13 bp inverted repeats.[241] Therefore, by delivering a single recombinase and engineering loxP sites in desired chromosomal positions, it is possible to introduce several genome modifications such as insertions, deletions, and transversions.[241] Cre-lox recombination has been used to tailor the genomes of several organisms, and it has been a particular important genetic tool for fruit flies and mice.[241–243]

The Cre-lox system was recently leveraged for genome engineering in *E. coli* in a strategy named recombinase-assisted genome engineering (RAGE).[244] RAGE involves the incorporation of a "targeting cassette" containing loxP sites into a desired genomic position. Then, the gene of interest is introduced in a plasmid flanked by the same loxP sites. Upon expression of Cre-recombinase, the flanked gene is incorporated into the genome. With this approach, the group reported the introduction of a 34-kb alginate catabolic pathway. The strategy also has the advantage of testing multiple genetic backgrounds, integration loci, and copy number for best heterologous expression.

Other organisms that are readily transformable with linear pieces of DNA had an advantage over *E. coli* for genome engineering efforts. Gene replacements are easily performed in yeast using PCR products with 50 bp of homology arms.[212] Even the genomes of more complex eukaryotic organisms, like mice, have been engineered using this same idea.[245] Notably, initial methodology involved the use of very large homology arms (10–20 kb), and the frequency of homologous recombination in mice was very low (on the order of 0.000001) targeted insertions per treated cell. A major breakthrough for genome engineering in organisms like mice came with the realization that introducing a targeted double-strand break can dramatically increase homologous recombination efficiency.[246,247]

Double-strand breaks can be repaired by two pathways that are present in most organisms: nonhomologous end-joining (NHEJ) and homology-directed repair (HDR). NHEJ combines the exposed ends of the double-strand break together, often introducing small insertions/deletions (indels) that change the reading frame of a coding sequence. HDR uses a complementary template DNA to repair the lesion, providing a way to effectively introduce desired sequences through recombination with a supplied template DNA. Either by using NHEJ or HDR, introducing a precise double-strand break in the genome is a crucial step for targeted genome editing. However, the ability to introduce double-strand breaks in precise genomic positions has not always been an easy task.

Initial studies relied on the rare-cutting homing meganuclease I-SceI from yeast. This idea was recently explored in *E. coli* and allowed the insertion of 7 kb constructs.[234] The approach consists of initially integrating I-SceI sites into the genome using lambda red recombination followed by the introduction of a plasmid containing the desired gene flanked by homology arms and I-SceI sites. Introduction of I-SceI endonuclease cleaves both the genome and the plasmid and allows recombination to efficiently occur and insert the targeted gene. I-SceI-based strategies have also been used in other organisms such as mice, increasing homologous recombination efficiency by 10,000-fold.[248]

The search for methods to introduce double-strand breaks in precise manners led to the discovery of user-targetable DNA-binding proteins that could be fused to nuclease domains. The resulting zinc-finger nucleases (ZFNs) and transcription activator-like effector nucleases (TALENs) could be engineered to target user-defined sequences. This approach allowed genome engineering in a multitude of organisms, including human cells, mice, pigs, yeast, and several plants.[248] Despite the remarkable breakthroughs allowed by these engineered nucleases, each application required reengineering the whole enzyme at the protein level to allow different site recognition. Additionally, the design rules are still not completely elucidated as several designed ZFNs or TALENs fail to perform as expected. Therefore, the genome engineering field is still in need of an easier and reliable tool to introduce precise double-strand breaks in the genome.

The most recent addition to the genome engineering toolbox is the CRISPR/Cas system. This system revolutionized the field and allowed scientists to introduce precise double-strand breaks with an easy, efficient, and broadly applicable method. CRISPR (clustered regularly interspaced short palindromic repeats) is a bacterial adaptive immune system that works as a defense against genetic parasites such as viruses and plasmids.[249] The native system works by integrating short fragments (typically 20-50 bp long) of foreign nucleic acid into the chromosome separated by repeats of similar length. These incorporated fragments can target RNA-guided nucleases to the complementary sites to cleave the invading DNA upon further invasions. Cas (CRISPR-associated) genes are located adjacent to the CRISPR loci and code for the required machinery for adaptive immunity: incorporation of foreign DNA fragments, processing of the transcribed loci (crRNA), and the RNA-guided targeted interference.

Based on the sequence comparisons between *cas* genes as well as between CRISPR repeats, the system can be classified into three major types. The type II system from *Streptococcus pyogenes* is the most extensively studied and has been leveraged for several genome engineering applications.[250,251] This system comprises a single nuclease (Cas9) whose specificity can be determined by an engineered guide RNA (gRNA). Therefore, changing the gRNA sequence can easily reprogram Cas9 to target distinct sites (Figure 9.3c), offering an unprecedented advantage over engineered nucleases like ZFNs or TALENs. Essential for cleavage is the presence of a 5′-NGG-3′ consensus sequence immediately downstream of the target site, called protospacer adjacent motif (PAM). Consequently, any mutations or deletions of the target PAM sequence abrogate Cas9 cleavage.[252]

CRISPR is a broadly applicable tool allowing genome editing of multiple cell lines and organisms, from bacteria to humans.[251] In addition to the common usage of Cas9 to introduce double-strand breaks and leverage repair mechanisms such as NHEJ or HDR, Cas9 has also been engineered for regulatory applications. Mutating the nuclease domains RuvC and HNH rendered a catalytically inactive "dead" Cas9 (dCas9) that is able to bind to complementary targets but not to introduce double-strand breaks.[253,254] Therefore, dCas9 can be used for repressing gene expressions. In addition, other groups demonstrated that it is also possible to induce gene expression by fusing dCas9 with activation domains to efficiently tune gene expression in several organisms.[255]

A major concern with the CRISPR/Cas system regards off-target effects, which is critical for gene therapy applications. Recent studies revealed that the amount of off-target binding by dCas9 can vary substantially in human cell lines depending on the gRNA, ranging from 10 to more than a 1000.[256] Even though there are strategies that alleviate this issue by utilizing double nickase versions of Cas9,[257] the full extent of off-target effects remains to be addressed in order to establish CRISPR/Cas as a platform for gene therapy.

9.5 Concluding Remarks

The energy crisis of the 1970s sparked intense research efforts in renewable fuel sources that have continued more than four decades. Today major efforts are focused on engineering biological systems to redirect carbon trapped in recalcitrant plant material back into the energy sector using microbial cell factories. Since the 1980s, several microbial chassis strains have been selectively bred for their ability to produce biogasoline candidates (alcohols and alkanes), biodiesels (fatty acids and their derivatives), jet fuels (hydrogenated terpenes), and advanced fuel additives for improving the fuel economy of petroleum-based blends. Although the success of these ventures is clear, productivity of biological systems is inherently limited by the chassis strain's susceptibility to toxic components found in processed feedstocks as well as the biofuel itself. Some alleviation can be achieved through process optimization (media components, growth temperatures, pH buffers, etc.) and improvements in product recovery such as gas stripping, adsorption, and liquid-liquid extraction concomitant with microbial product production. However, engineers have also found that optimization of the chassis strain genome is a fruitful means for enhancing tolerance phenotypes. As methods for tracking and manipulating production strain genetics become more readily available, engineers will have the opportunity to very rapidly screen for optimal product production as well as tolerance to industrial production conditions, allowing cost-effective biofuel production to become a much more tenable prospect.

Acknowledgments

The authors would like to thank Carrie Eckert, Andrew Garst, Sean Lynch, Ramsey Zeitoun, Gur Pines, and Emily Freed for thoughtful discussions during preparation of this manuscript. This work was funded by US Department of Energy, Office of Science, Office of Biological and Environmental Research, Genomic Science program, under Award Number DE-SC0008812. We would also like to acknowledge the Colorado Center for Biorefining and Biofuels and National Science Foundation REU Program (NSF#1261303).

References

1. US EPA OAR OTAQ. Renewable Fuel Standard (RFS). http://www.epa.gov/otaq/fuels/renewablefuels/index.htm; 2011.
2. EIA. Weekly petroleum status report. http://www.eia.gov/petroleum/supply/weekly/; 2014.
3. EPA Proposes 2014 Renewable Fuel Standards. http://www.epa.gov/otaq/fuels/renewablefuels/documents/420f13048.pdf; 2015.
4. The Renewable Fuel Standard: Issues for 2014 and Beyond. http://www.cbo.gov/sites/default/files/45477-Biofuels2.pdf.
5. Murray J, King D. Climate policy: oil's tipping point has passed. *Nature* 2012;**481**:433–5.
6. Murphy DJ. Fossil fuels: peak oil is affecting the economy already. *Nature* 2012;**483**:541.
7. Howarth RW, Ingraffea A, Engelder T. Natural gas: should fracking stop? *Nature* 2011;**477**:271–5.
8. The Upstream Survivors of 2009 How to Live to Tell the Tale. http://www.strategyand.pwc.com/media/file/Upstream_Survivors_2009.pdf.
9. Downing M, Eaton LM, Graham RL, Langholtz MH. *US billion-ton update: biomass supply for a bioenergy and bioproducts industry*; 2011. http://www.osti.gov/bridge/product.biblio.jsp?osti_id=1023318.
10. Fairley P. Introduction: next generation biofuels. *Nature* 2011;**474**:S2–5.
11. IEA Bioenergy. *From 1st-to 2nd-generation biofuel technologies. An overview of current industry and RD&D*; 2008. http://environmentportal.in/files/2nd_Biofuel_Gen.pdf.
12. Doeden M. *Crazy cars*. Minneapolis, MN: Lerner Publications Company; 2007.
13. CRC Handbook of Chemistry and Physics Online Connect. http://liblearn.osu.edu/guides/CRC_Online.pdf.
14. Knothe G. "Designer" biodiesel: optimizing fatty ester composition to improve fuel properties†. *Energy Fuel* 2008;**22**:1358–64. http://pubs.acs.org/doi/full/10.1021/ef700639e.
15. Rude MA, Schirmer A. New microbial fuels: a biotech perspective. *Curr Opin Microbiol* 2009;**12**:274–81.
16. Hirshfeld DS, Kolb JA, Anderson JE, Studzinski W, Frusti J. Refining economics of U.S. gasoline: octane ratings and ethanol content. *Environ Sci Technol* 2014;**48**:11064–71.
17. Speth RL, Chow EW, Malina R, Barrett SRH, Heywood JB, Green WH. Economic and environmental benefits of higher-octane gasoline. *Environ Sci Technol* 2014;**48**:6561–8.
18. Fischer CR, Klein-Marcuschamer D, Stephanopoulos G. Selection and optimization of microbial hosts for biofuels production. *Metab Eng* 2008;**10**:295–304.
19. Demirel Y. *Energy*. London: Springer; 2012, pp. 27-70.

20. Lam FH, Ghaderi A, Fink GR, Stephanopoulos G. Biofuels. Engineering alcohol tolerance in yeast. *Science* 2014;**346**:71–5.
21. Dunlop MJ, Dossani ZY, Szmidt HL, Chu HC, Lee TS, Keasling JD, et al. Engineering microbial biofuel tolerance and export using efflux pumps. *Mol Syst Biol* 2011;**7**:487.
22. Rogers PL, Jeon YJ, Lee KJ, Lawford HG. Zymomonas mobilis for fuel ethanol and higher value products. *Adv Biochem Eng Biotechnol* 2007;**108**:263–88.
23. Coton M, Laplace JM, Auffray Y, Coton E. "Framboisé" spoilage in French ciders: Zymomonas mobilis implication and characterization. *LWT- Food Sci Technol* 2006;**39**:972–9.
24. Jarboe LR, Grabar TB, Yomano LP, Shanmugan KT, Ingram LO. Development of ethanologenic bacteria. *Adv Biochem Eng Biotechnol* 2007;**108**:237–61.
25. York SW, Ingram LO. Ethanol production by recombinant Escherichia coli KO11 using crude yeast autolysate as a nutrient supplement. *Biotechnol Lett* 1996;**18**:683–8.
26. Ankarloo J, Wikman S, Nicholls IA. Escherichia coli mar and acrAB mutants display no tolerance to simple alcohols. *Int J Mol Sci* 2010;**11**:1403–12.
27. Lamsen EN, Atsumi S. Recent progress in synthetic biology for microbial production of C3-C10 alcohols. *Front Microbiol* 2012;**3**:196.
28. Srirangan K, Liu X, Westbrook A, Akawi L, Pyne ME, Moo-Young M, et al. Biochemical, genetic, and metabolic engineering strategies to enhance coproduction of 1-propanol and ethanol in engineered Escherichia coli. *Appl Microbiol Biotechnol* 2014;**98**:9499–515.
29. Collas F, Kuit W, Clément B, Marchal R, López-Contreras AM, Monot F. Simultaneous production of isopropanol, butanol, ethanol and 2,3-butanediol by Clostridium acetobutylicum ATCC 824 engineered strains. *AMB Express* 2012;**2**:45.
30. Inokuma K, Liao JC, Okamoto M, Hanai T. Improvement of isopropanol production by metabolically engineered Escherichia coli using gas stripping. *J Biosci Bioeng* 2010;**110**:696–701.
31. Jin C, Yao M, Liu H, Lee C-FF, Ji J. Progress in the production and application of n-butanol as a biofuel. *Renew Sustain Energy Rev* 2011;**15**:4080–106.
32. Peralta-Yahya PP, Keasling JD. Advanced biofuel production in microbes. *Biotechnol J* 2010;**5**:147–62.
33. Krivoruchko A, Serrano-Amatriain C, Chen Y, Siewers V, Nielsen J. Improving biobutanol production in engineered Saccharomyces cerevisiae by manipulation of acetyl-CoA metabolism. *J Ind Microbiol Biotechnol* 2013;**40**:1051–6.
34. Xue C, Zhao X-Q, Liu C-G, Chen L-J, Bai F-W. Prospective and development of butanol as an advanced biofuel. *Biotechnol Adv* 2013;**31**:1575–84.
35. Yu M, Du Y, Jiang W, Chang WL, Yang ST, Tang IC. Effects of different replicons in conjugative plasmids on transformation efficiency, plasmid stability, gene expression and n-butanol biosynthesis in Clostridium tyrobutyricum. *Appl Microbiol Biotechnol* 2012;**93**:881–9.
36. Lin YL, Blaschek HP. Butanol production by a butanol-tolerant strain of Clostridium acetobutylicum in extruded corn broth. *Appl Environ Microbiol* 1983;**45**:966–73.
37. Baez A, Cho K-M, Liao JC. High-flux isobutanol production using engineered Escherichia coli: a bioreactor study with in situ product removal. *Appl Microbiol Biotechnol* 2011;**90**:1681–90.
38. Atsumi S, Hanai T, Liao JC. Non-fermentative pathways for synthesis of branched-chain higher alcohols as biofuels. *Nature* 2008;**451**:86–9.
39. Matsuda F, Ishii J, Kondo T, Ida K, Tezuka H, Kondo A. Increased isobutanol production in Saccharomyces cerevisiae by eliminating competing pathways and resolving cofactor imbalance. *Microb Cell Fact* 2013;**12**:119.
40. Yamamoto S, Suda M, Niimi S, Inui M, Yukawa H. Strain optimization for efficient isobutanol production using Corynebacterium glutamicum under oxygen deprivation. *Biotechnol Bioeng* 2013;**110**:2938–48.
41. Smith KM, Cho K-M, Liao JC. Engineering Corynebacterium glutamicum for isobutanol production. *Appl Microbiol Biotechnol* 2010;**87**:1045–55.
42. Chou HH, Keasling JD. Synthetic pathway for production of five-carbon alcohols from isopentenyl diphosphate. *Appl Environ Microbiol* 2012;**78**:7849–55.

43. Connor MR, Cann AF, Liao JC. 3-Methyl-1-butanol production in Escherichia coli: random mutagenesis and two-phase fermentation. *Appl Microbiol Biotechnol* 2010;**86**:1155–64.
44. Renninger NS, Mcphee DJ. Fuel compositions comprising farnesane and farnesane derivatives and method of making and using same. EP Patent; 2011.
45. Peralta-Yahya PP, Ouellet M, Chan R, Mukhopadhyay A, Keasling JD, Lee TS. Identification and microbial production of a terpene-based advanced biofuel. *Nat Commun* 2011;**2**:483.
46. Sarria S, Wong B, García Martín H, Keasling JD, Peralta-Yahya P. Microbial synthesis of pinene. *ACS Synth Biol* 2014;**3**:466–75.
47. Pourbafrani M, McKechnie J, MacLean HL, Saville BA. Life cycle greenhouse gas impacts of ethanol, biomethane and limonene production from citrus waste. *Environ Res Lett* 2013;**8**:015007.
48. Akhtar MK, Dandapani H, Thiel K, Jones PR. Microbial production of 1-octanol: a naturally excreted biofuel with diesel-like properties. *Metab Eng Commun* 2015;**2**:1–5. http://dx.doi.org/10.1016/j.meteno.2014.11.001.
49. Youngquist JT, Schumacher MH, Rose JP, Raines TC, Politz MC, Copeland MF, et al. Production of medium chain length fatty alcohols from glucose in Escherichia coli. *Metab Eng* 2013;**20**:177–86.
50. Liu A, Tan X, Yao L, Lu X. Fatty alcohol production in engineered E. coli expressing Marinobacter fatty acyl-CoA reductases. *Appl Microbiol Biotechnol* 2013;**97**:7061–71.
51. Lennen RM, Kruziki MA, Kumar K, Zinkel RA, Burnum KE, Lipton MS, et al. Membrane stresses induced by overproduction of free fatty acids in Escherichia coli. *Appl Environ Microbiol* 2011;**77**:8114–28.
52. Choi YJ, Lee SY. Microbial production of short-chain alkanes. *Nature* 2013;**502**:571–4.
53. Demain AL, Báez-Vásquez MA. *New and future developments in catalysis*. In: Suib SL, editor. Chapter 14: Biofuels of the Present and the Future Elsevier; 2013. p. 325–70.
54. Eyidogan M, Ozsezen AN, Canakci M, Turkcan A. Impact of alcohol–gasoline fuel blends on the performance and combustion characteristics of an SI engine. *Fuel* 2010;**89**:2713–20.
55. Clark DP. The fermentation pathways of Escherichia coli. *FEMS Microbiol Rev* 1989;**5**:223–34.
56. Chandrakant P, Bisaria VS. Simultaneous bioconversion of cellulose and hemicellulose to ethanol. *Crit Rev Biotechnol* 1998;**18**:295–331.
57. De Graef MR, Alexeeva S, Snoep JL, Teixeira de Mattos MJ. The steady-state internal redox state (NADH/NAD) reflects the external redox state and is correlated with catabolic adaptation in Escherichia coli. *J Bacteriol* 1999;**181**:2351–7.
58. Richards M, Corbey DA. Isolation of Zymomonas mobilis from primed beer*. *J Inst Brew* 1974;**80**:241–4.
59. Swings J, De Ley J. The biology of Zymomonas. *Bacteriol Rev* 1977;**41**:1–46.
60. Conway T. The Entner-Doudoroff pathway: history, physiology and molecular biology. *FEMS Microbiol Lett* 1992;**103**:1–27.
61. Abreu-Cavalheiro A, Monteiro G. Solving ethanol production problems with genetically modified yeast strains. *Braz J Microbiol* 2013;**44**:665–71.
62. Wheals AE, Basso LC, Alves DM, Amorim HV. Fuel ethanol after 25 years. *Trends Biotechnol* 1999;**17**:482–7.
63. Basso LC, de Amorim HV, de Oliveira AJ, Lopes ML. Yeast selection for fuel ethanol production in Brazil. *FEMS Yeast Res* 2008;**8**:1155–63.
64. NREL: News—NREL Industry Partners Move Cellulosic Ethanol Technology Forward. http://www.nrel.gov/news/press/2008/600.html.
65. Buchholz SE, Dooley MM, Eveleigh DE. Zymomonas—an alcoholic enigma. *Trends Biotechnol* 1987;**5**:199–204.
66. Rogers PL, Lee KJ, Skotnicki ML, Tribe DE. *Microbial reactions*. Berlin, Heidelberg: Springer; 1982, 37-84.
67. O'Mullan PJ, Buchholz SE, Chase Jr T, Eveleigh DE. Roles of alcohol dehydrogenases of Zymomonas mobilis (ZADH): characterization of a ZADH-2-negative mutant. *Appl Microbiol Biotechnol* 1995;**43**:675–8.

68. D'Amore T, Celotto G, Russell I, Stewart GG. Selection and optimization of yeast suitable for ethanol production at 40 °C. *Enzyme Microb Technol* 1989;**11**:411–6.
69. da Silva-Filho EA, Brito dos Santos SK, Resende Ado M, de Morais JO, de Morais MA Jr., Ardaillon Simões D. Yeast population dynamics of industrial fuel-ethanol fermentation process assessed by PCR-fingerprinting. *Antonie Van Leeuwenhoek* 2005;**88**:13–23.
70. Argueso JL, Carazzolle MF, Mieczkowski PA, Duarte FM, Netto OV, Missawa SK, et al. Genome structure of a Saccharomyces cerevisiae strain widely used in bioethanol production. *Genome Res* 2009;**19**:2258–70.
71. Wyman CE, Hinman ND. Ethanol. *Appl Biochem Biotechnol* 1990;**24-25**:735–53.
72. Global Reach Internet Productions, LLC-Ames, IA—globalreach.com. Ethanol, gasoline, crude oil and corn prices: are the relationships changing? Agricultural Marketing Resource Center. http://www.agmrc.org/renewable_energy/ethanol/ethanol-gasoline-crude-oil-and-corn-prices-are-the-relationships-changing/; 2014.
73. Anwar Z, Gulfraz M, Irshad M. Agro-industrial lignocellulosic biomass a key to unlock the future bio-energy: a brief review. *J Radiat Res Appl Sci* 2014;**7**:163–73.
74. Osborn D, Chen LF. Corn hull hydrolysis using glucoamylase and sulfuric acid. *Starch - Stärke* 1984;**36**:393–5.
75. Asghari A, Bothast RJ, Doran JB, Ingram LO. Ethanol production from hemicellulose hydrolysates of agricultural residues using genetically engineered Escherichia coli strain KO11. *J Ind Microbiol* 1996;**16**:42–7.
76. Kim SR, Park Y-C, Jin Y-S, Seo J-H. Strain engineering of Saccharomyces cerevisiae for enhanced xylose metabolism. *Biotechnol Adv* 2013;**31**:851–61.
77. Brat D, Boles E, Wiedemann B. Functional expression of a bacterial xylose isomerase in Saccharomyces cerevisiae. *Appl Environ Microbiol* 2009;**75**:2304–11.
78. Ha SJ, Wei Q, Kim SR, Galazka JM, Cate JH, Jin YS. Cofermentation of cellobiose and galactose by an engineered Saccharomyces cerevisiae strain. *Appl Environ Microbiol* 2011;**77**:5822–5.
79. Karhumaa K, Garcia Sanchez R, Hahn-Hägerdal B, Gorwa-Grauslund M-F. Comparison of the xylose reductase-xylitol dehydrogenase and the xylose isomerase pathways for xylose fermentation by recombinant Saccharomyces cerevisiae. *Microb Cell Fact* 2007;**6**:5.
80. Kuyper M, Hartog MM, Toirkens MJ, Almering MJ, Winkler AA, van Dijken JP, et al. Metabolic engineering of a xylose-isomerase-expressing Saccharomyces cerevisiae strain for rapid anaerobic xylose fermentation. *FEMS Yeast Res* 2005;**5**:399–409.
81. Madhavan A, Tamalampudi S, Ushida K, Kanai D, Katahira S, Srivastava A, et al. Xylose isomerase from polycentric fungus Orpinomyces: gene sequencing, cloning, and expression in Saccharomyces cerevisiae for bioconversion of xylose to ethanol. *Appl Microbiol Biotechnol* 2009;**82**:1067–78.
82. Ho NW, Chen Z, Brainard AP. Genetically engineered Saccharomyces yeast capable of effective cofermentation of glucose and xylose. *Appl Environ Microbiol* 1998;**64**:1852–9.
83. Kötter P, Ciriacy M. Xylose fermentation by Saccharomyces cerevisiae. *Appl Microbiol Biotechnol* 1993;**38**:776–83.
84. Walfridsson M, Hallborn J, Penttilä M, Keränen S, Hahn-Hägerdal B. Xylose-metabolizing Saccharomyces cerevisiae strains overexpressing the TKL1 and TAL1 genes encoding the pentose phosphate pathway enzymes transketolase and transaldolase. *Appl Environ Microbiol* 1995;**61**:4184–90.
85. Matsushika A, Inoue H, Murakami K, Takimura O, Sawayama S. Bioethanol production performance of five recombinant strains of laboratory and industrial xylose-fermenting Saccharomyces cerevisiae. *Bioresour Technol* 2009;**100**:2392–8.
86. Panesar PS, Marwaha SS, Kennedy JF. Zymomonas mobilis: an alternative ethanol producer. *J Chem Technol Biotechnol* 2006;**81**:623–35.
87. Skotnicki ML, Lee KJ, Tribe DE, Rogers PL. Genetic alteration of Zymomonas mobilis for ethanol production. *Basic Life Sci* 1982;**19**:271–90.
88. Carey VC, Walia SK, Ingram LO. Expression of a lactose transposon (Tn951) in Zymomonas mobilis. *Appl Environ Microbiol* 1983;**46**:1163–8.
89. Buchholz SE, Eveleigh DE. Genetic modification of Zymomonas mobilis. *Biotechnol Adv* 1990;**8**:547–81.

90. Chatterjee R, Yuan L. Directed evolution of metabolic pathways. *Trends Biotechnol* 2006;**24**:28–38.
91. Pál C, Papp B, Lercher MJ. Adaptive evolution of bacterial metabolic networks by horizontal gene transfer. *Nat Genet* 2005;**37**:1372–5.
92. Portnoy VA, Bezdan D, Zengler K. Adaptive laboratory evolution—harnessing the power of biology for metabolic engineering. *Curr Opin Biotechnol* 2011;**22**:590–4.
93. Conrad TM, Lewis NE, Palsson BØ. Microbial laboratory evolution in the era of genome-scale science. *Mol Syst Biol* 2011;**7**:509.
94. Dragosits M, Mattanovich D. Adaptive laboratory evolution—principles and applications for biotechnology. *Microb Cell Fact* 2013;**12**:64.
95. Zhang M, Eddy C, Deanda K, Finkelstein M, Picataggio S. Metabolic engineering of a pentose metabolism pathway in ethanologenic Zymomonas mobilis. *Science* 1995;**267**:240–3.
96. Deanda K, Zhang M, Eddy C, Picataggio S. Development of an arabinose-fermenting Zymomonas mobilis strain by metabolic pathway engineering. *Appl Environ Microbiol* 1996;**62**:4465–70.
97. Mohagheghi A, Evans K, Chou Y-C, Zhang M. Cofermentation of glucose, xylose, and arabinose by genomic DNA-integrated xylose/arabinose fermenting strain of Zymomonas mobilis AX101. *Appl Biochem Biotechnol* 2002;**98-100**:885–98.
98. DuPont Cellulosic Ethanol Biorefinery Project—Chemicals Technology. http://www.chemicals-technology.com/projects/dupont-cellulosic-ethanol-biorefinery-project/.
99. He MX, Wu B, Qin H, Ruan ZY, Tan FR, Wang JL, et al. Zymomonas mobilis: a novel platform for future biorefineries. *Biotechnol Biofuels* 2014;**7**:101.
100. Belll DS, Ingram LO, Ben-Bassat A, Doran JB, Fowler DE, Hall RG, et al. Conversion of hydrolysates of corn cobs and hulls into ethanol by recombinant Escherichia coli B containing integrated genes for ethanol production. *Biotechnol Lett* 1992;**14**:857–62.
101. Moniruzzaman M, Ingram LO. Ethanol production from dilute acid hydrolysate of rice hulls using genetically engineered Escherichia coli. *Biotechnol Lett* 1998;**20**:943–7.
102. Richter K, Gescher J. The molecular toolbox for chromosomal heterologous multiprotein expression in Escherichia coli. *Biochem Soc Trans* 2012;**40**:1222–6.
103. Lin Y, Tanaka S. Ethanol fermentation from biomass resources: current state and prospects. *Appl Microbiol Biotechnol* 2006;**69**:627–42.
104. Dien BS, Cotta MA, Jeffries TW. Bacteria engineered for fuel ethanol production: current status. *Appl Microbiol Biotechnol* 2003;**63**:258–66.
105. Zaldivar J, Nielsen J, Olsson L. Fuel ethanol production from lignocellulose: a challenge for metabolic engineering and process integration. *Appl Microbiol Biotechnol* 2001;**56**:17–34.
106. Ingram LO, Aldrich HC, Borges AC, Causey TB, Martinez A, Morales F, et al. Enteric bacterial catalysts for fuel ethanol production. *Biotechnol Prog* 1999;**15**:855–66.
107. Ingram LO, Conway T, Clark DP, Sewell GW, Preston JF. Genetic engineering of ethanol production in Escherichia coli. *Appl Environ Microbiol* 1987;**53**:2420–5.
108. Raj KC, Talarico LA, Ingram LO, Maupin-Furlow JA. Cloning and characterization of the Zymobacter palmae pyruvate decarboxylase gene (pdc) and comparison to bacterial homologues. *Appl Environ Microbiol* 2002;**68**:2869–76.
109. Ohta K, Beall DS, Mejia JP, Shanmugam KT, Ingram LO. Genetic improvement of Escherichia coli for ethanol production: chromosomal integration of Zymomonas mobilis genes encoding pyruvate decarboxylase and alcohol dehydrogenase II. *Appl Environ Microbiol* 1991;**57**:893–900.
110. Turner PC, Yomano LP, Jarboe LR, York SW, Baggett CL, Moritz BE, et al. Optical mapping and sequencing of the Escherichia coli KO11 genome reveal extensive chromosomal rearrangements, and multiple tandem copies of the Zymomonas mobilis pdc and adhB genes. *J Ind Microbiol Biotechnol* 2012;**39**:629–39.
111. Hamilton CM, Aldea M, Washburn BK, Babitzke P, Kushner SR. New method for generating deletions and gene replacements in Escherichia coli. *J Bacteriol* 1989;**171**:4617–22.
112. Yomano LP, York SW, Ingram LO. Isolation and characterization of ethanol-tolerant mutants of Escherichia coli KO11 for fuel ethanol production. *J Ind Microbiol Biotechnol* 1998;**20**:132–8.

113. Searle S, Sanchez FP, Malins C, German J. *Technical barriers to the consumption of higher blends of ethanol*; 2014. http://theicct.org/sites/default/files/publications/ICCT_ethanol_revised_02_03_format.pdf.
114. Stephanopoulos G. Challenges in engineering microbes for biofuels production. *Science* 2007;**315**:801–4.
115. Jones DT, Woods DR. Acetone-butanol fermentation revisited. *Microbiol Rev* 1986;**50**:484–524.
116. Jang YS, Lee J, Malaviya A, Seung do Y, Cho JH, Lee SY. Butanol production from renewable biomass: rediscovery of metabolic pathways and metabolic engineering. *Biotechnol J* 2012;**7**:186–98.
117. Dürre P. Fermentative production of butanol—the academic perspective. *Curr Opin Biotechnol* 2011;**22**:331–6.
118. Lee SY, Park JH, Jang SH, Nielsen LK, Kim J, Jung KS. Fermentative butanol production by Clostridia. *Biotechnol Bioeng* 2008;**101**:209–28.
119. Zverlov VV, Berezina O, Velikodvorskaya GA, Schwarz WH. Bacterial acetone and butanol production by industrial fermentation in the Soviet Union: use of hydrolyzed agricultural waste for biorefinery. *Appl Microbiol Biotechnol* 2006;**71**:587–97.
120. Mitchell WJ. Physiology of carbohydrate to solvent conversion by clostridia. *Adv Microb Physiol* 1998;**39**:31–130.
121. Qureshi N, Ezeji TC, Ebener J, Dien BS, Cotta MA, Blaschek HP. Butanol production by Clostridium beijerinckii. Part I: use of acid and enzyme hydrolyzed corn fiber. *Bioresour Technol* 2008;**99**:5915–22.
122. Andersch W, Bahl H, Gottschalk G. Level of enzymes involved in acetate, butyrate, acetone and butanol formation by Clostridium acetobutylicum. *Eur J Appl Microbiol Biotechnol* 1983;**18**:327–32.
123. Hartmanis MG, Gatenbeck S. Intermediary metabolism in Clostridium acetobutylicum: levels of enzymes involved in the formation of acetate and butyrate. *Appl Environ Microbiol* 1984;**47**:1277–83.
124. Tummala SB, Welker NE, Papoutsakis ET. Design of antisense RNA constructs for downregulation of the acetone formation pathway of Clostridium acetobutylicum. *J Bacteriol* 2003;**185**:1923–34.
125. Jiang Y, Xu C, Dong F, Yang Y, Jiang W, Yang S. Disruption of the acetoacetate decarboxylase gene in solvent-producing Clostridium acetobutylicum increases the butanol ratio. *Metab Eng* 2009;**11**:284–91.
126. Han B, Gopalan V, Ezeji TC. Acetone production in solventogenic Clostridium species: new insights from non-enzymatic decarboxylation of acetoacetate. *Appl Microbiol Biotechnol* 2011;**91**:565–76.
127. Jang YS, Lee JY, Lee J, Park JH, Im JA, Eom MH, et al. Enhanced butanol production obtained by reinforcing the direct butanol-forming route in Clostridium acetobutylicum. *MBio* 2012;**3**.
128. Shen CR, Lan EI, Dekishima Y, Baez A, Cho KM, Liao JC. Driving forces enable high-titer anaerobic 1-butanol synthesis in Escherichia coli. *Appl Environ Microbiol* 2011;**77**:2905–15.
129. Bond-Watts BB, Bellerose RJ, Chang MCY. Enzyme mechanism as a kinetic control element for designing synthetic biofuel pathways. *Nat Chem Biol* 2011;**7**:222–7.
130. Berríos-Rivera SJ, Bennett GN, San K-Y. Metabolic engineering of Escherichia coli: increase of NADH availability by overexpressing an NAD(+)-dependent formate dehydrogenase. *Metab Eng* 2002;**4**:217–29.
131. Choi YJ, Park JH, Kim TY, Lee SY. Metabolic engineering of Escherichia coli for the production of 1-propanol. *Metab Eng* 2012;**14**:477–86.
132. Atsumi S, Liao JC. Directed evolution of Methanococcus jannaschii citramalate synthase for biosynthesis of 1-propanol and 1-butanol by Escherichia coli. *Appl Environ Microbiol* 2008;**74**:7802–8.
133. Atsumi S, Wu TY, Eckl EM, Hawkins SD, Buelter T, Liao JC. Engineering the isobutanol biosynthetic pathway in Escherichia coli by comparison of three aldehyde reductase/alcohol dehydrogenase genes. *Appl Microbiol Biotechnol* 2010;**85**:651–7.
134. Cann AF, Liao JC. Production of 2-methyl-1-butanol in engineered Escherichia coli. *Appl Microbiol Biotechnol* 2008;**81**:89–98.
135. Zhang K, Sawaya MR, Eisenberg DS, Liao JC. Expanding metabolism for biosynthesis of nonnatural alcohols. *Proc Natl Acad Sci U S A* 2008;**105**:20653–8.
136. Chen X, Nielsen KF, Borodina I, Kielland-Brandt MC, Karhumaa K. Increased isobutanol production in Saccharomyces cerevisiae by overexpression of genes in valine metabolism. *Biotechnol Biofuels* 2011;**4**:21.
137. Li S, Wen J, Jia X. Engineering Bacillus subtilis for isobutanol production by heterologous Ehrlich pathway construction and the biosynthetic 2-ketoisovalerate precursor pathway overexpression. *Appl Microbiol Biotechnol* 2011;**91**:577–89.

138. Li S, Jia X, Wen J. Improved 2-methyl-1-propanol production in an engineered Bacillus subtilis by constructing inducible pathways. *Biotechnol Lett* 2012;**34**:2253–8.
139. Higashide W, Li Y, Yang Y, Liao JC. Metabolic engineering of Clostridium cellulolyticum for production of isobutanol from cellulose. *Appl Environ Microbiol* 2011;**77**:2727–33.
140. Blombach B, Riester T, Wieschalka S, Ziert C, Youn JW, Wendisch VF, et al. Corynebacterium glutamicum tailored for efficient isobutanol production. *Appl Environ Microbiol* 2011;**77**:3300–10.
141. Lu J, Brigham CJ, Gai CS, Sinskey AJ. Studies on the production of branched-chain alcohols in engineered Ralstonia eutropha. *Appl Microbiol Biotechnol* 2012;**96**:283–97.
142. Bastian S, Liu X, Meyerowitz JT, Snow CD, Chen MM, Arnold FH. Engineered ketol-acid reductoisomerase and alcohol dehydrogenase enable anaerobic 2-methylpropan-1-ol production at theoretical yield in Escherichia coli. *Metab Eng* 2011;**13**:345–52.
143. Wallner T, Miers SA, McConnell S. A comparison of ethanol and butanol as oxygenates using a direct-injection, spark-ignition engine. *J Eng Gas Turbines Power* 2009;**131**:032802.
144. Patzak J, Nesvadba V, Krofta K, Henychova A, Marzoev AI, Richards K. Evaluation of genetic variability of wild hops (Humulus lupulus L.) in Canada and the Caucasus region by chemical and molecular methods. *Genome* 2010;**53**:545–57.
145. Enserink M. Infectious diseases. Source of new hope against malaria is in short supply. *Science* 2005;**307**(33).
146. Ro D-K, Paradise EM, Ouellet M, Fisher KJ, Newman KL, Ndungu JM, et al. Production of the antimalarial drug precursor artemisinic acid in engineered yeast. *Nature* 2006;**440**:940–3.
147. Zeng Q, Qiu F, Yuan L. Production of artemisinin by genetically-modified microbes. *Biotechnol Lett* 2008;**30**:581–92.
148. Chang W-C, Song H, Liu H-W, Liu P. Current development in isoprenoid precursor biosynthesis and regulation. *Curr Opin Chem Biol* 2013;**17**:571–9.
149. Withers ST, Keasling JD. Biosynthesis and engineering of isoprenoid small molecules. *Appl Microbiol Biotechnol* 2007;**73**:980–90.
150. Kung Y, Runguphan W, Keasling JD. From fields to fuels: recent advances in the microbial production of biofuels. *ACS Synth Biol* 2012;**1**:498–513.
151. Withers ST, Gottlieb SS, Lieu B, Newman JD, Keasling JD. Identification of isopentenol biosynthetic genes from Bacillus subtilis by a screening method based on isoprenoid precursor toxicity. *Appl Environ Microbiol* 2007;**73**:6277–83.
152. Renninger NS, Ryder JA, Fisher KJ. Jet fuel compositions and methods of making and using same. U.S. Pat. Off. Gaz.; 2011.
153. Harvey BG, Wright ME, Quintana RL. High-density renewable fuels based on the selective dimerization of pinenes. *Energy Fuel* 2010;**24**:267–73.
154. Ignea C, Cvetkovic I, Loupassaki S, Kefalas P, Johnson CB, Kampranis SC, et al. Improving yeast strains using recyclable integration cassettes, for the production of plant terpenoids. *Microb Cell Fact* 2011;**10**:4.
155. Reiling KK, Yoshikuni Y, Martin VJ, Newman J, Bohlmann J, Keasling JD. Mono and diterpene production in Escherichia coli. *Biotechnol Bioeng* 2004;**87**:200–12.
156. Murugesan A, Umarani C, Chinnusamy TR, Krishnan M, Subramanian R, Neduzchezhain N. Production and analysis of bio-diesel from non-edible oils—a review. *Renew Sustain Energy Rev* 2009;**13**:825–34.
157. Murugesan A, Umarani C, Subramanian R, Neduzchezhain N. Bio-diesel as an alternative fuel for diesel engines—a review. *Renew Sustain Energy Rev* 2009;**13**:653–62. http://www.sciencedirect.com/science/article/pii/S1364032107001438.
158. Rakopoulos CD, Antonopoulos KA, Rakopoulos DC, Hountalas DT, Giakoumis EG. Comparative performance and emissions study of a direct injection diesel engine using blends of diesel fuel with vegetable oils or bio-diesels of various origins. *Energ Convers Manage* 2006;**47**:3272–87.
159. Qi DH, Geng LM, Chen H, Bian YZH, Liu J, Ren XCH. Combustion and performance evaluation of a diesel engine fueled with biodiesel produced from soybean crude oil. *Renew Energy* 2009;**34**:2706–13.
160. Monthly Biodiesel Production Report—Energy Information Administration. http://www.eia.gov/biofuels/biodiesel/production/.

161. Narwal SK, Gupta R. Biodiesel production by transesterification using immobilized lipase. *Biotechnol Lett* 2013;**35**:479–90.
162. Basha SA, Gopal KR, Jebaraj S. A review on biodiesel production, combustion, emissions and performance. *Renew Sustain Energy Rev* 2009;**13**:1628–34.
163. Aktas DF, Lee JS, Little BJ, Ray RI, Davidova IA, Lyles CN, et al. Anaerobic metabolism of biodiesel and its impact on metal corrosion. *Energy Fuel* 2010;**24**:2924–8.
164. Singh B, Korstad J, Sharma YC. A critical review on corrosion of compression ignition (CI) engine parts by biodiesel and biodiesel blends and its inhibition. *Renew Sustain Energy Rev* 2012;**16**:3401–8.
165. Wang C, Tian Z, Wang L, Xu R, Liu Q, Qu W, et al. One-step hydrotreatment of vegetable oil to produce high quality diesel-range alkanes. *ChemSusChem* 2012;**5**:1974–83.
166. Beopoulos A, Cescut J, Haddouche R, Uribelarrea JL, Molina-Jouve C, Nicaud JM. Yarrowia lipolytica as a model for bio-oil production. *Prog Lipid Res* 2009;**48**:375–87.
167. Maier T, Leibundgut M, Ban N. The crystal structure of a mammalian fatty acid synthase. *Science* 2008;**321**:1315–22.
168. Edwards AL, Matsui T, Weiss TM, Khosla C. Architectures of whole-module and bimodular proteins from the 6-deoxyerythronolide B synthase. *J Mol Biol* 2014;**426**:2229–45.
169. Boehringer D, Ban N, Leibundgut M. 7.5-Å cryo-em structure of the mycobacterial fatty acid synthase. *J Mol Biol* 2013;**425**:841–9.
170. Lomakin IB, Xiong Y, Steitz TA. The crystal structure of yeast fatty acid synthase, a cellular machine with eight active sites working together. *Cell* 2007;**129**:319–32.
171. Jenni S, Leibundgut M, Maier T, Ban N. Architecture of a fungal fatty acid synthase at 5 A resolution. *Science* 2006;**311**:1263–7.
172. Tehlivets O, Scheuringer K, Kohlwein SD. Fatty acid synthesis and elongation in yeast. *Biochim Biophys Acta* 2007;**1771**:255–70.
173. Liu T, Khosla C. Genetic engineering of Escherichia coli for biofuel production. *Annu Rev Genet* 2010;**44**:53–69.
174. Steen EJ, Kang Y, Bokinsky G, Hu Z, Schirmer A, McClure A, et al. Microbial production of fatty-acid-derived fuels and chemicals from plant biomass. *Nature* 2010;**463**:559–62.
175. Zheng Y-N, Li L-L, Liu Q, Yang J-M, Wang X-W, Liu W, et al. Optimization of fatty alcohol biosynthesis pathway for selectively enhanced production of C12/14 and C16/18 fatty alcohols in engineered Escherichia coli. *Microb Cell Fact* 2012;**11**:65.
176. Schirmer A, Rude MA, Li X, Popova E, del Cardayre SB. Microbial biosynthesis of alkanes. *Science* 2010;**329**:559–62.
177. Mills T, Sandoval N, Gill R. Cellulosic hydrolysate toxicity and tolerance mechanisms in Escherichia coli. *Biotechnol Biofuels* 2009;**2**:26.
178. Jin H, Chen L, Wang J, Zhang W. Engineering biofuel tolerance in non-native producing microorganisms. *Biotechnol Adv* 2014;**32**:541–8.
179. Dunlop MJ. Engineering microbes for tolerance to next-generation biofuels. *Biotechnol Biofuels* 2011;**4**:32.
180. Mosier N, Wyman C, Dale B, Elander R, Lee YY, Holtzapple M, et al. Features of promising technologies for pretreatment of lignocellulosic biomass. *Bioresour Technol* 2005;**96**:673–86.
181. Sainz MB. Commercial cellulosic ethanol: the role of plant-expressed enzymes. *In Vitro Cell Dev Biol Plant* 2009;**45**:314–29.
182. Olsson L, Hahn-Hägerdal B. Fermentation of lignocellulosic hydrolysates for ethanol production. *Enzyme Microb Technol* 1996;**18**:312–31.
183. Axe DD, Bailey JE. Transport of lactate and acetate through the energized cytoplasmic membrane of Escherichia coli. *Biotechnol Bioeng* 1995;**47**:8–19.
184. Cherrington CA, Hinton M, Chopra I. Effect of short-chain organic acids on macromolecular synthesis in Escherichia coli. *J Appl Bacteriol* 1990;**68**:69–74.
185. Haselkorn R, Doty P. The reaction of formaldehyde with polynucleotides. *J Biol Chem* 1961;**236**:2738–45.

186. Modig T, Lidén G, Taherzadeh MJ. Inhibition effects of furfural on alcohol dehydrogenase, aldehyde dehydrogenase and pyruvate dehydrogenase. *Biochem J* 2002;**363**:769–76.
187. Bowles LK, Ellefson WL. Effects of butanol on Clostridium acetobutylicum. *Appl Environ Microbiol* 1985;**50**:1165–70.
188. Ding J, Huang X, Zhang L, Zhao N, Yang D, Zhang K. Tolerance and stress response to ethanol in the yeast Saccharomyces cerevisiae. *Appl Microbiol Biotechnol* 2009;**85**:253–63.
189. Ramos JL, Duque E, Gallegos MT, Godoy P, Ramos-Gonzalez MI, Rojas A, et al. Mechanisms of solvent tolerance in gram-negative bacteria. *Annu Rev Microbiol* 2002;**56**:743–68.
190. Stambuk BU, Dunn B, Alves Jr SL, Duval EH, Sherlock G. Industrial fuel ethanol yeasts contain adaptive copy number changes in genes involved in vitamin B1 and B6 biosynthesis. *Genome Res* 2009;**19**:2271–8.
191. Mohagheghi A, Linger J, Smith H, Yang S, Dowe N, Pienkos PT. Improving xylose utilization by recombinant Zymomonas mobilis strain 8b through adaptation using 2-deoxyglucose. *Biotechnol Biofuels* 2014;**7**:19.
192. Zhang YX, Perry K, Vinci VA, Powell K, Stemmer WP, del Cardayré SB. Genome shuffling leads to rapid phenotypic improvement in bacteria. *Nature* 2002;**415**:644–6.
193. Zhu B-F, Xu Y, Fan W-L. High-yield fermentative preparation of tetramethylpyrazine by Bacillus sp. using an endogenous precursor approach. *J Ind Microbiol Biotechnol* 2010;**37**:179–86.
194. Shi D-J, Wang C-L, Wang K-M. Genome shuffling to improve thermotolerance, ethanol tolerance and ethanol productivity of Saccharomyces cerevisiae. *J Ind Microbiol Biotechnol* 2009;**36**:139–47.
195. Winkler JD, Garcia C, Olson M, Callaway E, Kao KC. Evolved osmotolerant Escherichia coli mutants frequently exhibit defective N-acetylglucosamine catabolism and point mutations in cell shape-regulating protein MreB. *Appl Environ Microbiol* 2014;**80**:3729–40.
196. Patnaik R, Louie S, Gavrilovic V, Perry K, Stemmer WP, Ryan CM, et al. Genome shuffling of Lactobacillus for improved acid tolerance. *Nat Biotechnol* 2002;**20**:707–12.
197. Cooper TF. Recombination speeds adaptation by reducing competition between beneficial mutations in populations of Escherichia coli. *PLoS Biol* 2007;**5**, e225.
198. Winkler J, Kao KC. Harnessing recombination to speed adaptive evolution in Escherichia coli. *Metab Eng* 2012;**14**:487–95.
199. Gonzalez R, Tao H, Purvis JE, York SW, Shanmugam KT, Ingram LO. Gene array-based identification of changes that contribute to ethanol tolerance in ethanologenic Escherichia coli: comparison of KO11 (parent) to LY01 (resistant mutant). *Biotechnol Prog* 2003;**19**:612–23.
200. Minty JJ, Lesnefsky AA, Lin F, Chen Y, Zaroff TA, Veloso AB, et al. Evolution combined with genomic study elucidates genetic bases of isobutanol tolerance in Escherichia coli. *Microb Cell Fact* 2011;**10**:18.
201. Alper H, Moxley J, Nevoigt E, Fink GR, Stephanopoulos G. Engineering yeast transcription machinery for improved ethanol tolerance and production. *Science* 2006;**314**:1565–8.
202. Gall S, Lynch MD, Sandoval NR, Gill RT. Parallel mapping of genotypes to phenotypes contributing to overall biological fitness. *Metab Eng* 2008;**10**:382–93.
203. Warnecke TE, Lynch MD, Karimpour-Fard A, Lipscomb ML, Handke P, Mills T, et al. Rapid dissection of a complex phenotype through genomic-scale mapping of fitness altering genes. *Metab Eng* 2010;**12**:241–50.
204. Woodruff LB, Pandhal J, Ow SY, Karimpour-Fard A, Weiss SJ, Wright PC, et al. Genome-scale identification and characterization of ethanol tolerance genes in Escherichia coli. *Metab Eng* 2013;**15**:124–33.
205. Lynch MD, Warnecke T, Gill RT. SCALEs: multiscale analysis of library enrichment. *Nat Methods* 2007;**4**:87–93.
206. Fralick JA. Evidence that TolC is required for functioning of the Mar/AcrAB efflux pump of Escherichia coli. *J Bacteriol* 1996;**178**:5803–5.
207. Ma D, Cook DN, Alberti M, Pon NG, Nikaido H, Hearst JE. Genes acrA and acrB encode a stress-induced efflux system of Escherichia coli. *Mol Microbiol* 1995;**16**:45–55.

208. White DG, Goldman JD, Demple B, Levy SB. Role of the acrAB locus in organic solvent tolerance mediated by expression of marA, soxS, or robA in Escherichia coli. *J Bacteriol* 1997;**179**:6122–6.
209. Kobayashi H, Yamamoto M, Aono R. Appearance of a stress-response protein, phage-shock protein A, in Escherichia coli exposed to hydrophobic organic solvents. *Microbiology* 1998;**144**(Pt 2):353–9.
210. Woodruff LBA, Boyle NR, Gill RT. Engineering improved ethanol production in Escherichia coli with a genome-wide approach. *Metab Eng* 2013;**17**:1–11.
211. Warner JR, Reeder PJ, Karimpour-Fard A, Woodruff LBA, Gill RT. Rapid profiling of a microbial genome using mixtures of barcoded oligonucleotides. *Nat Biotechnol* 2010;**28**:856–62.
212. Shoemaker DD, Lashkari DA, Morris D, Mittmann M, Davis RW. Quantitative phenotypic analysis of yeast deletion mutants using a highly parallel molecular bar-coding strategy. *Nat Genet* 1996;**14**:450–6.
213. Sandoval NR, Kim JY, Glebes TY, Reeder PJ, Aucoin HR, Warner JR, et al. Strategy for directing combinatorial genome engineering in Escherichia coli. *Proc Natl Acad Sci U S A* 2012;**109**:10540–5.
214. Ferrante AA, Augliera J, Lewis K, Klibanov AM. Cloning of an organic solvent-resistance gene in Escherichia coli: the unexpected role of alkylhydroperoxide reductase. *Proc Natl Acad Sci U S A* 1995;**92**:7617–21.
215. Poole LB. Bacterial defenses against oxidants: mechanistic features of cysteine-based peroxidases and their flavoprotein reductases. *Arch Biochem Biophys* 2005;**433**:240–54.
216. Seaver LC, Imlay JA. Alkyl hydroperoxide reductase is the primary scavenger of endogenous hydrogen peroxide in Escherichia coli. *J Bacteriol* 2001;**183**:7173–81.
217. Stephanopoulos G. Synthetic biology and metabolic engineering. *ACS Synth Biol* 2012;**1**:514–25.
218. Ellis T, Adie T, Baldwin GS. DNA assembly for synthetic biology: from parts to pathways and beyond. *Integr Biol* 2011;**3**:109–18.
219. Tsai C-S, Kwak S, Turner TL, Jin Y-S. Yeast synthetic biology toolbox and applications for biofuel production. *FEMS Yeast Res* 2015;**15**:1–15. http://dx.doi.org/10.1111/1567-1364.12206.
220. Avitabile A, Carlomagno-Cerillo S, Favvre R, Blasi F. Isolation of transducing bacteriophages for the histidine and isoleucine-valine operons in Escherichia coli K-12. *J Bacteriol* 1972;**112**:40–7.
221. Lennox ES. Transduction of linked genetic characters of the host by bacteriophage P1. *Virology* 1955;**1**:190–206.
222. Groth AC, Olivares EC, Thyagarajan B, Calos MP. A phage integrase directs efficient site-specific integration in human cells. *Proc Natl Acad Sci U S A* 2000;**97**:5995–6000.
223. Haldimann A, Wanner BL. Conditional-replication, integration, excision, and retrieval plasmid-host systems for gene structure-function studies of bacteria. *J Bacteriol* 2001;**183**:6384–93.
224. Boyd D, Weiss DS, Chen JC, Beckwith J. Towards single-copy gene expression systems making gene cloning physiologically relevant: lambda InCh, a simple Escherichia coli plasmid-chromosome shuttle system. *J Bacteriol* 2000;**182**:842–7.
225. Hoekstra WP, Bergmans JE, Zuidweg EM. Role of recBC nuclease in Escherichia coli transformation. *J Bacteriol* 1980;**143**:1031–2.
226. Jasin M, Schimmel P. Deletion of an essential gene in Escherichia coli by site-specific recombination with linear DNA fragments. *J Bacteriol* 1984;**159**(2):783–6. http://europepmc.org/backend/ptpmcrender.fcgi?accid=PMC215717&blobtype=pdf.
227. Russell CB, Thaler DS, Dahlquist FW. Chromosomal transformation of Escherichia coli recD strains with linearized plasmids. *J Bacteriol* 1989;**171**:2609–13.
228. Dillingham MS, Kowalczykowski SC. RecBCD enzyme and the repair of double-stranded DNA breaks. *Microbiol Mol Biol Rev* 2008;**72**:642–71.
229. Cox MM. Regulation of bacterial RecA protein function. *Crit Rev Biochem Mol Biol* 2007;**42**:41–63.
230. Murphy KC. Use of bacteriophage lambda recombination functions to promote gene replacement in Escherichia coli. *J Bacteriol* 1998;**180**:2063–71.
231. Mosberg JA, Lajoie MJ, Church GM. Lambda red recombineering in Escherichia coli occurs through a fully single-stranded intermediate. *Genetics* 2010;**186**:791–9.
232. Datsenko KA, Wanner BL. One-step inactivation of chromosomal genes in Escherichia coli K-12 using PCR products. *Proc Natl Acad Sci U S A* 2000;**97**:6640–5.

233. Ryu YS, Biswas RK, Shin K, Parisutham V, Kim SM, Lee SK. A simple and effective method for construction of Escherichia coli strains proficient for genome engineering. *PLoS One* 2014;**9**, e94266.
234. Kuhlman TE, Cox EC. Site-specific chromosomal integration of large synthetic constructs. *Nucleic Acids Res* 2010;**38**, e92.
235. Zhang Y, Buchholz F, Muyrers JP, Stewart AF. A new logic for DNA engineering using recombination in Escherichia coli. *Nat Genet* 1998;**20**:123–8.
236. Zhang Y, Muyrers JP, Testa G, Stewart AF. DNA cloning by homologous recombination in Escherichia coli. *Nat Biotechnol* 2000;**18**:1314–7.
237. Fu J, Bian X, Hu S, Wang H, Huang F, Seibert PM, et al. Full-length RecE enhances linear-linear homologous recombination and facilitates direct cloning for bioprospecting. *Nat Biotechnol* 2012;**30**:440–6.
238. Wang HH, Isaacs FJ, Carr PA, Sun ZZ, Xu G, Forest CR, et al. Programming cells by multiplex genome engineering and accelerated evolution. *Nature* 2009;**460**:894–8.
239. Isaacs FJ, Carr PA, Wang HH, Lajoie MJ, Sterling B, Kraal L, et al. Precise manipulation of chromosomes in vivo enables genome-wide codon replacement. *Science* 2011;**333**:348–53.
240. Wang HH, Kim H, Cong L, Jeong J, Bang D, Church GM. Genome-scale promoter engineering by coselection MAGE. *Nat Methods* 2012;**9**:591–3.
241. Nagy A. Cre recombinase: the universal reagent for genome tailoring. *Genesis* 2000;**26**:99–109.
242. Birling M-C, Gofflot F, Warot X. Site-specific recombinases for manipulation of the mouse genome. *Methods Mol Biol* 2009;**561**:245–63.
243. García-Otín AL, Guillou F. Mammalian genome targeting using site-specific recombinases. *Front Biosci* 2006;**11**:1108–36.
244. Santos CNS, Regitsky DD, Yoshikuni Y. Implementation of stable and complex biological systems through recombinase-assisted genome engineering. *Nat Commun* 2013;**4**:2503.
245. Capecchi MR. Gene targeting in mice: functional analysis of the mammalian genome for the twenty-first century. *Nat Rev Genet* 2005;**6**:507–12.
246. Rouet P, Smih F, Jasin M. Expression of a site-specific endonuclease stimulates homologous recombination in mammalian cells. *Proc Natl Acad Sci U S A* 1994;**91**:6064–8.
247. Choulika A, Perrin A, Dujon B, Nicolas JF. Induction of homologous recombination in mammalian chromosomes by using the I-SceI system of Saccharomyces cerevisiae. *Mol Cell Biol* 1995;**15**:1968–73.
248. Segal DJ, Meckler JF. Genome engineering at the dawn of the golden age. *Annu Rev Genomics Hum Genet* 2013;**14**:135–58.
249. Sorek R, Lawrence CM, Wiedenheft B. CRISPR-mediated adaptive immune systems in bacteria and archaea. *Annu Rev Biochem* 2013;**82**:237–66.
250. Cong L, Ran FA, Cox D, Lin S, Barretto R, Habib N, et al. Multiplex genome engineering using CRISPR/Cas systems. *Science* 2013;**339**:819–23.
251. Sander JD, Joung JK. CRISPR-Cas systems for editing, regulating and targeting genomes. *Nat Biotechnol* 2014;**32**:347–55.
252. Jiang W, Bikard D, Cox D, Zhang F, Marraffini LA. RNA-guided editing of bacterial genomes using CRISPR-Cas systems. *Nat Biotechnol* 2013;**31**:233–9.
253. Bikard D, Jiang W, Samai P, Hochschild A, Zhang F, Marraffini LA. Programmable repression and activation of bacterial gene expression using an engineered CRISPR-Cas system. *Nucleic Acids Res* 2013;**41**:7429–37.
254. Qi LS, Larson MH, Gilbert LA, Doudna JA, Weissman JS, Arkin AP, et al. Repurposing CRISPR as an RNA-guided platform for sequence-specific control of gene expression. *Cell* 2013;**152**:1173–83.
255. Cheng AW, Wang H, Yang H, Shi L, Katz Y, Theunissen TW, et al. Multiplexed activation of endogenous genes by CRISPR-on, an RNA-guided transcriptional activator system. *Cell Res* 2013;**23**:1163–71.
256. Kuscu C, Arslan S, Singh R, Thorpe J, Adli M. Genome-wide analysis reveals characteristics of off-target sites bound by the Cas9 endonuclease. *Nat Biotechnol* 2014;**32**:677–83.
257. Ran FA, Hsu PD, Lin CY, Gootenberg JS, Konermann S, Trevino AE, et al. Double nicking by RNA-guided CRISPR Cas9 for enhanced genome editing specificity. *Cell* 2013;**154**:1380–9.

CHAPTER 10

Evolutionary Methods for Improving the Production of Biorenewable Fuels and Chemicals

Tao Jin, Yingxi Chen, Laura R. Jarboe

Department of Chemical and Biological Engineering, Iowa State University, Ames, Iowa, USA

10.1 Introduction

Metabolic engineering is, by definition, a rational and predictive field.[1,2] This type of deliberate, planned approach has enabled the production of a wide variety of biorenewable fuels and chemicals, including but not limited to ethanol, butanol, styrene, and carboxylic acids.[3–6] Similarly, an engineering approach has proven useful in enabling the utilization of renewable biomass-derived carbon sources, such as pentose sugars,[7–9] anhydrosugars,[10] and syngas.[11,12] However, our incomplete and imperfect knowledge of microbial metabolism, physiology, and regulatory networks hinders our ability to consistently and efficiently obtain strains with sufficiently high titer, yield, and/or productivity through rational engineering alone.

One of the challenges frequently associated with the production of biorenewable compounds at a high yield and titer from biomass-derived substrates is biocatalyst inhibition.[13–19] This inhibition can either be from the desired product or from nonsugar components of the "dirty" feedstock. This inhibition can sometimes be mitigated via *in situ* product removal[20–25] or treatment of the feedstock,[26–35] though this may incur an unacceptable increase in cost.

Another option for dealing with this inhibition is to modify the biocatalyst for increased tolerance of the inhibitory compounds. Metabolic engineering for increased tolerance has included strategies such as provision of the appropriate exporter[36,37] and engineering of stronger membranes.[38,39] However, this rational approach to increasing tolerance can be limited when either the mechanism of toxicity is not known and/or there is not a clear engineering strategy to mitigate the identified inhibition mechanism. Omics analysis can prove quite useful in identifying the mechanism of inhibition,[40–43] though interpretation of omics data is dependent upon the quality and thoroughness of enzyme annotation and regulatory network characterization.

Biotechnology for Biofuel Production and Optimization. http://dx.doi.org/10.1016/B978-0-444-63475-7.00010-8

Evolutionary strategies for improving biocatalyst performance provide an excellent counterpoint to the rational design strategy. Evolution can be used to improve tolerance to exogenous inhibitors, improve utilization of a target substrate, or improve production of the target compound. Such strategies are not hindered by erroneous or incomplete enzyme annotation or lack of regulatory network characterization. When the appropriate selective pressure is applied and the key mutations supporting the evolved phenotype are identified, novel design strategies can be distilled and our understanding of the biocatalyst can increase. As we increase the pool of available design strategies and our understanding of the organism, our ability to apply rational, predictive engineering strategies also increases (Figure 10.1).

The goal of this chapter is to describe the use of evolution to improve the production of biorenewable fuels and chemicals. Other similar works have been published recently on this topic.[44–53] Here we give an overview of how evolution works, the development of evolution-based strain design tools, evolution to increase tolerance to exogenously supplied inhibitors, evolution to improve strain performance, and reverse engineering to identify design principles. Note that evolutionary efforts that target specific enzymes have recently been reviewed elsewhere[54–56] and are not discussed here.

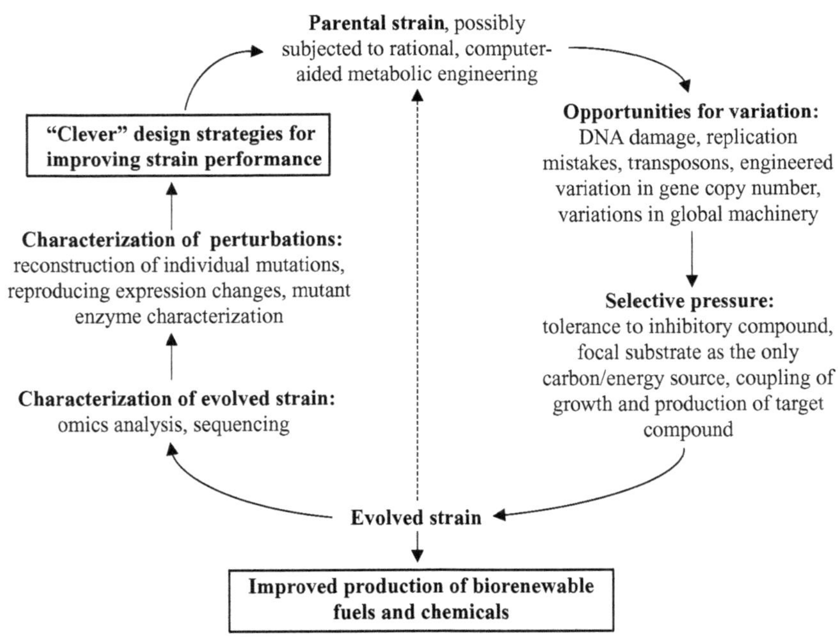

Figure 10.1

Evolutionary strain improvement can play an important role in the strain design cycle. Characterization of the evolved strain and its sequence and expression changes can yield novel design strategies for future improvement.

10.2 Mechanisms of Evolution

The power of natural selection is promising and very useful for the modification of microorganisms for a variety of properties. Put more succinctly as Orgel's Second Rule, "evolution is cleverer than you are".[57,58] In this work, we are interested in the use of adaptive evolution to improve microbial production of biorenewable fuels and chemicals, such as biocatalyst tolerance to toxic compounds or operating conditions, or to otherwise improve strain performance.

10.2.1 How Does Evolution Work?

The essence of evolution is perturbation of the DNA sequence. These changes can occur within coding regions (genes) or noncoding regions. Changes within coding regions can either impact the resulting amino acid sequence, or be "silent" mutations that change the codon identity but not the resulting amino acid. Changes to the genome sequence can occur through a variety of events, as described elsewhere.[59–61] These events can involve changes of single nucleotides or insertion or deletion of various-sized segments.

Single-base-pair (point) mutations can arise either through damage to the DNA or during DNA replication. The physical structure of the nucleotides can be impacted by mutagens, such as radiation or ethidium bromide.[62,63] Mutations can also arise via incorporation of the wrong nucleotide during DNA replication. Though proofreading systems exist to detect and repair these types of errors, they are not infallible.[64,65]

As opposed to the single-base-pair changes described above, many observed mutations involve the insertion or deletion of more than just one base pair. These types of insertions and deletions can possibly arise via transposons and retrotransposons. Transposons are a class of genetic elements which can duplicate themselves and move to different positions within a genome. Retrotransposons are a subclass of transposons, which need RNA to be intermediates. Transposons can have dual impacts on their host, in that they can confer mutations that are either beneficial or harmful. However, in adaptive evolution, transposons are considered to be an accelerator of the evolutionary process.[66] Large-scale genomic rearrangements have also been observed.[67] Another possible mutation type is the insertion or deletion of a small number of nucleotides, that is, in-frame deletion of a single amino acid.[68]

Most microorganisms have a relatively low and stable mutation rate. For example, *Escherichia coli* averages 0.0025 mutations per genome per replication and *Saccharomyces cerevisiae* maintains 0.0027 mutations per generation.[69] Mutations to the proofreading systems, general stress response, or DNA repair systems can increase this mutation rate.[60,61,70] For example, RpoS is a well-characterized regulator of the general stress response and two mechanisms of the RpoS-mediated response can increase the mutation rate of stressed cells.

One mechanism is the increased expression of the error-prone DNA polymerase, Pol IV; the other is the decreased expression of the mismatch repair enzymes.[60,61,71]

Thus, there are a variety of natural mechanisms for generating diversity in microbial genomes (Table 10.1). These mechanisms include single-base-pair changes, either within a coding or noncoding region; short insertions or deletions that can possibly truncate a gene or change the number of amino acids; gene silencing or increased expression via transposon insertion; or large-scale rearrangements in genome organization. One can envision that these changes can

Table 10.1: Representative mutations within the *E. coli* genome

Location	Type	Possible impacts	Example	Ref.
Coding	"Silent" single nucleotide change where the amino acid sequence is not changed	Transcription and translation rate, protein folding	Silent mutations in *ompA* impacted transcription and translation. Silent mutations in recombinant EgFABP1 gene affected translation and protein folding.	72,73
Coding	Nonsynonymous; single nucleotide change where amino acid sequence is altered	Transcription and translation rate, enzyme activity	Single point mutation in *rho* decreased the activity of Rho-dependent terminators. Three single point mutations in *lpdA* eliminated NADH sensitivity.	68,74
Noncoding	Single nucleotide changes	Transcription and translation of nearby genes	Four different single point mutations in the *mgl* operator (*mglO*) increased the glucose transport rate.	75
Coding	Insertion	Silencing or increased expression of the affected gene and possible downstream genes	An insertion mutation in *hfq* led to significant pleiotropic phenotypes, such as increased cell size and decreased growth rates. An IS10 insertion mutation in *yqhC* affected the transcription of downstream genes *yqhD* and *dkgA*.	76,77
Noncoding	Insertion or deletion	Perturbed expression of nearby genes	Insertion mutations upstream of the *bgl* promoter activated the silent *bgl* operon. Deletion mutation in the *mgl* operator (*mglO*) increased the glucose transport rate.	75,78
Coding	Deletion	In-frame deletions can alter enzyme activity. Out-of-frame deletions can result in translation termination and altered or absent enzyme activity.	The single amino acid deletion in *metJ* altered MetJ activity, resulting in increased expression of the methionine biosynthesis pathway.	68

influence the strain behavior in a variety of ways, including but not limited to perturbation of transcript abundance, transcript stability, translation initiation, translation efficiency, and enzyme amino acid sequence. This change in amino acid sequence could impact protein function, protein stability, and interaction with other cellular components. In this manner, directed evolution enables screening of these variants and selection of those that confer the desired behavior.

10.2.2 Directed Evolution

Directed evolution performed in the laboratory seeks to exploit the natural evolution process in order to obtain strains with desired properties. There are a variety of techniques for guided laboratory evolution. Each of these essentially aims to obtain evolved strain(s) by propagating the parent strain in a controlled environment. The specifics of the propagation methods and the controlled environment depend on the experimental purpose.

A traditional example of this propagation is the sequential transfer method.[79–81] Essentially, cells are grown in a condition which is stressful enough to slow growth, but not stressful enough to halt DNA replication. As the cells grow in this stressful condition, some will acquire mutations, as described above. Those that have acquired a mutation that enhances growth in the condition of interest will grow faster than their cohorts. As additional beneficial mutations are acquired, one strain will eventually become the dominant strain within the population. During this time, the selective pressure can be increased. Since mutations are largely acquired during the growth phase and since growth is the selective marker, it is desirable to maintain the cells in a state of near-continual growth. Ideally, such growth would involve the use of a carefully maintained chemostat culture. However, in practice this is often approximated through the use of serial batch cultures. Since the culture at the end of the serial transfer process presumably contains a variety of strains, it is very important to isolate and characterize single colonies before proceeding. Briefly, the culture is diluted into new growth medium on a regular basis, that is, every 24 h, and the magnitude of the selective pressure is gradually increased. Finally, the evolved strain is obtained when the cells display the desired behavior. Here we briefly describe several cases where this type of approach was used.

In order to increase thermal tolerance, *E. coli* cells were sequentially transferred in minimal medium for 523 days. During the course of this evolution, the temperature was increased from 36.9 to 44.8 °C. The final evolved strain was able to grow at 45.9 °C, while the original parent strain cannot grow at 43.2 °C.[79]

A variety of studies have aimed to increase ethanol tolerance. For example, ethanologenic *E. coli* KC01 was sequentially transferred in rich medium with 50 g L^{-1} xylose for 6 weeks while the concentration of ethanol was increased from 10 to 40 g L^{-1}. The evolved strain SZ470 could successfully grow anaerobically in the presence of 40 g L^{-1} ethanol. This represents a twofold increase in ethanol tolerance relative to the parent strain.[80] Yomano et al. took a slightly different approach to select for ethanol tolerance. Ethanologenic *E. coli* strain KO11 was subjected to alternating rounds of growth in liquid and on solid media. The liquid media contained selective concentrations of ethanol. At the end of the growth period on liquid media, cells were plated onto solid media and colonies with high ethanol production capability were identified based on size and morphology.[82] This alternation between liquid and solid media represents a useful deviation from the standard liquid-to-liquid serial transfer method.

Recently, a new sequential transfer method was created and demonstrated by increasing n-butanol tolerance in *E. coli*. This new method is named "visualizing evolution in real time" (VERT)[83,84] and is based on traditional evolution, but can track distinct improvements in fitness by fluorescence-based techniques.[83,84] In the VERT scheme, subpopulations of the parental strain are generated. These subpopulations are isogenic except for the recombinant expression of distinct fluorescence proteins. These fluorescence reporters enable rapid determination of the population distribution. When the various subpopulations are growing at similar rates, the population distribution is roughly equal between the various groups. When a cell in one subpopulation acquires a useful mutation that enables faster growth, this subpopulation will become dominant. This alerts researchers that a useful mutation has occurred.

10.3 Engineered Methods of Evolution

Modification of microbial biocatalysts for some complex functionalities, such as increased tolerance or production of desired compounds, can require perturbation of multiple genes. However, our ability to rationally plan and then introduce these modifications can be limited by our knowledge of microbial physiology and metabolism and resource availability. These combinations can also require unacceptably lengthy directed evolution schemes.

10.3.1 Global Transcription Machinery Engineering

Global transcription machinery engineering (gTME) aims to introduce these multiple perturbations to the transcriptome by altering the transcription machinery. This method has proven to be an appealing and efficient tool to improve the performance of biocatalysts.[85]

One of the first examples of gTME was the improvement of ethanol tolerance and production by *S. cerevisiae*.[86] A library of mutated versions of the *SPT15* gene, which encodes the TATA-binding protein, was generated by error-prone PCR. This led to increased ethanol tolerance and yield, as described below. *SPT15* was also selected for mutation in a gTME project that aimed to improve xylose tolerance and consumption by *S. cerevisiae*.[87] The mutant library was screened in xylose defined medium; the resulting strain showed a conversion efficiency of 93.5% for $50\,g\,L^{-1}$ xylose. This conversion efficiency is comparable to the 98.3% observed for $50\,g\,L^{-1}$ glucose. Even when glucose and xylose were supplied at a ratio of 1:1, the utilization was 97.3% and 90.8%, respectively.

The power of gTME has also been demonstrated in *E. coli*.[85] The *rpoD* gene, which encodes the main sigma factor σ^{70}, was subjected to random mutagenesis, along with its promoter region, for three rounds of error-prone PCR. After screening and selection from the library in the presence of $50\,g\,L^{-1}$ ethanol, the best mutant showed better growth in the presence of 20-70 $g\,L^{-1}$ ethanol relative to the strain harboring only the wild-type *rpoD*. gTME was also demonstrated to increase the production of lycopene by 50% and increase tolerance to multiple inhibitors, such as sodium dodecyl sulfate.

Global transcription factor cyclic AMP receptor protein (CRP) of *E. coli* was subjected to gTME for increasing the tolerance to 1-butanol.[88] Four mutants with improved performance were selected from the library. DNA shuffling was then used to combine these beneficial mutations. The final mutant MT5 showed a growth rate of $0.18\,h^{-1}$ in the presence of $9.7\,g\,L^{-1}$ 1-butanol, which was twice as high as that observed for the parent strain.

Thus, gTME has proven to be an effective method for enabling complex phenotypes in a variety of organisms. These complex phenotypes include tolerance of inhibitory compounds and production of target compounds.

10.3.2 SCALEs

gTME aims to perturb the expression of a large number of genes by changing the sequence, and therefore behavior, of proteins involved in controlling gene expression. SCALEs aims to perturb the expression of multiple genes by increasing the copy number of genomic DNA fragments of various sizes. Libraries of cells containing these fragments are then enriched for their ability to confer a phenotype of interest. The cells isolated by this enrichment procedure are typically then subjected to a genome-wide, microarray-type multiscale analysis.[89]

For example, *E. coli* K12 was used to construct five genomic libraries for the demonstration and validation of SCALEs.[89] Transformants carrying the plasmid-based DNA fragments of different sizes were cultured in a continuous fermentation system for selection and sampled periodically. Microarray and multiscale analysis were then conducted on the enriched plasmids to identify the position and size of the inserts contributing to the increased fitness.

This approach was used to identify *E. coli* K12 genes responsible for the increased tolerance of *E. coli* to acetate. This interest in acetate tolerance was motivated by the fact that acetate is a common inhibitor found in biomass hydrolysate.[90] The populations surviving in the presence of $1.75\,g\,L^{-1}$ acetate were selected and subjected to microarray and multiscale analysis. In this manner, a list of genes whose overexpression contributed to high acetate tolerance was obtained. This list includes such genes as *murC* for peptidoglycan biosynthesis and *yjdL* for dipeptide transporter.

The SCALEs method was also employed to improve the growth of *E. coli* NZN111.[91] Previous metabolic engineering of NZN111 had resulted in a growth deficiency due to a high NADH/NAD^+ ratio. The SCALEs analysis led to the identification of 9 *E. coli* K12 genes that can increase the growth of NZN111 approximately fivefold when their expression was increased. The enriched population also showed a 20% increase in succinate production.

Other methods for improving the relative fitness of biocatalysts include COMPACTER[92] and TRMR.[93]

10.4 Evolution for Increased Tolerance

As described in many recent reviews,[17,94–96] the metabolic engineering of biocatalysts for economically viable production of desired compounds is often restricted by product-mediated inhibition, such as that imposed by organic solvents[97,98] and carboxylic acids.[18] Additionally, low-cost, renewable feedstocks, such as those derived from biomass or industrial by-products, often contain compounds that can inhibit biocatalyst growth and metabolism.[99] One approach for dealing with this inhibition is selective removal of the inhibitory compound as it is formed.[100–102] Another approach is to evolve the microbes for improved tolerance of the inhibitory compounds[19,103] (Table 10.2). Performing reverse engineering to reproduce the phenotype or trait of interest can aid in the identification of design principles and increased characterization of the host microbe, as described below.

Table 10.2: The use of directed evolution to increase tolerance to inhibitory products or inhibitors in biomass-derived sugars

Strain	Inhibitory compound	Tolerance	Duration of evolution	% increase compared to parent strain	Ref.
E. coli SA481	Isobutanol	0.8% (w/v)	45 transfers	100	104
E. coli G3.2	Isobutanol	2% (w/v) (glucose)	500 generations	60	105
E. coli X3.5	Isobutanol	1.75% (w/v) (xylose)	430 generations	40	105
S. cerevisiae	Isobutanol	1.9% (v/v)	30 transfers	58	106
JBA-mut	2-Butanol	3% (v/v)	30 transfers		106
	1-Butanol	1.9% (v/v)	30 transfers		106
P. putida JUCT1	Cyclohexane	60% (v/v)	12 transfers	1100	107
	Decalin	60% (v/v)	12 transfers		107
	Methyl cyclohexane	60% (v/v)	12 transfers		107
	Toluene	60% (v/v)	12 transfers		107
E. coli EMFR9	Furfural	0.13% (w/v)	54 transfers	160	108
	5-HMF	0.25% (w/v)	54 transfers		109
S. cerevisiae RK60-5	Simulated spruce hydrolysate	60% (w/v)	429 generations	200	110
S. cerevisiae KE1-17	Spruce hydrolysate	50% (v/v)	97 generations	150	110

Organic solvents, including next-generation biofuels such as butanol,[19] are highly toxic to microorganisms. This toxicity is largely attributed to their hydrophobicity and the resulting damage to the cell membrane.[111,112] A variety of rational engineering strategies have been applied to increase the tolerance of biocatalysts to these inhibitory products. Here we briefly summarize two of these. As mentioned above, provision of biocatalysts with appropriate efflux pumps has been shown to increase tolerance to inhibitory products.[19] For example, *marA*, which is responsible for activation of the AcrAB-TolC efflux pump, was overexpressed in *E. coli* leading to the increased tolerance to geraniol.[113] On solid media with 1.0% (v/v) geraniol, the strain harboring overexpressed *marA* gene showed 10^4-fold colony-forming efficiency compared to that of the parent strain. Additionally, the intracellular accumulation of geraniol was only half of the parent strain. Organic solvents can interfere with appropriate protein folding, and thus the GroESL chaperone system, which facilitates protein folding, was overexpressed in *E. coli*. The resulting strain had increased tolerance to organic solvents such as ethanol, *n*-, 2-, *i*-butanol and 1,2,4-butanetriol.[114]

The standard sequential transfer method has been used in a variety of cases to adapt microbes to inhibitory environments (Table 10.3).[125] Here we summarize a few of these instances. Uses of modified evolution techniques are described at the end of this section.

Directed evolution was conducted in batch fermentors to acquire a strain of *E. coli* with increased isobutanol tolerance. After 45 sequential transfers, tolerance increased from $4\,g\,L^{-1}$ to $8\,g\,L^{-1}$.[104] This demonstrates the ability of the sequential transfer method to

Table 10.3: The use of directed evolution to increase production of biorenewable fuels and chemicals

Strain	Product	Medium	Production by evolved strain	Ref.
E. coli KO12	Ethanol	Luria Broth + $100\,g\,L^{-1}$ glucose	1.2 M	115
E. coli KO11	Ethanol	Luria Broth + $80\,g\,L^{-1}$ xylose	0.90 M	115
E. coli TCS009 e50rep1/pLOI297	Ethanol	Defined minimal medium + $40\,g\,L^{-1}$ glycerol	0.39 M	116
S. cerevisiae GS1.11-26	Ethanol	YP + $35\,g\,L^{-1}$ xylose	$0.46\,g\,g^{-1}$	117
S. cerevisiae T1-E	Ethanol	Defined minimal medium + $25\,g\,L^{-1}$ lactose	0.22 M	118
E. coli SZ63	D-Lactic acid	M9 + $50\,g\,L^{-1}$ glucose	0.54 M	119
E. coli SZ132	D-Lactic acid	Luria Broth + $100\,g\,L^{-1}$ glucose	1.0 M	120
E. coli SZ132	D-Lactic acid	NBS + $100\,g\,L^{-1}$ glucose	0.70 M	120
E. coli SZ194	D-Lactic acid	NBS + $120\,g\,L^{-1}$ glucose	1.2 M	121
E. coli SZ85	L-Lactic acid	M9 + $50\,g\,L^{-1}$ glucose	0.51 M	122
E. coli HX024	Succinate	NBS + $120\,g\,L^{-1}$ glucose	0.81 M	74
E. coli KJ060	Succinate	AM1 + $100\,g\,L^{-1}$ glucose	~0.70 M	123
E. coli KJ071	Malate	AM1 + $100\,g\,L^{-1}$ glucose	0.52 M	123
E. coli XZ132	L-Alanine	AM1 + $120\,g\,L^{-1}$ glucose	1.3 M	124

obtain significant gains in tolerance in a relatively short time. In another isobutanol-focused project, parallel directed evolution was applied to multiple populations of *E. coli* simultaneously in minimal medium containing isobutanol with either glucose or xylose as carbon source.[105] After approximately 500 generations in glucose medium and 430 generations in xylose medium, the tolerance of the mutants was improved from 0.75% (w/v) to 2% (w/v) and 1.75% (w/v), respectively. However, the strain evolved on glucose also showed increased isobutanol tolerance in xylose minimal media and the strain evolved on xylose showed increased isobutanol tolerance in glucose minimal media. Thus, the specificity of the adaptation in the two carbon sources was relatively low. When these evolved strains were further characterized in rich media with isobutanol, there was no significant gain in fitness relative to the parent. These results highlight the fact that selected mutations can be specific to the experimental conditions (media type) while other aspects of the experimental condition can be less impactful (carbon source).

Another evolutionary approach to improve isobutanol production by *E. coli* is focused on tolerance of the valine analog norvaline.[126] It should be noted that the isobutanol production pathway in this case is this same pathway that is used to produce valine. Through a combination of mutagenic treatment, selection for growth in the presence of norvaline, and repair of a mutation within *rpoS*, the final strain NV3r1 was obtained. This strain produced 21.2 $g\,L^{-1}$ isobutanol at a yield of 0.31 $g\,g^{-1}$, relative to the 12.0 $g\,L^{-1}$ and 0.24 $g\,g^{-1}$ observed for the parent strain.

Directed evolution has also been applied to microorganisms other than *E. coli*. A butanol-tolerant *S. cerevisiae* strain was obtained after 30 serial transfers in the presence of 2-butanol.[106] The resulting mutant was able to maintain growth in 3% (v/v) 2-butanol, a concentration which completely inhibited growth of the parent strain. The evolved strain also showed increased tolerance to other butanol isomers. Specifically, the evolved strain was able to grow in the presence of 1.9% (v/v) 1-butanol and 1.9% (v/v) isobutanol, while no growth was observed for the parent strain in these concentrations. Similarly, when *Pseudomonas putida* was evolved for cyclohexane tolerance, it also acquired enhanced tolerance to the organic solvents decalin, methyl cyclohexane, and toluene.[107] This evolved strain was acquired by only 12 sequential transfers with cyclohexane content increased from an original value of 5% (v/v) to a final concentration of 60% (v/v). Note that evolved strain was evaluated in 60% (v/v) of each of the solvents listed. These two studies demonstrate that adaptive mutations that confer tolerance one inhibitor can also be beneficial in conferring tolerance to other, chemically similar, inhibitors.

Carboxylic acids are very appealing biorenewable chemicals for synthesizing a variety of commodity chemicals.[127,128] However, as with organic solvents, the fact that they hinder the performance of the biocatalyst is still an issue that needs to be addressed for economically viable production.[129,130] Many rational engineering efforts have been taken to improve the resistance of microbes to the free fatty acids. For example, the A as pathway for incorporating medium-chain fatty acids into membrane phospholipids was deleted in *E. coli* with the expression of the gene encoding cytosolic acyl-ACP thioesterase from *Umbellularia californica* to alleviate the toxicity and restore the strength of cell membrane.[131] The colony-forming units of the engineered strain were twice as high as those of the parent strain in the presence of 1 g L^{-1} of 12:0 and 14:0 fatty acids. By employing an alternative thioesterase from *Geobacillus* sp. Y412MC10, the membrane content of unsaturated fatty acids was decreased, thus increasing the tolerance to the free fatty acids.[38] An evolutionary approach has also been used to improve tolerance to octanoic acid in *E. coli.*[132] Specifically, 17 transfers were performed in glucose minimal media over the course of approximately 700 h. Initially, 10 mM octanoic acid was used, though the concentration was gradually increased to 30 mM. The resulting evolved strain showed increased tolerance not only to octanoic acid, but also to hexanoic acid, decanoic acid, and somewhat surprisingly, isobutanol and n-butanol.

In addition to dealing with inhibitory products, biocatalysts can also be limited by the inhibitory compounds contained in the substrates. Biomass-derived sugars have been utilized as substrate for the production of biorenewable chemicals and fuels.[133,134] However, inhibitory compounds are often formed during the pretreatment process.[16,135] Some of these compounds, including furan derivatives, aldehydes, and organic acids, have been under investigation regarding their impact on microbes.[16,43,136–138] Furfural is the most thoroughly characterized of these biomass-associated inhibitors.[108,139–144]

A furfural-resistant *E. coli* strain EMFR9 was derived from ethanologenic LY180 based on growth in batch fermentors with gradually increased furfural concentration from 0.5 to 1.3 g L^{-1} during 54 serial transfers.[108] In minimal medium with 100 g L^{-1} xylose, this resulting strain could produce approximately 40 g L^{-1} ethanol in the presence of 1.0 g L^{-1} furfural; the parent LY180 could not even maintain growth in such conditions. The increased tolerance of EMFR9 to furfural also resulted in increased tolerance to 5-hydroxymethyl furfural (5-HMF).[109] Specifically, EMFR9 produced more than 40 g L^{-1} ethanol after 96 h in minimal medium with 100 g L^{-1} xylose and 2.5 g L^{-1} 5-HMF. Note that 1.0 g L^{-1} 5-HMF completely inhibited growth of the parent strain.

Directed evolution has also been applied to directly improve the tolerance of biocatalysts to lignocellulosic biomass hydrolysate. *S. cerevisiae* strain TMB3400 was evolved in batch cultures with 20 g L^{-1} glucose and xylose, respectively, and 12 inhibitory compounds. Specifically, these cultures included HMF, furfural, acetic acid, formic acid, levulinic acid, vanillin, syringaldehyde, coniferyl aldehyde, hydroquinone, cinnamic acid, 4-hydroxybenzoic acid, and guaiacyl acetone, each of which have been detected in spruce hydrolysate.[110] When characterized in the presence of 20 wt.% of this simulated spruce hydrolysate "cocktail," the evolved strain's maximum specific growth rate ($0.33\,h^{-1}$) was double that of the parent and the length of the lag phase was similarly reduced twofold. The original parent strain was also evolved in a chemostat for 97 generations using actual spruce hydrolysate, as opposed to the simulated cocktail. Initially, 20% (v/v) hydrolysate was used and the concentration was gradually increased to 50% (v/v). Mutants were selected from both evolutionary procedures and further characterized. Strains from both groups demonstrated a 25-38% increase in specific sugar consumption rates and a 32-50% increase in specific ethanol productivity compared to the parent strain using actual spruce hydrolysate.

Modified evolutionary strategies have also been used to increase tolerance to organic solvents. gTME was used on the global transcription factor CRP for improving the tolerance of *E. coli* to isobutanol.[145] One of the resulting mutants showed growth of $0.18\,h^{-1}$, compared to $0.05\,h^{-1}$ of the parent strain, in $9.6\,g\,L^{-1}$ of isobutanol.

VERT was applied to *E. coli* BW25113 for improving its tolerance to *n*-butanol in chemostat with M9 minimal medium and $5\,g\,L^{-1}$ glucose.[146] After 144 generations, the resistance of the evolved strain was increased from 0.5% (v/v) to 1.3% (v/v). VERT was also used to identify the adaptive events of *S. cerevisiae* derived from FY2 during corn stover hydrolysate utilization.[147] The growth rate of the best mutant was increased by 57% in 25% (v/v) hydrolysate relative to the parent strain.

Here we have described the use of serial transfer-based evolution, gTME and VERT to increase the tolerance of a microbial biocatalyst to specific inhibitory compounds. It can be seen that these are effective tools and sometimes the acquired tolerance also applies to other compounds. However, as described in the next section, acquiring tolerance with the goal of increasing production of the focal compound, has proven challenging.

10.5 Evolution for Increased Production

Besides enhancing biocatalyst tolerance to both substrates and products, metabolic engineering of biocatalysts for improving their performance of production is also of great importance. In some cases, evolution-based strategies can be designed so that selecting for cells with faster growth co-selects for strains with improved production abilities (Table 10.3).

E. coli W was engineered for ethanol production by chromosomal insertion of the *pdc* and *adhB* genes from *Zymomonas mobilis*.[115] Successive dilutions were transferred to solid media for selecting large colonies with higher expression of these two genes, enabling higher ethanol production. Serial transfers in rich media with 10% glucose were then used to stabilize both the alcohol dehydrogenase activity and the high resistance to Cm, where Cm resistance is indicative of *pdc*/*adhB* expression. These serial transfers were performed for more than 60 generations. The resulting strains were able to produce more than $50\,g\,L^{-1}$ ethanol from $100\,g\,L^{-1}$ glucose and more than $40\,g\,L^{-1}$ ethanol from $80\,g\,L^{-1}$ xylose.

E. coli strain TCS009, a derivative of MG1655 engineered for the production of ethanol from glycerol, was also improved through co-selection of growth and ethanol production.[116] After 500 serial dilutions in minimal media with glycerol as sole carbon source, the specific growth rate had increased 1.8-fold. Ethanol production by this evolved strain was also improved both in terms of yield and rate. Specifically, the evolved strain produced ethanol from glycerol at a yield of $0.45 \pm 0.01\ g/g^{-1}$ over 48 h, while its parent had a yield of $0.34 \pm 0.00\ g/g^{-1}$ over 72 h.

This growth-based strain improvement was also used as part of the engineering of *E. coli* W3110 for D-lactic acid production. Alternating periods of rational pathway engineering and serial transfer-based metabolic evolution were used to fine-tune strain performance. In this case, the engineered central metabolic pathway enabled co-selection for growth and lactic acid production because the engineered strain was only able to regenerate NAD^+ through the production of lactic acid. The final strain was able to produce more than 500 mM D-lactic acid with 98% purity from mineral salts medium containing 5% glucose.[119]

E. coli KO11 was modified for D-lactate production by alternating rounds of rationally selected genetic manipulations and directed evolution. This reengineering included elimination of the ethanol pathway, along with other competing pathways. This pathway elimination left the lactate-producing pathway as the sole route for regenerating NAD^+. Then, the engineered strain was subjected to sequential transfer-based evolution. By co-selecting growth and lactate production, the resulting strain SZ132 was able to produce 1000 ± 20 mmol D-lactate in rich medium with 10% (w/v) glucose and 700 ± 11 mmol D-lactate from defined NBS medium, also with 10% (w/v) glucose.[120] SZ132 was further modified to eliminate loss of carbon due to coproduct formation. The poor performance of the resulting strain SZ186 in mineral salts medium motivated further rounds of serial transfer-based metabolic evolution to improve the ability to completely utilize glucose.[121] 1.23 ± 0.03 M D-lactate could be produced by the final strain SZ194 from NBS medium with 12% (w/v) glucose.

A similar strategy was used for succinate production. *E. coli* ATCC 8739 was rationally engineered for succinate production by the traditional elimination of competing pathways. In this case, cell growth was coupled with succinate production since the succinate-synthesizing pathway was the sole route for regeneration of NAD^+. With the alternating use of metabolic engineering and evolution for more than 1000 generations, the resulting strain HX024 was able to produce 813 ± 28 mM succinate from NBS mineral salts medium with 12% (w/v) glucose, which was 4.5-fold more than the parent strain.[74]

Similarly, deleting the alternative NAD^+-generating pathways resulted in succinate production as the only route for oxidizing NADH. After conducting 2000 generations of metabolic evolution, the selected mutants produced 622-733 mM succinate from AM1 medium with $100\ g\ L^{-1}$ glucose and 516 mM malate from NBS medium with $100\ g\ L^{-1}$ glucose.[123]

The examples described above were all able to redesign the central metabolic pathways so that the biocatalyst could only maintain redox balance by producing the target compound. In this manner, cells that acquired mutations that enabled increased production of the target compound would also have increased ability to turn over NADH, resulting in increased biomass production and growth. There are also examples of metabolic pathway redesigns so that production of the target compound is the only way to produce the ATP needed for growth.

For example, the L-lactic acid producer SZ194 described above was reengineered for alanine production by replacing the *ldhA* gene with the gene encoding an alanine dehydrogenase from *Geobacillus stearothermophilus*.[124] In the redesigned metabolic pathways, cells could produce ATP by producing alanine. Through a similar alternating strategy of rational engineering and serial transfer-based evolution, the resulting strain could convert 120 g L^{-1} of glucose to 1279 mmol L-alanine within 48 h.

A similar strategy was used to engineer *E. coli* for the production of optically pure L-lactic acid. *E. coli* SZ63,[119] initially engineered for D-lactate production, was modified for L-lactate production by replacing the native *ldhA* gene with the *ldhL* gene from *Pediococcus acidilactici*.[122] Metabolic evolution was then required to improve the expression of the foreign gene, as determined by comparison of growth and lactate dehydrogenase activity to the L-lactic acid producing parent strain. After a period of evolutionary selection, the selected strain SZ85 was able to produce 505 ± 12 mM L-lactate from M9 medium with 5% (w/v) glucose within 24 h.

Computational strain design via the OptKnock optimization algorithm was used to predict manipulations needed for lactic acid production by *E. coli* strain MG1655.[148] Growth of this engineered strain was coupled to lactate production and thus approximately 1000 generations of adaptive evolution in minimal media followed the strain engineering. Eleven different strains were characterized by this method. The best lactate producer attained a titer of 19 mM lactate from 0.2% glucose.

Directed evolution was also applied to biocatalysts in conjunction with targeted genetic manipulations to enhance strain performance. An industrial *S. cerevisiae* strain was modified for the production of ethanol from pentose sugars by introducing the D-xylose-utilization cassette from *Clostridium phytofermentans*. This engineered strain was then subjected to both treatment with the chemical mutagen ethyl methanesulfonate (EMS) and genome shuffling. The EMS treatment aimed to increase the mutation rate and the genome shuffling provided large-scale rearrangements that were unlikely to occur naturally. The resulting strain was then subjected to directed evolution in the form of 11 sequential batch cultures in the presence of increasing concentrations of xylose. The initial xylose concentration was 40 g L^{-1}, the final concentration was 100 g L^{-1}. The final strain produced ethanol at a yield of 0.46 g g^{-1} in YP medium from a mixture of xylose and glucose. This yield was double that observed for the parent strain.[117]

Another example of evolutionary improvement of yeast behavior is the improvement of lactose utilization by *S. cerevisiae* T1.[118] After serial transfers in batch fermenters, the ethanol production from lactose was increased from 7.1±0.8 to 10.52±0.04 $g\,L^{-1}$.

10.6 Enabling Increased Production by Increasing Tolerance

As described above, it is sometimes possible to co-select for increased production of a target compound and growth. This enables evolution-based selection of strains with the ability to produce the target compound at increased yield, titer, and/or productivity. But such selection schemes are not always feasible. In such cases, strains are often evolved for increased tolerance of an exogenously supplied target compound, with the goal of increasing the strain's ability to produce the target compound. This has met with mixed results (Table 10.4).

Various projects have aimed to increase the ethanol tolerance of ethanologenic *E. coli* strains through directed evolution. Recombinant KO11 grew poorly in the presence of 35 $g\,L^{-1}$ ethanol and had a maximum observed ethanol titer of more than 50 $g\,L^{-1}$. Its evolved derivative LY01 could tolerate up to 50 $g\,L^{-1}$ of ethanol and was able to produce 60 $g\,L^{-1}$ ethanol from 140 $g\,L^{-1}$ of either glucose or xylose within 72 h.[82] Note that the evolution of strain LY01 included selection for high ethanol-producing strains, as evaluated by colony size and morphology on solid media. Strain SZ470, the evolved derivative of nontransgenic *E. coli* KC01,[80] not only had increased ethanol tolerance but also had improved ethanol production. Specifically, within 72 h SZ470 produced 23.5 $g\,L^{-1}$ ethanol from 50 $g\,L^{-1}$ xylose; KC01 only produced 16 $g\,L^{-1}$.

A noncarboxylic acid producing derivative of *E. coli* strain MG1655 was evolved for tolerance to exogenously supplied octanoic acid. When the evolved strain LAR was provided with the thioesterase for octanoic acid production, the observed octanoic acid titer was approximately twice as high as that observed for the parent strain, also encoding the thioesterase.[132]

Table 10.4: Increased tolerance leading to increased production

Strain	Medium	Production	% increase relative to parent strain	Ref.
E. coli LY01	Luria Broth + 140 $g\,L^{-1}$ glucose	61±3 $g\,L^{-1}$ ethanol	42.8	82
E. coli LY01	Luria Broth + 140 $g\,L^{-1}$ xylose	63±1 $g\,L^{-1}$ ethanol	42.8	82
E. coli SZ470	Luria Broth + 50 $g\,L^{-1}$ xylose	23.5 $g\,L^{-1}$ ethanol	100	80
E. coli LAR1 + pJMY-EEI82564	Luria Broth + 15 $g\,L^{-1}$ glucose	540±50 $mg\,L^{-1}$ carboxylic acid	400	132
S. cerevisiae spt15-300	Minimal medium + 100 $g\,L^{-1}$ glucose	0.40 g ethanol g^{-1} glucose	n/a	86

In one of the first demonstrations of gTME, TATA-binding protein Spt15 of *S. cerevisiae* was targeted, with the goal of improving tolerance to 20% (v/v) ethanol and up to 120 g L^{-1} glucose.[86] The selected mutant strain had an ethanol yield of 0.40 g/g, an increase of 15% relative to the parent.

However, increased tolerance does not always contribute to improved production. This disconnects between tolerance of exogenously supplied inhibitor and increased ability to produce the inhibitor has also been observed in some rational engineering-based design schemes.[38] *E. coli* strain SA481 was obtained by evolution of an isobutanol-producing strain for tolerance to exogenously supplied isobutanol.[104] However, the isobutanol titer of 20 g L^{-1} achieved by this evolved strain in standard fermentation conditions was similar to the parent.

10.7 Reverse Engineering of Evolved Strains

As is evident from the cases described above, evolution-based strain improvement is a valuable partner for rational engineering strategies. Our reliance on evolutionary techniques is partially due to incomplete understanding of our biocatalysts. Investing the time and effort to fully characterize the evolved strains and identify and understand the mutations that confer evolved strain behavior can increase our understanding of these organisms (Table 10.5). This

Table 10.5: Examples of reverse engineering projects involving *E. coli*

Motivation	Characterization method	Results	Ref.
Increasing isobutanol tolerance and production	Whole-genome sequencing	Silencing of *acrA*, *gatY*, *tnaA*, *yhbJ* and *marCRAB* enables increased isobutanol tolerance.	104
Increasing furfural tolerance	Transcriptome analysis, targeted sequencing	YqhD is the primary NADPH-dependent furfural reductase; YqhC is a regulator of *yqhD*	76,108
Understand the impact of ethanol on gene expression machinery	Genome sequencing, Transcriptome analysis, Ribosome profiling	RpsQ [H31P]: decreased the effect of ethanol on miscoding errors during translation; Rho [L270M]: decreased activity of Rho-dependent terminators; MetJ [ΔE91]: increased expression of methionine biosynthesis gene	68
Increasing succinate production	Genome sequencing, Targeted enzyme assay, Real-time reverse transcription-PCR	A point mutation in *ptsI* allows more PEP to be directed towards succinate; Increased usage of Pck pathway instead of Ppc pathway enables increased ATP production.	149
Increasing succinate production	Genome sequencing; Transcriptome sequencing; Targeted enzyme assay;	Mutations in *lpdA* eliminated the NADH sensitivity of the pyruvate dehydrogenase enzyme complex, which enabled increased succinate production.	74

increased understanding can in turn lead to decreased dependence on evolutionary methods. We refer to this characterization of evolved strains as reverse engineering.

Recently, characterization tools such as genome sequencing, transcriptome analysis, and metabolic flux analysis, have become faster and cheaper, all of which support reverse engineering efforts. Reverse engineering aims not to identify all mutations that occurred during evolution, but to instead identify the crucial mutations which enable the evolved phenotype, and to understand how these mutations are beneficial. Reverse engineering is a tool to expose the story behind the evolution experiments, which can in turn accelerate the pace of future biorenewable fuels and chemicals design processes.

Identifying and understanding the key mutations that support the evolved phenotype requires the knowledge of what mutations occurred during the evolutionary engineering. Whole-genome sequencing is invaluable to this goal.[150] As described above, isobutanol-producing *E. coli* was subjected to directed evolution to increase both isobutanol tolerance and isobutanol production.[104] The evolved strain, SA481, was found to have acquired one single nucleotide change, 25 insertion sequences and a genomic deletion involving 62 genes. Twenty-eight of these mutations were individually reconstructed in the parent strain and tested for their contribution to the evolved phenotype. Five of these mutations were found to contribute to the evolved phenotype. These five mutations, essentially deletion of *acrA*, *gatY*, *tnaA*, *yhbJ*, and *marCRAB*, were then combined into a single strain TW306 and the resulting phenotype was assessed. The engineered strain TW306 reproduced the evolved strain phenotype.[104]

After finding the important mutations, the next step is to explore which mutations promote fitness, the mechanisms of how the tolerance to inhibitors has increased, and the functions of poorly characterized enzymes and pathways involved in the evolved phenotype. In addition to the use of genome sequence as described above, transcriptome analysis and metabolic flux analysis have proven quite useful in such efforts.[43,108]

For the ethanologenic *E. coli* strain evolved for furfural tolerance, transcriptome analysis led to the discovery that YqhD is the primary NADPH-dependent furfural reductase in *E. coli*.[108,151] Sequence analysis of the region surrounding *yqhD* identified an insertion mutation within *yqhC* in the evolved strain. This mutation was found to be responsible for the decreased expression of YqhD because YqhC is a regulator of *yqhD*.[76]

E. coli strain MG1655 was subjected to evolution for increased ethanol tolerance and then characterized via a variety of techniques, including genome sequencing, transcriptome analysis and ribosome profiling. Six mutated genes (*metJ*, *rho*, *rpsQ*, *ispB*, *lptF*, *topA*) were identified in evolved strain MTA156. Only one of these mutations, H122L within *topA*, was found to have no impact on ethanol tolerance. The main goal of this study was to understand the impact of ethanol on gene expression machinery, and thus these researchers focused the rest of their characterization on the MetJ transcriptional repressor, Rho transcription termination factor, and RpsQ ribosomal subunit protein. The H31P mutation in RpsQ was found to decrease the effect of ethanol on miscoding errors during translation. This decrease in translational misreading was due to enhanced ribosome accuracy. The *rho* [L270M] mutation led to decreased activity of Rho-dependent terminators, where these terminators can cause abnormal transcription termination under ethanol stress. The ΔE91mutation within MetJ results in increased expression of methionine biosynthesis genes; this presumably leads to increased methionine availability, preventing some ribosomal stalling.[68] The thorough reverse engineering efforts employed here have provided multiple new mechanisms of ethanol toxicity, as well as design strategies for mitigating this toxicity.

As described above, a metabolic pathway was designed in *E. coli* C so that production of succinate was linked to growth. It was found that in this evolved strain, the glucose uptake system had switched from the traditional PEP-based phosphotransferase system to the GalP permease. This change occurred via a mutation in *ptsI* and allowed more PEP to be directed towards succinate, instead of being converted to pyruvate via glucose import. A second change in the evolved strain was the increased usage of the PEP carboxykinase (Pck) pathway to convert PEP to OAA instead of the PEP carboxylase (Ppc) pathway. The Pck pathway produces a molecule of ATP for every conversion of PEP to OAA while the PPC pathway does not. This results in the production of more ATP in the evolved strain. This change in pathway usage was evidenced by both increased activity of the Ppc pathway and increased expression of the genes encoding this pathway. This increased expression was attributed to at least three distinct mutations. These results demonstrate the ability to identify important mutations and pathway changes even in the absence of whole-genome sequence data.[149]

A distinct strain of *E. coli* engineered for succinate production, NZ-037, was also subjected to metabolic evolution to improve its succinate production. The evolved strain, HX024, was found to have three distinct single base pair changes within the *lpdA* gene. Each of these nucleotide changes resulted in a change in the corresponding amino acid. This alteration to the LpdA protein sequence eliminated the NADH sensitivity of the pyruvate dehydrogenase enzyme complex, of which LpdA is a component. This decreased NADH sensitivity enabled increased succinate production.[74]

Eleven *E. coli* strains subjected to computationally designed engineering strategies and then evolved by co-selecting for growth and lactate production were subjected to whole-genome sequencing and characterization.[152] Seven of these sequenced strains were found to contain a 82-bp deletion mutation in the *rph-pyrE* region. *rph* encodes an RNase and *pyrE* encodes an enzyme involved in biosynthesis of pyrimidine nucleotides. Introduction of this mutation into the parent strain was shown to increase growth by approximately 15%.

These examples demonstrate how reverse engineering efforts can reveal novel design strategies and increase our understanding of microbial physiology and metabolism.

10.8 Modeling of Evolution

There has been much work performed recently in modeling the evolutionary process *in silico*. This work is discussed here only briefly. A thermodynamic perspective has also been applied to these models and, as supported by experimental data of *Thermoanaerobacterium saccharolyticum* evolution, shown that the rate of entropy production increases as a strain is allowed to evolve.[153] This entropy increase was shown to be asymptotically approaching a maximum rate of entropy production predicted according to the Boltzmann distribution. Further consideration of this evolved organism in terms of metabolic control analysis successfully predicted the enzyme likely to be altered during the evolution of *T. saccharolyticum* in serial cultures.[154] Specifically, the metabolic reaction performed by phosphoglucose isomerase had the large flux control coefficient of the various metabolic reactions and when sequenced in 10 evolved clones, the gene encoding phosphoglucose isomerase was found to contain a variety of mutations. These results demonstrate the ability of a predictive model to prioritize targets for engineering, targeted enzyme evolution, or analysis when characterizing evolved isolates.

10.9 Concluding Statement

We have summarized the use of serial dilution and engineering evolutionary methods to improve the production of biorenewable fuels and chemicals. These evolutionary tools have proven quite effective and as reverse engineering tools become more accessible, additional design strategies can be distilled from these evolved strains.

References

1. Bailey JE. Toward a science of metabolic engineering. *Science* 1991;**252**(5013):1668–75.
2. Stephanopoulos G, Vallino JJ. Network rigidity and metabolic engineering in metabolite overproduction. *Science* 1991;**252**(5013):1675–81.
3. McKenna R, Nielsen DR. Styrene biosynthesis from glucose by engineered *E. coli*. *Metab Eng* 2011;**13**(5):544–54.
4. Ranganathan S, Tee TW, Chowdhury A, Zomorrodi AR, Yoon JM, Fu Y, et al. An integrated computational and experimental study for overproducing fatty acids in *Escherichia coli*. *Metab Eng* 2012;**14**(6):687–704.
5. Bankar SB, Survase SA, Ojamo H, Granstrom T. Biobutanol: the outlook of an academic and industrialist. *RSC Adv* 2013;**3**(47):24734–57.
6. Liang Y, Zhao X, Chi Z, Rover M, Johnston P, Brown R, et al. Utilization of acetic acid-rich pyrolytic bio-oil by microalga *Chlamydomonas reinhardtii*: reducing bio-oil toxicity and enhancing algal toxicity tolerance. *Bioresour Technol* 2013;**133**:500–6.
7. Kim SR, Ha S-J, Wei N, Oh EJ, Jin Y-S. Simultaneous co-fermentation of mixed sugars: a promising strategy for producing cellulosic ethanol. *Trends Biotechnol* 2012;**30**(5):274–82.
8. Kim SR, Park Y-C, Jin Y-S, Seo J-H. Strain engineering of Saccharomyces cerevisiae for enhanced xylose metabolism. *Biotechnol Adv* 2013;**31**(6):851–61.
9. Madhavan A, Srivastava A, Kondo A, Bisaria VS. Bioconversion of lignocellulose-derived sugars to ethanol by engineered *Saccharomyces cerevisiae*. *Crit Rev Biotechnol* 2012;**32**(1):22–48.

10. Layton DS, Ajjarapu A, Choi DW, Jarboe LR. Engineering ethanologenic *Escherichia coli* for levoglucosan utilization. *Bioresour Technol* 2011;**102**(17):8318–22.
11. Bengelsdorf FR, Straub M, Duerre P. Bacterial synthesis gas (syngas) fermentation. *Environ Technol* 2013;**34**(13-14):1639–51.
12. Daniell J, Koepke M, Simpson SD. Commercial biomass syngas fermentation. *Energies* 2012;**5**(12):5372–417.
13. Jarboe LR, Wen Z, Choi D, Brown RC. Hybrid thermochemical processing: fermentation of pyrolysis-derived bio-oil. *Appl Microbiol Biotechnol* 2011;**91**(6):1519–23.
14. Jonsson LJ, Alriksson B, Nilvebrant N-O. Bioconversion of lignocellulose: inhibitors and detoxification. *Biotechnol Biofuels* 2013;**6**:13.
15. Lo T-M, Teo WS, Ling H, Chen B, Kang A, Chang MW. Microbial engineering strategies to improve cell viability for biochemical production. *Biotechnol Adv* 2013;**31**(6):903–14.
16. Mills TY, Sandoval NR, Gill RT. Cellulosic hydrolysate toxicity and tolerance mechanisms in *Escherichia coli*. *Biotechnol Biofuels* 2009;**2**:26.
17. Nicolaou SA, Gaida SM, Papoutsakis ET. A comparative view of metabolite and substrate stress and tolerance in microbial bioprocessing: from biofuels and chemicals, to biocatalysis and bioremediation. *Metab Eng* 2010;**12**(4):307–31.
18. Jarboe LR, Royce LA, Liu P. Understanding biocatalyst inhibition by carboxylic acids. *Front Microbiol* 2013;**4**:272.
19. Dunlop MJ. Engineering microbes for tolerance to next-generation biofuels. *Biotechnol Biofuels* 2011;**4**:32.
20. Chen Y, Ren H, Liu D, Zhao T, Shi X, Cheng H, et al. Enhancement of n-butanol production by in situ butanol removal using permeating-heating-gas stripping in acetone-butanol-ethanol fermentation. *Bioresour Technol* 2014;**164**:276–84.
21. Hemavathy RV, Sankaran K, Vadanasundari V, Rangabhashiyam S. in situ separation of ethanol with aqueous two-phase system and assessment of K(L)a for yeast growth in batch cultivation. *Prep Biochem Biotechnol* 2014;**44**(6):633–44.
22. Huang H-J, Ramaswamy S, Liu Y. Separation and purification of biobutanol during bioconversion of biomass. *Sep Purif Technol* 2014;**132**:513–40.
23. Teh AL, Layton DS, Hyduke DR, Jarboe LR, Rollins DK. Data mining based on principal component analysis: application to the nitric oxide response in *Escherichia coli*. *J Stat Sci Appl* 2014;**2**:1–18.
24. Lopez-Garzon CS, Straathof AJJ. Recovery of carboxylic acids produced by fermentation. *Biotechnol Adv* 2014;**32**(5):873–904.
25. Xu K, Xu P. Efficient calcium lactate production by fermentation coupled with crystallization-based in situ product removal. *Bioresour Technol* 2014;**163**:33–9.
26. Chan JKS, Duff SJB. Methods for mitigation of bio-oil extract toxicity. *Bioresour Technol* 2010;**101**(10):3755–9.
27. Chi Z, Rover M, Jun E, Deaton M, Johnston P, Brown RC, et al. Overliming detoxification of pyrolytic sugar syrup for direct fermentation of levoglucosan to ethanol. *Bioresour Technol* 2013;**150**:220–7.
28. Datta S, Lin YJ, Schell DJ, Millard CS, Ahmad SF, Henry MP, et al. Removal of acidic impurities from corn stover hydrolysate liquor by resin wafer based electrodeionization. *Ind Eng Chem Res* 2013;**52**(38):13777–84.
29. Fakhrudin J, Setyaningsih D, Rahayuningsih M. Bioethanol production from seaweed *Eucheuma cottonii* by neutralization and detoxification of acidic catalyzed hydrolysate. *Int J Environ Sci Dev* 2014;**5**(5):455–8.
30. Grzenia DL, Wickramasinghe SR, Schell DJ. Fermentation of reactive-membrane-extracted and ammonium-hydroxide-conditioned dilute-acid-pretreated corn stover. *Appl Biochem Biotechnol* 2012;**166**(2):470–8.
31. Guo X, Cavka A, Jonsson LJ, Hong F. Comparison of methods for detoxification of spruce hydrolysate for bacterial cellulose production. *Microb Cell Factories* 2013;**12**:93.

32. Hodge DB, Andersson C, Berglund KA, Rova U. Detoxification requirements for bioconversion of softwood dilute acid hydrolyzates to succinic acid. *Enzym Microb Technol* 2009;**44**(5):309–16.
33. Mateo S, Roberto IC, Sanchez S, Moya AJ. Detoxification of hemicellulosic hydrolyzate from olive tree pruning residue. *Ind Crop Prod* 2013;**49**:196–203.
34. Rover MR, Johnston PA, Jin T, Smith RG, Brown RC, Jarboe L. Production of clean pyrolytic sugars for fermentation. *Chemsuschem* 2014;**7**(6):1662–8.
35. Stoutenburg RM, Perrotta JA, Nakas JP. Overcoming inhibitors in a hemicellulosic hydrolysate: improving fermentability by feedstock detoxification and adaptation of *Pichia stipitis*. *J Ind Microbiol Biotechnol* 2011;**38**(12):1939–45.
36. Dunlop MJ, Dossani ZY, Szmidt HL, Chu HC, Lee TS, Keasling JD, et al. Engineering microbial biofuel tolerance and export using efflux pumps. *Mol Syst Biol* 2011;**7**:487.
37. Frederix M, Huetter K, Leu J, Batth TS, Turner WJ, Rueegg TL, et al. Development of a native *Escherichia coli* induction system for ionic liquid tolerance. *PLoS ONE* 2014;**9**(7):e101115.
38. Lennen RM, Pfleger BF. Modulating membrane composition alters free fatty acid tolerance in *Escherichia coli*. *PLoS ONE* 2013;**8**(1):e54031.
39. Luo LH, Seo P-S, Seo J-W, Heo S-Y, Kim D-H, Kim CH. Improved ethanol tolerance in *Escherichia coli* by changing the cellular fatty acids composition through genetic manipulation. *Biotechnol Lett* 2009;**31**(12):1867–71.
40. Anfelt J, Hallstrom B, Nielsen J, Uhlen M, Hudson EP. Using transcriptomics to improve butanol tolerance of *Synechocystis* sp strain PCC 6803. *Appl Environ Microbiol* 2013;**79**(23):7419–27.
41. He M-X, Wu B, Shui Z-X, Hu Q-C, Wang W-G, Tan F-R, et al. Transcriptome profiling of *Zymomonas mobilis* under furfural stress. *Appl Microbiol Biotechnol* 2012;**95**(1):189–99.
42. Brynildsen MP, Liao JC. An integrated network approach identifies the isobutanol response network of *Escherichia coli*. *Mol Syst Biol* 2009;**5**:277.
43. Miller EN, Jarboe LR, Turner PC, Pharkya P, Yomano LP, York SW, et al. Furfural inhibits growth by limiting sulfur assimilation in ethanologenic *Escherichia coli* strain LY180. *Appl Environ Microbiol* 2009;**75**(19):6132–41.
44. Abatemarco J, Hill A, Alper HS. Expanding the metabolic engineering toolbox with directed evolution. *Biotechnol J* 2013;**8**(12):1397–410.
45. Cakar ZP, Turanli-Yildiz B, Alkim C, Yilmaz U. Evolutionary engineering of *Saccharomyces cerevisiae* for improved industrially important properties. *FEMS Yeast Res* 2012;**12**(2):171–82.
46. Conrad TM, Lewis NE, Palsson BO. Microbial laboratory evolution in the era of genome-scale science. *Mol Syst Biol* 2011;**7**:509.
47. Dragosits M, Mattanovich D. Adaptive laboratory evolution—principles and applications for biotechnology. *Microb Cell Factories* 2013;**12**:64.
48. Furusawa C, Horinouchi T, Hirasawa T, Shimizu H. Systems Metabolic Engineering: The Creation of Microbial Cell Factories by Rational Metabolic Design and Evolution. *Future Trends Biotechnol* 2013;**131**:1–23.
49. Kumar A, Singh S. Directed evolution: tailoring biocatalysts for industrial applications. *Crit Rev Biotechnol* 2013;**33**(4):365–78.
50. Pal C, Papp B, Posfai G. The dawn of evolutionary genome engineering. *Nat Rev Genet* 2014;**15**(7):504–12.
51. Portnoy VA, Bezdan D, Zengler K. Adaptive laboratory evolution—harnessing the power of biology for metabolic engineering. *Curr Opin Biotechnol* 2011;**22**(4):590–4.
52. Soyer OS, O'Malley MA. Evolutionary systems biology: what it is and why it matters. *Bioessays* 2013;**35**(8):696–705.
53. Tee KL, Wong TS. Polishing the craft of genetic diversity creation in directed evolution. *Biotechnol Adv* 2013;**31**(8):1707–21.
54. Allewell NM. Thematic minireview series on enzyme evolution in the post-genomic era. *J Biol Chem* 2012;**287**(1):1–2.

55. Goldsmith M, Tawfik DS. Directed enzyme evolution: beyond the low-hanging fruit. *Curr Opin Struct Biol* 2012;**22**(4):406–12.
56. Jarboe LR, Liu P, Kautharapu KB, Ingram LO. Optimization of enzyme parameters for fermentative production of biorenewable fuels and chemicals. *Comput Struct Biotechnol J* 2012;**3**:e201210005.
57. Austad SN. An experimental paradigm for the study of slowly aging organisms. *Exp Gerontol* 2001;**36**(4-6):599–605.
58. Dennett DC. *Darwin's dangerous idea: evolution and meanings of life*. New York: Touchstone; 1996.
59. Wright BE. Stress-directed adaptive mutations and evolution. *Mol Microbiol* 2004;**52**(3):643–50.
60. Galhardo RS, Hastings PJ, Rosenberg SM. Mutation as a stress response and the regulation of evolvability. *Crit Rev Biochem Mol Biol* 2007;**42**(5):399–435.
61. Foster PL. Stress-induced mutagenesis in bacteria. *Crit Rev Biochem Mol Biol* 2007;**42**(5):373–97.
62. Steinborn G. Uvm mutants of *Escherichia coli* K 12 deficient in UV mutagenesis. *Mol Gen Genet MGG* 1978;**165**(1):87–93.
63. Ohta T, Nakamura N, Moriya M, Shirasu Y, Kada T. The SOS-function-inducing activity of chemical mutangens in *Escherichia coli*. *Mutat Res* 1984;**131**(3-4):101–9.
64. Radman M. Fidelity and infidelity. *Nature* 2001;**413**(6852):115.
65. Loeb LA, Cheng KC. Errors in DNA-synthesis—a source of spontaneous mutations. *Mutat Res* 1990;**238**(3):297–304.
66. Zhang Z, Saier Jr MH. Transposon-mediated adaptive and directed mutations and their potential evolutionary benefits. *J Mol Microbiol Biotechnol* 2011;**21**(1-2):59–70.
67. Turner PC, Yomano LP, Jarboe LR, York SW, Baggett CL, Moritz BE, et al. Optical mapping and sequencing of the *Escherichia coli* KO11 genome reveal extensive chromosomal rearrangements, and multiple tandem copies of the *Zymomonas mobilis* pdc and adhB genes. *J Ind Microbiol Biotechnol* 2012;**39**(4):629–39.
68. Haft RJF, Keating DH, Schwaegler T, Schwalbach MS, Vinokur J, Tremaine M, et al. Correcting direct effects of ethanol on translation and transcription machinery confers ethanol tolerance in bacteria. *Proc Natl Acad Sci U S A* 2014;**111**(25):E2576–85.
69. Drake JW, Charlesworth B, Charlesworth D, Crow JF. Rates of spontaneous mutation. *Genetics* 1998;**148**(4):1667–86.
70. Foster PL, Gudmundsson G, Trimarchi JM, Cai H, Goodman MF. Proofreading-defective DNA-polymerase-II increases adaptive mutation in *Escherichia coli*. *Proc Natl Acad Sci U S A* 1995;**92**(17):7951–5.
71. Weber H, Polen T, Heuveling J, Wendisch VF, Hengge R. Genome-wide analysis of the general stress response network in *Escherichia coli*: sigma(S)-dependent genes, promoters, and sigma factor selectivity. *J Bacteriol* 2005;**187**(5):1591–603.
72. Cortazzo P, Cervenansky C, Marin M, Reiss C, Ehrlich R, Deana A. Silent mutations affect in vivo protein folding in *Escherichia coli*. *Biochem Biophys Res Commun* 2002;**293**(1):537–41.
73. Deana A, Ehrlich R, Reiss C. Silent mutations in the *Escherichia coli* ompA leader peptide region strongly affect transcription and translation in vivo. *Nucleic Acids Res* 1998;**26**(20):4778–82.
74. Zhu X, Tan Z, Xu H, Chen J, Tang J, Zhang X. Metabolic evolution of two reducing equivalent-conserving pathways for high-yield succinate production in *Escherichia coli*. *Metab Eng* 2014;**24**:87–96.
75. Notley-McRobb L, Ferenci T. Adaptive mgl-regulatory mutations and genetic diversity evolving in glucose-limited *Escherichia coli* populations. *Environ Microbiol* 1999;**1**(1):33–43.
76. Turner PC, Miller EN, Jarboe LR, Baggett CL, Shanmugam KT, Ingram LO. YqhC regulates transcription of the adjacent *Escherichia coli* genes yqhD and dkgA that are involved in furfural tolerance. *J Ind Microbiol Biotechnol* 2011;**38**(3):431–9.
77. Tsui HCT, Leung HCE, Winkler ME. Characterization of broadly pleiotropic phenotypes caused by an hfq insertion mutation in *Escherichia coli* K12. *Mol Microbiol* 1994;**13**(1):35–49.
78. Giel M, Desnoyer M, Lopilato J. A mutation in a new gene, bglJ, activates the bgl operon in *Escherichia coli* K-12. *Genetics* 1996;**143**(2):627–35.

79. Kishimoto T, Iijima L, Tatsumi M, Ono N, Oyake A, Hashimoto T, et al. Transition from positive to neutral in mutation fixation along with continuing rising fitness in thermal adaptive evolution. *PLoS Genet* 2010;**6**(10):e1001164.
80. Wang Y, Manow R, Finan C, Wang J, Garza E, Zhou S. Adaptive evolution of nontransgenic *Escherichia coli* KC01 for improved ethanol tolerance and homoethanol fermentation from xylose. *J Ind Microbiol Biotechnol* 2011;**38**(9):1371–7.
81. Lenski RE, Travisano M. Dynamics of adaptation and diversifcation—a 10,000-generation experiment with bacterial-populations. *Proc Natl Acad Sci U S A* 1994;**91**(15):6808–14.
82. Yomano LP, York SW, Ingram LO. Isolation and characterization of ethanol-tolerant mutants of *Escherichia coli* KO11 for fuel ethanol production. *J Ind Microbiol Biotechnol* 1998;**20**(2):132–8.
83. Reyes LH, Winkler J, Kao KC. Visualizing evolution in real-time method for strain engineering. *Front Microbiol* 2012;**3**:198.
84. Huang M, McClellan M, Berman J, Kao KC. Evolutionary dynamics of candida albicans during in vitro evolution. *Eukaryot Cell* 2011;**10**(11):1413–21.
85. Alper H, Stephanopoulos G. Global transcription machinery engineering: a new approach for improving cellular phenotype. *Metab Eng* 2007;**9**(3):258–67.
86. Alper H, Moxley J, Nevoigt E, Fink GR, Stephanopoulos G. Engineering yeast transcription machinery for improved ethanol tolerance and production. *Science* 2006;**314**(5805):1565–8.
87. Liu HM, Yan M, Lai CG, Xu L, Ouyang PK. gTME for improved xylose fermentation of *Saccharomyces cerevisiae*. *Appl Biochem Biotechnol* 2010;**160**(2):574–82.
88. Zhang H, Chong H, Ching CB, Song H, Jiang R. Engineering global transcription factor cyclic AMP receptor protein of *Escherichia coli* for improved 1-butanol tolerance. *Appl Microbiol Biotechnol* 2012;**94**(4):1107–17.
89. Lynch MD, Warnecke T, Gill RT. SCALEs: multiscale analysis of library enrichment. *Nat Methods* 2007;**4**(1):87–93.
90. Sandoval NR, Mills TY, Zhang M, Gill RT. Elucidating acetate tolerance in *E. coli* using a genome-wide approach. *Metab Eng* 2011;**13**(2):214–24.
91. Singh A, Lynch MD, Gill RT. Genes restoring redox balance in fermentation-deficient *E. coli* NZN111. *Metab Eng* 2009;**11**(6):347–54.
92. Yuan Y, Du J, Zhao H. Customized optimization of metabolic pathways by combinatorial transcriptional engineering. *Methods Mol Biol (Clifton, NJ)* 2013;**985**:177–209.
93. Sandoval NR, Kim JYH, Glebes TY, Reeder PJ, Aucoin HR, Warner JR, et al. Strategy for directing combinatorial genome engineering in *Escherichia coli*. *Proc Natl Acad Sci U S A* 2012;**109**(26):10540–5.
94. Garcia V, Pakkila J, Ojamo H, Muurinen E, Keiski RL. Challenges in biobutanol production: how to improve the efficiency? *Renew Sustain Energy Rev* 2011;**15**(2):964–80.
95. Ezeji T, Milne C, Price ND, Blaschek HP. Achievements and perspectives to overcome the poor solvent resistance in acetone and butanol-producing microorganisms. *Appl Microbiol Biotechnol* 2010;**85**(6):1697–712.
96. Warnecke T, Gill RT. Organic acid toxicity, tolerance, and production in *Escherichia coli* biorefining applications. *Microb Cell Factories* 2005;**4**:25.
97. Ingram LO, Buttke TM. Effects of alcohols on microorganisms. *Adv Microb Physiol* 1984;**25**:253–300.
98. Ramos JL, Duque E, Gallegos MT, Godoy P, Ramos-Gonzalez MI, Rojas A, et al. Mechanisms of solvent tolerance in gram-negative bacteria. *Annu Rev Microbiol* 2002;**56**:743–68.
99. Klinke HB, Thomsen AB, Ahring BK. Inhibition of ethanol-producing yeast and bacteria by degradation products produced during pre-treatment of biomass. *Appl Microbiol Biotechnol* 2004;**66**(1):10–26.
100. McKenna R, Moya L, McDaniel M, Nielsen DR. Comparing in situ removal strategies for improving styrene bioproduction. *Bioprocess Biosyst Eng* 2014;**38**:1–10.
101. Roffler SR, Blanch HW, Wilke CR. In situ extractive fermentation of acetone and butanol. *Biotechnol Bioeng* 1988;**31**(2):135–43.
102. Inokuma K, Liao JC, Okamoto M, Hanai T. Improvement of isopropanol production by metabolically engineered *Escherichia coli* using gas stripping. *J Biosci Bioeng* 2010;**110**(6):696–701.

103. Jarboe LR, Liu P, Royce LA. Engineering inhibitor tolerance for the production of biorenewable fuels and chemicals. *Curr Opin Chem Eng* 2011;**1**(1):38–42.
104. Atsumi S, Wu T-Y, Machado IMP, Huang W-C, Chen P-Y, Pellegrini M, et al. Evolution, genomic analysis, and reconstruction of isobutanol tolerance in *Escherichia coli*. *Mol Syst Biol* 2010;**6**:449.
105. Ward RJ, McConnell CA, Minty JJ, Rouillard J-M, Lin F, Chen Y, et al. Evolution combined with genomic study elucidates genetic bases of isobutanol tolerance in *Escherichia coli*. ArrayExpress Archive; 2012.
106. Ghiaci P, Norbeck J, Larsson C. Physiological adaptations of *Saccharomyces cerevisiae* evolved for improved butanol tolerance. *Biotechnol Biofuels* 2013;**6**:101.
107. Ni Y, Song L, Qian X, Sun Z. Proteomic analysis of *Pseudomonas putida* reveals an organic solvent tolerance-related gene mmsB. *PLoS ONE* 2013;**8**(2):e55858.
108. Miller EN, Jarboe LR, Yomano LP, York SW, Shanmugam KT, Ingram LO. Silencing of NADPH-dependent oxidoreductase genes (yqhD and dkgA) in furfural-resistant ethanologenic *Escherichia coli*. *Appl Environ Microbiol* 2009;**75**(13):4315–23.
109. Miller EN, Turner PC, Jarboe LR, Ingram LO. Genetic changes that increase 5-hydroxymethyl furfural resistance in ethanol-producing *Escherichia coli* LY180. *Biotechnol Lett* 2010;**32**(5):661–7.
110. Koppram R, Albers E, Olsson L. Evolutionary engineering strategies to enhance tolerance of xylose utilizing recombinant yeast to inhibitors derived from spruce biomass. *Biotechnol Biofuels* 2012;**5**:32.
111. Isken S, de Bont JAM. Bacteria tolerant to organic solvents. *Extremophiles* 1998;**2**(3):229–38.
112. Sardessai Y, Bhosle S. Tolerance of bacteria to organic solvents. *Res Microbiol* 2002;**153**(5):263–8.
113. Shah AA, Wang C, Chung Y-R, Kim J-Y, Choi E-S, Kim S-W. Enhancement of geraniol resistance of *Escherichia coli* by MarA overexpression. *J Biosci Bioeng* 2013;**115**(3):253–8.
114. Zingaro KA, Papoutsakis ET. GroESL overexpression imparts *Escherichia coli* tolerance to i-, n-, and 2-butanol, 1,2,4-butanetriol and ethanol with complex and unpredictable patterns. *Metab Eng* 2013;**15**:196–205.
115. Ohta K, Beall DS, Mejia JP, Shanmugam KT, Ingram LO. Genetic improvement of *Escherichia coli* for ethanol-production—chromosomal integration of *Zymomonas mobilis* genes encoding pyruvate decarboxylase and alcohol dehydrogenase-II. *Appl Environ Microbiol* 1991;**57**(4):893–900.
116. Trinh CT, Srienc F. Metabolic engineering of *Escherichia coli* for efficient conversion of glycerol to ethanol. *Appl Environ Microbiol* 2009;**75**(21):6696–705.
117. Demeke MM, Dietz H, Li Y, Foulquie-Moreno MR, Mutturi S, Deprez S, et al. Development of a D-xylose fermenting and inhibitor tolerant industrial Saccharomyces cerevisiae strain with high performance in lignocellulose hydrolysates using metabolic and evolutionary engineering. *Biotechnol Biofuels* 2013;**6**:89.
118. Guimaraes PMR, Francois J, Parrou JL, Teixeira JA, Domingues L. Adaptive evolution of a lactose-consuming *Saccharomyces cerevisiae* recombinant. *Appl Environ Microbiol* 2008;**74**(6):1748–56.
119. Zhou SD, Causey TB, Hasona A, Shanmugam KT, Ingram LO. Production of optically pure D-lactic acid in mineral salts medium by metabolically engineered *Escherichia coli* W3110. *Appl Environ Microbiol* 2003;**69**(1):399–407.
120. Zhou S, Yomano LP, Shanmugam KT, Ingram LO. Fermentation of 10% (w/v) sugar to D(−)-lactate by engineered *Escherichia coli* B. *Biotechnol Lett* 2005;**27**(23-24):1891–6.
121. Zhou S, Shanmugam KT, Yomano LP, Grabar TB, Ingram LO. Fermentation of 12% (w/v) glucose to 1.2 M lactate by *Escherichia coli* strain SZ194 using mineral salts medium. *Biotechnol Lett* 2006;**28**(9):663–70.
122. Zhou SD, Shanmugam KT, Ingram LO. Functional replacement of the *Escherichia coli* D-(−)-lactate dehydrogenase gene (ldhA) with the L-(+)-lactate dehydrogenase gene (ldhL) from *Pediococcus acidilactici*. *Appl Environ Microbiol* 2003;**69**(4):2237–44.
123. Jantama K, Haupt MJ, Svoronos SA, Zhang X, Moore JC, Shanmugam KT, et al. Combining metabolic engineering and metabolic evolution to develop nonrecombinant strains of *Escherichia coli* C that produce succinate and malate. *Biotechnol Bioeng* 2008;**99**(5):1140–53.
124. Zhang X, Jantama K, Moore JC, Shanmugam KT, Ingram LO. Production of L-alanine by metabolically engineered *Escherichia coli*. *Appl Microbiol Biotechnol* 2007;**77**(2):355–66.

125. Cobb RE, Si T, Zhao H. Directed evolution: an evolving and enabling synthetic biology tool. *Curr Opin Chem Biol* 2012;**16**(3-4):285–91.
126. Smith KM, Liao JC. An evolutionary strategy for isobutanol production strain development in *Escherichia coli*. *Metab Eng* 2011;**13**(6):674–81.
127. Kabara JJ, Marshall DL. Medium-chain fatty acids and esters. *Antimicrobials in food*. 3rd ed. BocaRaton: CRC Press; 2010, pp. 327–60.
128. Russell AD. Mechanisms of bacterial-resistance to non-antibiotics- food-additives and food and pharmaceutical preservatives. *J Appl Bacteriol* 1991;**71**(3):191–201.
129. Lennen RM, Kruziki MA, Kumar K, Zinkel RA, Burnum KE, Lipton MS, et al. Membrane stresses induced by overproduction of free fatty acids in *Escherichia coli*. *Appl Environ Microbiol* 2011;**77**(22):8114–28.
130. Royce LA, Liu P, Stebbins MJ, Hanson BC, Jarboe LR. The damaging effects of short chain fatty acids on *Escherichia coli* membranes. *Appl Microbiol Biotechnol* 2013;**97**(18):8317–27.
131. Sherkhanov S, Korman TP, Bowie JU. Improving the tolerance of *Escherichia coli* to medium-chain fatty acid production. *Metab Eng* 2014;**25**:1–7.
132. Royce LA, Yoon JM, Chen Y, Rickenbach E, Shanks JV, Jarboe LR. *Evolution for exogenous octanoic acid tolerance improves carboxylic acid production and membrane integrity. Metab Eng* 2015;**29**:180–8.
133. Stephanopoulos G. Challenges in engineering microbes for biofuels production. *Science* 2007;**315**(5813):801–4.
134. Jorgensen H, Kristensen JB, Felby C. Enzymatic conversion of lignocellulose into fermentable sugars: challenges and opportunities. *Biofuels Bioprod Bior* 2007;**1**(2):119–34.
135. Almeida JRM, Modig T, Petersson A, Hahn-Hagerdal B, Liden G, Gorwa-Grauslund MF. Increased tolerance and conversion of inhibitors in lignocellulosic hydrolysates by *Saccharomyces cerevisiae*. *J Chem Technol Biotechnol* 2007;**82**(4):340–9.
136. Sanchez B, Bautista J. Effects of furfural and 5-hydroxymethylfurfural on the fermentation of *Saccharomyces cerevisiae* and biomass production from *Candida guilliermondii*. *Enzym Microb Technol* 1988;**10**(5):315–8.
137. Zaldivar J, Ingram LO. Effect of organic acids on the growth and fermentation of ethanologenic *Escherichia coli* LY01. *Biotechnol Bioeng* 1999;**66**(4):203–10.
138. Zaldivar J, Martinez A, Ingram LO. Effect of selected aldehydes on the growth and fermentation of ethanologenic *Escherichia coli*. *Biotechnol Bioeng* 1999;**65**(1):24–33.
139. Glebes TY, Sandoval NR, Reeder PJ, Schilling KD, Zhang M, Gill RT. Genome-wide mapping of furfural tolerance genes in *Escherichia coli*. *PLoS ONE* 2014;**9**(1), e87540.
140. Glebes TY, Sandoval NR, Gillis JH, Gill RT. Comparison of genome-wide selection strategies to identify furfural tolerance genes in *Escherichia coli*. *Biotechnol Bioeng* 2014;**112**:129–40.
141. Luhe AL, Lim CY, Gerken H, Wu J, Zhao H. Furfural and HMF tolerance in *Escherichia coli* ΔacrR regulatory mutants. *Biotechnol Appl Biochem* 2014;**62**:32–6.
142. Miller EN, Jarboe LR, Turner PC, Pharkya P, Yomano LP, York SW, et al. Furfural inhibits growth by limiting sulfur assimilation in ethanologenic *Escherichia coli* strain LY180. *Appl Environ Microbiol* 2009;**75**(19):6132–41.
143. Wang X, Miller EN, Yomano LP, Zhang X, Shanmugam KT, Ingram LO. Increased furfural tolerance due to overexpression of NADH-dependent oxidoreductase FucO in *Escherichia coli* strains engineered for the production of ethanol and lactate. *Appl Environ Microbiol* 2011;**77**(15):5132–40.
144. Geddes RD, Wang X, Yomano LP, Miller EN, Zheng H, Shanmugam KT, et al. Polyamine transporters and polyamines increase furfural tolerance during xylose fermentation with ethanologenic *Escherichia coli* strain LY180. *Appl Environ Microbiol* 2014;**80**(19):5955–64, p. AEM. 01913-14.
145. Chong H, Geng H, Zhang H, Song H, Huang L, Jiang R. Enhancing *E. coli* isobutanol tolerance through engineering its global transcription factor cAMP Receptor Protein (CRP). *Biotechnol Bioeng* 2014;**111**(4):700–8.

146. Reyes LH, Almario MP, Winkler J, Orozco MM, Kao KC. Visualizing evolution in real time to determine the molecular mechanisms of n-butanol tolerance in *Escherichia coli*. *Metab Eng* 2012;**14**(5):579–90.
147. Almario MP, Reyes LH, Kao KC. Evolutionary engineering of saccharomyces cerevisiae for enhanced tolerance to hydrolysates of lignocellulosic biomass. *Biotechnol Bioeng* 2013;**110**(10):2616–23.
148. Fong SS, Burgard AP, Herring CD, Knight EM, Blattner FR, Maranas CD, et al. In silico design and adaptive evolution of *Escherichia coli* for production of lactic acid. *Biotechnol Bioeng* 2005;**91**(5):643–8.
149. Zhang X, Jantama K, Moore JC, Jarboe LR, Shanmugam KT, Ingram LO. Metabolic evolution of energy-conserving pathways for succinate production in *Escherichia coli*. *Proc Natl Acad Sci U S A* 2009;**106**(48):20180–5.
150. Royce L, Boggess E, Jin T, Dickerson J, Jarboe L. Identification of mutations in evolved bacterial genomes. *Methods Mol Biol (Clifton, NJ)* 2013;**985**:p. 249–67.
151. Jarboe LR. YqhD: a broad-substrate range aldehyde reductase with various applications in production of biorenewable fuels and chemicals. *Appl Microbiol Biotechnol* 2011;**89**(2):249–57.
152. Conrad TM, Joyce AR, Applebee MK, Barrett CL, Xie B, Gao Y, et al. Whole-genome resequencing of *Escherichia coli* K-12 MG1655 undergoing short-term laboratory evolution in lactate minimal media reveals flexible selection of adaptive mutations. *Genome Biol* 2009;**10**(10):R118.
153. Unrean P, Srienc F. Metabolic networks evolve towards states of maximum entropy production. *Metab Eng* 2011;**13**(6):666–73.
154. Unrean P, Srienc F. Predicting the adaptive evolution of *Thermoanaerobacterium saccharolyticum*. *J Biotechnol* 2012;**158**(4):259–66.

CHAPTER 11

Biomass Utilization

K. Merklein[*], **S.S. Fong**[†], **Y. Deng**[‡]
[*]Department of Biological and Agricultural Engineering, Kansas State University-Olathe, Olathe, Kansas, USA [†]Department of Chemical and Life Science Engineering, Virginia Commonwealth University, Richmond, Virginia, USA [‡]National Engineering Laboratory for Cereal Fermentation Technology (NELCF), Jiangnan University, Wuxi, China

11.1 Introduction

This chapter focuses on the utilization of biomass for ethanol production. The characteristics of biomass are introduced as well as the main sources of biomass. Following that, the different pretreatment methods that have been investigated to date will be explored to understand the pretreatment goals as well as evaluate the methods for potential industrial viability. The pretreatment products are then used in the enzymatic hydrolysis process, which is also explored. Finally, an alternative to enzymatic hydrolysis called consolidated bioprocessing is examined as well as the information and the tools that will be used for strain development in consolidated bioprocessing microorganisms.

This chapter evaluates these technologies for use in the industrial production of ethanol. The industrial production means that it could efficiently be scaled up to become economically feasible in large volumes. Some technologies are promising in the lab scale but, are very resource intensive, with either monetary or time resources as the limiting factor. Additionally, the production of ethanol is evaluated because of its existing use as a petroleum additive. The infrastructure is in place to handle ethanol, it is well known in its characteristics, and has already been accepted by governing agencies around the world for its use as a transportation fuel. There are a number of alternatives to ethanol that have been proposed, such as butanol or other higher level alcohols. Because of numerous other challenges not related to biomass utilization, those alternatives are outside the scope of this chapter; however, the majority of this chapter would be applicable in the production of these other chemicals. Additionally, utilization of biomass through the routes of pyrolysis and liquefaction are not considered in this chapter. This is because those processes do not rely on biotechnology to the same extent as microbial utilization, as well as not being able to produce ethanol selectively as an end product.

Biotechnology for Biofuel Production and Optimization. http://dx.doi.org/10.1016/B978-0-444-63475-7.00011-X

11.2 Introduction to Biomass

The term biomass refers to the general organic matter created by plants during the photosynthesis cycle, which converts and stores the energy from sunlight into chemical bonds. Biomass serves as a potential energy source for human use by breaking these chemical bonds. Utilization of this potential energy source is a major focus of research, as biomass is regarded as a sustainable energy source. The simplest method, and historically most common, is to use this energy through a simple combustion reaction. Current desires focus on converting biomass into a form with higher energy value as a substitute for petroleum fuel.

Numerous forms of biomass have been investigated, and first-generation biofuels have even seen success in industrial operations. The production of first-generation biofuels includes biodiesel, ethanol, and small amounts of biogas. However, the current biofuel market is composed of fuels derived from food sources, which has sparked extensive debate regarding impact on biodiversity and land use as well as competition with food crops.[1] Additionally, there are claims that first-generation biofuels are not maximizing the abatement of greenhouse gas emissions.[2]

Because of the reasons identified above, this chapter will focus exclusively on the context of fuels derived from processing lignocellulosic biomass (LCB), which is also known as the second-generation biofuels.[2] Among the sources of LCB are forest residues such as woody plant species, energy crops like switchgrass which is bred specifically to produce bioenergy, agricultural residues such as corn stover or rice or wheat straw, and cellulosic waste such as food or municipal solid waste. The production of second-generation biofuels is currently not cost competitive and currently not commercially produced. The production of second-generation biofuels seems to alleviate many of the problems raised with first-generation fuels.

11.2.1 Structure of LCB

Plant cells have a cell wall (0.1 to 10 μm) that provides rigidity to the cell, as well as preventing attack by pathogens.[3] The cell walls are typically composed of three layers—the middle lamella, the primary cell wall, and the secondary cell wall. Further sublayers of the secondary cell wall are present only in certain tissues.[4] The chemical composition of cell walls is different between monocots, like corn stover or switchgrass, and dicots, like poplar. Major components of the cell wall are cellulose, hemicellulose, and lignin with minor components of proteins, lipids, soluble sugars, and minerals.[5]

The use of LCB to produce a liquid biofuel, specifically ethanol, is currently unable to proceed commercially because the LCB is not broken down into fermentable sugars by current ethanol-producing microorganisms, such as *Saccharomyces cerevisiae* and *Escherichia coli*. In fact, the biochemical utilization of LCB is extremely difficult due to its recalcitrant structure, and one of the major hurdles in production of biofuels from LCB is the development of an economical process to break down the LCB into fermentable sugars.

As mentioned above, the primary LCB components are cellulose, hemicellulose, and lignin. The structure of woody plant species is characterized by tightly bound fibers while the structure of herbaceous plants, such as corn stover, sugarcane bagasse, and switchgrass, are more loosely bound. The binding level of the fibers is associated with the lignin content of each crop.[6,7]

Cellulose consists of linear chains of (1,4)-D-glucopyranose units, and those units are linked 1-4 in the β-configuration. The average molecular weight of cellulose is around 100,000 atomic mass units (amu).[6]

The hemicellulose is a mixture of polysaccharides, composed of sugars such as glucose, mannose, xylose, and arabinose, as well as methyglucoronic and galaturonic acids. Hemicellulose is a heterogeneous branched polysaccharide that binds tightly to the surface of each cellulose microfibril. Hemicellulose binding to cellulose is noncovalent bonding, and the average molecular weight of hemicellulose is <30,000 amu.[6]

Lignin is a high molecular weight group of amorphous compounds. Lignin is composed of three-carbon chain attached to rings of phenyl propanes, which is a six-carbon ring. There are three possible structures of lignin, termed I, II, and III which correspond to zero, one, or two methoxy groups attached to the rings. The structures vary based on the source of the polymer; structure I is found in herbaceous plants, structure II is in the wood of conifers, and structure III is found in deciduous wood.[6]

Biomass is not only composed of varying amounts of cellulose, hemicellulose, and lignin, but also is composed of different types of lignin. Therefore, any attempt to utilize biomass in a processing environment must be robust enough to accommodate the differing structures of biomass.

11.3 Pretreatment Technologies for LCB

The main goal of pretreatment of LCB is to disrupt the recalcitrant structure for enzyme access in further downstream processing as outlined in its most basic form in Figure 11.1. There are many ways to accomplish this, and all of the methods do this with different final results. For example, enzymatic hydrolysis may require the removal of hemicellulose since no hemicellulases are typically present, while consolidated bioprocessing utilizes organisms that naturally secrete hemicellulases and can effectively handle hemicellulose. Each of those processes could use biomass treated from completely different pretreatment strategies. Because of this, it is virtually impossible to identify one strategy that will be the industrial "gold standard" for pretreatment. However, the pros and cons of each will be identified, and this information can be used for a system-wide approach to biomass utilization.

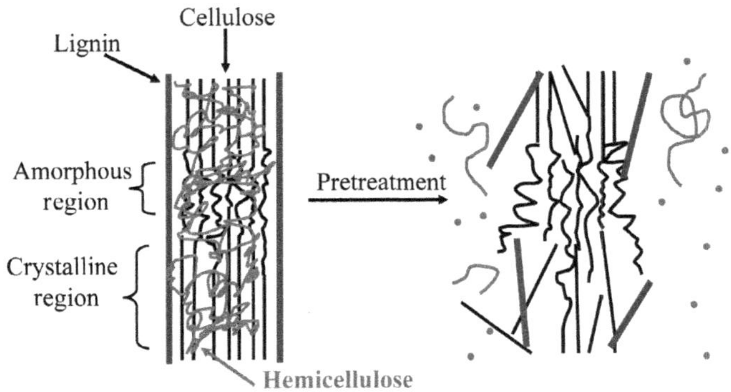

Figure 11.1
The ultimate goal of pretreatment methods is to break LCB down from recalcitrant structure to its components of lignin, cellulose, and hemicellulose.[8]

11.3.1 Physical Pretreatment

Physical pretreatment of LCB consists of physically altering the structure of cellulose through one of two primary methods: the mechanical method that consists of chipping, grinding, and milling; and the extrusion method.[9]

Mechanical method

The goal of mechanical pretreatment is to reduce the crystallinity and particle size of the LCB so that the surface area is increased and the degree of polymerization is reduced.[9] The technologies available for mechanical pretreatment are very mature, and are already used in a variety of bioprocessing areas, such as corn milling for grain ethanol production. It is well known that a combination of chipping, grinding, and milling can be used to obtain a variety of final particle sizes. After chipping, the size of particle is typically in the range of 10-30 mm. After milling or grinding, the particle size is in the range of 0.2-2 mm.[10] Many different types of milling exist and are suitable for different applications. For wet material, the colloid mill, fibrillator, and dissolver are suitable. For dry material, the roller mill, cryogenic mill, and hammer mill are used.[11] The ball mill can be used for either dry or wet material.

However, despite being economically feasible for some industries, it is unlikely that this technology will be used for LCB, as it takes a large amount of energy to get LCB into a small enough particle size for enzymatic hydrolysis. The energy consumption for mechanical size reduction has been calculated in most cases (depending on final particle size and size reduction ratio) to be higher than the theoretical energy content available in the biomass.[12] It should be noted that use of milling as the sole pretreatment strategy is not economically feasible, it can be used to reduce particle size for other pretreatment methods. In terms of a practical point of view, it makes sense that wood chips, rather than an entire tree, would be subjected to a chemical or physicochemical pretreatment method.

Extrusion method

The extrusion process of physical pretreatment is a means to subject LCB to a variety of heating, mixing, and shearing forces during passage through the extruder.[9] All of these forces produce desirable physical and chemical treatment methods for enzymatic hydrolysis. Extrusion processing was first evaluated as a pretreatment method in the 1980s for crop residues, sawdust, and municipal waste.[13] It is believed that the screw speed and the barrel temperature are able to disrupt the LCB structure, which causes defibrillation, fibrillation, and shortening of fibers, all of which allows higher accessibility of enzymatic depolymerization.[9] Studies recently have focused on the many variables that can be altered in extrusion such as single or twin extrusion, barrel temperature, screw speed, grinding for extrusion pretreatment, addition of chemicals, as well as solids loading.[13,14] While extrusion can be a continuous process, it has the same downfall of grinding; the energy required for this method to work properly does not seem like this method will easily scale from lab to industry.

11.3.2 Chemical Pretreatment

Chemical pretreatment methods work by using additional chemicals to break down biomass. All have been proven to be effective at improving the enzymatic hydrolysis of LCB in a laboratory scale; as can be expected, each has its own advantages and disadvantages that will be discussed with each method.

Alkali pretreatment

Some bases can effectively solubilize lignin, as well as have minor cellulose and hemicellulose solubilization. It is believed that this occurs by breaking the ester bond between lignin, hemicellulose, and cellulose.[9,15] This can cause less sugar degradation than acid pretreatment. Additionally, alkali pretreatment is believed to be more effective on agricultural residues than on wood materials.[16]

Alkali pretreatment occurs at room temperature, and the duration of pretreatment can range from just seconds up to days. Ammonium, calcium, potassium, and sodium hydroxides are effective bases for alkali pretreatment. Sodium hydroxide (NaOH) is effective as a pretreatment strategy because it causes swelling which increases the internal surface of cellulose as well as decreasing the polymerization and crystallinity.[17] This further causes lignin structure disruption. NaOH has been used to increase hardwood digestibility from 14% to 55% by reducing the lignin content from 24-25% to 20%.[16] Alkali treatment with NaOH has been compared to other pretreatment strategies. In comparing NaOH pretreatment of cotton stalks with sulfuric acid, hydrogen peroxide, and ozone pretreatment, the NaOH-pretreated stalks resulted in the highest cellulose conversion and highest level of delignification.[18]

Lime, $Ca(OH)_2$, is a well-studied hydroxide for alkali pretreatment. Compared to other hydroxides, lime has a lower cost as well as fewer safety concerns. It also can be easily recovered from hydrolysates through a reaction with CO_2.[8] Lime pretreatment has been effective at reducing the lignin fraction in LCB and also removes the acetyl group from hemicellulose.[8,19] Lime pretreatment of corn stover[19] and poplar wood[20] at a temperature range from 85 to 150°C for 3-13 h has been successfully demonstrated.

The use of NaOH has been more studied than other hydroxides, and the effectiveness is well known. The use of lime is desirable because of the process safety aspect as well as cost, but is a little less studied in terms of its effectiveness. However, as will be discussed later, the use of ammonia can combine both aspects of alkali pretreatment as well as the added ability of physical disruption. Because of this combined process, it is unlikely that an alkaline only pretreatment will reach industrial viability.

Acid pretreatment

The primary objective through acid pretreatment is to solubilize the hemicellulosic portion of the LCB. Treatment can occur either at low-acid concentration and high temperature (dilute acid pretreatment) or high-acid concentration and low temperatures (concentrated acid pretreatment).[9] While both dilute or concentrated acids could be used for pretreatment, using concentrated acids is less desirable because it produces more inhibiting compounds, such as HMF and furfural, than diluted acids. It also can cause greater corrosion and acid recovery problems compared to diluted acid, which increase the cost on a commercial scale. Diluted acid pretreatments have been studied for a variety of reactor types. The pretreatment can occur at high temperature (180°C) for a short period of time, or a lower temperature (120°C) for a longer time period.

The most studied acid for pretreatment is diluted H_2SO_4, which may in fact be the most common chemical pretreatment method.[11] Experiments have shown the ability to remove nearly 100% of the hemicellulose with dilute sulfuric acid although it is not effective at dissolving lignin. It does disrupt lignin as well as increase the cellulose susceptibility for enzymatic hydrolysis.[21,22] The conditions that are optimal for hemicellulose removal may not be optimal for the pretreatment for enzymatic hydrolysis. Studies testing various parameters of the pretreatment of olive tree biomass showed different process conditions for highest overall sugar yield, highest hemicellulose removal, and highest enzymatic hydrolysis yield.[23] The major drawback of this method is formation of inhibitors. Compounds such as carboxylic acid, furfural, and phenolic compounds were formed, which may affect enzymatic hydrolysis and have a negative impact on microbial growth.[17,24]

Additional acids used for dilute acid pretreatment are the organic acids, such as fumaric or maleic acids. Initial studies show that less furfural was formed when using these organic acids, as opposed to sulfuric acid.[25] However, as fewer studies have been performed with organic acids, it is unknown if they can achieve the same results as dilute sulfuric acid.

Ozonolysis

Ozone is an oxidant that has been shown to delignify LCB. This pretreatment method is performed at room temperature, and does not lead to the formation of inhibitory compounds.[10] However, it requires a large amount of ozone, which likely makes the process economically nonviable.

Organosolv

Organosolv pretreatment occurs with numerous organic or aqueous solvent mixtures, such as methanol, ethanol, acetone, ethylene glycol, and tetrahydrofurfuryl alcohol. These mixtures work to solubilize lignin.[26] A potential advantage of organosolv compared to other pretreatment methods is the recovery of relatively pure lignin as a by-product.

The removal of solvents is necessary with this pretreatment method because they could be inhibitory to enzymatic hydrolysis and fermentation.[10] Because of this, low-molecular-weight solvents such as ethanol and methanol are favored because of their low boiling points and ease of removal.

Others have suggested combining organosolv with acid hydrolysis to separate hemicellulose and lignin in a two-stage fractionation.[27] First, the biomass would be treated with acid to selectively hydrolyze the hemicellulose; next the process of delignification would occur through organosolv. Through this method, high lignin removal was seen with negligible cellulose loss.[27] However, this process also would require a large amount of chemicals to be effective which increase both operation cost as well as process risk.

Ionic liquids

Ionic liquids (ILs) used as solvents are a relatively new pretreatment method that is receiving a large amount of research. These solvents are salts composed of large organic cations and small inorganic anions, and are typically defined as molten salts with melting points less than 100°C. They are chemically and thermally stable, nonflammable, relatively nontoxic, and have low vapor pressures.[9]

The pretreatment using ILs is carried out by the dissolution of biomass in the solvent at temperatures between 90 and 130°C, ambient pressures, and varying amounts of time (typically 1-24 h). Water is then added to precipitate the biomass, and is washed prior to enzymatic hydrolysis.[28,29] The IL mechanism is the formation of a hydrogen bond between the anion of the IL and the cellulose, which breaks up the cellulose crystalline structure.[30] Additionally, the chemistry of certain ILs lends to the ability of selectively removing lignin and hemicellulose from biomass.[31–33]

The research into ILs as a pretreatment strategy was first reported in 2002,[4,34] so there are still numerous problems to solve before this is an industrially viable pretreatment strategy.

Specifically, a means to efficiently recycle the IL is a prerequisite, as is the toxicity of ILs to enzymes and fermentation microorganisms. Initial studies have shown that ILs can have a significant negative impact on cellulase activity.[9] Additionally, ILs are currently too expensive for industrial use because they are derived from multiple synthesis of nonrenewable resources, so a means to produce ILs at a much lower cost is required for widespread use as a pretreatment technology.[35]

A strategy that could accomplish the production of ILs at a lower cost is the synthesis of ILs from carbohydrates. One promising approach in the formation of ILs from renewable resources is to use the lignin and hemicellulose fractions of biomass as a starting point to derive ILs.[35] This is described as a hypothetical closed-loop biorefinery where the cellulose portion of the LCB would be used for fermentation, while the lignin and hemicellulose portions would be depolymerized and converted to ILs, which would be used by the refinery as the pretreatment method. Compared to the "gold standard" IL, which could remove 44.5% xylan and 52.4% lignin, the biomass-derived ILs had the ability to remove xylan (33.9-51.4%) and lignin (3.9-43.0%) through pretreatment at rates close enough to warrant further research. Additionally, the initial technoeconomic analysis showed that the raw materials cost of the biomass-derived ILs is in the range of 60-70% the cost of the "gold standard" used in the study, with potential to be much cheaper. This study is very promising in its potential industrial applications; however, it is the first known study of ILs derived from biomass so, much like ILs in general, there is still much to be investigated before this is an industrially viable pretreatment strategy.

11.3.3 Physicochemical Pretreatment

As the name implies, physicochemical pretreatment methods combine the advantages of physical disruption to cellulose as well as chemical processing. Similar to the purely physical or purely chemical methods, all have been proven to be effective at improving the enzymatic hydrolysis of LCB. Which emerge as an industrial standard will depend on many additional factors such as feedstock used or the desired end products of pretreatment.

Steam explosion

Steam explosion is a treatment method where the biomass is subjected to pressurized steam for a time period of several seconds to several minutes, followed by sudden depressurization. The mechanism of pretreatment involves both mechanical and chemical effects. The chemical effects occur through a hydrolysis (autohydrolysis) of acetyl groups that are present in the hemicellulose.[9] This autohydrolysis takes place in high temperatures, which promote the formation of acetic acid from acetyl groups; additionally, the water from steam can act as an acid at high temperatures.[9] Mechanically, the reduced pressure causes the fibers of biomass to separate as a result of explosive decompression.[9]

Multiple factors can affect the effectiveness of steam explosion, including particle size, temperature (T), residence time (t), as well as the severity factor. The severity factor (R_0) is characterized by the equation "$\log R_0 = \log(t * e^{((T-100)/14.75)})$" where "$t$" is minutes and "$T$" is degrees Celsius.[36] It should be noted, that the severity factor R_0 does not take into account any acids used as a catalyst in the process. Combining dilute acids with steam explosion have been shown to effectively pretreat softwoods, improve hemicellulose hydrolysis during the pretreatment, as well as further decrease retention time and temperature of the steam explosion. Therefore, a lower severity factor will be required for the process when using dilute acids.[30] A decrease in the retention time and temperature has been shown to reduce inhibitory compounds formed, which helps improve the later enzymatic hydrolysis and fermentation steps.[37] With all treatment strategies, addition of diluted acid comes with potential downfalls such as added processing cost, added steps of neutralizing the acid and removal of salts, as well as additional concerns when it comes to equipment requirements.[30]

Liquid hot water

Liquid hot water pretreatment is carried out with only water and no catalysts or additional chemicals. The water is both heated to elevated temperatures (160°C-240°C) while applying a high enough pressure to keep the water in a liquid state. This process causes alterations in the structure of lignocellulose, primarily through the removal of hemicellulose. After pretreatment, the slurry formed can be filtered into a solid-cellulose-enriched fraction and a hemicellulose-rich liquid fraction.[9] This process is a hydrothermal pretreatment method that has been used for several decades as part of the pulp industry.[11]

The liquid fraction consists primarily of oligosaccharides derived from hemicellulose, lignin (35-60% of starting material), with a minor amount of cellulose (4-15% of starting material).[30] To prevent inhibitory compounds from effecting enzymatic hydrolysis and fermentation, it is desired to have as much lignin as possible be solubilized and removed with this liquid fraction as well. It has been shown that the amount of lignin solubilized is related to lower process temperatures for liquid hot water treatment.[38]

Condition optimization of the liquid hot water method for pretreating corn stover found that conditions of 190°C for 15 min resulted in a 90% conversion of the cellulose in the subsequent enzymatic hydrolysis step.[39] A different study compared the performance of liquid hot water and steam pretreatment of sugarcane bagasse.[40] Both methods were shown to improve enzymatic hydrolysis, but the liquid hot water method had a better xylan recovery. It was also found that the liquid hot water was comparable to dilute acid pretreatment.

The main advantage of the liquid hot water system is the potential cost savings involved. Because no catalyst is required, as well as low amounts of corrosion occurring during the process, it can have both lower initial and maintenance costs when compared to some other pretreatment methods that involve harsh chemicals. However, this pretreatment method is very energy intensive which will increase the operating costs.

Ammonia fiber explosion (AFEX)

The AFEX pretreatment process involves biomass being treated with liquid anhydrous ammonia at high pressures and a temperature of 60-100°C. After a period of time, ranging from a low of 5 min to a moderate time of 30 min, the pressure is released which causes a rapid expansion of the ammonia gas. This rapid expansion causes swelling and physical disruption of the LCB, as well as partial decrystallization of cellulose. The ammonia loading for AFEX is typically a 1:1 ratio of ammonia to dry weight biomass.[41] The AFEX process differs from other processes like steam explosion as it only produces a pretreated solid material as opposed to separate liquid and solid fractions.

AFEX results in a decrease in cellulose crystallinity as well as the disruption in lignin-carbohydrate linkages.[42] Because both cellulose and hemicellulose remains in the pretreated material, both cellulases and hemicellulases will be required during the enzymatic hydrolysis step. Since the pretreated material contains full amounts of lignin, it has been shown that AFEX is more effective on agricultural residues and herbaceous crops as compared to high-lignin feedstocks like woody biomass.[21] An advantage of AFEX pretreatment is the lack of inhibitor formation, which as has been previously discussed, helps the steps of enzymatic hydrolysis and fermentation more efficiently.

CO_2 explosion

Carbon dioxide explosion uses CO_2 as a supercritical fluid, which means the CO_2 is compressed at temperatures above its critical point to form a liquid-like density. In an aqueous solution, CO_2 forms carbonic acid, which favors polymers hydrolysis. The CO_2 molecules are comparable in size to water and ammonia, so they penetrate the pores of LCB in the same way as water and ammonia. After the CO_2 molecule penetrates LCB, the pressure is decreased for an explosive release of CO_2. This release disrupts cellulose and hemicellulose structure, which increases the accessible surface area.[9] While both hardwood and softwood biomass that was subjected to supercritical CO_2 showed efficient enzymatic hydrolysis,[43] the supercritical CO_2 process may be too expensive for industrial scale-up.[11]

Wet oxidation

Wet oxidation is a pretreatment method that occurs at relatively low temperatures and short reactor times, and uses oxygen in air as a catalyst. The process is performed for 10-15 min at temperatures of 170-200°C and pressures of 10-12 bar O_2.[44] Oxygen addition at temperatures above 170°C makes the process an exothermic reaction, which reduces the energy demand. Hydrolytic processes and oxidative reactions occur, which results in the formation of acids. Wet oxidation solubilizes hemicellulose and lignin and can increase the digestibility of cellulose.[45] However, because this uses a large amount of oxygen, the cost is a major disadvantage in the commercialization of this technology.

Microwave pretreatment

Microwave pretreatment is a process that immerses the biomass in dilute chemical reagents and exposes this slurry to a microwave radiation for 5-20 min. Experiments have identified alkalis as a suitable chemical reagent, and a further evaluation of alkalis showed NaOH as most effective.[46] Microwave pretreatment is unlikely to be industrially applied, as it is an energy-intense method that will not scale up efficiently enough to be economically feasible.

11.4 Enzymatic Hydrolysis

After pretreatment and enzymatic hydrolysis, the biomass is broken down into monomeric sugars that can be used through standard fermentation pathways. Because the use of these sugars is well understood in terms of fermentation, it is not an emphasis of this chapter. However, it should be noted that the order of enzymatic hydrolysis, fermentation of six-carbon sugars and fermentation of five-carbon sugars, is variable. Three schematics exist, and each contains a different order for these processes.

The first scheme that exists is called separate saccharification and fermentation (SSF). This scheme separates enzymatic hydrolysis, fermentation of six-carbon sugars, and fermentation of five-carbon sugars into separate reaction vessels. The second scheme is called simultaneous hydrolysis and fermentation (SHF). This scheme will combine the process of enzymatic hydrolysis and fermentation of six-carbon sugars in a single vessel. However, SHF still performs fermentation of five-carbon sugars in a separate vessel. The final schematic available is simultaneous saccharification and co-fermentation (SSCF). The SSCF process combines the enzymatic hydrolysis, fermentation of six-carbon sugars, and fermentation of five-carbon sugars into one single vessel. One common element for all schemes is that they all require production of cellulase systems in a separate system. All of these schemes are illustrated in Figure 11.4.

The focus of the remainder of this section will be on the factors affecting enzymatic hydrolysis, as well as an overview of the enzymes used to break down the LCB recalcitrant structure.

11.4.1 Factors Effecting Enzymatic Hydrolysis of LCB

The effectiveness of enzymatic hydrolysis depends on breaking down the recalcitrant structure of LCB materials into a more manageable substrate for enzymatic hydrolysis. Therefore, pretreatment of LCB material is necessary prior to performing any hydrolysis or fermentation steps. Specifically, the areas pretreatment should address are the following[9]:

- Cellulose crystallinity.
- Cellulose degree of polymerization.
- Substrate's available surface area.

- Lignin barrier.
- Hemicellulose content.
- Feedstock particle size.
- Porosity.
- Cell wall thickness.
- Change in accessibility with conversion.

11.4.2 Enzymes Used for LCB Hydrolysis

Hydrolysis of LCB is accomplished through the use of at least different types of enzymes[48]:

I. Endoglucanase (EG) or 1,4-β-D-glucan-4-glucanohydrolase
II. Exoglucanase, including 1,4-β-D-glucan glucanohydrolase (also known as cellodextrinase) and 1,4-β-D-glucan cellobiohydrolase (cellobiohydrolase; CBH)
III. β-Glucosidase (BGL) or β-glucoside glucohydrolase

The process of degrading cellulose begins with the EG hydrolyzing the β-glycoside linkages of the internal regions of cellulose, which yields oligosaccharides of varying degrees of polymerization. Additionally, this hydrolysis step generates new chain ends. Exoglucanases hydrolyze in a processive manner, starting at the reducing or nonreducing ends of cellulose to yield either glucose or cellobiose. Additionally, exoglucanases can hydrolyze microcrystalline cellulose by peeling the cellulose chains from the microcrystalline structure. Finally, BGL cleaves the soluble cellodextrins and cellobiose into glucose.[48]

Fungi belonging to the genus Trichoderma, specifically *Trichoderma reesei*, are the primary microorganisms for commercial production of these cellulases. *T. Reesei* is known to produce at least two exoglucanases (CBHI and CBHII), five EGs (EGI, EGII, EGIII, EGIV, and EGV), and two BGLs (BGLI and BGLII).[49] The cellulolytic system of *T. Reesei* is a typical noncomplexed cellulase system where cellulolytic enzymes are secreted and act in an uncoordinated, diffusion-driven manner.[48] This can be seen in Figure 11.2.

A complexed cellulase system, or cellulosome, is a multienzymatic complex that is produced by some anaerobic bacteria. These multienzymatic complexes can efficiently degrade plant biomass, by assembling different cellulose degrading enzymes on a structural scaffoldin subunit. These enzymes are connected to the scaffoldin subunit through strong noncovalent protein-protein interactions between the docking modules (dockerin) and complementary modules (cohesins). Additionally, scaffoldin contains a carbohydrate-binding module, which allows the entire complex to bind to the cellulose surface,[48] shown in Figure 11.3.

Pretreatment strategies have the goal of solubilizing hemicellulose from the biomass, thus only leaving cellulose for hydrolysis. However, as the previous review of the pretreatment strategies has shown, they are not all perfect at doing this so it is important to identify

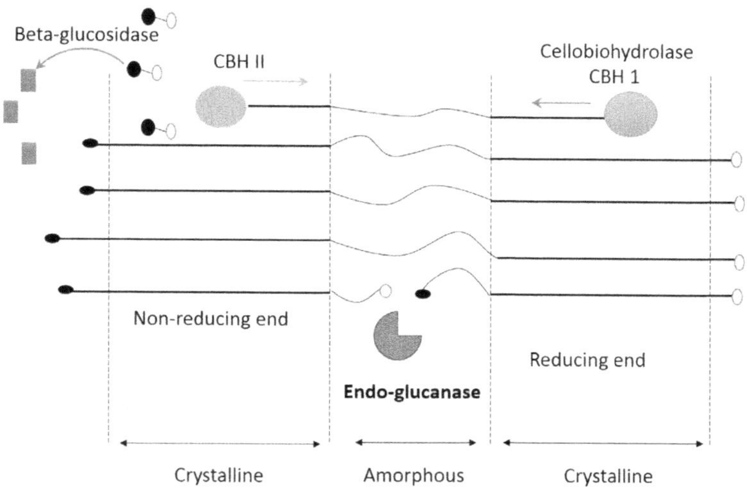

Figure 11.2
Example of a noncomplexed cellulase system where enzymes are secreted and act in an uncoordinated, diffusion-driven manner.[49]

the enzymes that can efficiently hydrolyze hemicellulose. In fact, because of the tangled nature of biomass, the hemicelluloses can liberate cellulose for easier access by cellulases. The common hemicelluloses of β-glucan, xylan, xyloglucan, arabinoxylan, mannan, galactomannan, arabinan, galactan, and polygalacturonan have hemicellulases that target them called β-glucanase, xylanase, xyloglucanase, arabinase, mannanase, galactanase, polygalacturonase, glucuronidase, and acetyl xylan esterase.[51,52] For enzymes that work specifically on the structure of hemicellulose, glycoside hydrolases hydrolyze glycosidic bonds, carbohydrate esterases hydrolyze ester bonds, and polysaccharide lyases cleave glycosidic bonds. *endo*-Hemicellulases cleave internal/backbone glycosidic bonds, and other glycosidases mainly remove substituent groups or side chains.[53] Different hemicellulase combinations would be required for different feedstocks as the makeup of hemicellulose varies widely across plant type. Taking this a step further, synergisms have been found amongst hemicellulases as well as synergism between hemicellulases and cellulases.[54–59] This suggests that complex data processing and bioinformatics tools must be applied to correctly apply these synergisms to an industrial process. Detailed mechanistic models have been created for enzymatic hydrolysis of cellulose.[60–63] Some mechanistic models have been created, although less advanced than the cellulose modeling, to understand the enzymatic hydrolysis of hemicelluloses.[64–67] However, only one such study to date was found for a mechanistic model that considers simultaneous enzymatic hydrolysis of cellulose and hemicellulose.[68] Developing the synergism of cellulases and hemicellulases could be extremely valuable in enzymatic hydrolysis, or through consolidated bioprocessing.

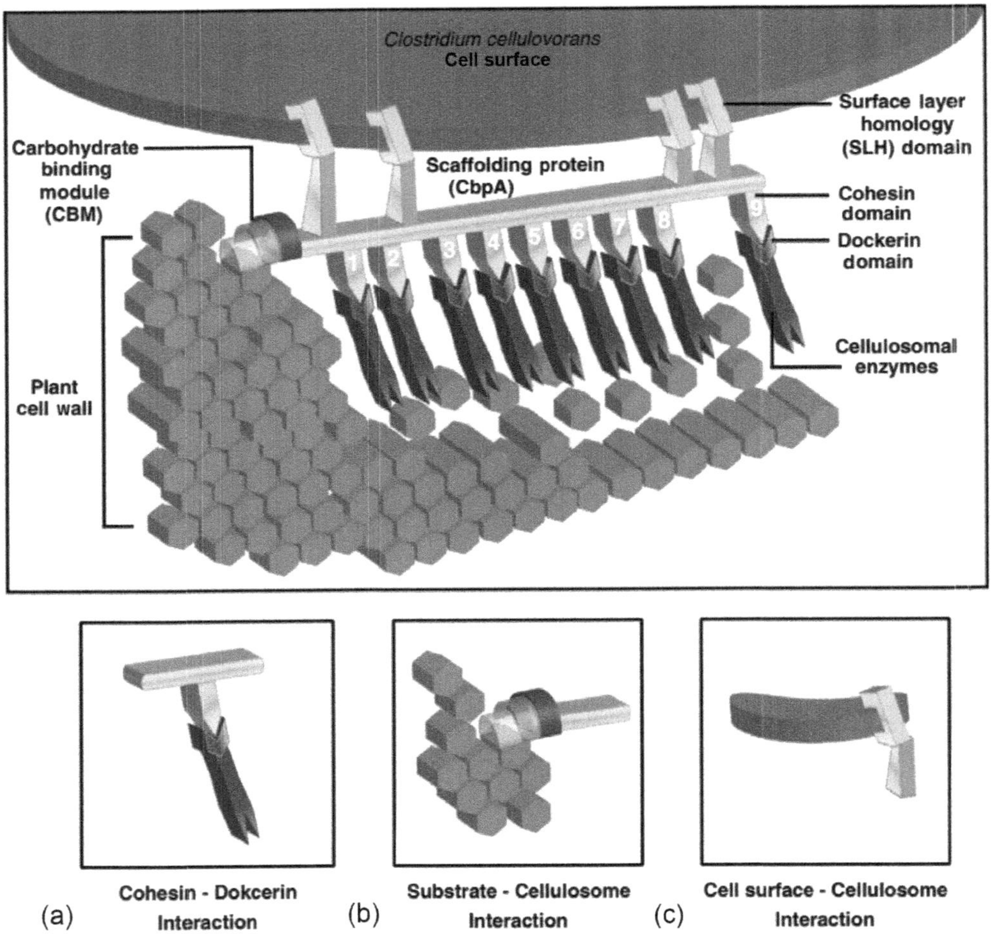

Figure 11.3
Example of a complexed cellulase system, or cellulosome with the necessary parts required to bind to a plant cell wall for biomass degradation.[50]

11.5 Consolidated Bioprocessing (CBP)

The previous sections have described the processes involved in biomass processing: pretreatment of the biomass to help remove lignin and solubilize hemicellulose, saccharolytic enzymes (both cellulases and hemicellulases) and their use in the enzymatic hydrolysis process, and the fermentation of the sugars produced. While these processes all are highly connected, they exist as separate steps and require separate process parameters and process vessels. Combining the production of cellulases, the enzymatic hydrolysis process, and fermentation into a single step is consolidated bioprocessing (CBP).[49] Because cellulolytic

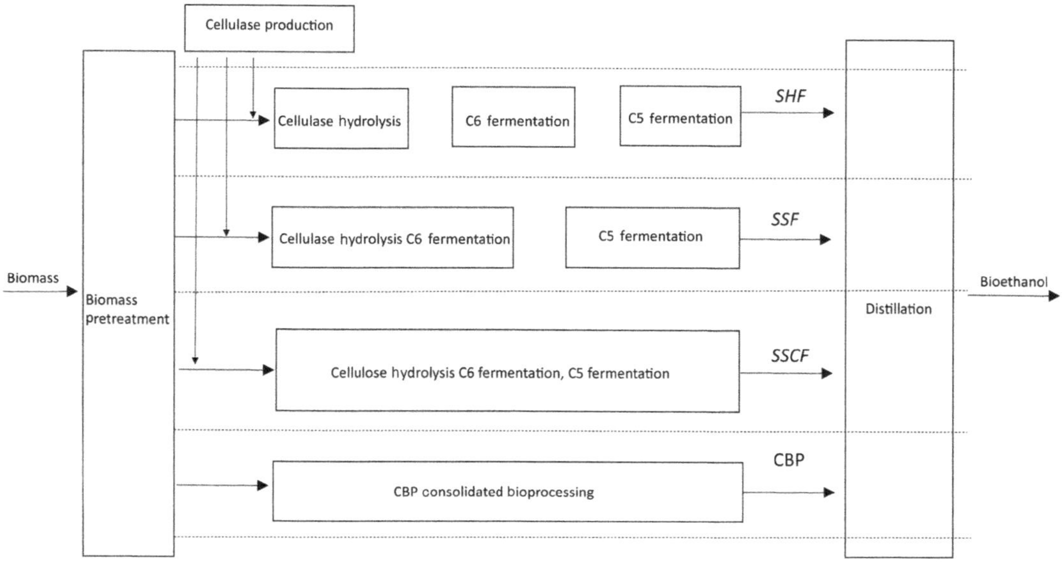

Figure 11.4
Overview of schemes for conversion of LCB to ethanol. The first three routes of SHF, SSF, and SSCF all require external production of cellulases, whereas CBP does not. Also, SSCF and CBP are the only schemes that can have hydrolysis, fermentation of six-carbon sugars, and fermentation of five-carbon sugars all occur in the same vessel.[47]

microorganisms can produce their own cellulases, the commercial production of cellulases is eliminated in CBP. Additionally, the cellulolytic bacteria can excrete their cellulases and ferment sugars in a single step rather than in the multiple steps of enzymatic hydrolysis and fermentation. A diagram showing the differences between the schemes for biomass utilization is illustrated in Figure 11.4.

The primary advantage of CBP is the potential to both decrease the cost and raise the efficiency of biomass processing. Cellulase costs have remained relatively unchanged over the years, and still is a large cost in the biological conversion to biofuels.[69] As mentioned earlier, a synergy between cellulases and hemicellulases has been show to occur. However, as the cellulase cost is already a large cost in the enzymatic hydrolysis step, it would not seem feasible to also add cost by using hemicellulases so this synergism is not utilized in standalone hydrolysis processes. Also having CBP occur in one step eliminates the need for additional process equipment, as well as the utilities involved for enzymatic hydrolysis. The increase in efficiency can be realized through enzyme-microbe synergy as well as eliminate potential contaminants to the process, as noncellulolytic organisms will compete with cellulose-adherent cellulolytic microorganisms during the hydrolysis step.[70]

11.5.1 Cellulose Utilization by Microorganisms in CBP

There are two mechanisms for crystalline cellulose utilization in cellulolytic organisms—either directly adhering to cellulose or by producing noncomplexed cellulases. Anaerobic bacteria typically follow the former mechanism, while aerobic bacteria and fungi tend to follow the later mechanism.[49] Some microbial species however, exhibit different combinations of complexed and noncomplexed cellulases.[49]

After primary hydrolysis, which breaks cellulose down into soluble oligomers, the strictly anaerobic, cellulolytic bacteria utilize cellobiose and cellodextrin phosphorylases (CbP and CdP) to catalyze an ATP-independent phosphorolysis reaction. Further intracellular metabolism creates glucose-6-phosphate, which is the entry point in the Embden-Meyerhoff pathway. From the Embden-Meyerhoff pathway, there are a wide variety of end products that can be produced. The end product distribution appears to be a result of metabolic flux at several levels: mass action effects involving the concentrations of intermediates, fermentation products, and electron carriers; enzyme activity and enzyme synthesis.[49]

11.5.2 Substrate Pretreatment for CBP

All of the pretreatment methods reviewed earlier were initially designed for separate enzymatic hydrolysis and fermentation steps. However, since CBP combines these steps into one, the pretreatment methods are still applicable for CBP. Microorganism development for CBP should keep in mind the strengths and weaknesses of a pretreatment method, and vice versa. Specifically, the following points should be analyzed when comparing pretreatment/CBP compatibility[49]:

I. Cellulose crystallinity is not affected by some pretreatment methods, so it is important that the cellulolytic enzyme system of the CBP microorganism is capable of utilizing crystalline cellulose.

II. The ability of the CBP microorganism to hydrolyze soluble or insoluble hemicellulose should be examined. This is a minor consideration, as most processing methods offer a significant solubilization. However, strong alkali pretreatment methods may have low solubilization levels, which may pose challenges for microorganism utilization. It has been shown that residual insoluble hemicellulose present after pretreatment have been hydrolyzed by commercial cellulase preparations without any specific enzyme for hemicellulose hydrolysis. However, the synergy of hemicellulases and cellulases has also been shown. Therefore, it may or may not be necessary for CBP organisms to produce hemicellulose-specific enzymes; this should be examined further for each microorganism and each process case.

III. The ability to convert all sugars derived from hemicellulose is an objective for organism development. Therefore, it is important for the microorganism to utilize both monomeric sugars, such as xylose, as well as soluble oligomers.

IV. A microorganism should remain metabolically active in the presence of inhibitory compounds created during pretreatment. Because detoxification measures would add to overall cost of processing, it is desired to keep this to a minimum. The amount and types of inhibitors produced depend on the process conditions and configurations.

Degradation of lignin is not necessary in organism development, although decrease in cellulose binding to lignin could be potentially valuable.

11.5.3 Organism Development for CBP

The ideal microbial culture for CBP combines the properties of both substrate utilization as well as product formation. Additionally, the CBP process must occur in an amount of time that is industrially feasible. Therefore, slow growth organism will be ruled out, or the growth must be improved to a sufficient level. The substrate utilization means production of a hydrolytic enzyme system that allows high rates of hydrolysis as well as using the hydrolysis products for growth. Product formation indicates high product selectivity, as well as ability to withstand high product concentration in the growth medium.

Developing an organism for CBP is separated into two distinct strategies. The first strategy is to identify a native cellulolytic organism and develop its ability to produce ethanol (or other fuel product). A second strategy is to find a natural ethanol-producing organism and develop a cellulase system for it to degrade cellulose. This strategy is called the recombinant cellulolytic strategy.[49]

11.5.4 Native Cellulolytic Organisms

One possible strategy in CBP is to make use of naturally occurring cellulolytic microorganisms as a starting point for organism development. These cellulolytic organisms are isolated from nature in areas where cellulose must be used by microorganisms to survive, such as in the soil or in the rumens of animals. Because these organisms have developed the ability to use cellulose and/or hemicellulose, they are a good starting point for strain development for CBP. Additional bioprospecting and characterization of cellulolytic organisms will lead to improvements in the CBP field and provide flexibility for targeting different compounds. Table 11.1 shows a (noninclusive) list of cellulolytic organisms that have been identified to date, relevant growth information, and the genome accession number if applicable.

Metabolic engineering of cellulolytic microorganisms

A necessity to make process CBP economical is to have an organism that is able to produce a high yield of saleable product. The most promising methods to develop a high-yield organism is through metabolic engineering. While native cellulolytic bacteria typically produce large amounts of mixed organic end products, it is believed that these bacteria will

Table 11.1: A noninclusive list of cellulolytic organims that have been studied to date. The organisms source is listed, as well as the genome accession number. If the genome accession number is not available, it will contain an N.[71]

Phylogeny	Species	Source	Genome accession number
Class Clostridia	*Caldicellulosiruptor bescii*		NC_012034
Order Thermoanaerobacterales			
Family Incertae Sedis	*Caldicellulosiruptor hydrothermalis*	Hot spring	NC_014652
	Caldicellulosiruptor kristjanssonii	Hot spring	NC_014721
	Caldicellulosiruptor kronotskyensis	Hot spring	NC_014720
	Caldicellulosiruptor lactoaceticus		NC_015949
	Caldicellulosiruptor obsidiansis	Hot spring	NC_014392
	Caldicellulosiruptor saccharolyticus	Hot spring	NC_009437
	Thermoanaerobacter cellulolyticus	Hot spring	N
	Thermoanaerobacter thermocopriae	Hot spring	PRJNA224116
Class Clostridia	*Butyrivibrio fibrisolvens*	Rumen	NC_021031
Order Clostridiales			
Family Lachnospiraceae	*Cellulosilyticum lentocellum*	Rumen	NC_015275
	Cellulosilyticum ruminicola	Rumen	N
Class Clostridia	*Eubacterium cellulosolvens*	Rumen	NZ_CM001487
Order Clostridiales			
Family Eubacteriaceae			
Class Clostridia	*Clostridium aldrichii*	Wood fermenter	N
Order Clostridiales	*Clostridium alkalicellulosi*	Soda lake	PRJNA65293
Family Clostridiaceae	*Clostridium caenicola*	Sludge	N
	Clostridium celerescens	Manure	N
	Clostridium cellobioparum	Rumen	N
	Clostridium cellulofermentans	Manure	N
	Clostridium cellulolyticum	Compost	NC_011898
	Clostridium cellulosi	Manure	PRJNA212730
	Clostridium cellulovorans	Wood fermenter	NC_014393
	Clostridium chartatabidum	Rumen	N
	Clostridium clariflavum	Sludge	NC_016627
	Clostridium herbivorans	Pig intestine	N
	Clostridium hungatei	Soil	N
	Clostridium josui	Compost	PRJNA195880

	Clostridium longisporum	Rumen	N
	Clostridium papyrosolvens	Paper mill	PRJNA55815
	Clostridium phytofermentans	Soil	NC_010001
	Clostridium populeti	Wood fermenter	N
	Clostridium sp. *C7*	Mud	N
	Clostridium stercorarium	Compost	NC_020134
	Clostridium straminisolvens	Rice straw	PRJDB821
	Clostridium sufflavum	Cattle waste	N
	Clostridium termitidis	Termite	PRJNA196409
	Clostridium thermocellum	Sewage+soil	NC_009012
	Clostridium thermopapyrolyticum	Mud	N
Class Clostridia	*Acetivibrio cellulolyticus*	Sewage	PRJNA51533
Order Clostridiales	*Acetivibrio cellulosolvens*	Sewage	N
Family Ruminococcaceae	*Ruminococcus albus*	Rumen	NC_014833
	Ruminococcus flavefaciens	Rumen	PRJNA224116
Class Clostridia	*Hallocella cellulosilytica*	Saline lake	N
Order Halanaerobiales			
Family Halanaerobiaceae			
Class Bacilli	*Thermoactinomyces* sp. *YX*		N
Order Bacillales			
Family Thermoactinomycetaceae			
Class Bacilli	*Caldibacillus cellulovorans*		N
Order Bacillales			
Family Alicyclobacillaceae			
Class Bacilli	*Bacillus circulans*		PRJDB1314
Order Bacillales	*Bacillus pumilis*		NC_009848
Family Bacillaceae			
Class Actinobacteria	*Acidothermus cellulolyticus*	Acidic hot spring	NC_008578
Subclass Actinobacteridae			
Order Actinomycetales			
Suborder Frankineae			
Family Acidothermaceae			
Class Actinobacteria	*Cellulomonas biazotea*		N
Subclass Actinobacteridae	*Cellulomonas cartae*		N
Order Actinomycetales	*Cellulomonas cellasea*		N
Suborder Micrococcineae	*Cellulomonas fimi*	Soil	NC_015514

(Continued)

Table 11.1: A noninclusive list of cellulolytic organims that have been studied to date. The organisms source is listed, as well as the genome accession number. If the genome accession number is not available, it will contain an N.[71]—cont'd

Phylogeny	Species	Source	Genome accession number
Family Cellulomonadaceae	*Cellulomonas flavigena*	Soil	NC_014151
	Cellulomonas gelida		N
	Cellulomonas iranensis	Forest soil	PRJDB435
	Cellulomonas persica	Forest soil	PRJDB980
	Cellulomonas terrae	Soil	N
	Cellulomonas uda	Sewage	N
Class Actinobacteria	*Cellulosimicrobium cellulans*	Soil	PRJEB571
Subclass Actinobacteridae			
Order Actinomycetales			
Suborder Micrococcineae			
Family Promicromonosporaceae			
Class Actinobacteria	*Curtobacterium flaccumfaciens*	Soil	PRJNA199964
Subclass Actinobacteridae			
Order Actinomycetales			
Suborder Micrococcineae			
Family Microbacteriaceae			
Class Actinobacteria	*Xylanimonas cellulosilytica*	Decayed tree	NC_013530
Subclass Actinobacteridae			
Order Actinomycetales			
Suborder Micrococcineae			
Family Promicromonosporaceae			
Class Actinobacteria	*Actinoplanes aurantiaca*	Soil	N
Subclass Actinobacteridae			
Order Actinomycetales	*Micromonospora melonosporea*	Compost	N
Suborder Micromonosporineae	*Micromonospora chalcae*	Soil	N
Family Micromonosporaceae	*Micromonospora propionici*	Termite	N
	Micromonospora ruminantium	Rumen	N
Class Actinobacteria	*Actinosynnema mirum*	Grass blade	NC_013093
Subclass Actinobacteridae			
Order Actinomycetales			
Suborder Pseudonocardineae			
Family Pseudonocardiaceae			

Class Actinobacteria	*Streptomyces albogriseolus*		PRJNA73805
Subclass Actinobacteridae	*Streptomyces aureofaciens*	Compost	N
Order Actinomycetales	*Streptomyces cellulolyticus*		N
Suborder Streptomycineae	*Streptomyces flavogriseus*	Soil	N
Family Streptomycetaceae	*Streptomyces lividans*		NZ_CM001889
	Streptomyces nitrosporeus		N
	Streptomyces olivochromogenes		N
	Streptomyces reticuli	Soil	N
	Streptomyces rochei	Termite gut	N
	Streptomyces thermovulgaris		N
	Streptomyces viridosporus		PRJNA200053
Class Actinobacteria	*Thermobifida alba*		N
Subclass Actinobacteridae			
Order Actinomycetales			
Suborder Streptosporangineae			
Family Nocardiopsaceae	*Thermobifida cellulolytica*	Compost	N
	Thermobifida fusca	Soil	NC_007333
Class Actinobacteria	*Thermomonospora curvata*		NC_013510
Subclass Actinobacteridae			
Order Actinomycetales			
Suborder Streptosporangineae			
Family Thermomonosporaceae			
Class Actinobacteria	*Thermobispora bispora*	Soil	NC_014165
Subclass Actinobacteridae	*Streptosporangium subroseum*	Soil	*N*
Order Actinomycetales			
Suborder			
StreptosporangineaeFamily			
Streptosporangiaceae			
Class Fibrobacteria	*Fibrobacter succinogenes*	Rumen	NC_013410
Order Fibrobacterales			
Family Fibrobacteraceae			
Class Cytophagia	*Cytophaga aurantiaca*	Soil	PRJNA199184
Order Cytophagales	*Cytophaga haloflava*	Soil	N
Family Cytophagaceae	*Cytophaga hutchinsonii*	Soil	NC_008255
	Cytophaga krzemieniewskae	Soil	N
	Cytophaga rosea	Soil	N
	Sporocytophaga myxococcoides	Soil	PRJNA224116

(Continued)

Table 11.1: A noninclusive list of cellulolytic organims that have been studied to date. The organisms source is listed, as well as the genome accession number. If the genome accession number is not available, it will contain an N.[71]—cont'd

Phylogeny	Species	Source	Genome accession number
Class Flavobacteriia	*Flavobacterium johnsoniae*	Soil	NC_009441
Order Flavobacteriales			
Family Flavobacteriaceae			
Class Bacteroidia	*Bacteroides cellulosilyticus*	Human faecal sample	PRJNA55279
Order Bacteroidales	*Bacteroides cellulosolvens*	Sewage	N
Family Bacteroidaceae	*Bacteroides* sp. *P-1*	Rotting biomass	N
Class Bacteroidia	*Rhodothermus marinus*	Hot spring	NC_013501
Order Bacteroidetes			
Order II. Incertae sedis			
Family Rhodothermaceae			
Class Betaproteobacteria	*Achromobacter piechaudii*	Soil	PRJNA47029
Order Burkholderiales			
Family Alcaligenaceae			
Class Gammaproteobacteria	*Dickeya dadantii*	Plant biomass	NC_012880
Order Enterobacteriales			
Family Enterobacteriaceae			
Class Gammaproteobacteria	*Xanthomonas* sp.	Brack water	N
Order Xanthomonadales	*Rudaea cellulosilytica*	Soil	PRJNA199387
Family Xanthomonadaceae			
Class Gammaproteobacteria	*Cellvibrio gilvus*	Soil	NC_015671
Order Pseudomonadales	*Cellvibrio mixtus*	Soil	N
Family Pseudomonadaceae	*Cellvibrio vulgaris, fulvus*	Soil	N
	Cellvibrio japonicus	Plant biomass	NC_010995
Subphylum delta/epsilon subdivisions	*Myxobacter* sp. *AL-1*	Soil	N
Class Deltaproteobacteria			
Order Myxococcales			
Family Incertae Sedis			
Class Thermotogae	*Fervidobacterium islandicum*	Hot spring	N
Order Thermotogales	*Thermotoga maritima*	Hot spring	NC_000853
Family Thermotogaceae	*Thermotoga neopolitana*	Hot spring	N

be able to metabolize cellulose to a near exclusive end product.[49] This is supported by initial experiments with cellulolytic bacteria, as well as the large amounts of experience producing an exclusive end product through metabolic engineering of noncellulolytic microorganisms.

A major challenge in the development of native cellulolytic organisms is the lack of universal genetransfer and genetic engineering techniques. However, much work in recent years has gone into developing these techniques for some organisms, including *Thermobifida fusca*[72,73] and *Clostridium thermocellum*.[74,75] While these gene transfer techniques are generally more challenging than for noncellulolytic bacteria, it does still provide the ability to further develop these cellulolytic bacteria. Using the knowledge gained from noncellulolytic bacteria, metabolic flux can be altered by blocking undesirable pathways by generating gene deletions (knockout strains) or through overexpression of genes associated with desirable pathways. When performing initial attempts to redirect flux, it is common for unintended consequences to occur. These are a result of either a metabolic imbalance of organic intermediates,[76,77] electron carriers,[78] or interactions with metabolic control systems.[79] In the goal of engineering a microorganism for a certain function, like producing ethanol, it is important to fully evaluate the consequences of the changes that were made. Some of the changes may provide no value other than knowledge gained and will require a change in theory. Still other changes may not initially improve product formation, but paired with an additional compounding change, there may be significant improvement in product formation.

Because of the challenges facing metabolic engineering, including the lack of accuracy and a slow process, systems biology tools have been developed to rationally design microorganisms. Rational design strain optimization protocols have been developed, and they include many computational tools that help to model the cell metabolism and help to predict strategies for metabolic engineering.[80–83] These models help process all available knowledge in a way that are intuitively difficult to analyze, and they take into account many factors including experimental data, all known biochemical pathways, genomic information including enzyme functionality, stoichiometry, and much more possible information. By employing these models for metabolic engineering, the genetic manipulations that do occur have a much higher chance of being successful. Additionally, the advantage is that experimental results (even failed designs) can continue to improve modeling predictions to further increase the likelihood of successful metabolic engineering. As more tools are developed for synthetic biology, transcriptomics, metabolomics and, systems biology, these models should become ever more robust.

Having computational models for rational design of microorganisms is only part of the battle in metabolic engineering. Modeling an organism and identifying possible gene targets is useless if the organism under development is unable to be genetically modified. The genetic manipulation platform for *C. thermocellum*, a gram-positive, anaerobic, thermophilic bacterium, was noted in a preliminary report with protoplast transformation.[74] However, the first reported platform was reported through the use of an electrotransformation (ET) protocol using a custom-designed pulse generating system as well as custom-designed cuvettes for electroporation.[74] The protocol for this ET was optimized. Among the variables that were optimized, it was determined that chilling during cell collection and washing prior to electric-pulse application as well as the duration of the postelectropulsing recovery time had the largest effects. Additionally, the electric-pulse parameters had a large impact of transformation efficiency. However, as this was the first round of genetic manipulation via ET, it was expected that this method would become obsolete with a better understanding of the bacteria, as well as better technology. This proved to be the case as a new protocol for ET of *C. thermocellum* was developed through creation of a gene deletion system based on the *pyrF* gene.[75] This new platform reduced the need to have custom cuvettes, as well as simplifying the process to obtain competent cells. The custom pulse generator still worked best for this system; however, it was noted that commercial generators could obtain successfully modified cells.

The microbe *T. fusca* has also had a genetic manipulation platform developed for future metabolic engineering efforts. *T. fusca* had been studied and characterized for its cellulolytic capabilities, and has also seen its genome sequenced. Initial genetic manipulation for this bacteria occurred through a nonrational design protocol. These efforts were made by continually altering the growth conditions of *T. fusca*, which induced stress, leading to natural evolution of the bacteria over time. Desired traits were imparted to the bacteria, but it was clear that this procedure would have little utility in future strain development. The efforts were made for rational design of the bacteria through chromosomal DNA manipulation. The high exonuclease activity in *T. fusca* required atypically long flanking regions of homology in the disruption cassette. This disruption cassette was introduced as a plasmid; however, since *T. fusca* does not possess a functional methyl-directed mismatch repair system, the plasmid preparation and propagation were conducted using a nonmethylating host strain. For transformation, *T. fusca* protoplasts (cell wall removed) were prepared. After transformation, because of the fragility of the protoplast, there was a period for protoplast regeneration in a nonselective medium. After regeneration, a selective media was used to screen for positive transformants. In the metabolic engineering work using the platform, *T. fusca* was manipulated to replace the celR gene with the adhE2 gene in the chromosome of *T. fusca* in an attempt to improve the production of 1-propanol, which is a higher alcohol that could be compatible with existing infrastructure for ethanol. Attempts were successful at higher production, although maximum titers were still low with a maximum titer of 0.48 g/L on raw switchgrass. However, having the platform for a thermophilic bacteria, combined with systems biology tools could lead to the production of much higher yields of various chemicals.

The information above described the overall strategy as well as the information needed to fully understand metabolic engineering, which can be applied at a basic level for ethanol production. Based on the branched catabolism of a typical ethanol-forming cellulolytic anaerobe in both the carbon- and electron-centered perspective, a few possible targets for gene knockout arise. Some specific knockout targets are hydrogenase, acetate kinase, and phosphotransacetylase to redirect flux away from acetic acid formation. It has been noted that the basic requirements of fermentative metabolism, mainly the generation of net ATP and regeneration of reduced electron carriers, can be met by producing ethanol, acetic acid, or lactic acid either singly or in combination. All of this information is helpful to understand, but these are just starting points for metabolic engineering efforts. The systems biology-based model described previously should be employed for this purpose in order to achieve full optimization of metabolic engineering efforts.

Product inhibition of cellulolytic bacteria

An additional consideration when selecting and developing a microorganism for CBP is the ability to maintain growth in high concentrations of product. While the microorganism should continue to grow in high concentrations of final product, in this case ethanol, it is also necessary for the organism to be able to withstand the accumulation of other products and inhibitory compounds that can arise in the industrial process.

It is believed that microorganisms are inhibited by ethanol through end product inhibition of glycolytic enzymes, as well as the damage to the cell membrane.[84,85] Among cellulolytic organisms, *C. thermocellum* has been investigated extensively in terms of end product tolerance. In *C. thermocellum,* the inhibition is believed to be a blockage in glycolysis associated with ethanol-induced changes in the cell membrane.[86–88] Additionally, increased temperature has been observed to markedly decrease ethanol tolerance in *C. thermocellum* and other organisms.[89,90] This occurs because both increased temperature and ethanol result in increased membrane fluidity.

C. thermocellum has been shown to grow at added ethanol concentrations up to 50 g/L.[91,92] However, *C. thermocellum* cannot natively produce ethanol at this high concentration, and the maximum titer was about 25 g/L.[91,93] This discrepancy is known as a titer gap, which is the difference between the maximum concentrations of a compound tolerated by a microorganism when added to the culture, versus the maximum concentration produced by the organism.[69] This titer gap has been decreased in noncellulolytic microorganisms, so it could be assumed that this will occur for cellulolytic bacteria as well. However, just like the challenges of metabolic engineering, this progress will initially be slow as the molecular platforms are less developed.

In additional to ethanol inhibition, various other compounds are likely to occur in an industrial setting that could cause an inhibition to growth with the most studies to date have also occurred with *C. thermocellum.* In multiple studies, *C. thermocellum* was shown to have the ability to grow on pretreated LCB substrates that contained most or all of the lignin present prior to pretreatment, which suggests that soluble lignin does not have inhibitory effects.[94–96] However, two other common inhibitors have been shown to have a negative effect on *C. thermocellum.* The presence of hydrolysates that resulted from autoclaved corn stover[97] and steam-exploded aspen[98] were shown to inhibit growth. Additionally, the presence of acetate, which is a by-product of pretreatment as well as a fermentation product were shown to decrease the growth of unacclimated cells by 50% in a near-neutral pH medium with an acetate concentration of 0.28 M.[88] The suggestion behind the acetate inhibition is that the magnitude of the transmembrane proton-motive force is decreased. It is also suggested that anion accumulation is the explanation for the toxicity of cells to organic acids.[99]

Similar to strain development, the reasons for product inhibition are extremely complex, and the development of bioinformatics and systems biology tools could be useful. Ever increasing amounts of data can be accumulated with transcriptomics and proteomics tools. This data could feedback to strain development, and genetic engineering efforts could be made to improve product tolerance. Genetic engineering has been used to create a mutant of *E. coli* that can produce ethanol in the presence of 15 mM of furfural.[100,101] This mutant was produced using known experimental information; it is possible that, just like metabolic engineering, a model could improve furfural tolerance further.

11.5.5 Recombinant Cellulolytic Microorganisms

The recombinant cellulolytic strategy starts with a microorganism that is a model organism for product formation, and engineer a cellulolytic system for this microorganism that would enable it to grow on cellulose through heterologous gene expression. This cross species gene expression can be applied to a number of different process goals, but in this case it is specifically discussed in terms of utilizing cellulose as a carbon source.

The heterologous expression of cellulase systems has seen research in the yeasts *S. cerevisiae,*[102] *Pachysolen tannophilus,*[103] *Candida shehatae,*[104] and *Pichia stipitis.*[105] Recent research efforts in bacteria include *Zymomonas mobilis,*[106,107] *E. coli,*[108,109] and *Klebsiella oxytoca.*[110,111] This section will focus almost exclusively on *S. cerevisae*, as that has both extensive research performed, as well as being the standard microorganism in current industrial ethanol production.

When designing a cellulase enzyme system, it is first necessary to identify what components are necessary to be added to the noncellulolytic organism. For a noncomplexed cellulase system to properly function there must be a cellobiohydrolase attacking reducing ends, a cellobiohydrolase attacking nonreducing ends, and EG, and a BGL. Of these, the cellobiohydrolases should have a cellulose-binding module (CBM). For complexed cellulases, the following are required: a scaffolding protein with a CBM; two cohesins and a domain that binds to a cell wall-anchoring protein; a cell wall-anchoring protein; and at least one exoglucanase and an endoglucanse, both containing dockerins capable of binding to the scaffolding protein. Additionally, either a BGL or cellodextrin and cellobiose phosphorylases, together with the appropriate permeases, would be required.[49,70] From a practical standpoint, the engineering of a noncomplexed cellulase system appears to be much easier than that of the complexed systems.

Cellulase expression in S. cerevisiae

The expression levels of cellulases in *S. cerevisiae* has improved significantly in recent years. It has been reported that two critical cellulases, Cel6A (CBH1) and Cel7a (CBH2) have achieved a maximum titer at values corresponding to 4-5% of total cell protein, which meets the calculated levels for growth on cellulose in an industrial process.[70]

An approach that has been used for cellulase expression is to closely resemble the naturally occurring systems. *T. reesei,* has been analyzed extensively for its application in cellulase production for enzymatic hydrolysis. The functional enzyme classes in *T. reesei* from the Herpoel-Gimbert study[112] have been 100% expressed in *S. cerevisiae* while the enzymes identified in Nagendran et al. have been 80% expressed in *S. cerevisiae.*[113] Because of the intricacy of complexed cellulase systems, the expression has not been as successful with only 20% of *C. thermocellum* cellulases functionally expressed in *S. cerevisiae.*[69]

CBP process and challenges using S. cerevisiae

After designing and expressing the cellulase systems, the next step is to test the microbial system on phosphoric acid swollen cellulose (PASC). Recent reports have achieved 75% of theoretical ethanol yield by optimizing the level and ratios of cellulase enzymes in *S. cerevisiae,*[114] which shows a large improvement over the approximately 27% conversion of PASC first published in 2007.[115] This report is promising in the improvements made, however, much more improvement needs to occur for industrial success. Additionally, PASC is more of a model substrate due to its extreme reactivity. Therefore, these organisms would likely need to undergo further refinement to achieve high yields on industrially pretreated cellulose, which will be considerably more recalcitrant than PASC.

In addition to designing appropriate cellulase systems as well as achieving high cellulase expression levels in yeast, there are a number of other challenges facing the recombinant strategy. By adding a large amount of heterologous genes to yeast, long-term gene stability is an issue that should be addressed before industrial acceptance. Gene mutations over time could reduce the expression levels of the cellulases, which obviously would decrease recombinant organisms effectiveness in CBP.

11.6 Concluding Remarks

LCB is a promising inexpensive and abundant raw material that can be replenished in a sustainable manner. Potential sources of biomass include agricultural waste (corn stover), energy crops (switchgrass), woody residue (poplar), and human waste. The attributes of LCB make it attractive as a bulk raw material for chemical and biotechnology industries; however, there are a variety of technical challenges to industrial use of LCB.

The initial challenge to using LCB is to develop efficient, cost-effective methods to separate the major components of LCB (lignin, hemicellulose, and cellulose) and convert fractionated LCB to monomeric substrates that can be directly utilized for chemical production. Combinations of chemical and mechanical approaches have been studied to address this aspect of LCB utilization, with some process flexibility required to address the fact that different sources of LCB have different specific lignin, hemicellulose, and cellulose content. Successes in these pretreatment approaches have facilitated the development of processes that utilize LCB as a starting raw material, but there remain technical issues (e.g., inhibitory effects associated with carryover components of the pretreatment) and economic constraints.

While it is currently possible to utilize LCB as a raw material following a pretreatment, enzymatic hydrolysis, and fermentation methodology, it has been found that inefficiencies in each step coupled with associated costs make this LCB processing paradigm almost infeasible from an economic standpoint. Thus, there are a number of approaches that have arisen to improve the overall prospect of LCB-derived industries. The first is to continue to improve the efficiency of each of the individual steps of pretreatment, enzymatic hydrolysis, and fermentation and the linkages between these steps. The second approach is look to eliminate some of the individual steps in this process through the development of a consolidated bioprocessing approach. The third approach is to identify and develop secondary high-value products that can be made in the same process parallel to production of the biofuel.

Acknowledgments

The authors thank Xiongjun Shao, Ph.D of Dartmouth College for valuable discussions. This work was supported by The Key Laboratory of Industrial Biotechnology, Ministry of Education, Jiangnan University, China (KLIB-KF201403).

References

1. Anonymous US billion ton update: biomass supply for a bioenergy and bioproducts industry (executive summary). *Industrial Biotechnology* 7 pp. 375+. 2011. Available: http://go.galegroup.com/ps/i.do?id=GALE%7CA272510991&v=2.1&u=ksu&it=r&p=AONE&sw=w&asid=bed0da8bceada9e5a961ab60f40258f0.

2. Naik SN, Goud VV, Rout PK, Dalai AK. Production of first and second generation biofuels: a comprehensive review. *Renew Sust Energ Rev* 2010;**14**(2):578–97. http://dx.doi.org/10.1016/j.rser.2009.10.003.
3. McCann MC, Buckeridge MS, Carpita NC. Plants and bioenergy. New York: Springer; 2014, p. 153–181.
4. Chundawat SPS, Beckham GT, Himmel ME, Dale BE. Deconstruction of lignocellulosic biomass to fuels and chemicals. *Annu Rev Chem Biomol Eng* 2011;**2**(1):121–45. http://dx.doi.org/10.1146/annurev-chembioeng-061010-114205.
5. Pauly M, Keegstra K. Cell-wall carbohydrates and their modification as a resource for biofuels. *Plant J* 2008;**54**(4):559–68. http://dx.doi.org/10.1111/j.1365-313X.2008.03463.x.
6. McKendry P. Energy production from biomass (part 1): overview of biomass. *Bioresour Technol* 2002;**83**(1):37–46. http://dx.doi.org/10.1016/S0960-8524(01)00118-3.
7. (). *Biomass Feedstock Composition and Property Database*. Available: http://www.afdc.energy.gov/biomass/progs/search3.cgi?13124.
8. Mosier N, Wyman C, Dale B, Elander R, Lee YY, Holtzapple M, et al. Features of promising technologies for pretreatment of lignocellulosic biomass. *Bioresour Technol* 2005;**96**(6):673–86. http://dx.doi.org/10.1016/j.biortech.2004.06.02.
9. Alvira P, Tomás-Pejó E, Ballesteros M, Negro MJ. Pretreatment technologies for an efficient bioethanol production process based on enzymatic hydrolysis: a review. *Bioresour Technol* 2010;**101**(13):4851–61. http://dx.doi.org/10.1016/j.biortech.2009.11.09.
10. Sun Y, Cheng J. Hydrolysis of lignocellulosic materials for ethanol production: a review. *Bioresour Technol* 2002;**83**(1):1–11. http://dx.doi.org/10.1016/S0960-8524(01)00212-7.
11. Taherzadeh MJ, Karimi K. Pretreatment of lignocellulosic wastes to improve ethanol and biogas production: a review. *Int J Mol Sci* 2008;**9**(9):1621–51. http://dx.doi.org/10.3390/ijms9091621.
12. Cadoche L, López GD. Assessment of size reduction as a preliminary step in the production of ethanol from lignocellulosic wastes. *Biol Wastes* 1989;**30**(2):153–7. http://dx.doi.org/10.1016/0269-7483(89)90069-4.
13. Lamsal B, Yoo J, Brijwani K, Alavi S. Extrusion as a thermo-mechanical pre-treatment for lignocellulosic ethanol. *Biomass Bioenergy* 2010;**34**(12):1703–10. http://dx.doi.org/10.1016/j.biombioe.2010.06.009.
14. Yoo J, Alavi S, Vadlani P, Amanor-Boadu V. Thermo-mechanical extrusion pretreatment for conversion of soybean hulls to fermentable sugars. *Bioresour Technol* 2011;**102**(16):7583–90. http://dx.doi.org/10.1016/j.biortech.2011.04.092.
15. Gáspár M, Kálmán G, Réczey K. Corn fiber as a raw material for hemicellulose and ethanol production. *Process Biochem* 2007;**42**(7):1135–9. http://dx.doi.org/10.1016/j.procbio.2007.04.003.
16. Kumar P, Barrett DM, Delwiche MJ, Stroeve P. Methods for pretreatment of lignocellulosic biomass for efficient hydrolysis and biofuel production. *Ind Eng Chem Res* 2009;**48**(8):3713–29. http://dx.doi.org/10.1021/ie801542g.
17. Taherzadeh M, Karimi K. Enzymatic-based hydrolysis processes for ethanol. *BioResources* 2007;**2**(4):707–38.
18. Silverstein RA, Chen Y, Sharma-Shivappa RR, Boyette MD, Osborne J. A comparison of chemical pretreatment methods for improving saccharification of cotton stalks. *Bioresour Technol* 2007;**98**(16):3000–11, S0960-8524(06)00578-5 [pii].
19. Kim S, Holtzapple MT. Delignification kinetics of corn stover in lime pretreatment. *Bioresour Technol* 2006;**97**(5):778–85. http://dx.doi.org/10.1016/j.biortech.2005.04.002.
20. Chang V, Nagwani M, Kim C, Holtzapple M. Oxidative lime pretreatment of high-lignin biomass. *Appl Biochem Biotechnol* 2001;**94**(1):1–28. http://dx.doi.org/10.1385/ABAB:94:1:01.
21. Wyman CE, Dale BE, Elander RT, Holtzapple M, Ladisch MR, Lee YY. Coordinated development of leading biomass pretreatment technologies. *Bioresour Technol* 2005;**96**(18):1959–66. http://dx.doi.org/10.1016/j.biortech.2005.01.01.
22. Yang B, Wyman CE. Effect of xylan and lignin removal by batch and flowthrough pretreatment on the enzymatic digestibility of corn stover cellulose. *Biotechnol Bioeng* 2004;**86**(1):88–98. http://dx.doi.org/10.1002/bit.20043.

23. Cara C, Ruiz E, Oliva JM, Sáez F, Castro E. Conversion of olive tree biomass into fermentable sugars by dilute acid pretreatment and enzymatic saccharification. *Bioresour Technol* 2008;**99**(6):1869–76.
24. Taherzadeh MJ. Ethanol from lignocellulose: physiological effects of inhibitors and fermentation strategies. Ph.D thesis, Gothenburg, Sweden: Chalmers University of Technology; 1999.
25. Kootstra AMJ, Beeftink HH, Scott EL, Sanders JP. Comparison of dilute mineral and organic acid pretreatment for enzymatic hydrolysis of wheat straw. *Biochem Eng J* 2009;**46**(2):126–31.
26. Zhao X, Cheng K, Liu D. Organosolv pretreatment of lignocellulosic biomass for enzymatic hydrolysis. *Appl Microbiol Biotechnol* 2009;**82**(5):815–27. http://dx.doi.org/10.1007/s00253-009-1883-1.
27. Papatheofanous MG, Billa E, Koullas DP, Monties B, Koukios EG. Two-stage acid-catalyzed fractionation of lignocellulosic biomass in aqueous ethanol systems at low temperatures. *Bioresour Technol* 1995;**54**(3):305–10. http://dx.doi.org/10.1016/0960-8524(95)00152-.
28. Samayam IP, Schall CA. Saccharification of ionic liquid pretreated biomass with commercial enzyme mixtures. *Bioresour Technol* 2010;**101**(10):3561–6. http://dx.doi.org/10.1016/j.biortech.2009.12.066.
29. Zhu S, Wu Y, Chen Q, Yu Z, Wang C, Jin S, et al. Dissolution of cellulose with ionic liquids and its application: a mini-review. *Green Chem* 2006;**8**(4):325–7. http://dx.doi.org/10.1039/B601395C.
30. Brodeur G, Yau E, Badal K, Collier J, Ramachandran KB, Ramakrishnan S. Chemical and physicochemical pretreatment of lignocellulosic biomass: a review. *Enzyme Res* 2011;**2011**:17. http://dx.doi.org/10.4061/2011/787532.
31. Pu Y, Pu N, Jiang A. Ionic liquid as a green solvent for lignin. *J Wood Chem Technol* 2007;**27**(1):23–33. http://dx.doi.org/10.1080/02773810701282330.
32. Binder JB, Raines RT. Simple chemical transformation of lignocellulosic biomass into furans for fuels and chemicals. *J Am Chem Soc* 2009;**131**(5):1979–85. http://dx.doi.org/10.1021/ja808537j [doi].
33. Lee SH, Doherty TV, Linhardt RJ, Dordick JS. Ionic liquid-mediated selective extraction of lignin from wood leading to enhanced enzymatic cellulose hydrolysis. *Biotechnol Bioeng* 2009;**102**(5):1368–76. http://dx.doi.org/10.1002/bit.22179.
34. Swatloski RP, Spear SK, Holbrey JD, Rogers RD. Dissolution of cellose with ionic liquids. *J Am Chem Soc* 2002;**124**(18):4974–5. http://dx.doi.org/10.1021/ja025790m.
35. Socha AM, Parthasarathi R, Shi J, Pattathil S, Whyte D, Bergeron M, et al. Efficient biomass pretreatment using ionic liquids derived from lignin and hemicellulose. *Proc Natl Acad Sci* 2014;**111**(35):E3587–95. http://dx.doi.org/10.1073/pnas.1405685111.
36. Overend RP, Chornet E, Gascoigne JA. Fractionation of lignocellulosics by steam-aqueous pretreatments and discussion. *Philos Trans R Soc, A* 1987;**321**(1561):523–36.
37. Ballesteros I, Negro M, Oliva J, Cabañas A, Manzanares P, Ballesteros M. Ethanol production from steam-explosion pretreated wheat straw. *Appl Biochem Biotechnol* 2006;**130**(1–3):496–508. http://dx.doi.org/10.1385/ABAB:130:1:496.
38. Yu Q, Zhuang X, Yuan Z, Wang Q, Qi W, Wang W, et al. Two-step liquid hot water pretreatment of eucalyptus grandis to enhance sugar recovery and enzymatic digestibility of cellulose. *Bioresour Technol* 2010;**101**(13):4895–9. http://dx.doi.org/10.1016/j.biortech.2009.11.05.
39. Mosier N, Hendrickson R, Ho N, Sedlak M, Ladisch MR. Optimization of pH controlled liquid hot water pretreatment of corn stover. *Bioresour Technol* 2005;**96**(18):1986–93. http://dx.doi.org/10.1016/j.biortech.2005.01.013.
40. Laser M, Schulman D, Allen SG, Lichwa J, Antal Jr MJ, Lynd LR. A comparison of liquid hot water and steam pretreatments of sugar cane bagasse for bioconversion to ethanol. *Bioresour Technol* 2002;**81**(1):33–44.
41. Alizadeh H, Teymouri F, Gilbert T, Dale B. Pretreatment of switchgrass by ammonia fiber explosion (AFEX). *Appl Biochem Biotechnol* 2005;**124**(1–3):1133–41. http://dx.doi.org/10.1385/ABAB:124:1-3:1133.
42. Laureano-Perez L, Teymouri F, Alizadeh H, Dale B. Understanding factors that limit enzymatic hydrolysis of biomass. *Appl Biochem Biotechnol* 2005;**124**(1–3):1081–99. http://dx.doi.org/10.1385/ABAB:124:1-3:1081.

43. Kim KH, Hong J. Supercritical CO2 pretreatment of lignocellulose enhances enzymatic cellulose hydrolysis. *Bioresour Technol* 2001;**77**(2):139–44. http://dx.doi.org/10.1016/S0960-8524(00)00147-4.
44. Dumitriu S. Polysaccharides: structural diversity and functional versatility. New York: Marcel Dekker; 2005.
45. Martín C, Thomsen MH, Hauggaard-Nielsen H, BelindaThomsen A. Wet oxidation pretreatment, enzymatic hydrolysis and simultaneous saccharification and fermentation of clover–ryegrass mixtures. *Bioresour Technol* 2008;**99**(18):8777–82. http://dx.doi.org/10.1016/j.biortech.2008.04.03.
46. Keshwani DR. *Microwave pretreatment of switchgrass for bioethanol production*. Ph.D thesis. Raleigh, NC, USA: North Carolina State University; 2009.
47. Chiaramonti D, Prussi M, Ferrero S, Oriani L, Ottonello P, Torre P, et al. Review of pretreatment processes for lignocellulosic ethanol production, and development of an innovative method. *Biomass Bioenergy* 2012;**46**:25–35. http://dx.doi.org/10.1016/j.biombioe.2012.04.02.
48. Hasunuma T, Okazaki F, Okai N, Hara KY, Ishii J, Kondo A. A review of enzymes and microbes for lignocellulosic biorefinery and the possibility of their application to consolidated bioprocessing technology. *Bioresour Technol* 2013;**135**:513–22. http://dx.doi.org/10.1016/j.biortech.2012.10.04.
49. Lynd LR, Weimer PJ, van Zyl WH, Pretorius IS. Microbial cellulose utilization: fundamentals and biotechnology. *Microbiol Mol Biol Rev* 2002;**66**:506–77.
50. Hyeon JE, Jeon SD, Han SO. Cellulosome-based, clostridium-derived multi-functional enzyme complexes for advanced biotechnology tool development: advances and applications. *Biotechnol Adv* 2013;**31**(6):936–44. http://dx.doi.org/10.1016/j.biotechadv.2013.03.00.
51. van DB, de Vries R. Fungal enzyme sets for plant polysaccharide degradation. *Appl Microbiol Biotechnol* 2011;**91**(6):1477–92. http://dx.doi.org/10.1007/s00253-011-3473-2.
52. Scheller HV, Ulvskov P. Hemicelluloses. *Annu Rev Plant Biol* 2010;**61**(1):263–89. http://dx.doi.org/10.1146/annurev-arplant-042809-112315.
53. Sweeney MD, Xu F. Biomass converting enzymes as industrial biocatalysts for fuels and chemicals: recent developments. *Catalysts* 2012;**2**(2):244–63.
54. Kumar R, Singh S, Singh O. Bioconversion of lignocellulosic biomass: biochemical and molecular perspectives. *J Ind Microbiol Biotechnol* 2008;**35**(5):377–91. http://dx.doi.org/10.1007/s10295-008-0327-8.
55. Gao D, Uppugundla N, Chundawat SP, Yu X, Hermanson S, Gowda K, et al. Hemicellulases and auxiliary enzymes for improved conversion of lignocellulosic biomass to monosaccharides. *Biotechnol Biofuels* 2011;**4**:5–6834. http://dx.doi.org/10.1186/1754-6834-4-5.
56. Couturier M, Haon M, Coutinho PM, Henrissat B, Lesage-Meessen L, Berrin JG. Podospora anserina hemicellulases potentiate the trichoderma reesei secretome for saccharification of lignocellulosic biomass. *Appl Environ Microbiol* 2011;**77**(1):237–46. http://dx.doi.org/10.1128/AEM.01761-10.
57. Kumar R, Wyman CE. Effect of xylanase supplementation of cellulase on digestion of corn stover solids prepared by leading pretreatment technologies. *Bioresour Technol* 2009;**100**(18):4203–13. http://dx.doi.org/10.1016/j.biortech.2008.11.05.
58. Banerjee G, Scott-Craig JS, Walton JD. Improving enzymes for biomass conversion: a basic research perspective. *BioEnergy Res* 2010;**3**(1):82–92.
59. Banerjee G, Car S, Scott-Craig JS, Borrusch MS, Walton JD. Rapid optimization of enzyme mixtures for deconstruction of diverse pretreatment/biomass feedstock combinations. *Biotechnol Biofuels* 2010;**3**:22–6834. http://dx.doi.org/10.1186/1754-6834-3-22.
60. Griggs AJ, Stickel JJ, Lischeske JJ. A mechanistic model for enzymatic saccharification of cellulose using continuous distribution kinetics I: depolymerization by EGI and CBHI. *Biotechnol Bioeng* 2012;**109**(3):665–75. http://dx.doi.org/10.1002/bit.23355.
61. Levine SE, Fox JM, Blanch HW, Clark DS. A mechanistic model of the enzymatic hydrolysis of cellulose. *Biotechnol Bioeng* 2010;**107**(1):37–51. http://dx.doi.org/10.1002/bit.22789.
62. Okazaki M. Kinetics of enzymatic hydrolysis of cellulose: analytical description of a mechanistic model. *Biotechnol Bioeng* 1978;**20**(5):637–63. http://dx.doi.org/10.1002/bit.260200503.

63. Bansal P, Hall M, Realff MJ, Lee JH, Bommarius AS. Modeling cellulase kinetics on lignocellulosic substrates. *Biotechnol Adv* 2009;**27**(6):833–48. http://dx.doi.org/10.1016/j.biotechadv.2009.06.005.
64. Belkacemi K, Hamoudi S. Enzymatic hydrolysis of dissolved corn stalk hemicelluloses: reaction kinetics and modeling. *J Chem Technol Biotechnol* 2003;**78**(7):802–8. http://dx.doi.org/10.1002/jctb.865.
65. Feng YY, He ZM, Song LF, Ong SL, Hu JY, Zhang ZG, Ng WJ. Kinetics of Î^{2}-mannanase fermentation by bacillus licheniformis. *Biotechnol Lett* 2003;**25**(14):1143–6. http://dx.doi.org/10.1023/A:1024517228270.
66. HarjunpÄÄ V, Teleman A, Siika-Aho M, Drakenberg T. Kinetic and stereochemical studies of manno-oligosaccharide hydrolysis catalysed by?-mannanases from trichoderma reesei. *Eur J Biochem* 1995;**234**(1):278–83. http://dx.doi.org/10.1111/j.1432-1033.1995.278_c.x.
67. Liu X, Lu M, Ai N, Yu F, Ji J. Kinetic model analysis of dilute sulfuric acid-catalyzed hemicellulose hydrolysis in sweet sorghum bagasse for xylose production. *Ind Crop Prod* 2012;**38**:81–6. http://dx.doi.org/10.1016/j.indcrop.2012.01.01.
68. Zhang Y, Xu B, Zhou W. On a novel mechanistic model for simultaneous enzymatic hydrolysis of cellulose and hemicellulose considering morphology. *Biotechnol Bioeng* 2014;**111**(9):1767–81. http://dx.doi.org/10.1002/bit.25244.
69. Olson DG, McBride JE, Joe Shaw A, Lynd LR. Recent progress in consolidated bioprocessing. *Curr Opin Biotechnol* 2012;**23**(3):396–405.
70. Lynd LR, van Zyl WH, McBride JE, Laser M. Consolidated bioprocessing of cellulosic biomass: an update. *Curr Opin Biotechnol* 2005;**16**:577–83.
71. Koeck DE, Pechtl A, Zverlov VV, Schwarz WH. Genomics of cellulolytic bacteria. *Curr Opin Biotechnol* 2014;**29**:171–83. http://dx.doi.org/10.1016/j.copbio.2014.07.002.
72. Deng Y, Fong SS. Development and application of a PCR-targeted gene disruption method for studying CelR function in thermobifida fusca. *Appl Environ Microbiol* 2010;**76**(7):2098–106. http://dx.doi.org/10.1128/AEM.02626-09.
73. Deng Y, Fong SS. Metabolic engineering of<i>thermobifida fusca for direct aerobic bioconversion of untreated lignocellulosic biomass to 1-propanol. *Metab Eng* 2011;**13**(5):570–7.
74. Tyurin MV, Desai SG, Lynd LR. Electrotransformation of clostridium thermocellum. *Appl Environ Microbiol* 2004;**70**(2):883–90.
75. Tripathi SA, Olson DG, Argyros DA, Miller BB, Barrett TF, Murphy DM, et al. Development of pyrF-based genetic system for targeted gene deletion in clostridium thermocellum and creation of a pta mutant. *Appl Environ Microbiol* 2010;**76**(19):6591–9. http://dx.doi.org/10.1128/AEM.01484-10.
76. Diaz-Ricci JC, Regan L, Bailey JE. Effect of alteration of the acetic acid synthesis pathway on the fermentation pattern of escherichia coli. *Biotechnol Bioeng* 1991;**38**(11):1318–24. http://dx.doi.org/10.1002/bit.260381109.
77. Ingram LO, Clark DP. Ethanol production using engineered mutant *E. coli*. Available: http://www.google.com/patents/US5028539, 1991.
78. Hols P, Ramos A, Hugenholtz J, Delcour J, de Vos WM, Santos H, et al. Acetate utilization in lactococcus lactis deficient in lactate dehydrogenase: a rescue pathway for maintaining redox balance. *J Bacteriol* 1999;**181**(17):5521–6.
79. Desai RP, Papoutsakis ET. Antisense RNA strategies for metabolic engineering of clostridium acetobutylicum. *Appl Environ Microbiol* 1999;**65**(3):936–45.
80. Burgard AP, Maranas CD. Optimization-based framework for inferring and testing hypothesized metabolic objective functions. *Biotechnol Bioeng* 2003;**82**(6):670–7.
81. Herrgard MJ, Fong SS, Palsson BØ. Identification of genome-scale metabolic network models using experimentally measured flux profiles. *PLoS Comput Biol* 2006;**2**(7):e72.
82. Milne CB, Eddy JA, Raju R, Ardekani S, Kim PJ, Senger RS, et al. Metabolic network reconstruction and genome-scale model of butanol-producing strain clostridium beijerinckii NCIMB 8052. *BMC Syst Biol* 2011;**5**:130–0509. http://dx.doi.org/10.1186/1752-0509-5-130.

83. Feist AM, Zielinski DC, Orth JD, Schellenberger J, Herrgard MJ, Palsson BØ. Model-driven evaluation of the production potential for growth-coupled products of escherichia coli. *Metab Eng* 2010;**12**(3):173–86. http://dx.doi.org/10.1016/j.ymben.2009.10.00.
84. Ingram LO. Ethanol tolerance in bacteria. *Crit Rev Biotechnol* 1990;**9**(4):305–19.
85. Jung S, Zeikus JG, Hollingsworth RI. A new family of very long chain alpha, omega-dicarboxylic acids is a major structural fatty acyl component of the membrane lipids of thermoanaerobacter ethanolicus 39E. *J Lipid Res* 1994;**35**(6):1057–65.
86. Herrero AA, Gomez RF, Roberts MF. Ethanol-induced changes in the membrane lipid composition of clostridium thermocellum. *Biochim Biophys Acta* 1982;**693**(1):195–204, 0005-2736(82)90487-4 [pii].
87. Herrero AA, Gomez RF, Roberts MF. 31P NMR studies of clostridium thermocellum. mechanism of end product inhibition by ethanol. *J Biol Chem* 1985;**260**(12):7442–51.
88. Herrero A, Herrero R, Gomez B, Snedecor C, Tolman M. Growth inhibition of clostridium thermocellum by carboxylic acids: a mechanism based on uncoupling by weak acids. *Appl Microbiol Biotechnol* 1985;**22**(1):53–62. http://dx.doi.org/10.1007/BF00252157.
89. Herrero AA, Gomez RF. Development of ethanol tolerance in clostridium thermocellum: effect of growth temperature. *Appl Environ Microbiol* 1980;**40**(3):571–7.
90. Kundu S, Ghose TK, Mukhopadhyay SN. Bioconversion of cellulose into ethanol by clostridium thermocellum–product inhibition. *Biotechnol Bioeng* 1983;**25**(4):1109–26. http://dx.doi.org/10.1002/bit.260250418.
91. Brown SD, Guss AM, Karpinets TV, Parks JM, Smolin N, Yang S, et al. Mutant alcohol dehydrogenase leads to improved ethanol tolerance in clostridium thermocellum. *Proc Natl Acad Sci* 2011;**108**(33):13752–7. http://dx.doi.org/10.1073/pnas.1102444108.
92. Williams T, Combs J, Lynn B, Strobel H. Proteomic profile changes in membranes of ethanol-tolerant clostridium thermocellum. *Appl Microbiol Biotechnol* 2007;**74**(2):422–32. http://dx.doi.org/10.1007/s00253-006-0689-7.
93. Demain AL, Newcomb M, Wu JH. Cellulase, clostridia, and ethanol. *Microbiol Mol Biol Rev* 2005;**69**(1):124–54, 69/1/124 [pii].
94. Lynd LR, Grethlein HE, Wolkin RH. Fermentation of cellulosic substrates in batch and continuous culture by clostridium thermocellum. *Appl Environ Microbiol* 1989;**55**(12):3131–9.
95. Saddler J, Chan M. Conversion of pretreated lignocellulosic substrates to ethanol by clostridium thermocellum in mono-and co-culture with clostridium thermosaccharolyticum and clostridium thermohydrosulphuricum. *Can J Microbiol* 1984;**30**(2):212–20.
96. Hörmeyer HF, Hrmeyer P, Tailliez J, Millet H, Girard G, Bonn O, et al. Ethanol production by clostridium thermocellum grown on hydrothermally and organosolv-pretreated lignocellulosic materials. *Appl Microbiol Biotechnol* 1988;**29**(6):528–35. http://dx.doi.org/10.1007/BF0026098.
97. Avgerinos GC, Wang DI. Selective solvent delignification for fermentation enhancement. *Biotechnol Bioeng* 1983;**25**(1):67–83. http://dx.doi.org/10.1002/bit.260250107.
98. Khan AW. Factors affecting the utilization of steam- and explosion-decompressed aspen wood by cellulolytic anaerobes. *Biomass* 1988;**15**(4):269–79. http://dx.doi.org/10.1016/0144-4565(88)90062-5.
99. Russell J. Another explanation for the toxicity of fermentation acids at low pH: anion accumulation versus uncoupling. *J Appl Bacteriol* 1992;**73**(5):363–70.
100. Wang X, Miller EN, Yomano LP, Zhang X, Shanmugam KT, Ingram LO. Increased furfural tolerance due to overexpression of NADH-dependent oxidoreductase FucO in escherichia coli strains engineered for the production of ethanol and lactate. *Appl Environ Microbiol* 2011;**77**(15):5132–40. http://dx.doi.org/10.1128/AEM.05008-11.
101. Wang X, Yomano LP, Lee JY, York SW, Zheng H, Mullinnix MT, Shanmugam KT, Ingram LO. Engineering furfural tolerance in escherichia coli improves the fermentation of lignocellulosic sugars into renewable chemicals. *Proc Natl Acad Sci U S A* 2013;**110**(10):4021–6. http://dx.doi.org/10.1073/pnas.1217958110.
102. van Zyl W, Lynd L, den Haan R, McBride J. Consolidated bioprocessing for bioethanol production using saccharomyces cerevisiae. In: Olsson L, editor. 2007. doi: 10.1007/10_2007_061.

103. Slininger PJ, Bolen PL, Kurtzman CP. Pachysolen tannophilus: properties and process considerations for ethanol production from d-xylose. *Enzym Microb Technol* 1987;**9**(1):5–15. http://dx.doi.org/10.1016/0141-0229(87)90043-3.

104. Prior B, Kilian S, Preez JDu. Fermentation of D-xylose by the yeasts candida shehatae and pichia stipitis. *Process Biochem* 1989;**24**(1):21–32.

105. Xu Q, Singh A, Himmel ME. Perspectives and new directions for the production of bioethanol using consolidated bioprocessing of lignocellulose. *Curr Opin Biotechnol* 2009;**20**(3):364–71. http://dx.doi.org/10.1016/j.copbio.2009.05.00.

106. Lawford H, Rousseau J. Performance testing of zymomonas mobilis metabolically engineered for cofermentation of glucose, xylose, and arabinose. *Appl Biochem Biotechnol* 2002;**98–100**(1–9):429–48. http://dx.doi.org/10.1385/ABAB:98-100:1-9:429.

107. Mohagheghi A, Evans K, Chou Y, Zhang M. Cofermentation of glucose, xylose, and arabinose by genomic DNA-Integrated xylose/arabinose fermenting strain of zymomonas mobilis AX101. In: *Biotechnology for Fuels and Chemicals*Anonymous, 2002.

108. Ingram LO, Aldrich HC, Borges ACC, Causey TB, Martinez A, Morales F, Saleh A, Underwood SA, Yomano LP, York SW, Zaldivar J, Zhou S. Enteric bacterial catalysts for fuel ethanol production. *Biotechnol Prog* 1999;**15**(5):855–66. http://dx.doi.org/10.1021/bp9901062.

109. Tao H, Gonzalez R, Martinez A, Rodriguez M, Ingram LO, Preston JF, et al. Engineering a homo-ethanol pathway in escherichia coli: increased glycolytic flux and levels of expression of glycolytic genes during xylose fermentation. *J Bacteriol* 2001;**183**(10):2979–88. http://dx.doi.org/10.1128/JB.183.10.2979-2988.2001.

110. Dien BS, Cotta MA, Jeffries TW. Bacteria engineered for fuel ethanol production: current status. *Appl Microbiol Biotechnol* 2003;**63**(3):258–66. http://dx.doi.org/10.1007/s00253-003-1444-y.

111. Wood BE, Yomano LP, York SW, Ingram LO. Development of industrial-medium-required elimination of the 2,3-butanediol fermentation pathway to maintain ethanol yield in an ethanologenic strain of klebsiellaoxytoca. *Biotechnol Prog* 2005;**21**(5):1366–72. http://dx.doi.org/10.1021/bp050100e.

112. Herpoel-Gimbert I, Margeot A, Dolla A, Jan G, Molle D, Lignon S, et al. Comparative secretome analyses of two trichoderma reesei RUT-C30 and CL847 hypersecretory strains. *Biotechnol Biofuels* 2008;**1**(1):18–6834. http://dx.doi.org/10.1186/1754-6834-1-18.

113. Nagendran S, Hallen-Adams HE, Paper JM, Aslam N, Walton JD. Reduced genomic potential for secreted plant cell-wall-degrading enzymes in the ectomycorrhizal fungus amanita bisporigera, based on the secretome of trichoderma reesei. *Fungal Genet Biol* 2009;**46**(5):427–35. http://dx.doi.org/10.1016/j.fgb.2009.02.00.

114. Yamada R, Yamakawa S, Tanaka T, Ogino C, Fukuda H, Kondo A. Direct and efficient ethanol production from high-yielding rice using a saccharomyces cerevisiae strain that express amylases. *Enzym Microb Technol* 2011;**48**(4–5):393–6. http://dx.doi.org/10.1016/j.enzmictec.2011.01.00.

115. Den Haan R, Rose SH, Lynd LR, van Zyl WH. Hydrolysis and fermentation of amorphous cellulose by recombinant saccharomyces cerevisiae. *Metab Eng* 2007;**9**(1):87–94. http://dx.doi.org/10.1016/j.ymben.2006.08.005.

CHAPTER 12

Ralstonia eutropha H16 as a Platform for the Production of Biofuels, Biodegradable Plastics, and Fine Chemicals from Diverse Carbon Resources

Jingnan Lu*, Christopher J. Brigham†, Sophia Li‡, Anthony J. Sinskey§,¶

**Department of Chemistry, Massachusetts Institute of Technology, Cambridge, Massachusetts, USA †Department of Bioengineering, University of Massachusetts Dartmouth, North Dartmouth, Massachusetts, USA ‡Department of Biological Engineering, Massachusetts Institute of Technology, Cambridge, Massachusetts, USA §Department of Biology, Massachusetts Institute of Technology, Cambridge, Massachusetts, USA ¶Engineering Systems Division, Massachusetts Institute of Technology, Cambridge, Massachusetts, USA*

12.1 Introduction

The Gram-negative betaproteobacterium *Ralstonia eutropha* is found in soil and freshwater environments. It has been previously named and renamed *Hydrogenomonas eutropha, Alcaligenes eutrophus,* and *Wautersia eutropha*, in part due to its ability to utilize many different compounds as sources of nutrients and energy. The current official name of the organism is *Cupriavidus necator*, which was a result of genome sequence comparison with other *C. necator* strains.[1] However, many prominent researchers in the field have opted to continue using the *R. eutropha* name for the sake of continuity.

Given the ability of *R. eutropha* to grow autotrophically in the presence of hydrogen, carbon dioxide, and oxygen, it was originally examined for the application of CO_2 mitigation for astronauts in space flight,[2] as well as single-cell protein for animals.[3,4] For the past 3 decades, *R. eutropha* has been studied globally as a model industrial organism for the production of polyhydroxyalkanoate (PHA) bioplastic. Mechanisms for intracellular PHA homeostasis have been largely elucidated in *R. eutropha* strains, revealing a complex physiological pathway.

R. eutropha can grow and produce PHA from a wide variety of carbon substrates including sugars,[5] short-chain organic acids and other compounds,[6] fatty acids and lipids,[7,8] and CO_2.[9,10] Because it is capable of metabolizing such diverse carbon sources, it is an excellent platform organism for the production of value-added compounds. First and foremost, PHA

Biotechnology for Biofuel Production and Optimization. http://dx.doi.org/10.1016/B978-0-444-63475-7.00012-1

has been synthesized by wild-type and engineered *R. eutropha* strains on a variety of carbon compounds in the laboratory and at pilot manufacturing scale. Due to the genetic tractability of the organism, the availability of genetic tools, and the advent of metabolic engineering, the range of compounds *R. eutropha* can produce is being expanded to include molecules like fatty acids, alcohols, and esters. This chapter discusses the biomanufacturing capabilities of *R. eutropha*, including the current state of the research in value-added product biosynthesis.

12.2 Advantages in Substrate and Metabolite Diversity and Flexibility

12.2.1 Lithoautotrophic Metabolism

R. eutropha assimilates CO_2 as a carbon source through the Calvin-Benson-Bassham (CBB) cycle. As a facultative chemoautotroph, *R. eutropha* is able to induce the CBB cycle when needed, even under heterotrophic conditions.[11,12] Carbon fixation enzymes in the CBB cycle are encoded in two *cbb* operons located on chromosome 2 and on the megaplasmid pHG1.[11] A LysR-type transcriptional regulator CbbR on chromosome 2 activates both operons, while the metabolite phosphoenolpyruvate (PEP) acts as a negative effector of the operon.[11] While the mechanism of the *cbb* operon in *R. eutropha* has been well-studied, the regulator of the *cbb* operon and the reduced metabolite that acts as a positive effector remain to be identified.

Two important carbon fixation enzymes, carbonic anhydrase (CA) and ribulose-1,5-bisphosphate carboxylase/oxygenase (RuBisCO) are found in *R. eutropha*. CA interconverts carbon dioxide and bicarbonate, and also has a role in controlling intracellular pH. These CAs operate through a zinc-dependent mechanism and are classified into five different classes (α, β, γ, δ, and ζ), which have evolved convergently.[12a] *R. eutropha* has four CA enzymes that are necessary for proper function. The genes *can* and *can2* encode β-CAs, *caa* encodes a periplasmic α-CA, and *cag* encodes a γ-CA.[13] Each CA has a different role in *R. eutropha* (Figure 12.1). Can is required for growth under atmospheric CO_2 levels, and supplies CO_2 to RuBisCO.[14,15] Caa differs from the other CA enzymes by its location, in the periplasm, and its preference for CO_2 as a substrate.[14] These differences are necessary for Caa's role in converting CO_2 that has diffused through the membrane to the more soluble bicarbonate, effectively trapping the carbon inside the cell. The roles of Cag and Can2 are less clear, but Cag may also provide CO_2 to RuBisCO, while Can2 has a possible role in pH homeostasis[14] (Figure 12.1). More research is required to elucidate the exact roles of these four CAs in the metabolism of CO_2.

RuBisCO, the second key enzyme, is responsible for the carbon fixation step where ribulose-1,5,-bisphosphate is carboxylated with CO_2 to form two molecules of 3-phosphoglycerate. *R. eutropha*, similar to many other facultative autotrophs, has Form 1C of RuBisCO, which is more suited to O_2-containing environments with medium to high levels of CO_2.[16] RuBisCO, however, is an inefficient enzyme with a slow catalytic turnover rate, a low affinity for carbon dioxide, and O_2 acts as a competing substrate.[16] The oxygenation reaction results in the

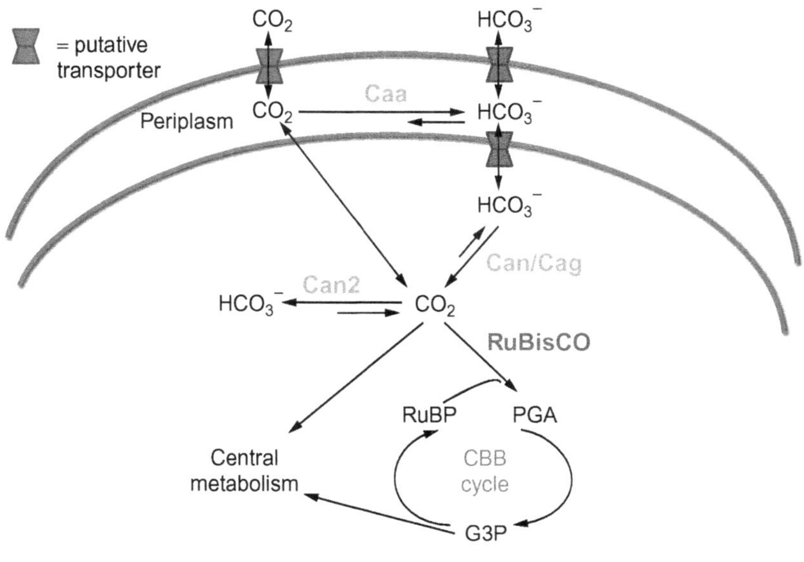

Figure 12.1

Proposed roles of carbonic anhydrases (CAs) in *R. eutropha*. Periplasmic Caa preferentially converts CO_2 into bicarbonate. Cytosolic Can supplies CO_2 to RuBisCO in the Calvin-Benson-Bassham (CBB) cycle. The roles of Cag and Can2 are less clear; however, Cag could play a similar role to Can, while Can2 could be involved with intracellular pH homeostasis. Abbreviations: Ribulose-1,5-bisphosphate carboxylase oxygenase (RuBisCO), ribulose-1,5-bisphosphate (RuBP), 3-phosphoglycerate (PGA), glyceraldehyde 3-phosphate (G3P).

formation of toxic phosphoglycolate, further inhibiting carbon fixation.[17] The efficiency of RuBisCO is a challenge to be addressed when maximizing the efficiency of carbon fixation.

12.2.2 Improvements on Carbon Fixation

Improving carbon fixation in *R. eutropha* is an active area of research. A metabolic model of *R. eutropha* H16 has been used to study the effect of different feeding ratios of H_2, O_2, and CO_2 on lithoautotrophic growth.[18] Electricity has also been used to reduce CO_2 to formate, providing the reducing equivalents needed for carbon fixation, growth, and biofuel production in *R. eutropha.*[9,19] However, few studies have attempted to improve the efficiency of carbon metabolism and the CBB cycle in *R. eutropha.*

There has been an increasing interest in the engineering of CBB cycle genes, alternative carbon fixation pathways, and carbon-concentrating mechanisms in bacteria in order to promote carbon fixation.[20] The *cbb* operon from *R. eutropha* H16 has been expressed in *E. coli*, leading to an increase in 2-phosphoglycolate phosphatase and transketolase activities in the normally heterotrophic bacteria.[21] However, growth of *E. coli* on CO_2 has not been successful to date. Attempts have been made to engineer RuBisCO with limited success, since improvements in velocity often decrease the selectivity of the enzyme.[22] The 3-hydroxypropionate (3-HOP) CO_2 fixation cycle in the phototrophic bacterium *Chloroflexus aurantiacus* may be an advantageous alternative pathway, since its RuBisCO enzymes incorporate bicarbonate, instead of CO_2, as a substrate, and have no known oxygenation side reactions.[23] Finally, the expression of the carbon-concentrating mechanism of carboxysomes in nonnative hosts may maximize RuBisCO's efficiency and improve carbon fixation.[17] Functional carboxysomes have been expressed and assembled heterologously in *E. coli,*[24] demonstrating the applicability of these bacterial microcompartments in nonnative hosts. However, carboxysomes have not yet been demonstrated in *R. eutropha*.

A carboxysome is a bacterial microcompartment, similar to a synthetic organelle, where an icosahedral protein shell encases the enzymes CA and RuBisCO (Figure 12.2).[25] Carboxysomes are natively found in cyanobacteria and some chemoautotrophic bacteria.[26] These bacteria often grow in environments with CO_2 concentrations lower than that of the K_M of RuBisCO, but they are still able to fix CO_2 efficiently by concentrating it.[27] It has been hypothesized that the carboxysome increases the efficiency of carbon fixation through two

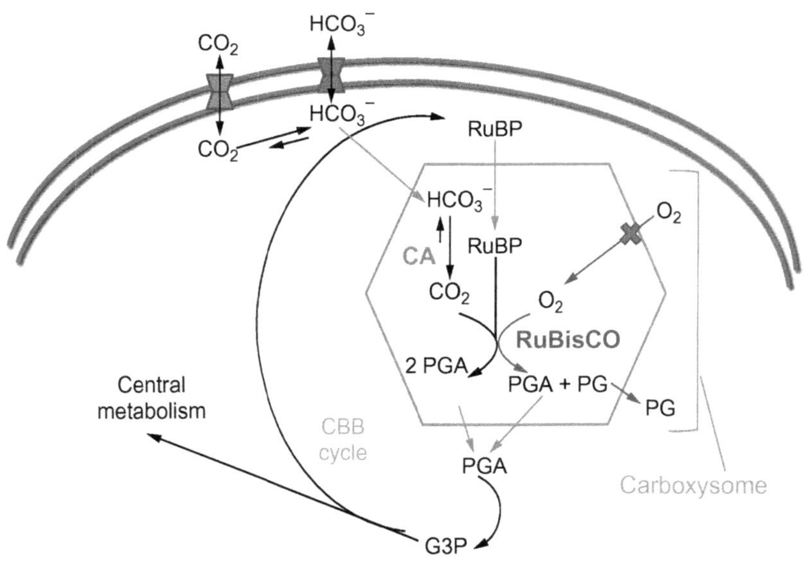

Figure 12.2

Carbon-concentrating mechanism of the carboxysome. Carbonic anhydrase (CA) and ribulose-1, 5-bisphosphate carboxylase oxygenase (RuBisCO) are inside the protein shell of the carboxysome. Bicarbonate enters the carboxysome through pores and is converted to CO_2. RuBisCO then catalyzes the carbon fixation reaction where ribulose-1,5-bisphosphate (RuBP) and CO_2 are converted to two molecules of 3-phosphoglycerate (PGA). The carboxysome shell restricts oxygen from entering and interacting with RuBisCO to form the toxic side product, phosphoglycolate (PG). Abbreviation: Calvin-Benson-Bassham cycle (CBB cycle).

main mechanisms: saturating RuBisCO with CO_2 and blocking RuBisCO from taking O_2 as a substrate for the production of wasteful byproducts.[17] In native organisms, bicarbonate is able to diffuse through pores in the protein shell into the carboxysome, where CA converts it to CO_2 (Figure 12.2). The proximity of CA and RuBisCO and the trapping of CO_2 inside the carboxysome increase the specificity and efficiency of RuBisCO. The protein shell also shields RuBisCO from O_2, minimizing the substrate competition between O_2 and CO_2 for RuBisCO and decreasing the amount of the toxic side product phosphoglycolate in the cytoplasm.[17] Through these two mechanisms,[27,28] the carboxysome allows the bacteria to more efficiently fix CO_2 to form 3-phosphoglycerate, a precursor for cell growth and the production of valuable products.

Improving *R. eutropha*'s ability to utilize CO_2 as a substrate has become an attractive way to address problems associated with rising CO_2 emissions, but many challenges lie ahead. While the general roles of *R. eutropha*'s various CAs[14] and the genetics and regulation of the CBB cycle have been studied,[11] the exact mechanisms and regulation of carbon fixation in *R. eutropha* need to be elucidated. Expressing a more efficient natural variant of RuBisCO, implementing a carbon-concentrating mechanism through carboxysomes, engineering alternative carbon fixation pathways, and finding novel ways to supply the reducing power required for the CBB cycle are all future directions to explore in improving carbon fixation in this lithoautotrophic organism.

12.2.3 Oxygen-Tolerant Hydrogenases

R. eutropha is capable of growing on an explosive gas mixture of 70% H_2, 10% CO_2, and 20% O_2.[11,21] Cellular metabolism relies entirely on CO_2 fixation, which is an extremely energy intensive process. In order to keep up with the energy supply via oxidation of H_2, *R. eutropha* possesses three hydrogenases.[5,11] The membrane-bound hydrogenase (MBH), soluble hydrogenase (SH), and regulatory hydrogenase (RH) all belong to the [NiFe] hydrogenase family. [NiFe] hydrogenases catalyze the reversible conversion of molecular hydrogen to protons and electrons.[29,30] Besides the [NiFe] cofactor, this type of hydrogenase also contains [FeS] clusters ~14 Å apart from the [NiFe] active site, which facilitate electron transfer. Both MBH and SH produce energy from the reducing equivalents harvested via H_2 oxidation to support cell growth under autotrophic conditions. In most organisms, oxygen and carbon monoxide can bind at the [NiFe] active site and inhibit or inactivate the hydrogenase. All three *R. eutropha* hydrogenases function differently but are tolerant to oxygen and carbon monoxide, which makes *R. eutropha* the model organism for the study of mechanisms of tolerance.[29,31]

The MBH is a heterotrimer consisting of HoxG, HoxK, and HoxZ subunits for H_2 catalysis, electron transfer, and cytoplasmic membrane localization, respectively. Besides the [NiFe] active site, it also contains two [4Fe-4S] clusters and one [3Fe-4S] cluster, in addition to

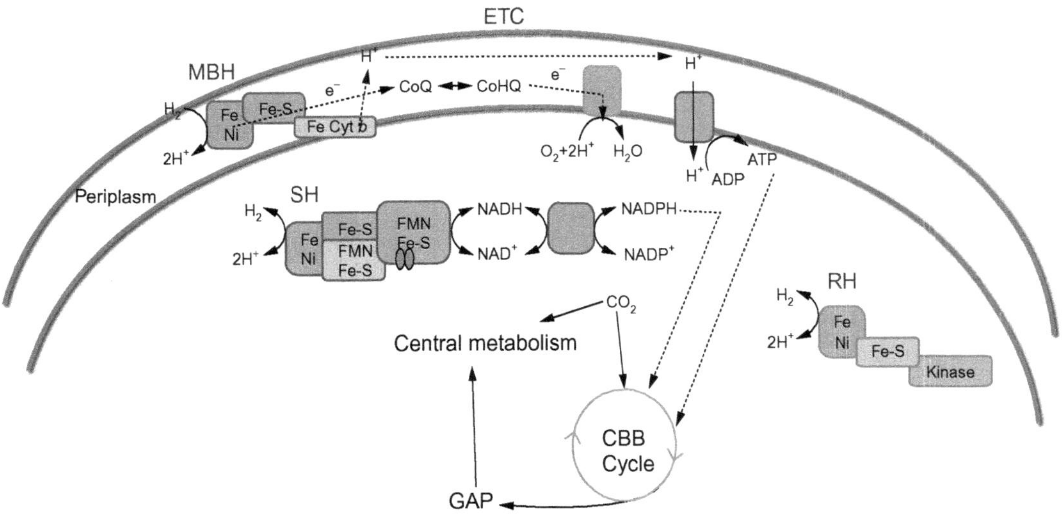

Figure 12.3

Depiction of *R. eutropha* lithoautotrophic growth. The membrane-bound hydrogenase (MBH) complex transfers electrons from hydrogen (H_2) oxidation to coenzyme Q and subsequently passes to cytochrome *a* oxidase of the electron transport chain (ETC), which results in the reduction of molecular O_2 to H_2O. Protons are translocated to the periplasm as electrons pass through the ETC and return into the cell by the ATPase to produce ATP. The soluble hydrogenase (SH) catalyzes H_2 oxidation and transfers electron potential to the production of NADH. The presence of a transhydrogenase can subsequently reduce $NADP^+$ to NADPH via NADH oxidation. The ATP and NADPH generated through hydrogen oxidation by MBH and SH are utilized in the Calvin-Benson-Bassham cycle for carbon fixation. Regulatory hydrogenase (RH) could activate transcription of hydrogenase genes through phosphorylation, but is inactivated in wild-type *R. eutropha* H16.

two *b*-type hemes. The heterotrimer of MBH is fused with a cytochrome *b*, which transfers electrons from H_2 oxidation into the electron transport chain for the production of ATP (Figure 12.3). MBH allows *R. eutropha* to grow aerobically using H_2 as an energy source and O_2 as the terminal electron acceptor. MBH is considered a slow oxidase and possesses a strong bias toward H_2 oxidation.[29,30] The active site and proximal cluster surrounding the active site of MBH are significantly different from other [NiFe] hydrogenases. As revealed by biochemical and spectroscopic studies, six cysteine residues coordinate the [4Fe-4S] cluster instead of only four found in other [NiFe] hydrogenases. Mutation of the two extra cysteine residues to glycines abolished the MBH's ability to tolerate O_2 and caused lithoautotrophic growth retardation. Such findings suggest that the two additional cysteine residues located around the [4Fe-4S] cluster of MBH allow for two sequential one-electron transfers during hydrogenase reactivation in O_2-containing atmosphere.[32] An O_2-sensitive hydrogenase from *Desulfovibrio gigas* cannot be engineered to tolerate O_2 by mutating the two glycine residues to cysteine, indicating that other factors in concert contribute to O_2 tolerance in *R. eutropha* MBH.[30,32,33]

SH is a bidirectional cytoplasmic hydrogenase and acts as a redox valve to prevent overreduction of the electron transport chain. SH consists of a heterodimer with subunits HoxH and HoxY. HoxH contains the [NiFe] catalysis site, while HoxY contains one [4Fe-4S] site and one FMN cofactor for electron transfer. The HoxHY heterodimer is bound to a diaphorase dimer HoxF and HoxU, which contains two [4Fe-4S] clusters, one on each subunit, two [2Fe-2S] on HoxU, and an FMN and NAD^+ binding site on HoxF. A homodimer of the HoxI subunit is bound to HoxY and contains an $NADP^+$ binding pocket. Such complex structure allows first for the oxidation of H_2, then the harvesting of electron potential into the reduction of NAD^+ to NADH, and lastly the subsequential reduction of $NADP^+$ to NADPH.[34,35] The ATP generated via the MBH, in addition to the NADPH reducing equivalents generated through SH, together support the high energy-demanding carbon fixation that occurs with the CBB cycle (Figure 12.3). The presence of a hydrophobic gas channel into the active site of SH allows it to favor H_2 entrance over O_2, thus reducing inhibition due to O_2 exposure. Furthermore, the [NiFe] active site of SH contains two extra CN ligands coordinated to Ni; removal of such CN groups resulted in a hydrogenase that is sensitive to O_2.[30,34,35]

RH of *R. eutropha* is based on a bacterial two-component regulatory system. The RH complex exists in a dimeric state. Each holoenzyme contains a H_2-sensing HoxBC unit, which contains the [NiFe] active site in addition to a [4Fe-4S] cluster (Figure 12.3). The H_2-signaling regulator, HoxA, is similar to a NtrC response regulator. HoxA is inactive in the phosphorylated form, and phosphorylation is mediated by HoxJ, a Histidine-protein kinase. Nonphosphorylated HoxA is capable of initiating transcription of both MBH and SH genes. It was hypothesized that in an environment with H_2, HoxBC oxidizes H_2 and passes along the electrons as a signal to the HoxJ transmitter, which is dependent on the G422 residue to autophosphorylate. The dephosphorylated HoxJ is no longer able to phosphorylate HoxA, resulting in an increase in unphosphorylated HoxA and activation of MBH and SH gene transcription and maturation proteins.[30,34,36] In wild-type *R. eutropha* H16, the expression of MBH and SH solely depends on the global energy status of the cell. An increase in the expression is independent of H_2 availability, but solely occurs under energy-limiting conditions.

Studies revealed that the H_2 sensor protein in H16 was silenced as a result of a single residue mutation in HoxJ G422S. When such mutation is repaired, SH is capable of responding to both a global energy signal and H_2 accessibility.[37] Such silencing and inactivation of the regulatory cascade of SH could be explained by evolutionary adaptation under low and unsteady availability of H_2 in the environment. The transcription, translation, and maturation of MBH and SH take time; therefore, in order to not miss the opportunity to access H_2 and survive under limited and unreliable amounts of H_2, *R. eutropha* adapted to dismiss the H_2-sensing signal transduction chain and instead expresses both hydrogenases under low energy state.

O_2-tolerance is critical for the survival of cells under aerobic lithoautotrophic growth mode. Recently, additional structural model analysis and mutant comparisons support a multipartite approach to combat O_2 in both MBH and SH of *R. eutropha*. The diverse factors are summarized here: (1) specific chaperone proteins are produced to shield the active site [NiFe] from O_2-inactivation during hydrogenase maturation process; (2) the presence of hydrophobic cavities in the [NiFe] active sites limits O_2 passage and allows only H_2 to enter the active site; (3) rapid and spontaneous reactivation of the active site to its EPR Ni-B ready state when exposed to O_2; (4) [NiFe] active site nearby [4Fe-4S] cluster is distorted with a distal Fe, which can assist [NiFe] site reactivation; (5) two additional cysteine residues located around the [4Fe-4S] cluster of O_2-tolerant hydrogenases that are glycines in O_2-sensitive hydrogenases allow two sequential one-electron transfers during reactivation, thus eliminating the formation of reactive oxygen species and reducing O_2 rapidly to H_2O.[30–33,35] *R. eutropha* hydrogenases' ability to overcome O_2 toxicity is very complex, unique, and still not well understood.

12.2.4 Hydrogenase-Based Technologies

R. eutropha oxygen-tolerant hydrogenases not only play crucial roles in microbial energy metabolism but also hold significant potential in H_2-based biotechnological applications. Recent developments with *R. eutropha* hydrogenases include platinum-free hydrogenase electrodes for fuel cells and light-driven H_2 production. Platinum-based fuel cells present three major drawbacks, which include the high cost and low availability of platinum electrocatalysts, significant and irreversible inhibition by CO and H_2S impurities, and unspecific selectivity toward both H_2 oxidation and O_2 reduction. Electrocatalysis by a hydrogenase is an alternative to a platinum-based fuel cell. Hydrogenase enzymes can be immobilized on an electrode and serve as an electrocatalyst for hydrogen oxidation. Since the *R. eutropha* hydrogenase is renewable, able to tolerate CO, and is highly selective toward H_2 oxidation, energy conversion was able to reach 100% theoretical yield in neutral media.[38,39] The second promising technology is the photosynthetic conversion of sunlight and water directly into H_2 and O_2, which bypasses the current natural gas-based H_2 generation. Such solar H_2 production has been achieved by coupling photosystem I (PS I) with a hydrogenase. In one study, *R. eutropha* hydrogenase subunit HoxK was directly coupled to PS I as a protein fusion. Unfortunately H_2 production from solar energy by this protein fusion system is incredibly low and inefficient. Studies to improve efficiency included the addition of a gold electroplate *in vitro* to the fusion system and chemically linking electron transfer [Fe-S] clusters. *R. eutropha* hydrogenases are highly valued because both MBH and SH are capable of H_2 production in the presence of O_2. Additionally, such a system is renewable and carbon neutral.[39,40] Efficiency and sustainability are the challenges faced by both hydrogenase-based technologies. In the case of hydrogenase fuel cells, it is difficult to achieve hydrogen equilibrium, especially in a fuel cell setting. On the other hand, the solar

H_2 production system is limited because proton reduction at low potential is not favored and H_2 is a potent inhibitor of MBH-mediated H_2 production.

12.2.5 Lithoorganotropic Metabolism

Formate (HCO_2^-) acts both as a carbon and energy source to support cell growth under organoautotrophic conditions. Once inside the cell, formate is irreversibly broken down to CO_2 and enters the central metabolism via the CBB cycle. The electrons from this oxidization are transferred to quinol and into the electron transport chain or utilized for the reduction of NAD^+.[11,41] *R. eutropha* processes two types of formate dehydrogenases, the soluble formate dehydrogenase (S-FDH) and the membrane-bound formate dehydrogenase (M-FDH). S-FDH is Molybdenum (Mo)-dependent with subunits FdsGBACD encoded on chromosome 1. Transcription of S-FDH genes is controlled by formate concentration, although studies have shown oxalate can also induce S-FDH gene transcription. FdsA is the catalytic subunit, which contains Mo and four [4Fe-4S] electron transfer sites. Surprisingly, unlike other S-FDH, tungsten (W) cannot replace the Mo cofactor. Electrons released from formate oxidation are transferred to FdsB via the iron-sulfur clusters and FMN. NAD^+ on FdsB then gets reduced to NADH. Other subunits FdsG, FdsC, and FdsD all facilitate the transfer of electrons. *R. eutropha* was hypothesized to contain one Mo-dependent M-FDH and two W-dependent M-FDHs. The expression of M-FDH is constitutive and well understood, since FDH activity was detected at various growth conditions. The M-FDH also contains a catalytic subunit and various iron-sulfur clusters, in addition to a transmembrane cytochrome *b* subunit that transfers the electrons to quinone and lastly reduces NAD^+ to NADH.[5,42,43]

Studying autotrophic growth on CO_2 can be challenging, since it requires an exact control of CO_2, O_2, and H_2 gas compositions. The mixture of H_2 and O_2 is considered unsafe, even though preventative strategies for mixed-gas explosion do exist. Utilization of formate to study autotrophic growth and production in *R. eutropha* has been routinely utilized as a safe alternative.[9] The downside of formate utilization is that formate can be toxic to the cells at concentrations as low as 0.2 g/L. A recent study combined stoichiometric-modeling data with well-controlled fermentation cultivation to achieve up to 85% theoretical yield of biomass in a formate-limited continuous fermentation culture. The maximum growth capacities can also be reached in a chemostat or a pH-controlled fed-batch culture at null residual formic acid concentration to avoid toxic effect.[44]

Since the S-FDH is capable of catalyzing the reversible conversion of formate to CO_2, various research has focused on the production of formate from CO_2. One group was able to use electrochemical conversion for the production of formate, and subsequently bioconvert it into branched-chain alcohol biofuel in engineered *R. eutropha.*[19] Another idea includes coupling formate dehydrogenase with PS I, so the energy harvested from sunlight could be utilized for the conversion of CO_2 to formate. The main challenge facing all these engineered systems is energy balance. In other words, the energy output should outweigh the energy input.

12.2.6 Respiratory Metabolism

R. eutropha transitions between heterotrophic and lithotrophic growth conditions due to environmental variations; therefore, its energetic metabolism is capable of rapid adaptation to carbon source fluctuations. The respiratory chain is highly flexible in terms of sources and destinations of electrons. NADH dehydrogenases (NDH-1 and NDH-2) transfer electrons from NADH to the membrane-soluble transporters, ubiquinone, or menaquinone, and then subsequently pass on the electrons, one at a time, to cytochrome *c* oxydoreductase (complex bc_1). Electron paramagnetic resonance spectroscopy analysis identified three [FeS] clusters in NDH-1, and that NDH-2 contains a FAD, both of which assist in the transfer of electrons. Electrons can also enter the electron transport chain via oxidation of succinate by succinate dehydrogenase.[5,41]

Proteomic analysis, and redox titration, in addition to inhibition studies by HQNO, antimycin A, rotenone, and myxothiazol revealed at least three different oxidases in both lithotrophic and heterotrophic growth conditions. Two oxidases belong to the heme-copper oxidase family, which contains a binuclear heme a_3-copper center and an additional heme a_3 subunit II with CuA. The third oxidase is a quinol oxidase, which does not contain heme or CuA, nor does it pump protons.[5] The abundance of high-affinity oxidases increases during lithoautotrophic growth conditions, which can be explained by the low availability of O_2 inside the cell during H_2 oxidation. Oxidative phosphorylation coupled with various substrates resulted in H^+/O (number of protons pumped acrossed the membrane per each oxygen molecule reduced) ratio of 6 and 8 under lithoautotrophic and heterotrophic growth settings, respectively. The chemiosmotic theory suggests that approximately three protons are required to generate the chemical energy to make a single ATP. Therefore, the *R. eutropha* phosphorylation coefficient (P/O), also known as the number of ATP per oxygen can be determined from the H^+/O ratio. For NADH under lithoautotrophic growth conditions, 2 ATPs are generated, while close to 3 ATPs per NADH are made during heterotrophic growth.[45]

R. eutropha is capable of anaerobic metabolism by reducing nitrogen compounds instead of oxygen. Four reductase enzymes carry out this denitrification process. Nitrate (NO_3^-) is sequentially reduced to nitrite (NO_2^-), nitric oxide (NO), and nitrous oxide (N_2O), respectively, by nitrate reductase (NAR), nitrite reductase (NIR), nitric oxide reductase (NOR), and nitrous oxide reductase (NOS).[46–48] All four enzymes are encoded on the megaplasmid (pHG1), which is required for anaerobic growth on nitrate, even though a second copy of genes encoding for all key denitrification enzymes were also found on chromosome 2. Stoichiometric modeling indicates that 1.42 ATP is produced per NADH and 0.75 ATP per $FADH_2$ when nitrate is used as the terminal electron acceptor.[45]

The genome sequence of *R. eutropha* strain H16 revealed various genes encoding an ArsR-type transcription regulator, sulfur dehydrogenase, sulfur-chelating/binding complex, thiosulfate-oxidizing complex, and sulfate thioesterase.[13,49] These enzymes make up a sulfur reduction pathway and suggest *R. eutropha* could use sulfur as an alternative terminal electron acceptor during anaerobic growth. Although the isolated sulfur dehydrogenase is active, growth on sulfur has not been established to date. Additional research is needed to elucidate the expression of sulfur reduction genes, cofactor requirements, and growth conditions.

The presence of alternative electron accepting pathways improves *R. eutropha*'s likelihood of surviving under harsh environmental growth conditions. It also presents an alternative strategy to eliminating O_2-induced metabolic inhibition in industrial production settings. Lastly, growth on CO_2, H_2, and an alternative electron acceptor could be utilized to allow for safe and easy fermentations, since the use of nitrate or sulfur eliminates the possibility of explosion caused by the interaction of H_2 and O_2, in addition to difficulties associated with gas diffusion.

12.2.7 Growth of *R. eutropha* on C2-C3 Waste Carbon Substrates

Lactate is often produced in anaerobic treatment of waste carbon, and is a suitable carbon source for fermentations to synthesize value-added products. PHA production in fed batch culture using lactate and acetate as carbon sources has been studied using a novel carbon substrate feeding technique. Since lactate and acetate are both acids and could potentially play a role in culture pH regulation, feedstock supply was linked to pH-stat control.[50] Fed batch co-culture of *R. eutropha* and *Lactobacillus plantarum* has been attempted, and lactate concentration (produced by *L. plantarum*) in the culture plays a key role in PHA productivity.[51] Furthermore, lactate-containing PHA copolymers have been produced in engineered *R. eutropha* with 3-hydroxybutyrate (3HB) as the co-monomer.[52] Like lactate, the C2 α-hydroxy acid glyoxylate can be used as a carbon source. Recently, glycolate and other 2-hydroxyalkanoic acids have been incorporated into PHA in *E. coli*.[53] Similar polymers have yet to be produced in *R. eutropha*, but a broad-specificity propionate CoA-transferase has been characterized in *R. eutropha* that could be used to produce the precursor molecule, glycolyl-CoA, although relative activity is low with this substrate compared to propionate.[54] Glyoxylate has not been demonstrated as an extracellular carbon source, but *R. eutropha* expresses glyoxylate bypass enzymes malate synthase and isocitrate lyase. A functional glyoxylate bypass has been shown to be of importance in *R. eutropha*, especially during growth on acetate and compounds that are metabolized to an acetyl-CoA intermediate, such as fatty acids.[55,56] Disruption of the glyoxylate bypass has been shown to severely inhibit growth on acetate as the main carbon source.[57]

Wild-type *R. eutropha* is not capable of producing ethanol or utilizing it as a carbon source. Engineered strains have been constructed that are capable of synthesizing short-chain alcohols on a variety of substrates, including CO_2 and H_2. *R. eutropha* harbors multiple alcohol dehydrogenase genes in its genome. The constitutive activity of one particular alcohol dehydrogenase gene, *adh*, in *R. eutropha* resulted in its ability to grow on alcohols like ethanol or 2,3-butanediol.[58,59] PHA could be produced by these strains from ethanol in co-culture with the brewer's yeast *Saccharomyces cerevisiae*.

Pure glycerol is an important industrial feedstock in the food, drug, and cosmetic industries; however, crude glycerol, the main byproduct of biodiesel production from the transesterification of plant or animal fats and oils, has very low value due to the presence of methanol, salt, and fatty acid impurities.[60] Crude glycerol treatment involves filtration, addition of chemicals, and fractional vacuum distillation, which can be very costly and time-consuming. There is a high incentive to redirect these waste crude glycerols into other industries.[61] Cellular utilization of glycerol can be achieved by the conversion of glycerol into glyceraldehyde 3-phosphate via glycerol kinase, glycerol-3-phosphate dehydrogenase, and triosephosphate isomerase. Glyceraldehyde 3-phosphate then enters the central metabolism via the Entner-Doudoroff pathway and can be converted to acetyl-CoA for energy and growth. Since *R. eutropha* is able to tolerate growth conditions with high salts, in addition to utilizing glycerol, methanol, and fatty acids for growth, 1.5 g/L/h PHA productivity has been achieved using unprocessed crude glycerol directly from biodiesel production.[60–62] Such a process can be directly coupled with biodiesel production and achieve environmentally friendly coproduction of two high value products.

12.2.8 Growth of *R. eutropha* on Organic Acids

In many industries, carbon-rich waste streams are produced, which require treatment to mitigate Chemical Oxygen Demand and Biological Oxygen Demand. Agricultural and municipal waste streams are often treated by anaerobic digestion, resulting in high concentrations of mixed organic acids, containing compounds like acetate, propionate, lactate, and butyrate.[63] These carbon substrates, if suitably purified, can serve as feedstocks for *R. eutropha* growth and production.

Innovative processes have been investigated to utilize organic acids in *R. eutropha* growth media. Since these organic acids are toxic and inhibitory to the bacterium, the mixed acids concentration is kept at a low level.[6,64,65] Organic acids can penetrate freely through the cell membrane, so therefore dissociate and acidify the cytoplasm. In order to maintain intracellular pH and proton gradient, cells must rapidly pump out protons, which results in an increase in the release of free energy via ATPase.[66] Such a process decouples the energy production and electron transport system, and therefore affects the overall wellness of *R. eutropha*, including organic acid utilization, growth, and PHA production. Various mixture analysis models were implemented to study the effect of mixed organic acids ratio on cell growth, PHA production, and PHA monomer composition.[65,66] These studies have demonstrated *R. eutropha* utilizing short-chain organic acids for PHA production. Upon feeding different mixtures of acetic, propionic, and butyric acids, carbon flux was studied in *R. eutropha* during growth and PHA biosynthesis. With acetate, a C2 carbon source, present in the growth media, *R. eutropha* can produce polyhydroxybutyrate (PHB). When propionic acid (C3) is present in the growth medium, the biopolymer poly(hydroxybutyrate-*co*-hydroxyvalerate) [P(HB-*co*-HV)] is produced. Flux analysis demonstrates that, even in the presence of propionate, the majority of the carbon that is routed to PHA in *R. eutropha* is going to the hydroxybutyrate (HB) monomer.[67]

Organic acids from treated oil palm milling waste have been used to produce P(HB-*co*-HV).[68,69] Organic acid solutions made to simulate treated waste streams have also been used for demonstrating optimal acid mixtures for PHB and P(HB-*co*-HV) production.[65] Valerate (C5) has also been supplemented to *R. eutropha* growth media to achieve increased HV monomer content in P(HB-*co*-HV) copolymer. Recently, the incorporation of 3-HOP into *R. eutropha* PHA has been investigated. For the production of a poly(hydroxybutyrate-*co*-hydroxypropionate) [P(HB-*co*-HP)] copolymer from fructose and other unrelated carbon source, portions of the 3-HOP cycle from *C. aurantiacus* were inserted into *R. eutropha.*[70] At the current time, P(HB-*co*-HP) has not been produced from propionate in engineered *R. eutropha*.

R. eutropha is also capable of utilizing organic acids such as fumarate, succinate, aspartate, malate, and glutamate, from the TCA (tricarboxylic acid) cycle by mediating between C3 and C4 molecules. Proteomic studies revealed active TCA cycle enzymes in both autotrophically and heterotrophically grown bacteria.[71] Additionally, three enzymes, PEP carboxykinase, malic enzyme (malate dehydrogenase or NAD-malic enzyme), and PEP synthase were found to be active under heterotrophic growth conditions on organic carbon sources. These enzymes allow for the flow of metabolites between the Entner-Doudoroff pathway and the TCA cycle. In PHA production phase, the gene encoding malic enzyme became upregulated. Malic enzyme catalyzes the reversible conversion of malate to pyruvate and is part of a metabolic shunt for the production of NADPH.[66,71] Since *R. eutropha* does not contain 6-phosphogluconate dehydrogenase, an important enzyme in the pentose-phosphate pathway, production of NADPH for cellular metabolism and PHA production could be from this malic shunt (Figure 12.4).

12.2.9 Carbohydrate Utilization

Unlike many commensal and pathogenic microorganisms, *R. eutropha* does not express a wide variety of glycohydrolases and sugar-degrading enzymes and thus is unable to metabolize many different carbohydrates. In physiological studies of the bacterium, fructose[45] and gluconate[72] are often used as carbon sources. It has also been shown that *R. eutropha* can grow on xylose and *N*-acetyl-D-glucosamine (NAG), although fructose is a more preferred carbon source.[73] A wild-type strain of *R. eutropha* H16 is not able to utilize glucose as a main carbon source. However, glucose-utilizing variants have been engineered.[74] Also, other glucose-utilizing *R. eutropha* strains have been uncovered in nature. The addition of

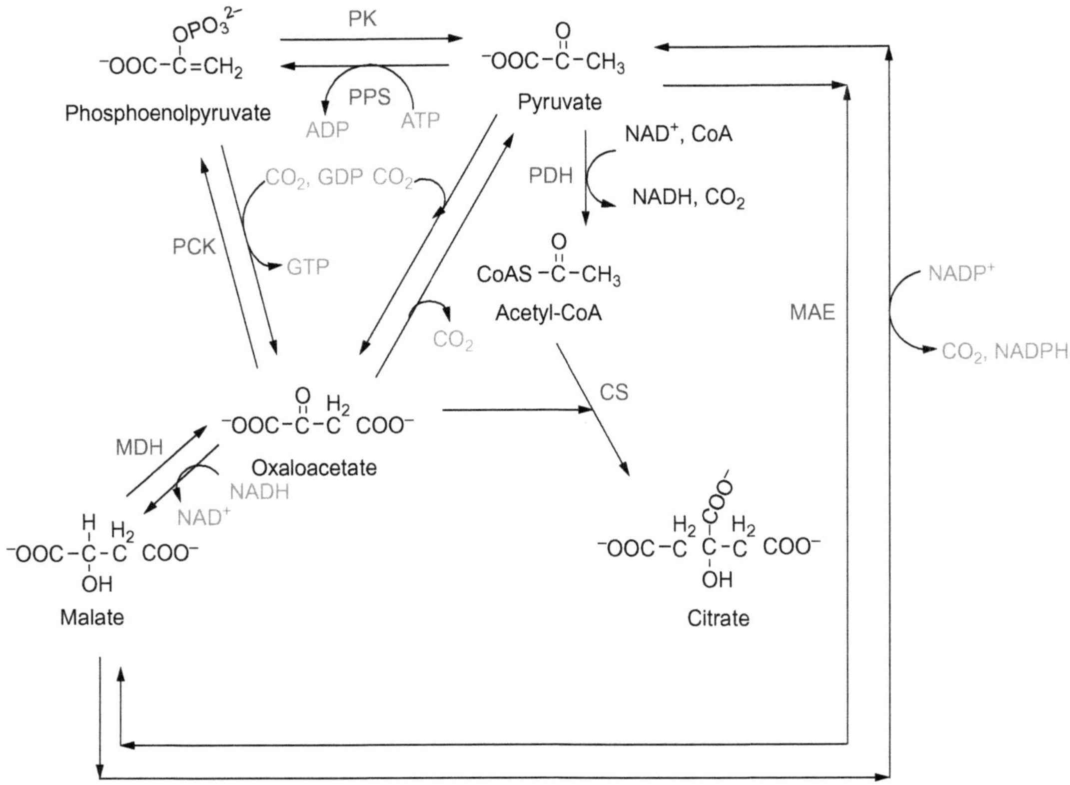

Figure 12.4

Pathway mediating between C3 and C4 intermediates. Malic enzyme (MAE) produces NADPH at the expense of ATP and NADH. Abbreviations: pyruvate kinase (PK), phosphoenolpyruvate synthase (PPS), pyruvate dehydrogenase (PDH), citrate synthase (CS), malate dehydrogenase (MDH), phosphoenolpyruvate carboxykinase (PCK).

molasses has been shown to enhance PHB production in solid-state fermentations featuring *R. eutropha*[75]; however, it has not been characterized whether the specific strain utilized both glucose and fructose from the crude carbon source. PHB has also been synthesized by *R. eutropha* from a crude sugar solution extracted from oil palm fronds. In this study, both glucose and fructose were shown to be consumed.[76]

12.2.10 Cellulosic Biomass Utilization

R. eutropha is also able to utilize cellulosic biomass, a renewable feedstock, to generate products. PHB has been produced from wheat-based feedstock streams, where wheat was converted into glucose-rich wheat hydrolysate and nitrogen-rich fungal extract.[77] *R. eutropha* grown on levulinic acid, a chemical made from refining cellulosic biomass, produced P(HB-*co*-HV)[78] and a terpolyester comprising 3-hydroxybutyrate and 3-hydroxyvalerate as major monomers, and 4-hydroxyvalerate as the minor monomer.[79] Fed-batch cultivation on hydrolysed sago starch led to PHA production,[80] and up to 57% cell dry weight PHA was produced from pretreated sugarcane bagasse,[81] further demonstrating *R. eutropha's* substrate diversity.

12.2.11 Aromatics Utilization

R. eutropha is also unique in its ability to use aromatic compounds such as benzoate and phenol as sole carbon and energy sources.[82] The catabolism of these aromatic compounds is inducible and is achieved through various pathways. *R. eutropha* consumes phenols through the meta pathway encoded in chromosome 1,[13] where catechol-2,3-oxygenase catalyzes the first reaction.[82] *R. eutropha* strain E2 has a phenol hydroxylase that allows phenol to be converted to catechol.[83] *p*-Nitrophenol, a pollutant from pesticide and explosives production,[84] is degraded by *R. eutropha* JMP134 through 3-nitrophenol nitroreductase.[85] JMP134 can also degrade substituted aromatic pollutants such as 2,4-dichlorophenoxyacetic acid (2,4-D) through *tfd* genes on plasmid pJP4[86] and is chemotactic toward 2,4-D.[87] Lastly, benzoate is degraded through the beta-ketoadipate pathway,[82] and phenol catabolism is repressed when this pathway is activated.[88] The ability of *R. eutropha* to utilize aromatics, a large source of environmental pollutants, demonstrates its potential for bioremediation applications.

12.2.12 Utilization of Lipids and Fats

Fatty acids and lipids pack in more energy per mole of carbon than any other carbon source; therefore, they are a more efficient group of carbon source for growth and production of value-added products such as PHAs.[89] Increasing research has been focused on utilizing plant oil or animal fats in *R. eutropha* fermentations. Triacylglycerols in plant oils and animal fats consist of glycerol triesters of fatty acids. The length and saturation of fatty acids may vary depending on the source of triacylglycerol, whether it is plant oil or animal fats. Triacylglycerols cannot be utilized directly by the cells and must be broken down to free fatty acids by the digestive enzyme, lipase. *R. eutropha* has one extracellular lipase, LipA that is responsible for the breaking down of triacylglycerols. LipA is a nonspecific lipase, which is able to completely digest triacylglycerols into glycerols and free fatty acids. Triacylglycerols are water insoluble, whereas lipases are water soluble, so digestion of triacylglycerol takes place at the lipid-water interface. The rate of LipA catalysis depends on the surface area of the interface, which increases up to 500% with enhancement of interfacial surface.[8] In various fermentation settings, surfactants were used to emulsify oils and fatty acids, in order to increase the interfacial surface area and allowing fatty acids to be uptaken by cells.[7] Free fatty acids released from LipA catalyzed triacylglycerol breakdown can act as natural surfactants to emulsify the lipids and fatty acids in *R. eutropha* growth culture, thus increasing the bioavailability of the fatty acids.[89]

Once inside the cells, fatty acids are broken down two carbons at a time in a cyclic manner to produce one molecule each of acetyl-CoA, $FADH_2$, and NADH. $FADH_2$ and NADH are reoxidized by electron-transport chain through the intermediacy of a series of electron-transfer reactions to produce ATP. Genome sequencing of *R. eutropha* indicated the presence of two sets of β-oxidation pathway genes. A previous study also revealed that both β-oxidation pathways are active and essential for growth on oils and fats.[57] Fatty acyl-CoA intermediates from the β-oxidation pathway can serve as PHA precursors. Expression of a gene encoding for enoyl-CoA hydratase (*phaJ*) from *Pseudomonas aeruginosa* in *R. eutropha* allowed for conversion of enoyl-CoA to (*R*)-3-hydroxyacyl-CoA, a PHA precursor. Similarly, 3-hydroxyacyl-ACP:CoA transferase gene (*phaG*) from *Pseudomonas putida* when expressed in *R. eutropha* is capable of redirecting 3-ketoacyl-ACP from the fatty acid biosynthesis pathway to (*R*)-3-hydroxyacyl-CoA. Expression of either *phaJ* or *phaG* gene allowed for the production of copolymer poly(3-hydroxybutyrate-*co*-hydroxyhexanoate).[7,89]

12.3 Diverse Pathways for Biosynthesis of Value-Added Compounds

12.3.1 R. eutropha *PHA Production*

As mentioned, *R. eutropha* was proposed to be used as single cell protein (SCP) for animal feeds. However, the large amounts of intracellular PHAs that are synthesized under nitrogen limiting conditions were demonstrated to be poorly digestible by rats.[4] PHA is produced by *R. eutropha* under stress conditions, such as noncarbon nutrient starvation. PHA synthesis as a stress response has been demonstrated to be regulated in part by the bacterial stringent response.[90] In the past couple of decades, utility for PHA was, indeed, firmly established as a biodegradable, biocompatible replacement for petroleum-based polymer. Many applications for PHA have been suggested, including disposable household products,[91] packaging,[92] agricultural,[93] and waste treatment.[94] However, the most successful application for PHA to date is in medical devices and consumable items, including sutures, implants, surgical meshes, scaffolds, and timed-release drug delivery devices.[95] In particular, sutures and surgical meshes have been marketed and are being produced and used today.

A schematic of native and engineered PHA biosynthesis pathways in *R. eutropha* is shown in Figure 12.5. PHA can be synthesized enzymatically or chemically by incorporating many different monomer types. In nature, short-chain length (*scl*) PHA is a polymerization of C3-C5 monomers, with the primary example being the PHB homopolymer, which is comprised of C4 monomers. Medium-chain length (*mcl*) PHA incorporates C6-C12 monomers, and is typically produced by members of the genus *Pseudomonas*, among some others. Some bacteria can produce PHA containing a combination of *scl* and *mcl* monomers (i.e., *scl-co-mcl* PHA). The PHA synthases from these microbes, such as *Aeromonas caviae* and *Rhodococcus aetherivorans*, are highly sought after for *R. eutropha* strain engineering to produce biodegradable polymer with characteristics very similar to petroleum-based plastics.

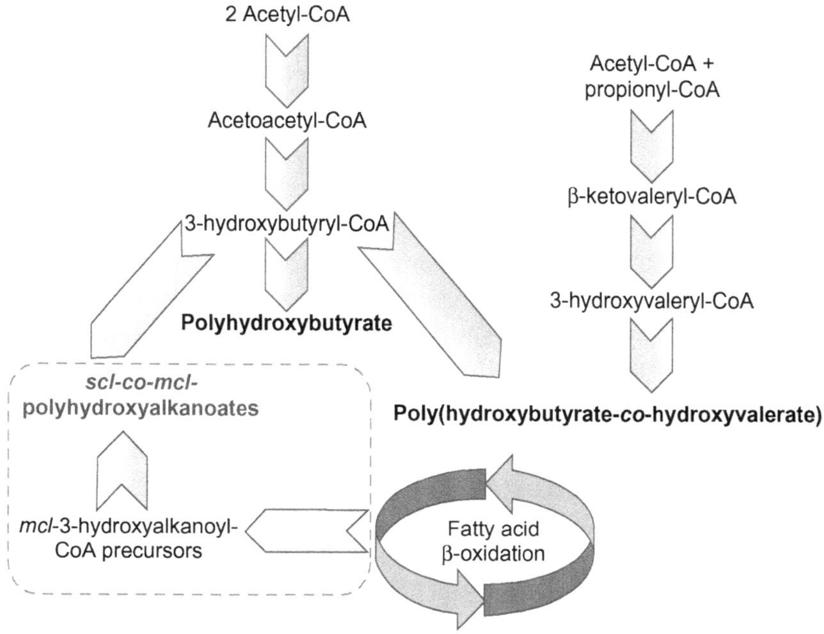

Figure 12.5

Wild-type *R. eutropha* produces polyhydroxybutyrate from condensation of two acetyl-CoA molecules. When the cells are fed with odd-chain length precursors, poly(hydroxybutyrate-*co*-hydroxyvalerate) can be produced. Engineered *R. eutropha* can incorporate hydroxyalkanoyl-CoA precursors from β-oxidation of fatty acids and produce PHA copolymers with both SCL and MCL monomers.

The versatile metabolism and genetic tractability of *R. eutropha* allow for this organism to be used to produce many different types of PHA biopolymer, each with different thermal and mechanical properties. Wild-type *R. eutropha* is capable of producing *scl*-PHA, typically PHB on most carbon sources. Odd-chain length (C3 or C5) precursors must be fed, in order to synthesize the copolymer P(HB-*co*-HV).

Given their advantageous properties, *scl-co-mcl* PHAs are often the goal of production. This requires strain engineering of *R. eutropha* to confer the capability of synthesizing and polymerizing *mcl* 3-hydroxyacyl-CoA (*mcl*-3HA-CoA) precursor molecules. The *mcl*-3HA-CoA precursors can be produced via intermediates of fatty acid metabolism. Typically, fatty acid β-oxidation intermediates are employed to synthesize these precursors using an (*R*)-specific-enoyl-CoA reductase enzyme, PhaJ.[96] In most strain engineering studies available in the literature, the *phaJ* gene from *Pseudomonas* is expressed in *R. eutropha* heterologously.[97] However, *phaJ* homologs have been characterized in *R. eutropha*.[98]

R. eutropha is considered to be the model organism of PHA homeostasis. It can produce up to 90% of its cell dry weight as polymer,[96] its PHB biosynthetic enzymes were among the first to be studied extensively,[98a] and many different PHA granule-associated proteins (GAPs) have been initially characterized in *R. eutropha*. Pilot scale PHA production operations have been established in areas around the world using *R. eutropha* as the biocatalyst. This bacterium is efficient at moving carbon from feedstock to product and is often considered in many biotechnological applications.

Polythioesters (PTEs) are another group of bio-based polymers. PTEs, although non-biodegradable in nature, are thought to be a durable, bio-based alternative to petrochemical plastics. The main difference between PTEs and PHAs is the thioester group in the backbone, resulting in the polymer's non-biodegradability. *R. eutropha* was first demonstrated to produce a PTE copolymer of poly(3-hydroxybutyrate-*co*-3-mercaptopropionate) [P(HB-*co*-3MP)], when 3MP was fed to cells. Other PTE copolymers could be synthesized by *R. eutropha* if the corresponding mercapto-acid (MA) was presented in the growth media. Other precursor substrates that promote PTE biosynthesis in *R. eutropha* are 3,3′-dithiodipropionic acid (DTDP) and 3,3′-thiodipropionic acid (TDP). The feeding of the TDP and DTDP precursors to *R. eutropha* cells resulted in synthesis of P(HB-co-3MP) with >50 mole% 3MP monomer content. The main issue with production of PTEs is that specialized precursors are needed, and many potential thiol-containing precursors are toxic to the cells.[99] PTEs have been examined for their thermal and mechanical properties, and have been found to be generally more heat stable than PHAs. Also, the exact characteristics observed depended on the chain length of the MA monomer incorporated into the PTE.[100]

12.3.2 *Biofuel Production by* R. eutropha

R. eutropha has also been examined for biosynthesis of non-alcohol fuel molecules. PHA can be converted by an acid-catalyzed hydrolysis to produce biofuel molecules, i.e., methyl esters.[92] Engineered strains have been constructed to produce free fatty acids,[101] which can also be converted to methyl esters. A strain of *R. eutropha* with inactivated PHB biosynthesis and fatty acid β-oxidation pathways also produced free fatty acids. This strain, however, also produced methyl ketones, which are another potential biofuel molecule. The addition of a heterologous methyl ketone biosynthesis pathway to this strain resulted in a strain with even higher ketone production capabilities.[102] It should be noted that these biofuel molecules could be synthesized from most of the carbon feedstocks discussed in this chapter, including CO_2.

Isobutanol, *n*-butanol, and isopropanol have attracted research and industrial attentions as promising next generation alcohol biofuel alternatives to ethanol and biodiesel. Each alcohol has a similar research octane number as gasoline and a higher energy density than ethanol. Most importantly, each alcohol can be used directly in the current fuel delivery infrastructure due to its low vapor pressure, low hygroscopicity, and low water solubility.[103,104] When the native PHA production genes were eliminated from *R. eutropha*, the bacteria secrete large amounts of pyruvate and stored more energy in the form of NADH than wild type.[9,105] The secreted pyruvate and extra energy in these PHA-negative strains were redirected for the production of isobutanol and isopropanol. Production of isobutanol relies on the valine biosynthesis pathway, in which the intermediate 2-ketoisovalerate is decarboxylated and reduced to form isobutanol. Expression of heterologous ketoisovalerate decarboxylase from *Lactococcus lactis* in a mutant *R. eutropha* strain with constitutively expressed alcohol dehydrogenase enabled the cells to produce isobutanol. Two studies utilizing *R. eutropha* native or heterologous valine biosynthesis enzymes in additional to ketoisovalerate decarboxylase and alcohol dehydrogenase were able to produce isobutanol from fructose, formate, and CO_2 in *R. eutropha.*[19,105] Another study redirected acetoacetyl-CoA, the precursor for the production of PHB, for the production of isopropanol via expression of acetoacetate decarboxylase and alcohol dehydrogenase from *Clostridium.*[106] Since the production of isopropanol and PHB share common precursors, isopropanol production titer reached near theoretical maximum in fed-batch fermentation. Although the *de novo* production of *n*-butanol has not been demonstrated in *R. eutropha*, one study utilized genes from the *R. eutropha* PHB production pathway for *n*-butanol production in *E. coli.*[107]

12.3.3 Production of Fine-Chemicals and Pharmaceutical Precursors

Hydroxyalkanoate monomers incorporated into PHA polymers are all chiral molecules. Each monomer has a chiral center on the carbon with the hydroxyl group and possesses the (*R*)-configuration (Figure 12.6). These chiral molecules also contain two functional groups, a hydroxyl group and a carboxyl group, which promote further modification to produce fine chemicals such as antibiotics, vitamins, perfumes, and pheromones.[108] The most widely produced polymer by *R. eutropha*, PHB, contains monomer units of (*R*)-3-hydroxybutyric

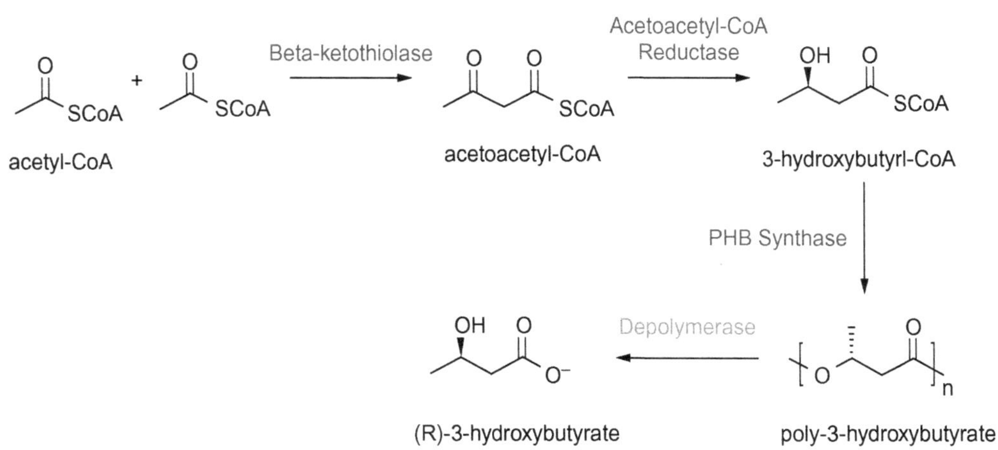

Figure 12.6
Synthesis and degradation of poly-3-hydroxybutyrate. (*R*)-3-hydroxybutyrate, the chiral molecule that can be used as pharmaceutical precursors can be generated from depolymerization of poly-3-hydroxybutyrate.

acid, which is an important precursor for the production of 4-acetoxyazetidinone for carbapenem antibiotics. (*R*)-3-hydroxybutyric acid has also shown medical potential for the activation of Ca ion channels and memory enhancement. Production of (*R*)-hydroxyalkanoates can be achieved through chemical synthesis or chemical hydrolysis, although both methods are very complicated and expensive. As discussed above, *R. eutropha* naturally produces PHAs as a way to store excess carbon. When the cells undergo starvation, PHA depolymerases are capable of breaking down PHAs into (*R*)-hydroxyalkanoate monomers. This is a well-controlled and utilized process for *R. eutropha* to reserve and reuse excess carbon and energy sources. *R. eutropha* has seven identified PHA depolymerases, and in carbon-limited cultivation condition, these depolymerases are able to hydrolyze PHB to (*R*)-3-hydroxybutyric acid monomers (Figure 12.6). This allows for the production of (*R*)-hydroxyalkanoates without any *in vitro* hydrolysis step.[109,110] Since engineered *R. eutropha* is capable of producing PHAs with various monomer lengths, a variety of (*R*)-hydroxyalkanoates chiral compounds can be produced. Additionally, the gene encoding the PHA synthase can be eliminated from the *R. eutropha* genome, which could potentially allow for the accumulation and direct isolation of (*R*)-hydroxyalkanoates precursors.

R. eutropha has also been engineered to produce (*R*)-1,2-propanediol, a pharmaceutical precursor for the production of antibacterial agents. Alcohol dehydrogenase from *Kluyveromyces lactis* was used to convert hydroxyacetone, the product from reduction and dehydration of glycerol, to (*R*)-1,2-propanediol.[111] This study coupled hydrogen oxidation as the reduction energy source with the synthesis of (*R*)-1,2-propanediol, which increased production yield and minimized carbon dioxide generation.

12.3.4 Recombinant Protein Expression System

R. eutropha has been engineered to harbor an efficient recombinant protein expression system. This recombinant protein expression system contains a T7 RNA polymerase under the control of *R. eutropha phaP* inducible promoter. As a proof of principle, organophosphohydrolase, a protein that is prone to form inclusion bodies in *E. coli* overexpression systems, was overexpressed in this *R. eutropha* system. Using this novel expression system, *R. eutropha* produced greater than 10 g/L soluble and active organophosphohydrolase,[112–114] which is 200 times higher yield than *E. coli*. *R. eutropha* recombinant protein expression system has great potential in the expression and production of commodity enzymes, therapeutic proteins, vaccines, and peptides.

12.3.5 Application in Bioremediation

The metabolically versatile *R. eutropha* is capable of surviving in harsh environmental conditions. It is tolerant to various environmental pollutants including aromatics and heavy metals. As described earlier, *R. eutropha* can break down aromatic compounds and utilize them for growth. Soils contaminated with *p*-nitrophenol or 2,4-dichlorophenoxyacetic acid can be cleaned up using *R. eutropha.*[115,116] A study also demonstrated that heavy metal-polluted soil had decreased toxic effect on plants when pre-inoculated with *R. eutropha*, which suggests that *R. eutropha* also contains a detoxification machinery for heavy metal bioremediation.[116,117]

12.4 Conclusion and Outlook

Studies of *R. eutropha* metabolic processes have demonstrated that it is a model microbial factory. Specifically, *R. eutropha* is very versatile in the metabolic processes that it can run. Taking advantage of the organism's ability to move large quantities of carbon for PHA biosynthesis, metabolism can be rerouted to produce a variety of useful bio-based commodity products. Figure 12.7 gives an overview of the many production pathways that were demonstrated in wild-type and engineered *R. eutropha* strains and discussed in this chapter. Undoubtedly, more useful pathways remain to be constructed and streamlined in *R. eutropha*.

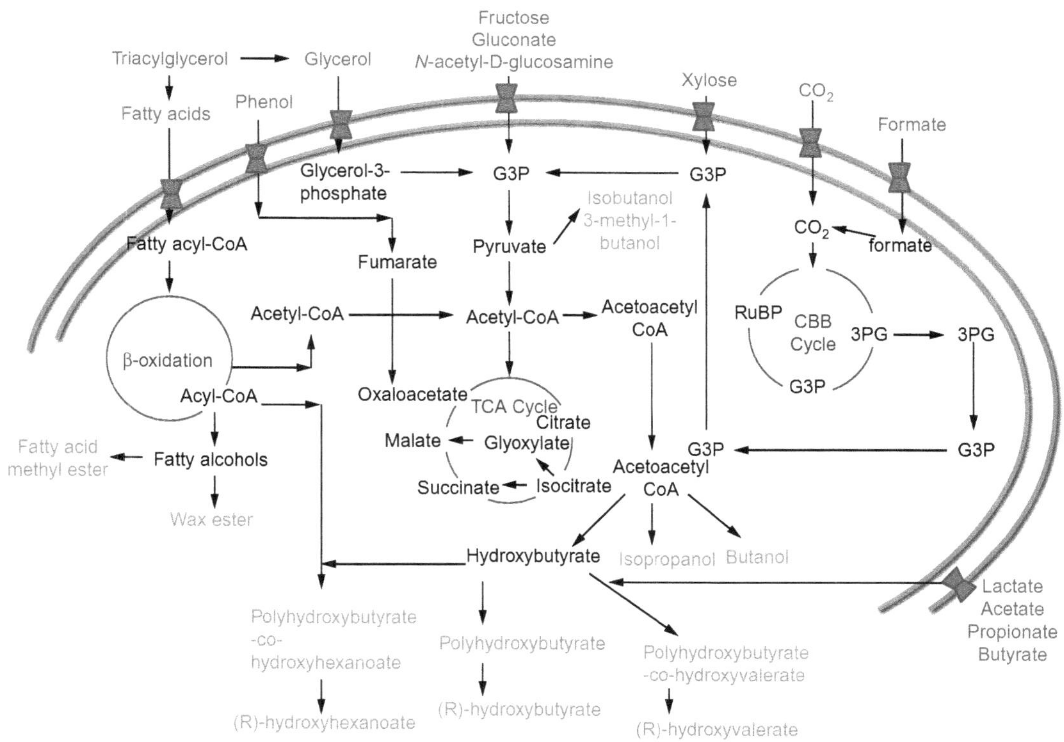

Figure 12.7
Schematic demonstrating *R. eutropha*'s metabolic versatility and its ability to produce value-added materials.

Much of the focus in *R. eutropha* bioengineering has been on optimizing the yields and productivities of compounds the organism can synthesize. Future directions include optimization of utilization and conversion of different carbon sources to be used in the synthesis of value-added products. This aspect of cellular metabolism is especially critical given the role of *R. eutropha* as a bioremediator and the trend toward usage of unrefined, waste carbon as feedstock for production of value added compounds. As much of the metabolic potential of *R. eutropha* is untapped, there may be no limit in sight of which natural products can be synthesized and subsequently marketed.

Acknowledgments

The authors thank Mr. John W. Quimby for critical review of this manuscript. J.L. is supported by the Malaysian Palm Oil Board. S.L. is supported by the MIT Undergraduate Research Opportunities Program. C.B. is supported by the Office of the Provost, University of Massachusetts Dartmouth. The authors would like to thank the following funding sources: U.S. Department of Energy-Advanced Research Projects Agency-Energy, The Malaysia MIT Biotechnology Partnership Program, and The Mega-Grant of Russian Federation. The authors also wish to thank Professor Cong Trinh at University of Tennessee and Professor Carrie Eckert at NREL/University of Colorado for giving us the opportunity to participate in the making of this book.

References

1. Vandamme P, Coenye T. Taxonomy of the genus *Cupriavidus*: a tale of lost and found. *Int J Syst Evol Microbiol* 2004;**54**:2285–9.
2. Schlegel HG, Lafferty RM. Novel energy and carbon sources A. The production of biomass from hydrogen and carbon dioxide. *Advances in biochemical engineering*, vol. 1. Berlin Heidelberg: Springer; 1971. p. 143–68.
3. Calloway DH, Kumar AM. Protein quality of the bacterium Hydrogenomonas eutropha. *Appl Microbiol* 1969;**17**:176–8.
4. Waslien CI, Calloway DH. Nutritional value of lipids in *Hydrogenomonas eutropha* as measured in the rat. *Appl Microbiol* 1969;**18**:152–5.
5. Cramm R. Genomic view of energy metabolism in *Ralstonia eutropha* H16. *J Mol Microbiol Biotechnol* 2009;**16**:38–52.
6. Yan Q, Du G, Chen J. Biosynthesis of polyhydroxyalkanoates (PHAs) with continuous feeding of mixed organic acids as carbon sources by *Ralstonia eutropha. Process Biochem* 2003;**39**:387–91.
7. Budde CF, Riedel SL, Willis LB, Rha C, Sinskey AJ. Production of poly(3-hydroxybutyrate-co-3-hydroxyhexanoate) from plant oil by engineered *Ralstonia eutropha* strains. *Appl Environ Microbiol* 2011;**77**:2847–54.
8. Lu J, Brigham CJ, Rha C, Sinskey AJ. Characterization of an extracellular lipase and its chaperone from *Ralstonia eutropha* H16. *Appl Microbiol Biotechnol* 2013;**97**:2443–54.
9. Brigham CJ, Gai CS, Lu J, Speth DR, Worden RM, Sinskey AJ. Engineering *Ralstonia eutropha* for production of isobutanol from CO_2, H_2, and O_2. In: Lee JW, editor. *Advanced biofuels and bioproducts*. New York: Springer; 2013. p. 1065–90.
10. Jeffke T, Gropp NH, Kaiser C, Grzeszik C, Kusian B, Bowien B. Mutational analysis of the cbb operon (CO_2 assimilation) promoter of *Ralstonia eutropha. J Bacteriol* 1999;**181**:4374–80.
11. Bowien B, Kusian B. Genetics and control of CO2 assimilation in the chemoautotroph *Ralstonia eutropha. Arch Microbiol* 2002;**178**:85–93.

12. Friedrich CG, Friedrich B, Bowien B. Formation of enzymes of autotrophic metabolism during heterotrophic growth of *Alcaligenes eutrophus*. *J Gen Microbiol* 1981;**122**:69–78.
12a. Smith KS, Ferry JG. Prokaryotic carbonic anhydrases. *FEMS Microbiol Rev* 2000;**24**(4):335–66.
13. Pohlmann A, Fricke WF, Reinecke F, Kusian B, Liesegang H, Cramm R, et al. Genome sequence of the bioplastic-producing "Knallgas" bacterium R*alstonia eutropha* H16. *Nat Biotechnol* 2006;**24**:1257–62.
14. Gai CS, Lu J, Brigham CJ, Bernardi AC, Sinskey AJ. Insights into bacterial CO_2 metabolism revealed by the characterization of four carbonic anhydrases in *Ralstonia eutropha* H16. *AMB Express* 2014;**4**:2.
15. Kusian B, Sültemeyer D, Bowien B. Carbonic anhydrase is essential for growth of *Ralstonia eutropha* at ambient CO_2 concentrations. *J Bacteriol* 2002;**184**:5018–26.
16. Badger MR, Bek EJ. Multiple Rubisco forms in proteobacteria: their functional significance in relation to CO_2 acquisition by the CBB cycle. *J Exp Bot* 2008;**59**:1525–41.
17. Yeates TO, Kerfeld CA, Heinhorst S, Cannon GC, Shively JM. Protein-based organelles in bacteria: carboxysomes and related microcompartments. *Nat Rev Microbiol* 2008;**6**:681–91.
18. Park JM, Kim TY, Lee SY. Genome-scale reconstruction and in silico analysis of the *Ralstonia eutropha* H16 for polyhydroxyalkanoate synthesis, lithoautotrophic growth, and 2-methyl citric acid production. *BMC Syst Biol* 2011;**5**:101.
19. Li H, Opgenorth PH, Wernick DG, Rogers S, Wu TY, Higashide W, et al. Integrated electromicrobial conversion of CO_2 to higher alcohols. *Science* 2012;**335**:1596.
20. Ducat DC, Silver PA. Improving carbon fixation pathways. *Curr Opin Chem Biol* 2012;**16**:337–44.
21. Schäferjohann J, Yoo JG, Kusian B, Bowien B. The cbb operons of the facultative chemoautotroph Alcaligenes eutrophus encode phosphoglycolate phosphatase. *J Bacteriol* 1993;**175**:7329–40.
22. Tcherkez GGB, Farquhar GD, Andrews TJ. Despite slow catalysis and confused substrate specificity, all ribulose bisphosphate carboxylases may be nearly perfectly optimized. *Proc Natl Acad Sci U S A* 2006;**103**:7246–51.
23. Zarzycki J, Brecht V, Müller M, Fuchs G. Identifying the missing steps of the autotrophic 3-hydroxypropionate CO_2 fixation cycle in *Chloroflexus aurantiacus*. *Proc Natl Acad Sci U S A* 2009;**106**:21317–22.
24. Bonacci W, Teng PK, Afonso B, Niederholtmeyer H, Grob P, Silver PA, et al. Modularity of a carbon-fixing protein organelle. *Proc Natl Acad Sci U S A* 2012;**109**:478–83.
25. Frank S, Lawrence AD, Prentice MB, Warren MJ. Bacterial microcompartments moving into a synthetic biological world. *J Biotechnol* 2013;**163**:273–9.
26. Price GD, Badger MR. Evidence for the role of carboxysomes in the cyanobacterial CO_2-concentrating mechanism. *Can J Bot* 1991;**69**:963–73.
27. Cannon GC, Bradburne CE, Aldrich HC, Baker SH, Heinhorst S, Shively JM. Microcompartments in prokaryotes: carboxysomes and related polyhedra. *Appl Environ Microbiol* 2001;**67**:5351–61.
28. Rae BD, Long BM, Whitehead LF, Förster B, Badger MR, Price GD. Cyanobacterial carboxysomes: microcompartments that facilitate CO_2 fixation. *J Mol Microbiol Biotechnol* 2013;**23**:300–7.
29. Burgdorf T, Lenz O, Buhrke T, van der Linden E, Jones AK, Albracht SPJ, et al. [NiFe]-Hydrogenases of *Ralstonia eutropha* H16: modular enzymes for oxygen-tolerant biological hydrogen oxidation. *J Mol Microbiol Biotechnol* 2005;**10**:181–96.
30. Lenz O, Ludwig M, Schubert T, Bürstel I, Ganskow S, Goris T, et al. H_2 conversion in the presence of O_2 as performed by the membrane-bound [NiFe]-hydrogenase of *Ralstonia eutropha*. *Chemphyschem* 2010;**11**:1107–19.
31. Buhrke T, Lenz O, Krauss N, Friedrich B. Oxygen tolerance of the H2-sensing [NiFe] hydrogenase from *Ralstonia eutropha* H16 is based on limited access of oxygen to the active site. *J Biol Chem* 2005;**280**:23791–6.
32. Saggu M, Zebger I, Ludwig M, Lenz O, Friedrich B, Hildebrandt P, et al. Spectroscopic insights into the oxygen-tolerant membrane-associated [NiFe] hydrogenase of *Ralstonia eutropha* H16. *J Biol Chem* 2009;**284**:16264–76.
33. Fritsch J, Lenz O, Friedrich B. The maturation factors HoxR and HoxT contribute to oxygen tolerance of membrane-bound [NiFe] hydrogenase in *Ralstonia eutropha* H16. *J Bacteriol* 2011;**193**:2487–97.

34. Burgdorf T, van der Linden E, Bernhard M, Yin QY, Back JW, Hartog AF, et al. The soluble NAD+-reducing [NiFe]-hydrogenase from *Ralstonia eutropha* H16 consists of six subunits and can be specifically activated by NADPH. *J Bacteriol* 2005;**187**:3122–32.
35. Linden EV, der Burgdorf T, Bernhard M, Bleijlevens B, Friedrich B, Albracht SPJ. The soluble [NiFe]-hydrogenase from *Ralstonia eutropha* contains four cyanides in its active site, one of which is responsible for the insensitivity towards oxygen. *J Biol Inorg Chem* 2004;**9**:616–26.
36. Kleihues L, Lenz O, Bernhard M, Buhrke T, Friedrich B. The H2 sensor of *Ralstonia eutropha* is a member of the subclass of regulatory [NiFe] hydrogenases. *J Bacteriol* 2000;**182**:2716–24.
37. Lenz O, Friedrich B. A novel multicomponent regulatory system mediates H_2 sensing in *Alcaligenes eutrophus*. *PNAS* 1998;**95**:12474–9.
38. Kim DH, Kim MS. Hydrogenases for biological hydrogen production. *Bioresour Technol* 2011;**102**:8423–31.
39. Winkler M, Kawelke S, Happe T. Light driven hydrogen production in protein based semi-artificial systems. *Bioresour Technol* 2011;**102**:8493–500.
40. Ihara M, Nishihara H, Yoon KS, Lenz O, Friedrich B, Nakamoto H, et al. Light-driven hydrogen production by a hybrid complex of a [NiFe]-hydrogenase and the cyanobacterial photosystem I. *Photochem Photobiol* 2006;**82**:676–82.
41. Schwartz E, Voigt B, Zühlke D, Pohlmann A, Lenz O, Albrecht D, et al. A proteomic view of the facultatively chemolithoautotrophic lifestyle of *Ralstonia eutropha* H16. *Proteomics* 2009;**9**:5132–42.
42. Lee SE, Li QX, Yu J. Proteomic examination of *Ralstonia eutropha* in cellular responses to formic acid. *Proteomics* 2006;**6**:4259–68.
43. Oh JI, Bowien B. Dual control by regulatory gene fdsR of the fds operon encoding the NAD+-linked formate dehydrogenase of *Ralstonia eutropha*. *Mol Microbiol* 1999;**34**:365–76.
44. Grunwald S, Mottet A, Grousseau E, Plassmeier JK, Popovic MK, Uribelarrea JL, et al. Kinetic and stoichiometric characterization of organoautotrophic growth of *Ralstonia eutropha* on formic acid in fed-batch and continuous cultures. *Microbiol Biotechnol* 2014;**8**:155–63.
45. Grousseau E, Blanchet E, Déléris S, Albuquerque MGE, Paul E, Uribelarrea JL. Impact of sustaining a controlled residual growth on polyhydroxybutyrate yield and production kinetics in *Cupriavidus necator*. *Bioresour Technol* 2013;**148**:30–8.
46. Bernhard M, Friedrich B, Siddiqui RA. *Ralstonia eutropha* TF93 is blocked in tat-mediated protein export. *J Bacteriol* 2000;**182**:581–8.
47. Kohlmann Y, Pohlmann A, Otto A, Becher D, Cramm R, Lu S, et al. Analyses of soluble and membrane proteomes of *Ralstonia eutropha* H16 reveal major changes in the protein complement in adaptation to lithoautotrophy. *J Proteome Res* 2011;**10**:2767–76.
48. Ye RW, Thomas SM. Microbial nitrogen cycles: physiology, genomics and applications. *Curr Opin Microbiol* 2001;**4**:307–12.
49. Schwartz E, Henne A, Cramm R, Eitinger T, Friedrich B, Gottschalk G. Complete nucleotide sequence of pHG1: a *Ralstonia eutropha* H16 megaplasmid encoding key enzymes of H_2-based lithoautotrophy and anaerobiosis. *J Mol Biol* 2003;**332**:369–83.
50. Tsuge T, Tanaka K, Ishizaki A. Development of a novel method for feeding a mixture of L-lactic acid and acetic acid in fed-batch culture of *Ralstonia eutropha* for poly-D-3-hydroxybutyrate production. *J Biosci Bioeng* 2001;**91**:545–50.
51. Patnaik PR. Response coefficient analysis of a fed-batch bioreactor to dissolved oxygen perturbation in complementary cultures during PHB production. *J Biol Eng* 2008;**2**:4.
52. Park SJ, Jang YA, Lee H, Park AR, Yang JE, Shin J, et al. Metabolic engineering of Ralstonia eutropha for the biosynthesis of 2-hydroxyacid-containing polyhydroxyalkanoates. *Metab Eng* 2013;**20**:20–8.
53. Matsumoto K, Taguchi S. Biosynthetic polyesters consisting of 2-hydroxyalkanoic acids: current challenges and unresolved questions. *Appl Microbiol Biotechnol* 2013;**97**:8011–21.
54. Volodina E, Schürmann M, Lindenkamp N, Steinbüchel A. Characterization of propionate CoA-transferase from *Ralstonia eutropha* H16. *Appl Microbiol Biotechnol* 2014;**98**:3579–89.

55. Brigham CJ, Budde CF, Holder JW, Zeng Q, Mahan AE, Rha C, et al. Elucidation of β-oxidation pathways in *Ralstonia eutropha* H16 by examination of global gene expression. *J Bacteriol* 2010;**192**:5454–64.
56. Wang ZX, Brämer CO, Steinbüchel A. The glyoxylate bypass of *Ralstonia eutropha*. *FEMS Microbiol Lett* 2003;**228**:63–71.
57. Brigham CJ, Budde CF, Holder JW, Zeng Q, Mahan AE, Rha C, et al. Elucidation of beta-oxidation pathways in *Ralstonia eutropha* H16 by examination of global gene expression. *J Bacteriol* 2010;**192**:5454–64.
58. Jendrossek D, Steinbüchel A, Schlegel HG. Alcohol dehydrogenase gene from *Alcaligenes eutrophus*: subcloning, heterologous expression in Escherichia coli, sequencing, and location of Tn5 insertions. *J Bacteriol* 1988;**170**:5248–56.
59. Jendrossek D, Krüger N, Steinbüchel A. Characterization of alcohol dehydrogenase genes of derepressible wild-type *Alcaligenes eutrophus* H16 and constitutive mutants. *J Bacteriol* 1990;**172**:4844–51.
60. Bormann EJ, Roth M. The production of polyhydroxybutyrate by Methylobacterium rhodesianum and *Ralstonia eutropha* in media containing glycerol and casein hydrolysates. *Biotechnol Lett* 1999;**21**:1059–63.
61. Solaiman DKY, Ashby RD, Foglia TA, Marmer WN. Conversion of agricultural feedstock and coproducts into poly(hydroxyalkanoates). *Appl Microbiol Biotechnol* 2006;**71**:783–9.
62. Cavalheiro JMBT, de Almeida MCMD, Grandfils C, da Fonseca MMR. Poly(3-hydroxybutyrate) production by *Cupriavidus necator* using waste glycerol. *Process Biochem* 2009;**44**:509–15.
63. Hassan AN, Nelson BK. Invited review: Anaerobic fermentation of dairy food wastewater. *J Dairy Sci* 2012;**95**:6188–203.
64. Du G, Si Y, Yu J. Inhibitory effect of medium-chain-length fatty acids on synthesis of polyhydroxyalkanoates from volatile fatty acids by *Ralstonia eutropha*. *Biotechnol Lett* 2001;**23**:1613–7.
65. Yang YH, Brigham CJ, Budde CF, Boccazzi P, Willis LB, Hassan MA, et al. Optimization of growth media components for polyhydroxyalkanoate (PHA) production from organic acids by *Ralstonia eutropha*. *Appl Microbiol Biotechnol* 2010;**87**:2037–45.
66. Jian Yu LXLC. Biopolyester synthesis and protein regulations in *Ralstonia eutropha* on levulinic acid and its derivatives from biomass refining. *J Biobased Mater Bioenergy* 2009;**3**:113–22.
67. Yu J, Si Y. Metabolic carbon fluxes and biosynthesis of polyhydroxyalkanoates in *Ralstonia eutropha* on short chain fatty acids. *Biotechnol Prog* 2004;**20**:1015–24.
68. Hassan MA, Nawata O, Shirai Y, Rahman NAA, Yee PL, Ariff AB, et al. A proposal for zero emission from palm oil industry incorporating the production of polyhydroxyalkanoates from palm oil mill effluent. *J Chem Eng Jpn* 2002;**35**:9–14.
69. Ali Hassan M, Shirai Y, Kusubayashi N, Ismail Abdul Karim M, Nakanishi K, Hashimoto K. Effect of organic acid profiles during anaerobic treatment of palm oil mill effluent on the production of polyhydroxyalkanoates by *Rhodobacter sphaeroides*. *J Ferment Bioeng* 1996;**82**:151–6.
70. Fukui T, Suzuki M, Tsuge T, Nakamura S. Microbial synthesis of poly((R)-3-hydroxybutyrate-co-3-hydroxypropionate) from unrelated carbon sources by engineered *Cupriavidus necator*. *Biomacromolecules* 2009;**10**:700–6.
71. Bruland N, Voß I, Brämer C, Steinbüchel A. Unravelling the C3/C4 carbon metabolism in *Ralstonia eutropha* H16. *J Appl Microbiol* 2010;**109**:79–90.
72. Brandt U, Raberg M, Voigt B, Hecker M, Steinbüchel A. Elevated poly(3-hydroxybutyrate) synthesis in mutants of *Ralstonia eutropha* H16 defective in lipopolysaccharide biosynthesis. *Appl Microbiol Biotechnol* 2012;**95**:471–83.
73. Holder JW, Ulrich JC, DeBono AC, Godfrey PA, Desjardins CA, Zucker J, et al. Comparative and functional genomics of *Rhodococcus opacus* PD630 for biofuels development. *PLoS Genet* 2011;**7**:e1002219.
74. Raberg M, Kaddor C, Kusian B, Stahlhut G, Budinova R, Kolev N, et al. Impact of each individual component of the mutated PTS(Nag) on glucose uptake and phosphorylation in *Ralstonia eutropha* $G^{+}1$.
Appl Microbiol Biotechnol 2012;**95**:735–44.
75. Oliveira FC, Dias ML, Castilho LR, Freire DMG. Characterization of poly(3-hydroxybutyrate) produced by *Cupriavidus necator* in solid-state fermentation. *Bioresour Technol* 2007;**98**:633–8.

76. Mohd Zahari MAK, Ariffin H, Mokhtar MN, Salihon J, Shirai Y, Hassan MA. Factors affecting poly(3-hydroxybutyrate) production from oil palm frond juice by *Cupriavidus necator* (CCUG52238(T)). *J Biomed Biotechnol* 2012;**2012**:125865.
77. Koutinas AA, Xu Y, Wang R, Webb C. Polyhydroxybutyrate production from a novel feedstock derived from a wheat-based biorefinery. *Enzyme Microbiol Technol* 2007;**40**:1035–44.
78. Wang Y, Chen R, Cai J, Liu Z, Zheng Y, Wang H, et al. Biosynthesis and thermal properties of PHBV produced from levulinic acid by *Ralstonia eutropha*. *PLoS ONE* 2013;**8**:e60318.
79. Yu J, Chen LXL, Sato S. Biopolyester synthesis and protein regulations in *Ralstonia eutropha* on levulinic acid and its derivatives from biomass refining. *J Biobased Mater Bioenergy* 2009;**3**:113–22.
80. Syamsu KM, Fauzi A, Hartoto L, Ani Suryani AS, Nur Atifah NA. Production of Pha (poly hydroxy alkanoates) by *Ralstonia eutropha* on hydrolysed sago starch as main substrate using fed-batch cultivation method. (Putrajaya, Malaysia); 2006, pp. 153–157.
81. Yu J, Stahl H. Microbial utilization and biopolyester synthesis of bagasse hydrolysates. *Bioresour Technol* 2008;**99**:8042–8.
82. Johnson BF, Stanier RY. Dissimilation of aromatic compounds by *Alcaligenes eutrophus*. *J Bacteriol* 1971;**107**:468–75.
83. Hino S, Watanabe K, Takahashi N. Phenol hydroxylase cloned from *Ralstonia eutropha* strain E2 exhibits novel kinetic properties. *Microbiol Read Engl* 1998;**144**:1765–72.
84. Spain JC. Biodegradation of nitroaromatic compounds. *Annu Rev Microbiol* 1995;**49**:523–55.
85. Schenzle A, Lenke H, Fischer P, Williams PA, Knackmuss H. Catabolism of 3-Nitrophenol by *Ralstonia eutropha* JMP 134. *Appl Environ Microbiol* 1997;**63**:1421–7.
86. Trefault N, De la Iglesia R, Molina AM, Manzano M, Ledger T, Pérez-Pantoja D, et al. Genetic organization of the catabolic plasmid pJP4 from *Ralstonia eutropha* JMP134 (pJP4) reveals mechanisms of adaptation to chloroaromatic pollutants and evolution of specialized chloroaromatic degradation pathways. *Environ Microbiol* 2004;**6**:655–68.
87. Hawkins AC, Harwood CS. Chemotaxis of *Ralstonia eutropha* JMP134(pJP4) to the herbicide 2,4-dichlorophenoxyacetate. *Appl Environ Microbiol* 2002;**68**:968–72.
88. Ampe F, Léonard D, Lindley ND. Repression of phenol catabolism by organic acids in Ralstonia eutropha. *Appl Environ Microbiol* 1998;**64**:1–6.
89. Riedel SL, Lu J, Stahl U, Brigham CJ. Lipid and fatty acid metabolism in *Ralstonia eutropha*: relevance for the biotechnological production of value-added products. *Appl Microbiol Biotechnol* 2014;**98**:1469–83.
90. Brigham CJ, Speth DR, Rha C, Sinskey AJ. Whole-genome microarray and gene deletion studies reveal regulation of the polyhydroxyalkanoate production cycle by the stringent response in *Ralstonia eutropha* H16. *Appl Environ Microbiol* 2012;**78**:8033–44.
91. Philip S, Keshavarz T, Roy I. Polyhydroxyalkanoates: biodegradable polymers with a range of applications. *J Chem Technol Biotechnol* 2007;**82**:233–47.
92. Chen GQ. A microbial polyhydroxyalkanoates (PHA) based bio- and materials industry. *Chem Soc Rev* 2009;**38**:2434–46.
93. Voinova ON, Kalacheva GS, Grodnitskaya ID, Volova TG. Microbial polymers as a degradable carrier for pesticide delivery. *Appl Biochem Microbiol* 2009;**45**:384–8.
94. Sudesh K, Bhubalan K, Chuah JA, Kek YK, Kamilah H, Sridewi N, et al. Synthesis of polyhydroxyalkanoate from palm oil and some new applications. *Appl Microbiol Biotechnol* 2011;**89**:1373–86.
95. Brigham CJ, Sinskey AJ. Applications of polyhydroxyalkanoates in the medical industry. *Int J Biotechnol Wellness Ind* 2012;**1**:52–60.
96. Brigham CJ, Zhila N, Shishatskaya E, Volova TG, Sinskey AJ. Manipulation of *Ralstonia eutropha* carbon storage pathways to produce useful bio-based products. In: Wang X, Chen J, Quinn P, editors. *Reprogramming microbial metabolic pathways*. Netherlands: Springer; 2012. p. 343–66.
97. Davis R, Chandrashekar A, Shamala TR. Role of (R)-specific enoyl coenzyme A hydratases of *Pseudomonas sp* in the production of polyhydroxyalkanoates. *Antonie Van Leeuwenhoek* 2008;**93**:285–96.

98. Kawashima Y, Cheng W, Mifune J, Orita I, Nakamura S, Fukui T. Characterization and functional analyses of R-specific enoyl coenzyme A hydratases in polyhydroxyalkanoate-producing *Ralstonia eutropha. Appl Environ Microbiol* 2012;**78**:493–502.

98a. Budde CF, Mahan AE, Lu J, Rha C, Sinskey AJ. Role of multiple acetoacetyl coenzyme A reductases in polyhydroxybutyrate biosynthesis in *Ralstonia eutrohpa* H16. *J Bacteriol* 2010;**192**:5319–28.

99. Lütke-Eversloh T, Steinbüchel A. Novel precursor substrates for polythioesters (PTE) and limits of PTE biosynthesis in *Ralstonia eutropha. FEMS Microbiol Lett* 2003;**221**:191–6.

100. Kawada J, Lütke-Eversloh T, Steinbüchel A, Marchessault RH. Physical properties of microbial polythioesters: characterization of poly(3-mercaptoalkanoates) synthesized by engineered *Escherichia coli. Biomacromolecules* 2003;**4**:1698–702.

101. Torella JP, Ford TJ, Kim SN, Chen AM, Way JC, Silver PA. Tailored fatty acid synthesis via dynamic control of fatty acid elongation. *Proc Natl Acad Sci U S A* 2013;**110**:11290–5.

102. Müller J, MacEachran D, Burd H, Sathitsuksanoh N, Bi C, Yeh YC, et al. Engineering of *Ralstonia eutropha* H16 for autotrophic and heterotrophic production of methyl ketones. *Appl Environ Microbiol* 2013;**79**:4433–9.

103. Connor MR, Atsumi S. Synthetic biology guides biofuel production. *J Biomed Biotechnol* 2010;1–9.

104. Lan EI, Liao JC. Microbial synthesis of n-butanol, isobutanol, and other higher alcohols from diverse resources. *Bioresour Technol* 2013;**135**:339–49.

105. Lu J, Brigham CJ, Gai CS, Sinskey AJ. Studies on the production of branched-chain alcohols in engineered *Ralstonia eutropha. Appl Microbiol Biotechnol* 2012;**96**:283–97.

106. Grousseau E, Lu J, Gorret N, Guillouet SE, Sinskey AJ. Isopropanol production with engineered *Cupriavidus necator* as bioproduction platform. *Appl Microbiol Biotechnol* 2014;**98**:4277–90.

107. Bond-Watts BB, Bellerose RJ, Chang MCY. Enzyme mechanism as a kinetic control element for designing synthetic biofuel pathways. *Nat Chem Biol* 2011;**7**:222–7.

108. Gao X, Chen JC, Wu Q, Chen GQ. Polyhydroxyalkanoates as a source of chemicals, polymers, and biofuels. *Curr Opin Biotechnol* 2011;**22**:768–74.

109. Lee SY, Lee Y. Metabolic engineering of *Escherichia coli* for production of enantiomerically pure (R)-hydroxycarboxylic acids. *Appl Environ Microbiol* 2003;**69**:3421–6.

110. Shiraki M, Endo T, Saito T. Fermentative production of (R)-3-hydroxybutyrate using 3-hydroxybutyrate dehydrogenase null mutant of *Ralstonia eutropha* and recombinant *Escherichia coli. J Biosci Bioeng* 2006;**102**:529–34.

111. Oda T, Oda K, Yamamoto H, Matsuyama A, Ishii M, Igarashi Y, et al. Hydrogen-driven asymmetric reduction of hydroxyacetone to (R)-1,2-propanediol by *Ralstonia eutropha* transformant expressing alcohol dehydrogenase from Kluyveromyces lactis. *Microb Cell Factories* 2013;**12**:2.

112. Barnard GC, McCool JD, Wood DW, Gerngross TU. Integrated recombinant protein expression and purification platform based on Ralstonia eutropha. *Appl Environ Microbiol* 2005;**71**:5735–42.

113. Schmidt FR. Recombinant expression systems in the pharmaceutical industry. *Appl Microbiol Biotechnol* 2004;**65**:363–72.

114. Srinivasan S, Barnard GC, Gerngross TU. A novel high-cell-density protein expression system based on *Ralstonia eutropha. Appl Environ Microbiol* 2002;**68**:5925–32.

115. Chen WM, Chang JS, Wu CH, Chang SC. Characterization of phenol and trichloroethene degradation by the rhizobium *Ralstonia taiwanensis. Res Microbiol* 2004;**155**:672–80.

116. Watanabe K. Microorganisms relevant to bioremediation. *Curr Opin Biotechnol* 2001;**12**:237–41.

117. Pandey G, Jain RK. Bacterial chemotaxis toward environmental pollutants: role in bioremediation. *Appl Environ Microbiol* 2002;**68**:5789–95.

CHAPTER 13

Methane Biocatalysis: Selecting the Right Microbe

M.G. Kalyuzhnaya
Biology Department, Viral Information Institute, San Diego State University, San Diego, California, USA

Abbreviations

AMO	arobic methane oxidation
ANME	anaerobic methanotrophic archaea
AnMO	anaerobic methane oxidation
BS	bifidobacterium shunt
CBB	Calvin-Benson-Bassham cycle
CCE	carbon conversion efficiency
DCW	dry cell weight
EDD pathway	Entner-Doudoroff pathway
EMP pathway	Embden-Meyerhof-Parnas pathway
MCC	methanol condensation cycle
PHB	polyhydroxybutyrate
RuMP pathway	ribulose monophosphate pathway
SC	serine cycle
pSC	partial serine cycle
SCP	single-cell protein
TCA	tricarboxylic acid cycle
rTCA	reverse tricarboxylic acid cycle
WLP	Wood-Ljungdahl pathway

13.1 Introduction

Methane-oxidizing bacteria (methanotrophs) are natural systems for the attenuation of methane emission. These microbes can use methane as their sole source of energy and carbon and thus are unique systems for methane-based bioconversions. The industrial potential of methane metabolism has actively been discussed for almost a half a century. Methanotrophic cultures have been tested for the production of single-cell protein (SCP),[1,2] poly-β-hydroxybutyrate (PHB),[3] components for nanotechnological applications,[4] lipids,[5] and carotenoids[6] and as a carbohydrate feedstock for fermentation.[7,8] A number of methane-driven transformations, such as epoxidation, calcium carbonate precipitation, bioleaching, biofiltration, and bioremediation, have also been explored.[9–15] However, only the production of methanotrophic SCP (such as BioProtein/Norferm, Norway,

Biotechnology for Biofuel Production and Optimization. http://dx.doi.org/10.1016/B978-0-444-63475-7.00013-3

http://calysta.com; UniProtein/Unibio, Denmark http://www.unibio.dk, and Gaprin, Russia, http://atelgroup.org/en/biotechnology) using a mixed culture with *Methylococcus capsulatus* has remained a commercially established product. A number of factors contributed to the slow progress in methane-based biotechnology, including the increased energy prices in the late 1970s to early 1980s, which reduced the attractiveness of natural gas as a carbon feedstock. By the early 1990s it had already been concluded that the future of C_1 biocatalysis lays in the production of high-cost products,[16] although difficulties in the genetic reprogramming of methanotrophic bacteria together with fairly cheap chemical fossil-based synthesis almost completely eliminated the attention given to bioconversions.

Resurgence of interest in the industrial application of methane biocatalysis is occurring now. The energy market landscape (cheap natural gas *vs* expensive oil) and energy boom in the United States are most likely the main drivers of the new wave of attention being given to methane.[4,5,17–19] However, history shows that this interest can easily decline in parallel with oil prices. Experts in the field believe that new methane-based technology must go beyond short-term economic considerations and focus on long-term sustainability. It is important to remember that methane is a very controversial source of energy—one which comes with an exceptionally large environmental footprint.[20,21] Anthropogenic methane emission accounts for more than 60% of the greenhouse gas budget and it is the number one target for global climate regulation.[22] Fossil fuel production (e.g., underground coal mining, oil-associated gas, and natural gas pipelines/compressors) is the first target for new technologies. However, the technology can also benefit other areas, such as agriculture (e.g., enteric fermentation in livestock, manure management, and rice cultivation), landfills, municipal wastewater, and rising biofuel production.[23,24] Up to 116 million tons of oil-associated methane and 40 million tons of biogas (equivalent to 27% of the total United States transportation fuel) are flared each year. Any gas flare, from small to large, represents lost energy, lost revenues, and unnecessary CO_2 emissions. Gas flares are also a dangerous source of air pollution. A search for new environment-friendly approaches for methane capture is an ongoing journey, and it is not yet apparent which innovation might represent the best solution. If developed, methane biocatalysis could fill this niche and the technology would certainly withstand future oil-price drops and rises.

Microbial biotechnology is grounded on the fundamental knowledge of related systems. Our understanding of methane utilization has changed dramatically in recent years. A number of novel microbial strains with novel metabolic arrangements of methane utilization have been described. The new microbes inspire conceptually novel ideas for the intensification of methane conversion and stimulate new elaborations in the area. This chapter summarizes the most recent development in microbial methane biocatalysis. Several strategies are outlined below for streamlining the selection of microbial systems for biotechnological alterations. These include metabolic considerations and the physiological advantages or constraints of specific microbial cultures.

13.2 Methane Biocatalysis: Basic Concepts

Two fundamentally different mechanisms of methane utilization are known: oxygen dependent (mostly aerobic) and oxygen independent (anaerobic). Aerobic methane oxidation (AMO) relies on oxygen for methane activation and oxidation and has thus far been described for Proteobacteria, Verrucomicrobia, and NC10 phyla. Oxygen-independent methane oxidation is based on reverse methanogenesis. A number of anaerobic methane oxidation (AnMO) couplings have been proposed.[25,26]

So far only aerobic fermentation with methane has been proven at the large scale, demonstrating that safe and efficient operation is achievable. However, when realized at the industrial scale, methane catalysis appeared to be more expensive then was initially projected.[16] One of the overlooked economic aspects of methane-based conversion could have been the cost of oxygen, the second compound required for the process. Today a number of alternate strategies for methane conversion, which engage nitrate, sulfate, or three metal oxides to fulfill oxygen's functions, are being discussed. Below, the cost of each conversion is estimated based on the current prices of the nutrients required for methane oxidation.

Aerobic methane conversion requires an oxygen input with at least a 1:1 methane to oxygen ratio. Oxygen is widely available, but it is a pricy feedstock. The cost of oxygen is directly linked to energy cost (mostly electricity) and might vary from \$100 to \$125 per ton.[27] That estimation would bring the cost of aerobic methane conversion to at least \$200-\$250 per one ton of methane (Figure 13.1). In fact, high cell growth would actually require higher oxygen input (at 1:1.5-2 methane to oxygen ratio) making the cost of AMO conversion even higher, up to \$375-\$500 per ton of methane.

The cheapest AnMO process is the one linked to sulfate reduction (SR). With the price of sulfuric acid ranging from \$25 to 100 per ton (www.icis.com), the AnMO-SR conversion would cost \$30-\$600 per ton of methane; the price would be much higher (\$1000-\$1500) if sulfates are used. AnMO supported by denitrification (which requires 4 mol of nitrate per 1 mol of methane oxidized) is about \$3400 per ton of methane if nitric acid is used [at the price of \$215-\$225 per ton (www.icis.com)] and at least \$7600 per ton if sodium nitrate is used. AnMO supported by metal oxides comes with a remarkably high price tag (Figure 13.1).

Other factors challenging the applicability of AnMO-based technology which are rarely discussed in the literature are its overall process sustainability and possible environmental impact. The current markets for sulfuric acid/sulfates, nitric acid/nitrates, and metal oxides are quite narrow, and they would have to be expanded significantly to support methane catalysis. For example, the conversion of flared oil-associated methane (estimated to be at least 116 million tons) would require 696 million tons of sulfate, but the global annual production of sulfuric acid is only 200 million tons (http://www.essentialchemicalindustry.org) while sodium sulfate is only 16-17 million tons (http://www.cefic.org).

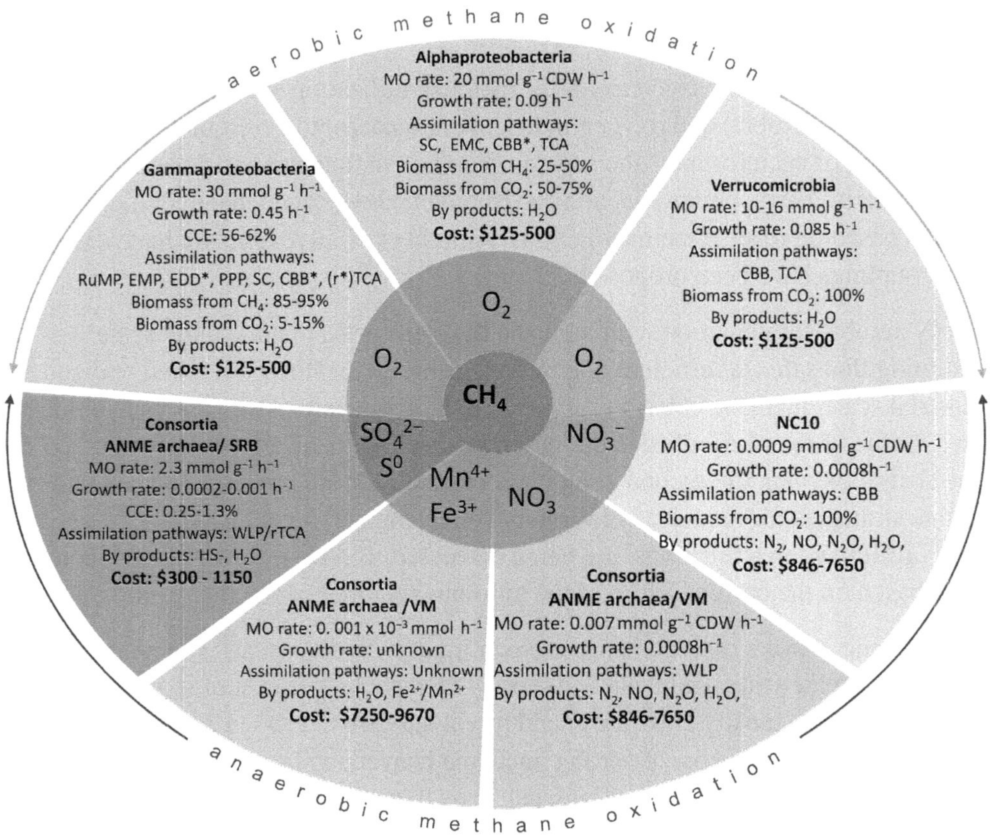

Figure 13.1

Overview of aerobic and anaerobic methane oxidation: microbes, parameters, by-products, and economics. DCW, dry cell weight; MO, methane oxidation; RuMP, assimilatory ribulose monophosphate pathway; SC, serine cycle; EMP, ethylmalonyl-CoA pathway; CBB, Calvin cycle; WLP, the reductive acetyl-CoA (Wood-Ljungdahl) pathway; *, pathway is present only in some species. Cost of methane conversion was estimated based on the methane conversion equations (25-26, and below) and current price of electron acceptors (taking from www.icis.com).

$$CH_4 + 2O_2 \rightarrow CO_2 + 2H_2O$$

$$CH_4 + SO_4^{2-} + 2H^+ \rightarrow CO_2 + H_2S + 2H_2O$$

$$CH_4 + 4NO^{3-} \rightarrow CO_2 + 4NO^{2-} + 2H_2O$$

$$5CH_4 + 8MnO^{4-} + 24H^+ \rightarrow 5CO_2 + 8Mn^{2+} + 22H_2O$$

$$CH4 + 8Fe^{3+} + 2H_2O \rightarrow CO_2 + 8Fe^{2+} + 8H^+$$

Another aspect that challenges the AnMO process is that intensive oxidation of methane requires significant input of sulfate or nitrate into the cultivation environment. Assuming that the rate of AnMO is in the same range at AMO, or about 20 mmol h^{-1} g^{-1} dry cell weight (DCW), a bioreactor with about 30 g DCW per liter would consume sodium sulfate at the rate 58.8 g h^{-1} L^{-1} and sodium nitrate at the rate 37.2 g h^{-1} L^{-1}. That indicates that the AnMO-based process has to be built using strains capable of tolerating high salt (halophiles) or, if electron acceptors are added as acids, extremely low pH (acidophiles). That would remain true even for an alternate host of the AnMO modules.

Even with a very cursory look, AnMO-based technology might have dramatic environmental impacts. The ANMO-SR mode is directly linked with the formation of volatile sulfur compounds, such as H_2S. This mode of methane oxidation would not be feasible for technological applications unless a sustainable solution for conversion of H_2S back to sulfate or elemental sulfur is found. In theory, that could also be accomplished via biological pathways, such as photosynthetic sulfur oxidation linked to light, oxygen, or nitrate.[28] The AnMO-DR system might leak toxic intermediates and dangerous greenhouse gases, such as NO, N_2O, and ammonia. Like SR, some impacts of the denitrification process could be complimented by additional (bio)technological modules, such as coupling to ammonium oxidation.[29] That, however, also requires additional breakthrough developments in areas beyond methane biocatalysis.

Taking into account all three parameters discussed for AnMO-based technologies (cost, sustainability, and environmental impact), AMO remains the more feasible solution for a methane-based technology if it is considered for a wide deployment. However, the AnMO process might be beneficial for some areas. For example, methane-activated denitrification might find its way to the market as an innovative solution to wastewater treatment, and perhaps some biomass produced in the process could be converted into value-added chemicals. In the long term, feasibility of AnMO will depend on breakthrough developments in synthetic biology. While it is hard to predict what innovation might bring advanced methane biocatalysts, one aspect of the AnMO-based technology is apparent—it has to be accomplished only as a closed, self-sustained cycle.

13.3 Selecting the Right Microbe: C_1-Oxidation Options

13.3.1 Methane Oxidation—Aerobic

Methane oxidation is driven by mixed function oxidases, in which one atom from O_2 goes to methanol and the other to water, requiring the input of two electrons and two protons[30]; Figure 13.2a).

Two isoforms of the enzyme are described: cytosolic soluble methane monooxygenase (sMMO) and membrane bound, or particulate methane monooxygenase (pMMO).[31] sMMO uses NADH as its electron donor[32]; the physiological electron donor to pMMO is still not known, but it is widely accepted that reducing equivalents produced by formaldehyde and/or formate oxidation (NADH or QH_2) are required for its enzyme activity.[30] In all known methanotrophs, the second step of the AMO pathway is catalyzed by a periplasmic pyrroloquinoline quinone (PQQ)-linked methanol dehydrogenase (MDH). Two systems are described: a two subunit enzyme, known as Mxa, encoded by *mxaFI* genes[33,34] and a

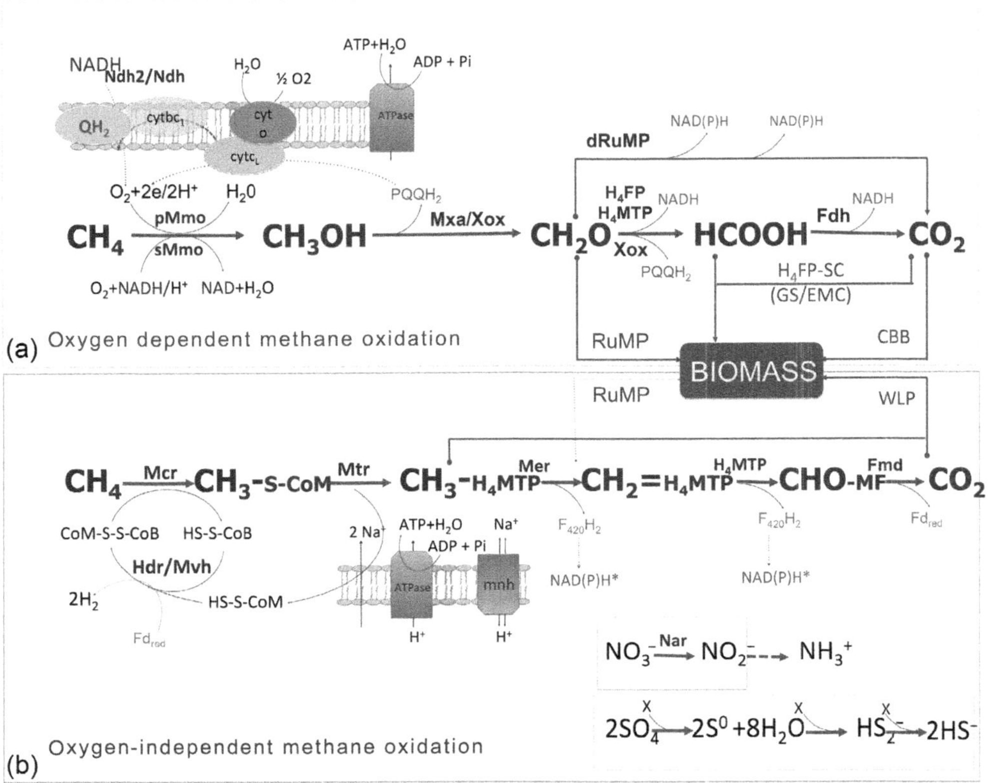

Figure 13.2

Schematic illustration of methane metabolism. (a) Aerobic methane utilization; (b) anaerobic methane utilization. pMMO, membrane-bound methane monooxygenase; sMMO, soluble methane monooxygenase; Mxa, PQQ-linked methanol dehydrogenases; Xox, PQQ-linked methanol and formaldehyde dehydrogenases; H_4MTP, methanopterin-linked C_1 transfer; H_4FP, folate-linked C_1 transfer; Fdh, formate dehydrogenases; NDH2, NADH:quinone oxidoreductase; dPPP, dissimilatory pentose phosphate cycle; RuMP, assimilatory ribulose monophosphate pathway; SC, serine cycle; EMP, ethylmalonyl-CoA pathway; CBB, Calvin-Benson-Bassham cycle.

monosubunit enzyme, encoded by *xoxF* genes.[35] Both systems are associated with cytochrome *c* L (*mxaG* or *xoxG*) and a protein of unknown function protein (*mxaJ* or *xoxJ*), which enhances the activity of the enzyme.[36,37] The methanol oxidation machinery also includes a number of proteins for the enzyme assembly and biosynthesis of PQQ.[34,38]

While several formaldehyde dehydrogenases were reported in early studies of methanotrophs,[39–42] so far, only the presence of the broad-specificity housekeeping aldehyde dehydrogenases has been confirmed by genomics.[43–46] The aldehyde dehydrogenases show quite low expression levels upon growth on methane and their role in primary C_1 oxidation remains to be validated.[47,48] However, it has been shown that the Xox-type MDH also contributes to oxidation of formaldehyde in Verrucomicrobial methanotrophs.[49] As *xox* genes are quite widespread among bacteria,[49,50] it is possible that the Xox enzyme plays at least some role in formaldehyde oxidation in all methanotrophs. Among the pathways commonly predicted for formaldehyde conversion in methanotrophic bacteria are the oxidation through H_4folate or H_4methanopterin intermediates[1,51] and the dissimilatory ribulose monophosphate phosphate (dRuMP) pathways.[52] The cofactor-linked pathways generate 1 mol of NADH per 1 mol of formaldehyde converted to formate. The dRuMP pathway leads to the production of 2 mol of NAD(P)H and 1 mol of CO_2 per 1 mol of formaldehyde. Formate oxidation is typically carried out by NAD-dependent formate dehydrogenases containing molybdenum or tungsten.[45,48,53]

13.3.2 Methane Oxidation—Anaerobic

The energetic and environmental parameters of ANMO have recently been reviewed by Caldwell et al.[25] ANMO coupled with the reduction of nitrate,[54] sulfate or elemental sulfur,[26,55–57] or metals[58] are discussed in this chapter, as they are the only naturally occurring reactions that have been at least partially defined on a biological level. AnMO is typically driven by a mix culture (or consortia) of anaerobic methanotrophic archaea (ANME) and bacteria, rather than a pure culture. Because none of the anaerobic methanotrophs have been isolated in pure culture or coculture, not much is known about pathways and enzymes involved in primary methane oxidation and assimilation. The key kinetic parameters of the most typical arrangements of AnMO are summarized in Figure 13.1. The proposed reverse methanogenesis is supported by remarkable enzymatic studies on methyl-CoA reductase (MCR), which also carries out biological methane formation.[59] The arrangement of metabolic reactions downstream from the MCR step remains elusive. It is assumed that methane-oxidation products follow the reverse methanogenic pathways all the way back to CO_2 or acetate (Figure 13.2b).[54,60]

13.3.3 Methane Oxidation—Summary

A summary of the core enzymes involved in methane oxidation is shown in Figure 13.2. Based on available data, complete AMO can be achieved by only three enzymes: MMO, Xox-type methanol/formaldehyde dehydrogenase, and formate dehydrogenase. This most simplistic configuration of methane utilization would require three genes for pMMO, at least three genes for methanol and formaldehyde oxidation—assuming that the host has the PQQ biosynthesis inventory (nine genes if it does not)—and at least two genes to complete the conversion of formate into CO_2 and to generate reducing power.

The rates and efficiencies of methane oxidation might vary quite significantly even among closely related species; however, some common trends can be noted. For AMO, typically cells expressing pMMO show higher growth rates and higher carbon conversion efficiencies (CCEs) than cells using sMMO.[61–64] Interestingly, the activity of the enzyme in archaea/sulfate-reducing bacteria consortia aligns with the lowest parameters observed for aerobic methane oxidizers. The rates of AnMO linked to reduction of sulfates are almost three orders of magnitude higher than the rates of AnMO coupled to denitrification or reduction of metal oxides (Figure 13.1).[54–58,65,66] Since the exact energy requirements for pMMO-based methane utilization are not well established, any *in silico* predictions of methane oxidation efficiencies are to some degree arbitrary. All comparisons provided below are based entirely on assimilation, with the assumption that the energetic inputs and outputs of the primary oxidation machinery are identical among all methanotrophs.

13.4 Selecting the Right Microbe: C_1-Assimilation Options

Natural pathways for single carbon assimilation come in a great variety of metabolic arrangements, which fundamentally can be split into three main categories: *CH_2O to biomass conversion*, which occurs via the assimilatory ribulose monophosphate (aRuMP) pathway (Figure 13.3)[50,67]; *[CH_2] to biomass conversion*, which involves the serine cycle

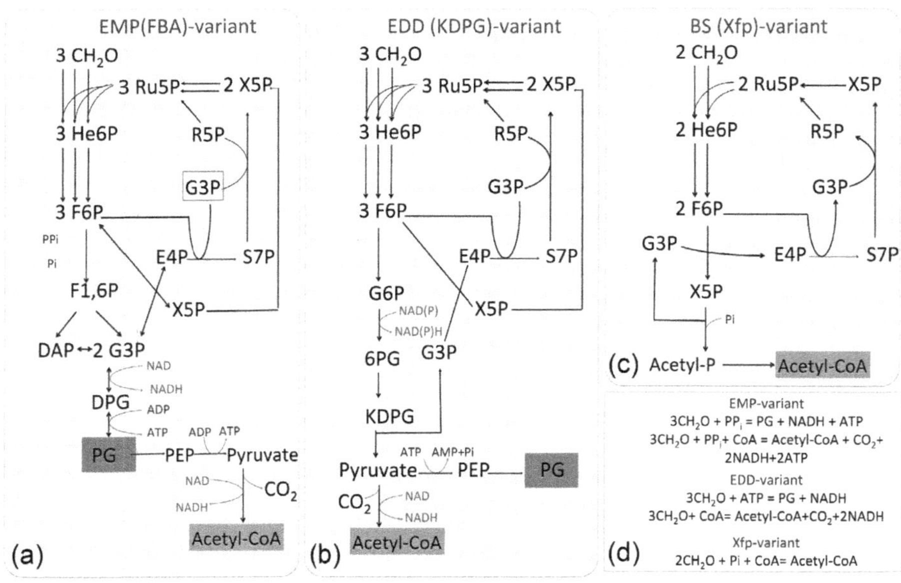

Figure 13.3

The ribulose monophosphate pathway (RuMP) in methanotrophic bacteria. (a) EMP (FBA) variant; (b) EDD (KDPG) variant; and (c) BS (Xfp) variant. (d) Summary equations for the production of phosphoglycerate or acetyl-CoA from formaldehyde. DAP, dihydroxyacetone phosphate; DPG, 2,3-diphosphoglycerate; E4P, erythrose-4-phosphate; F6P, fructose 6-phosphate; F1,6P, fructose 1,6-bisphosphate; G3P, glyceraldehyde 3-phosphate; G6P, glucose 6-phosphate; He6P, 3-hexulose 6-phosphate; KDPG, 2-keto-3-deoxy 6-phosphogluconate; 6PG, 6-phosphogluconate; PEP, phosphoenolpyruvate; PGA, 3-phosphoglycerate; Ru5P, ribulose 5-phosphate; R5P, ribose 5-phosphate; S7P, sedoheptulose 7-phosphate; X5P, xylulose 5-phosphate; PPi, diphosphate; Pi, inorganic phosphate.

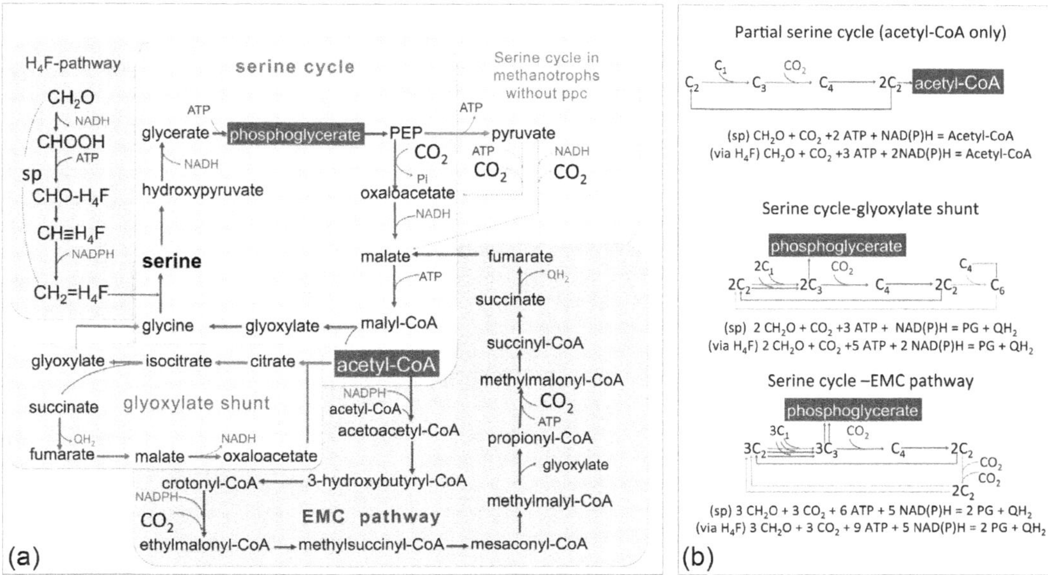

Figure 13.4
Overview of the metabolic arrangement for $[CH_2]$- to biomass conversion. (a) The serine cycle and associated reactions in methanotrophic bacteria. (b) Summary equations for the production of phosphoglycerate or acetyl-CoA from formaldehyde. sp, spontaneous condensation of formaldehyde on H_4-folate.

(SC), at least one of the glyoxylate regeneration pathways, and, under certain circumstances, tetrahydrofolate (H_4F)-linked C_1 transfer (Figure 13.4)[68–71]; and *CO_2 to biomass conversion*, which theoretically could cover all the described CO_2-fixation pathways.[72,73] Different microbes show different inventories for assimilation, and some of them might have pathways for carbon fixation at all three levels (see Table 13.1). The details of methylotrophic assimilatory pathways will be described below.

13.4.1 CH_2O to Biomass Conversion

The two key enzymes of the CH_2O to biomass conversion, 3-hexulose 6-phosphate synthase (EC 4.1.2.43, Hps) and 3-hexulose 6-phosphote isomerase (EC 5.3.1.27, Hpi), are responsible for condensation of formaldehyde on ribulose 5-phosphate (R5P) to form hexulose 6-phosphate and its conversion into fructose 6-phosphate. Hps and Hpi are core enzymes for C_1 assimilation in methanotrophic Gammaproteobacteria.[117] Theoretically, three variants of the RuMP pathway could be predicted, including the Entner-Doudoroff pathway (EMP) (glycolytic) variant (also known as FBP variant); the Entner-Doudoroff (EDD) pathway (KDPG variant); and the bifidobacterium shunt (BS) variant (also known as phosphoketolase, Xfp variant, also proposed as a biocatalytic methanol condensation cycle (Figure13.3)).[1,118,119]

Table 13.1: General summary for publicly available methanotrophic cultures

Name	Species*	Growth parameters					Culture collection ID [Genome Accession Number (NCBI)]	Reference
		Growth rate (h^{-1})	Metabolic pathways	pH	T °C	Other		
Group I								
Methylobacter	*M. luteus*	0.12	pMMO, Mxa, H_4MTP, H_4F, RuMP (EMP,EDD,BS)pSC	NA (6.8)	5-55 (37)	Produces yellow water-soluble pigment and bacteriocin-like compounds	**89/VKM B-2111** [NZ_ATYJ00000000.1]	74
	M. marinus	0.15	pMMO, Mxa, H_4MTP, H_4F, RuMP (EMP,EDD,BS)pSC	NA (7.2)	5-35 (30)	Grows at modest salinity (0.1-5%)	**A45**, Dr. Lidstrom culture collections (University of Washington) [NZ_ARVS00000000.1]	
	M. tundripaludum	0.02	pMMO, Mxa, H_4MTP, H_4F, RuMP (EMP,EDD,BS)pSC	5.5-8 (6.8)	5-30 (23)	Does not grow well on methanol, might be capable of nitrogen fixation	**SV96**^{T} (= DSM 17260^T = ATCC BAA-1195^T) [NZ_AEGW00000000.2]	75,76
	M. whittenburyi	0.1	pMMO, Mxa, H_4MTP, H_4F, RuMP (EMP,EDD,BS)pSC	5-8 (6.8)	5-40 (25)		**UCM-B-3033**, Dr. Romanovskaya V.A. culture collection (Institute of Microbiology, Ukraine) [JQNS01000000]	
	Methylobacter sp.BBA5.1	0.06	pMMO, Mxa, H_4MTP, H_4F, RuMP (EMP,EDD,BS)pSC	NA (7.2)	5-35 (30)	Grows at modest salinity (0.1-5%)	**BBA5.1**, Dr. Lidstrom culture collections (University of Washington) [JQKS01000000.1]	
Methylocaldum	*M. szegediense*	0.03	pMMO, Mxa, H_4MTP, H_4F, RuMP (EMP,EDD,BS) CBB, pSC	NA (6.8)	30-61 (55)	Produces melanin-like pigments, and can excrete 4-hydroxyphenyl acetic acid. Might be capable of PHB production	**O-12**, Dr. Trotsenko Yu.A. culture collection (IBPM) [ATXX01000000]	77,78

Methylococcus	*M. capsulatus* Bath	0.3-0.4	pMMO, sMMO, Mxa, H_4MTP, H_4F, RuMP (EMP,EDD,BS) CBB, pSC	6-8 (7)	30-55 (42)	Well-established catalyst for SCP production	**Bath**^{T} (= ACM 3302^T = ATCC 3309^T = NCIMB 11132^T) [NC_002977]	46,79
	M. capsulatus Texas	0.1	pMMO, sMMO, Mxa, H_4MTP, H_4F, RuMP (EMP,EDD,BS) CBB, pSC	6-8 (7)	30-50 (37)	Grows well on methanol	**Texas** (=ACM 1292^T = ATCC 19069^T = NCIMB 11853^T) [NZ_AMCE00000000.1]	80,81
Methylogaea	*M. oryzae*	NA	pMMO, Mxa, RuMP (EMP,EDD,BS) CBB, pSC	5-8 (7)	20-37 (30)	Grows at low salinity. Might be capable of nitrogen fixation and PHB production	**E10**^{T} (=JCM 16910^T = DSM 23452^T) [PRJDB848]**	82
Methyloglobulus	*M. morosus* KoM1	0.14	pMMO, RuMP (EMP,EDD,BS)pSC	5-8.5 (6-8)	4-30 (20)	Can fix nitrogen, prefers to grow at low O2	**KoM1**^{T} (=DSM 22980) [NZ_AYLO01000180]	83,84
Methylohalobius	*M. crimeensis* 10Ki	0.03-0.05	pMMO, Mxa, H_4MTP, H_4F, RuMP (EMP, BS), CBB, pSC	6.5-7.5	15-42 (30)	Grow optimally at 1-1.5 M NaCl	**10Ki** = DSM 16011^T = ATCC BAA-967^T [NZ_ATXB00000000]	[85]P. Dunfield personal communication
Methylomarinum	*M. vadi* IT-4	0.33	pMMO, Mxa, H_4MTP, H_4F, RuMP	4.5-7.0 (6.4)	20-44 (37)	Grows well at low pH and relatively high salinity. The strain was isolated from thermal vent and might be tolerant to short-chain alkanes	**IT-4**^{T} (=JCM 13665^T = DSM 18976^T) [NZ_JPON00000000]	86
Methylomarinovum	*M. caldicuralii*	0.29-0.33	pMMO, RuMP	5-7 (6-6.4)	30-55 (45-50)	Growth at salinity 1-5% (optimum 3%) and high temperature	**IT-9**^{T} (=JCM 13666(T) = DSM 19749(T)	87
Methylomicrobium	*M. agile* A30	0.06	pMMO, Mxa, H_4MTP, H_4F, RuMP (EMP,EDD,BS)pSC	5-8 (6.8)	5-30 (25)		**A30** (=ACM 3308^T = ATCC 35068^T = NCIMB 11124^T) [NZ_JPOJ00000000.1]	79
	M. album BG8	0.11	pMMO, Mxa, H_4MTP, H_4F, RuMP (EMP,EDD,BS)pSC	5-8 (6.8)	5-30 (25)		**BG8**^{T} (= ACM 3314^T = NCIMB 11123^T) [NZ_AFJF00000000.2]	88–90

(Continued)

Table 13.1: General summary for publicly available methanotrophic cultures—cont'd

Name	Species*	Growth parameters					Culture collection ID [Genome Accession Number (NCBI)]	Reference
		Growth rate (h^{-1})	Metabolic pathways	pH	T °C	Other		
	M. alcaliphilum 20Z	0.12	pMMO, Mxa, H_4MTP, H_4F, RuMP (EMP,EDD,BS)pSC	6-9 (8.5)	5-42 (30-37)	Grows well at 1 M NaCl. Resistant to desiccation	FO082060 **20Z**^{T}: VKM B-2133;5G Dr. Trotsenko Yu.A. culture collection (IBPM) [FO082060]	39,45,91,92
	M. buryatense 5G	0.23	pMMO, sMMO, Mxa, H_4MTP, H_4F, RuMP (EMP,EDD,BS)pSC	6-9 (8.5)	5-45 (30)	Robust growth	**5G** Dr. Trotsenko Yu.A. culture collection (IBPM) [NZ_AOTL00000000.1]	93,94
Methylomonas	*M. methanica* MC09	NA	pMMO, sMMO, Mxa, H_4MTP, H_4F, RuMP (EMP,EDD,BS)pSC	NA	NA	Strain was isolated from seawater, and can grow at 1 M salinity	Dr. Murell lab culture collection (University of East Anglia) [CP002738; NC_015572.1]	95
	Methylomonas LW13	0.16	pMMO, sMMO, Mxa, H_4MTP, H_4F, RuMP (EMP,EDD,BS)pSC	NA (6.8)	NA (30)	Strain shows good growth parameters	**LW13**, Dr. Lidstrom culture collections (University of Washington) [PRJNA225845]**	96
	M. paludis	NA	pMMO	3.8-7.3 (5.8-6.4)	8-30 (20-25)	Acid-tolerant methanotroph	**MG30**^{T} = DSM24973^T = VKM B-2745^T	97
	M. lenta	NA	pMMO	6.3-7.8 (6.8-7.3)	15-28 (20)	Strain was isolated from a slurry pit of a cow stable	**R-45377**^{T} = LMG 26260^T = JCM 19378^T	98
Methyloparacoccus	*M. murrellii*	NA	pMMO	5.8-9 (6.3-6.9)	20-37 (25-33)	Strain was isolated from pond water	**LMG 27482**^{T} = JCM 19379^T	99
Methyloprofundus	*M. sedimenti*	0.08	pMMO RuMP, pSC	6-8 (6.5-7.5)	4-26 (23)	Nitrogen fixation	Draft genome **WF1** = BCCM LMG 28393 = ATCC BAA-2619) [Draft genome]	100

Methylosarcina	*M. fibrata* AML-C10	0.08	pMMO, RuMP, pSC	5-9 (5)	10-37 (30)	Grows well at pH 5.0. Isolated from landfill soil	**AML-C10**^{T} (= ATCC 700909^{T} = DSM 13736^{T}) [NZ_ ARCU00000000.1]	101
	M. lacus LW14	0.05	pMMO, RuMP, pSC	NA (6.8)	NA (6.8)		**LW14**^{T} (=ATCC BAA-1047^{T} = JCM 13284^{T}) [NZ_ AZUN00000000.1]	96
Methylothermus	*M. subterraneus* HTM55	0.3	pMMO, RuMP	5.2-7.5 (6)	37-65 (55-60)	Grows well on methane and methanol. Moderately-acidophilic, thermophilic strain	**HTM55**^{T} (=JCM 13664^{T} = DSM 19750^{T})	102
Methylovulum	*M. miyakonense* HT12	0.009	pMMO, sMMO, RuMP	6-7.5	5-34 (24)	High sensitivity to NaCl	**HT12**^{T} (=NBRC 106162^{T} = DSM 23269^{T} = ATCC BAA-2070^{T}) 106162^{T} = DSM 23269^{T} = ATCC BAA-2070^{c} [NZ_ AQZU00000000.1]	103
Group II								
Methylosinus	*M. trichosporium OB3b*	0.12	pMMO, sMMO, SC-EMP, H4MTP, H4FP.	6-8 (6.8)	5-37 (30)	Well-established catalysis for PHB production, epoxidation. Are heat-resistant, form spores	NZ_ ADVE00000000.2 **OB3b**, Dr. Stein culture collection (University of Alberta) [NZ_ ADVE00000000.2] PRJNA199154	79
	*Methylosinus sp.*LW4	0.12-0.15	pMMO, SC-EMP, H_4MTP, H_4F	6-8 (6.8)	5-37 (30)	Strain shows very robust growth	**LW4**, Dr. Lidstrom culture collections (University of Washington) [PRJNA199154]	96,104

(Continued)

Table 13.1: General summary for publicly available methanotrophic cultures—cont'd

Name	Species*	Growth parameters					Culture collection ID [Genome Accession Number (NCBI)]	Reference
		Growth rate (h^{-1})	Metabolic pathways	pH	T °C	Other		
Methylocystis	*M. rosea* SV97	NA	pMMO, SC-EMP, H_4MTP, H_4F	5.5-9.0	4-40 (30)		**SV97**^{T} (=DSM 17261^T=ATCC BAA-1196^T) [NZ_ ARCT00000000.1]	74,105
	Methylocystis Rockwell	0.16	pMMO, SC-EMP, H_4MTP, H_4F	NA (6.8)	NA (30)	Highly tolerant of ammonium, capable of nitrogen fixation	**ATCC 49242** [NZ_ AJTV00000000.1]	44,88
	M. parvus OBBp	0.12	pMMO, SC-EMP, H_4MTP, H_4F	6-8 (6.8)	5-37 (30)	Does not form spores, resistant to desiccation, but not heat. Tested for calcium carbonate precipitation	**OBBP** (NCIMB 11129) [AJTV00000000]	79,106
	Methylocystis sp. SC2	NA	pMMO, SC-EMP, H_4MTP, H_4F	NA	NA	Has low affinity pMMO	**SC2**, Dr. Liesack culture collection (Max Planck Institute) [HE956757/ FO000001 and FO000002]	107
	Methylocystis sp. SB2		pMMO, sMMO, SC-EMP, H_4MTP, H_4F	6-9 (6.8)	10-30 (30)	Facultative methanotroph with EMP pathway. Low CCE (12-13%)	**SB2**. Dr. Semrau lab culture collection (University of Michigan) [NZ_ AYNA00000000]	108
Methylocella	*M. silvestris*	0.02	sMMO, Mxa, SC-GS	4-7.5 (5.5)	4-30 (15-20)	Facultative methanotroph, can use short-chain alkanes	**BL2**^{T} (=DSM 15510^T=NCIMB 13906^T) [NC_011666.1]	109

Methylocapsa	*M. acidiphila B2*	0.03	sMMO, Mxa, SC-GS		3-6 (4.5)	Facultative methanotroph	**B2**^{T} (= DSM 13967^T = NCIMB 13765^T) [NZ_ ATYA00000000.1]	110
	M. aurea KYG	0.018	pMMO, Mxa, SC-GS	5.2-7.2	2-33	Facultative methanotroph. Capable of nitrogen fixation	**KYG**^{T} (=DSM 22158^T =VKM B-2544^T) [JQKO00000000.1]	111
Methyloferula	*M. stellata* AR4	0.005	sMMO, Mxa, SC-GS/CBB	3.5-7.2			**AR4**^{T} = DSM 22108^T = LMG 25277^T = VKM B-2543^T [NZ_ ARWA00000000.1]	112
Group III								
Methylacidiphilum	*M. fumariolicum* SolV	0.07-0.08	pMMO, Xox, CBB	0.8-5.8 (2)	40-65 (55)	Grows at very low pH. Capable of nitrogen fixation.	**SolV**. Dr. Op den Camp culture collection (Radboud University Nijmegen) [PRJEA85607]	89,113,114
	M. kamchatkensis Kam1	0.01	pMMO, Xox, CBB	2-5 (3)	40-65 (55)	Thermoacidophilic methanotroph	**Kam1**. culture collection (University of Bergen) [PRJNA169117]*	115
	M. infernorum V4	0.03	pMMO, Xox, CBB	1-6 (2)	50	Thermoacidophilic methanotroph	**V4, Dr**. Dunfield culture collection (University of Calgary) [NC_010794.1]	12
Methylacidimicrobium	*M. cyclopophantes*	0.042	pMMO, Xox, CBB	0.6-6 (1.5-3)	20-49 (44)	Mesophilic acidophilic methanotrophs	**3B**, Dr. Op den Camp culture collection (Radboud University Nijmegen)	116

(Continued)

Table 13.1: General summary for publicly available methanotrophic cultures—cont'd

Name	Species*	Growth parameters					Culture collection ID [Genome Accession Number (NCBI)]	Reference
		Growth rate (h^{-1})	Metabolic pathways	pH	T °C	Other		
	M. tartarophylax	0.035	pMMO, Xox, CBB	0.5-5 (1-3)	20-43 (38)	Mesophilic acidophilic methanotrophs	**4 AC**, Dr. Op den Camp culture collection (Radboud University Nijmegen)	116
	M. fagopyrum	0.013	pMMO, Xox, CBB	0.6-6 (1.5-3)	20-39 (35)	Mesophilic acidophilic methanotrophs	**3C**, Dr. Op den Camp culture collection (Radboud University Nijmegen)	116

**NCBI BioProject Accession number; pSC, partial Serine cycle; NA, data not available.

All of these pathways can be identified in genomes of most methanotrophic Gammaproteobacteria (but S. and Kalyuzhnaya M.G., unpublished). However, it has been shown that most C_1 carbon (about 75%) enters the glycolytic route in two methanotrophic strains, *Methylomicrobium alcaliphilum* 20Z and *Methylomonas* sp. LW13.[47] The FBP variant of the RuMP is the most efficient pathway for C_1 assimilation (Figure 13.3a). It enables the conversion of 3 mol of formaldehyde into 1 mol of a three-carbon compound (assumed to be 2-phosphoglycerate for the convenience of balancing) with the generation of 1 mol NADH and 0.7 mol ATP.[47] In *M. alcaliphilum* 20Z and *Methylomonas* sp. LW13 part of the cell carbon (about 25%) is assimilated via the KDPG variant (Figure 13.3b), which requires 2 mol of ATP and generates 1 mol NADH upon assimilation of 3 mol of formaldehyde into 2-phosphoglycerate.

A number of phosphoketolases (EC 4.1.2.9, Xfp), a key enzyme of the bifidobacterial shunt, have been detected in all sequenced Gammaproteobacteria (but S. and Kalyuzhnaya M.G., unpublished). Their function in methanotrophs is not clear yet, but it could be predicted that methanotrophic phosphoketolases contribute to C_2-compound production (Figure 13.3c) and might play a role in C_1 assimilation and/or acetate production upon oxygen limitation. The biotechnological potential of the Xfp-based synthetic pathways has been highlighted in recent publications by Bogodar et al.[118,119]

Hps/Hpi-driven assimilation seems to be an ideal starting point for sugar-based technologies aiming to expand application to natural gas or methane resources. The key genes of the pathway have been identified in two groups of methanotrophs: Gammaproteobacteria and ANME.[1,52,54] It has been shown that in methanogenic archaea, the enzymes operate in reverse direction to produce R5P needed for nucleic acid synthesis.[120,121] Nothing is known about Hps/Hpi function in anaerobic methane utilizers, except that members of the ANME consortia express the enzymes.[54] It is tempting to speculate that these enzymes play a role in C_1 assimilation in AnMO.

13.4.2 [CH_2]- to Biomass Conversion

It has been suggested that direct nonenzymatic condensation of formaldehyde with tetrahydrofolate (H_4F) contributes to methylene-H_4F formation, an entry metabolite for the SC (Figure 13.4).

However, the kinetics of the chemical condensation is quite unfavorable.[69] It is predicted that the assimilation of C_1 units derived from methane or methanol via the SC involves the H_4F-linked transfer and starts from formate.[69] The SC in methanotrophs can be supported by the glyoxylate cycle (GC)[112,122] or the ethylmalonyl-CoA pathway (EMC)[48,123] for glyoxylate regeneration (Figure 13.4).

The biotechnological potential of the EMC pathway is well presented by Alber.[124] All benefits of the EMC conversion could be even further extending by the use of methanotrophs. According to genomic evidence, the SC-EMC pathway seems to be the most common arrangement of the C_1 assimilation in Alphaproteobacterial methanotrophs.[48,104,123] It has been shown that a significant portion of cell carbon in *Methylosinus thrichosporium* OB3b, a model system, comes from CO_2 (over 60%), suggesting that the EMC pathway plays a significant role in assimilation.[48] The overall balance of assimilation via the H_4F, SC, and EMC pathways is: 1.5 mol of formaldehyde and 1.5 mol of CO_2 to make 1 mol of 2-phosphoglycerate, using 2.5 mol of NADH and 4.5 mol of ATP and generating 0.5 mol of ubiquinol (QH_2).

In some microbes, such as *Methylocella*, *Methylocapsa*, and *Methyloferula*, the SC operates in combination with the glyoxylate shunt.[122,125]

The activity of two enzymes of the SC, serine glyoxylate aminotransferase (EC 2.6.1.45) and hydroxypyruvate reductase (EC 1.1.1.81), have been detected in some Gammaproteobacterial methanotrophs.[117] According to available genomic data, all Gammaproteobacterial methanotrophs possess a complete or partial SC; however, none of them have a glyoxylate regeneration pathway. The functional significance of the pathways in C_1 assimilation is not well established. The incomplete SC cannot function as a main assimilatory route and the pathway most likely contributes to biosynthesis of serine or to the conversion of formate into acetyl-CoA (Figure 13.4).

13.4.3 CO_2 to Biomass Conversion

So far only the Calvin-Benson-Bassham (CBB) cycle has been confirmed by experimental measurements as the main assimilatory route in Verrucomicrobial methanotrophs grown in aerobic/microaerobic environments[113] and NC10-phylum grown in anaerobic systems (Figure 13.3c).[126]

Some members of the Group I (*Methylococcus* and *Methylocaldum*) and Group II (*Methyloferula*) are also capable of CO_2 fixation via the CBB.[77,127,128] The CO_2-fixation reactions in these microbes complement other assimilatory routes, contributing to assimilation, energy balancing, and replenishment of some common intermediates of the pentose-phosphate pathway. The CBB cycle would require nine ATP and six NAD(P)H per three-carbon compound.

13.4.4 Other Possible Assimilation Arrangements

Among other possible arrangements of assimilation in methane-consuming bacteria and/or their metabolic partners are: the reductive acetyl-CoA (Wood-Ljungdahl) pathway, which is predicted for oxygen-independent methanotrophy[54]; the reductive tricarboxylic acid cycle (complete or partial), predicted for some Gammaproteobacterial methanotrophs[45]; and the dicarboxylate/4-hydroxybutyrate, the 3-hydroxypropionate/malyl-CoA, and the 3-hydroxypropionate/4-hydroxybutyrate cycles.[72,73] It remains to be seen whether these pathways are used by natural methanotrophs.

13.5 Selecting the Right Microbe: Available Cultures

This part of the chapter is focused primarily on aerobic methanotrophs, as unfortunately no pure microbial cultures capable of carrying out AnMO are available as laboratory strains. Examples of methanotrophic bacteria that could be considered for industrial applications are shown in Table 13.1.

The summary includes microbial cultures that are currently available from culture collections or laboratory stocks. Some specific parameters of different species incorporated in the description of strain are growth rate, methanotrophic network arrangement, and additional unique phenotypic characteristics. It should be remembered that growth characteristics vary significantly even among one species.

Aerobic methane utilization has been described for Gammaproteobacteria (Group I), Alphaproteobacteria (Group II), and Verrucomicrobia (Group III) and predicted for the Deltaproteobacteria (SAR324-clade, a yet-uncultivated marine bacterium).[129] Only members of Group I, II, and III are currently available in pure culture.

13.5.1 Group I: Gammaproteobacterial Methanotrophs

Gammaproteobacterial methanotrophs represent one order, Methylococcales, that includes three families: Crenotrichaceae, represented by one filamentous bacterium *Crenothrix polyspora* Cohn 1870, yet to be isolated in pure culture[130]; *Methylococcaceae*, which currently includes 14 genera (Table 13.1); and *Methylothermaceae*, which includes three genera.[131] Partial or complete genomes are currently available for cultivated representatives of at least 14 genera, with several additional strains in sequencing pipelines (Table 13.1). So far all characterized Gammaproteobacterial strains demonstrate high activity of the ribulose-monophosphate pathway enzymes (Figure 13.3).

Many members of the class are capable of nitrogen fixation[100,127] and almost all grow equally well on methane and methanol. Most methanotrophs of this group accumulate glycogen (up to 60% of cell biomass) as a carbon-storage compound.[132] There are some reports of PHB accumulation, which should be taken with some caution. The lipid-dye test typically used for the detection of PHB might also stain intracellular membrane compartments (ICMs) formed by many members of Group I. So far, PHB-biosynthesis genes have been found only in the genome of *Methylocaldum szegediense* O-12 (Dr. Khmelenina V.N, personal communication).

Representatives of Group I include acidophilic/acidotolerant,[133,134] alkaliphilic/alkalitolerant,[91–93] halophilic/halotolerant,[85,86,135] and thermophilic/thermotolerant[77,87,102,136] species. Two members of the class, *M. capsulatus* and *Methylomonas* spp. are well-established catalysts for SCP,[2,137,138] carotenoids,[6,7] and exopolysaccharides/glycogen production.[7] The biotechnological potential of members of the group has been broadly discussed.[4,5,139] It has been shown that Gammaproteobacterial methanotrophs are capable of methane fermentation, which expands the applicability of methane conversion from production of biomass-based products to synthesis of chemicals.[47] Many Group I methanotrophs are capable of producing commercially valuable osmoprotectors, such as ectoine and sucrose.[78,91,140] The *M. alcaliphilum* 20Z is the first nonphototrophic microbe to be investigated for direct conversion of methane into extractable fermentation-ready sucrose.[8]

The drawbacks of the group include unstable growth, spontaneous cell lysis, and sensitivity to ammonia.[88] Some species (such as *Methylomicrobium* and *Methylococcus*) seem to be more robust than others (*Methylomonas* and *Methylobacter*). The obligate methanotrophic nature of all members of the group also constrains modern metabolic engineering and requires the development of novel out-of-the-box approaches for the alteration of metabolite fluxes.

13.5.2 Group II: Alphaproteobacterial Methanotrophs

Methanotrophic Alphaproteobacteria are classified into two families: Beijerinckiaceae (genera *Methylocella*, *Methylocapsa*, and *Methyloferula*) and Methylocystaceae (genera *Methylocystis* and *Methylosinus*) in the order Rhizobiales. A summary of a few strains that show good growth is shown in Table 13.1. The drawbacks of the group include quite slow growth (6-8 h for the fastest cultures) and low CCEs (63-64). The main advantages of the group include high resistance to desiccation, more robust growth in the lab, and much greater metabolic flexibility. Many members of the class are capable of nitrogen fixation and some are facultative methanotrophs, utilizing ethanol, acetate, and short-chain alkanes.[141] Carbon assimilation occurs via the SC (GC variant is typical for Beijerinckiaceae; EMC variant is typical for Methylocystaceae members). Some members of the group show high activities of the Calvin cycle enzyme.[112] Two members of the group, *M. thrichosporium* OB3b and *Methylocystis* spp. have extensively been studied for PHB production,[3,142] epoxidation,[11] methanol,[143] bioleaching,[9] calcium carbonate precipitation,[10] and bioremediation.[15] The ability of *Methylocella silvestris* to use short-chain alkanes further expands its biotechnological potential.[141] The Group II methanotrophs are capable of fermenting methane and excreting acetate, succinate, acetone, 2,3-butanediol, and isopropanol.[144,145] It has been predicted that these products are derived from PHB, but the exact mechanisms and pathways are not yet understood.

13.5.3 Group III: Verrucomicrobial Methanotrophs

Verrucomicrobial methanotrophs are a relatively new clade of methane utilizers, which currently includes three genera: *Methylacidiphilum*,[12,146,147] *Methylacidimicrobium*,[116] and *Methyloacida*.[115] These microbes typically inhabit geothermal and acidic environments, where they thrive at pH 0.5-5 (optima 2) and temperature 30-65°C (optima between 35-50°C). Unlike proteobacterial methanotrophs, which prefer to assimilate carbon from reduced forms of C_1 carbon, all Group III members are autotrophs and use methane only as a source of energy, burning it all the way to CO_2. Despite all theoretical predictions of low yield and low CCE for CBB conversion, the Verrucomicrobial methanotrophs show exceptionally good growth parameters, comparable to the growth of other methanotrophs (Table 13.1).

Many members are capable of nitrogen fixation[148] and accumulate glycogen as a storage compound.[149] Methanotrophic Verrucomicrobia are the only methanotrophs capable of growth at low pH, relatively high osmolality, and high temperature. All these parameters are ideal for the fermentation of organic acids from methane. The biotechnological applications of the group are limited by the lack of appropriate tools for genetic alterations. As this relatively new area of methanotrophy is quickly developing,[89,115,116,146–149] significant progress can be expected in near future.

13.5.4 NC10

While the NC10-phylum members appear to thrive in anaerobic ecosystems with a substantial input of nitrite, at the metabolic level the microbes resemble canonical aerobic methanotrophs.[150] NC10 is predicted to assimilate carbon via the CBB cycle. The only exception of the methane network in NC10 is the presence of a yet-to-be characterized enzyme for intracellular production of oxygen from NO.[151] Despite the aerobic machinery for methane oxidation, the growth and methane oxidation kinetics of NC10 are well below AnMO-SR, and are similar to other AnMO-DR processes. Lack of pure strains and genetic tools, its requirement for NO_2, and its slow kinetics are obviously downsides of the group. However, the microbes can be a source of novel enzymes, such as its enzymatic system responsible for intracellular oxygen production. Once fully established, its enzymatic system could be incorporated into aerobic microbes to improve mass transfer upon gas fermentation or to increase the rate of oxygen-dependent conversion in other biotechnological applications.

13.5.5 ANME

Lack of pure cultures, slow kinetics, and exceptionally slow growth rates (typically below 0.001 h^{-1}) makes the anaerobic methanotrophic communities quite a challenging metabolic platform. Furthermore, anaerobic methanotrophic consortia show the lowest CCE, covering only 0.25-1.3 mol % of the methane oxidized.[54,152] Little to nothing is known about optimal cultivation parameters of ANME, and no genetic strategies for metabolic alterations exist. ANME cocultures have been established in a few laboratories, providing some background data to navigate cultivation attempts.[54,55] Nevertheless, it should be remembered that while it appears to be mission impossible for canonical industrial applications, the AnMO might be a breakthrough in C-H activation.[17] Metagenomic and metatranscriptomic data are available for some bioreactor studies, and those can represent *in silico* resources for synthetic applications and computational simulations for metabolic engineering of C_1 metabolism in alternative hosts.

13.6 Looking for a New Microbe: Isolation Strategies

It is always possible that the ideal microbe for biocatalysis still hides in nature. It is hard to cover all possible isolation strategies, as they are typically directed by the parameters relevant to technology of interest. However, a few general concepts can be outlined. As methanotrophic bacteria are widespread in nature, it is the best to find a natural environment which is similar to required cultivation conditions. Isolation of methanotrophic bacteria typically occurs in three stages: (1) enrichment culture; (2) active mixed culture; and (3) pure culture(s).

13.6.1 Enrichment and Mixed Cultures

The initial enrichment cultures should be set up using conditions close to the environmental setting. It is best to use water samples/extracts directly from the source. The enrichment culture should be set up in closed vials with inoculum (10-50% of the enrichment culture) and methane (10-20% of headspace). If an environmental water sample is not available or limited, key environmental parameters such as pH, temperature, and salinity should be precisely represented. It is also best to keep the concentration of oxygen as close to the environmental setting as possible. If parameters are not known for the isolation of aerobe microbes, the oxygen concentration should be kept at 5-10% (of headspace). It is best to monitor the enrichment process by tracing methane, oxygen, and CO_2 using GC-chromatography. When 20-50% of added methane is consumed, the culture should be transferred into a new vial containing higher amounts of nutrients. The cultivation medium should replicate environmental sample parameters, with an excess of main nutrients, such as nitrate, phosphate, sulfate, copper, and iron. This can be achieved by adding mineral medium (see Table 13.2) or similar media like Medium 632 and 1179 (http://www.dsmz.de) to the filtered water sample/extract from the environment at a 5:100 ratio.

The ratio can be increased to 50:50 and completely replaced with just medium in follow-up cultures. The growth of cultures should be monitored by optical density and gas consumption. This approach should lead to an active mixed culture.

13.6.2 Pure Culture Isolation

While getting an enrichment culture of methanotrophic bacteria is always quite easy, it is not always an easy task to separate a pure culture out of the well established, usually 3-10 species, consortia. A number of tricks have been used for identification and isolation of methanotrophs[153,154]; see also methanotroph.org. In some cases, the interactions between methanotrophs and satellite cultures are quite strong. One of the tricks for the separation of a methanotrophic strain is cultivation on a mixed-culture filtrate. To prepare the filtrate, the mixed culture should be grown to mid-exponential phase. The liquid phase should be separated from the cells by filtration via a 0.2-0.02 μm filter. The filtrate can then be used for serial dilutions. To make solid medium, the filtrate should be mixed with hot sterilized

Table 13.2: An example of growth medium for methanotrophic bacteria isolation and cultivation

Compounds	Amount	Comments
KNO_3	1 g	Autoclave at 121 °C for 30 min. Cool. Add phosphate solution
$MgSO_4 \times 7H_2O$	0.2 g	
$CaCl_2 \times 2H_2O$	0.02 g	
Trace solution	1 ml	
Phosphate solution	20 ml	
Distilled deionized water	1 L	
Trace solution	g/L	Autoclave at 121 °C for 20 min. After autoclaving the solution becomes pink
Na_2EDTA	5	
$FeSO_4 \times 7H_2O$	2	
$ZnSO_4 \times 7H_2O$	0.3	
$MnCl_2 \times 4H_2O$	0.03	
$CoCl_2 \times 6H_2O$	0.2	
$CuSO_4 \times 5H_2O$	0.6	
$NiCl_2 \times 6H_2O$	0.05	
$Na_2MoO_4 \times 2H_2O$	0.05	
H_3BO_3	0.03	
Phosphate solution	g/L	pH of the solution should be 6.8-7. Autoclave at 121 °C for 20 min
KH_2PO_4	5.44	
Na_2HPO_4	5.68	

medium containing 5% agar, at a 5:1 ratio. The isolation of a pure culture can be achieved by serial dilution and/or by spread plating on solid media.

13.7 Concluding Remarks

While aerobic methanotrophs are often seen as the most credible systems for methane biocatalysis, it should be kept in mind that natural microbes typically do not experience selective pressure for high rate/high CCE of methane utilization. In most environments, methanotrophic bacteria are part of a very complex community, and their functions realized under different environmental constrains such as nutrient limitation, competition, synergistic interactions, or predation. Such wild cultures possess specialized metabolic networks (which might rely on activities of other members of the community) and complex signaling mechanisms (which activate in response to high population density), and those might lead to granulation or lysis. Obviously, these factors might impact the performance of the selected strain in "man-made" settings or upon industrial scale-up.

When comparing all known clades of methane utilizers, it is not surprising that Gammaproteobacterial methanotrophs are the winners. They have the highest growth rate and the highest CCE of methane utilization, which is not surprising when their highly efficient arrangement of methane utilization is taken into account. However, both Alphaproteobacteria and Verrucomicrobia offer some additional advances that keep them on the consideration list: (1) only Verrucomicrobial methanotrophs can tolerate extremely low pH and they are an attractive system for the conversion of methane sources into organic acid and (2) so far only Alphaproteobacterial methanotrophs have shown metabolic flexibility. They are able to use other sources of carbon in addition to C_1 compounds. That makes them a unique system for the manipulation of C_1 machinary for biological production of methanol from methane. In addition, the ability of some members of the group to co-utilize short alkanes has obvious benefits for natural gas conversion. Finally, these microbes have a high flux of carbon into CoA-derivatives, such as acetyl-CoA and crotonyl-CoA.

A number of pathways have been described for anaerobic methane consumption, but none of them can be driven by a single pure culture. The methane oxidation in the NC10 phylum is driven by aerobic MMO, but mechanisms supporting how the cells generate oxygen remain unclear. The slow rate of methane oxidation, the lack of pure cultures, the need for expensive electron acceptors, the generation of toxic by-products (such as H_2S or ammonia), and the need for conceptually new approaches for genetic manipulation restrict the large-scale deployment of anaerobic methane conversion today. Nevertheless, these novel microbes represent a very valuable source of enzymes for the synthetic biology applications.

Acknowledgments

The development of this review was funded by a grant from ARPA-E (0670-5169). The author thanks Dr. Derek Griffin and Dr. Qiang Fei for their insightful suggestions on commodity industrial prices.

References

1. Anthony C. *The biochemistry of methylotrophs*. London, New York: Academic Press; 1982.
2. Winder R. Methane to biomass: Norferm, Norway, uses *Methylococcus capsulatus* bacteria to convert methane to bioprotein in a continuous loop fermentation process. *Chem Ind* 2004;. http://findarticles.com/p/articles/mi_hb5255/is_17/ai_n29124852/.
3. Doronina NV, Ezhov VA, Trotsenko YuA. Growth of *Methylosinus trichosporium* OB3b on methane and poly-beta-hydroxybutyrate biosynthesis. *Appl Biochem Microbiol* 2008;**44**:182–5.
4. Strong PJ, Xie S, Clarke WP. Methane as a resource: cam the methanotrophs ass value? *Environ Sci Technol* 2015;**49**(9):4001–18. http://dx.doi.org/10.1021/es504242n.
5. Fei Q, Guarnieri MT, Tao L, Laurens LM, Dowe N, Pienkos PT. Bioconversion of natural gas to liquid fuel: opportunities and challenges. *Biotechnol Adv* 2014;**32**:596–614.
6. Ye RW, Kelly K. Construction of carotenoid biosynthetic pathways through chromosomal integration in methane-utilizing bacterium *Methylomonas* sp. strain 16a. *Methods Mol Biol* 2012;**892**:185–95.
7. Koffas M, Odom JM, Schenzle A. Methane or methanol as sole carbon source; Embden-Meyerhof carbon pathway comprising a gene encoding a pyrophosphate dependent phosphofructokinase; biomass contains protein, carbohydrates and pigment on industrial scale. US6958222 B2, 2005.
8. Kalyuzhnaya MG, Lidstrom ME, But S, Khmelenina VN. Conversion of methane into chemicals via dry fermentation. US Provisional Patent Application 61/993,636, 2015.
9. DiSpirito AA, Choi DD, Semrau JD, Keeney D. Use of methanobactin. US7932052 B1, 2011.
10. Ganendra G, Muynck De, Ho A, Arvaniti EC, Hosseinkhani B, Ramos JA, et al. Formate oxidation driven calcium carbonate precipitation by *Methylocystis parvus* OBBP. *Appl Environ Microbiol* 2014;**80**:4659–67.

11. Hou CT. Oxidation of gaseous hydrocarbons by methanotrophs: heterogeneous bioreactor. *Ann N Y Acad Sci* 1984;**434**:541–8.
12. Hou S, Makarova KS, Saw JHW, Senin P, Ly BV, et al. Complete genome sequence of the extremely acidophilic methanotroph isolate V4, *Methylacidiphiluminfernorum*, a representative of the bacterial phylum Verrucomicrobia. *Biol Direct* 2008;**3**:26.
13. Jiang H, Chen Y, Jiang P, Zhang C, Smith TJ, Murrell JC, et al. Methanotrophs: multifunctional bacteria with promising applications in environmental bioengineering. *Biochem Eng J* 2010;**49**:277–88.
14. Strandberg GW, Donaldson TL, Farr LL. Degradation of trichloroethylene and trans-1,2-dichloroethylene by a methano tropic consortium in a fixed-film, packed-bed bioreactor. *Environ Sci Technol* 1989;**23**:1422–5.
15. Semrau JD. Bioremediation via methanotrophy: overview of recent findings and suggestions for future research. *Front Microbiol* 2011;**2**:209.
16. Quayle JR. Introduction. In: Goldberg I, Rokem JS, editors. *Biology of methylotrophs*. Madison: Butterworth-Heinemann, Science Tech Publishers; 1991.
17. Conrado RJ, Gonzales R. Envisioning the bioconversion of methane to liquid fuels. *Science* 2014;**343**:621–3.
18. Haynes CA, Gonzalez R. Rethinking biological activation of methane and conversion to liquid fuels. *Nat Chem Biol* 2014;**10**:331–9.
19. Silverman J. *BioGTL Platform for the conversion of natural gas to fuels and chemicals*. In: AIChE Spring meeting; 2014. http://calystaenergy.com/pdfs/AIChE_final_33114.pdf.
20. EPA reports. *Municipal solid waste generation, recycling, and disposal in the united states*; 2008. http://www.epa.gov/epawaste/nonhaz/municipal/index.htm.
21. EPA reports. *U.S. Greenhouse gas inventory report*; 2010. http://epa.gov/climatechange/emissions/usinventoryreport.html.
22. Shindell D, Kuylenstierna JCI, Vignati E, Dingenen R, Amann M, Klimont Z, et al. Simultaneously mitigating near-term climate change and improving human health and food security. *Science* 2012;**335**:183–9.
23. IPCC report. *Climate change 2013: the physical science basis*; 2013; http://www.ipcc.ch/report/ar5/wg1/.
24. IPCC report. *Climate change 2014: impacts, adaptation, and vulnerability*; 2014. http://www.ipcc.ch/report/ar5/wg2/.
25. Caldwell SL, Laidler JR, Brewer EA, Eberly JO, Sandborgh SC, Colwell FS. Anaerobic oxidation of methane: mechanisms, bioenergetics, and the ecology of associated microorganisms. *Environ Sci Technol* 2008;**42**:6791–9.
26. Knittel K, Boetius A. Anaerobic oxidation of methane: progress with an unknown process. *Annu Rev Microbiol* 2009;**63**:311–34.
27. Rao P, Muller M. Industrial oxygen: its generation and use. *ACEEE summer study on energy efficiency in industry*; 2007. http://aceee.org/files/proceedings/2007/data/papers/78_6_080.pdf.
28. Friedrich CG, Bardischewsky F, Rother D, Quentmeier A, Fischer J. Prokaryotic sulfur oxidation. *Curr Opin Microbiol* 2005;**8**:253–9.
29. Kartal B, de Almeida NM, Maalcke WJ, Op den Camp HJM, Jetten MSM, Keltjens JT. How to make a living from anaerobic ammonium oxidation. *FEMS Microbiol Rev* 2013;**37**:428–61.
30. Semrau JD, DiSpirito AA, Yoon S. Methanotrophs and copper. *FEMS Microbiol Rev* 2010;**34**:496–531.
31. Hakemian AS, Rosenzweig AC. The biochemistry of methane oxidation. *Annu Rev Biochem* 2007;**76**:223–41.
32. Dalton H. The Leeuwenhoek Lecture 2000 the natural and unnatural history of methane-oxidizing bacteria. *Philos Trans R Soc Lond B Biol Sci* 2005;**360**:1207–22.
33. Goodwin MG, Anthony C. Characterization of a novel methanol dehydrogenase containing a Ba^{2+} ion at the active site. *Biochem J* 1996;**318**:673–9.
34. Goodwin PM, Anthony C. The biochemistry, physiology and genetics of PQQ and PQQ-containing enzymes. *Adv Microb Physiol* 1998;**40**:1–80.
35. Pol A, Barends TR, Dietl A, Khadem AF, Eygensteyn J, Jetten MSM, et al. Rare earth metals are essential for methanotrophic life in volcanic mudpots. *Environ Microbiol* 2014;**16**:255–64.

36. Kim HG, Han GH, Kim D, Choi J-S, Kim SW. Comparative analysis of two types of methanol dehydrogenase from *Methylophaga aminisulfidivorans* MPT grown on methanol. *J Basic Microbiol* 2012;**52**:141–9.
37. Choi JM, Kang JH, Lee DW, Kim SW, Lee SH. Crystallization and preliminary X-ray crystallographic analysis of MxaJ, a component of the methanol-oxidizing system operon from the marine bacterium *Methylophaga aminisulfidivorans* MPT. *Acta Crystallogr* 2013;**F69**:902–5.
38. Van Spanning RJM, Wansell CW, De Boer T, Hanelaar MJ, Anazawa H, Harms N, et al. Isolation and characterization of the *moxJ*, *moxG*, *moxI*, and *moxR* genes of *Paracoccus denitrificans*: Inactivation of *moxJ*, *moxG*, and *moxR* and the resultant effect on methylotrophic growth. *J Bacteriol* 1991;**173**:6948–61.
39. Khmelenina VN, Kalyuzhnaya MG, Starostina NG, Suzina NE, Trotsenko YuA. Isolation and characterization of halotolerant alkaliphilic methanotrophic bacteria from Tuva soda lakes. *Curr Microbiol* 1997;**35**:257–61.
40. Patel RN, Hou CT, Derelanko P, Felix A. Purification and properties of a heme-containing aldehyde dehydrogenase from *Methylosinus trichosporium*. *Arch Biochem Biophys* 1980;**203**:654–62.
41. Speer BS, Chistoserdova L, Lidstrom ME. Sequence of the gene for a NAD(P)-dependent formaldehyde dehydrogenase (class III alcohol dehydrogenase) from a marine methanotroph *Methylobacter marinus* A45. *FEMS Microbiol Lett* 1994;**121**:349–55.
42. Zahn JA, Bergmann DJ, Boyd JM, Kunz RC, DiSpirito AA. Membrane-associated quinoprotein formaldehyde dehydrogenase from *Methylococcus capsulatus* Bath. *J Bacteriol* 2001;**183**:6832–40.
43. Stein LY, Yoon S, Semrau JD, Dispirito AA, Murrell JC, Vuilleumier S, et al. Genome sequence of the obligate methanotroph, *Methylosinus trichosporium* strain OB3b. *J Bacteriol* 2010;**192**:6497–8.
44. Stein LY, Bringel F, Dispirito AA, Han S, Jetten MS, Kalyuzhnaya MG, et al. Genome sequence of the methanotrophic Alphaproteobacterium, *Methylocystis* sp. Rockwell (ATCC 49242). *J Bacteriol* 2011;**193**:2668–9.
45. Vuilleumier S, Khmelenina VN, Bringel F, Reshetnikov AS, Lajus A, Mangenot S, et al. Genome sequence of the haloalkaliphilic methanotrophic bacterium *Methylomicrobium alcaliphilum* 20Z. *J Bacteriol* 2012;**194**:551–2.
46. Ward N, Larsen O, Sakwa J, Bruseth L, Khouri H, Durkin AS, et al. Genomic insights into methanotrophy: the complete genome sequence of *Methylococcus capsulatus* (Bath). *PLoS Biol* 2004;**2**:303.
47. Kalyuzhnaya MG, Yang S, Rozova ON, Smalley NE, Clubb J, Lamb A, et al. Highly efficient methane biocatalysis revealed in methanotrophic bacterium. *Nat Commun* 2013;**4**:2785.
48. Matsen JB, Yang S, Stein LY, Beck D, Kalyuzhnaya MG. Global molecular analyses of methane metabolism in methanotrophic Alphaproteobacterium, *Methylosinus trichosporium* OB3b. Part I. Transcriptomic study. *Front Microbiol* 2013;**7**:1–13.
49. Keltjens JT, Pol A, Reimann J, Op den Camp HJ. PQQ-dependent methanol dehydrogenases: rare-earth elements make a difference. *Appl Microbiol Biotechnol* 2014;**98**:6163–83.
50. Chistoserdova L. Modularity of methylotrophy, revisited. *Environ Microbiol* 2011;**13**:2603–22.
51. Vorholt JA, Chistoserdova L, Stolyar SM, Thauer RK, Lidstrom ME. Distribution of tetrahydromethanopterin-dependent enzymes in methylotrophic bacteria and phylogeny of methenyl tetrahydromethanopterin cyclohydrolases. *J Bacteriol* 1999;**181**:5750–7.
52. Trotsenko YuA, Murrell JC. Metabolic aspects of aerobic obligate methanotrophy. *Adv Appl Microbiol* 2008;**63**:183–229.
53. Jollie DR, Lipscomb JD. Formate dehydrogenase from Methylosinus trichosporium OB3b. Purification and spectroscopic characterization of the cofactors. *J Biol Chem* 1991;**266**:21853–63.
54. Haroon MF, Hu S, Shi Y, Imelfort M, Keller J, Hugenholtz P, et al. Anaerobic oxidation of methane coupled to nitrate reduction in a novel archaeal lineage. *Nature* 2013;**500**:567–70.
55. Milucka J, Ferdelman TG, Polerecky L, Franzke D, Wegener G, Schmid M, et al. Zero-valent sulphur is a key intermediate in marine methane oxidation. *Nature* 2012;**491**:541–6.
56. Nauhaus K, Boetius A, Krüger M, Widdel F. *In vitro* demonstration of anaerobic oxidation of methane coupled to sulphate reduction in sediment from a marine gas hydrate area. *Environ Microbiol* 2002;**4**:296–305.

57. Nauhaus K, Albrecht M, Elvert M, Boetius A, Widdel F. *In vitro* cell growth of marine archaeal-bacterial consortia during anaerobic oxidation of methane with sulfate. *Environ Microbiol* 2007;**9**:187–96.
58. Beal EJ, House CH, Orphan VJ. Manganese- and iron-dependent marine methane oxidation. *Science* 2009;**325**:184–7.
59. Scheller S, Goenrich M, Boecher R, Thauer RK, Jaun B. The key nickel enzyme of methanogenesis catalyses the anaerobic oxidation of methane. *Nature* 2010;**465**:606–8.
60. Chistoserdova L, Vorholt J, Lidstrom ME. A genomic view of methane oxidation by aerobic bacteria and anaerobic archaea. *Genome Biol* 2005;**6**:208.
61. Leak DJ, Dalton H. Growth yields of methanotrophs. *Appl Microbiol Biotechnol* 1986;**23**:470–6.
62. Leak DJ, Dalton H. Growth yields of methanotrophs 2. A theoretical analysis. *Appl Microbiol Biotechnol* 1986;**23**:477–81.
63. Park S, Hanna LM, Taylor RT, Droege MW. Batch cultivation of *Methylosinus trichosporium* OB3b. I: Production of soluble methane monooxygenase. *Biotechnol Bioeng* 1991;**38**:423–33.
64. Park S, Shah NN, Taylor RT, Droege MW. Batch cultivation of *Methylosinus trichosporium* OB3b: II. production of particulate methane monooxygenase. *Biotechnol Bioeng* 1992;**40**:151–7.
65. Girguis PR, Orphan VJ, Hallam SJ, DeLong EF. Growth and methane oxidation rates of anaerobic methane-oxidizing archaea in a continuous-flow bioreactor. *Appl Environ Microbiol* 2003;**69**:5472–82.
66. Girguis PR, Cozen AE, DeLong EF. Growth and population dynamics of anaerobic methane-oxidizing archaea and sulfate-reducing bacteria in a continuous-flow bioreactor. *Appl Environ Microbiol* 2005;**71**:3725–33.
67. Lawrence AJ, Quayle JR. Alternative carbon assimilation pathways in methane-utilizing bacteria. *J Gen Microbiol* 1970;**63**:371–4.
68. Anthony C. How half a century of research was required to understand bacterial growth on C_1 and C_2 compounds; the story of the serine cycle and the ethylmalonyl-CoA pathway. *Sci Prog* 2011;**94**:109–37.
69. Crowther GJ, Kosály G, Lidstrom ME. Formate as the main branch point for methylotrophic metabolism in *Methylobacterium extorquens* AM1. *J Bacteriol* 2008;**190**:5057–62.
70. Large PJ, Quayle JR. Microbial growth on C(1) compounds. 5. Enzyme activities in extracts of *Pseudomonas* AM1. *Biochem J* 1963;**87**:386–96.
71. Peyraud R, Kiefer P, Christen P, Massou S, Portais JC, Vorholt JA. Demonstration of the ethylmalonyl-CoA pathway by using ^{13}C metabolomics. *Proc Natl Acad Sci* 2009;**106**:4846–51.
72. Berg IA. Ecological aspects of the distribution of different autotrophic CO2 fixation pathways. *Appl Environ Microbiol* 2011;**77**:1925–36.
73. Sato T, Atomi H. Microbial inorganic carbon fixation. *Encyclopedia of life science (eLS)*, Chichester, UK: Wiley; 2010, http://dx.doi.org/10.1002/9780470015902.a0021900, p. 1–12.
74. Starostina NG, Pashkova NI, Tsiomenko AB. Detection and partial characterization of bacteriocin in the methanotrophic bacterium *Methylobacter bovis*. *Biochemistry* 1998;**63**:1122–5.
75. Svenning MM, Hestnes AG, Wartiainen I, Stein LY, Klotz MG, Kalyuzhnaya MG, et al. Genome sequence of the Arctic methanotroph *Methylobacter tundripaludum* SV96. *J Bacteriol* 2006;**193**:6418–9.
76. Wartiainen I, Hestnes AG, McDonald IR, Svenning MM. *Methylobacter tundripaludum* sp. nov., a methane-oxidizing bacterium from Arctic wetland soil on the Svalbard islands, Norway (78° N). *Int J Syst Evol Microbiol* 2006;**56**:109–13.
77. Eshinimaev BTs, Medvedkova KA, Khmelenina VN, Suzina NE, Osipov GA, Lysenko AM, et al. New thermophilic methanotrophs of the genus *Methylocaldum*. *Mikrobiologiia* 2004;**73**:448–56.
78. Medvedkova KA, Khmelenina VN, Trotsenko YuA. Sucrose as a factor of thermal adaptation of the thermophilic methanotroph *Methylocaldum szegediense* O-12. *Mikrobiologiia* 2007;**76**:500–2.
79. Whittenbury R, Phillips KC, Wilkinson JF. Enrichment, isolation and some properties of methane-utilizing bacteria. *J Gen Microbiol* 1970;**61**:205–18.
80. Foster JW, Davis RH. A methane-dependent coccus, with notes on classification and nomenclature of obligate, methane-utilizing bacteria. *J Bacteriol* 1966;**91**:1924–31.
81. Kleiveland CR, Hult LT, Kuczkowska K, Jacobsen M, Lea T, Pope PB. Draft genome sequence of the methane-oxidizing bacterium *Methylococcus capsulatus* (Texas). *J Bacteriol* 2012;**194**:6626.

82. Geymonat E, Ferrando L, Tarlera SE. *Methylogaea oryzae* gen. nov., sp. nov., a mesophilic methanotroph isolated from a rice paddy field. *Int J Syst Evol Microbiol* 2011;**61**:2568–72.
83. Poehleina A, Deutzmannb JS, Daniel R, Simeonova DD. Draft genome sequence of the methanotrophic gammaproteobacterium *Methyloglobulus morosus* DSM 22980 Strain KoM1. *Genome Announc* 2013;**1**:e01078-13.
84. Deutzmann JS, Hoppert M, Schink B. Characterization and phylogeny of a novel methanotroph, *Methyloglobulus morosus* gen. nov., spec. nov. *Syst Appl Microbiol* 2014;**37**:165–9.
85. Heyer J, Berger U, Hardt M, Dunfield PF. *Methylohalobius crimeensis* gen. nov., sp. nov., a moderately halophilic, methanotrophic bacterium isolated from hypersaline lakes of Crimea. *Int J Syst Evol Microbiol* 2005;**55**:1817–26.
86. Hirayama H, Fuse H, Abe M, Miyazaki M, Nakamura T, Nunoura T, et al. *Methylomarinum vadi* gen. nov., sp. nov., a methanotroph isolated from two distinct marine environments. *Int J Syst Evol Microbiol* 2013;**63**:1073–82.
87. Hirayama H, Abe M, Miyazaki M, Nunoura T, Furushima Y, Yamamoto H, et al. *Methylomarinovum caldicuralii* gen. nov., sp. nov., a moderately thermophilic methanotroph isolated from a shallow submarine hydrothermal system, and proposal of the family *Methylothermaceae* fam. nov. *Int J Syst Evol Microbiol* 2014;**64**:989–99.
88. Nyerges G, Han SK, Stein LY. Effects of ammonium and nitrite on growth and competitive fitness of cultivated methanotrophic bacteria. *Appl Environ Microbiol* 2010;**76**:5648–51.
89. Anvar SY, Frank J, Pol A, Schmitz A, Kraaijeveld K, den Dunnen JT, et al. The genomic landscape of the verrucomicrobial methanotroph *Methylacidiphilum fumariolicum* SolV. *BMC Genomics* 2014;**15**:914.
90. Kits KD, Kalyuzhnaya MG, Klotz MG, Jetten MS, Op den Camp HJ, Vuilleumier S, et al. Genome sequence of the obligate gammaproteobacterial methanotroph *Methylomicrobium album* Strain BG8. *Genome Announc* 2013;**11**:e0017013.
91. Khmelenina VN, Kalyuzhnaya MG, Sakharovsky VG, Suzina NE, Trotsenko YuA, Gottschalk G, Osmoadaptation in halophilic and alcaliphilic methanotrophs. *Arch Microbiol* 1999;**172**:321–9.
92. Kalyuzhnaya MG, Khmelenina V, Eshinimaev BT, Sorokin DYu, Fuse H, Lidstrom ME, et al. Reclassification and emended description of halo(alkali)philic and halo(alkali)tolerant methanotrophs of the genera *Methylomicrobium* and *Methylobacter*. *Int J Syst Evol Microbiol* 2008;**58**:591–6.
93. Kalyuzhnaya MG, Khmelenina VN, Lysenko AM, Suzina NE, Trotsenko YuA. New methanotrophic isolates from soda lakes of the southern Transbaikal region. *Mikrobiologiia* 1999;**68**:689–97.
94. Khmelenina VN, Beck D, Munk C, Davenport K, Daligault H, Erkkila T, et al. Draft genome sequence of *Methylomicrobium buryatense* 5G, a haloalkaline-tolerant methanotrophic bacterium. *Genome Announc* 2013;**1**:1–2.
95. Boden R, Cunliffe M, Scanlan J, Moussard H, Kits KD, Klotz MG, et al. Complete genome sequence of the aerobic marine methanotroph *Methylomonas methanica* MC09. *J Bacteriol* 2011;**193**:7001–2.
96. Kalyuzhnaya MG, Lamb AE, mcTaggart TL, Oshkin IY, Shapiro N, Woyke T, et al. Draft genome sequences of gammaproteobacterial methanotrophs isolated from Lake Washington sediment. *Genome Announc* 2015;**3**:e00103-15.
97. Hoefman S, Heylen K, De Vos P. *Methylomonas lenta* sp nov a methanotroph isolated from manure and a denitrification tank. *Int J Syst Evol Microbiol* 2014;**64**:1210–7.
98. Danilova OV, Kulichevskaya IS, Rozova ON, Detkova EN, Bodelier LE, Trotsenko PLE, et al. *Methylomonas paludis sp nov* the first acid tolerant member of the genus *Methylomonas* from an acidic wetland. *Int J Syst Evol Microbiol* 2013;**63**:2282–9.
99. Hoefman S, van der Ha D, Iguchi H, Yurimoto H, Sakai Y, Boon N, et al. *Methyloparacoccus murrellii* gen. nov., sp. nov., a methanotroph isolated from pond water. *Int J Syst Evol Microbiol* 2014;**64**:2100–7.
100. Tavormina PL, Hatzenpichler R, McGlynn S, Chadwick G, Dawson KS, Connon SA, et al. *Methyloprofundus sedimenti* gen. nov., sp. nov., an obligate methanotroph from ocean sediment belonging to the Deep Sea 1 clade of marine methanotrophs. *Int J Syst Evol Microbiol* 2014;**65**:251–9.
101. Wise MG, McArthur JV, Shimkets LJ. *Methylosarcina fibrata* gen. nov., sp. nov. and *Methylosarcina quisquiliarum* sp. nov., novel type I methanotrophs. *Int J Syst Evol Microbiol* 2001;**51**:611–21.

102. Hirayama H, Suzuki Y, Abe M, Miyazaki M, Makita H, Inagaki F, et al. *Methylothermus subterraneus* sp. nov., a moderately thermophilic methanotroph isolated from a terrestrial subsurface hot aquifer. *Int J Syst Evol Microbiol* 2011;**61**:2646–53.
103. Wartiainen I, Hestnes AG, McDonald IR, Svenning MM. *Methylocystis rosea* sp. nov., a novel methanotrophic bacterium from Arctic wetland soil, Svalbard, Norway (786 N). *Int J Syst Evol Microbiol* 2006;**56**:541–7.
104. Beck DAC, McTaggart TL, Setboonsarng U, Vorobev A, Goodwin L, Shapiro N, et al. Multiphyletic origins of methylotrophy in Alphaproteobacteria, exemplified by comparative genomics of Lake Washington isolates. *Environ Microbiol* 2015;**17**:547–54. http://dx.doi.org/10.1111/1462-2920.12736.
105. Iguchi H, Yurimoto H, Sakai Y. *Methylovulum miyakonense* gen. nov., sp. nov., a type I methanotroph isolated from forest soil. *Int J Syst Evol Microbiol* 2011;**61**:810–5.
106. del Cerro C, García JM, Rojas A, Tortajada M, Ramón D, Galán B, et al. Genome sequence of the methanotrophic poly-β-hydroxybutyrate producer *Methylocystis parvus* OBBP. *J Bacteriol* 2012;**194**:5709–10.
107. Dam B, Dam S, Kube M, Reinhardt R, Liesack W. Complete genome sequence of *Methylocystis* sp. strain SC2, an aerobic methanotroph with high-affinity methane oxidation potential. *J Bacteriol* 2012;**194**:6008–9.
108. Im J, Lee SW, Yoon S, DiSpirito AA, Semrau JD. Characterization of a novel facultative *Methylocystis* species capable of growth on methane, acetate and ethanol. *Environ Microbiol Rep* 2010;**3**:174–81.
109. Dunfield PF, Khmelenina VN, Suzina NE, Trotsenko YA, Dedysh SN. *Methylocella silvestris* sp. nov., a novel methanotroph isolated from an acidic forest cambisol. *Int J Syst Evol Microbiol* 2003;**53**:1231–9.
110. Dedysh SN, Panikov NS, Liesack W, Grobkopf R, Zhou J, Tiedje JM. Isolation of acidophilic methane-oxidizing bacteria from northern peat wetlands. *Science* 1998;**282**:281–4.
111. Dunfield PF, Belova SE, Vorob'ev AV, Cornish SL, Dedysh SN. *Methylocapsa aurea* sp. nov., a facultative methanotroph possessing a particulate methane monooxygenase, and emended description of the genus *Methylocapsa*. *Int J Syst Evol Microbiol* 2010;**60**:2659–64.
112. Vorobev AV, Baani M, Doronina NV, Brady AL, Liesack W, Dunfield PF, et al. *Methyloferula stellata* gen. nov., sp. nov., an acidophilic, obligately methanotrophic bacterium that possesses only a soluble methane monooxygenase. *Int J Syst Evol Microbiol* 2011;**61**:2456–63.
113. Khadem AF, Pol A, Wieczorek A, Mohammadi SS, Francoijs KJ, Stunnenberg HG, et al. Autotrophic methanotrophy in Verrucomicrobia: Methylacidiphilum fumariolicum SolV uses the Calvin-Benson-Bassham cycle for carbon dioxide fixation. *J Bacteriol* 2011;**193**:4438–46.
114. Pol A, Heijmans K, Harhangi HR, Tedesco D, Jetten MSM, Op den Camp HJM. Methanotrophy below pH 1 by a new Verrucomicrobia species. *Nature* 2007;**450**:874–8.
115. Islam T, Jensen S, Reigstad LJ, Larsen O, Birkeland NK. Methane oxidation at 55°C and pH 2 by a thermoacidophilic bacterium belonging to the Verrucomicrobia phylum. *Proc Natl Acad Sci U S A* 2008;**105**:300–4.
116. Van Teeseling MCF, Pol A, Harhangi HR, van der Zwart S, Jetten MSM, Op den Camp HJM, et al. Expanding the verrucomicrobial methanotrophic world: description of three novel species of *Methylacidimicrobium* gen. nov. *Appl Environ Microbiol* 2014;**80**:6782–91.
117. Lawrence AJ, Kemp MB, Quayle JR. Synthesis of cell constituents by methane-grown *Methylococcus capsulatus* and *Methanomonas methanooxidans*. *Biochem J* 1970;**116**:631–9.
118. Bogodar IW, Lin TS, Liao JC. Synthetic non-oxidative glycolysis enables complete carbon conservation. *Nature* 2013;**502**:693–7.
119. Bogodar IW, Chen CT, Theisen MK, Wu TY, Schlenz AR, Lam AT, et al. Building carbon-carbon bonds using a biocatalytic methanol condensation cycle. *Proc Natl Acad Sci* 2014;**111**:15928–33.
120. Goenrich M, Thauer RK, Yurimoto H, Kato N. Formaldehyde activating enzyme (Fae) and hexulose-6-phosphate synthase (Hps) in *Methanosarcina barkeri*: a possible function in ribose-5-phosphate biosynthesis. *Arch Microbiol* 2005;**184**:41–8.
121. Grochowski LL, Xu H, White RH. Ribose-5-phosphate biosynthesis in *Methanocaldococcus jannaschii* occurs in the absence of a pentose-phosphate pathway. *J Bacteriol* 2005;**187**:7382–9.

122. Crombie AT. *Metabolism of methane and propane and the role of the glyoxylate bypass enzymes in Methylocella silvestris BL2*. PhD thesis, University of Warwick; 2011. http://www.dart-europe.eu/full.php?id=585409.
123. Vorobev A, Jagadevan S, Jain S, Anantharaman K, Dick GJ, Vuilleumier S, et al. Genomic and transcriptomic analyses of the facultative methanotroph *Methylocystis* sp. strain SB2 grown on methane or ethanol. *Appl Environ Microbiol* 2014;**80**:3044–52.
124. Alber BE. Biotechnological potential of the ethylmalonyl-CoA pathway. *Appl Microbiol Biotechnol* 2011;**89**:17–25.
125. Dedysh SN, Naumoff DG, Vorobev AV, Kyrpides N, Woyke T, Shapiro N, et al. Draft genome sequence of Methyloferula stellata AR4, an obligate methanotroph possessing only a soluble methane monooxygenase. *Genome Announc* 2015;**3**:e01555-14 (in press).
126. Rasigraf O, Kool DM, Jetten MSM, Damsté JSS, Ettwig KF. Autotrophic carbon dioxide fixation via the Calvin-Benson-Bassham cycle by the denitrifying methanotroph "Candidatus *Methylomirabilis oxyfera*". *Appl Environ Microbiol* 2014;**80**:2451–60.
127. Taylor SC. Evidence for the presence of ribulose-1,5-bisphosphate carboxylase and phosphoribulokinase in *Methylococcus capsulatus* (Bath). *FEMS Microbiol Lett* 1977;**2**:305–7.
128. Taylor SC, Dalton H, Dow CS. Ribulose-1,5-bisphosphate carboxylase/oxygenase and carbon assimilation in *Methylococcus capsulatus* (Bath). *J Gen Microbiol* 1981;**122**:89–94.
129. Swan BK, Martinez-Garcia M, Preston CM, Sczyrba A, Woyke T, Lamy D, et al. Potential for chemolithoautotrophy among ubiquitous bacteria lineages in the dark ocean. *Science* 2011;**333**:1296–300.
130. Stoecker K, Bendinger B, Schoning B, Nielsen PH, Nielsen JL, Baranyi C, et al. Cohn's Crenothrix is a novel filamentous methane oxidizer with an unusual methane monooxygenase. *Proc Natl Acad Sci U S A* 2006;**103**:2363–7.
131. Hirayama H, Abe M, Miyazaki M, Nunoura T, Furushima Y, Yamamoto H, et al. *Methylomarinovum caldicuralii* gen. nov., sp. nov., a moderately thermophilic methanotroph isolated from a shallow submarine hydrothermal system, and proposal of the family *Methylothermaceae* fam. nov. *Int J Syst Evol Microbiol* 2013;**64**:989–99.
132. Koffas M, Odom JM, Schenzle A. High growth methanotrophic bacterial strain Methylomonas 16a. EP1320579 A2/WO2002020728 A2, 2001.
133. Kip N, Ouyang W, van Winden J, Raghoebarsing A, van Niftrik L, Pol A, et al. Detection, isolation and characterization of acidophilic methanotrophs from *Sphagnum* mosses. *Appl Environ Microbiol* 2011;**77**:5643–54.
134. Danilova OV, Kulichevskaya IS, Rozova ON, Detkova EN, Bodelier PLE, Trotsenko YA, et al. *Methylomonas paludis* sp. nov., the first acid-tolerant member of the genus *Methylomonas*, from an acidic wetland. *Int J Syst Evol Microbiol* 2013;**63**:2282–9.
135. Kalyuzhnaya MG, Khmelenina VN, Starostina NG, Baranova SV, Suzina NE, Trotsenko YuA. A new moderately halophilic methanotroph of the genus *Methylobacter*. *Mikrobiologiia* 1998;**67**:527–34.
136. Dunfield PF. Methanotrophy in extreme environments. *Encyclopedia of Life Sciences (eLS)*, Chichester: John Wiley & Sons Ltd; 2009. http://dx.doi.org/10.1002/9780470015902.a0021897. http://www.els.net.
137. Egorov I, Kupina L, Aksyuk I, Murtazaeva R. Gaprin—a protein source. *Ptitsevodstvo (Russian)* 1990;**8**:25–7.
138. Yazdian F, Hajizadeh S, Shojaosadati SA, Khalilzadeh R, Jahanshahi M, Nosrati M. Production of single cell protein from natural gas: parameter optimization and RNA evaluation. *Iran J Biotechnol* 2005;**3**:235–42.
139. Kacbelyuzhnaya MG, Puri A, Lidstrom ME. Metabolic engineering in methanotrophic bacteria. *Metab Eng* 2015;**29**:142–52 (in press).
140. But S, Khmelenina VN, Reshetnikov AS, Mustakhimov II, Kalyuzhnaya MG, Trotsenko YA. Sucrose metabolism in halotolerant methanotroph Methylomicrobium alcaliphilum 20Z. *Arch Microbiol* 2015;**197**(3):471–80 (in press).
141. Crombie AT, Murrell JC. Trace-gas metabolic versatility of the facultative methanotroph *Methylocella silvestris*. *Nature* 2014;**510**:148–51.

142. Criddle CS, Hart JR, Wu WM, Sundstrom ER, Morse MC, Billington SL, Rostkowski KH, Frank CW, Production of PHA using biogas as feedstock and power, source. US20130071890 A1, 2013.
143. Park D, Lee J. Biological conversion of methane to methanol. *Korean J Chem Eng* 2013;**30**:977–87.
144. Vecherskaya M, Dijkema C, Stams AJM. Intracellular PHB conversion in a Type II methanotroph studied by ^{13}C NMR. *J Ind Microbiol Biotechnol* 2001;**26**:15–21.
145. Vecherskaya M, Dijkeme C, Saad HR, Stams AJM. Microaerobic and anaerobic metabolism of a *Methylocystis parvus* strain isolated from a denitrifying bioreactor. *Environ Microbiol Rep* 2009;**1**:442–9.
146. Dunfield PF, Yuryev A, Senin P, Smirnova AV, Stott MB, Hou S, et al. Methane oxidation by an extremely acidophilic bacterium of the phylum Verrucomicrobia. *Nature* 2007;**450**:879–82.
147. Khadem AF, Wieczorek AS, Pol A, Vuilleumier S, Harhangi HR, Dunfield PF, et al. Draft genome sequence of the volcano-inhabiting thermoacidophilic methanotroph *Methylacidiphilum fumariolicum* strain SolV. *J Bacteriol* 2012;**194**:3729–30.
148. Khadem AF, Pol A, Jetten MS, Op den Camp HJ. Nitrogen fixation by the verrucomicrobial methanotroph '*Methylacidiphilum fumariolicum*' SolV. *Microbiology* 2010;**156**:1052–9.
149. Khadem AF, van Teeseling MC, van Niftrik L, Jetten MS, Op den Camp HJ, Pol A. Genomic and physiological analysis of carbon storage in the verrucomicrobial methanotroph "Ca. *Methylacidiphilum fumariolicum*" SolV. *Front Microbiol* 2012;**3**:345.
150. Ettwig KF, van Alen T, van de Pas-Schoonen KT, Jetten MSM, Strous M. Enrichment and molecular detection of denitrifying methanotrophic bacteria of the NC10 Phylum. *Appl Environ Microbiol* 2009;**75**:3656–62.
151. Ettwig KF, Butler MK, Le Paslier D, Pelletier E, Mangenit S, Kuypers MMM, et al. Nitrite-driven anaerobic methane oxidation by oxygenic bacteria. *Nature* 2010;**464**:543–8.
152. Wegener G, Niemann H, Elvert M, Hinrichs KU, Boetius A. Assimilation of methane and inorganic carbon by microbial communities mediating the anaerobic oxidation of methane. *Environ Microbiol* 2008;**10**:2287–98.
153. Svenning MM, Wartiainena I, Hestnesa AG, Binnerupb SJ. Isolation of methane oxidising bacteria from soil by use of a soil substrate membrane system. *FEMS Microbiol Ecol* 2003;**44**:347–57.
154. Dedysh SN, Dunfield PF. Facultative and obligate methanotrophs how to identify and differentiate them. *Methods Enzymol* 2011;**495**:31–44.

CHAPTER 14

Photosynthetic Platform Strain Selection: Strain Selection Considerations and Large-Scale Production Limitations

Whitney E. Sinclair, Nanette R. Boyle
Chemical and Biological Engineering, Colorado School of Mines, Golden, Colorado, USA

14.1 Introduction

In terms of long-term sustainability of biofuels, phototrophic microorganisms, such as cyanobacteria, diatoms, and algae, have a great potential to serve as a renewable source of next-generation biofuels. The major advantage of using photosynthetic microorganisms is their preferred substrate: carbon dioxide. Phototrophs are capable of directly converting carbon dioxide and light into biomass and useful chemical feedstocks and fuels.[1–14] This overcomes one of the major limitations of first-generation biofuels, which is the use of feedstocks that would otherwise be used in the food or feed supply. When this is coupled to the decline in the number of arable acres (383 million acres in 1949-340 million acres today[15]), the use of first-generation biofuels becomes decidedly unsustainable—we cannot feed the current global population AND produce feedstocks for biofuels on the same land using the same finite resources we have today. Luckily, phototrophic microorganisms can be grown on marginal land in waste or brackish water that is not typically used for high-yielding crop plants. In terms of photosynthetic efficiency, algae and cyanobacteria also have an advantage over traditional land plants; they are 3- to 9-fold more efficient at light harvesting than land plants,[13,16] which makes them even more attractive in terms of long-term sustainability.

Despite the numerous advantages of using phototrophs for biofuel or chemical feedstock production, using one in place a heterotrophic bacterium is not always a straight forward replacement. Model organisms like *Escherichia coli* and *Saccharomyces cerevisiae* have much more advanced genetic tools and more complete data. The genomes of both organisms are very well annotated and there is a plethora of information available about gene function, known mutations, and their effect on metabolism. Many phototrophs are now benefiting from larger research efforts to improve genetic engineering tools, genome annotations, and increased research in the field in general. In this chapter, we will discuss some of the

Biotechnology for Biofuel Production and Optimization. http://dx.doi.org/10.1016/B978-0-444-63475-7.00014-5

advantages and disadvantages of using phototrophs for biofuel and chemical production and how to overcome some of the limitations with large-scale production.

14.2 Platform Strain Selection

Selecting a platform strain for engineering new biofuel production strains requires careful consideration of the strengths and limitations of each possible organism. Here, we will discuss several aspects of cellular metabolism, genetic tools, and availability of information that must be weighed carefully before the platform strain is selected.

14.2.1 Genetic Tools/Genetic Tractability

One of the most important considerations for a platform strain is the availability and ease of use of genetic tools. Engineering any organism for biofuel production requires a multitude of genetic changes to optimize pathway flux, eliminate or minimize competing pathways, balance redox levels, and design optimal methods for end-product purification. Unlike the premiere model organisms, *E. coli* and *S. cerevisiae*, which have been used extensively for metabolic engineering, phototrophs lack the well-developed molecular tools available for model organisms which make it possible to implement changes and finely tune gene expression or protein translation. Fortunately, there has been a significant effort to develop tools specifically for phototrophs in recent years.

Many of the more widely used cyanobacteria (*Synechocystis* and *Synechococcus*) are naturally competent and capable of homologous recombination, which allows targeted replacement or introduction of genes anywhere on the chromosome. However, in one of the most widely studied green algae, *Chlamydomonas reinhardtii,* genetic changes to the nuclear genome can only be introduced by random insertion via electroporation.[17] There is no tool to target specific genes, which then requires screening thousands of mutants to find knockouts of the specific gene of interest. The same is true for other eukaryotic algae and diatoms (see Ref. 18 for a complete summary of transformation methods used in eukaryotic algae). Luckily, there are species which have naturally high homologous recombination rates, such as *Nannochloropsis*,[19] which do allow for targeted gene insertion/knockout. In algae without high homologous recombination rates, it is possible to create artificial microRNAs to specifically knock down the expression of nuclear encoded genes in *Chlamydomonas*,[20–23] *Dunaliella salina*,[24] and *Phaeodactylum tricornutum*.[25]

Counter-selection markers are useful tools for genetic engineering because they can be used to select for the desired mutation and against the wild-type allele in the same experiment, thus greatly reducing the number of clones which must be screened to identify the desired mutations (or eliminating the screening step all together). Recently, counter-selection markers have been developed for *Synechocystis* sp. PCC 6803 and *Synechococcus* sp. PCC 7002[26,27,7] which enable mutations to be introduced without leaving antibiotic resistance markers on the chromosome. A *sacB/aphX* counter-selection method was developed for a glucose-tolerant strain of *Synechocystis* sp. PCC 6803; loss of the *sacB* gene allows growth in the presence of sucrose.[7] Unfortunately, the *sacB/aphX* counter-selection marker cannot be used in cyanobacteria which do not consume glucose because they lack the transporters to uptake sucrose. To overcome this limitation, a *mazF/aphII* counter-selection marker was developed by Cheah et al.[26] This marker takes advantage of the *E. coli mazF* gene under the control of a nickel-inducible promoter. MazF is an endoribonuclease which cleaves mRNA at the ACA triplet, thus allowing the use of this marker in any cyanobacteria that is capable of homologous recombination. The *sacB/aphX* and *mazF/aphII* counter-selection markers were both developed specifically for use with *Synechocystis* sp. PCC 6803; Begemann et al.[27] developed a counter-selection marker (*acsA/aadA*) specifically for use with *Synechococcus* sp. PCC 7002, but it also has the potential to be used in *Synechocystis* sp. PCC 6803. This counter-selection is based on the sensitivity of cyanobacteria to organic acids (acrylate, 3-hydroxypropionate, and propionate) and the loss of *acsA* function, which enables growth in the presence of organic acids.

Another desirable trait for a model system for biofuel production is the availability of genetic elements which allow fine tuning of metabolic pathways. Although there are a number of endogenous inducible promoters in cyanobacteria (see Ref. 28 for a comprehensive list), most of them are induced by trace metals (Fe^{3+}, Zn^{2+}, Cu^{2+}, Co^{2+}, etc.), which can be problematic to implement without special considerations such as acid washing of glassware and using trace metal grade chemicals. There are also a handful of promoters from *E. coli* that have been shown to work in *Synechococcus* sp. PCC 7942 and *Synechocystis* sp. PCC 6803 with dynamic ranges reported from no significant difference up to 290-fold.[1,29–33] There are a few native promoters in cyanobacteria which are inducible and have been used for genetic engineering studies. One such promoter is *psbA*, which controls expression of the photosystem II reaction center D1 protein. This promoter drives high expression of the gene under its control when cells are grown in light (50-500 μmol photons $m^{-2} s^{-1}$).[4,34] Another inducible/repressible promoter is *nirA*, which controls the expression of nitrate reductase; in the presence of more desirable nitrogen sources (such as ammonia), the expression of *nirA* is repressed. In the three species it has been used to drive heterologous expression of a gene (*Synechocystis*, *Synechococcus*, and *Anabaena*), the suggested media recipes provide nitrogen in the form of nitrate; thus, *nirA* is expressed during growth in normal media and repressed in the presence of ammonium. There has recently been an effort to engineer and characterize additional promoters for cyanobacteria in order to expand the tools available for genetic engineering; ideally, metabolic engineers would have a suite of promoters available to them which span a large range of expression and whose regulation is orthogonal to both one another and to regulation in the cell. Huang and Lindblad[29] engineered a library of *tetR* promoters which were capable of a wide range of gene expression (repressed to 290-fold induction) in *Synechocystis* sp. PCC 6803. LacI repressible promoters have also been engineered for use in cyanobacteria[35,36]; however, neither was the dynamic range as large nor repression as effective as in *E. coli*. Markley et al. engineered two constitutive promoters (*Synechocystis* pc. PCC 6803 *cpcB* and *E. coli* BioBrick BBa_J23119) which resulted in a library of promoters which was expressed and tested in *Synechococcus* sp. PCC 7002 and resulted in 3 and 2.5 log dynamic ranges, respectively. The RBS calculator,[37,38] which is a computational tool that has been used to predict ribosome binding sites with desired strengths for modifying protein translation, has been used extensively in *E. coli* metabolic engineering studies[39–43] to tune metabolic pathways. It has also been used to aide in complete refactoring of the nitrogen fixation gene cluster in *Klebsiella oxytoca*.[44] Unfortunately, the predictions made by the RBS calculator cannot be directly translated into cyanobacteria; two recent studies which have tried to use RBS sequences report different levels of translations than predicted *in vivo*.[2,45] Therefore, protein translation in phototrophic prokaryotes must be controlled by a different mechanism and thus requires the development of cyanobacteria-specific prediction tools. There is a pressing need to develop and characterize more synthetic biology tools for phototrophs because the more sophisticated the genetic tools available, the faster and easier it will be to implement genetic changes and the faster we can design and create production strains.

A recent advance in genome engineering, CRISPR/Cas, has been shown to be an effective tool for a wide range of organisms, from bacteria[46,47] to mice[48,49] and human cells.[50–52] For photosynthetic microorganisms that are not currently genetically tractable, the CRISPR/Cas system holds great promise for enabling advanced genetic engineering. Although its use has not yet been reported in a photosynthetic microorganism, it is only a matter of time before it has been applied to these systems as well.

14.2.2 Substrate

For sustainability, the ideal substrate for phototrophic microorganisms is carbon dioxide. However, it is sometimes advantageous when introducing mutations into the genome to have the ability to use alternative substrates. For example, in some strains, such as *Chlamydomonas* and *Synechocystis*, mutants defective in important photosynthetic genes can be made by growing the mutants on alternative substrates (acetate and glucose, respectively). Without the option to provide an alternative substrate, these mutants could never have been made because defects in the photosynthetic apparatus would have prevented growth on carbon dioxide. Another advantage of providing alternative substrates is that heterotrophic or mixotrophic growth is typically faster than purely autotrophic growth, which means that the time required to introduce mutations and create high-yielding production strain is reduced significantly due to faster growth rates.

14.2.3 Availability of Information

Sequenced genome and quality of annotation

Advances in sequencing technology have brought down the cost and time associated with sequencing whole genomes dramatically (see Figure 14.1); however, one major limiting factor in the usability of these genomes is the quality of genome annotation.

Unfortunately, our ability to accurately annotate genomes has not kept pace with the ability to obtain genome sequences. The changes in technology which allows us to sequence more genomes faster and cheaper also contributes to the complexity of annotating new genomes; next-generation sequencing relies on much shorter read lengths, so the contiguity of genomes sequenced using classical techniques (i.e., human genome[53,54] and *Drosophila melanogaster*[55,56]) is rarely achieved. Another contributing factor is that many of the genomes currently being sequenced are not well-known organisms; therefore, there is less information available to use during annotation efforts. Unlike early genome annotation efforts, it is also much more likely that genome sequencing and annotation efforts are performed by small groups instead of the large consortia required for earlier efforts (for example, the Human Genome Project). Most genome sequencing projects follow the same general pipeline for assembly and annotation which can be roughly divided into two distinct phases: the computation phase and the annotation phase.[57] During the computation phase, expressed sequence tags (ESTs), known protein sequences, RNA-Seq data, etc., are aligned to the genome and *ab initio* or evidence-based gene predictions are made. The goal of the

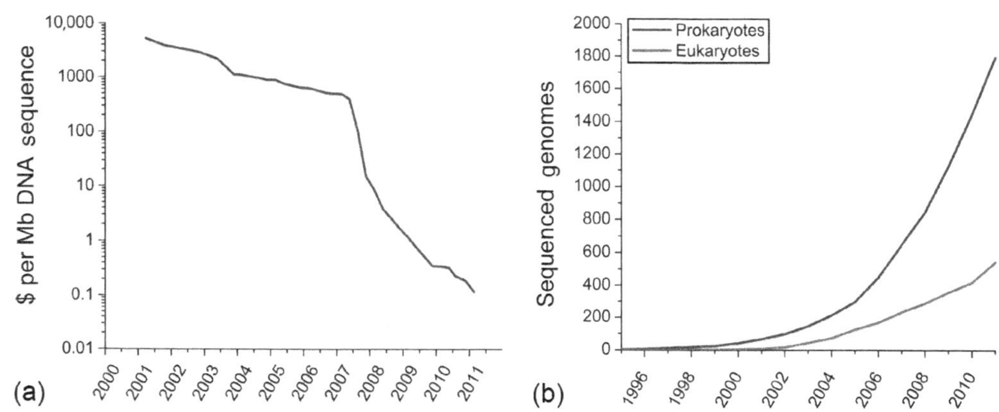

Figure 14.1

Rapidly decreasing sequence costs have enabled the rapid expansion of genome sequencing. (a) Cost of sequencing per MB DNA sequence reported by the National Human Genome Research Institute (http://www.genome.gov/sequencingcosts) has dropped from nearly $6000 in 2001 to slightly more than $0.10 in 2011. (b) Total number of completed sequence genomes for prokaryotes and eukaryotes (www.ncbi.nlm.nih.gov).

annotation phase is to combine the alignment-based data and *ab initio* data to produce the best possible gene predictions. Manually curated projects produce high-quality, reliable genome annotations[58,59]; however, this requires hundreds of man hours and is extremely costly so very few annotations have extensive manual curation. Annotating eukaryotic genomes is even more of a challenge because even the best annotation pipelines are only 80% accurate at predicting the correct intron/exon boundaries.[60] Due to the algorithm in which genomes are annotated, incorrect annotations can spread to other organisms very quickly, which sets the stage for any experiment that makes use of them to fail. For example, one of the most widely research algal species, *C. reinhardtii*, has been used as a model system for over 50 years. Its genome was sequenced in 2007[61] and had prolonged effort by a number of research groups to improve the annotation by revising annotations using more accurate gene prediction software[62] and manually curating gene annotations using previous research studies, ESTs, and extensive RNA-seq data. Despite the concerted effort of the *Chlamydomonas* research community, there are still mis-annotated genes. One example in particular is a gene which encodes a type-2 diacylglycerol acyltransferase, *DGTT1*; the function of the gene was predicted correctly based on homology to other organisms; however, researchers were unable to prove its function by complementation in a yeast mutant lacking all diacylglycerol acyltransferase activity.[63,64] In *Chlamydomonas*, *DGTT1* is only expressed highly in nitrogen-deficient conditions; therefore, it is difficult to get EST and RNA-Seq data for this gene when cells are grown in replete conditions. It was later discovered by comparing ESTs and RNA-Seq data from nitrogen-deficient cells that an exon on the 5′ end of the gene has been missed during previous annotation efforts (see Figure 14.2).

Using the newly predicted cDNA sequence of *DGTT1*, Boyle et al.[65] were able to prove the function of the gene by complementing a mutant yeast strain. This is just one example of how the quality of a genome annotation can affect the outcome of an experiment; the higher the quality of annotation, the more reliable the information, but even then, it can adversely affect even the most well-designed experiments. Unfortunately, genome annotations for photosynthetic microorganisms lag behind those for heterotrophic model systems like *E. coli* and *S. cerevisiae*. However, the larger the community that is working on a particular organism, the more likely it is that the annotation of the genome is being continuously improved and will become increasingly more reliable.

Metabolic model

Although not strictly necessary for a platform strain, the availability of a metabolic model can be a significant aide in strain design. Metabolic models have been used in several different organisms to predict the effect of genetic changes in a small fraction of the time that it would take to implement such changes in the laboratory.[66–69] An early example of the utility of metabolic models is the improvements made in lycopene production in *E. coli*; Alper et al.[66] used stoichiometric metabolic models to identify single and multiple gene knockouts which

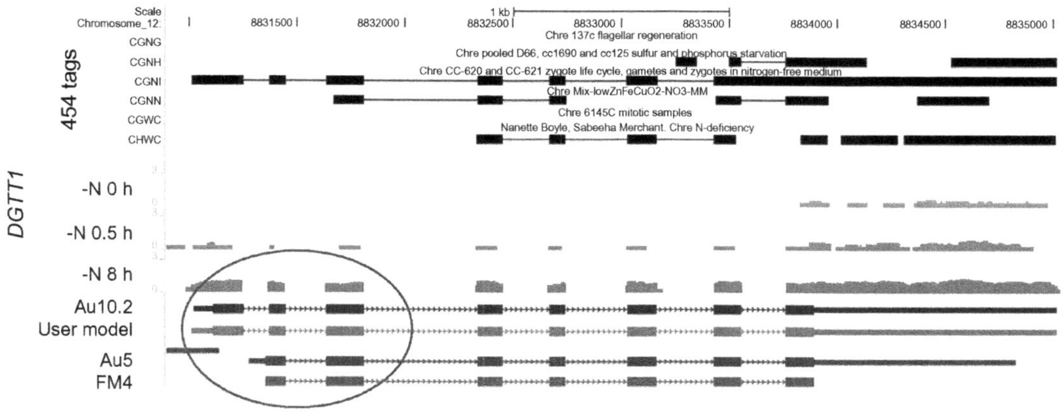

Figure 14.2

UCSC browser view of *Chlamydomonas reinhardtii DGTT1*. The UCSC browser was used to view RNA-Seq coverage; genome coordinates are shown at the *top*. The next six tracks labeled 454 ESTs represent sequences, collapsed into a single track, from the UCLA/JGI EST project (accession no. SRX038871). The tracks labelled -N 0 h, -N 0.5 h and -N 8 h represent RNA-Seq coverage, on a log scale, for the locus at three time points after *Chlamydomonas* cells were transferred to nitrogen-depleted medium. The JGI best gene model for the V4 assembly, FM4, and two different gene models predicted by the Augustus algorithm[62] (Au5 and Au10.2) are shown below as labeled. *Thick blocks* represent exons, and *thin blocks* represent UTRs, and *arrowed lines* represent introns. RNA-Seq, 454, and EST coverage were used to inform the manual construction of a single gene model at each locus, labeled user model. From this browser view, it is evident that the FM4 and Au5 models are missing the UTR and most 5′ exon of the gene. It was only by using conditions to induce higher expression for this gene (nitrogen starvation) that the actual gene structure could be determined from experimental evidence.

would lead to increased lycopene production. In order to have more accurate predictions of fluxes, Alper et al. utilized an algorithm called minimization of metabolic adjustment (MOMA)[70] which minimizes the difference in flux when comparing the wild-type fluxes to the mutant. By constructing the mutant strains, the authors were able to identify one strain (*ΔgdhA ΔaceE ΔfdhF*) which produced 37% more lycopene than the parent strain. Another excellent example of this is the design and creation of an *E. coli* strain which is capable of producing high titers of 1,4-butanediol (BDO), a product that was not previously known to be synthesized naturally in any known organism.[71] Researchers at Genomatica used an in-house modeling tool, Simpheny Biopathway Predictor, to enumerate all possible pathways from *E. coli* central metabolism to BDO. The algorithm which identifies possible pathways is similar to those reported previously and is not limited to known enzymes[72–74] and resulted in over 10,000 possibilities with 4-6 steps. These possible pathways were then evaluated and ranked on the basis of several different attributes: maximum theoretical yield, pathway length, number of nonnative steps, number of novel steps, and thermodynamic feasibility, ultimately narrowing down the options to two pathways. After functionally expressing the necessary genes for BDO production in *E. coli*, the team used OptKnock[75] to identify gene knockout strategies that would further improve the yield. Ultimately, combing *in silico* modeling with metabolic engineering strategies to further improve yield, the team were able to design a strain capable of producing almost 20 g/l BDO. This study illustrates how important the use of metabolic models and pathway identification algorithms are in reducing the time needed to produce a commercially viable production strain.

The time savings of using *in silico* models are even more pronounced in phototrophs, which have much slower growth rates and thus require much longer time periods to create homogenous mutants. As discussed above, *in silico* models allow investigators to evaluate hundreds (or thousands) of genetic or environmental changes which could lead to improved production titers in a fraction of the time it would take to create the corresponding mutants in a laboratory. Also, computational tools such as the Simpheny Biopathway Predictor speed up the identification of possible pathways from metabolites to a desired product. Currently, there are metabolic models available for the following photosynthetic microorganisms: *C. reinhardtii*,[76–78] *Synechocystis* sp. PCC 6803,[79,80] *Synechococcus* sp. PCC 7942,[81] *Synechococcus* sp PCC 7002,[82] and *Cyanothece* sp. ATCC 51142.[83] Unfortunately, these models have not yet been used in metabolic engineering studies which may be in part due to the smaller community that works with these types of organisms and the limited tools we have to perform sophisticated genetic engineering. Vu et al. used their genome-scale model of *Synechococcus* sp. PCC 7002 to evaluate theoretical yields of a variety of potential biofuels and feedstocks in different growth conditions (phototrophic, dark oxic, and dark anoxic) which illustrates the importance of choosing the appropriate growth conditions in order to optimize yields. Several metabolites of interest had the highest yields in phototrophic growth; however, 2-phenylethanol had the highest yield (g/g C) in dark oxic conditions.[82]

The models mentioned above are all *predictions* of the fluxes in the cell. In order to obtain estimates of the actual fluxes in the cell, ^{13}C-Metabolic Flux Analysis (^{13}C-MFA) must be used. ^{13}C-MFA is a technique used to track carbon throughout metabolism which provides additional data needed to fit fluxes to (instead of using linear programming methods to optimize an objective function). Traditional ^{13}C-MFA performed in heterotrophic organisms allows a variety of substrates which are ^{13}C-labeled on different carbon atoms to be used (glucose, amino acids, and organic acids), metabolic and isotopic steady states are achieved, and then the labeling pattern of metabolites are measured by gas chromatography/mass spectrometry or liquid chromatography/mass spectrometry. Unfortunately, with phototrophic organisms, if isotopic steady state is used, then all metabolites will be labeled because the substrate is a single carbon molecule ($^{13}CO_2$). Therefore, nonstationary or instationary ^{13}C-MFA must be used to determine *in vivo* fluxes.[84–86] This is computationally and experimentally more demanding, but it has been used to determine fluxes in cyanobacteria and plants.[84,87,88] Models that can distinguish between intracellular compartments have also been used for plant cells (which have similar structure to algae).[89–91] These models/flux determination techniques are extremely useful for metabolic engineering efforts, pointing out by-product carbon fluxes, bottle necks, and futile cycles that can be reengineered.

The metabolic models described above rely solely on stoichiometry of the reactions in the cell; therefore, they are unable to predict the effect of kinetics or regulation on the final yield of product. Metabolic models have seen continuous improvement and refinement since the first report of flux balance analysis. With advancing technologies in systems biology, the future of these models will incorporate additional information about regulation and protein expression and thus be more predictive.

Endogenous pathways

Photosynthetic microorganisms are a largely untapped well of natural products. They are extremely diverse, being found in every ecosystem on the planet[92] and because of this, have developed a wide and diverse set of metabolic pathways to outcompete other organisms in very nutrient-limited conditions. Many of the secondary metabolites produced by these phototrophs have properties that make them ideal as potential pharmaceuticals, nutraceuticals, fuels, or chemical feedstocks. Algae and cyanobacteria can be used for large-scale production of food additives, such as vitamin E, B-complex vitamins, and omega fatty acids.[92–96] Other metabolites isolated from cyanobacteria have been shown to be effective in treating cancer[97,98] and HIV[99] or to have anti-fungal, anti-bacterial, and immune modulatory properties. Triacylglycerols produced naturally in algae have been shown to be used as a source of renewable biodiesel.[13,64,65,100] Despite the great promise phototrophs have in terms of natural product development, only a handful of strains have been studied extensively in a laboratory setting and even fewer have been engineered for biofuel production. Part of this may be due to the challenges discussed above.

14.3 Limitations/Large-Scale Production Considerations

Of course, the use of phototrophs for large-scale biofuel production does not come without unique challenges. There are numerous considerations that must be taken into account in the design of any large-scale phototroph production facility. In order to maximize productivity, cells must be provided with the optimal amount of light, carbon dioxide, and other nutrients. There are a number of different types of cultivation methods, a selection of some of the most widely used methods are shown in Figure 14.3.

Below, we discuss the specific challenges with providing optimal growth conditions for these organisms below.

14.3.1 Light

Light provides the energy necessary to fix carbon in photosynthetic organisms; therefore, it is extremely important to provide them adequate light at the right wavelength(s) to have optimal growth rates and productivities. Without adequate light, cells will cease growth or

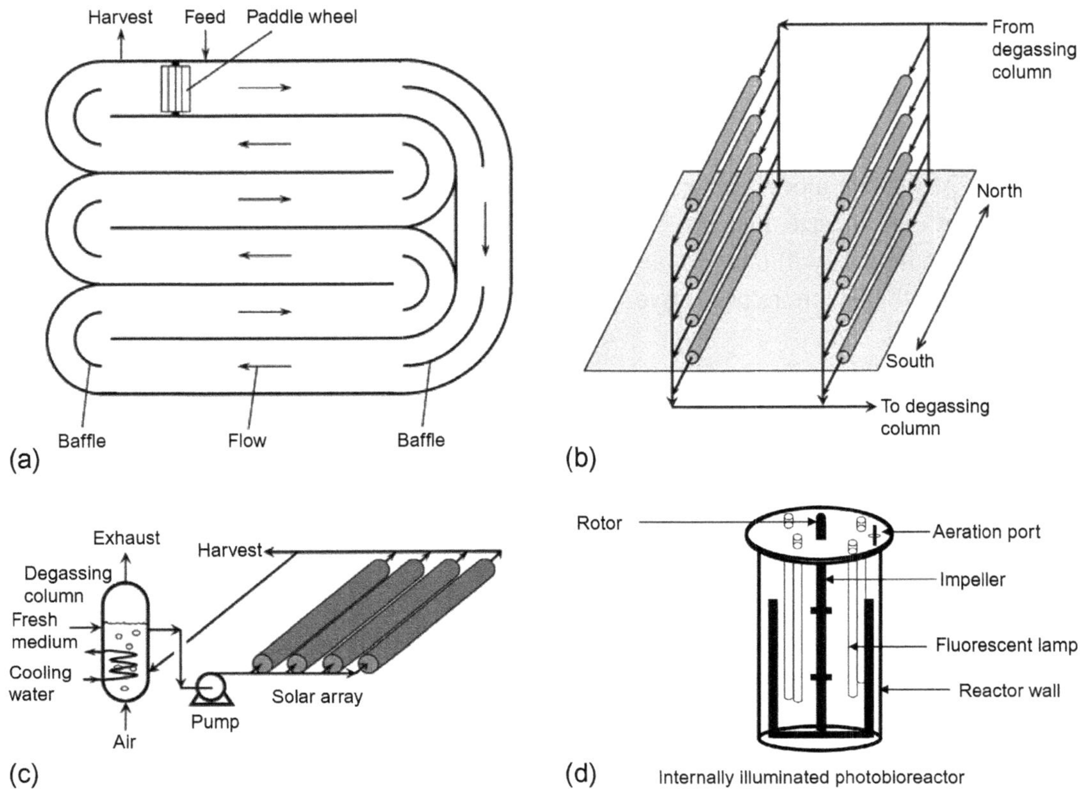

Figure 14.3

Methods for large-scale production of phototrophic microorganisms. (a) Arial view of a raceway pond. (b) Fence-like solar collector. (c) Tubular photobioreactor with parallel run horizontal tubes. (d) Internally illuminated photobioreactor. Panels a-c reprinted from Chisti Y. Biodiesel from microalgae. Biotechnol Adv 2007;25(3):294-306 and panel d reprinted from with permission from Elsevier. Ugwu CU, Aoyagi H, Uchiyama H. Photobioreactors for mass cultivation of algae. *Bioresour Technol* 2008;**99**(10):4021-8.

require a reduced carbon source which can supply both energy and carbon. In their natural environment, phototrophs receive the required energy to perform photosynthesis from sunlight. Ideally, for large-scale production of phototrophs, the necessary light would be supplied by the sun; however, this resource is inherently variable which makes its use for consistent production challenging. Not only is there the diurnal changes in light from day to night and seasonal variation in the length of day versus night hours, cloud cover and other weather-related changes also cause sunlight to vary. In nature, phototrophs grow in diurnal cycles, storing carbon as starch or glycogen during the day to use at night when metabolism shifts from carbon fixation to respiration. However, the overall growth rate for any organism grown in day/night cycles is significantly less than those for organisms grown in 24 h of light.[101] It is important to note however that some bacteria have higher growth rates in diurnal cycles (or will not tolerate continuous light); so, they may be more ideal platform strains for outdoor production. Any weather that brings cloud cover over the facility used for large-scale production also decreases the light energy available to the phototrophic organisms as well; this is an issue that is more easily averted by strategic siting of large-scale production facilities rather than supplying light energy powered by traditional energy sources (electricity, coal/natural gas-fired power plants, etc.). When considering siting for large-scale production facilities, it would also be advantageous to choose locations closer to the equator which have lower seasonal variations in day/night cycles. In most laboratory research studies, artificial lights are used to grow the phototrophs under constant light. Typically, scale-up studies take advantage of photobioreactors, which also supply light to the cells using artificial lights. The benefits of using artificial light sources include the ability to supply the optimal wavelength of light (by choosing the appropriate light source) and the ability to place light at different angles to illuminate cells from different directions.[102] Although this is an effective method to increase growth rates, it is expensive for commercial-scale application. The cost of light sources and energy to power them removes profit from the bottom line.[103] Advances in photobioreactors and photovoltaics may make artificial light a worthwhile investment to increase product output.[102]

Supplying the correct amount of light to a growing culture requires a fine balance; too much light at lower cell densities can cause photoinhibition, while too little light at higher cell densities causes self-shading effects. High light intensity can cause overheating or photoinhibition. Photoinhibition occurs when the photosynthetic apparatus is exposed to intense light for extended periods of time which causes damage to photosystem II and subsequently lowers photosynthetic capacity (and thus growth). This is evident by the naked eye by bleaching that occurs in photo-inhibited cells; this can also be coupled to a change in the color of pigments in the photosynthetic apparatus. An example of this is the production of astaxanthin by *Haematococcus pluvialis*; when it is shifted to high light conditions (coupled to other stress conditions), the algae accumulates the red-colored pigment astaxanthin as a response to photoinhibition.[104] High-density cultures also pose a challenge for large-scale production; high cell densities can lead to self-shading (when light is absorbed by the cells closest to the light supply and cells deeper in the culture are not able to harvest sufficient light for growth).[105] This leads to large variations in light availability depending on where a particular cell is located in the culture. This can lead to lowered output and large variations in the active metabolic pathways in the culture. In order to minimize this effect, cultures can be grown at lower densities or the surface area to volume ration can be increased (large, shallow open ponds are preferable to smaller, deeper ponds). Another option to minimize self-shading is to use thin film photobioreactors.[106] Since both of these conditions (photoinhibition and self-shading) can lead to decreased growth rates and lowered productivity, it is important that large-scale cultures are monitored closely for cell density and light intensity so that such issues can be mitigated.[103]

14.3.2 Carbon Dioxide

In August 2014, the current level of carbon dioxide in the atmosphere (measured at Mauna Loa Observatory in Hawaii) was 397.01 ppm, or slightly less than 0.04%.[107] The average atmospheric carbon dioxide concentration has been on a steady rise since the industrial revolution, rising from 315.97 ppm in 1958.[107] Due to the poor selectivity of the carbon fixing enzyme, Ribulose-1,5-bisphosphate carboxylase/oxygenase (RuBisCO), the increase in carbon dioxide actually improves the rate of carbon fixation in photosynthetic organisms; providing supplemental carbon dioxide (1-5% vvm) increases growth rate significantly, indicating that carbon dioxide supply limits growth rate. However, much the same way light must be supplied to cultures carefully, carbon dioxide concentrations that are too high can also have deleterious effects.

Using exhaust streams from fossil fuel-fired power plants is one way to supplement carbon dioxide supply to cells; however, it is not as easy as bubbling exhaust gas through the growth medium. First and foremost, end-of-pipe technologies must be utilized to capture the carbon dioxide from the flue gas, which also contains contaminants such as soot, carbon monoxide, nitrogen oxides, and sulfur oxides. Chemical scrubbing techniques can be used to obtain pure carbon dioxide which can then be mixed with air and bubbled into bioreactors or open ponds. The two most common methods of recovering carbon dioxide involve the use of solvents [monoethanolamine (MEB) or ammonia], but these methods can be expensive.[108] Storage of the purified carbon dioxide may also be a consideration, if the carbon dioxide is produced faster than it is needed in the production facility.[103] In some cases, additional compression may also be needed for storage or transport, which is a significant operational cost.

In the type of high-density cell cultures which are optimal for the large-scale production of biofuels and chemicals, it is necessary to bubble carbon dioxide into the growth medium to minimize the limitation of carbon dioxide. In addition to supplying more carbon dioxide to the system, bubbling gases throughout the pond or photobioreactor also increase mass transfer rates. Without bubbling, the only gas exchange that occurs in the system is located at the interface where the atmosphere and medium are in contact. Bubbling air that contains carbon dioxide increases the surface area of mass transfer, exposing cells throughout the medium to a continuous supply of carbon.[109] Bubbling can also increase flotation, which is a method for cell harvesting discussed later in the chapter. Stirring the cell culture medium along with bubbling increases the mass transfer of carbon dioxide into the growth medium and thus the transport of carbon dioxide into the cell. Higher stirring and bubbling rates result in more turbulent fluid flow and higher mass transfer rates; however, cells cannot withstand these high levels of shear stress without lysing. Therefore, bubbling and stirring rates must be controlled in order to minimize cell damage.

14.3.3 Mechanical Stirring

Stirring is essential during growth to maximize mass transfer and to minimize self-shading, nutrient gradients, and settling. This can be achieved by creating a mild flow in the medium using mechanical stirring and/or bubbling of gases. Another advantage of using mechanical stirring is that it minimizes temperature gradients within the pond and creates a more homogeneous environment for the cells.[110] Mixing of the medium can be accomplished through paddle wheels or a combination of pumps and gravity flow. Paddle wheels and pumps must be powered with electricity and therefore add additional operational expenses to the process. Hand paddle wheels and pumps do not require high financial costs but is only effective on a smaller scale.[110] When continuously mixing or pumping the medium, it is beneficial to work with cells that can tolerate a high shear force.[111] Mixing and pumping can be expensive, but having a cell that can withstand increased amount of force allows for the utilization of inexpensive methods and equipment.

14.3.4 Contamination

When growing any organism on a large scale, contamination is an issue that should be considered. These foreign organisms are commonly microbes or native phototrophs to the region or predators. In open pond systems, foreign organisms often enter open ponds alongside debris or rainfall; if the contaminating organisms have faster growth rates than the desired phototroph, they can quickly overrun the pond.[111] Therefore, when choosing the platform strain, it is beneficial to consider growth rate not only for productivity but also to outcompete invading species. There are a few options to protect production strains from contamination in open pond systems. The first advantage phototrophs have over possible invading species is that they can grow in media without a reduced carbon source; this minimizes contamination from other species in newer cultures, but once cells lyse, the medium contains free organic acids, nucleic acids, proteins, and sugars, which can serve as substrates for the invading species. Despite the lack of a reduced carbon source in the medium, predators can easily invade because the cells themselves serve as substrates. Selecting an organism that has a very high growth rate will help to decrease the likelihood that a foreign invader can compete in growth. Choosing an organism that can grow in extreme environments is another method to minimize issues with contamination. These extreme environments often include temperature or pH levels that are very low or very high.[111] Tolerance to high concentrations of salts and toxic metals are other valuable traits to consider when choosing a phototroph that will have low concern for contamination. It is also important to note that organisms that can grow or live in extreme environments oftentimes have little available research surrounding them. Unfortunately, minimal research has been completed on the differing pathogens that can contaminate algae and cyanobacteria farms, but there is a potential to use pesticides and antibiotics to minimize the threat of unwanted species.[112]

If the species of interest is not resistant to the pesticides and antibiotics that are applied to the medium, a specific strain of that species can be evolved or genetically engineered in a way that they can survive in the new medium, while the unwanted species will die off. Of course, closed ponds and photobioreactors have a lower likelihood of contamination because they are closed systems, but it does happen.

14.3.5 Evaporation

Appropriate siting of large-scale production facilities in order to minimize fluctuations in seasonal day/night lengths and to maximize the number of sun-filled days has already been discussed. Another consideration when choosing the location of a large-scale production facility is the supply of water; areas which are ideal in terms of sun are typically arid, which limit the supply of water and accelerate evaporation (low humidity). Evaporation causes an overall decrease in the volume of water suspending the cell culture and medium, resulting in changes in the concentration of salts and trace metals in solution.[101] The decrease in water also has the potential to change the pH of the liquid environment. Drastic changes in salt concentrations and pH levels can negatively impact the productivity and health of the chosen phototroph. To ensure ideal conditions for cell growth, water levels must be maintained at constant levels, which can be a challenge in arid environments. Minimizing evaporation is the preferred method to keep salt concentrations and pH levels consistent in a culture of cells, but can also be challenging or costly. For example, closed ponds or bioreactors have little to no evaporation, but the initial capital and operating are much higher than that of open ponds. Water is not a readily available resource; there is limited supply and it is always in high demand. To minimize the use of clean water when harvesting phototrophs, research has been performed to test the ability of these organisms to grow in waste waters.[113] Wastewater is rich in organic and inorganic nutrients that can be used for cellular growth. The main concern with the use of wastewater is that it carries many other undesired organisms that can compete with the phototroph, and nutrient levels are not always balanced to meet the demands of the desired organism. Seawater is also an alternative source of water for production[114]; in fact, many of the widely researched phototrophs are marine organisms and are grown in artificial seawater media. Seawater contains many of the essential salts that are needed in the medium depending on the phototroph or cyanobacteria; these salt ions include calcium, chloride, magnesium, potassium, sodium, and sulfate [104]. More detailed information about the use of waste water or sea water as a growth medium can be found in Refs. 113,114

14.3.6 Downstream Processing

Water removal

If the final product is not excreted by the cell, the cells must be harvested and processed further to collect the product. It is critical that the cell wall remains intact during harvesting to minimize product loss. Sedimentation, flotation, centrifugation, and filtration are currently the most commonly used forms of water removal.[103,110] Sedimentation and flotation are often used to harvest cells from an open or closed pond, whereas centrifugation and filtration are often used when cells are grown in closed reactor systems.[109] Flocculating the cells before preforming a separation processes can help to improve the quantity of biomass that is recovered. Although these are the most commonly used methods of water removal, they are only cost-effective on a small scale or for very valuable products, such as pharmaceuticals or beauty products. Alternative technologies which improve on these techniques or new technologies are required in order to produce alternative fuels which can compete with low petroleum prices.

Sedimentation is the simplest method to recover cells from their medium. Cells have a higher density than the surrounding liquid they are suspended in. This property promotes the cells to settle at the bottom of a pond because the force from gravity is greater for the cells that have a larger mass than the particles that are suspended in the liquid medium. The strain and cellular structure of cyanobacteria/algae or diatom determine the efficiency of the sedimentation process[102]; cells that are higher in mass often have a greater yield of settled culture. Sedimentation is a very simple process; typically, methods that have been used to keep cells suspended (bubbling or stirring) are turned off and cells are allowed to settle. Depending on the cellular properties, this can take minutes to hours (or even days). One way to speed up this process is to use flocculation. A flocculant is a chemical substance that promotes clumping of particles, thus speeding up natural sedimentation processes. Flocculants can be inorganic or organic molecules, carry a negative or positive charge, and work at various pH levels. Phototrophs that grow in water-based media have a negatively charged surface. With the addition of a cation to the medium, cells with a negative charge can be neutralized.[102] The flocculant can also create a molecular link between two organisms. This is extremely useful when working with smaller cyanobacteria and microalgae because it effectively increases the cell size and mass. Flocculants are oftentimes expensive and can increase the cost of production; however, some can be recovered after sedimentation and reused, which minimizes the cost of this method of water removal.[115]

Flotation is a valuable approach used to harvest organisms that are smaller and unicellular. Commonly used in laboratory settings, this method attaches gaseous bubbles to a solid. The solid particle will be compelled to float to the top of the mixture, making it easy for the phototrophs to be skimmed off the top of the medium. Charged bubbles are commonly used during floatation harvesting because the negative charge on the bubble interacts with the hydrophilic surface of a cell creating an ionic bond.[102] The bubble promotes the cell to float to the surface of the pond. In some strains of algae and cyanobacteria, oxygen that is naturally produced in the organisms gets captured in the cell and allows the cells to naturally accumulate at the top of the solution.[109] Due to the ease of skimming the accumulated cell layer from the surface of the pond, this method of harvesting cells is used in open ponds.[103] The use of flotation in open ponds has shown to increase contamination levels and can be an expensive procedure. The high expense is linked to the technology needed to supply bubbles to the medium and producing the charged bubbles that interact with the negative surface of the organism.

Centrifugation is often referred to as an accelerated version of the sedimentation process. This process takes advantage of the density of the biomass in comparison to the liquid medium. During centrifugation, rotating walls or fixed walls in a hydrocyclone add an angular velocity to the separation process.[109] This angular acceleration increases the rate at which the biomass is concentrated at the bottom of the previously homogeneous mixture. The cells collect at the bottom of the vessel because they have a higher density than the liquid medium it was suspended in. Centrifugation is able to separate biomass more effectively then sedimentation, but requires large capital investment and increases operation costs significantly.[116]

Filtration is another commonly performed method to separate solids from liquids. The homogeneous mixture is moved through a semipermeable membrane and the liquid flows through the membrane, while the cells are unable to do so. Gravity is commonly used as the central driving force for the filtration; however, centrifugal force or pressure-driven force can be applied as well. For larger algae, a vacuum is applied to the filtration process. For smaller algae or cyanobacteria, a micro- or ultra-filtration must be used due to the smaller size of the cell (0.5-60 μm).[116] Filtration systems are prone to plugging and require constant maintenance and monitoring. Similar to centrifugation, filtration requires large capital investment and increases operation costs. These operation costs come from the need for electricity to power vacuum, centrifugal, and pressure filtration; and if the membrane is not robust or experiences extreme use, it will need to be replaced regularly.

After water removal, further processing is needed to extract and purify the product from the cell. Current methods of extraction include freezing and thawing the cell, acid/solvent extraction, or homogenization of cells,[102,117], and the chosen method depends heavily on the cell type and structure, and so will not be discussed in depth here. Purification after extraction or to remove the desired product from the growth medium (if it is excreted) is also heavily dependent on the type of molecule that is being purified, but it is an important consideration both during the design and construction phase of the platform strain and designing the downstream processing steps. Ideally, the upstream design of the platform strain and the downstream processing would be integrated into the total design from the very beginning of the design of the production strain.

14.4 Current Industrial Production

There are a handful of companies that are currently pursuing large-scale production of phototrophic biomass/bioproducts. Algenol Biofuels, located in Fort Meyers, Florida, is focused on producing ethanol, gasoline, and jet fuel using cyanobacteria grown in their proprietary plastic film photobioreactors, which are hung vertically and bubbled with carbon dioxide. Algenol reports yields of: 125 gallons of ethanol/ton CO_2, 8 gallons of diesel/ton CO_2, 6 gallons gasoline/ton CO_2, and 5 gallons of jet fuel/ton CO_2. Joule Unlimited, located in Bedford, Massachusetts, produces Joule *Sunflow® E* (photosynthetically produced ethanol) and *Sunflow®-D* (photosynthetically produced diesel) using cyanobacteria in closed tubular photobioreactor.[118] They report that they have successfully completed pilot-scale production and are now looking to scale their technology to 10,000 acres. Solazyme is the only commercial venture to use algae for large-scale production, and the only company which actually has a commercial product on the market. Located in San Francisco, California, the company focuses mainly on tailor-made hydrocarbons for beauty, food, and fuel industries. Some of their commercial products can be found in Algenist® and EverDeep® beauty products, AlgaVia™ proteins and flour, and SolaDiesel™ and SolaJet™. Their SolaJet™ aviation fuel was used in a commercial flight from Washington, D.C. to Chicago O'Hare in 2011 (Continental Airlines). While Solazyme has been very successful in the production of hydrocarbons and other value-added products from algae, they do not produce their products photosynthetically. Their large-scale production facilities take advantage of typical heterotrophic growth conditions in large bioreactors providing their algae cells with sugars or other sources of reduced carbon.

14.5 Conclusion

Phototrophic microorganisms, such as algae, diatoms, and cyanobacteria, have great potential to serve as a source of renewable fuels and chemicals. In recent years, there has been more effort focused on developing transformation techniques and more sophisticated tools for genetic manipulation of these organisms. With advanced genetic tools and more complete knowledge of the genomes and gene regulation of phototrophs, it is becoming easier to engineer these organisms. There is still a pressing need for developing more sophisticated genetic tools, a larger suite of selection markers, and characterized constitutive and inducible promoters with large dynamic ranges in order to fully take advantage of the potential of these organisms. Large-scale cultivation techniques for phototrophs are also advancing. There are numerous commercial-scale production facilities for value-added products, such as nutraceuticals and beauty products, which can be grown with higher operational costs than commodities like chemicals and fuels. However, there are a number of successful pilot-scale studies which have been completed by start-up companies who are now poised to move to larger production facilities, such as Joule Unlimited. With the increased emphasis on research in this area, the ability to cultivate these organisms in an economical way will be achieved in the near future. Despite the challenges associated with engineering and growing phototrophs, they have characteristics which make them ideal platforms in terms of long-term sustainability which cannot be ignored, the most important of which is their ability to directly convert carbon dioxide into useful products.

References

1. Oliver JWK, Machado IMP, Yoneda H, Atsumi S. Cyanobacterial conversion of carbon dioxide to 2,3-butanediol. *Proc Natl Acad Sci U S A* 2013;**110**(4):1249–54.
2. Oliver JWK, Machado IMP, Yoneda H, Atsumi S. Combinatorial optimization of cyanobacterial 2,3-butanediol production. *Metab Eng* 2014;**22**:76–82.
3. Kallas T, Nelson M, Singsaas E. *Methods for isoprene and pinene production in cyanobacteria.* WISYS Technology Foundation, Inc; Madison, WI; 2014.
4. Lindberg P, Park S, Melis A. Engineering a platform for photosynthetic isoprene production in cyanobacteria, using Synechocystis as the model organism. *Metab Eng* 2010;**12**(1):70–9.
5. Melis A. Photosynthesis-to-fuels: from sunlight to hydrogen, isoprene, and botryococcene production. *Energ Environ Sci* 2012;**5**(2):5531–9.
6. Liu XY, Sheng J, Curtiss R. Fatty acid production in genetically modified cyanobacteria. *Proc Natl Acad Sci U S A* 2011;**108**(17):6899–904.
7. Lagarde D, Beuf L, Vermaas W. Increased production of zeaxanthin and other pigments by application of genetic engineering techniques to Synechocystis sp strain PCC 6803. *Appl Environ Microbiol* 2000;**66**(1):64–72.
8. Zhou J, Zhang H, Zhang Y, Li Y, Ma Y. Designing and creating a modularized synthetic pathway in cyanobacterium *Synechocystis* enables production of acetone from carbon dioxide. *Metab Eng* 2012;**14**(4):394–400.
9. Tan X, Yao L, Gao Q, Wang W, Qi F, Lu X. Photosynthesis driven conversion of carbon dioxide to fatty alcohols and hydrocarbons in cyanobacteria. *Metab Eng* 2011;**13**(2):169–76.
10. Ungerer J, Tao L, Davis M, Ghirardi M, Maness PC, Yu JP. Sustained photosynthetic conversion of CO_2 to ethylene in recombinant cyanobacterium Synechocystis 6803. *Energ Environ Sci* 2012;**5**(10):8998–9006.
11. Atsumi S, Higashide W, Liao JC. Direct photosynthetic recycling of carbon dioxide to isobutyraldehyde. *Nat Biotechnol* 2009;**27**(12):1177–80.
12. Li H, Liao JC. Engineering a cyanobacterium as the catalyst for the photosynthetic conversion of CO_2 to 1,2-propanediol. *Microb Cell Fact* 2013;**12**:4.
13. Dismukes GC, Carrieri D, Bennette N, Ananyev GM, Posewitz MC. Aquatic phototrophs: efficient alternatives to land-based crops for biofuels. *Curr Opin Biotechnol* 2008;**19**(3):235–40.
14. Hu Q, Sommerfeld M, Jarvis E, Ghirardi M, Posewitz M, Seibert M, et al. Microalgal triacylglycerols as feedstocks for biofuel production: perspectives and advances. *Plant J* 2008;**54**(4):621–39.
15. Lubowski RN, Vesterby M, Bucholtz S, Baez A, Roberts MJ. *Major uses of land in the United States, 2002* [U.S.D.o. Agriculutre, editor.]. Washington D.C: United States Department of Agriculture; 2006. p. 12–21.
16. Brenner M *Engineering microorganisms for energy production* [U.S.D.o. Energy, editor.]. 2006. Washington D.C.
17. Shimogawara K, Fujiwara S, Grossman A, Usuda H. High-efficiency transformation of chlamydomonas reinhardtii by electroporation. *Genetics* 1998;**148**(4):1821–8.
18. Radakovits R, Jinkerson RE, Darzins A, Posewitz MC. Genetic engineering of algae for enhanced biofuel production. *Eukaryot Cell* 2010;**9**(4):486–501.
19. Kilian O, Benemann CSE, Niyogi KK, Vick B. High-efficiency homologous recombination in the oil-producing alga Nannochloropsis sp. *Proc Natl Acad Sci* 2011;**108**(52):21265–9.
20. Schmollinger S, Strenkert D, Schroda M. An inducible artificial microRNA system for Chlamydomonas reinhardtii confirms a key role for heat shock factor 1 in regulating thermotolerance. *Curr Genet* 2010;**56**(4):383–9.
21. Molnar A, Bassett A, Thuenemann E, Schwach F, Karkare S, Ossowski S, et al. Highly specific gene silencing by artificial microRNAs in the unicellular alga Chlamydomonas reinhardtii. *Plant J* 2009;**58**(1):165–74.
22. Zhao T, Wang W, Bai X, Qi Y. Gene silencing by artificial microRNAs in Chlamydomonas. *Plant J* 2009;**58**(1):157–64.

23. Moellering ER, Benning C. RNA interference silencing of a major lipid droplet protein affects lipid droplet size in chlamydomonas reinhardtii. *Eukaryot Cell* 2010;**9**(1):97–106.
24. Sun G, Zhang X, Sui Z, Mao Y. Inhibition of pds gene expression via the RNA interference approach in Dunaliella salina (Chlorophyta). *Marine Biotechnol* 2008;**10**(3):219–26.
25. De Riso V, Raniello R, Maumus F, Rogato A, Bowler C, Falciatore A. Gene silencing in the marine diatom Phaeodactylum tricornutum. *Nucleic Acids Res* 2009;**37**(14):e96.
26. Cheah YE, Albers SC, Peebles CAM. A novel counter-selection method for markerless genetic modification in Synechocystis sp. PCC 6803. *Biotechnol Prog* 2013;**29**(1):23–30.
27. Begemann MB, Zess EK, Walters EM, Schmitt EF, Markley AL, Pfleger BF. An organic acid based counter selection system for cyanobacteria. *PLoS One* 2013;**8**(10).
28. Berla BM, Saha R, Immethun CM, Maranas CD, Moon TS, Pakrasi H. Synthetic biology of cyanobacteria: unique challenges and opportunities. *Front Microbiol* 2013;**4**.
29. Huang H-H, Lindblad P. Wide-dynamic-range promoters engineered for cyanobacteria. *J Biol Eng* 2013;**7**(1):10.
30. Geerts D, Bovy A, de Vrieze G, Borrias M, Weisbeek P. Inducible expression of heterologous genes targeted to a chromosomal platform in the cyanobacterium Synechococcus sp. PCC 7942. *Microbiology* 1995;**141**(4):831–41.
31. Huang HH, Camsund D, Lindblad P, Heidorn T. Design and characterization of molecular tools for a Synthetic Biology approach towards developing cyanobacterial biotechnology. *Nucleic Acids Res* 2010;**38**(8):2577–93.
32. Guerrero F, Carbonell V, Cossu M, Correddu D, Jones PR. Ethylene synthesis and regulated expression of recombinant protein in synechocystis sp. PCC 6803. *PLoS One* 2012;**7**(11):e50470.
33. Niederholtmeyer H, Wolfstädter BT, Savage DF, Silver PA, Way JC. Engineering cyanobacteria to synthesize and export hydrophilic products. *Appl Environ Microbiol* 2010;**76**(17):6023.
34. Berto P, D'Adamo S, Bergantino E, Vallese F, Giacometti GM, Costantini P. The cyanobacterium Synechocystis sp. PCC 6803 is able to express an active [FeFe]-hydrogenase without additional maturation proteins. *Biochem Biophys Res Commun* 2011;**405**(4):678–83.
35. Camsund D, Heidorn T, Lindblad P. Design and analysis of LacI-repressed promoters and DNA-looping in a cyanobacterium. *J Biol Eng* 2014;**8**(1):4.
36. Camsund D, Lindblad P. Engineered transcriptional systems for cyanobacterial biotechnology. *Front Bioeng Biotech* 2014;**2**.
37. Salis HM, Mirsky EA, Voigt CA. Automated design of synthetic ribosome binding sites to control protein expression. *Nat Biotech* 2009;**27**(10):946–50.
38. Salis HM. Chapter two - the ribosome binding site calculator. In: Voigt C, editor. *Methods in enzymology*. Academic Press; 2011. p. 19–42.
39. Lennen RM, Pfleger BF. Modulating membrane composition alters free fatty acid tolerance in Escherichia coli. *PLoS One* 2013;**8**(1):e54031.
40. Makino T, Skretas G, Kang T-H, Georgiou G. Comprehensive engineering of Escherichia coli for enhanced expression of IgG antibodies. *Metab Eng* 2011;**13**(2):241–51.
41. Zurbriggen A, Kirst H, Melis A. Isoprene production via the mevalonic acid pathway in Escherichia coli (Bacteria). *BioEnergy Res* 2012;**5**(4):814–28.
42. Minty J, Lesnefsky A, Lin F, Chen Y, Zaroff T, Veloso A, et al. Evolution combined with genomic study elucidates genetic bases of isobutanol tolerance in Escherichia coli. *Microb Cell Fact* 2011;**10**(1):18.
43. Xu P, Gu Q, Wang W, Wong L, Bower AGW, Collins CH, et al. Modular optimization of multi-gene pathways for fatty acids production in E.coli. *Nat Commun* 2013;**4**:1409.
44. Temme K, Zhao D, Voigt CA. Refactoring the nitrogen fixation gene cluster from Klebsiella oxytoca. *Proc Natl Acad Sci* 2012;**109**(18):7085–90.
45. Markley AL, Begemann MB, Clarke RE, Gordon GC, Pfleger BF. A synthetic biology toolbox for controlling gene expression in the cyanobacterium Synechococcus sp. PCC 7002. *ACS Synth Biol* 2015;**4**:595–603.

46. Cong L, Ran FA, Cox D, Lin S, Barretto R, Habib N, et al. Multiplex genome engineering using CRISPR/Cas systems. *Science* 2013;**339**(6121):819–23.
47. Jiang W, Bikard D, Cox D, Zhang F, Marraffini LA. RNA-guided editing of bacterial genomes using CRISPR-Cas systems. *Nat Biotech* 2013;**31**(3):233–9.
48. Wang H, Yang H, Shivalila CS, Dawlaty MM, Cheng AW, Zhang F, et al. One-step generation of mice carrying mutations in multiple genes by CRISPR/Cas-mediated genome engineering. *Cell* 2013;**153**(4):910–8.
49. Yang H, Wang H, Shivalila CS, Cheng AW, Shi L, Jaenisch R. One-step generation of mice carrying reporter and conditional alleles by CRISPR/Cas-mediated genome engineering. *Cell* 2013;**154**(6):1370–9.
50. Mali P, Yang L, Esvelt KM, Aach J, Guell M, DiCarlo JE, et al. RNA-guided human genome engineering via Cas9. *Science* 2013;**339**(6121):823–6.
51. Fu Y, Foden JA, Khayter C, Maeder ML, Reyon D, Joung JK, et al. High-frequency off-target mutagenesis induced by CRISPR-Cas nucleases in human cells. *Nat Biotech* 2013;**31**(9):822–6.
52. Cho SW, Kim S, Kim JM, Kim J-S. Targeted genome engineering in human cells with the Cas9 RNA-guided endonuclease. *Nat Biotech* 2013;**31**(3):230–2.
53. Venter JC, Adams MD, Myers EW, Li PW, Mural RJ, Sutton GG, et al. The sequence of the human genome. *Science* 2001;**291**(5507):1304–51.
54. Human Genome Sequencing C. Finishing the euchromatic sequence of the human genome. *Nature* 2004;**431**(7011):931–45.
55. Adams MD, Celniker SE, Holt RA, Evans CA, Gocayne JD, Amanatides PG, et al. The genome sequence of Drosophila melanogaster. *Science* 2000;**287**(5461):2185–95.
56. Celniker S, Wheeler D, Kronmiller B, Carlson J, Halpern A, Patel S, et al. Finishing a whole-genome shotgun: release 3 of the Drosophila melanogaster euchromatic genome sequence. *Genome Biol* 2002;**3**(12):1–14.
57. Yandell M, Ence D. A beginner's guide to eukaryotic genome annotation. *Nat Rev Genet* 2012;**13**:329–42.
58. Misra S, Crosby M, Mungall C, Matthews B, Campbell K, Hradecky P, et al. Annotation of the Drosophila melanogaster euchromatic genome: a systematic review. *Genome Biol* 2002;**3**(12). research0083.1–0083.22.
59. Yandell M, Bailey AM, Misra S, Shu S, Wiel C, Evans-Holm M, et al. A computational and experimental approach to validating annotations and gene predictions in the Drosophila melanogaster genome. *Proc Natl Acad Sci U S A* 2005;**102**(5):1566–71.
60. Reese M, Guigo R. EGASP: introduction. *Genome Biol* 2006;**7**(Suppl. 1):S1.
61. Merchant SS, Prochnik SE, Vallon O, Harris EH, Karpowicz SJ, Witman GB, et al. The chlamydomonas genome reveals the evolution of key animal and plant functions. *Science* 2007;**318**(5848):245–50.
62. Stanke M, Steinkamp R, Waack S, Morgenstern B. AUGUSTUS: a web server for gene finding in eukaryotes. *Nucleic Acids Res* 2004;**32**(Suppl. 2):W309–12.
63. Miller R, Wu G, Deshpande RR, Vieler A, Gärtner K, Li X, et al. Changes in transcript abundance in Chlamydomonas reinhardtii following nitrogen deprivation predict diversion of metabolism. *Plant Physiol* 2010;**154**(4):1737–52.
64. Benning C, Miller R, Moellering ER. *Enzyme directed oil biosynthesis in microalgae* [U.S.P. Office, editor.]. Boars of Trustees of Michigan State Univeristy United States; 2010. US patent number 20100192258-A1.
65. Boyle NR, Page MD, Liu B, Blaby IK, Casero D, Kropat J, et al. Three acyltransferases and nitrogen-responsive regulator are implicated in nitrogen starvation-induced triacylglycerol accumulation in Chlamydomonas. *J Biol Chem* 2012;**287**(19):15811–25.
66. Alper H, Jin Y-S, Moxley JF, Stephanopoulos G. Identifying gene targets for the metabolic engineering of lycopene biosynthesis in Escherichia coli. *Metab Eng* 2005;**7**(3):155–64.
67. Lee KH, Park JH, Kim TY, Kim HU, Lee SY. Systems metabolic engineering of Escherichia coli for L-threonine production. *Mol Syst Biol* 2007;**3**(1):n/a–.
68. Asadollahi MA, Maury J, Patil KR, Schalk M, Clark A, Nielsen J. Enhancing sesquiterpene production in Saccharomyces cerevisiae through in silico driven metabolic engineering. *Metab Eng* 2009;**11**(6):328–34.
69. Hong SH, Lee SY. Metabolic flux analysis for succinic acid production by recombinant Escherichia coli with amplified malic enzyme activity. *Biotechnol Bioeng* 2001;**74**(2):89–95.

70. Segrè D, Vitkup D, Church GM. Analysis of optimality in natural and perturbed metabolic networks. *Proc Natl Acad Sci* 2002;**99**(23):15112–7.
71. Yim H, Haselbeck R, Niu W, Pujol-Baxley C, Burgard A, Boldt J, et al. Metabolic engineering of Escherichia coli for direct production of 1,4-butanediol. *Nat Chem Biol* 2011;**7**(7):445–52.
72. Cho A, Yun H, Park J, Lee S, Park S. Prediction of novel synthetic pathways for the production of desired chemicals. *BMC Syst Biol* 2010;**4**(1):35.
73. Hatzimanikatis V, Li C, Ionita JA, Henry CS, Jankowski MD, Broadbelt LJ. Exploring the diversity of complex metabolic networks. *Bioinformatics* 2005;**21**(8):1603–9.
74. Henry CS, Broadbelt LJ, Hatzimanikatis V. Discovery and analysis of novel metabolic pathways for the biosynthesis of industrial chemicals: 3-hydroxypropanoate. *Biotechnol Bioeng* 2010;**106**(3):462–73.
75. Burgard AP, Pharkya P, Maranas CD. Optknock: a bilevel programming framework for identifying gene knockout strategies for microbial strain optimization. *Biotechnol Bioeng* 2003;**84**(6):647–57.
76. Boyle N, Morgan J. Flux balance analysis of primary metabolism in Chlamydomonas reinhardtii. *BMC Syst Biol* 2009;**3**(1):4.
77. Chang RL, Ghamsari L, Manichaikul A, Hom EFY, Balaji S, Fu W, et al. Metabolic network reconstruction of Chlamydomonas offers insight into light-driven algal metabolism. *Mol Syst Biol* 2011;**7**.
78. Gomes de Oliveira Dal'Molin C, Quek L-E, Palfreyman R, Nielsen L. AlgaGEM - a genome-scale metabolic reconstruction of algae based on the Chlamydomonas reinhardtii genome. *BMC Genomics* 2011;**12**(Suppl. 4):S5.
79. Montagud A, Navarro E, Fernandez de Cordoba P, Urchueguia J, Patil K. Reconstruction and analysis of genome-scale metabolic model of a photosynthetic bacterium. *BMC Syst Biol* 2010;**4**(1):156.
80. Fu P. Genome-scale modeling of Synechocystis sp. PCC 6803 and prediction of pathway insertion. *J Chem Technol Biotechnol* 2009;**84**(4):473–83.
81. Triana J, Montagud A, Siurana M, Fuente D, Urchueguía A, Gamermann D, et al. Generation and evaluation of a genome-scale metabolic network model of Synechococcus elongatus PCC7942. *Metabolites* 2014;**4**(3):680–98.
82. Vu TT, Hill EA, Kucek LA, Konopka AE, Beliaev AS, Reed JL. Computational evaluation of Synechococcus sp. PCC 7002 metabolism for chemical production. *Biotechnol J* 2013;**8**(5):619–30.
83. Vu TT, Stolyar SM, Pinchuk GE, Hill EA, Kucek LA, Brown RN, et al. Genome-scale modeling of light-driven reductant partitioning and carbon fluxes in diazotrophic unicellular cyanobacterium Cyanothece sp. ATCC 51142. *PLoS Comput Biol* 2012;**8**(4):e1002460.
84. Young JD, Shastri AA, Stephanopoulos G, Morgan JA. Mapping photoautotrophic metabolism with isotopically nonstationary ^{13}C flux analysis. *Metab Eng* 2011;**13**(6):656–65.
85. Nöh K, Wiechert W. Experimental design principles for isotopically instationary ^{13}C labeling experiments. *Biotechnol Bioeng* 2006;**94**(2):234–51.
86. Zheng Y, Sriram G. Steady-state and instationary modeling of proteinogenic and free amino acid isotopomers for flux quantification. In: Dieuaide-Noubhani M, Alonso AP, editors. *Plant metabolic flux analysis*. Humana Press; 2014. p. 155–79.
87. Martzolff A, Cahoreau E, Cogne G, Peyriga L, Portais J-C, Dechandol E, et al. Photobioreactor design for isotopic non-stationary ^{13}C-metabolic flux analysis (INST 13C-MFA) under photoautotrophic conditions. *Biotechnol Bioeng* 2012;**109**(12):3030–40.
88. Chen X, Alonso AP, Shachar-Hill Y. Dynamic metabolic flux analysis of plant cell wall synthesis. *Metab Eng* 2013;**18**:78–85.
89. Allen DK, Evans BS, Libourel IGL. Analysis of isotopic labeling in peptide fragments by tandem mass spectrometry. *PLoS One* 2014;**9**(3).
90. Allen DK, Shachar-Hill Y, Ohlrogge JB. Compartment-specific labeling information in ^{13}C metabolic flux analysis of plants. *Phytochemistry* 2007;**68**(16–18):2197–210.
91. Mandy DE, Goldford JE, Yang H, Allen DK, Libourel IGL. Metabolic flux analysis using ^{13}C peptide label measurements. *Plant J* 2014;**77**(3):476–86.
92. Thajuddin N, Subramanian G. Cyanobacterial biodiversity and potential applications in biotechnology. *Curr Sci* 2005;**89**(1):47–57.

93. Hirschberg J. Production of high-value compounds: carotenoids and vitamin E. *Curr Opin Biotechnol* 1999;**10**(2):186–91.
94. Carballo-Cárdenas EC, Tuan PM, Janssen M, Wijffels RH. Vitamin E (α-tocopherol) production by the marine microalgae Dunaliella tertiolecta and Tetraselmis suecica in batch cultivation. *Biomol Eng* 2003;**20**(4–6):139–47.
95. Ward OP, Singh A. Omega-3/6 fatty acids: alternative sources of production. *Process Biochem* 2005;**40**(12):3627–52.
96. Tani Y. Algal and microbial production of vitamin E. In: Vandamme E, editor. *Biotechnology of vitamins, pigments and growth factors*. Netherlands: Springer; 1989. p. 95.
97. Simmons TL, Andrianasolo E, McPhail K, Flatt P, Gerwick WH. Marine natural products as anticancer drugs. *Mol Cancer Ther* 2005;**4**(2):333–42.
98. Gerwick WH, Sitachitta N. Chapter 4 nitrogen-containing metabolites from marine bacteria. In: Geoffrey AC, editor. *The alkaloids: chemistry and biology*. Academic Press; 1999. p. 239–85.
99. Gustafson KR, Cardellina JH, Fuller RW, Weislow OS, Kiser RF, Snader KM, et al. AIDS-antiviral sulfolipids from cyanobacteria (Blue-Green Algae). *J Natl Cancer Inst* 1989;**81**(16):1254–8.
100. Chisti Y. Biodiesel from microalgae. *Biotech Adv* 2007;**25**(3):294–306.
101. Brennan L, Owende P. Biofuels from microalgae—a review of technologies for production, processing, and extractions of biofuels and co-products. *Renew Sustain Energy Rev* 2010;**14**(2):557–77.
102. Rawat I, Ranjith Kumar R, Mutanda T, Bux F. Dual role of microalgae: phycoremediation of domestic wastewater and biomass production for sustainable biofuels production. *Appl Energ* 2011;**88**(10):3411–24.
103. Parmar A, Singh NK, Pandey A, Gnansounou E, Madamwar D. Cyanobacteria and microalgae: a positive prospect for biofuels. *Bioresour Technol* 2011;**102**(22):10163–72.
104. Boussiba S, Vonshak A. Astaxanthin accumulation in the green alga Haematococcus pluvialis1. *Plant Cell Physiol* 1991;**32**(7):1077–82.
105. Chen C-Y, Yeh K-L, Aisyah R, Lee D-J, Chang J-S. Cultivation, photobioreactor design and harvesting of microalgae for biodiesel production: A critical review. *Bioresour Technol* 2011;**102**(1):71–81.
106. Nakada E, Asada Y, Arai T, Miyake J. Light penetration into cell suspensions of photosynthetic bacteria and relation to hydrogen production. *J Ferment Bioeng* 1995;**80**(1):53–7.
107. Tans P, Keeling R. *Recent monthly average Muana Loa CO_2*. Global Greenhouse Gas Reference Netowrk; 2014. Available from: http://www.esrl.noaa.gov/gmd/ccgg/trends/index.html.
108. Yeh AC, Bai H. Comparison of ammonia and monoethanolamine solvents to reduce CO_2 greenhouse gas emissions. *Sci Total Environ* 1999;**228**(2–3):121–33.
109. Singh A, Nigam PS, Murphy JD. Mechanism and challenges in commercialisation of algal biofuels. *Bioresour Technol* 2011;**102**(1):26–34.
110. Saeid A, Chojnacka K. Toward production of microalgae in photobioreactors under temperate climate. *Chem Eng Res Des* (0).
111. Griffiths MJ, Harrison STJ. Lipid productivity as a key characteristic for choosing algal species for biodiesel production. *J Appl Physiol* 2008;**21**(5):493–507.
112. Geordianna DR, Mayfield SP. Exploiting diversity and synthetic biology for the production of algal biofuels. *Nature* 2012;**488**:329–35.
113. Bhatnagar A, Chinnasamy S, Singh M, Das KC. Renewable biomass production by mixotrophic algae in the presence of various carbon sources and wastewaters. *Appl Energ* 2011;**88**(10):3425–31.
114. Yang J, Xu M, Zhang X, Hu Q, Sommerfeld M, Chen Y. Life-cycle analysis on biodiesel production from microalgae: water footprint and nutrients balance. *Bioresour Technol* 2011;**102**(1):159–65.
115. Milledge JJ, Heaven S. A reveiw of the harvesting of micro-algae for biofuel production. *Rev Environ Sci Biotechnol* 2012;**12**:165–78.
116. Molina Grima E, Belarbi EH, Acién Fernández FG, Robles Medina A, Chisti Y. Recovery of microalgal biomass and metabolites: process options and economics. *Biotech Adv* 2003;**20**(7–8):491–515.
117. Lee AK, Lewis DM, Ashman PJ. Disruption of microalgal cells for the extraction of lipids for biofuels: processes and specific energy requirements. *Biomass Bioenerg* 2012;**46**:89–101.
118. Robertson D, Jacobson S, Morgan F, Berry D, Church G, Afeyan N. A new dawn for industrial photosynthesis. *Photosynth Res* 2011;**107**(3):269–77.

CHAPTER 15

Interpreting and Designing Microbial Communities for Bioprocess Applications, from Components to Interactions to Emergent Properties

A.E. Beck[*,†,1], **K.A. Hunt**[*,‡,1], **H.C. Bernstein**[§], **R.P. Carlson**[*,‡]
*Center for Biofilm Engineering, Montana State University, Bozeman, Montana, USA †Department of Microbiology and Immunology, Montana State University, Bozeman, Montana, USA ‡Department of Chemical and Biological Engineering, Montana State University, Bozeman, Montana, USA §Chemical and Biological Signature Science, Pacific Northwest National Laboratory, Richland, Washington, USA
[1]Contributed equally

15.1 Introduction

Most natural microbial ecosystems are the result of millions of years of natural selection in spatially and temporally dynamic landscapes. These ecosystems possess effective, highly evolved functions and are almost exclusively organized as polymicrobial communities. The study of natural and engineered microbial communities has benefited from new technologies such as increased resolution and throughput of omics measurements, development of new genetic systems for establishing model organisms, advanced cell isolation methods such as flow cytometry and cell sorting,[1] comprehensive databases such as KBase (http://kbase.us), and ever-growing computational power for performing *in silico* experiments such as community-scale metabolic network modeling.[2,3] Interest in using polymicrobial systems for applied bioprocesses stems largely from an effort to mimic and ultimately control the beneficial emergent properties that are often observed in natural ecosystems. These attractive attributes of communities have a potential to enable superior catalytic function compared to traditional monocultures based on simultaneous optimization of multiple tasks, increased productivity, and greater stability.

While the study of applied microbial communities is growing in popularity, the appreciation and use of communities for societal purposes is not new. In fact, applications of microbial communities date back at least 5000 years to early food preservation via lactic acid-producing bacterial communities used for yogurt production,[4] and evidence suggests that directed biogas production was practiced in Assyria and China going back at least 3000 years.[5] The significance of polymicrobial systems was also observed and tested by the earliest pioneers

Biotechnology for Biofuel Production and Optimization. http://dx.doi.org/10.1016/B978-0-444-63475-7.00015-7

of modern microbiology. In 1683, Antoine van Leeuwenhoek recorded observations of morphologically distinct "animalcules" collected from oral scrapings and, in 1877, Louis Pasteur tested antagonistic interactions between medically relevant bacteria.[6–9]

Numerous reviews reiterate, with overlapping content, the relevance and use of polymicrobial systems for human health, biological circuit synthesis, microbial computing, synthetic ecology, biomass degradation, and a myriad of other biological applications (Table 15.1). This review provides a generalized discussion of ecological foundations useful in understanding polymicrobial systems and highlights classical theories applicable to future microbial community engineering efforts. No attempt is made here to exhaustively outline every relevant study in this rapidly growing field; rather, case studies are selected to epitomize ecological themes and design motifs within the context of the current and future state of polymicrobial bioprocessing.

15.2 Definitions

The study of microbial communities has expanded from traditional biological disciplines to include a wide cross section of applied sciences. This broad expansion has resulted in the merging of concepts and terms from classical disciplines including biochemistry, computational biology, ecology, engineering, genetics, and microbiology. These fields use vernacular with varying connotations; hence, a list of terms is provided with the definitions used here to facilitate a unified discussion.

Table 15.1: Recent community-relevant review articles categorized by focus

Focus	Reference
Artificial symbiosis	[100]
Biodegradation	[101]
Bioenergy, biomaterials	[102]
Bioengineering	[46]
Bioprocess	[103, 104]
Bioprocess, experimental and theoretical	[11]
Biotechnology, algal biofuels	[105]
Cellulose degradation, bioprocess	[106]
Cyanobacteria/microalgae and bacteria	[107]
Ecological interactions, symbiosis	[108]
Engineering, bioprocess	[109]
Food fermentations	[110]
Industrial bioprocessing	[111]
Microbiome	[112]
Mining	[113]
Polymicrobial infections	[114]
Synthetic biology	[115]
Synthetic biology, bioprocess	[116]
Synthetic communities	[117]
Synthetic ecosystems	[22]
Viral interactions	[12]

A **microbiological community** is a collection of populations that may be comprised of prokaryotic, eukaryotic, or viral components; definable interactions are not a requirement. A **microbial community** is a microbiological community limited to prokaryotic and eukaryotic populations with no explicit accounting for viruses and no constraints on interactions. However, the ubiquity of viruses suggests that the vast majority of microbial species hosts viruses, and the existence of virus-free microbial communities is therefore thought to be exceedingly rare. A microbial community exhibiting positive interactions is referred to as a **consortium** (plural consortia)[10]; some uses of the term are more generic, referring to any interacting community. The origin of community populations can be used to further categorize the system. A **natural community** is defined as a collection of wild-type populations that have interacted in nature on evolutionary time scales. An **artificial community** is a collection of unmodified populations that have been assembled through manual intervention and are not thought to interact in natural habitats. A **synthetic community** is defined as a collection of genetically modified populations, whereas a **semisynthetic community** is a collection of populations with at least one wild-type population and at least one genetically modified population. A review of recent case studies exemplifying these different organizations can be found in Bernstein and Carlson.[11]

15.2.1 Community Components

Communities are ensembles of populations that serve as the system components. Populations can be classified by phylogenetic and/or phenotypic distinction. **Phylogenetically distinct** community populations are different species that can range across the domains of microbial life and can include viruses. Table 15.2 provides examples of components comprising interacting communities ranging from virus-virus systems to archaeon-eukaryote systems. Virus-virus and host-virus interactions are often overlooked in reviews of applied microbial communities despite their common distribution across most natural systems and their potential to impact engineered systems.

Table 15.2: Representative microbial communities organized by system components

Community members	Description	Reference
Virus-virus	Enhanced pathogenicity	[118]
Virus-bacterium	Enhanced infectivity of human cells	[119]
Virus-bacterium	Lysis of competitors	[120]
Virus-eukaryote	Suppression of additional virus infection (HIV)	[121, 122]
Virus-eukaryote	Prevention of diabetes	[123]
Virus-eukaryote	Thermal stress tolerance	[124]
Bacterium-bacterium	Differentiation of cyanobacteria to fix nitrogen	[125]
Archaeon-bacterium	Enhanced function via hydrogen exchange and methanogenesis	[21]
Bacterium-eukaryote	Cellulose conversion to value-added biochemicals	[126]
Archaeon-eukaryote	Enhanced cellulose degradation via hydrogen exchange and methanogenesis	[97]

These community interactions are not limited to pathogenicity and can represent a spectrum of interaction outcomes, including mutually beneficial effects as reviewed in Roossinck.[12]

Phenotypically distinct populations are not differentiable on the species level but exhibit separate expression patterns that often lead to niche differentiation. An **ecological niche** is the function or location of an organism within an ecosystem. The cyanobacterial strategy of cellular differentiation into specialized nitrogen-fixing heterocysts and vegetative cells is a well-studied example of a single-species population functioning as a community.[13] Another example is synthetic or adapted communities of *Escherichia coli* strains with different substrate specificities which can lead to niche partitioning in biofilms (e.g., oxic or anoxic separation).[14–17] **Biofilms** are polymer-encapsulated microbial populations attached to biological or abiotic surfaces.[18] Mass transfer typically limits the availability of resources within a biofilm, resulting in spatial heterogeneity which, in turn, leads to phenotypic differentiation.[16,18,19]

15.2.2 Interaction Outcomes

Community interactions can be classified as one of six outcomes. These interactions may be unidirectional, bidirectional, or of higher order; they can also be obligatory or facultative. Each of the interaction outcome categories presented here is symbolically summarized for a two-population community using "+" to indicate a population benefit, "–" to indicate an adverse effect, and "0" to indicate no effect. In practice, these interaction outcomes are often observed in communities comprised of more than two populations.[20]

(1) **Mutualism** (+/+): both populations benefit from the interaction(s). Syntrophy is a specific instance of mutualism associated with the cross-feeding of essential resources between populations.[21] (2) **Commensalism** (+/0): one population benefits from the interaction(s) while the other is unaffected. (3) **Ammensalism** (–/0): one population is adversely affected by the interaction(s) while the other is unaffected. (4) **Competition/antagonism** (–/–): both populations are negatively affected by the interaction(s), which can be due to functional redundancy or antagonistic interactions. (5) **Parasitism** (+/–): one population benefits and one population is adversely affected by the interaction(s). (6) **Neutralism** (0/0): neither population is affected by the interaction(s); alternatively, there is no interaction between populations. Synthetic ecology, another scientific field witnessing rapid growth, often attempts to assemble tractable, albeit constrained, systems to test these interaction outcomes.[22]

15.2.3 Interaction Mechanisms

The mechanisms that mediate known interactions can be divided into three generalized categories: metabolite exchange, physical interaction, and environmental modification. These mechanisms are typically combinatorial and interrelated within natural and engineered microbial communities. **Metabolite exchange** is any transfer of material and/or chemical energy between community populations and requires some combination of active transport (e.g., ABC-type transporters)

Table 15.3: Representative microbial communities organized by system interaction mechanisms

Interaction mechanism	Community	Description	Reference
Anabolic metabolite exchange	Filamentous anoxygenic phototrophs/cyanobacteria	Vitamin	[127]
Anabolic metabolite exchange	Uncultured marine bacteria	Siderophore	[128]
Anabolic and catabolic metabolite exchange	Sulfate-reducing bacteria/ methanogens	Alanine and hydrogen/ formate	[95]
Catabolic metabolite exchange	Sulfate-reducing bacteria/ methanogens	Hydrogen/formate	[129]
Catabolic metabolite exchange	Colon microbiota	Acetate cross-feeding	[130]
Quorum sensing exchange	Sludge community	Granulation and EPS production	[131]
Antibiotic/antimicrobial peptide exchange	*K. pneumoniae*/*E. coli*	Microcin-mediated ammensalism	[132]
Environment sensing exchange	*Vibrio* pathogen/crustacean hosts	Differential virulence impacts	[133]
Direct physical interaction	Cellulose-fed soil biofilm community	Filamentous structures for electron transfer	[134]
Environment modulation	Marine phytoplankton community	Hydrogen peroxide scavenging	[135]

and/or passive transport (e.g., diffusion). Exchanges of **anabolic resources,** resources used in biosynthetic processes (e.g., amino acids, nucleotides, vitamins, cofactors, and siderophores), are widely distributed among natural systems and represent targets for metabolically engineered polymicrobial systems (Table 15.3).[23,24] Exchanged metabolites can also serve as **catabolic resources**, resources used to produce cellular energy. Examples include electron donors or acceptors (e.g., hydrogen or oxygen, respectively), which are used in biologically mediated redox reactions to facilitate production of energetic molecules like NAD(P)H or ATP.

Exchanged material can serve functions other than anabolic or catabolic roles, such as modulating the behavior of community populations. **Quorum sensing**, a process by which organisms secrete and receive specific soluble metabolites that act as regulatory signals, is associated with a wide variety of natural multicellular functions, including coordinated biofilm formation, microbial pathogenicity, and culture bioluminescence.[25] In addition, quorum sensing has been proposed as a means of interrogating local environments, permitting a feedback mechanism for regulating phenotype (e.g., resource-intensive strategies such as enzyme secretion).[26,27] Modulating community population activity can also be realized through competitive strategies like the exchange of inhibitory or toxic metabolites.[28,29] Secretion of antibiotics and antimicrobial peptides is widely distributed and provides both offensive and defensive mechanisms within some communities.

Direct **physical interactions** between community populations can dictate community structure and function. These interactions are a hallmark mechanism of biofilms[19,30]; this physical association can be controlled or constrained by metabolite exchange.[31] Evidence suggests that direct physical contact between community populations can also be used for the transfer of electrons through materials such as cytochrome-rich extracellular structures, which have been characterized as outer membrane and periplasmic extensions.[32] Similarly, filamentous cable bacteria in microbial sediment communities permit electron transfer across centimeter-length scales.[33] Direct interpopulation electron transfer remains an active area of investigation.

Environmental modification is a mechanism of interaction in which a community population influences the local environment and thereby alters the niche(s) that other populations can inhabit. Modulation often occurs when one population consumes a chemical species (e.g., organic acids, oxygen, or hydrogen sulfide) that inhibits other populations.[14,34,35] Table 15.3 provides examples of natural and engineered systems organized by mechanism of interaction.

15.2.4 Emergent Properties

A major advantage of polymicrobial systems over traditional monoculture biotechnology is the potential for emergence of higher order properties. **Emergent properties** are attributes that are either not present or increased in magnitude from those properties characteristic of the individual system components. Contrary to many connotations, emergent properties are not always beneficial to a community; however, the discussions here are limited to positive attributes. **Stability** is defined by a community's response to perturbations,[36] although many other interpretations have been used for the concept of ecological stability and related properties such as robustness.[37] Generally, a stable community returns to its initial state after a small perturbation, while an unstable community does not. Stable community behavior is attributed to the two quantifiable metrics, resistance and resilience.[38] **Resistance** is defined as the degree to which community behavior is insensitive to a perturbation, and **resilience** is the rate at which community behavior returns to its original condition.[39] An increase in any combination of resistance and resilience is desirable for bioprocessing and often for **ecological fitness**, defined as the ability of an organism to survive and reproduce in an environment. Another set of properties that can emerge from community interactions is productivity and uptake. **Productivity** is defined by the rate at which material is produced; **uptake** is the material consumption rate. Many ecological studies have observed correlations between productivity and diversity (or species richness),[40] although a universally accepted mechanism has not been identified.

15.3 Ecological Theories for Interpreting and Designing Communities

Ecologists have historically studied multiscale material and energy flows between populations and their environments. Resource scarcity has influenced fundamental aspects of biological organization, including the elemental and macromolecular composition of microorganisms.[41–43] Hence, theories focused on themes of resource acquisition, resource investment, energetic efficiencies, and tradeoffs have been developed to explain observations of natural phenomena. Fundamental theories help organize observations, describe community behaviors, and enable strategic engineering of community composition or manipulation of environmental factors to improve natural and biotechnological processes.[44–46] The current section details three distinct but complementary ecological theories that provide explanations for the competitive basis of component organization and interaction outcomes found in many natural and engineered communities. These classical theories will remain relevant to ecologists and bioengineers as their respective fields further mature. Table 15.4 outlines some existing studies of microbial communities employing these theoretical concepts.

15.3.1 Maximum Power Principle

The maximum power principle describes community interactions based on acquisition of available energy. The maximum power principle asserts that the fitness of biological systems will increase with the rate of available energy harvest, resulting in a maximization of metabolic power (units of J/s).[47–49] From an evolutionary viewpoint, the principle predicts selection of systems that capture previously unutilized energy sources.[50] The ability to acquire available energy from the environment at faster rates leads to enhanced fitness by enabling more energy to be apportioned to survival and reproduction, while reducing available energy for competitors. Interactions between populations that increase the overall metabolic power of a system can lead to coexistence via cooperation[49] or niche differentiation such as the use of different substrates.[50] Additional related postulates have been proposed over the years, including maximization or minimization of entropy.[51,52] Box 15.1 graphically demonstrates interaction outcomes for two populations with different metabolic powers.

15.3.2 Resource Ratio Theory

Resource ratio theory is an ecological theory describing resource consumption, competition, and niche partitioning in which interpopulation interactions are defined with respect to shared resources.[53,54] These resources are often essential, but hemi-essential and substitutable resources

Table 15.4: Ecological theory studies

Theory	Community	Description	Reference
Maximum power	Aquatic planktonic microcosms	Effect of pH-controlled light on power acquisition	[50]
Maximum power	Two closely related picocyanobacteria	Stable coexistence due to light partitioning	[136]
Resource ratio	Competition studies with bacteria, phytoplankton, and zooplankton	Meta-analysis of prediction consistency	[137]
Resource ratio	Nine phytoplankton species	Competition for nitrogen, phosphorus, and light	[60]
Pareto/Tradeoff	Three-member hot springs mat community	Community productivity and inhibitory byproduct tradeoffs	[3]
Pareto/Tradeoff	Syntrophic *Geobacter* species	Interspecies electron transfer	[138]

BOX 15.1 Maximum Power Principle

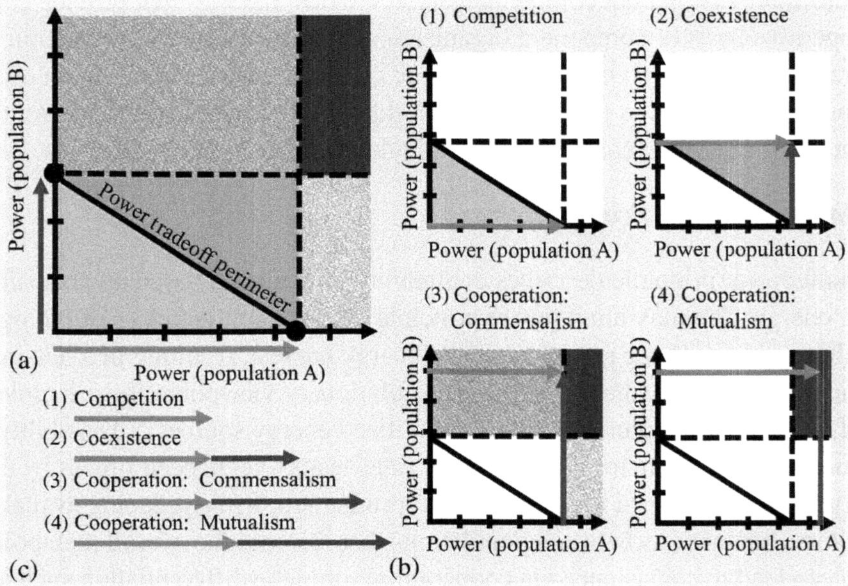

(a) *Graphical representation of key concepts from the maximum power principle.* Populations A and B are each characterized by a steady-state metabolic power when growing in monoculture (black dots on axes; red and blue arrows represent magnitude of power for A and B, respectively). The line connecting these two points, the **power tradeoff perimeter**, represents the partitioning of metabolic power between the two populations. Any point within the plot represents a cumulative metabolic power of the community through summing the individual metabolic powers for populations A and B.

(b) *Interaction outcomes interpreted with maximum power principle.*

(b1) Populations A and B are competing for resources within the same ecological niche. According to the maximum power principle, the fittest population has the largest metabolic power. Therefore, population A will outcompete population B. The total metabolic power of any mixture of two competing populations in the same niche is a nonnegative linear combination of the individual metabolic powers and by definition cannot exceed the power tradeoff perimeter at steady state. For competition-based unsteady state scenarios, the community power will shift toward the population capable of the most metabolic power, which will eventually dominate (population A here).

(b2) Coexistence can occur when populations obtain energy from separate resources, allowing each population the potential to achieve its respective maximum power. Total community power may exceed the power tradeoff perimeter but is bounded by the magnitude of the individual population maximum powers. The gray region represents antagonism, as neither population is able to achieve its maximum power. A neutralistic community exists at the upper right corner (both populations achieve their maximum powers), while the edges bounding this region indicate an ammensal community in which one population realizes its maximum power and one does not.

BOX 15.1 Maximum Power Principle—cont'd

(b3) Cooperating populations can exceed the maximum power of neutralistic populations if one population facilitates the acquisition of energy by the other population (e.g., commensal interaction) (boundary of the green region). Parasitic interactions may reduce such gains in community power because the increased power acquisition of one population occurs at the expense of the other population (red and blue patterned regions indicate parasitism favoring populations A and B, respectively).
(b4) The total community power can exceed the maximum power of neutralistic populations through mutualistic interactions. The two populations facilitate acquisition of metabolic power by each other, placing the community power in the green region. The final magnitude of an unsteady state trajectory is unknown without further information about the interaction.
(c) *Comparing community metabolic powers.* The example metabolic powers of populations A and B shown in (b1)–(b4) are summed and aligned to visualize the increase in metabolic power that may occur through community interactions.
Figure adapted from DeLong.[51]

have also been studied.[55] Resource ratio theory postulates that the population best capable of depleting a limiting resource while maintaining a positive growth rate will be most competitive. The resource ratio theory has been used to assess outcomes (competitive exclusion or coexistence) between populations competing for shared limiting resources and can predict resource levels that will permit coexistence of multiple populations.[53,55–57] The theory can be consistent with classic chemostat theory, which states that it is possible to sustain multiple stable populations when there are multiple limiting substrates, provided that each population is limited by a different resource.[58–60]

Resource ratio theory can describe cooperative populations by accounting for mutualistic resource exchange. Cooperating populations that exchange limiting resources can exist in a wider range of resource environments than is possible for either population individually. This scenario describes what has been termed a super-competitor unit,[54] a community with the emergent property of enhanced resource utilization that can deplete limiting resources more effectively than the individual monocultures. This ability to survive at reduced resource availabilities highlights an evolutionary advantage of cooperation that has been observed in natural ecosystems.[54] Resource ratio theory extended to cooperation is analogous to the economic concept of comparative advantage, where specialization and resource exchange enable enhanced community function.[61] Box 15.2 describes a generalized example of resource ratio theory applied to two different populations, each of which is more effective at depleting a different essential resource.[54]

15.3.3 Resource Allocation Theories: Pareto Surfaces and Metabolic Tradeoff Analysis

Biological systems from enzymes to communities represent competitive resource allocation.[62] Phenotypic plasticity, the ability to change phenotype with changing environment, can enable different relationships between cellular function and resource allocation. However,

BOX 15.2 Resource Ratio Theory

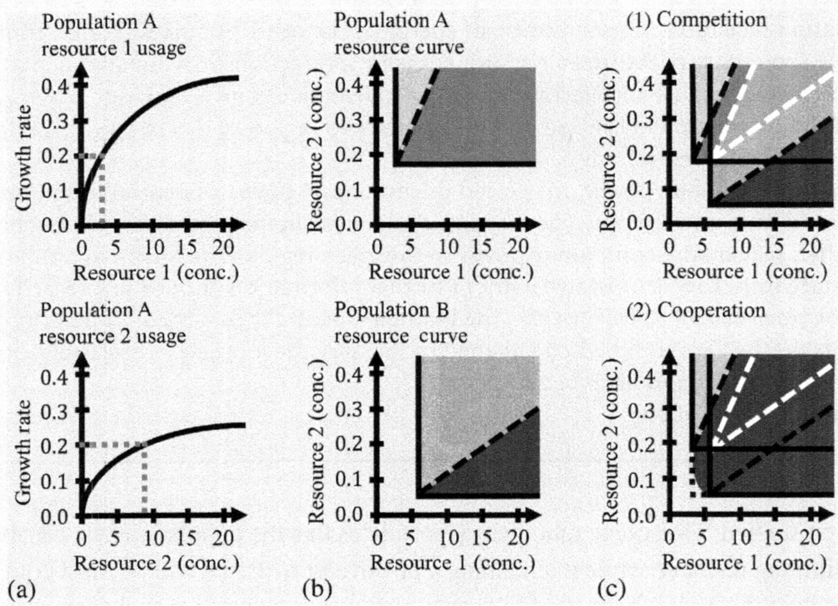

Graphical representation of key concepts from resource ratio theory.

(a) *Determination of resource requirements.* The resource concentrations required to support steady state growth of population A at different growth rates are tabulated using Monod expressions; an analogous set of data for population B is not shown. Resources 1 and 2 are essential for both species (e.g., nitrogen or phosphorus sources). The red dotted lines denote a particular growth rate selected for study and the corresponding resource concentrations required to support steady state growth.

(b) *Population growth responses to limiting resources.* The minimum requirements for resources 1 and 2 to maintain a steady state growth rate are indicated by the solid black edges for populations A (top) and B (bottom). Population A has a higher affinity for resource 1 and a lower affinity for resource 2 as compared to population B. Thus, populations A and B are more effective at depleting resources 1 and 2, respectively. The shaded areas represent conditions sustaining positive growth for each population. The slope of the dashed line separating the lightly shaded region (resource 1-limited) from the darkly shaded region (resource 2-limited) represents the resource ratio requirement (resource 2 per resource 1 necessary for growth). If both resources are in excess, the population will maintain positive growth and consume resources as governed by its resource ratio requirement, until its consumption reaches the steady-state resource level indicated by the solid lines.

(c) *Interaction outcomes interpreted with resource ratio theory.*

(c1) Resource ratio theory describes community interactions by combining the growth responses of the two populations and examining the resulting regions. If populations A and B are competing for resources 1 and 2, the population that can drive a resource concentration below the acquisition capability of the other will dominate the ecological niche. Population A will dominate the community in the red region due to its superior ability to deplete resource 1,

BOX 15.2 Resource Ratio Theory—cont'd

and population B will dominate the community in the blue region due to its superior ability to deplete resource 2. These regions are bounded by intersecting the resource ratio requirements of the two populations (white dashed lines). The gray region (coexistence), bounded by the white dashed lines, represents an excess of resource 1 for population A and an excess of resource 2 for population B. The ratio of resources permits coexistence as both populations are limited by their noncompetitive resources.
(c2) Resource ratio theory can describe cooperating populations utilizing mutualistic resource exchange. The region of coexistence (green region) is expanded (c1 gray region) due to resource trading. In the light red region, population A is resource 1-limited, leaving no excess resource 1 to trade with population B. Therefore, population A will dominate the community since it is more competitive not to trade with population B. In contrast, to the right of the light red region, population A is resource 2-limited and has excess resource 1, making it more competitive to exchange resource 1 with population B for resource 2. Analogous reasoning explains the dark blue region. Additionally, through resource trading, populations A and B are able to exist in a wider range of resource environments than is possible for either population individually (dotted box, lower left). In this region, both resources 1 and 2 are limiting for both populations. Thus, resource exchange is more competitive since neither population can maintain growth under these resource conditions without the other. The rectangular boundary around this extended resource availability region represents an ideal one-to-one trading scenario; however, the shape of this region may take on an ellipsoidal form depending on the costs and relative fluxes of resource exchange.
Figure adapted from de Mazancourt and Schwartz.[54]

phenotypic plasticity requires additional resources for the genes and regulatory systems to express different phenotypes under different environments. This investment cost for phenotypic plasticity may be ecologically justified for populations that persist in spatially and/or temporally dynamic environments.[63]

Tradeoff theories analogous to those used in economics have been applied to metabolic systems to assess the competitive use of phenotypic plasticity.[64] For many cellular behaviors, a benefit in one objective is realized only at the detriment of another objective. These tradeoffs are postulated to have tamed the "Darwinian demon," a superspecies that can optimize all objectives simultaneously and which has never been observed in nature or the laboratory.[65] The boundary that describes the tradeoff between resource uses is referred to as the Pareto front, efficiency frontier, or tradeoff surface.[64,66–68] Analysis of these resource allocation strategies is often paired with mathematical analysis such as stoichiometric modeling approaches (e.g., elementary flux mode and flux balance analyses) and/or kinetic models (Table 15.5).[66,69] Phenotypic phase plane analysis and multi-objective optimization have been used in stoichiometric modeling approaches to examine the effects of resource allocation on competitive network function.[70–75] The tradeoff concept can be represented graphically via rates or efficiencies of resource utilization, a measure of cost. Box 15.3 demonstrates a generalized

Table 15.5: Microbial community *in silico* analysis studies

Focus	Description	Reference
Review of modeling	Application of genome-scale models to microbial communities	[2]
Cyanobacterial mat	Three EFMA modeling approaches	[3]
Geobacter electron transfer	Multi-omic modeling	[138]
Synthetic microbial ecosystems	Prediction of symbiotic communities via environmental constraints	[139]
SRB-methanogen interactions	FBA	[140]
Spatial community dynamics	Reaction-diffusion FBA	[141]
Three case studies of OptCom algorithm	Community FBA	[142]
Consortial studies	Dynamic community FBA	[143-145]

EFMA, elementary flux mode analysis; FBA, flux balance analysis

BOX 15.3 Resource Allocation Theory: Pareto/Tradeoff Analysis

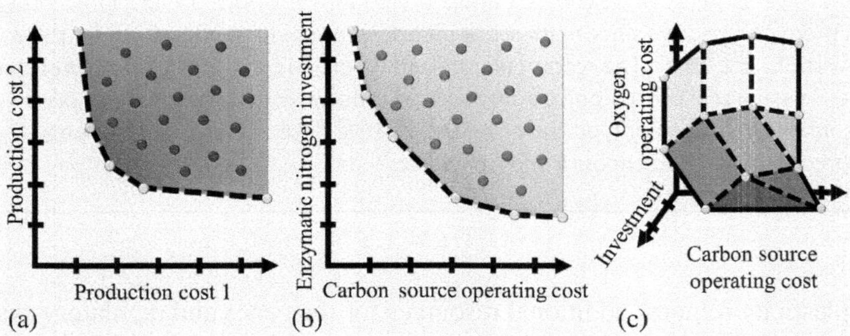

(A) *A generalized Pareto tradeoff curve* between two economic costs associated with the production of a good (for instance, the cost of fertilizer and pesticide required to produce a bushel of corn). The plot quantifies the tradeoff between minimizing the two different costs. In the example, the yellow points and connecting line segments represent optimal cost-minimizing relationships between the variables, while green interior points represent higher cost alternative strategies.
(B) *A metabolic tradeoff curve* analogous to a generalized Pareto tradeoff curve between two metabolic costs: (1) the carbon source operating cost, or the amount of carbon source required to synthesize a unit of bioproduct, and (2) enzymatic nitrogen investment, or the amount of nitrogen required to synthesize the enzymes in the utilized pathways. The yellow points represent the continuum of optimal pathways for producing the bioproduct, given the objectives of minimizing the cost of carbon source or nitrogen investment; interior red points represent less efficient alternative pathways for producing the bioproduct.
(C) *Three-dimensional metabolic tradeoff surface* considering the requirements for three resources (carbon source, nitrogen, and oxygen). The resulting Pareto surface represents the optimal relationship between these three costs for synthesizing a bioproduct. For ease of visualization, the interior volume of the tradeoff surface is omitted.

example of a Pareto optimization tradeoff curve for economic costs, with an analogous scenario translated into metabolic terms.

Single population examples of the predictions from resource allocation tradeoffs include the predicted use of the glyoxylate shunt and the Entner-Doudoroff glycolysis pathway during *E. coli* nutrient-limited growth. These alternative pathways produce less ATP than the citric acid cycle and Embden-Meyerhof-Parnas glycolysis pathway, respectively, but also require fewer anabolic resources, such as nitrogen and carbon, to synthesize the pathway enzymes as compared to the higher ATP-yielding pathways.[74,76,77]

In addition to modulating metabolic pathways, resource investment transitions are predicted to occur based on allocation of resources to either enzyme or substrate pools. Classic Michaelis-Menten kinetics describe the driving of flux via enzyme ($v_{max} = k_{cat} \cdot [E]$) and substrate pools. These two pools create a continuum of drivers for a single flux, ranging from high relative substrate concentration to high relative enzyme concentrations (Figure 15.1a); these different flux-driving mechanisms are referred to as push- or pull-based mechanisms, respectively.[78–80] The optimal distribution of resources between substrate and enzyme pools can vary depending on the required flux (Figure 15.1a, dotted line). Taking the FumC enzyme as an example, relatively small fluxes are best driven by a substrate-based push mechanism which minimizes overall resource requirements, while higher fluxes are best driven by a pull mechanism in which elevated enzyme concentrations represent a more efficient use of anabolic carbon.

Resource allocation between enzyme and substrate pools can also benefit microbial communities when populations exchange metabolites such as amino acids, vitamins, or nucleotides. Maximization of flux per investment of a valuable resource can be applied to a community exchanging metabolites such as cyanobacterial heterocyst-vegetative cell interactions.[13] As a numerical illustration, five interacting populations can each individually support a flux of 50 μM/s through enzyme X, which would require a total substrate and enzyme carbon investment of 1.51 CmM (millimolar concentration of carbon). Alternatively, one population can specialize in that function and drive a flux of 250 μM/s with a total carbon investment of 4.43 CmM and then exchange the product with the community. The division of labor strategy with product exchange reduces the total community carbon investment requirement by ~41%, neglecting the cost of metabolite exchange. Figure 15.1b demonstrates the general principle of flux as a function of total carbon investment into both enzyme and substrate pools. The net community investment savings continue to improve as the flux magnitude increases. Extrapolating to entire pathways will likely offset the involvement of transport proteins while maintaining resource investment savings. While this economy of scale holds from a purely reaction kinetics basis, additional factors such as diffusion rates and protein synthesis machinery could play substantial roles.

15.4 Case Studies of Communities with Interpretation

The maximum power principle, resource ratio theory, and resource allocation theory are useful for developing and interrogating the design principles of microbial communities. The current section provides an analysis of natural and engineered communities with dissection of relevant components, interactions, and theories.

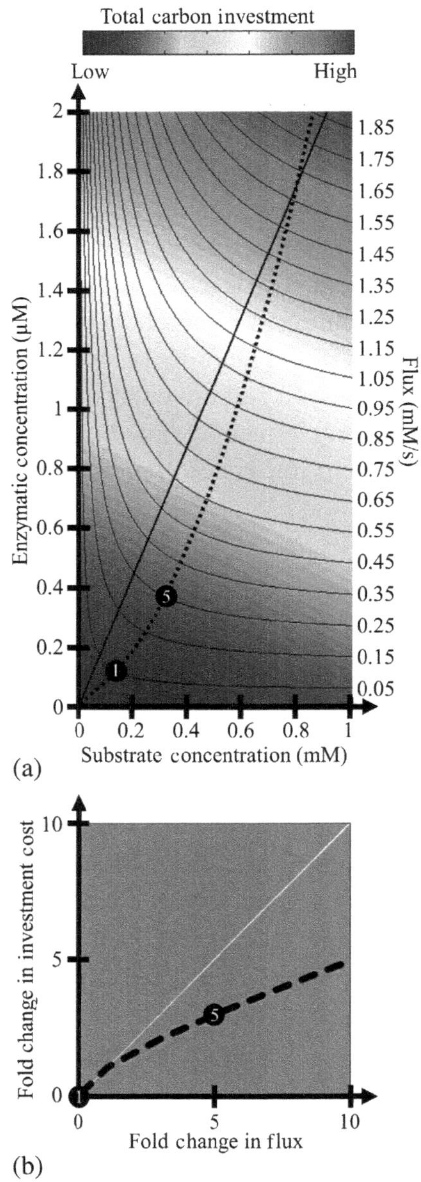

Figure 15.1

See figure legend on next page

15.4.1 Case Study: Multilevel Interactions in a Naturally Occurring Phototrophic Community

Community Description

The naturally occurring phototrophic community "*Chlorochromatium aggregatum*" is comprised of the phototrophic green sulfur bacterium *Chlorobium chlorochromatii* physically attached to the heterotrophic β-proteobacterium *Candidatus* Symbiobacter mobilis (Figure 15.2). This community is typically arranged as 13 green sulfur bacteria surrounding a single central β-proteobacterium and is found in sulfide-containing, oxic-anoxic interfaces of stratified lakes.[81,82] The green sulfur bacteria are nonmotile anoxygenic phototrophs which use light energy and electrons from sulfide to fix carbon and nitrogen. The heterotrophic β-proteobacterium is motile and possesses genetic evidence of phototactic and chemotactic capabilities but cannot harvest light energy and has limited respiratory capacities.[83] The green sulfur bacterium, although culturable in the laboratory, is not found free-living in nature, and the β-proteobacterium is unculturable independent of the green sulfur bacteria.[83,84]

Interaction Mechanisms

"*C. aggregatum*" interacts via a number of mechanisms. First, the green sulfur bacteria exchange anabolic and catabolic metabolites (reduced carbon and nitrogen) with the β-proteobacterium. Metabolite exchange is hypothesized to be bidirectional; the β-proteobacterium may synthesize compounds such as acetate that could be assimilated by the green sulfur bacteria due to limited respiratory capabilities.[83] Additionally, the β-proteobacterium and green sulfur bacteria are connected by periplasmic tubules. The

Figure 15.1
Resource investment and economies of scale for enzymatic flux. (a) A target enzymatic flux described by Michaelis-Menten kinetics can be achieved by a continuum of substrate or enzyme concentrations. Thin solid curves depict reaction isoflux lines that increase in 100 μM/s increments moving up and to the right. The substrate and enzyme concentrations each have associated resource costs related to the size and composition of the molecules. The total resource investment cost, depicted in color, sums the carbon investment from both the substrate and enzyme pools. The heavy solid line marks equal carbon investment contributions from substrate and enzyme pools. The dotted line represents the minimum total carbon investment for each isoflux line. Enzyme and substrate calculations model FumC from *E. coli* using $K_{cat} = 1150\,s^{-1}$, $K_m = 0.207\,mM$, assuming fumarate, and 8800 carbon atoms per functional enzyme. This principle can be extrapolated to entire pathways, posing a mechanism for improved resource efficiency with increased specialization, as represented by the economies of scale. (b) The relationship between increasing flux and resource investment can be described based on the optimal minimization of investment. Total carbon investment per flux can be higher (red), lower (green), or the same (boundary) at larger fluxes as compared to smaller fluxes. In the example depicted, an increase in flux from 50 μM/s (point 1) to 250 μM/s (point 5) requires only threefold more resources.

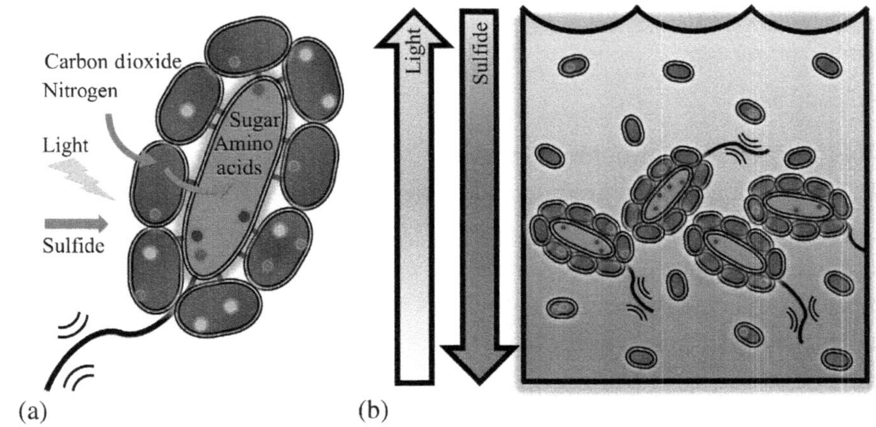

Figure 15.2
Phototrophic community interactions. (a) Green sulfur bacteria fix carbon dioxide and nitrogen from the environment and exchange reduced products with the heterotrophic β-proteobacterium. In return, the β-proteobacterium maintains a flagellum and provides taxis toward light and sulfide, which the green sulfur bacteria require for photosynthesis. (b) In a stratified lake environment, light is more intense at the top of the lake, and sulfide concentration increases toward the bottom. The motile community is able to take advantage of ideal concentrations of both resources. Free-living green sulfur bacteria cannot actively direct their movement toward resources and the β-proteobacterium is not able to survive independently.

direct physical interaction between the bacteria enables the β-proteobacterium to provide community motility toward light and sulfide.[85,86] These intercellular contacts could also permit exchange of electrons via soluble carriers,[87] enabling community-wide energy transfer and redox balancing.[83]

Theory Applications

Resource ratio and resource allocation theories provide an evolutionary rationale for the interactions of these two bacteria. In an environment with scarce resources (e.g., light, sulfide, reduced carbon, reduced nitrogen), this cooperative strategy confers a selective advantage.[54] The physical interactions and metabolite exchange are consistent with the resource ratio theory concept of a super-competitor unit. The motility contributed by the β-proteobacterium enables a competitive advantage for growth at low sulfide concentrations and lower light intensities via directed movement toward essential resources.[88] Additionally, the community configuration minimizes the associated resource investment costs. The direct physical connection between the bacteria permits 14 cells to benefit from the ability to move actively toward light and sulfide while only one cell needs to maintain the sensing and motility genes, synthesize flagellar proteins, and power flagellar operation.[83]

15.4.2 Case Study: Anaerobic Syntrophy in Methanogenic Communities

Community Description

Methane is a major focus of renewable energy efforts and plays an essential role in ecological food webs. Anaerobic communities produce methane by catabolizing organic feedstocks like biomass via a cascade of cross-feeding microorganisms. Sulfate-reducing bacteria (SRBs) and methanogenic archaea are key members of these anaerobic communities found in aquatic sediments and anaerobic digesters.[89] The SRBs oxidize organic acids and produce hydrogen or formate in the absence of alternative electron acceptors. The methanogens catabolize hydrogen and formate for cellular energy, ultimately producing methane as a by-product (Figure 15.3).[90,91]

Interaction Mechanisms

Methanogenic communities have a strong requirement for metabolite exchange due to the thermodynamic constraints of hydrogen synthesis.[92–94] Hydrogen partial pressures above a very small critical threshold (~10 Pa) shift the chemical reaction equilibrium from hydrogen synthesis to hydrogen consumption, stalling SRB metabolism in the absence of sulfate. The consumption of hydrogen by methanogens reduces local hydrogen partial pressures below the critical threshold, enabling further SRB catabolism.[94] The mutualistic interactions can extend further. In addition to providing the methanogens with substrate, the SRBs can also modify the local environment by reducing sulfate to sulfide rather than forming hydrogen. Sulfide is highly reactive with oxygen and reduces the environmental concentrations of oxygen,

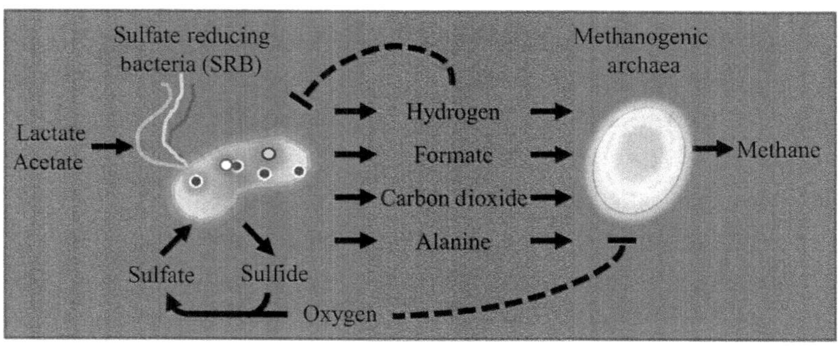

Figure 15.3
Sulfate-reducing bacteria (SRBs)—methanogen anaerobic syntrophy. SRBs create a reducing environment conducive to methanogenesis by reducing sulfate to sulfide, which can then abiotically react with oxygen. In the absence of sulfate, SRBs produce hydrogen; this metabolism becomes unfavorable without a hydrogen sink. Methanogens consume the hydrogen, facilitating SRB oxidation of carbon. SRBs cross-feed anabolic resources like amino acids to facilitate methanogen growth, resulting in larger hydrogen sink.

benefiting the oxygen-inhibited methanogens. In addition, research suggests that SRBs enhance methanogen consumption of hydrogen by exchanging anabolic resources like amino acids.[95]

Theory Applications

SRB-methanogen communities are a well-studied example of anaerobic syntrophy.[21] The cross-feeding of hydrogen, which can also be viewed as an environmental modification interaction, represents a classic example of the maximum power principle by collectively increasing the potential for energy acquisition from the environment. Metabolite exchange allows the SRBs to consume substrate at higher rates, thus increasing the collective community metabolic power. The exchange of anabolic resources such as amino acids is proposed to facilitate methanogen growth by providing an energetically expensive metabolite, thus increasing methanogen consumption of hydrogen.[95] The amino acid cross-feeding strategy represents an example of improved function per anabolic resource investment according to resource allocation theory (Figure 15.1b).

The syntrophic interactions of methanogens are not strictly limited to hydrogen-producing SRBs. Hydrogen-producing fermenters, which provide the substrate for SRBs, also benefit from methanogens.[96] Hydrogen transfer has also been observed in communities of hydrogen-producing anaerobic fungi and methanogens, which demonstrate the emergent property of more complete degradation of cellulosic biomass.[97]

15.4.3 Case Study: Cross-Feeding Chemostat Communities

Community Description

Culturing a clonal population of *E. coli* in a glucose-limited chemostat for hundreds of generations results in the reproducible formation of multiple, phenotypically distinct *E. coli* populations which cross-feed secondary metabolites (Figure 15.4).[17,98,99] The interacting community consists of three distinct functional populations: a glucose-catabolizing glycerol- and acetate-producing specialist, an acetate-catabolizing specialist, and a glycerol-catabolizing specialist. Both the acetate- and glycerol-catabolizing specialists completely oxidize their respective substrates. The population functions are based primarily on mutations in gene regulatory sequences, resulting in altered gene expression.[98] For instance, the glucose-catabolizing specialist expresses the high-affinity Mgl sugar transporter and has reduced expression of citric acid cycle enzymes, resulting in acetate and glycerol secretion. This strain, which represents 80% of the community, grown in isolation has a lower specific growth rate and a lower biomass per glucose yield than the parent monoculture; yet the community demonstrates a 15% improvement in biomass per glucose yield. The reproducibility of the interacting consortium suggests that the organization of components and interactions is more competitive and stable than an *E. coli* monoculture in a homogeneous environment like a chemostat.

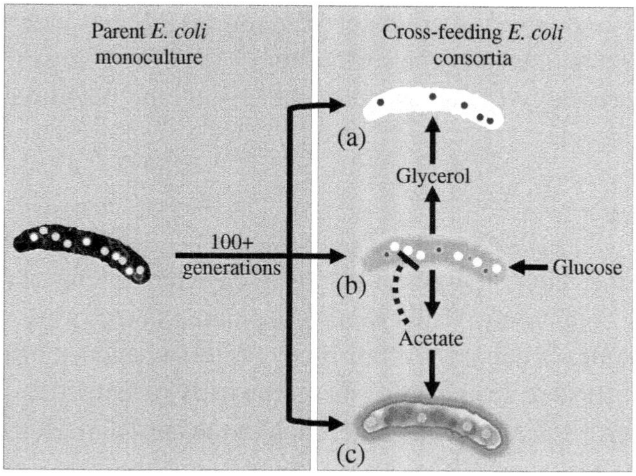

Figure 15.4
Naturally occurring *E. coli* crossing-feeding communities. *E. coli* monocultures grown long-term in glucose-limited chemostats evolve into cross-feeding communities. Community members include a glucose-catabolizing specialist that secretes acetate and glycerol (b), an acetate-catabolizing specialist (c), and a glycerol-catabolizing specialist (a). The cross-feeding community has an ~15% improvement on biomass per glucose yield as compared to the parent monoculture and depletes available substrates more completely than the parent monoculture.

Interaction Mechanisms

The community populations interact through the cross-feeding of catabolic resources. Additionally, consumption of acetate represents an environmental modification interaction that reduces the inhibition of the glucose-catabolizing strain.

Theory Applications

The cross-feeding *E. coli* community highlights key features of the maximum power principle, resource ratio theory, and resource allocation theory. The consortium has a higher metabolic power than the monoculture because inhibitory by-products (e.g., acetate) are removed during growth, allowing for a higher community metabolic rate. The community, as a whole, demonstrates the emergent property of a super-competitor unit defined in resource ratio theory; through metabolite exchange, the community drives the concentrations of carbon-based resources to levels unattainable by the parent monoculture. The populations also demonstrate an enhanced resource allocation configuration through division of metabolic labor. Each respective population increases expression of enzymes associated with a portion of central metabolism. The glucose-consuming specialist has higher glycolytic expression but lower citric acid cycle fluxes, resulting in the secretion of acetate and glycerol.[17,98,99] The strategy is representative of the tradeoff between substrate-driven and enzyme-driven fluxes; driving a reduced number of higher fluxes can be less resource intensive than driving a greater number of smaller fluxes.

A synthetic biology analog of the cross-feeding community has been created based on gene deletion rather than altered gene regulation.[14] The synthetic system demonstrated improved biomass productivity under batch, chemostat, and biofilm culturing conditions.

15.5 Conclusions

The ebbs and flows of scientific efforts build largely on achievements of the predecessors of the modern scientific community. While gaining popularity in recent years, the importance of microbial communities is not new. At this time of renewed interest in a systems-focused future, it is critical to keep a broad view. The early reductionist scientists are to be commended for laying a solid foundation that enables the current trajectory toward systems studies. In fact, reductionist approaches have a bright future because systems cannot be constructed solely on noble postulations but must be grounded in well-described components. It is not possible to define an emergent property until the basic components and interactions are appropriately catalogued.

A paramount goal and current challenge facing microbial community design is to mechanistically understand the design principles that govern higher order attributes of polymicrobial communities. The definitions presented here lay a foundation for a common language to bridge gaps across varying disciplines and enable greater collaboration between fields studying polymicrobial communities. The three ecological theories discussed—maximum power principle, resource ratio theory, and resource allocation theory—distill a broad range of literature regarding ecological theory into a concise synthesis and highlight the relevance of these theories to community interaction outcomes. The usefulness of these theories is illustrated through application to specific case studies in both natural and engineered environments. The tenets of resource acquisition, concentration, and allocation will remain useful for developing and examining the design principles of microbial communities and guide the future of polymicrobial bioprocess engineering.

Acknowledgments

The development of this work was supported by the National Institutes of Health awards (1U01EB019416-01, P20GM103474) and the Genomic Science Program (GSP), Office of Biological and Environmental Research (OBER), U.S. Department of Energy (DOE), and is a contribution of the PNNL Foundational Scientific Focus Area (Principles of Microbial Community Design) subcontracted to Montana State University. Ashley E. Beck was supported by the Office of the Provost at Montana State University through the Molecular Biosciences Program. Hans C. Bernstein was supported by the Linus Pauling Distinguished Postdoctoral Fellowship which is a Laboratory Directed Research and Development program at PNNL, operated for the DOE by Battelle Memorial Institute under Contract DE-AC05-76RLO 1830. The content is solely the responsibility of the authors and does not necessarily represent the official views of the National Institutes of Health.

References

1. Muller S, Nebe-von-Caron G. Functional single-cell analyses: flow cytometry and cell sorting of microbial populations and communities. *FEMS Microbiol Rev* 2010;**34**(4):554–87.
2. Mahadevan R, Henson MA. Genome-based modeling and design of metabolic interactions in microbial communities. *Comput Struct Biotechnol J* 2012;**3**:e201210008.
3. Taffs R, et al. In silico approaches to study mass and energy flows in microbial consortia: a syntrophic case study. *BMC Syst Biol* 2009;**3**:114.
4. Tamime AY, Robinson RK. *Yoghurt: science and technology*. Cambridge: Woodhead; 1999.
5. Bond T, Templeton MR. History and future of domestic biogas plants in the developing world. *Energy Sustainable Dev* 2011;**15**(4):347–54.
6. Dobell C. *Antony van Leeuwenhoek and his "Little Animals." Being some account of the father of protozoology and bacteriology and his multifarious discoveries in these disciplines*. Collected, translated, and edited, from his printed works, unpublished manuscripts, and contemporary records. Published on the 300th anniversary of his birth, Great Titchfield Street, London W.C.1: John Bale, Sons & Danielsson, Ltd; 1932, p. 83–91.
7. Korgaonkar A, et al. Community surveillance enhances Pseudomonas aeruginosa virulence during polymicrobial infection. *Proc Natl Acad Sci U S A* 2013;**110**(3):1059–64.
8. Florey HW. The use of micro-organisms for therapeutic purposes. *Yale J Biol Med* 1946;**19**(1):101–18.
9. Pasteur L, Joubert J. Charbon et septicemie. *C R Acad Sci* 1877;**85**:101–5.
10. Madigan MT, Martinko JM, Bender K, Bender KS, Clark DP, Buckley DP, et al. *Brock biology of microorganisms*. 13 ed. San Francisco: Benjamin Cummings; 2010.
11. Bernstein HC, Carlson RP. Microbial consortia engineering for cellular factories: in vitro to in silico systems. *Comput Struct Biotechnol J* 2012;**3**(4):e201210017.
12. Roossinck MJ. The good viruses: viral mutualistic symbioses. *Nat Rev Microbiol* 2011;**9**(2):99–108.
13. Kumar K, Mella-Herrera RA, Golden JW. Cyanobacterial heterocysts. *Cold Spring Harbor Perspect Biol* 2010;**2**(4):a000315.
14. Bernstein HC, Paulson SD, Carlson RP. Synthetic Escherichia coli consortia engineered for syntrophy demonstrate enhanced biomass productivity. *J Biotechnol* 2012;**157**(1):159–66.
15. Eiteman MA, Lee SA, Altman E. A co-fermentation strategy to consume sugar mixtures effectively. *J Biol Eng* 2008;**2**:3.
16. Poltak SR, Cooper VS. Ecological succession in long-term experimentally evolved biofilms produces synergistic communities. *ISME J* 2011;**5**(3):369–78.
17. Rosenzweig RF, et al. Microbial evolution in a simple unstructured environment—genetic differentiation in Escherichia-Coli. *Genetics* 1994;**137**(4):903–17.
18. Hall-Stoodley L, Costerton JW, Stoodley P. Bacterial biofilms: from the natural environment to infectious diseases. *Nat Rev Microbiol* 2004;**2**(2):95–108.
19. Stoodley P, et al. Biofilms as complex differentiated communities. *Annu Rev Microbiol* 2002;**56**:187–209.
20. Holland JN, DeAngelis DL. Consumer-resource theory predicts dynamic transitions between outcomes of interspecific interactions. *Ecol Lett* 2009;**12**(12):1357–66.
21. McInerney MJ, Sieber JR, Gunsalus RP. Syntrophy in anaerobic global carbon cycles. *Curr Opin Biotechnol* 2009;**20**(6):623–32.
22. De Roy K, et al. Synthetic microbial ecosystems: an exciting tool to understand and apply microbial communities. *Environ Microbiol* 2014;**16**(6):1472–81.
23. Wintermute EH, Silver PA. Emergent cooperation in microbial metabolism. *Mol Syst Biol* 2010;**6**:407.
24. Phelan VV, et al. Microbial metabolic exchange—the chemotype-to-phenotype link. *Nat Chem Biol* 2012;**8**(1):26–35.
25. Waters CM, Bassler BL. Quorum sensing: cell-to-cell communication in bacteria. *Annu Rev Cell Dev Biol* 2005;**21**:319–46.
26. Hense BA, et al. Does efficiency sensing unify diffusion and quorum sensing? *Nat Rev Microbiol* 2007;**5**(3):230–9.

27. Decho AW, Norman RS, Visscher PT. Quorum sensing in natural environments: emerging views from microbial mats. *Trends Microbiol* 2010;**18**(2):73–80.
28. Hibbing ME, et al. Bacterial competition: surviving and thriving in the microbial jungle. *Nat Rev Microbiol* 2010;**8**(1):15–25.
29. Fajardo A, Martinez JL. Antibiotics as signals that trigger specific bacterial responses. *Curr Opin Microbiol* 2008;**11**(2):161–7.
30. Paerl HW, Pinckney JL. A mini-review of microbial consortia: their roles in aquatic production and biogeochemical cycling. *Microb Ecol* 1996;**31**(3):225–47.
31. Brenner K, Arnold FH. Self-organization, layered structure, and aggregation enhance persistence of a synthetic biofilm consortium. *PLoS One* 2011;**6**(2):e16791.
32. Pirbadian S, et al. Shewanella oneidensis MR-1 nanowires are outer membrane and periplasmic extensions of the extracellular electron transport components. *Proc Natl Acad Sci U S A* 2014;**111**(35):12883–8.
33. Pfeffer C, et al. Filamentous bacteria transport electrons over centimetre distances. *Nature* 2012;**491**(7423):218–21.
34. Carlton RG, Richardson LL. Oxygen and sulfide dynamics in a horizontally migrating cyanobacterial mat: black band disease of corals. *FEMS Microbiol Ecol* 1995;**18**(2):155–62.
35. Kuhl M, Jorgensen BB. Microsensor measurements of sulfate reduction and sulfide oxidation in compact microbial communities of aerobic biofilms. *Appl Environ Microbiol* 1992;**58**(4):1164–74.
36. Rykiel EJ. Towards a definition of ecological disturbance. *Aust J Ecol* 1985;**10**(3):361–5.
37. Grimm V, Wissel C. Babel, or the ecological stability discussions: an inventory and analysis of terminology and a guide for avoiding confusion. *Oecologia* 1997;**109**(3):323–34.
38. Pimm SL. The complexity and stability of ecosystems. *Nature* 1984;**307**(5949):321–6.
39. Shade A, et al. Fundamentals of microbial community resistance and resilience. *Front Microbiol* 2012;**3**:417.
40. Waide RB, et al. The relationship between productivity and species richness. *Annu Rev Ecol Syst* 1999;**30**:257–300.
41. Dekel E, Alon U. Optimality and evolutionary tuning of the expression level of a protein. *Nature* 2005;**436**(7050):588–92.
42. Elser JJ, Acquisti C, Kumar S. Stoichiogenomics: the evolutionary ecology of macromolecular elemental composition. *Trends Ecol Evol* 2011;**26**(1):38–44.
43. Zinn M, Witholt B, Egli T. Dual nutrient limited growth: models, experimental observations, and applications. *J Biotechnol* 2004;**113**(1–3):263–79.
44. Konopka A. What is microbial community ecology? *ISME J* 2009;**3**(11):1223–30.
45. Prosser JI, et al. The role of ecological theory in microbial ecology. *Nat Rev Microbiol* 2007;**5**(5):384–92.
46. Brenner K, You L, Arnold FH. Engineering microbial consortia: a new frontier in synthetic biology. *Trends Biotechnol* 2008;**26**(9):483–9.
47. Lotka AJ. Contribution to the energetics of evolution. *Proc Natl Acad Sci U S A* 1922;**8**(6):147–51.
48. Lotka AJ. Natural selection as a physical principle. *Proc Natl Acad Sci U S A* 1922;**8**:151–4.
49. Sciubba E. What did Lotka really say? A critical reassessment of the "maximum power principle". *Ecol Model* 2011;**222**(8):1347–53.
50. Cai TT, Montague CL, Davis JS. The maximum power principle: an empirical investigation. *Ecol Model* 2006;**190**(3–4):317–35.
51. DeLong JP. The maximum power principle predicts the outcomes of two-species competition experiments. *Oikos* 2008;**117**(9):1329–36.
52. Martyushev L. Entropy and entropy production: old misconceptions and new breakthroughs. *Entropy* 2013;**15**(4):1152–70.
53. Tilman D. *Resource competition and community structure*. Princeton: Princeton University Press; 1982.
54. de Mazancourt C, Schwartz MW. A resource ratio theory of cooperation. *Ecol Lett* 2010;**13**(3):349–59.
55. Tilman D. Resources—a graphical-mechanistic approach to competition and predation. *Am Nat* 1980;**116**(3):362–93.
56. Tilman D. Tests of resource competition theory using four species of Lake Michigan algae. *Ecology* 1981;**62**(3):802.

57. Passarge J, et al. Competition for nutrients and light: Stable coexistence, alternative stable states, or competitive exclusion? *Ecol Monogr* 2006;**76**(1):57–72.
58. Bull AT. The renaissance of continuous culture in the post-genomics age. *J Ind Microbiol Biotechnol* 2010;**37**(10):993–1021.
59. Taylor PA, le BWPJ. Theoretical studies on the coexistence of competing species under continuous-flow conditions. *Can J Microbiol* 1975;**21**(1):90–8.
60. Brauer VS, Stomp M, Huisman J. The nutrient-load hypothesis: patterns of resource limitation and community structure driven by competition for nutrients and light. *Am Nat* 2012;**179**(6):721–40.
61. Enyeart PJ, Simpson ZB, Ellington AD. A microbial model of economic trading and comparative advantage. *J Theor Biol* 2014;**364**:326–43.
62. Sterner RW, Elser JJ. *Ecological stoichiometry: the biology of elements from molecules to the biosphere.* Princeton, NJ: Princeton University Press; 2002.
63. Agrawal AA. Phenotypic plasticity in the interactions and evolution of species. *Science* 2001;**294**(5541):321–6.
64. Kitano H. Violations of robustness trade-offs. *Mol Syst Biol* 2010;**6**:384.
65. Law R. Optimal life histories under age-specific predation. *Am Nat* 1979;**114**(3):399–417.
66. Molenaar D, et al. Shifts in growth strategies reflect tradeoffs in cellular economics. *Mol Syst Biol* 2009;**5**:323.
67. Carlson RP, Taffs RL. Molecular-level tradeoffs and metabolic adaptation to simultaneous stressors. *Curr Opin Biotechnol* 2010;**21**(5):670–6.
68. Shoval O, et al. Evolutionary trade-offs, Pareto optimality, and the geometry of phenotype space. *Science* 2012;**336**(6085):1157–60.
69. Song H-S, et al. Mathematical modeling of microbial community dynamics: a methodological review. *Processes* 2014;**2**(4):711–52.
70. Edwards JS, Ramakrishna R, Palsson BO. Characterizing the metabolic phenotype: a phenotype phase plane analysis. *Biotechnol Bioeng* 2002;**77**(1):27–36.
71. Oh YG, et al. Multiobjective flux balancing using the NISE method for metabolic network analysis. *Biotechnol Prog* 2009;**25**(4):999–1008.
72. Schuetz R, et al. Multidimensional optimality of microbial metabolism. *Science* 2012;**336**(6081):601–4.
73. Carlson R, Srienc F. Fundamental Escherichia coli biochemical pathways for biomass and energy production: identification of reactions. *Biotechnol Bioeng* 2004;**85**(1):1–19.
74. Carlson RP. Metabolic systems cost-benefit analysis for interpreting network structure and regulation. *Bioinformatics* 2007;**23**(10):1258–64.
75. Carlson RP. Decomposition of complex microbial behaviors into resource-based stress responses. *Bioinformatics* 2009;**25**(1):90–7.
76. Flamholz A, et al. Glycolytic strategy as a tradeoff between energy yield and protein cost. *Proc Natl Acad Sci U S A* 2013;**110**(24):10039–44.
77. Fischer E, Sauer U. A novel metabolic cycle catalyzes glucose oxidation and anaplerosis in hungry Escherichia coli. *J Biol Chem* 2003;**278**(47):46446–51.
78. Carlson RP, Oshota OJ, Taffs RL. Systems analysis of microbial adaptations to simultaneous stresses. *Subcell Biochem* 2012;**64**:139–57.
79. Tepper N, et al. Steady-state metabolite concentrations reflect a balance between maximizing enzyme efficiency and minimizing total metabolite load. *PLoS One* 2013;**8**(9):e75370.
80. Valenzuela J, et al. Potential role of multiple carbon fixation pathways during lipid accumulation in Phaeodactylum tricornutum. *Biotechnol Biofuels* 2012;**5**(1):40.
81. Overmann J. The phototrophic consortium "Chlorochromatium aggregatum"—a model for bacterial heterologous multicellularity. *Adv Exp Med Biol* 2010;**675**:15–29.
82. Overmann J, et al. The ecological niche of the consortium "Pelochromatium roseum". *Arch Microbiol* 1998;**169**(2):120–8.
83. Liu Z, et al. Genomic analysis reveals key aspects of prokaryotic symbiosis in the phototrophic consortium "Chlorochromatium aggregatum". *Genome Biol* 2013;**14**(11):R127.

84. Vogl K, et al. Chlorobium chlorochromatii sp. nov., a symbiotic green sulfur bacterium isolated from the phototrophic consortium "Chlorochromatium aggregatum". *Arch Microbiol* 2006;**185**(5):363–72.
85. Bryant DA, et al. Comparative and functional genomics of anoxygenic green bacteria from the taxa chlorobi, chloroflexi, and acidobacteria. In: Burnap RL, Vermaas WFJ, editors. *Functional Genomics and Evolution of Photosynthetic Systems*, 33. New York: Springer; 2012. p. 47–102.
86. Muller J, Overmann J. Close interspecies interactions between prokaryotes from sulfureous environments. *Front Microbiol* 2011;**2**:146.
87. Wanner G, Vogl K, Overmann J. Ultrastructural characterization of the prokaryotic symbiosis in "Chlorochromatium aggregatum". *J Bacteriol* 2008;**190**(10):3721–30.
88. Frostl JM, Overmann J. Physiology and tactic response of the phototrophic consortium "Chlorochromatium aggregatum". *Arch Microbiol* 1998;**169**(2):129–35.
89. Offre P, Spang A, Schleper C. Archaea in biogeochemical cycles. *Annu Rev Microbiol* 2013;**67**:437–57.
90. Li X, et al. Metabolism of H2 by Desulfovibrio alaskensis G20 during syntrophic growth on lactate. *Microbiology* 2011;**157**(Pt 10):2912–21.
91. Meyer B, et al. Variation among Desulfovibrio species in electron transfer systems used for syntrophic growth. *J Bacteriol* 2013;**195**(5):990–1004.
92. Thauer RK, et al. Methanogenic archaea: ecologically relevant differences in energy conservation. *Nat Rev Microbiol* 2008;**6**(8):579–91.
93. Schink B. Energetics of syntrophic cooperation in methanogenic degradation. *Microbiol Mol Biol Rev* 1997;**61**(2):262–80.
94. Stams AJ, Plugge CM. Electron transfer in syntrophic communities of anaerobic bacteria and archaea. *Nat Rev Microbiol* 2009;**7**(8):568–77.
95. Walker CB, et al. Functional responses of methanogenic archaea to syntrophic growth. *ISME J* 2012;**6**(11):2045–55.
96. Bizukojc M, et al. Metabolic modelling of syntrophic-like growth of a 1,3-propanediol producer, Clostridium butyricum, and a methanogenic archeon, Methanosarcina mazei, under anaerobic conditions. *Bioprocess Biosyst Eng* 2010;**33**(4):507–23.
97. Cheng YF, et al. Diversity and activity of enriched ruminal cultures of anaerobic fungi and methanogens grown together on lignocellulose in consecutive batch culture. *Bioresour Technol* 2009;**100**(20):4821–8.
98. Kinnersley M, et al. Ex uno plures: clonal reinforcement drives evolution of a simple microbial community. *PLoS Genet* 2014;**10**(6):e1004430.
99. Treves DS, Manning S, Adams J. Repeated evolution of an acetate-crossfeeding polymorphism in long-term populations of Escherichia coli. *Mol Biol Evol* 1998;**15**(7):789–97.
100. Momeni B, et al. Correction: strong inter-population cooperation leads to partner intermixing in microbial communities. *Elife* 2014;**3**:e02945.
101. Mikeskova H, Novotny C, Svobodova K. Interspecific interactions in mixed microbial cultures in a biodegradation perspective. *Appl Microbiol Biotechnol* 2012;**95**(4):861–70.
102. Ortiz-Marquez JC, et al. Genetic engineering of multispecies microbial cell factories as an alternative for bioenergy production. *Trends Biotechnol* 2013;**31**(9):521–9.
103. Jagmann N, Philipp B. Design of synthetic microbial communities for biotechnological production processes. *J Biotechnol* 2014;**184**:209–18.
104. Bader J, et al. Relevance of microbial coculture fermentations in biotechnology. *J Appl Microbiol* 2010;**109**(2):371–87.
105. Pandhal J, Noirel J. Synthetic microbial ecosystems for biotechnology. *Biotechnol Lett* 2014;**36**(6):1141–51.
106. Zuroff TR, Curtis WR. Developing symbiotic consortia for lignocellulosic biofuel production. *Appl Microbiol Biotechnol* 2012;**93**(4):1423–35.
107. Subashchandrabose SR, et al. Consortia of cyanobacteria/microalgae and bacteria: biotechnological potential. *Biotechnol Adv* 2011;**29**(6):896–907.

108. Hussa EA, Goodrich-Blair H. It takes a village: ecological and fitness impacts of multipartite mutualism. *Annu Rev Microbiol* 2013;**67**:161–78.
109. Shong J, Jimenez Diaz MR, Collins CH. Towards synthetic microbial consortia for bioprocessing. *Curr Opin Biotechnol* 2012;**23**(5):798–802.
110. Smid EJ, Lacroix C. Microbe-microbe interactions in mixed culture food fermentations. *Curr Opin Biotechnol* 2013;**24**(2):148–54.
111. Sabra W, et al. Biosystems analysis and engineering of microbial consortia for industrial biotechnology. *Eng Life Sci* 2010;**10**(5):407–21.
112. Thiele I, Heinken A, Fleming RM. A systems biology approach to studying the role of microbes in human health. *Curr Opin Biotechnol* 2013;**24**(1):4–12.
113. Brune KD, Bayer TS. Engineering microbial consortia to enhance biomining and bioremediation. *Front Microbiol* 2012;**3**:203.
114. Murray JL, et al. Mechanisms of synergy in polymicrobial infections. *J Microbiol* 2014;**52**(3):188–99.
115. Goers L, Freemont P, Polizzi KM. Co-culture systems and technologies: taking synthetic biology to the next level. *J R Soc Interface* 2014;**11**(96).
116. Song H, et al. Synthetic microbial consortia: from systematic analysis to construction and applications. *Chem Soc Rev* 2014.
117. Grosskopf T, Soyer OS. Synthetic microbial communities. *Curr Opin Microbiol* 2014;**18**:72–7.
118. Lopez-Ferber M, et al. Defective or effective? Mutualistic interactions between virus genotypes. *Proc Biol Sci* 2003;**270**(1530):2249–55.
119. Kuss SK, et al. Intestinal microbiota promote enteric virus replication and systemic pathogenesis. *Science* 2011;**334**(6053):249–52.
120. Sandaa RA, et al. Viral control of bacterial biodiversity–evidence from a nutrient-enriched marine mesocosm experiment. *Environ Microbiol* 2009;**11**(10):2585–97.
121. Tillmann HL, et al. Infection with GB virus C and reduced mortality among HIV-infected patients. *N Engl J Med* 2001;**345**(10):715–24.
122. Ernst D, et al. Impact of GB virus C viraemia on clinical outcome in HIV-1-infected patients: a 20-year follow-up study. *HIV Med* 2014;**15**(4):245–50.
123. Oldstone M. Prevention of type I diabetes in nonobese diabetic mice by virus infection. *Science* 1988;**239**(4839):500–2.
124. Marquez LM, et al. A virus in a fungus in a plant: three-way symbiosis required for thermal tolerance. *Science* 2007;**315**(5811):513–5.
125. Zhang CC, et al. Heterocyst differentiation and pattern formation in cyanobacteria: a chorus of signals. *Mol Microbiol* 2006;**59**(2):367–75.
126. Minty JJ, et al. Design and characterization of synthetic fungal-bacterial consortia for direct production of isobutanol from cellulosic biomass. *Proc Natl Acad Sci U S A* 2013;**110**(36):14592–7.
127. Rodionova IA, et al. Genomic distribution of B-vitamin auxotrophy and uptake transporters in environmental bacteria from the Chloroflexi phylum. *Environ Microbiol Rep* 2014.
128. D'Onofrio A, et al. Siderophores from neighboring organisms promote the growth of uncultured bacteria. *Chem Biol* 2010;**17**(3):254–64.
129. Sieber JR, McInerney MJ, Gunsalus RP. Genomic insights into syntrophy: the paradigm for anaerobic metabolic cooperation. *Annu Rev Microbiol* 2012;**66**:429–52.
130. Falony G, De Vuyst L. *Ecological interactions of bacteria in the human gut*. 2009 p. 639–79.
131. Tan CH, et al. The role of quorum sensing signalling in EPS production and the assembly of a sludge community into aerobic granules. *ISME J* 2014;**8**(6):1186–97.
132. De Lorenzo V, Martinez JL, Asensio C. Microcin-mediated interactions between Klebsiella pneumoniae and Escherichia coli strains. *Microbiology* 1984;**130**(2):391–400.
133. Pande GS, et al. The Vibrio campbellii quorum sensing signals have a different impact on virulence of the bacterium towards different crustacean hosts. *Vet Microbiol* 2013;**167**(3–4):540–5.

134. Ishii S, et al. Characterization of a filamentous biofilm community established in a cellulose-fed microbial fuel cell. *BMC Microbiol* 2008;**8**:6.
135. Morris JJ, et al. Dependence of the cyanobacterium Prochlorococcus on hydrogen peroxide scavenging microbes for growth at the ocean's surface. *PLoS One* 2011;**6**(2):e16805.
136. Stomp M, et al. Adaptive divergence in pigment composition promotes phytoplankton biodiversity. *Nature* 2004;**432**(7013):104–7.
137. Wilson JB, Spijkerman E, Huisman J. Is there really insufficient support for Tilman's R* concept? A comment on Miller et al. *Am Nat* 2007;**169**(5):700–6.
138. Nagarajan H, et al. Characterization and modelling of interspecies electron transfer mechanisms and microbial community dynamics of a syntrophic association. *Nat Commun* 2013;**4**:2809.
139. Klitgord N, Segre D. Environments that induce synthetic microbial ecosystems. *PLoS Comput Biol* 2010;**6**(11):e1001002.
140. Stolyar S, et al. Metabolic modeling of a mutualistic microbial community. *Mol Syst Biol* 2007;**3**:92.
141. Harcombe WR, et al. Metabolic resource allocation in individual microbes determines ecosystem interactions and spatial dynamics. *Cell Rep* 2014;**7**(4):1104–15.
142. Zomorrodi AR, Maranas CD. OptCom: a multi-level optimization framework for the metabolic modeling and analysis of microbial communities. *PLoS Comput Biol* 2012;**8**(2):e1002363.
143. Zomorrodi AR, Islam MM, Maranas CD. d-OptCom: dynamic multi-level and multi-objective metabolic modeling of microbial communities. *ACS Synth Biol* 2014;**3**(4):247–57.
144. Hanly TJ, Henson MA. Dynamic flux balance modeling of microbial co-cultures for efficient batch fermentation of glucose and xylose mixtures. *Biotechnol Bioeng* 2011;**108**(2):376–85.
145. Hanly TJ, Urello M, Henson MA. Dynamic flux balance modeling of S. cerevisiae and E. coli co-cultures for efficient consumption of glucose/xylose mixtures. *Appl Microbiol Biotechnol* 2012;**93**(6):2529–41.

CHAPTER 16

Cell-Free Biotechnologies

S.D. Minteer

Departments of Chemistry and Materials Science & Engineering, University of Utah, Salt Lake City, Utah, USA

Over the last decade, there has been a great deal of progress in microbial biotechnologies, including synthetic biology for producing value-added products,[1] synthetic biology for producing biofuels,[1] microbial fuel cells for electricity generation,[2] microbial bioelectrochemical systems for fuel production,[3] and even microbial fuel cells for remediation. These technologies have been built on the ability of microbes to grow and reproduce providing regeneration of catalysts and long term stability. There are examples of microbial biotechnology lifetimes greater than 5 years with a single inoculation.[4] However, the rate of production is typically low due to slow transport across cellular membranes and low volumetric catalytic activity in the microbes.[2a] There is also concern about microbes evolving and other microbial contamination. Finally, many microbial biotechnologies are limited in efficiency, because they contain a variety of enzymatic cascades that can produce alternative products to the desired product (fuel, electricity, etc.). Therefore, there has also been a push for research in cell-free biotechnologies, which utilize isolated enzymes and enzyme cascades. These systems improve the volumetric catalytic activity and have no membrane transport issues. These enzymes and enzyme cascades can come in the form of a crude extract of the microbial cell or as purified enzyme, depending on the application. However, they require enzyme stabilization for long term function and viability and they frequently require enzyme immobilization for easy separation of the catalyst from the solution for enzyme re-use.

Cell-free biotechnology goes back over a century when Buchner first produced ethanol and carbon dioxide from yeast extract.[5] This first system utilized a natural system (yeast) and the natural metabolic enzymes and cofactors in the yeast cells. This *in vitro* fermentation process resulted in Buchner winning the Nobel Prize in Chemistry in 1907. From that point forward, there have been countless examples of cell-free biotechnology processes ranging from laboratory assays to polymerase chain reaction (PCR) systems to sugar transformations for food (i.e. invertase converting table sugar (sucrose) to fructose and glucose or glucose isomerase converting glucose to fructose). Today, cell-free biotechnology can be broken up into single enzyme systems, multi-enzyme cascades, and cellular extracts. We will detail each of these cell-free biotechnologies for value added products (food, medicine, etc.), biofuels, and electricity generation.

Biotechnology for Biofuel Production and Optimization. http://dx.doi.org/10.1016/B978-0-444-63475-7.00016-9

16.1 Cell Free Biotechnology for Value Added Products

Cell-free biotechnology for producing value added products is a commercial reality and has been one for decades. One of the simplest industrial examples of cell-free biotechnology is the use of glucose isomerase to produce fructose from glucose for food applications. In this case, glucose isomerase is very stable and doesn't require purification before use, so unpurified enzyme is immobilized on a solid support as a low-cost and re-useable catalyst. Another example of a single enzyme system is PCR for DNA replication. This is used in biology and biological chemistry labs across the world. This is an example where the enzyme requires purification and therefore is more costly then the glucose isomerase case. However, this enzyme is a thermophilic enzyme and therefore is quite stable.

When it comes to multi-enzyme systems for value added products, there are two strategies that are commonly discussed in the literature: crude extracts and synthetic enzymatic pathways. Cell extracts of *E. coli*, wheat germ, and other cells can be used to form polypeptides,[6] as shown in Figure 16.1. In this case, the cell lysate contains the enzymes, while the amino acids and other components can be added to the bioreactor. This type of technology can utilize post-processing of the peptide or protein for glycosylation, immobilization, or the incorporation of non-natural amino acids.[6] This strategy has been considered for hydrogenase production, therapeutics, and other medicines. Figure 16.2 shows a schematic for the use of cell extract technology for the production of a medical protein.[6] One issue with cell extracts is that proteases are present in the cell extract that can limit lifetime and activity. Therefore, many procedures include the addition of a protease inhibitor to prolong lifetime and activity. Another issue is that similar to intact microbes, the cell-extract has a variety of enzyme cascades present that might limit the purity of the product being formed and the efficiency of the process.

Purified enzymes have the ability to separate the proteases and all other enzymes that might cause side reactions, although there is additional cost to production for the purification process. The additional production costs vary dramatically with the enzyme and the degree of purification. If the enzyme is an expressed enzyme with an affinity tag (i.e. His-tagged

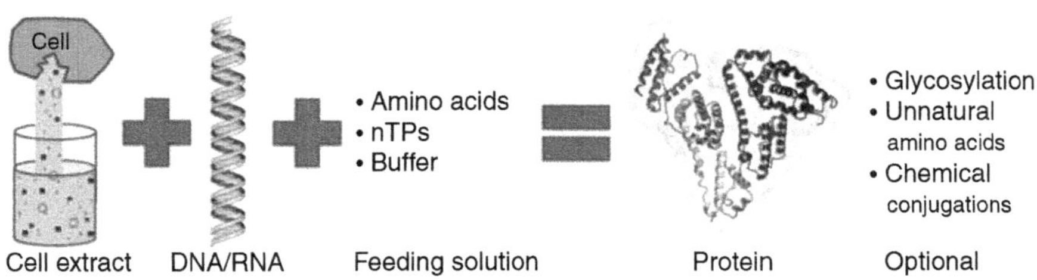

Figure 16.1
Schematic of cell extract based production of polypeptides or proteins. *Reproduced with permission from Ref. 6.*

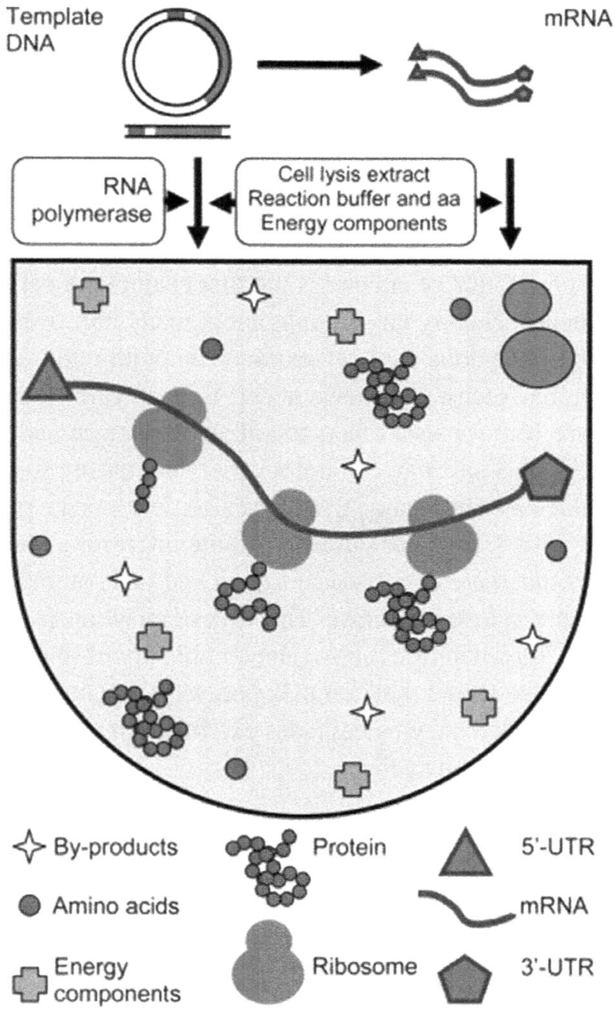

Figure 16.2
Detailed schematic of cell extract-based protein production. *Reproduced with permission from Ref. 6.*

enzymes), then purification can be relatively simple. However, purification of native enzymes via size exclusion and ion exchange chromatography can be quite time consuming and expensive, because frequently multiple columns are required. Enzyme precipitation and heat treatment based purification procedures are also becoming quite common in the field.

It is important to note that this technique also allows the user to combine different enzymes from different organisms to maximize flux through the synthetic enzyme pathway. One good example of this is the 12 enzyme cascade for producing Vitamin B12.[7] Another example is the production of dihydroxyacetone phosphate, which combines cell-free extract techniques with rabbit muscle aldolase to produce larger concentration of dihydroxyacetone phosphate than viable *in vivo* and then converting it to a more stable product.[8] One of the other advantages of enzyme cascades is that the coupling of multiple step reactions together allows for driving the overall reaction toward the value-added product even if certain steps have unfavorable equilibria.[9]

16.2 Cell Free Biotechnology for Fuel Production

Buchner's yeast extract production of ethanol is the first example of cell-free fuel production. However, the last century has brought about many improvements. Most cell-free biotechnology today is not done with cell extracts, but with multi-enzyme cascades. The Zhang group at Virginia Tech are pioneers at cell-free enzyme cascades for production of hydrogen fuel.[10] Figure 16.3 shows a schematic of an enzyme cascade for producing hydrogen from sucrose.[10a] This pathway can use a variety of starting sugars, including sucrose, fructose, and glucose. The Zhang group has also developed a pathway for producing hydrogen from xylose.[11] This 13 enzyme cascade is quite interesting, because it uses a series of thermophilic enzymes that were expressed in *E. coli* and one commercially available and naturally occurring enzyme xylose isomerase. This mix of enzymes from different organisms and with different temperature stabilities shows the complexity of multi-enzyme cascades for fuel production, but it also shows that complex pathways can result in very high product yields (>90%), which shows that enzyme cascades can function very efficiently outside of the living cell as well as inside the living cell.

Although hydrogen is a commonly discussed fuel for alternative energy applications (i.e. fuel cell powered cars), it is not the only fuel that can be produced from a multi-enzyme cascade. Figure 16.4 shows a schematic of the enzyme cascade for producing butanol (an alcoholic fuel) from glucose.[12] These type of cascades have been designed for a variety of

Figure 16.3

Enzymatic cascade for production of hydrogen from sucrose. (a) The scheme of the entire synthetic enzymatic pathway for dihydrogen production from sucrose, glucose or fructose, and water. The enzymes are no. 1 SP, sucrose phosphorlylase; no. 2 XI, xylose isomerase; no. 3 PPGK, polyphosphate glucokinase (PPGK); no. 4 PGM, phosphoglucomutase; no. 5, G6PDH, glucose-6-phosphate dehydrogenase; no. 6 6PGDH, 6-phosphogluconate dehydrogenase; no. 7, RPI, ribose 5-phosphate isomerase; no. 8, RPE, ribulose-5-phosphate 3-epimerase; no. 9 TK, transketolase; no. 10 TAL, transaldolase; no. 11 TIM, triose phosphate isomerase; no. 12 ALD, (fructose-bisphosphate) aldolase; no. 13 FBP, fructose bisphosphatase; no. 14 PGI, phosphoglucose isomerase; and no. 15 H_2ase, hydrogenase. *Pi* and (*Pi*)*n* are inorganic phosphate and polyphosphate with a degree of polymerization of *n*. (b) The partial synthetic pathway that regenerate five molecules of glucose 6-phosphate from six molecules of ribulose 5-phosphate. The metabolites are: g1p, glucose-1-phosphate; g6p, glucose-6-phosphate; ru5p, ribulose-5-phosphate; x5p, xylulose-5-phosphate; r5p, ribose-5-phosphate; s7p, sedoheptulose-7-phosphate; g3p, glyceraldehyde-3-phosphate; e4p, erythrose-4-phosphate; dhap, dihydroxyacetone phosphate; fdp, fructose-1,6-diphosphate; f6p, fructose-6-phosphate. *Reproduced with permission from Ref. 10a.*

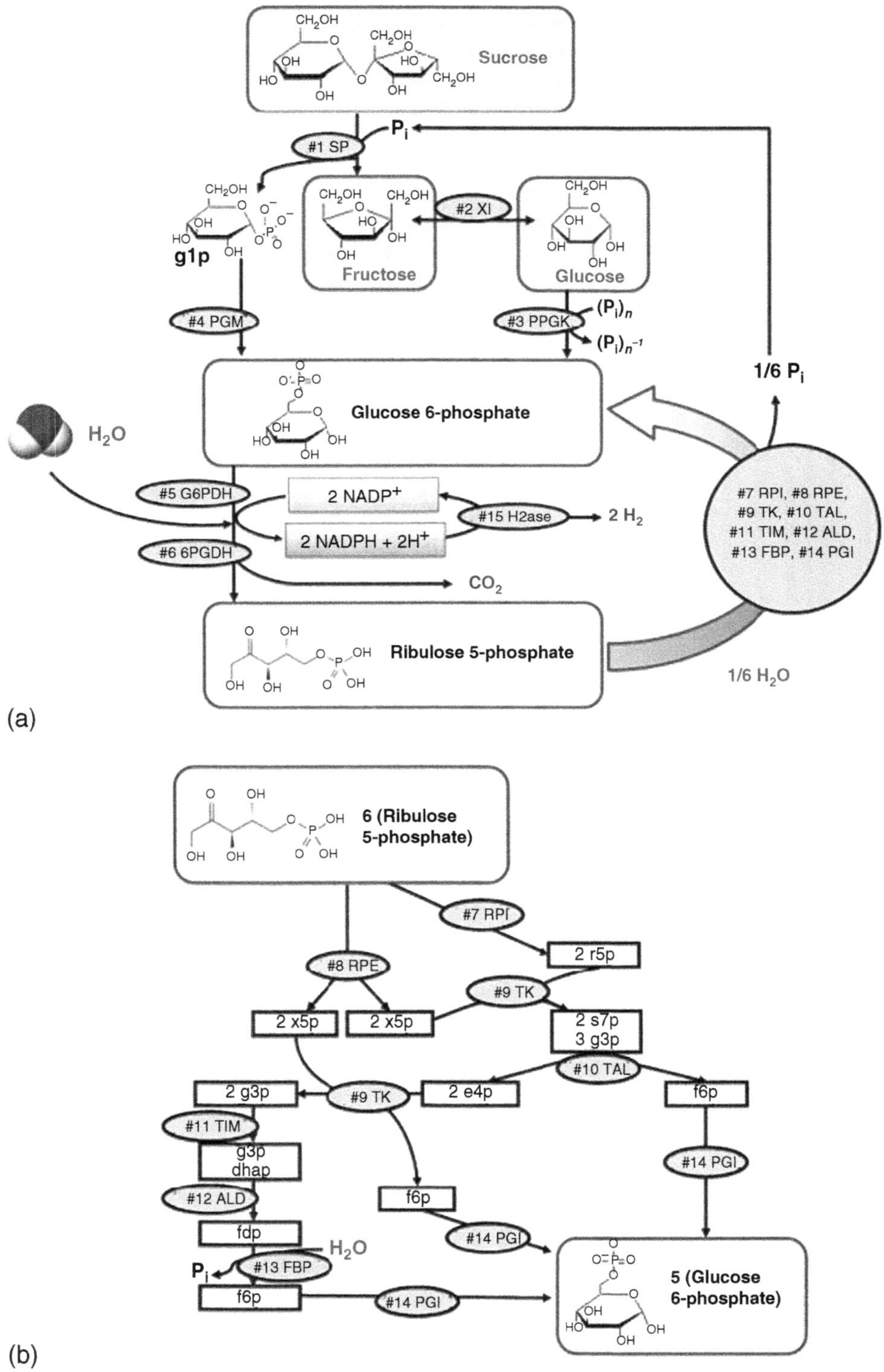

Figure 16.3

See figure Legend on Previous page

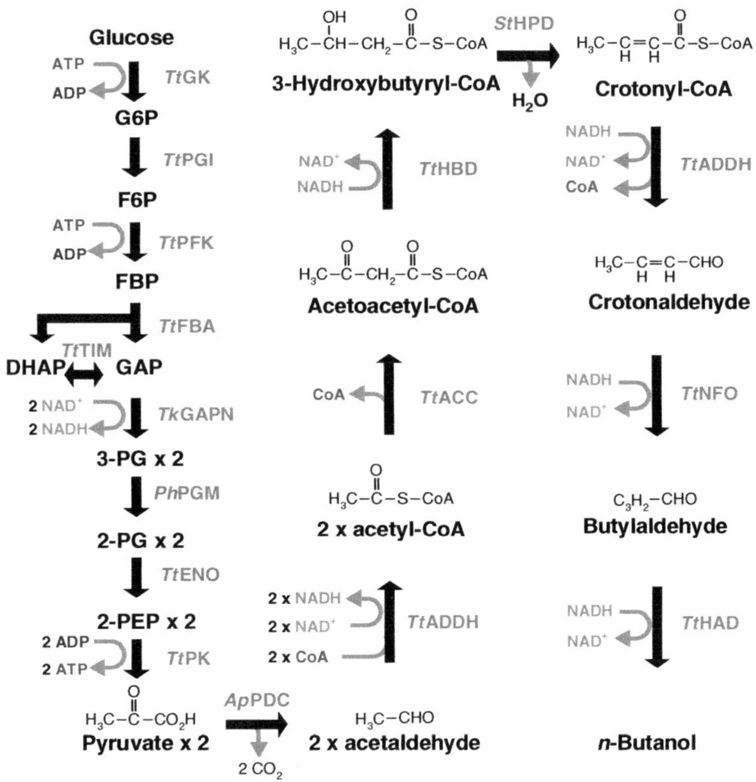

(a)

Enzyme	Source	Abbreviation
Glucokinase	*Thermus thermophilus* HBB	*Tt*GK
G6P isomerase	*Thermus thermophilus* HBB	*Tt*PGI
6-Phosphofructokinase	*Thermus thermophilus* HBB	*Tt*PFK
Fructose-bisphosphate aldolase	*Thermus thermophilus* HBB	*Tt*FBA
Triosephosphate isomerase	*Thermus thermophilus* HBB	*Tt*TIM
Non-phosphorylating GAP dehydrogenase	*Thermococcus kodakarensis* KOD1	*Tk*GAPN
Cofactor-independent phosphoglycerate mutase	*Pyrococcus horikoshii* OT3	*Ph*PGM
Enolase	*Thermus thermophilus* HBB	*Tt*ENO
Pyruvate kinase	*Thermus thermophilus* HBB	*Tt*PK
Pyruvate decarboxylase	*Acetobacter pasteurianus* NCIB8618	*Ap*PDC
CoA-acylating aldehyde dehydrogenase	*Thermus thermophilus* HBB	*Tt*ADDH
Acetyl-CoA acetyltransferase	*Thermus thermophilus* HBB	*Tt*ACC
Hydroxybutyryl-CoA dehydrogenase	*Thermus thermophilus* HBB	*Tt*HBD
3-Hydroxypropionyl-CoA dehydratase	*Sulfolobus tokodaii* strain 7	*St*HPD
NADH-dependent flavinoxidoreductase	*Thermus thermophilus* HBB	*Tt*NFO
3-Hydroxyacyl-CoA dehydrogenase	*Thermus thermophilus* HBB	*Tt*HAD

(b)

Figure 16.4

Enzymatic cascade for production of butanol from glucose. The in vitro metabolic pathway for conversion of glucose to *n*-butanol. (a) Schematic illustration of the designed pathway. Abbreviations used: G6P, glucose-6-phosphate; F6P, fructose-6-phosphate; FBP, fructose-1,6-bisphosphate;GAP, glyceraldehyde-3-phosphate; DHAP, dihydroxyacetonephosphate; 3-PG, 3-phosphoglycerate; 2-PG, 2-phosphoglycerate; and PEP, phosphoenolpyruvate. (b) List of the enzymes used for pathway construction. *Reproduced with permission from Ref. 12.*

alcoholic fuels, but mostly for combustion applications. The primary fuels produced by multi-enzyme cascades have been ethanol from a variety of sugars (predominantly glucose), isobutanol and *n*-butanol. It is important to note that for most of these multi-enzyme cascades for fuel production, it is necessary to consider co-factor regeneration due to the large cost of the labile co-factors. There are a number of strategies to consider. The first is to replace the costly co-factor with a cheaper or more stable, but natural, co-factor. A common example of this is replacing NADP/NADPH systems with NAD/NADH systems.[13] Another option is to engineer an enzyme to work with a minimal cofactor.[14] Another approach is to consider a synthetic analog.[15] Finally, there has been focus on immobilization of the cofactors for ease of regeneration and re-use.[16]

It is important to note that the multi-enzyme cascade systems of recent years are not the only application of cell-free biotechnology for fuel production. A single enzyme system for biodiesel production has been studied for years. For instance, biodiesel can be produced via lipase catalyzed production.[17] A variety of lipases from different organisms have been studied from bacteria and fungi, but in order to be useful industrially, lipases need to be immobilized and natural lipases do not have all of the properties that are desired, so genetic engineering of lipases is still on going to improve their specificity, stability, and tolerance to methanol.[17]

16.3 Cell Free Biotechnology for Energy Conversion

Over a century ago, researchers realized that microbes could be used at the anode of a fuel cell for energy conversion to electricity.[18] It took another 50 years before researchers realized that the enzymes responsible for oxidizing the fuel in the microbial fuel cell could be isolated and used directly at the bioanode of a biofuel cell.[19] This new type of biofuel cell called an enzymatic fuel cell (EFC) utilizes an enzyme catalyst at the anode and/or the cathode of the fuel cell. One of the positive features of bioanodes in biofuel cells is they regenerate the cofactor electrochemically at the bioanode and therefore decrease the need for an additional co-factor regeneration system.

Over the last decade, the field of EFCs has expanded for a number of reasons, including new methods for improved stability (see also section on enzyme stabilization)[20] and incorporation of nanomaterials for improving the biotic-abiotic interface and increasing the active surface area.[4,21] As researchers started developing biofuel cells and biobatteries, they started realizing the limitations of the technology for the anode. Biocathodes are typically fabricated from a single multi-copper oxidase (e.g. laccase or bilirubin oxidase) that performs a four electron reduction of oxygen to water,[22] but bioanodes are more complicated, because most enzymes catalyze a two electron oxidation, but most biofuels have many more electrons than two. Therefore, enzyme cascades are needed to harness all of the energy in the biofuel.

The first enzyme cascade was for the oxidation of methanol. This three enzyme cascade (alcohol dehydrogenase, formaldehyde dehydrogenase, and formate dehydrogenase) is responsible for the oxidation of methanol to formaldehyde to formate to carbon dioxide, but is limited by the fact that these enzymes are not stable at high methanol concentration, so biofuel cells can only operate in low fuel concentrations which lead to low energy density batteries. More recently, enzyme cascades for the oxidation of ethanol,[23] lactate,[24] pyruvate[23c,25] sucrose,[26] glycerol,[27] glucose,[28] and maltodextrin[29] have been developed. These enzyme cascades vary from natural metabolic pathways (ethanol, lactate, pyruvate, and sucrose) to engineered enzyme cascades. Figure 16.5 shows a natural enzyme cascade for the oxidation of ethanol to carbon dioxide using the Krebs cycle metabolic pathway. This figure shows the cascade of enzymes, as well as the power curve showing that the power and current density produced are a function of the degree of oxidation of the ethanol fuel. Figure 16.6 is an artificial enzyme cascade for the oxidation of glucose to carbon dioxide. In this cascade, promiscuous oxidoreductase enzymes were used to catalyze multiple steps of the oxidation reaction. Some engineered enzyme cascades are engineered to be minimal enzyme cascades (individual enzymes can catalyze more than one particular reaction (i.e. glucose and glycerol)) and others are engineered for flux through the pathway or stability. Figure 16.7 shows an enzyme cascade for the oxidation of sucrose, which is an example of a cascade engineered for stability, but limited in flux due to the need for glucose to mutorotate before oxidation.

Recently, there has been a great deal of work focused on forming metabolons from enzyme cascades. Metabolons are supercomplexes of metabolic pathway enzymes that form to channel substrate from one enzyme to the other. The Minteer Research Group showed the importance of metabolon formation with naturally forming Kreb's cycle metabolons, as shown in Figure 16.8. Since then, there has been a wealth of research in forming artificial metabolons. This has been done by *in vitro* crosslinking of enzymes,[30] use of DNA as a scaffold,[31] and also use of protein hydrogels for scaffolding.[31,32] This strategy can be important for improving the electricity generation of enzyme scaffolds.

16.4 Enzyme Stability

The main concern with cell free biotechnologies are the stability of the enzymes, since isolated and purified enzymes can be quite costly even with large scale production. There are several strategies that have been employed to improve stability. One approach is to utilize enzymes from thermophilic or extremophilic organisms.[33] These organisms typically function at elevated temperatures and/or high ionic strength and/or presence of organic solvent. Therefore, their enzymes tend to be stable in those environments. The negative aspect of this approach is that many thermophilic enzymes have minimal activity at moderate temperatures, so although they may have improved stability, they typically have

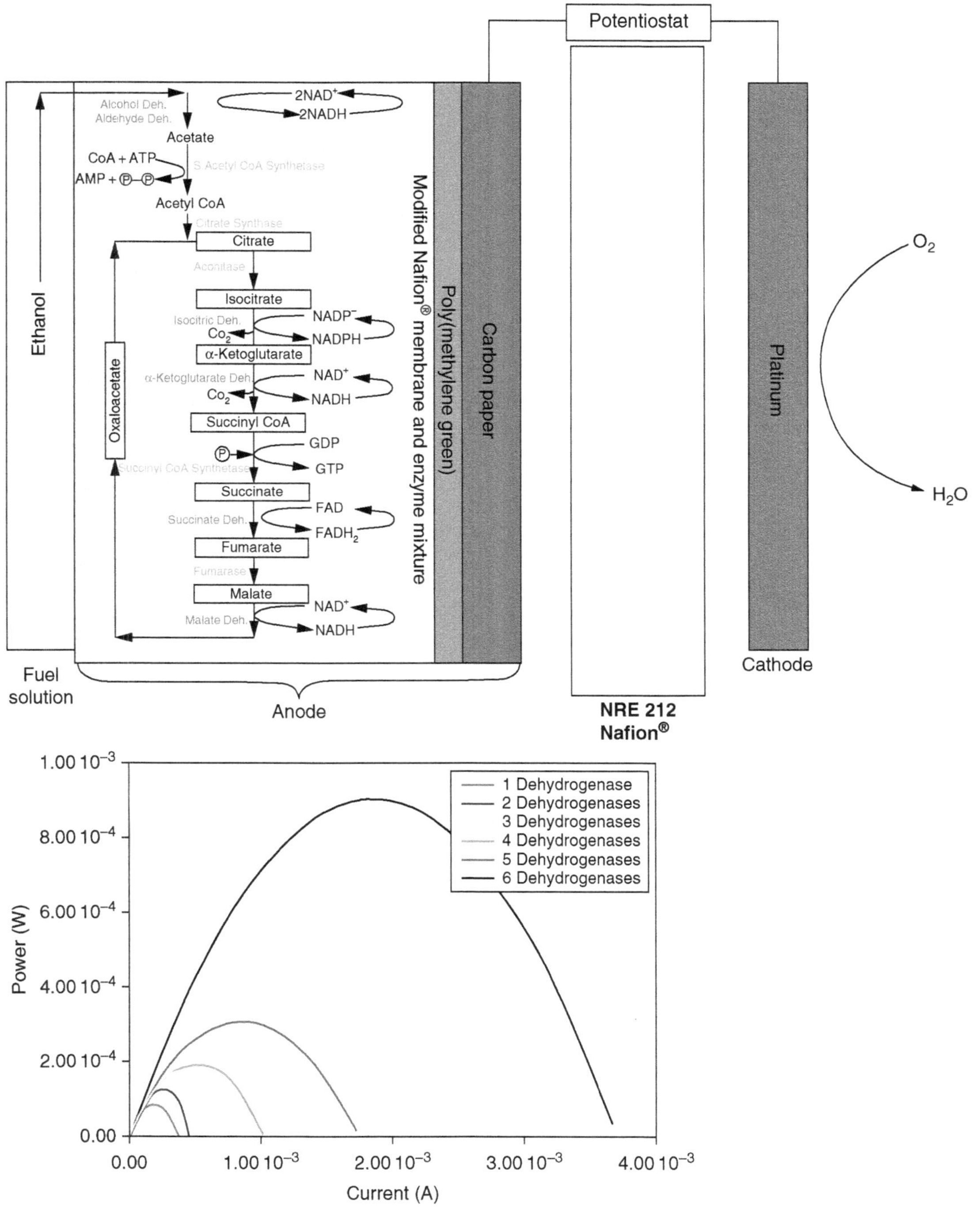

Figure 16.5

Electricity production from an enzyme cascade to oxidize ethanol to carbon dioxide. (Top) Enzyme cascade in a biofuel cell. (Bottom) Power curves for electricity production with different degrees of oxidation. *Reproduced with permission from Ref. 23b.*

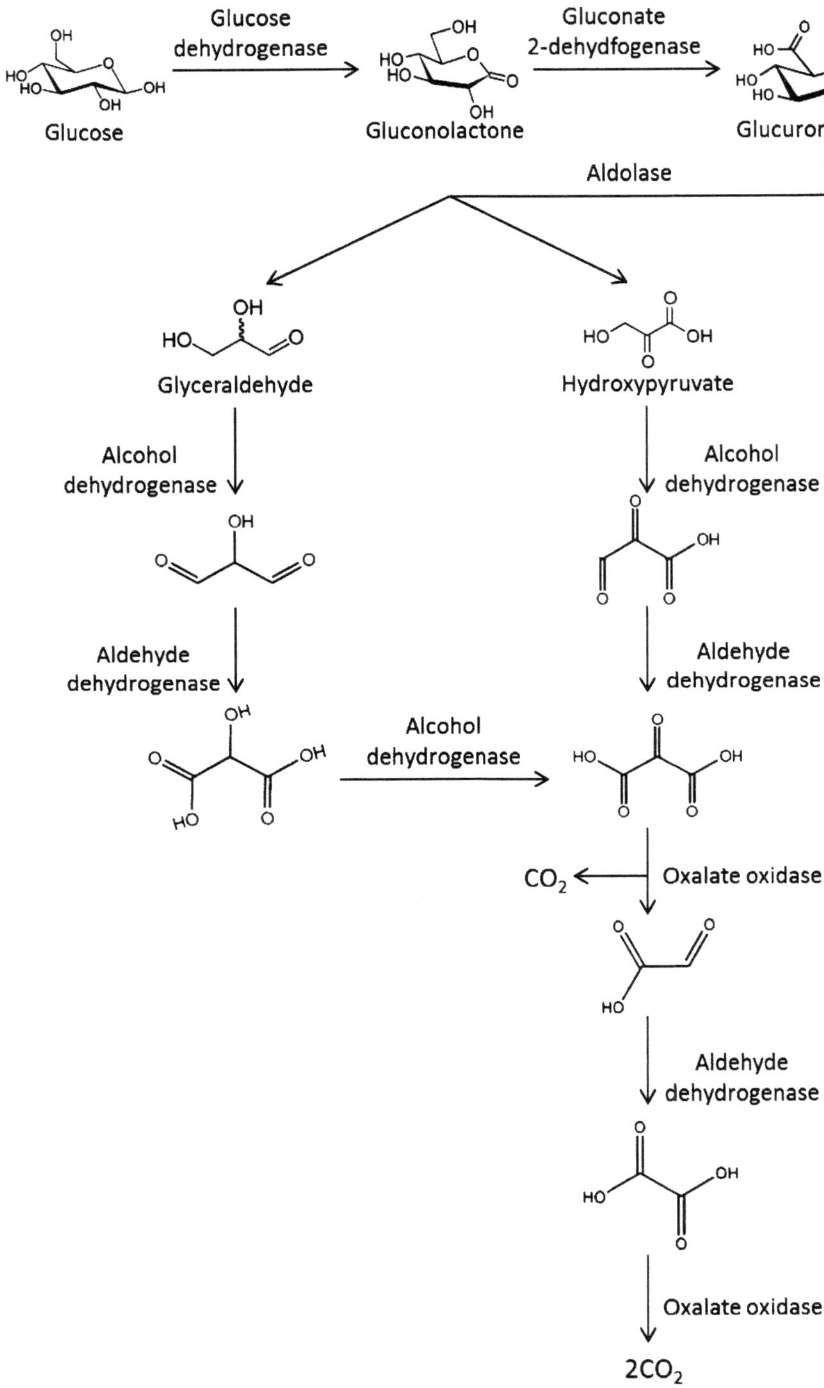

Figure 16.6

Electricity generation from a non-natural enzyme cascade for oxidation of glucose to carbon dioxide utilizing promiscuous enzymes (aldolase, PQQ-dependent alcohol dehydrogenase, PQQ-dependent aldehyde dehydrogenase, and oxalate oxidase). *Reproduced with permission from Ref. 28a.*

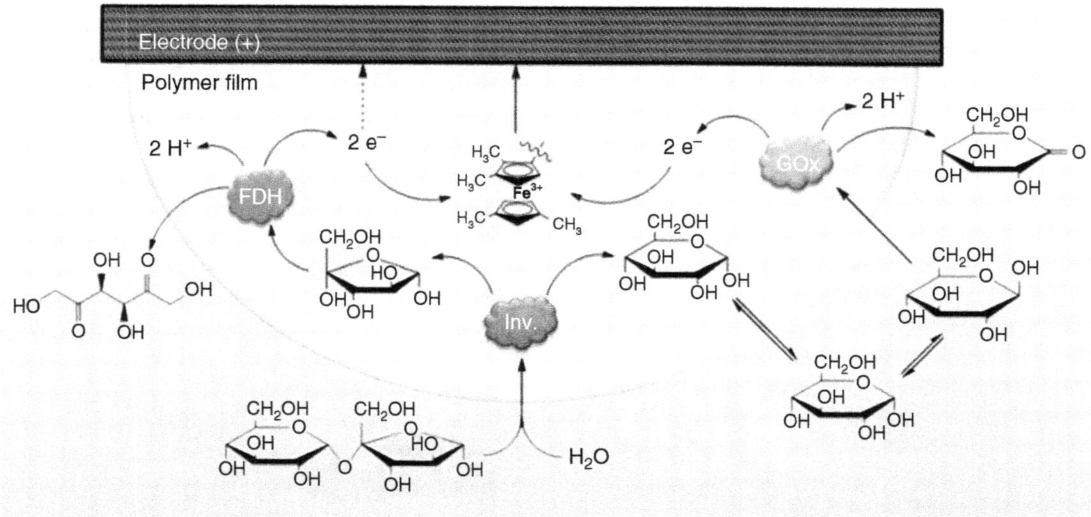

Figure 16.7
Enzyme cascade for oxidation of sucrose. *Reproduced with permission from Ref. 26.*

lower activity at room temperature operating conditions. Another approach has been enzyme engineering to improve stability.[6,34] Enzyme engineering can be used to improve long term stability, thermal stability, and/or stability in organic solvents. All three are important for a variety of the applications discussed above. Therefore, there has been a great deal of effort by academics and industrial enzyme companies to enzyme engineer proteins for improved stability. The final common approach to improve the stability of an enzyme or cascade of enzymes is immobilization of the enzyme(s). This will be discussed in the next section.

16.5 Enzyme Immobilization

Enzyme immobilization is a strategy to improve stability of an enzyme, but also a strategy for easily re-using enzymes in a bioreactor system. There are a variety of strategies for enzyme immobilization. The simplest strategies is adsorption of the protein to the surface of a packing material[35] (or electrode in the case of biofuel cells). Proteins non-specifically adsorb to a wide variety of materials ranging from metal and metal oxide surfaces to polymer surfaces, allowing for a wide variety of support materials. There are two concerns with adsorption as an enzyme immobilization strategy: (1) adsorbed enzymes can readily leach from surfaces decreasing the active lifetime and re-use of the enzymes and (2) depending on the surface preparation of the substrate/support particle, the enzyme can denature on the surface. Therefore, adsorption is commonly used for proof-of-concept biotechnologies, but not for industrial biotechnologies.

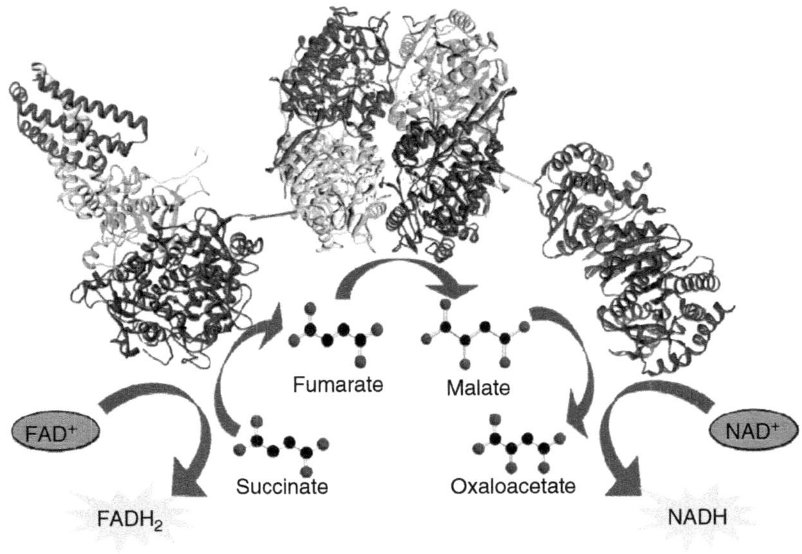

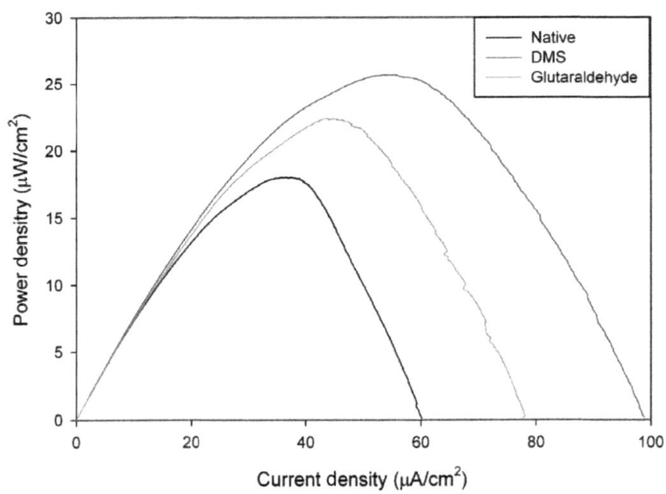

Figure 16.8
Metabolon-based electricity production. (Top) Schematic of the Krebs cycle metabolon. (Bottom) Representative power curves obtained for native and cross-linked mitochondrial electrodes in a pyruvate/air biofuel cell containing 100 mM pyruvate fuel at room temperature. *Reproduced with permission from Ref. 39.*

The second strategy for immobilization is probably the most common and involves crosslinking of the enzyme to the support particle[36] (or conductive material in the case of biofuel cells). This is done industrially for glucose isomerase and several other enzymes. Sometimes the enzyme is crosslinked to itself instead of to a particle and that is referred to as a crosslinked enzyme aggregate (CLEA).[37] There are a variety of other crosslinking techniques similar to CLEAs (i.e. CLECS, etc.) where enzyme is crosslinked into materials and to other enzymes. This technique minimizes leaching and has been shown to result in active lifetimes greater than 2 years, but crosslinking frequently decrease enzyme activity, because it decreases the degrees of freedom of the active site, which decreases the ability of the enzyme to turnover substrate. Therefore, it is typically used for very fast enzymes (i.e. carbonic anhydrase) where small decreases in enzyme activity don't massively affect throughput of the bioreactor.

Sandwich techniques are another type of enzyme immobilization technique. The sandwich technique involves coating the support material with enzyme and then coating the enzyme/support material with a polymer layer to trap the enzyme(s) at the surface of the support material.[35] If the polymer layer is deposited without organic solvent or crosslinker, then it can be a mild method of immobilization. However, transport of substrates through the thick polymer layer typically results in a lower apparent enzymatic activity.[38] Frequently, the sandwich technique is also used to provide some selectivity to the electrode. For instance, if the sandwich membrane is a gas-permeable membrane then it might be used for providing selectivity to only gas reactants or if it is an ion exchange membrane, then it might be used to preconcentrate an ionic analyte or to eliminate the response of a ionic interferent.

The final techniques for immobilization are entrapment and encapsulation. The entrapment process is typically performed by mixing enzyme, monomer, and crosslinker and casting on the support material. Many of the polymers (or hydrogels) that are formed this way have facile transport of substrate and the method can load a large quantity of enzyme, but the crosslinker can also crosslink the enzyme, which sometimes results in a decrease in enzyme activity. The encapsulation method typically involves co-casting enzyme with polymer. This technique does not involve crosslinker, so the enzyme can be immobilized while maintaining its structure; however, the polymer does retard transport of substrate to the enzyme, so frequently the apparent enzyme activity is lower due to the slow transport. Entrapment and encapsulation methods are typically good at minimizing leaching and promoting long term stability with reports in the literature of >2 years for these type of immobilized enzymes.

It is important to note with enzyme immobilization that the choice of enzyme does have an impact on the choice of immobilization strategy. For instance, if an enzyme has very low activity initially, then typically methods are chosen that do not affect the enzyme activity (i.e. avoid crosslinking) if an enzyme has a large substrate (polymeric substrate like cellulose), then entrapment and encapsulation are avoided, because it is too difficult for the substrate to diffuse through the polymer matrix.

16.6 Conclusions

In conclusions, cell-free biotransformations are an alternative to microbial biotechnology. They have the advantages of high volumetric catalytic activity and ability to incorporate catalysts from multiple organisms, but they are plagued with issues of instability. Instability can result from proteases degrading the enzymes, enzyme stability, co-factor stability, or low yield of co-factor regeneration. Enzyme stabilization is frequently improved by immobilization on surfaces (i.e. reactor support material), which also allow for re-use of the enzyme, but these immobilization strategies have to be combined with co-factor immobilization, stabilization, and regeneration strategies for industrial relevance. Cell-free biotechnologies range from value added products (i.e. fructose and therapeutics) to fuels (i.e. ethanol and hydrogen) to electricity via enzymatic biofuel cells.

References

1. Chemier JA, Fowler ZL, Koffas MAG, Leonard E. Trends in microbial synthesis of natural products and biofuels. *Adv Enzymol Relat Areas Mol Biol* 2009;**76**:151–217.
2. (a) Palmore G, Whitesides GM. Microbial and enzymatic biofuel cells. *ACS Symp Ser* 1994;**566**. 271–90. (b) Bond DR, Lovely DR. Electricity production by Geobacter sulfurreducens attached to electrodes. *Appl Environ Microbiol* 2003;**69**1548–55. (c) Rabaey K, Lissens G, Siciliano SD, Verstraete W. A microbial fuel cell capable of converting glucose to electricity at high rate and efficiency. *Biotechnol Lett* 2003;**25**:1531–5. (d) Schroeder U, Niessen J, Scholz F. A generation of microbial fuel cells with current outputs boosted by more than one order of magnitude. *Angew Chem Int Ed* 2003;**42**(25):2880–3. (e) Kim BH, et al. Enrichment of microbial community generating electricity using a fuel-cell-type electrochemical cell. *Appl Microbiol Biotechnol* 2004;**63**:672–81. (f) Logan B, Min B, Kim J, Heilmann J, Oh S, Liu H. Microbial power: electricity generation from domestic and agricultural wastewaters using microbial fuel cells. In: Technical Exhibition & Conference, 78th, Washington, DC, United States, Water Environment Federation: Washington, DC, United States; 2005. p. 93–9. (g) Rabaey K, Verstraete W. *Microbial fuel cells: novel biotechnology for energy generation. Trends Biotechnol* 2005;**23**(6):291–8. (h) Higgins SR, Foerster D, Cheung A, Lau C, Bretschger O, Minteer SD, et al. Fabrication of macroporous chitosan scaffolds doped with carbon nanotubes and their characterization in microbial fuel cell operation. *Enzym Microb Technol* 2011;**48**:458–65. (i) Higgins SR, Lau C, Atanassov P, Minteer SD, Cooney MJ. Standardized characterization of a flow through microbial fuel cell. *Electroanalysis* 2011;**23**:2174–81. (j) Higgins SR, Lau C, Atanassov P, Minteer SD, Cooney MJ. Hybrid biofuel cell: microbial fuel cell with an enzymatic air-breathing cathode. *ACS Catal* 2011;**1**:994–7.
3. Logan BE. Scaling up microbial fuel cells and other bioelectrochemical systems. *Appl Microbiol Biotechnol* 2010;**85**:1665–71.
4. Arechederra R, Minteer SD. Nanomaterials in biofuel cells. American Scientific Publishers, Stevenson Ranch, CA; 2009, pp. 287-299.
5. Buchner E. Alkoholische gahrun ohne hefezellen. *Ber Dtsch Chem Ges* 1897;**30**:117–24.
6. Casteleijn MG, Urtti A, Sarkhel S. Expression without boundaries: cell-free protein synthesis in pharmaceutical research. *Int J Pharm* 2013;**440**(1):39–47.
7. CA R, AI S. Genetically engineered synthesis of natural products: from alkaloids to corrins. *Annu Rev Microbiol* 1996;**50**:467–90.
8. Bujara M, Schümperli M, Billerbeck S, Heinemann M, Panke S. Exploiting cell-free systems: implementation and debuggin of a system of biotranformations. *Biotechnol Bioeng* 2010;**106**:376–89.
9. Daines A, Maltman B, Flitsch S. Synthesis and modifications of carbohydrates using biotransformations. *Curr Opin Chem Biol* 2004;**8**:106–13.
10. (a) Myung S, Rollin J, You C, Sun F, Chandrayan S, Adams MWW, et al. In vitro metabolic engineering of hydrogen production at theoretical yield from sucrose. *Metab Eng* 2014;**24**:70–7. (b) Zhang YHP, Evans BR, Mielenz JR, Hopkins RC, Adams MWW. High-yield hydrogen production from starch and water by a synthetic enzymatic pathway. *PLoS One* 2007;**2**(5):e456.
11. JS, MdC, Rollin J, Myung S, Chun Y, Chandrayan S, Patino R, et al. High-yield production of dihydrogen from xylose by using a synthetic enzyme cascade in a cell-free system. *Angew Chem Int Ed* 2013;**52**:1–5.

12. Krutsakorn B, Honda K, Ye X, Imagawa T, Bei X, Okano K, et al. In vitro production of n-butanol from glucose. *Metab Eng* 2013;**20**:84–91.
13. Banta S, Swanson BA, Wu S, Jarnagin A, Anderson S. Alteration of the specificity of the cofactor-binding pocket of Corynebacterium 2,5-diketo-D-gluconic acid reductase A. *Protein Eng* 2002;**15**(2):131–40.
14. Campbell E, Meredith M, Minteer SD, Banta S. Enzymatic biofuel cells utilizing a biomimetic cofactor. *Chem Commun* 2012;**48**(13):1898–900.
15. Ansell RJ, Lowe CR. Artificial redox coenzymes: biomimetic analogues of NAD^+. *Appl Microbiol Biotechnol* 1999;**51**(6):703–10.
16. Wichmann R, Vasic-Racki D. Cofactor regeneration at the lab scale. In: Kragl U, editor. *Technology transfer in biotechnology*, vol. 92. New York: Springer Berlin Heidelberg; 2005. p. 225–60.
17. Al-Zuhair S. Biochemical catalytic production of biodiesel. In: Luque R, Campelo J, Clark J, editors. *Handbook of biofuels production*. Cambridge: Woodhead Publishing Limited; 2011. p. 134–59.
18. Potter MC. Electrical effects accompanying the decomposition of organic compounds. *Proc R Soc London Ser B* 1912;**84**:260–78.
19. Yahiro AT, Lee SM, Kimble DO. Bioelectrochemistry. I. Enzyme utilizing biofuel cell studies. *Biochim Biophys Acta* 1964;**88**(2):375–83.
20. (a) Kim J, Grate JW, Wang P. Nanostructures for enzyme stabilization. *Chem Eng Sci* 2006;**61**:1017–26. (b) Moore CM, Akers NL, Hill AD, Johnson ZC, Minteer SD. Improving the environment for immobilized dehydrogenase enzymes by modifying Nafion with tetraalkylammonium bromides. *Biomacromolecules* 2004;**5**(4):1241–7.
21. (a) Wen D, Deng L, Zhou M, Guo S, Shang L, Xu G, et al. A biofuel cell with a single-walled carbon nanohorn-based bioanode operating at physiological condition. *Biosens Bioelectron* 2010;**25**:1544–7. (b) Zheng W, Zhao HY, Zhang JX, Zhou HM, Xu XX, Zheng YF, et al. A glucose/O_2 biofuel cell base on nanographene platelet-modified electrodes. *Electrochem Commun* 2010;**12**:869–71.(c) Kim J, Jia H, Wang P. Challenges in biocatalysis for enzyme-based biofuel cells. *Biotechnol Adv* 2006;**24**(3):296–308.
22. (a) Barriere F, Kavanaugh P, Leech D. A laccase-glucose oxidase biofuel cell prototype operating in a physiological buffer. *Electrochim Acta* 2006;**51**(24):5187–92. (b) Barton SC, Pickard M, Vazquez-Duhalt R, Heller A. Electroreduction of O_2 to water at 0.6 V (SHE) at pH 7 on the "wired" Pleurotus ostreatus laccase cathode. *Biosens Bioelectron* 2002;**17**(11–12):1071–4. (c) Calabrese Barton S, Kim H-H, Binyamin G, Zhang Y, Heller A. The "Wired" laccase cathode: high current density electroreduction of O_2 to Water at +0.7 V (NHE) at pH 5. *J Am Chem Soc* 2001;**123**(24):5802–3. (d) Meredith MT, Minson M, Hickey D, Artyushkova K, Glatzhofer DT, Minteer SD. Anthracene-modified multi-walled carbon nanotubes as direct electron transfer scaffolds for enzymatic oxygen reduction. *ACS Catal* 2011;**1**:1683–90.
23. (a) Akers NL, Moore CM, Minteer SD. Development of alcohol/O_2 biofuel cells using salt-extracted tetrabutylammonium bromide/Nafion membranes to immobilize dehydrogenases enzymes. *Electrochim Acta* 2005;**50**(12):2521–5. (b) Sokic-Lazic D, Minteer SD. Citric acid cycle biomimic on a carbon electrode. *Biosens Bioelectron* 2008;**24**:945–50. (c) Sokic-Lazic D, Arechederra RL, Treu BL, Minteer SD. Oxidation of biofuels: fuel diversity and effectiveness of fuel oxidation through multiple enzyme cascades. *Electroanalysis* 2010;**22**:757–64.
24. Sokic-Lazic D, de Andrade A, Minteer SD. Utilization of enzyme cascades for complete oxidation of lactate in an enzymatic biofuel cell. *Electrochim Acta* 2011;**56**:10772–5.
25. Sokic-Lazic D, Minteer SD. Pyruvate/air enzymatic biofuel cell capable of complete oxidation. *Electrochem Solid State Lett* 2009;**12**(9):F26–8.
26. Hickey DP, Giroud F, Schmidtke DW, Glatzhofer DT, Minteer SD. Enzyme cascade for catalyzing sucrose oxidation in a biofuel cell. *ACS Catal* 2013;**3**(12):2729–37.
27. (a) Arechederra RL, Treu BL, Minteer SD. Development of glycerol/O_2 biofuel cell. *J Power Sources* 2007;**173**156–61. (b) Arechederra RL, Minteer SD. Complete oxidation of glycerol in an enzymatic biofuel cell. *Fuel Cells* 2009;**9**(1):63–9.
28. (a) Xu S, Minteer SD. Enzymatic biofuel cell for oxidation of glucose to CO_2. *ACS Catal* 2012;**2012**(2):91. (b) Shao M, Nadeem Zafar M, Sygmund C, Guschin DA, Ludwig R, Peterbauer CK, et al. Mutual

enhancement of the current density and the coulombic efficiency for a bioanode by entrapping bi-enzymes with Os-complex modified electrodeposition paints. *Biosens Bioelectron* 2013;**40**(1):308–14. (c) Tasca F, Gorton L, Kujawa M, Patel I, Harreither W, Peterbauer CK, et al. Increasing the coulombic efficiency of glucose biofuel cell anodes by combination of redox enzymes. *Biosens Bioelectron* 2010;**25**:1710–6.

29. (a) Zhu Z, Wang Y, Minteer SD, Percival Zhang YH. Maltodextrin-powered enzymatic fuel cell through a non-natural enzymatic pathway. *J Power Sources* 2011;**196**(18):7505–9. (b) Zhu Z, Tam TK, Sun F, You C, Zhang Y-HP. A high-energy-density sugar biobattery via a synthetic enzymatic pathway. *Nat Commun* 2014;**5**:3026.
30. Moehlenbrock MJ, Meredith M, Minteer SD. Bioelectrocatalytic oxidation of glucose in CNT impregnated hydrogels: Advantages of synthetic enzymatic metabolon formation. *ACS Catal* 2012;**2**:17–25.
31. Lin J-L, Palomec L, Wheeldon I. Design and analysis of enhanced catalysis in scaffolded multienzyme cascade reactions. *ACS Catal* 2013;**4**(2):505–11.
32. Kim YH, Campbell E, Yu J, Minteer SD, Banta S. Complete oxidation of methanol in an enzymatic biofuel cell by a self-assembling hydrogel created from three modified dehydrogenase. *Angew Chem* 2013;**52**:1437–40.
33. (a) Beneyton T, El HA, Griffiths AD, Hellwig P, Taly V. Immobilization of CotA, an extremophilic laccase from Bacillus subtilis, on glassy carbon electrodes for biofuel cell applications. *Electrochem Commun* 2011;**13**:24–7. (b) Lojou E. Hydrogenases as catalysts for fuel cells: strategies for efficient immobilization at electrode interfaces. *Electrochim Acta* 2011;**56**:10385–97.
34. Asako H, Shimizu M, Itoh N. Engineering of NADPH-dependent aldo-keto reductase from Penicillium citrinum by directed evolution to improve thermostability and enantioselectivity. *Appl Microbiol Biotechnol* 2008;**80**:805–12.
35. Cooney MJ, Svoboda V, Lau C, Martin GP, Minteer SD. Enzyme catalysed biofuel cells. *Energy Environ Sci* 2008;**1**:320–37.
36. (a) Fernandez-Lafuente R, Rosell CM, Rodriguez V, Guisan JM. Strategies for enzyme stabilisation by intramolecular crosslinking with bifunctional reagents. *Enzym Microb Technol* 1995;**17**:517–23. (b) Szamocki R, Velichko A, Muecklich F, Reculusa S, Ravaine S, Neugebauer S, et al. Improved enzyme immobilization for enhanced bioelectrocatalytic activity of porous electrodes. *Electrochem Commun* 2007;**9**:2121–7. (c) Mena ML, Yanez-Sedeno P, Pingarron JM. A comparison of different strategies for the construction of amperometric enzyme biosensors using gold nanoparticle-modified electrodes. *Anal Biochem* 2005;**336**:20–7.
37. (a) Kim BC, Zhao X, Ahn H-K, Kim JH, Lee H-J, Kim KW, et al. Highly stable enzyme precipitate coatings and their electrochemical applications. *Biosens Bioelectron* 2011;**26**:1980–6. (b) Kim MI, Kim J, Lee J, Jia H, Na HB, Youn JK, Kwak JH, Dohnalkova A, Grate JW, Wang P, Hyeon T, Park HG, Chang HM. Crosslinked enzyme aggregates in hierarchically-ordered mesoporous silica: a simple and effective method for enzyme stabilization. *Biotechnol Bioeng* 2006;**96**(2):210–8.
38. Sahney R, Anand S, Puri BK, Srivastava AK. A comparative study of immobilization techniques for urease on glass-pH-electrode and its application in urea detection in blood serum. *Anal Chim Acta* 2006;**578**(2):156–61.
39. Moehlenbrock MJ, Toby TK, Waheed A, Minteer SD. Metabolon catalyzed pyruvate/air biofuel cell. *J Am Chem Soc* 2010;**132**:6288–9.

CHAPTER 17

Microbial Electrochemical Cells and Biorefinery Energy Efficiency*

Abhijeet P. Borole

Biosciences Division, Oak Ridge National Laboratory, Oak Ridge, Tennessee, USA; Energy Science and Engineering, Bredesen Center for Interdisciplinary Research and Graduate Education; Chemical and Biomolecular Engineering Department, University of Tennessee, Knoxville, Tennessee, USA

17.1 Introduction

Microbial fuel cells (MFCs) and bioelectrochemical systems (BESs) offer a novel method of energy recovery from wastewater and have evolved to derive energy and products from biomass and organic waste. Originally, MFCs were developed to produce electricity from degradable organic matter present in wastewaters.[1] Subsequently, microbial electrolysis emerged as a means of producing hydrogen, methane, and other products from the aqueous wastes.[2,3] In a biorefinery, hydrogen has many uses, and its production from renewable resources derived from biomass can have significant environmental benefits. It can be used as a reagent to upgrade the biofuel to a higher octane fuel or to hydrogenate intermediates to higher value chemicals or as an alternate fuel itself.[4]

Biorefinery wastewater contains a number of compounds generated during the various steps of biomass processing. The pretreatment step, which uses high temperature and an acid or a base to break down the polymeric carbohydrates, produces fermentation inhibitors such as 2-furfural and 5-hydroxymethyl-2-furfural (HMF), lignin degradation products such as phenolic acids, alcohols and ketones, plus acetate from deacetylation of hemicellulose.[5,6] These inhibitors can be detrimental to ethanol production if generated at high concentrations.[7] The need for high biomass solids loading of >17% (w/w) to enable higher biofuel yields generates higher concentrations of inhibitors in the hydrolyzate stream.[8] Inhibitor concentrations in the range of 5-20 mM can impact fermentation, depending on the ethanologen used.[9] Removal of these inhibitors has been studied via membrane-based technologies,[10] ion exchange, polymeric adsorbents,[11] overliming,[12] and solvent extraction.[13] However, these approaches are only

---

* This chapter has been authored by a contractor of the U.S. Government under contract No. DE-AC05-00OR22725. Accordingly, the U.S. Government retains a nonexclusive, royalty-free license to publish or reproduce the published form of this contribution, or allow others to do so, for U.S. Government purposes.

Biotechnology for Biofuel Production and Optimization. http://dx.doi.org/10.1016/B978-0-444-63475-7.00017-0

partially effective and add considerable costs to the fermentation process. Water recycling and reuse is significantly limited due to the build-up of these inhibitors.

Besides the by-products of pretreatment, the fermentation step itself generates many by-products. Production of organic acids such as acetate, propionate, lactate, and succinate is common during fermentation. Conversion of sugars in the fermentation process is usually not complete, leaving at least 5% of the sugars in the finished fermentation broth. Thus, the biorefinery process water, post biofuel recovery, contains sugar- and lignin-degradation products, acetate, fermentation by-products, and any unused carbohydrates. Removal of unused carbohydrates and sugars to low concentrations has been demonstrated in MFCs.[14,15] Conversion of the other components into electricity and hydrogen using an MFC/microbial electrolysis cell (MEC) has received attention in recent years.[3] Conversion of the inhibitory contaminants including 2-furfural, HMF, 4-hydroxybenzaldehyde (HB), hydroxyacetophenone (HAP), and vanillic acid (VA), which affect growth and activity of fermentation organisms, has been reported. The ability to generate additional energy from the biorefinery wastewater offers a way to optimize the conversion of biomass into bioenergy and increase process conversion efficiency and energy efficiency in the biorefinery. Production of coproducts besides the biofuel has the potential to provide an economic edge and may become a necessity as costs of biofuel production increase. This chapter compiles the reports from studies targeting conversion of biorefinery wastewater substrates and describes how the change from existing pathways to schemes including BES systems can improve energy efficiency in the biorefinery. Finally, the current state of the BES technology is discussed. The development of process options using microbial fuel and electrolysis cells for production of additional bioenergy from residual organics present in the biorefinery wastewater enables water clean-up and opens the door to water use minimization via recycle of the treated water.

17.2 Principles of BES Technology

BESs convert degradable organic matter in wastewater and other process waters into useful products such as electricity, hydrogen, methane, etc. This is done via extraction of electrons from the organic matter using microbes as the catalyst in an anode chamber. The electrons are then transferred to the cathode via an external circuit where the electrons are donated to an electron acceptor. Figure 17.1 shows a schematic of the BES system. In MFCs, the electrons are used to combine with oxygen at the cathode generating a potential difference of up to 0.5 V between anode and cathode which can be harvested as electricity. In MECs, the electrons are used to reduce protons and generate hydrogen, but an added external potential of $\geq$0.3 V is required to overcome the thermodynamic barrier for hydrogen production. The electrons can also be used to reduce partially oxygenated species such as acetate, glycerol, etc., producing valuable products such as ethanol and 1,3-propanediol, respectively.[16,17]

Production of hydrogen peroxide as well as caustic has been shown to be feasible via operation of the cathode under specific conditions.[18,19] The ability of microbes to efficiently harvest electrons from a variety of waste organics and to transfer them at low overpotentials to electrodes has made it possible to explore application of this technology in biorefinery.[20–22] The high efficiency

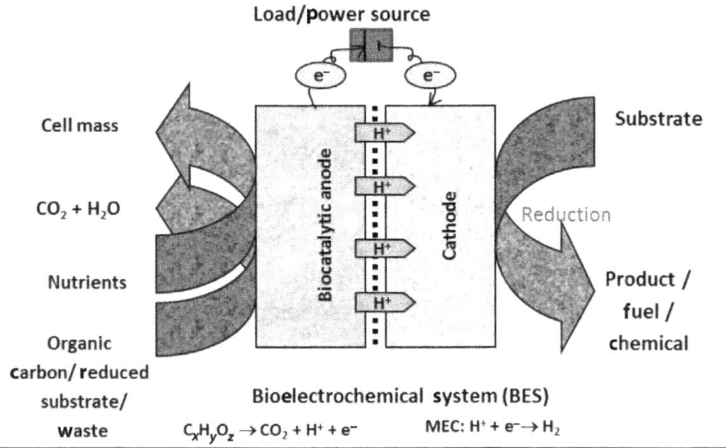

Type of BES	Cathode substrate	Product
MFC	Oxygen	Electricity
MEC	Protons	Biohydrogen
BES	Acetate	Ethanol/biofuel
BES	Oxygen	Hydrogen peroxide
BES	Carbon dioxide	Electrofuels
BES	Other/sunlight	Photo/biofuels

Figure 17.1
Schematic of a bioelectrochemical system.

is afforded by novel electrogenic microbes which use nanowires[23,24] and other mechanisms to facilitate direct electron transport, many of which are not yet fully understood.[25–28]

17.3 Conversion of Substrates Present in Biorefinery Process Water Using BES Systems

This section discusses the electrogenic conversion of the organic constituents present in biorefinery process water. The conversion of model compounds representing the organic compounds typically present in biorefinery process water is discussed first, followed by a discussion on the treatment of actual process water derived from corn stover processing. This is followed by estimates of hydrogen production from the corn stover process water and streams relevant to conversion of other feedstocks to biofuels, including grain to ethanol processing via wet-mill and dry-mill processes, conversion of sugar-beet and sugarcane juice to fuel, softwood, hardwood, and other lignocellulosic conversion processes.

17.3.1 Electrogenic Conversion of Biomass-Derived By-products

Electricity production from simulated biorefinery process streams

The removal of model fermentation inhibitor compounds typically present in biorefinery process water with simultaneous production of electricity has been demonstrated in MFCs.[29]

The model compounds included 2-furfural, HMF, VA, HAP, and HB. The use of MFCs for removal of acetate has been reported previously.[30–32] Development of a microbial consortium capable of handling a range of compounds including sugar-degradation products, lignin-degradation products, as well as hemicellulose-derived products such as acetate was reported.[29] In order to enable application of MFCs in the biorefinery process, a number of factors related to conversion of the individual components in the biorefinery process water need to be considered. These includes electrochemical performance, degree of mineralization, inhibitor levels after treatment, by-product formation, stability of current production, maintenance, contamination, cost, effect on ethanol yield, etc. These factors and others important for determining the suitability of applying MFCs for recovering energy from biorefinery process streams and controlling inhibitor concentration in the recycle water are discussed.

Electrogenic conversion. Conversion of 2-furfural was reported in MFCs demonstrating >99% removal of the substrate.[29] Similar results were reported with the other model compounds including HMF, VA, HAP, and HB. The MFC was investigated for its ability to handle streams with multiple substrates present at the same time and concentrations representative of biorefinery streams. The concentration of fermentation inhibitors in a pretreated biomass slurry or a typical biorefinery recycle stream ranges from a few mM to 20 mM or more.[9,33] The effect of the fermentation inhibitors on alcohol production in a cellulosic fermentation process increases with the concentration.[27] The effect of 2-furfural on electricity production was studied at a concentration of 0.1-2 g/L (20.8 mM). Figure 17.2 shows the effect of concentration on electricity production.[29] The concentrations of other substrates that were studied in the report were: HMF (0.1-2 g/L), HB (0.1-1 g/L), and VA (0.1-4 g/L). Experiments

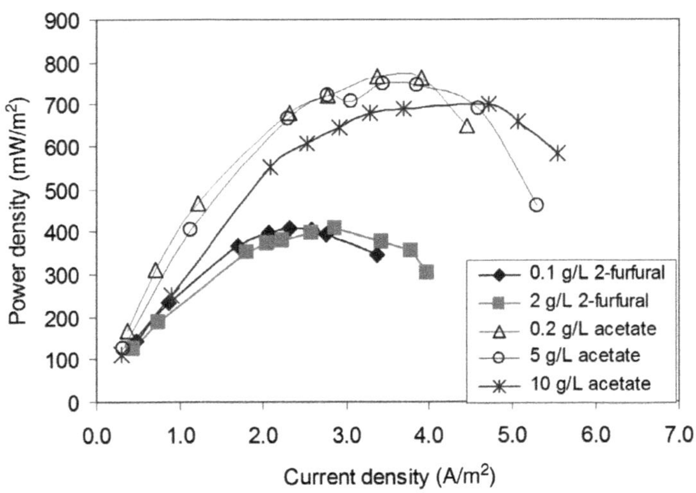

Figure 17.2

Effect of concentration on current production for two inhibitory molecules, 2-furfural and acetate present in biorefinery process water. *Figure extracted with permission from Borole et al.*[29]

with substrate concentration up to 20 mM did not reduce electrical output, indicating that the model compounds tested were not inhibitory to the anodic microbial biofilm consortium. The concentration of the fermentation inhibitor did not have any adverse effect on the power density. The increase in the inhibitor concentration actually resulted in an increase in the current density. The higher current densities at the high concentration indicate a high level of substrate tolerance by the microbial consortium. The ability to handle multisubstrate mixtures containing 2-furfural, HB, VA, and HAP was also reported.[29] The effects of by-products generated during MFC treatment is also important since it can have unexpected impact on the fermentation process. Analysis of the product stream via HPLC and the high recovery of the electrons in the form of electricity indicated near-complete mineralization of 2-furfural and HMF, thus implying minimal potential for any adverse effects on the fermentation process if the process water is recycled.

Coulombic efficiency (CE). CE is defined as the efficiency of capture of the electrons derived from a given substrate into electricity. Conversion in an MFC can result in transformation of the substrates into one of the following three products: carbon dioxide and water (complete mineralization), cellular biomass, or intermediate by-products. The CE reported for conversion of the model compounds ranged from 60% to 69%.[29] The balance 31-40% of the inhibitor was presumed to be assimilated into cells or biofilm since no persistent by-products were observed.

Power density. Power density is an important parameter for determining the cost-effectiveness of MFCs if electricity production is the main target. Water recycle and ethanol yield (using recycled water) are, however, more important in the case of the biorefinery application since the increase in revenue from electricity may not justify its use entirely. Nevertheless, energy recovery from the process has to be considered in the overall economic feasibility. The maximum power density reported for individual substrates was as follows: 3490 mW/m^2 (336 W/m^3) for 2-furfural, 2510 mW/m^2 (238 w/m^3) for HMF, and 630 mW/m^2 (62 W/m^3) for HB.[29]

MFC stability. The MFC investigating biorefinery process water model substrates has been reported to be operated for a total period of 10 months. A stable output of 0.15 ± 0.02 V (at a 50 Ω load) was reported using 2-furfural as the substrate. Operation of the MFC with an air cathode indicated cathode as the limiting factor for electricity production. Operating the MFC with a ferricyanide cathode for a period of 4 months, subsequent to its operation with an air cathode for 6 months, was shown to increase power density up to 3700 mW/m^2. This indicated potential for high output and the relative stability of the current output of the MFC.

Anode biocatalyst. Characterization of the microbial community capable of processing the biorefinery process water substrates in an MFC has been reported.[29] The consortium was reported to be diverse containing at least 42 different strains. It was reported that the community was dominated by δ-proteobacteria constituting 50% of the clone library population (Figure 17.3).

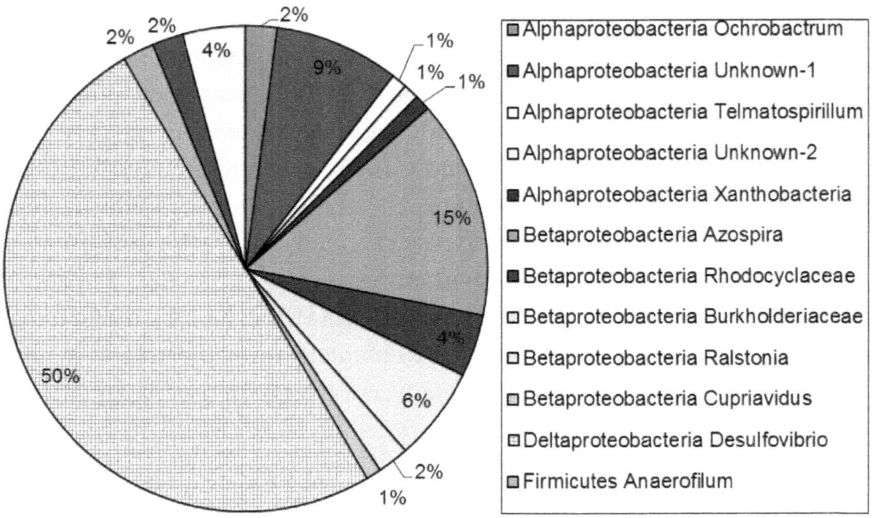

Figure 17.3
Distribution of microbial population in MFC anode consortium enriched using biorefinery process water model substrates.

Many strains identified in the anodic biofilm consortium were reported to be related to members of communities identified in other MFC/MEC systems. The closest match to the δ-proteobacteria was a previously uncultured bacterium Proteobacterium Core-3 enriched in a consortium in an electrochemically assisted bioreactor using iron for respiration. The closest known genus similar to the group was *Desulfovibrio* (92-99% similarity). The remaining 41 clones in the library represented more than 14 genera. The other two most dominant groups were reported to belong to *Azospira* (89-97% similarity) and a second type of proteobacterium. The phylogenetic tree for the clone library developed for the biorefinery-relevant MFC community reported by Borole et al. is shown in Figure 17.4.

Treatment of a process stream derived from corn stover

Biorefinery processes generate many by-products besides biofuels which end up in the process water generated after biofuel recovery. Treatment of a process stream obtained post-fermentation from dilute sulfuric acid-pretreated corn stover in an MFC has been reported.[34] Conversion of the inhibitory molecules in the presence of unused sugars present in the corn stover process water to electricity was reported. While the studies with model substrates indicated that substrate concentration did not have an effect on the current density,[29] its effect on CE was not investigated in detail. The biorefinery process water can contain a total organic concentration from <10 g/L to over 65 g/L.[29,33] The effect of substrate loading on conversion of the range of compounds present in the process water derived from corn stover, the CE, and the power density have been reported and are discussed in sections below.

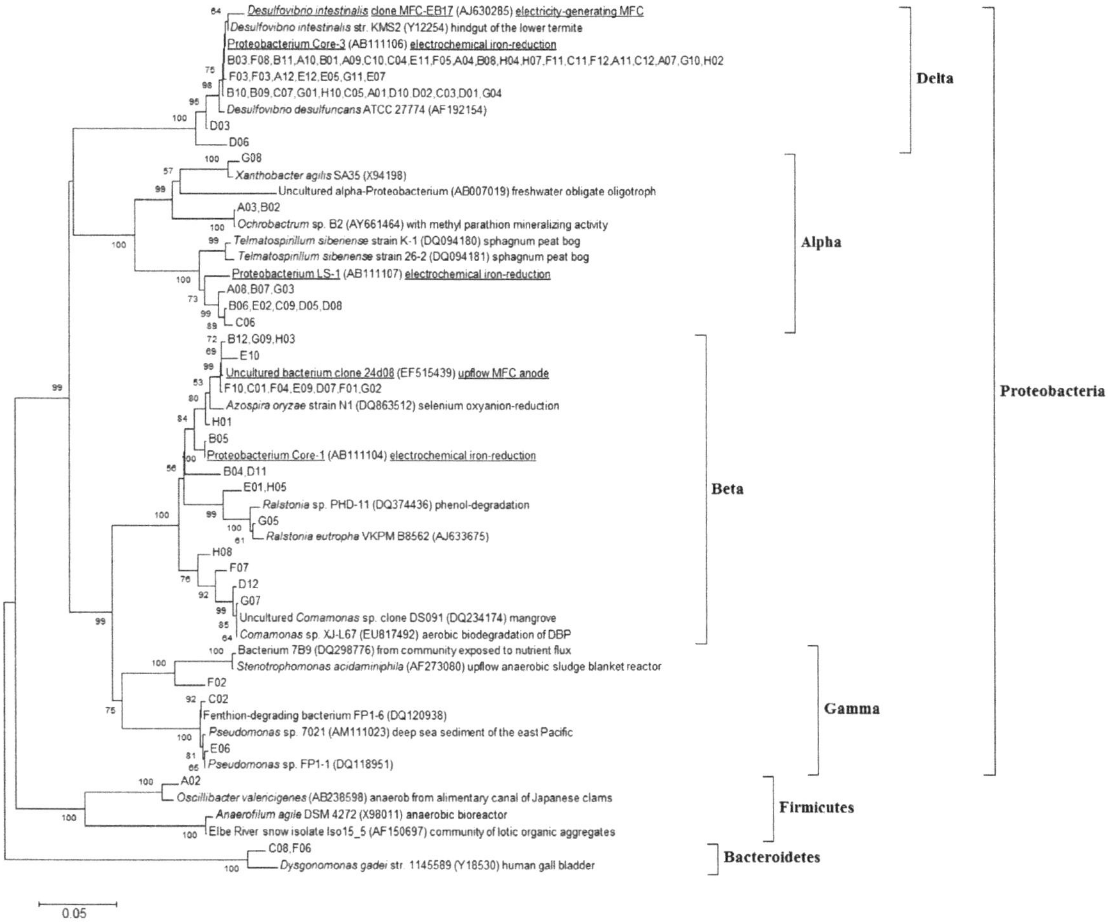

Figure 17.4

Phylogenetic tree of anode consortium capable of conversion of sugar- and lignin-degradation compounds in biorefinery process waters. *Figure extracted with permission from Borole et al.*[29]

Effect of loading on conversion and CE. The study was conducted with different loadings of the corn stover process water in batch experiments ranging from 1% (diluted with nutrient medium) to 100%.[34] The process water sample used in the study contained a total of 66.7 g/L of organic substrates. A 1% sample contained 0.67 g/L of the organic substrates. Transient increase in levels of some peaks was reported during the degradation process; however, experiments at lower loading levels indicated subsequent removal by the anodic consortium.[34] Complete removal of nearly all compounds was reported in the 1% loading experiment in a 49-h batch study.[34] A CE between 40% and 25% was reported for a loading between 1% and 4%, which dropped to 10% or lower at the higher loadings. Interestingly, the rate of removal of the organic compounds was reported to increase with the increase in organic loading. The electrical efficiency of the process, however, decreased with increasing loading.

A CE of 40% for the biorefinery process stream, which was dominated by sugars, primarily xylose, is comparable or higher than the CE reported in the literature for sugar-containing aqueous streams processed in MFCs.[14] A CE of 21-37% was reported for conversion of pentoses and hexoses used in an MFC.[14] When xylose is used as a single substrate, a CE of 13-50% was reported.[35,36] Thus, the CE depends to a large degree on the consortium and the enrichment process used to develop it. Borole et al. concluded that the organic loading in an MFC treating corn stover process water can have a significant effect on the efficiency of conversion of the compounds to electricity. Similar observations have been made by others when using xylose as a single substrate in MFCs.[36]

Conversion of sugar and lignin degradation products. The process stream representative of a biofuel producing biorefinery process contained over 20 inhibitory molecules. 2-Furfural, HMF, HB, VA, and several other phenolics were removed from the biorefinery process water sample at all organic loadings studied. It was reported that removal of these compounds occurred even in the presence of high concentrations of sugars and organic acids. Simultaneous removal of the furan aldehyde and phenolic derivatives with the sugars can be potentially advantageous for water recycle in the biorefinery.

Current density. Electricity production from the corn stover process stream was reported at a current density of 4 A/m^2 at a loading of 1%, which increased to 8 A/m^2 at a loading of 8%. At loading levels higher than 8%, the power and current density were both reported to decrease. The decreasing trend in the power density at higher loadings was reported to be due to factors such as higher biomass yield, mass transfer limitations, use of alternate electron acceptors, or a combination of these factors.[34] Formation of thick biofilms resulting from increased substrate availability at higher loadings was hypothesized to limit substrate delivery to or proton mass transfer from the electroactive biofilms growing on the electrode, limiting current production.[20] A study evaluating the stoichiometry of fermentative xylose conversion in an MFC reported rapid uptake at higher substrate concentrations but reduced power production.[35] Use of high substrate concentrations in MFCs can result in oversupply of the substrate compared to the metabolic capability of the exoelectrogenic consortium, promoting growth of fermentative organisms, methanogens, and other non-exoelectrogenic organisms.[37,38] Excessive biofilm growth can potentially result in mass transfer limitations, leading to poor electrogenic performance. A balance between substrate loading and electricity production capacity of the MFC is necessary for optimal performance. Therefore, control of organic loading is important for improving conversion efficiency and electricity production.

Implications for application in a biorefinery

The ability of an anode harboring an exoelectrogenic consortium to utilize the complete range of phenolics and furan aldehyde compounds generated during dilute acid pretreatment of corn stover in the presence of sugars was demonstrated by Borole et al.[34] A typical biorefinery wastewater stream is expected to contain about 15-65 g/L of organic matter.[39,40]

The organic content of the process water used in the study by Borole et al. was on the higher end due to incomplete conversion of the sugars to ethanol during fermentation. Even for streams with low organic strength, the concentration is too high for direct feeding to anode. In order to control the organic loading in the MFC, use of a recirculation stream is needed with controlled feeding of the wastewater stream at the anode inlet. This will enable efficient energy recovery. Thus, it is possible to treat a wastewater stream generated from a corn stover to ethanol process plant using an MFC. The energy harvested from the process in the form of electricity can potentially improve the overall energy recovery from biomass in the biorefinery.[4] Preliminary analysis suggests considerable potential for water savings as well if the water can be cleaned sufficiently for recycle.[4] The removal of the major portion of the compounds inhibitory to fermentation process via MFCs, in the presence of two to three orders of magnitude higher sugar concentration to nondetectable levels, demonstrates the potential for application of the process in a biorefinery for water recycle and energy recovery. Preliminary estimates project generation of 2.5 MW electricity in a commercial-scale biofuel production plant converting 2000 ton/day biomass.[29] Energy recovery in the form of hydrogen can result up to 7200 m^3/h of hydrogen production using MECs in similar sized plants.[41] Utilization of the biorefinery waste stream for biogas production via anaerobic digestion is an alternative; however, hydrogen production may result in a higher value proposition for the biorefinery.[4] For example, hydrogen may be used for upgrading of bio-oil or increasing yields of Fischer-Tropsch and other liquid fuels in hybrid thermo-biochem refineries. Further work is needed to accurately determine overall process energy efficiency, water savings, and environmental and economic benefits from incorporation of BESs into a biorefinery process.

17.3.2 Estimating Hydrogen Production via Microbial Electrolysis

The anode microbial catalyst in an MFC performs the primary function of substrate breakdown and electron and proton production. If an anaerobic cathode is employed instead of an aerobic cathode, hydrogen production can be achieved with application of an additional potential of about 0.3 V or higher. The excess potential is necessary to overcome the thermodynamic potential for hydrogen production. Thus, hydrogen production from biorefinery process streams is possible using an MEC. Production of hydrogen instead of electricity from the biorefinery process stream may be a more economically viable alternative due to the higher price of hydrogen and the need for hydrogen in a biorefinery. The hydrogen can potentially be used as a reagent for generation of high octane liquid fuels or production of higher value chemicals. The amount of hydrogen that can be produced via an MEC depends on a number of factors including the chemical oxygen demand (COD) of the process stream, flow rate, biodegradability of the organic matter, and efficiency of catalytic hydrogen production reactions. A method for determining hydrogen production rate from wastewater streams has been reported previously.[41,42] Estimation of hydrogen production potential from various stillage and biorefinery streams is possible using this methodology.

Theory/calculation

The data required for estimation of hydrogen production include the following: COD/concentration of organic content and the flow rate of the process stream[42]: Equation (17.1) below can then be used to determine the rate of hydrogen production[41]:

$$H_{\mathrm{MEC},i} = \frac{\mathrm{COD}_i}{M_{\mathrm{COD}}} \times Q_i b_i \times \frac{n_{\mathrm{H}_2}}{n_{\mathrm{COD}}} \times M_{\mathrm{H}_2}\,\eta_{\mathrm{H}} \tag{17.1}$$

where,

$H_{\mathrm{MEC},i}$, hydrogen production in L per gallon of biofuel/biochemical, *i*.
COD, chemical oxygen demand for wastewater stream *i*, g/L.
M_{COD}, molecular weight of oxygen (COD) = 32 g/mol.
Q_i, wastewater flow rate per gallon of biofuel/biochemical *i*, L/gal.
b_i, biodegradability of organic carbon in wastewater type *i*.
$n_{\mathrm{H}_2}/n_{\mathrm{COD}}$, ratio of number of moles of hydrogen obtained per mole of COD = 2.
M_{H_2}, molar volume of hydrogen, L/mol.
η_{H}, efficiency of hydrogen production or hydrogen yield.

The anodic and cathodic half reactions for the degradation of any organic compound present in process water can be given as:

$$\text{Anode:}\quad CH_2O + H_2O \rightarrow CO_2 + 4H^+ + 4e^- \tag{17.2}$$

$$\text{Cathode:}\quad 4H^+ + 4e^- \rightarrow 2H_2 \tag{17.3}$$

The following assumptions were used in the calculation of hydrogen production:

The biodegradability of the residual organic carbon (b) remaining after fermentation can be lower due to the presence of lignin and other components, which are slower to degrade. An average Biological Oxygen Demand (BOD) to COD ratio of these streams is about 0.54.[43] An estimate of $b_i = 0.5$ can be used for biorefinery streams. A higher conversion of pretreatment by-products to hydrogen is possible; therefore, b_i for these molecules can be assumed to be higher.[29] The ratio $n_{\mathrm{H}_2}/n_{\mathrm{COD}} = 2$ is based on Equation (17.2). While this is a theoretical estimate, it is close to the experimental data reported by Cheng and Logan.[44] Hydrogen production per mole of substrate ranging from 3.65 to 8.77 has been reported, which is equivalent to an average of 1.6 mol of hydrogen/mol of BOD. Ohmic and mass transfer losses can reduce the efficiency of hydrogen production, which are accounted via use of an efficiency factor and a biodegradability factor. The molar volume of hydrogen can be assumed to be 22.4 L/mol based on standard temperature (25 °C) and pressure (1 atm) conditions. An efficiency factor, η_{H} of 50%, was used. Higher hydrogen yields (up to 95%) have been reported with soluble model substrates[45]; however, hydrogen production from complex carbohydrates and pretreatment by-products present in the biorefinery process streams may be lower.

Hydrogen production estimates

In a typical biorefinery design using 2000 ton/day of lignocellulosic biomass as feedstock, the flow rate of the stillage stream is of the order of 350 m^3/h.[33] The total ethanol production for a biorefinery process using 2000 ton/day biomass is 265,000 m^3/year (70 million gallons/year). For such a biorefinery, the amount of hydrogen that can be potentially produced from the process water stream, based on its average COD, is about 1260 m^3/h. Table 17.1 provides the estimates for various other feedstocks used in the biorefinery and for other biofuel producing operations. This is based on a 10% solids loading; however, typical biomass loading in a biorefinery is expected to be in the 17% range (Personal communication, Daniel Schell), which will result in higher amounts of hydrogen production. Based on COD of stillage streams in a full-scale lignocellulosic biorefinery, the maximum hydrogen production rate can be 7200 m^3/h.

In addition to the calculation of hydrogen production from the average COD of the process water, it is possible to estimate hydrogen production from the individual compounds present in the process stream as well. Table 17.2 reports the estimates for hydrogen production from the hydrolyzate streams obtained from various feedstocks based on individual compounds: acetate, sugar-degradation products, and lignin-degradation products. It is observed that the amount of by-products present in the hydrolyzate and ultimately in the distillation bottoms depends on a number of factors including the biomass/solids loading, type of biomass, and the pretreatment process. The hardwoods can generate the highest amount of these by-products, while the herbaceous crops the least.

The amount of total hydrogen production in a lignocellulosic biorefinery from the process water stream was estimated to be about 7200 m^3/h,[41] which is equivalent to 91 GJ/h (based on a High Heating Value (HHV) value of hydrogen = 141.9 kJ/g)). It is worthwhile to assess the energy content in comparison to the energy needs of the biorefinery. Fuel distillation requires about 160 GJ/h of energy, typically provided by steam generated from lignin and residual organics.[33] The amount of hydrogen generated from residual organics via MEC alone is

Table 17.1: Potential hydrogen production from process water or stillage streams in sugar, starch, and lignocellulosic biorefineries

	L H_2/L stillage	m^3 H_2/h
Sugar beets	31.9	753
Cane molasses	29.8	2021
Cane juice	10.5	559
Corn: average plant 242,000 m^3/yr (64 million gal/yr) plant	21.0	8908
Biomass@265,000 m^3/yr (70 million gal/yr) plant	21.4	7200

Table extracted with permission from Borole et al.[41]

Table 17.2: Estimates of hydrogen production from three primary substrate groups present in biorefinery process water[41]

	Biomass	Pretreatment/biomass loading (wt%)	Potential for hydrogen production from specific substrate groups, L/L process stream[a]		Ref.
			Acetate, 2-furfural and HMF	Phenolic substrates	
Hardwood	Aspen	ASE[b]/33%	13.9		[46]
	Birch	ASE/33%	11.2		[46]
	Birch	SHE[c]/NA[d]		0.4	[47]
	Yellow poplar	AH[e]/10.6%	8.5	17.6	[48]
Softwood	Pine	ASE/33%	9.0		[46]
	Pine	AH/NA		4.5	[49]
	Spruce	Acid steam/NA	6.2	0.6	[50]
	Spruce	SO_2 steam/10%	6.8		[51]
Herbaceous	Wheat straw	AWO[f]/33%	4.3	0.3	[9]
	Switchgrass	Acid steam/10%		0.2	[52]
	Bagasse	SP[g]/NA		1.1	[9][h]
	Corn stover	Acid steam/10%		0.1	[52]
	Corn stover	AH[e]/10%	3.3	0.3	[53]

[a]Average concentration of each substrate was used to determine hydrogen production potential and then added together to obtain total hydrogen production. The biodegradability of each of the substrate was assumed to be 100%, while the hydrogen yield was assumed to be 50%.
[b]ASE, acid steam explosion.
[c]SHE, steaming and enzymatic hydrolysis.
[d]NA, not available.
[e]AH, acid hydrolysis.
[f]AWO, alkaline wet oxidation.
[g]SP, soda pulping.
[h]Energy recovery method (such as MECs).

equivalent to 91 GJ/h (57% of the energy needed for ethanol distillation). While it is unlikely that the hydrogen will be used as a fuel to generate heat, since significant amount of lignin is present, use of hydrogen for biofuel upgrading is a potential option for in-house utilization in the biorefinery.

17.4 Optimization of energy conversion in biorefinery

The energy efficiency of a biochemical conversion route for deriving liquid fuels such as ethanol from lignocellulosic biomass in a typical process is about 53.5%.[33] This constitutes 48.7% ethanol and 4.8% electricity (Figure 17.5). Typical process schemes derive energy from residual organics and by-products via combustion in a boiler or remove them from water via wastewater treatment. Significant portion of the biomass energy (46.5%) is lost in the form of heat via flue gases, water evaporation, and ambient heat when the thermochemical recovery

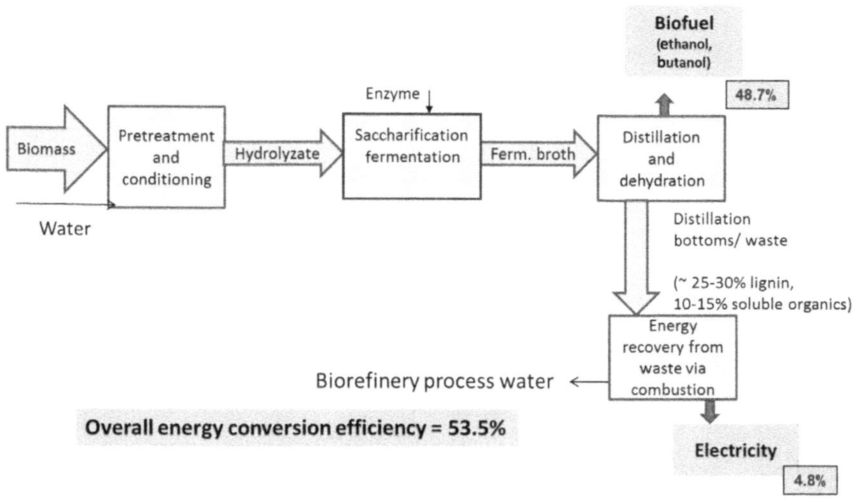

Figure 17.5
Simplified flow diagram and energy efficiency for biomass to ethanol production in biorefinery.[33] The energy recovery from biorefinery process stream post fuel recovery is usually via combustion of the lignin and concentrated organic stream, which is the main source of energy loss in the process.

options are used. Wastewater treatment can generate biogas which can be used as a fuel, as a boiler feed, or as a reagent in thermochemical fuel production, which can also result in heat losses. In addition to the energy loss, a significant amount of water is also lost via evaporation and release of flue gas to the environment. Recent work has revealed the importance of water recycle in a typical biorefinery process, particularly in the biochemical conversion process.[8]

17.4.1 Quantifying Energy and Water Losses in a Lignocellulosic Biorefinery

A process design report describing biochemical conversion of corn stover to ethanol reported by National Renewable Energy Laboratory (NREL) has reported details of the mass and energy balance in a typical lignocellulosic biorefinery.[33] An alternative process scheme using MEC has been described by Borole et al.[4] Figure 17.6 shows the energy and water flow in the original process, which uses a combination of evaporator and burner/boiler to derive energy from the unused carbohydrates and by-products present in the stillage stream. About 11.7% of the total biomass energy is vented as water vapor in the cooling tower and 7.6% is lost as ambient heat in the evaporator. Most of the incoming water, including the 88.5% make-up water, is lost via the evaporators, cooling tower, and burner effluent (flue gases). If the aqueous stream is treated using MECs, significant portion of the water and energy can be conserved. The MEC can be operated under ambient conditions or at operational temperature of the process stream, minimizing radiant heat loss. Also, the cooling requirement is minimized reducing evaporation losses. Replacement of the evaporators, wastewater treatment system, and cooling towers with MEC/MFCs can result in energy savings (e.g., reduction in the ambient heat loss, which is 7.6%

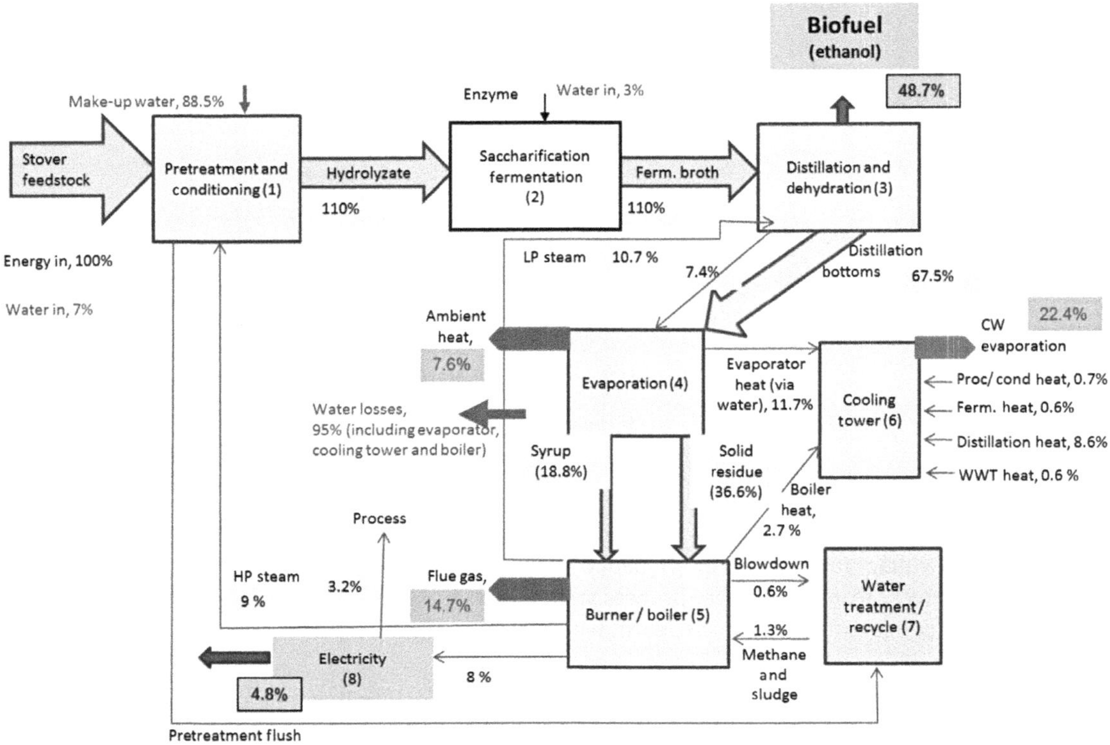

Figure 17.6

Energy and water losses in biomass to ethanol production in biorefinery.[4,33]

in the original process as well as cooling water evaporation loss (22.4%)). Alternately, a suitable combination of MFC/MEC and other treatment technologies may be implemented to improve process efficiency. The need for make-up water may be reduced due to water recycle after treatment of the process water using an MEC or an MFC.

17.5 High-efficiency Biorefinery Process Schemes Implementing BES

A number of advances have been made in biomass conversion to biofuels and bioenergy. Complex process scenarios, which target improved overall energy efficiency and recovery, have been reported via integration of multiple projected mature technologies into the biorefinery process.[54] For example, coproduction of fuels (ethanol, F-T liquids, gasoline and diesel blendstocks, methane, and hydrogen), power, and animal feed protein have been considered. The use of an ammonia fiber expansion (AFEX) pretreatment and consolidated bioprocessing for producing biofuel was implemented in one of the process scenarios. The downstream steps included gasification of lignin to generate syngas, F-T synthesis reactor to produce hydrocarbons (primarily alkanes), gas turbine combined cycle (GTCC) to generate electricity, anaerobic digestor to treat wastewater to generate biogas, pressure-swing adsorption for hydrogen separation hydrogen, partial oxidation (POX) of biogas, carbon dioxide separation and removal of hydrogen sulfide and carbon dioxide from syngas or water gas shift reactor to produce hydrogen from carbon monoxide. Optimization of the process train by inclusion of specific biological and thermochemical processes has been shown to result in improvements in the overall energy efficiency of the process up to 70.4%.[54] This shows the potential to reduce the energy loss from 46.5% in a process primarily based on the biochemical route[33] to under 30% in a mature, integrated bio-thermo-chemical processes.[54]

Integration of MFCs or MECs into these mature biorefinery process schemes has been reported by Borole et al.[4] Figure 17.7, Scheme A shows the process scenario generating power (via GTCC) as a coproduct along with biofuels reported by Laser et al.[54] The syngas obtained from gasification is used as feedstock for production of F-T hydrocarbons. The

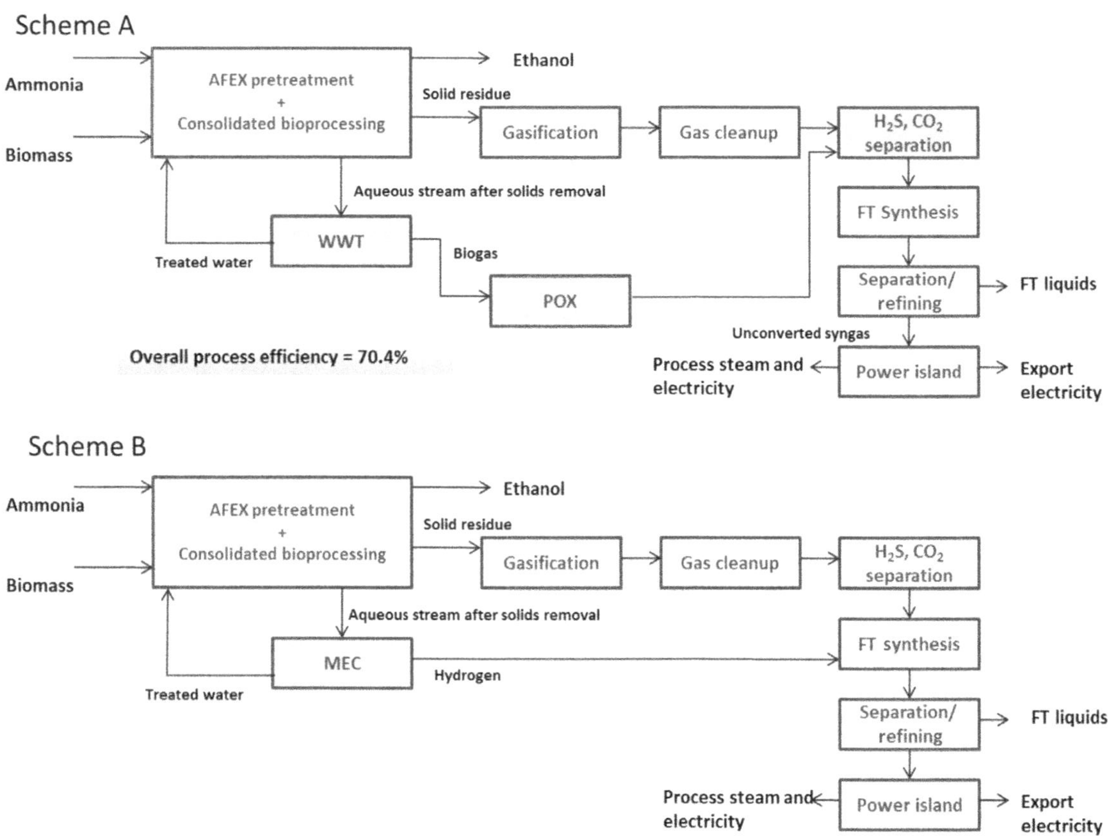

Figure 17.7

Mature biorefinery process schemes for improving energy efficiency. Scheme A was reported by Laser et al.,[54] while scheme B was reported by Borole,[4] incorporating MEC and eliminating the need for wastewater treatment via anaerobic digestion and partial oxidation and gas purification steps. *Figure extracted with permission from Borole et al.[41]*

syngas, however, does not have an optimal ratio for F-T synthesis which is 2.2. Conversion of biogas to syngas followed by POX and gas separation enables production of hydrogen which can be used to increase the H_2:CO ratio from 1.7 to 2.2. An alternate scheme using MECs is shown in Figure 17.7, Scheme B. In this scheme, use of the biorefinery process stream to produce hydrogen via an MEC was included,[4] which can be fed directly into the F-T synthesis reactor. This eliminates the need for POX as well as the separator for removal of H_2S and CO_2. This alternative process can therefore lead to reduced capital costs, while potentially maintaining the process yields around 70%.

17.6 Comparison with Alternate Technologies for Hydrogen Production

Several technologies are under development for hydrogen production from biomass and its derivatives. The unconverted sugars in the process stream after lignin and fuel separation can be converted to hydrogen via methods such as dark fermentation, photo fermentation, chemical conversion, or via a methane route, followed by steam reforming. Conversion of sugars to hydrogen via these alternatives has been considered by Westermann et al.[55] (Table 17.3). An autothermal route with two chemical catalytic steps can lead to hydrogen yields up to 10 mol of hydrogen per mole of hexose.[56] However, an expensive catalyst using Rhodium is needed for the reaction, making the process practically nonfeasible. The energy efficiency of the process at industrial scale can be lower than that observed in the laboratory (83%) due to heat loss from the exothermic chemical reaction

Table 17.3: Efficiencies of hydrogen production via various routes reported in the literature[4]

	Process scheme	Theoretical yield	Observed yield	Free energy change (for H_2-producing step)	Overall observed energy yield	Comments
1	Hypothetical H_2 production	12				
2	Hexose to ethanol to H_2 via autothermal reforming	10	9.5	−265[a] kJ/mol	~83%	Prohibitive catalyst (Rh) cost[56]
3	Dark-light fermentation: Glucose → acetate → H_2	8	7.1	+164 kJ/mol	59.2%	Limited by light penetration and cost[57]
4	Methanogenesis-steam reforming	8	6.0	+261 kJ/mol	50.5%	Mature technology components[55,58]
5	MEC	12	8.2	+104.6 kJ/mol	64%	Nascent technology[44,45]

[a]Processes 3-5 require energy input for the hydrogen-producing step, but this step is energy yielding in process 2. While the hydrogen producing reaction is energy yielding, energy input is required for the production of ethanol from hexose.
Figure extracted with permission from Borole et al.[4]

coupled to endothermic biological reaction. Another alternative, using dark fermentation, typically results in low hydrogen yields ranging from 1 to 3 mol H_2/mol hexose (the theoretical maximum being 4). The route via methanogenesis coupled to steam reforming has potential to yield up to 8 mol H_2/mol of hexose; however, the endothermic steam reforming step requires additional energy input, reducing the overall energy yield by at least a third.

Conversion of dead-end fermentation by-products to hydrogen has been reported by Ren et al.[59] Coupling of dark fermentation to photo fermentation is known to result in improved yields of hydrogen due to conversion of the dead end by-products to hydrogen by phototrophic organisms. Up to 8.3 mol H_2/mol of starch-glucose has been demonstrated[60] via this method. However, transfer of light to the phototrophs may not be efficient, which results in low rates of hydrogen production. Comparison of the various options for hydrogen production from hexose sugars indicates that MEC may be the most attractive option considering the yield and energy recovery efficiency.

17.7 Advantages of Hydrogen in Comparison to Methane as a Coproduct

Anaerobic digestion is a mature technology, which can be applied for treatment of biorefinery process water or wastewater. However, the MEC technology may offer certain advantages for producing hydrogen versus methane (biogas). These include higher revenue due to higher price of hydrogen, ability to use directly for upgrading of biorefinery products such as F-T liquids, potential for use as a fuel in fuel cells with higher efficiency, and as a valuable reagent for conversion of various other biorefinery intermediates to higher value products.[4] Hydrogen price can range from $12 to $26/GJ (price includes delivery costs)[61] versus $4 to $8/GJ for methane. Upgrading of biofuels and other fuel products generated in biorefinery to high octane or high decane fuels will be a necessity under the new renewable fuel standards for production of transportation fuels. While use of methane for upgrading F-T liquids is possible,[54] the additional steps in such a process compared to MECs can potentially lead to higher costs and lower energy efficiency. Use of hydrogen as a fuel in Polymer Electrolyte Membrane (PEM) and other fuel cells can also result in higher efficiencies compared to using methane as a fuel in high temperature fuel cells[62] or in turbines, etc.

A number of other processes in biorefineries can benefit from hydrogen availability on-site. Product trees from each of the three fractions derived from biomass, i.e., cellulose, hemicelluloses, and lignin can benefit from hydrogen. Higher value products can be generated via existing routes as shown in Figure 17.8. Some examples are discussed below. Glucose produced from cellulose can be converted to sorbitol via an industrially established hydrogenation process using a Raney Ni catalyst at 100-150 atm and 100-150 °C.[63] Sorbitol can then be converted to products including vitamin C, alkyd resins, and surfactants.[64] Fructose can be produced from glucose via glucosone using elevated pressure and temperature conditions in the presence of hydrogen. Besides its use in food and pharmaceutical industry, it can

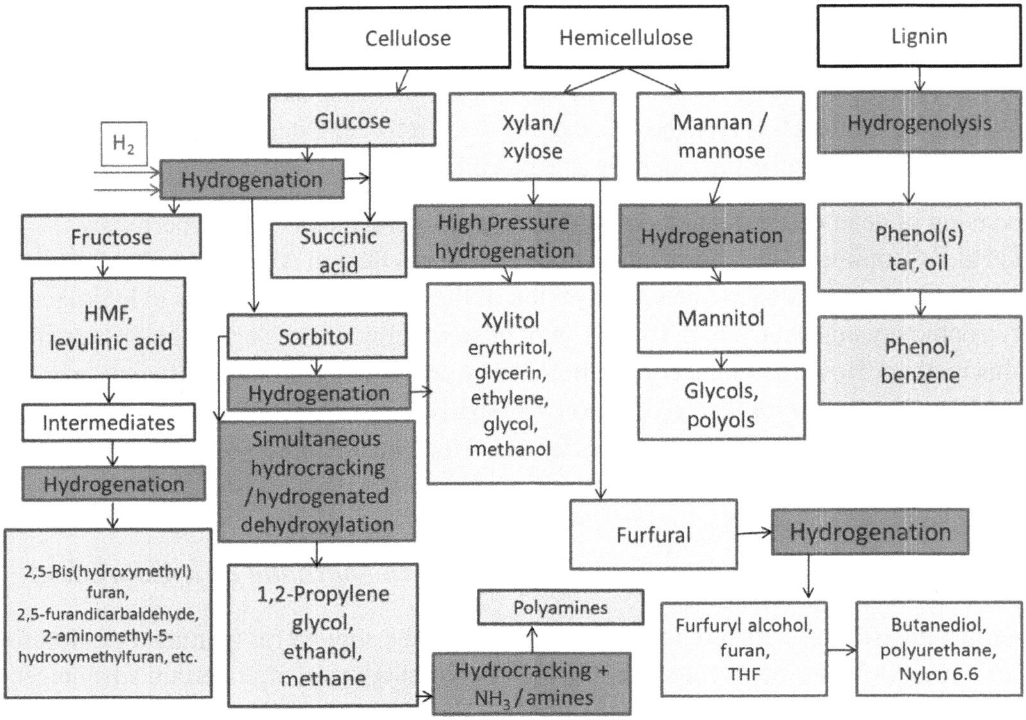

Figure 17.8
Potential hydrogen uses and product trees generated from biomass in biorefinery.[4] *Figure extracted with permission from Borole et al.*[4]

be converted into HMF and levulinic acid, which are important feedstocks for many other industrial chemicals. Hydrogen is used in several of these processes for converting HMF and levulinic acid into furan and levulinate derivatives as shown in Figure 17.8.[63] Production of succinic acid from glucose is in the commercialization stages already. Supplementation of the substrate glucose with hydrogen has been shown to increase succinic acid yields significantly.[65]

Hemicellulase is typically not available as a pure fraction, but as a partially or wholly depolymerized fraction, its oligomers form useful building blocks for reduction reactions. Mannose can be converted to mannitol via hydrogenation.[66] Subsequent catalytic hydrogenation via C–C bond cleavage can produce glycols and polyalcohols for polymer synthesis (e.g., polyurethane). Xylose hydrogenation via Raney Ni at high pressure results in formation of xylitol, which can be used to make resins, vitamin C, surfactants, plasticizers, etc.[67] Another product line resulting from xylose is 2-furfural, which can be converted to tetrahydrofuran and Nylon 6.6 via the following hydrogen-requiring steps: 2-furfural-furan-tetrahydrofuran-adiponitrile-1,6-diaminohexane-Nylon 6.6. Many processes using 2-furfural listed above are practiced commercially and use pentosan-containing raw materials. Furfuryl alcohol is another major intermediate produced via catalytic hydrogenation employing CuCrO or Ni catalyst.[68] Other products generated via hydrogenation include methyl furan, methyltetrahydrofuran, and 1,4-butanediol.[69] Catalytic hydrogenation/hydrocracking of lignin, the third fraction of biomass, via the Noguchi process[70] can produce ring products such as phenols, cresols, and ultimately aromatics or phenolic polymers, and resins.[63]

17.8 Status of the MEC Technology

The MEC technology was invented in 2005, and since then it has made rapid strides with potential applications in the municipal wastewater industry, food industry, biorefinery as well as the oil and gas industry.[1,2,21,42,71–73] Compared to conventional electrolysis, it has much higher electrical energy conversion efficiencies (up to 288%), primarily due to use of reduced organic substrates as additional energy input.[44] The overall energy efficiency of hydrogen production from organic molecules, based on the energy content of the organic substrate, ranges from 64% to 82% with the stoichiometric yields as high as 91%.[44] The rate of hydrogen production in the laboratory varies from 0.1 m^3/m^3 of reactor volume-day up to 3.1 m^3/m^3 of reactor volume-day.[45] The potential options for integration of BES systems into biorefinery are potentially advantageous and therefore, further investment of time and effort into development and scale-up of this technology can be beneficial to the biorefinery stakeholders. In order to evaluate the performance within the biorefinery and their impact on process efficiency, a detailed techno-economic analysis and performance assessment at larger scale with biorefinery streams as feedstock is essential.

17.9 Scale-up Issues and Future Directions

Scale-up of MFCs and MECs has shown significant loss of efficiency in larger scale systems, indicating that a better understanding of the scale-up parameters is necessary.[74] The parameters affecting losses can vary[75]; however, the growth of unwanted biofilms due to the use of anode effluent (not depleted of organic carbon) in the cathode and due to the presence of dissolved oxygen (in anode feed), higher inter-electrode distance at larger scale, heterogeneity in biofilm population and distribution in anode, inefficient distribution of the substrate, and channeling are probable causes for much of the loss in the systems where scale-up was studied. The reactor scale-up requires appropriate consideration of bioelectrochemical losses,[76] flow effects,[77] pH gradients,[78] long-term effects,[79] and polarization effects[80] so as to target adequate BOD or substrate distribution, conversion, and optimum proton and electron flow for achieving high conversion efficiency into bioenergy. Identification of the parameters for scale-up and a systematic optimization of these parameters are necessary in order to achieve higher performance at the pilot and larger scales.

Substrate organic loading and flow regime to maintain a mature biofilm may be two important parameters for optimal functioning of an electro-active biofilm. All the parameters affecting MFC or BES performance were compiled into three types of parameters: system design parameters, process parameters, and biology parameters.[20] The optimal region for MFC functioning is the intersection of these three sets of parameters. In terms of the system design, minimization of dead volume in the anode is an important consideration for performance improvement even at laboratory scale.[81] Additionally, the use of hybrid electrodes can potentially reduce current losses due to flow of electrons between the two electrodes, which can increase with scale. Potential process parameters that may be of importance for scale-up include retention time or flow of fluid through the anode, which directly impacts proton transfer between the anode and the cathode. A second parameter for improved performance at larger scale that needs further investigation is substrate loading or substrate delivery. Controlled substrate delivery can reduce growth of non-electrogens and reduce planktonic growth.[82,83]

High performance of bioanode to support commercial consideration of MECs has already been reported in the literature.[84,85] Current densities above 20 A/m^2 have been reported.[84,86] In addition to experimental investigations, modeling the bioelectrochemical process can potentially expedite progress in this field. Modeling effort at multiple scales can help with the design of the systems at larger scale. The biofilm-scale model, which simulates substrate, electron, and proton transfer within the biofilm, is necessary to understand the optimal biofilm thickness that would lead to high performance over the long term.[87–89] A flow model or process model simulating the substrate and product distribution along the length scale of the reactor can help determine substrate loading and flow rates and the potential need for recirculation.[80,90–92] An electrochemical model is needed to determine the electron yield and effect of pH and other parameters.[91,93,94] A combination of all these three models into one would be ideal, but can be very demanding on the computational needs. Solving them separately can help design teams identify suitable range of parameters and help integrate a better design and process moving further along toward optimally performing MECs and BESs at a larger scale.

Acknowledgments

This manuscript has been authored by UT-Battelle, LLC, under Contract No. DE-AC0500OR22725 with the U.S. Department of Energy. The United States Government retains and the publisher, by accepting the article for publication, acknowledges that the United States Government retains a non-exclusive, paid-up, irrevocable, world-wide license to publish or reproduce the published form of this manuscript, or allow others to do so, for the United States Government purposes.

This work was supported by funding from DOE Office of Energy Efficiency and Renewable Energy and the Laboratory Directed Research and Development Program of Oak Ridge National Laboratory (ORNL), managed by UT-Battelle, LLC.

References

1. Logan BE, Regan JM. Microbial fuel cells, challenges and applications. *Environ Sci Technol* 2006;**40**(17):5172–80.
2. Liu H, Grot S, Logan BE. Electrochemically assisted microbial production of hydrogen from acetate. *Environ Sci Technol* 2005;**39**(11):4317–20.
3. Cheng SA, Xing DF, Call DF, Logan BE. Direct biological conversion of electrical current into methane by electromethanogenesis. *Environ Sci Technol* 2009;**43**(10):3953–8.
4. Borole AP. Improving energy efficiency and enabling water recycle in biorefineries using bioelectrochemical cells. *Biofuels Bioprod Biorefin* 2011;**5**(1):28–36.
5. Yang B, Wyman CE. Pretreatment: the key to unlocking low-cost cellulosic ethanol. *Biofuels Bioprod Biorefin* 2008;**2**(1):26–40.
6. Mielenz J. Ethanol production from biomass: technology and commercialization status. *Curr Opin Microbiol* 2001;**4**(3):324–9.
7. Larsson S, Quintana-Sainz A, Reimann A, Nilvebrant NO, Jonsson LJ. Influence of lignocellulose-derived aromatic compounds on oxygen-limited growth and ethanolic fermentation by *Saccharomyces cerevisiae*. *Appl Biochem Biotechnol* 2000;**84–86**:617–32.
8. Schell D. *Biochemical processing integration*. In: Biochemical platform review meeting, Denver, Colorado; 2007.
9. Klinke HB, Thomsen AB, Ahring BK. Inhibition of ethanol-producing yeast and bacteria by degradation products produced during pre-treatment of biomass. *Appl Microbiol Biotechnol* 2004;**66**(1):10–26.
10. Nilvebrant NO, Reimann A, Larsson S, Jonsson LJ. Detoxification of lignocellulose hydrolysates with ion-exchange resins. *Appl Biochem Biotechnol* 2001;**91–93**:35–49.
11. Weil JR, Dien B, Bothast R, Hendrickson R, Mosier NS, Ladisch MR. Removal of fermentation inhibitors formed during pretreatment of biomass by polymeric adsorbents. *Ind Eng Chem Res* 2002;**41**(24):6132–8.
12. Martinez A, Rodriguez ME, York SW, Preston JF, Ingram LO. Effects of $Ca(OH)_2$ treatments ("overliming") on the composition and toxicity of bagasse hemicellulose hydrolysates. *Biotechnol Bioeng* 2000;**69**(5):526–36.
13. Palmqvist E, Hahn-Hagerdal B. Fermentation of lignocellulosic hydrolysates. I: inhibition and detoxification. *Bioresour Technol* 2000;**74**(1):17–24.
14. Catal T, Li K, Bermek H, Liu H. Electricity production from twelve monosaccharides using microbial fuel cells. *J Power Sources* 2008;**175**(1):196–200.
15. Niessen J, Schroder U, Scholz F. Exploiting complex carbohydrates for microbial electricity generation—a bacterial fuel cell operating on starch. *Electrochem Commun* 2004;**6**(9):955–8.
16. Rabaey K, Wise A, Johnstone AJ, Virdis B, Freguia S, Rozendal RA. Bioelectrochemical conversion of glycerol to 1,3-propanediol. *Abstr Pap Am Chem Soc* 2011;**241**.
17. Steinbusch KJJ, Hamelers HVM, Schaap JD, Kampman C, Buisman CJN. Bioelectrochemical ethanol production through mediated acetate reduction by mixed cultures. *Environ Sci Technol* 2010;**44**(1):513–7.
18. Rabaey K, Butzer S, Brown S, Keller J, Rozendal RA. High current generation coupled to caustic production using a lamellar bioelectrochemical system. *Environ Sci Technol* 2010;**44**(11):4315–21.
19. Rozendal RA, Leone E, Keller J, Rabaey K. Efficient hydrogen peroxide generation from organic matter in a bioelectrochemical system. *Electrochem Commun* 2009;**11**(9):1752–5.
20. Borole AP, Reguera G, Ringeisen B, Wang Z-W, Feng Y, Kim BH. Electroactive biofilms: current status and future research needs. *Energy Environ Sci* 2011;**4**:4813–34.
21. Borole AP. Microbial fuel cell with improved anode. US patent 7,695,834; 2010.
22. Pham TH, Aelterman P, Verstraete W. Bioanode performance in bioelectrochemical systems: recent improvements and prospects. *Trends Biotechnol* 2009;**27**(3):168–78.
23. Gorby YA, Yanina S, McLean JS, Rosso KM, Moyles D, Dohnalkova A, et al. Electrically conductive bacterial nanowires produced by *Shewanella oneidensis* strain MR-1 and other microorganisms. *Proc Natl Acad Sci USA* 2006;**103**(30):11358–63.

24. Reguera G, McCarthy KD, Mehta T, Nicoll JS, Tuominen MT, Lovley DR. Extracellular electron transfer via microbial nanowires. *Nature* 2005;**435**(7045):1098–101.
25. Strycharz-Glaven SM, Snider RM, Guiseppi-Elie A, Tender LM. On the electrical conductivity of microbial nanowires and biofilms. *Energy Environ Sci* 2011;**4**:4366–479.
26. Lovley DR. Electromicrobiology. In: Gottesman S, Harwood CS, Schneewind O, editors. *Annual review of microbiology*, vol. 66; 2012. p. 391–409.
27. Malvankar NS, Lovley DR. Microbial nanowires: a new paradigm for biological electron transfer and bioelectronics. *ChemSusChem* 2012;**5**(6):1039–46.
28. Pirbadian S, El-Naggar MY. Multistep hopping and extracellular charge transfer in microbial redox chains. *Phys Chem Chem Phys* 2012;**14**(40):13802–8.
29. Borole AP, Mielenz J, Vishnivetskaya TA, Hamilton CY. Controlling accumulation of fermentation inhibitors in biorefinery water recycle using microbial fuel cells. *Biotechnol Biofuels* 2009;**2**(1):7.
30. Lee JY, Phung NT, Chang IS, Kim BH, Sung HC. Use of acetate for enrichment of electrochemically active microorganisms and their 16S rDNA analyses. *FEMS Microbiol Lett* 2003;**223**(2):185–91.
31. Liu H, Cheng SA, Logan BE. Production of electricity from acetate or butyrate using a single-chamber microbial fuel cell. *Environ Sci Technol* 2005;**39**(2):658–62.
32. Rabaey K, Clauwaert P, Aelterman P, Verstraete W. Tubular microbial fuel cells for efficient electricity generation. *Environ Sci Technol* 2005;**39**(20):8077–82.
33. Aden A, Ruth M, Ibsen K, Jechura J, Neeves K, Sheehan J, et al. *Lignocellulosic biomass ethanol process design and economics utilizing co-current dilute acid prehydrolysis and enzymatic hydrolysis for corn stover*. NREL/TP-510-32438, Golden, CO: National Renewable Energy Laboratory (NREL); 2002, p. 88.
34. Borole AP, Hamilton C, Schell D. Conversion of residual organics in corn stover-derived biorefinery stream to bioenergy via microbial fuel cells. *Environ Sci Technol* 2013;**47**(1):642–8.
35. Huang LP, Logan BE. Electricity production from xylose in fed-batch and continuous-flow microbial fuel cells. *Appl Microbiol Biotechnol* 2008;**80**(4):655–64.
36. Huang LP, Angelidaki I. Effect of humic acids on electricity generation integrated with xylose degradation in microbial fuel cells. *Biotechnol Bioeng* 2008;**100**(3):413–22.
37. Pinto RP, Srinivasan B, Escapa A, Tartakovsky B. Multi-population model of a microbial electrolysis cell. *Environ Sci Technol* 2011;**45**(11):5039–46.
38. Choo YF, Lee J, Chang IS, Kim BH. Bacterial communities in microbial fuel cells enriched with high concentrations of glucose and glutamate. *J Microbiol Biotechnol* 2006;**16**(9):1481–4.
39. Humbird D, David R, Tao L, Kinchin C, Hsu D, Aden A, et al. *Process design and economics for biochemical conversion of lignocellulosic biomass to ethanol—dilute-acid pretreatment and enzymatic hydrolysis of corn stover*. NREL/TP-5100-47764, Golden, CO: National Renewable Energy Laboratory (NREL); 2011.
40. Mohagheghi A, Schell D. Impact of recycling stillage on conversion of dilute sulfuric acid pretreated corn stover to ethanol. *Biotechnol Bioeng* 2010;**105**:992–6.
41. Borole AP, Mielenz J. Estimating hydrogen production potential in biorefineries using microbial electrolysis cell technology. *Int J Hydrogen Energy* 2011;**36**(22):14787–95.
42. Borole AP, Hamilton CY. Energy production from food industry wastewaters using bioelectrochemical cells. In: Shah V, editor. *Emerging environmental technologies*, vol. 2. Dordrecht: Springer Publications; 2009. p. 97–114.
43. Wilkie AC, Riedesel KJ, Owens JM. Stillage characterization and anaerobic treatment of ethanol stillage from conventional and cellulosic feedstocks. *Biomass Bioenergy* 2000;**19**(2):63–102.
44. Cheng S, Logan BE. Sustainable and efficient biohydrogen production via electrohydrogenesis. *Proc Natl Acad Sci* 2007;**104**(47):18871–3.
45. Logan BE, Call D, Cheng S, Hamelers HVM, Sleutels T, Jeremiasse AW, et al. Microbial electrolysis cells for high yield hydrogen gas production from organic matter. *Environ Sci Technol* 2008;**42**(23):8630–40.
46. Taherzadeh MJ, Eklund R, Gustafsson L, Niklasson C, Liden G. Characterization and fermentation of dilute-acid hydrolyzates from wood. *Ind Eng Chem Res* 1997;**36**(11):4659–65.

47. Buchert J, Niemela K, Puls J, Poutanen K. Improvement in the fermentability of steamed hemicellulose hydrolysate by ion exclusion. *Process Biochem* 1990;**25**(5):176–80.
48. Ranatunga TD, Jervis J, Helm RF, McMillan JD, Wooley RJ. The effect of overliming on the toxicity of dilute acid pretreated lignocellulosics: the role of inorganics, uronic acids and ether-soluble organics. *Enzyme Microb Technol* 2000;**27**(3–5):240–7.
49. Clark TA, Mackie KL. Fermentation inhibitors in wood hydrolysates derived from the softwood *Pinus radiata*. *J Chem Technol Biotechnol B Biotechnol* 1984;**34**(2):101–10.
50. Larsson S, Cassand P, Jonsson L. Development of *Saccharomyces cerevisiae* strain with enhanced resistance to phenolic fermentation inhibitors in lignocellulosic hydrolyzates by heterologous expression of laccase. *Appl Environ Microbiol* 2001;**67**:1163–70.
51. Stenberg K, Bollók M, Réczey K, Galbe M, Zacchi G. Effect of substrate and cellulase concentration on simultaneous saccharification and fermentation of steam-pretreated softwood for ethanol production. *Biotechnol Bioeng* 2000;**68**(2):204–10.
52. Fenske J, Griffin D, Penner M. Comparison of aromatic monomers in lignocellulosic biomass hydrolyzates. *J Ind Microbiol Biotechnol* 1998;**20**:364.
53. Mukhopadhyay A, He ZL, Alm EJ, Arkin AP, Baidoo EE, Borglin SC, et al. Salt stress in *Desulfovibrio vulgaris* Hildenborough: an integrated genomics approach. *J Bacteriol* 2006;**188**(11):4068–78.
54. Laser M, Jin HM, Jayawardhana K, Dale BE, Lynd LR. Projected mature technology scenarios for conversion of cellulosic biomass to ethanol with coproduction thermochemical fuels, power, and/or animal feed protein. *Biofuels Bioprod Biorefin* 2009;**3**(2):231–46.
55. Westermann P, Jorgensen B, Lange L, Ahring BK, Christensen CH. Maximizing renewable hydrogen production from biomass in a bio/catalytic refinery. *Int J Hydrogen Energy* 2007;**32**(17):4135–41.
56. Deluga GA, Salge JR, Schmidt LD, Verykios XE. Renewable hydrogen from ethanol by autothermal reforming. *Science* 2004;**303**(5660):993–7.
57. Asada Y, Tokumoto M, Aihara Y, Oku M, Ishimi K, Wakayama T, et al. Hydrogen production by co-cultures of *Lactobacillus* and a photosynthetic bacterium, *Rhodobacter sphaeroides* RV. *Int J Hydrogen Energy* 2006;**31**(11):1509–13.
58. Guro D. *Distributed hydrogen fueling systems analysis: cost and performance comparison of stationary hydrogen fueling appliances. DOE Hydrogen, Fuel Cells & Infrastructure Technologies Program, Progress Report, II.E.4*. Washington: US Department of Energy; 2002.
59. Ren N, Wang A, Cao G, Xu J, Gao L. Bioconversion of lignocellulosic biomass to hydrogen: potential and challenges. *Biotechnol Adv* 2009;**27**(6):1051–60.
60. Kim MS, Baek JS, Yun YS, Sim SJ, Park S, Kim SC. Hydrogen production from *Chlamydomonas reinhardtii* biomass using a two-step conversion process: anaerobic conversion and photosynthetic fermentation. *Int J Hydrogen Energy* 2006;**31**:812–6.
61. Ogden JM. Prospects for building a hydrogen energy infrastructure. *Ann Rev Energy Environ* 1999;**24**:227–79.
62. Piroonlerkgul P, Wiyaratn W, Soottitantawat A, Kiatkittipong W, Arpornwichanop A, Laosiripojana N, et al. Operation viability and performance of solid oxide fuel cell fuelled by different feeds. *Chem Eng J* 2009;**155**(1–2):411–8.
63. Kamm B, Gruber PR, Kamm M. *Biorefineries-industrial processes and products*. Weinheim: Wiley Interscience; 2008, p. 498.
64. Hans-Jurgen Arpe, Ed., 5th ed. Ullmann's encyclopedia of industrial chemistry, vol. A25. Starch and Other Polysaccharides to Polymers. Weinheim: VCH Verlagsgesellschaft mbH; 1985.
65. McKinlay JB, Vieille C. C-13-metabolic flux analysis of *Actinobacillus succinogenes* fermentative metabolism at different $NaHCO_3$ and H_2 concentrations. *Metab Eng* 2008;**10**(1):55–68.
66. Lichtenthaler FW. The key sugars of biomass: availability, present non-food uses and potential industrial development lines. In: Kamm B, Gruber PR, Kamm M, editors. *Biorefineries-industrial processes and products*. Weinheim: Wiley Interscience; 2008. p. 498.
67. Oshima M. *Wood chemistry—process engineering aspects*. New York: Noyes Development Corp.; 1965.
68. Wojcik BH. Catalytic hydrogenation of furan compounds. *Ind Eng Chem* 1948;**40**:210.

69. Pinkos R, Fischer R. *Process for production of 1,4-butanediol and tetrahydrofuran from furan*; 1996.
70. Groheen DW. *Lignin, structures and reactions*. Washington, DC: American Chemical Society; 1966.
71. Borole AP. Microbial fuel cell treatment of ethanol fermentation process water. US patent 8,192,854 B2; 2012.
72. Borole AP, Tsouris C. Microbial fuel cell treatment of fuel processing wastewater. US patent 8,597,513 B2, December 3, 2013; 2009.
73. Rabaey K, Verstraete W. Microbial fuel cells: novel biotechnology for energy generation. *Trends Biotechnol* 2005;**23**(6):291.
74. Keller J, Rabaey K. *Experiences from MFC pilot plant operation*. In: MFC symposium, 2008; 2008.
75. Logan BE. Scaling up microbial fuel cells and other bioelectrochemical systems. *Appl Microbiol Biotechnol* 2010;**85**:1665–71.
76. Clauwaert P, Aelterman P, Pham TH, De Schamphelaire L, Carballa M, Rabaey K, et al. Minimizing losses in bio-electrochemical systems: the road to applications. *Appl Microbiol Biotechnol* 2008;**79**(6):901–13.
77. Aaron DS, Tsouris C, Hamilton CY, Borole AP. Assessment of the impact of flow rate and ionic strength on performance of microbial fuel cells using electrochemical impedance spectroscopy. *Energies* 2010;**3**(4):592.
78. Torres CI, Marcus AK, Rittmann BE. Proton transport inside the biofilm limits electrical current generation by anode-respiring bacteria. *Biotechnol Bioeng* 2008;**100**(5):872–81.
79. Borole AP, Aaron DS, Tsouris C, Hamilton CY. Understanding long term changes in microbial fuel cells using electrochemical impedance spectroscopy. *Environ Sci Technol* 2010;**44**(7):2740–5.
80. Harnisch F, Warmbier R, Schneider R, Schroder U. Modeling the ion transfer and polarization of ion exchange membranes in bioelectrochemical systems. *Bioelectrochemistry* 2009;**75**(2):136–41.
81. Borole AP, Hamilton CY, Vishnivetskaya TA, Leak D, Andras C, Morrell-Falvey J, et al. Integrating engineering design improvements with exoelectrogen enrichment process to increase power output from microbial fuel cells. *J Power Sources* 2009;**191**(2):520–7.
82. Borole AP, Hamilton CY, Vishnivetskaya TA. Enhancement in energy conversion efficiency and current density of 3-dimensional MFC anodes using pre-enriched consortium and continuous supply of electron donors. *Bioresour Technol* 2011;**102**(8):5098–104.
83. Borole AP, Hamilton CY, Vishnivetskaya TA, Leak D, Andras C. Improving power production from acetate-fed microbial fuel cells via enrichment of exoelectrogenic organisms in continuous flow systems. *Biochem Eng J* 2009;**48**:71–80.
84. Ichihashi O, Vishnivetskaya T, Borole AP. High-performance bioanode development for fermentable substrates via controlled electroactive biofilm growth. *ChemElectroChem* 2014;**1**:1940–7.
85. Sleutels T, Ter Heijne A, Buisman CJN, Hamelers HVM. Bioelectrochemical systems: an outlook for practical applications. *ChemSusChem* 2012;**5**(6):1012–9.
86. Jeremiasse AW, Hamelers HVM, Saakes M, Buisman CJN. Ni foam cathode enables high volumetric H-2 production in a microbial electrolysis cell. *Int J Hydrogen Energy* 2010;**35**(23):12716–23.
87. Torres CI, Kato Marcus A, Rittmann BE. Kinetics of consumption of fermentation products by anode-respiring bacteria. *Appl Microbiol Biotechnol* 2007;**77**(3):689–97.
88. Marcus AK, Torres CI, Rittmann BE. Conduction-based modeling of the biofilm anode of a microbial fuel cell. *Biotechnol Bioeng* 2007;**98**(6):1171–82.
89. Picioreanu C, Kreft JU, van Loosdrecht MCM. Particle-based multidimensional multispecies biofilm model. *Appl Environ Microbiol* 2004;**70**(5):3024–40.
90. Merkey BV, Chopp DL. The performance of a microbial fuel cell depends strongly on anode geometry: a multidimensional modeling study. *Bull Math Biol* 2012;**74**(4):834–57.
91. Picioreanu C, van Loosdrecht MCM, Curtis TP, Scott K. Model based evaluation of the effect of pH and electrode geometry on microbial fuel cell performance. *Bioelectrochemistry* 2010;**78**(1):8–24.
92. Picioreanu C, Katuri KP, Head IM, van Loosdrecht MCM, Scott K. Mathematical model for microbial fuel cells with anodic biofilms and anaerobic digestion. *Water Sci Technol* 2008;**57**(7):965–71.
93. Marcus AK, Torres CI, Rittmann BE. Analysis of a microbial electrochemical cell using the proton condition in biofilm (PCBIOFILM) model. *Bioresour Technol* 2011;**102**(1):253–62.
94. Hamelers HVM, ter Heijne A, Stein N, Rozendal RA, Buisman CJN. Butler-Volmer-Monod model for describing bio-anode polarization curves. *Bioresour Technol* 2011;**102**(1):381–7.

CHAPTER 18

Photobiohybrid Solar Conversion with Metalloenzymes and Photosynthetic Reaction Centers

Katherine A. Brown, Paul W. King, Peter Ciesielski
National Renewable Energy Laboratory, Biosciences Center, Golden, Colorado, USA

18.1 Introduction

Sunlight provides an abundant and sustainable supply of energy to the Earth's surface, at levels that far exceed the yearly human global energy demand. However, the intermittency and geographic variability of solar irradiation, combined with the need for storage, limits the ability to provide practical alternatives to use of conventional fossil fuels.[1–3] To better utilize the available solar energy, current photovoltaic technologies must be integrated with conversion technologies that produce storable chemical energy (fuels) that are easily distributed to meet regional demand as needed. In biological photosynthesis, conversion of solar energy into chemical energy is accomplished by the water-splitting and CO_2 fixation reactions. The molecular machinery of the natural system provides ideal models for the design and development of artificial solar-to-fuel systems.[4–7] The theoretical limit of biological photosynthesis is ~12%, and under optimal conditions, efficiencies of 7% have been achieved; however, 1% is a more typical benchmark.[8] Photosynthetic reactions rely on four key components which are integrated to act in concert as a highly functional energy transduction network: (i) the antenna, where photons are absorbed; (ii) the charge separation site, where high-energy excitons (electron-hole pairs) are separated into positive and negative charge carriers; (iii) the reduction catalyst, where electrons are utilized in a fuel-forming reaction (e.g., $NAD^+ \rightarrow NADH$ formation used for CO_2 fixation in photosynthesis); and (iv) the oxidation catalyst, where holes are utilized to drive an oxidation reaction (e.g., water oxidation by photosystem II during photosynthesis). Efforts are underway to translate photobiological design principles to develop artificial systems for solar fuel generation that circumvent or eliminate unwanted side reactions and attain higher efficiencies.[9,10] These efforts include the development of photochemical devices, inorganic biomimetic and bioinspired catalysts and light-harvesting complexes, and organic hybrid materials for photo-driven fuel production. Here, we discuss research focused on the development of hybrid materials that incorporate artificial and natural molecular

Biotechnology for Biofuel Production and Optimization. http://dx.doi.org/10.1016/B978-0-444-63475-7.00018-2

components into unified functional systems for light-harvesting and conversion into reduced chemical fuels.[11,12]

The development of catalysts that can effectively utilize the energy of absorbed photons for reductive and oxidative chemistry is vital to the advancement of direct solar-to-fuel technologies.[2,3,13] In photosynthesis, fuel-forming reduction reactions are catalyzed by redox enzymes such as hydrogenases, dehydrogenases, and reductases.[14] These enzymes generally have high substrate specificity and turnover frequencies (TOFs), low overpotential requirements, and can serve as excellent models for the development of synthetic catalysts. Although cellulase enzymes play a central role in recently industrialized biomass-to-fuel processes, enzymes and proteins are unlikely to be practical components of direct solar-to-fuel technologies due to the high cost of production and a lack of long-term stability relative to nonbiological components of solar energy conversion devices. However, with active sites formed from benign, earth-abundant, and inexpensive materials, efficient systems for electron and proton transfer, and the ability to operate at or near the formal redox couple for a given reaction, enzymes represent a benchmark for catalyst design, and their study will inspire development of efficient artificial catalysts that incorporate these advantages.[14] A number of enzymes have been incorporated into biohybrid fuel cells for fuel consumption, including glucose oxidases, peroxidases, and various dehydrogenases, but the enzymes and proteins that can function in hybrids for fuel production are more limited. To date, only H_2 production and CO_2 reduction (to CO or formate) have been demonstrated in enzyme-based photobiohybrid systems. Successful integration of photosystem I (PSI) in hybrid systems as an antenna for light absorption and reaction center for charge separation has also been demonstrated.

18.2 Biological Catalysts for Production of Reduced Products and Fuels

18.2.1 H_2 Production

The efficient production of hydrogen (H_2) directly from solar energy is an important component in the development of a robust renewable energy economy. In fuel applications, H_2 can be used directly by reacting with O_2 in a fuel cell,[1,15–17] as a substrate for Fischer-Tropsch production of liquid fuels, or in reactions with CO (e.g., water gas shift) to generate hydrocarbons or methanol.[18,19] In addition, H_2 is an important industrial chemical for petroleum refining and the production of nitrogen-based fertilizers (Haber-Bosch process).[20] There is a need for H_2 production catalysts that are efficient, selective, and composed of economically feasible and environmentally friendly materials. Platinum is an extremely efficient catalyst for proton reduction to H_2; however, it is cost prohibitive, a limited feedstock,[21] and lacks substrate specificity.[22] The long-term development of viable alternative catalysts to Pt is therefore an important challenge to designing large-scale H_2 production technologies.

Hydrogenase enzymes

Biology provides models for efficient, nontoxic, and economical H_2 production catalysts in the form of hydrogenases. These metalloenzymes catalyze both proton reduction and H_2 oxidation at high TOFs and low overpotentials. There are two classes of hydrogenases that can catalyze H_2 production, differentiated by the composition of their active sites, [FeFe]- and [NiFe]-hydrogenases. Figure 18.1 shows examples of the crystal and active site structures of both classes of hydrogenase, [FeFe]-hydrogenase from *Clostridium pasteurianum*[23] and a [NiFe]-hydrogenase from *Desulfovibrio gigas*.[24] The catalytic domain of all [FeFe]-hydrogenases contains three conserved cysteine-rich motifs that coordinate the catalytic H-cluster, as well as conserved proton and gas transport pathways.[25–27] The H-cluster is a unique organometallic [6Fe-6S]-cluster composed of a cubane [4Fe-4S]-subcluster linked by cysteine to a di-iron, di-thiolate, [2Fe]-subcluster that possesses Fe-CO and Fe-CN ligation and an organic (dithiomethylamine) bridging ligand.[23,28] Although structurally diverse, a distinguishing characteristic of [NiFe]-hydrogenases is the presence of a [NiFe]-cluster in the catalytic domain.[29] The [NiFe]-cluster is coordinated through conserved cysteines that bridge the Ni and Fe atoms.[29] Like [FeFe]-hydrogenases, the [NiFe]-hydrogenase catalytic domain also contains conserved gas transport[30] and proton transfer[31] regions, and both Fe-CO and Fe-CN ligation of the [NiFe]-cluster.[32,33]

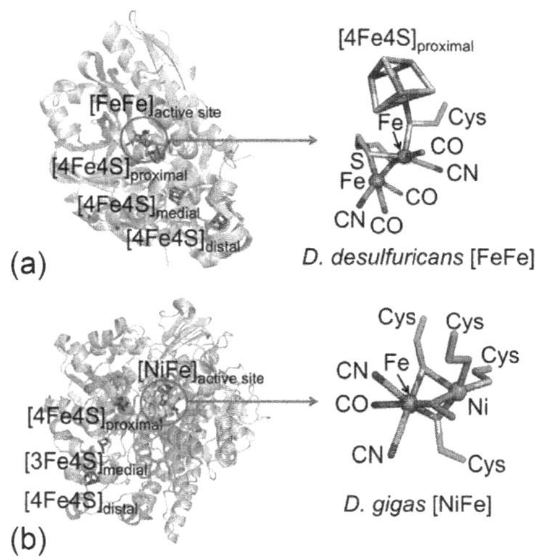

Figure 18.1

X-ray structure models of the multimeric hydrogenases from *Desulfovibrio*. (a) *D. desulfuricans* [FeFe]-hydrogenase (PDB code 1HFE)[24] and (b) *D. gigas* [NiFe]-hydrogenase (PDB code 1YQ9).[23] The small subunits are shown in gray and the large subunits in purple. *Illustration adapted with permission from Reisner E. Solar hydrogen evolution with hydrogenases: from natural to hybrid systems.* Eur J Inorg Chem *2011;7:1005-16 Copyright 2011, WILEY-VCH Verlag.*

Both [FeFe]- and [NiFe]-hydrogenases are functional models for the design of synthetic H_2 activation catalysts. When absorbed on an electrode, these hydrogenases produce high catalytic currents, and the transition from reduction to oxidation reactions occurs at the H^+/H_2 couple, meaning a minimal overpotential is required for catalysis.[34] TOFs of up to 50,000 s^{-1} have been measured on electrodes.[22] Both [FeFe]- and [NiFe]-hydrogenases are reversible though [FeFe]-hydrogenases are generally biased in favor of H_2 production and most [NiFe]-hydrogenases are biased for H_2 oxidation.[20] Both classes of hydrogenase can couple well to nonbiological electron donors due to the presence of [FeS] clusters which form an electron transport chain to efficiently move electrons to and from the active site.[35] The challenge of using hydrogenases, either as components of biohybrid fuel production systems or as enzymes in photobiological H_2 production, is their sensitivity to O_2.[36] Both classes of hydrogenases are sensitive to inhibition by O_2. The [FeFe]-hydrogenases are permanently inactivated as O_2 leads to degradation of the H-cluster,[26] whereas the inhibition of [NiFe]-hydrogenases can be reversible.[37,38] Details of the biological and physiological role of hydrogenases,[39,40] as well as their structural[41] and spectroscopic[42–44] properties, have been well reviewed elsewhere.

Electrochemical and photoelectrochemical hydrogenase systems

Photoelectrochemical cells offer a promising method of H_2 production driven directly by solar energy. Electrons produced at the light sensitive anode flow to a catalyst-modified cathode, where the proton reduction reaction takes place. For a hybrid photoelectrochemical cell, the H_2 production catalyst can be in the form of a hydrogenase enzyme. The design relies on effective electronic/vibronic coupling between the hydrogenase and the cathode in order for efficient electron transfer to take place.[45] Attachment of hydrogenases on electrodes has been investigated both in the context of a device for generating electricity from H_2 oxidation (e.g., fuel cell), for solar H_2 production (e.g., photoelectrochemical cell), or as a means to investigate the redox properties using electrochemical techniques.[34,46] Several methods are available for protein attachment, direct adsorption, covalent cross-linking, and entrapment in polymers.[47] Direct adsorption of [FeFe]- and [NiFe]-hydrogenases in electrochemically active confirmations has been achieved on pyrolytic graphite electrodes (PGE),[46,48–50] as well as electrodes modified with carbon nanotubes.[51–54] Inorganic electrode materials have also been used, with electronically active attachment to both TiO_2[55,56] and Au electrodes.[57–59] Covalent attachment to PGE[60] and carbon nanotubes[61–63] has been successful using small molecule linkers. Entrapment of hydrogenases within conductive polymers has advantages in terms of increasing enzyme stability.[20] Hydrogenases have been immobilized, along with the electron mediator methyl viologen, within polypyrrole[64–66] and clay films.[67,68] In the examples cited above, high rates of proton reduction and/or H_2 oxidation are achieved, sometimes comparable to Pt-based electrodes, and the overpotentials required are minimal.

Examples of incorporation of hydrogenases into fully functional photoelectrochemical cells are fewer in number than the electrode work detailed above. The [NiFe]-hydrogenase from

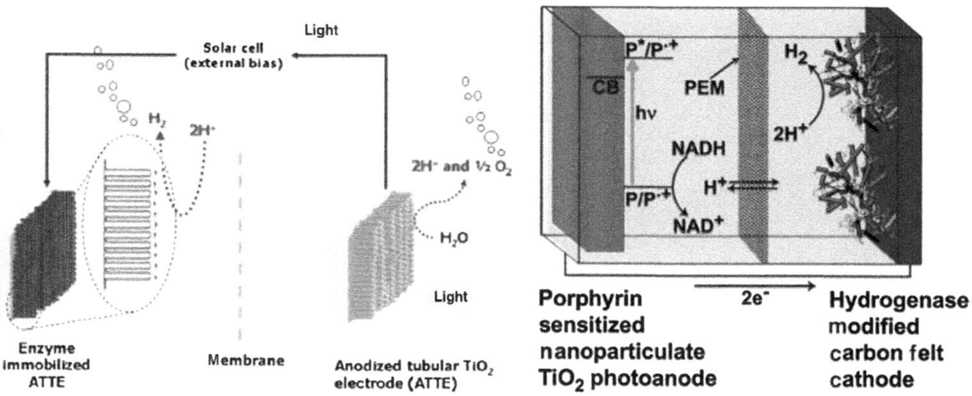

Figure 18.2

Schematic of a photoelectrochemical cell for enzymatic production of H_2. (Left) An anodized tubular TiO_2 (ATTE) photoanode is electrically coupled to a ATTE cathode that is modified with an adsorbed layer of [NiFe]-hydrogenase from *P. furiosus.*[69] *(Reprinted with permission from Yoon J, et al. Pyrococcus furiosus-immobilized anodized tubular titania cathode in a hydrogen production system.* J Power Sources *2009;189:1296-1301. Copyright 2009, Elsevier.)* (Right) A photoelectrochemical cell with a dye-sensitized TiO_2 photoanode electrically wired to a carbon felt anode modified with [FeFe]-hydrogenase from *C. acetobutylicum.*[70] *(Reprinted with permission from Hambourger M, et al. [FeFe]-hydrogenase-catalyzed H_2 production in a photoelectrochemical biofuel cell.* J Am Chem Soc *2008;* ***130****:2015-22. Copyright 2008, American Chemical Society.)*

Pyrococcus furiosus has been adsorbed and covalently linked to anodize tubular titania (TiO_2) cathodes for production of H_2 (Figure 18.2a), with seawater used as the reductant, requiring an external bias for successful H_2 production.[69] The covalently linked system showed 50% higher activity than the control system in which the hydrogenase was simply adsorbed. In an alternative system, Hambourger et al. constructed a photoelectrochemical fuel cell in which both half reactions involve enzymatic reactions (Figure 18.2b).[70] The cathode consisted of carbon felt electrode with an adsorbed [FeFe]-hydrogenase from *Clostridium acetobutylicum*, while at the TiO_2 photoanode, electrons were supplied by NADH or NADPH, which was regenerated by enzymatic oxidation of a biofuel (methanol, ethanol, glucose of glucose-6-phosphate). An identical fuel cell was also prepared with Pt rather than hydrogenase as the cathode catalyst. The two cells are similar to H_2 production rates under illumination.[70]

Nanomaterial-hydrogenase hybrids

An alternative approach to electrode immobilization of hydrogenase for H_2 production is to directly tether the enzyme to a light sensitive nanomaterial that can act as an antenna for photon absorption and exciton formation. Semiconducting nanomaterials generally have high extinction coefficients compared to organic dyes, can absorb large numbers of photons in solution, and are promising molecules for exceeding the quantum efficiencies

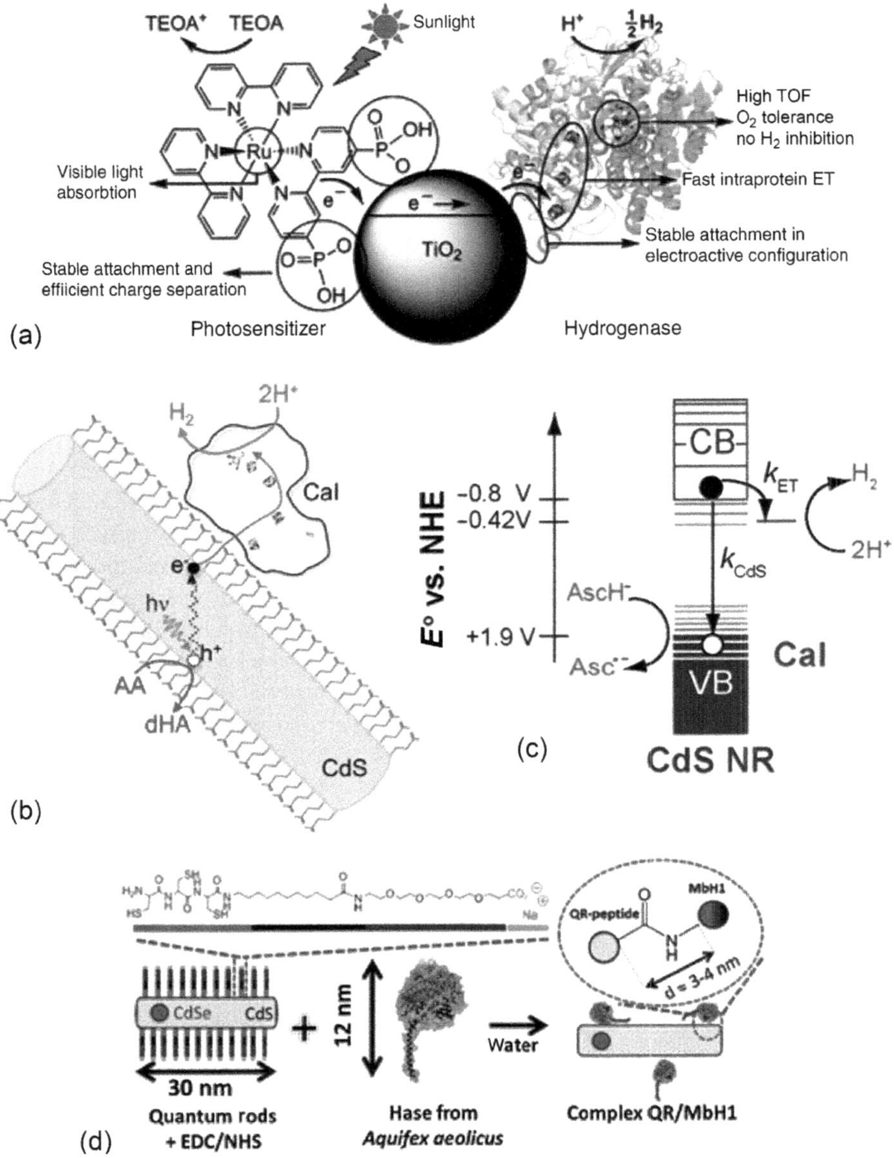

Figure 18.3

Schematics of nanoparticle-based photobiohybrids for H_2 production from sunlight. (a) Visible-light-driven H_2 evolution by *D. baculatum* [NiFeSe]-hydrogenase attached to RuP dye-sensitized TiO_2 nanoparticles.[72,73] *(Reprinted with permission from Reisner E, et al. Visible light-driven H_2 production by hydrogenases attached to dye-sensitized TiO_2 nanoparticles.* J Am Chem Soc *2009;**131**:18457-66. Copyright 2009, American Chemical Society.)* (b) Model of self-assembled complexes between CdS nanorods and *C. acetobutylicum* [FeFe]-hydrogenase CaI. Electron transfer [FeS]-clusters and the catalytic H-cluster of CaI are shown as yellow (sulfur) and orange (iron) spheres. CdS and CaI are drawn to scale, while MPA molecules are enlarged ~5×.[75] *(Reprinted with permission from Brown KA, et al. Characterization of photochemical processes for H_2 production by CdS nanorod-[FeFe] hydrogenase complexes.* J Am Chem Soc *2012;**134**:5627-36. Copyright 2012, American Chemical Society.)*

of biological absorbers.[71] Nanomaterials also have high surface areas available for binding due to their high surface to volume ratios.[35] This is an advantage when using enzymes for catalysis, since they have a large adsorption footprint. Semiconductor nanoparticles are of particular interest for these systems. Reisner et al. absorbed [NiFe]-hydrogenases to TiO_2 nanoparticles sensitized with a ruthenium dye (Figure 18.3a).[72,73] Photoexcitation of the dye produces an electron that transfers spontaneously into the TiO_2 conduction band and then to the enzyme. These complexes produce H_2 under illumination with visible light when in solution with a sacrificial electron donor to reduce the oxidized dye molecule. [NiFe]-hydrogenases from several organisms have been analyzed as enzymes for H_2 production, and the [NiFeSe]-hydrogenase from *Desulfomicrobium baculatum* was found to be particularly effective with a TOF of $50\,s^{-1}$.

Similar systems have been reported by King and coworkers using Cadmium nanoparticles and [FeFe]-hydrogenase from *C. acetobutylicum* (CaI) (Figure 18.3b).[74,75] The hydrogenase was shown to self-assemble on 3-mercaptopropionic acid (MPA)-capped CdTe and CdS nanoparticles by an electrostatic interaction between the negatively charged MPA and a region of positive charge on the hydrogenase, which corresponds to the binding site on the biological electron donor ferredoxin. The CdS-based complexes showed a TOF of $100\,s^{-1}$ under light intensity equivalent to AM 1.5 when ascorbic acid was used as a sacrificial electron donor.[75] Recently, the electron transfer kinetics of the CdS-CaI complexes were investigated by the Dukovic group (Figure 18.3c)[76] and found to be competitive with the internal recombination kinetics of the nanoparticle. This study highlights the potential for photobiohybrid systems to inform design of systems for solar-to-fuel systems. The authors propose methods of manipulating electron transfer and nanoparticle properties to improve efficiency based on the kinetic parameters measured.

The above examples rely on physisorption, or self-assembly, of the hydrogenase onto the nanoparticle. Recently, Hamon et al. demonstrated light-driven H_2 production by CdSe/CdS nanoparticles that were chemically linked to the [NiFe]-hydrogenase MbH1 from *Aquifex aeolicus* (Figure 18.3d).[77] The chemical reaction utilized the coupling agents ethylcarbodiimide and hydroxysuccinimide to covalently cross-link carboxylic acid groups on

Figure 18.3–Cont'd

(c) Energy level diagram and associated photochemical reaction processes leading to H_2 generation in CdS nanorod, CaI photocatalytic complexes. Under illumination, photon absorption by CdS leads to electron injection from CdS→CaI, with a rate constant of k_{ET}.[76] *(Reprinted with permission from Wilker MB, et al. Electron transfer kinetics in CdS nanorod-[FeFe]-hydrogenase complexes and implications for photochemical H_2 generation.* J Am Chem Soc *2014;***136***:4316-24. Copyright 2014, American Chemical Society.)* (d) Scheme for crosslinking of [NiFe]-hydrogenase MbH1 to the carboxylated surface of CdSe/CdS quantum rods (QRs).[77] *(Reprinted with permission from Hamon C, et al. Synthesis and enzymatic photoactivity of an O_2 tolerant hydrogenase—CdSe@CdS quantum rod bioconjugate.* Chem Commun *2014;***50***:4989-92. Copyright 2014, Royal Society of Chemistry.)*

the quantum rod (QR) and solvent exposed primary amines of MbH1, which resulted in stable solar-driven H_2 production activity observed up to 1 week.

18.2.2 CO_2 Reduction

The reduction of CO_2 for production of fuels is an important pathway for renewable production of energy carriers. Systems that can both capture CO_2 and reduce it into chemical products are an ideal solution for both mediating the over-abundance of atmospheric CO_2 and producing renewable fuels. The energetic requirements for the one-electron reduction of CO_2 and the two-electron reduction of CO_2 to either CO or formate are shown in Equations (18.1)-(18.3), respectively.

$$CO_2 + e^- \rightarrow CO_2^- \quad E^0 = -1.9\,V\,(pH\,7\,vs.\,NHE) \tag{18.1}$$

$$CO_2 + 2H^+ + 2e^- \rightarrow CO + H_2O \quad E^0 = -0.52\,V\,(pH\,7\,vs.\,NHE) \tag{18.2}$$

$$CO_2 + H^+ + 2e^- \rightarrow HCO_2 \quad E^0 = -0.43\,V\,(pH\,7\,vs.\,NHE) \tag{18.3}$$

While the one-electron reduction is highly energy intensive, due, in part, to the structural rearrangement of linear CO_2 to the bent $CO_2^{\bullet-}$,[78] the two-electron reduction reactions are feasible targets for fuel-producing reactions.[79] Both CO and formate can be used to generate reduced carbon compounds for fuel or as chemical feed stocks, and formate can be used directly in formic acids fuel cells.[79] Chemical catalysts for both two-electron CO_2 reduction reactions have been developed, but they suffer from instability and high overpotential requirements and tend to lack product selectivity.[80] In biology, CO_2 can be reduced to CO and eventually to formate in reactions catalyzed by CO-dehydrogenase (CODH) and formate dehydrogenases (FDH), respectively. Both enzymes contain highly evolved metal cofactors at their catalytic sites, which are coupled to accessory cofactors for electron transfer so that catalytic transformations of CO_2/CO occur at high rates and efficiencies compared to synthetic catalysts.[79]

CO_2 reduction enzymes

CO-dehydrogenases. CODH enzymes catalyze the reversible two-electron interconversion between CO and CO_2 shown in Equation (18.2). The rates of CO oxidation [reverse of the reaction sequence shown in Equation (18.2)] are generally significantly faster than the CO_2 reduction rates.[81,82] There are two principal types of CODHs that have been identified: Mo-[2Fe-2S]-FAD CODHs from aerobic bacteria[83] and Ni-[4Fe-4S] CODHs from anaerobic bacteria.[84–87] Crystal structures of Mo CODHs show they are composed of an FeS protein, containing two types of [2Fe-2S] clusters: a flavoprotein, which binds a FAD cofactor, and a Mo-containing protein, which carries the molybdopterin-cytosine dinucleotide-containing

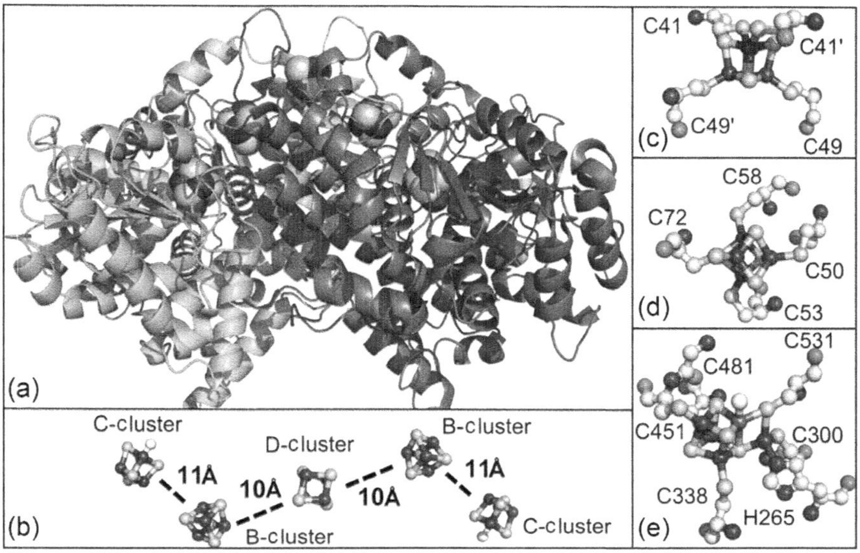

Figure 18.4
The atomic structure of *Rhodospirillum rubrum* CODH (a), with details of cluster coordination and structure. (b) Intramolecular cluster distances (Å). (c) Atomic structure of the [4Fe-4S] D-cluster. (d) Atomic structure of the [4Fe-4S] B-cluster. (e) Atomic structure of the catalytic [Ni-Fe-S] C-cluster. Atom colors: dark gray (iron), orange (sulfide), red (oxygen), blue (nitrogen), white (carbon), and dark green (nickel). Generated using Pymol from PDB 1JQK.[79] *Reprinted with permission from Can M, et al. Structure, function, and mechanism of the nickel metalloenzymes, CO-dehydrogenase, and acetyl-CoA synthase.* Chem Rev *2014;***114***:4149-74. Copyright 2014, American Chemical Society.*

CO oxidation site.[83,88,89] Crystal structures of several Ni-CODH enzymes have also been solved. These enzymes are generally homodimeric,[79] with multiple metal cofactor sites (between 5 and 7 per monomer) composed of two [Ni-Fe-S] clusters, with accessory [4Fe-4S] clusters that differ in coordination environments (Figure 18.4).[88,90] The reversible CO/CO_2 conversion occurs at the [Ni-Fe-S] cluster, or C-cluster, and numerous studies have investigated the spectroscopic and structural properties of these unique clusters.[91–99] The C-cluster is deeply buried in the enzyme, with hydrophobic channels for gas transport and transferring substrates to the catalytic Ni atom.[100] To date, only the Ni-dependent CODH enzymes have been successfully incorporated into hybrid systems.[101] The metal active sites are functional models for designing efficient CO_2 reduction catalysts due to the high turnover numbers, low overpotential requirements, and efficient CO_2 binding mechanisms. However, Ni-CODH enzymes suffer from the same limitations as hydrogenases, in that they are sensitive to O_2 and must be prepared and used under strictly anaerobic conditions.[79]

Formate dehydrogenases. FDH enzymes comprise a diverse group of prokaryotic and eukaryotic enzymes that catalyze the reversible two-electron oxidation of formate to CO_2 (Equation (18.3)).[102] In aerobic organisms, formate oxidation is a NAD^+-dependent process, with FDH catalyzing hydride transfer from formate to produce CO_2 and NADH.[103] In anoxic and anaerobic organisms, formate can serve as an electron donor for a variety of respiratory pathways.[104] Formate oxidation in these organisms is catalyzed by NAD^+-independent FDHs containing Mo- or W-active sites. Some of these FDHs are monodirectional, catalyzing only formate oxidation to CO_2, while others are reversible, primarily those with W as the active site metal.[102] Both NAD^+-dependent and -independent FDHs have been used as formate oxidation catalysts for biofuel cells, but due to the ability to reduce CO_2, the reversible NAD^+-independent FDHs are of interest for photobiohybrid fuel-producing systems. Crystal structures of three NAD^+-independent FDH enzymes have been reported, two Mo-active site FDHs from *Escherichia coli* K12[105,106] and one W-active site FDH from *D. gigas*.[107] These enzymes are structurally diverse, but the redox active sites show similarities, with the Mo or W metal ion coordinated by two pyranopterin guanosine dinucleotide molecules, a Selenium in the form of Selenocysteine, and Fe in either a FeS cluster or a heme (Figure 18.5).[105–108] These enzymes have potential both as CO_2 reduction catalysts and as models for synthetic catalyst design.

Electrochemical and photochemical CO_2 reduction systems

The first example of a photobiohybrid system for CO_2 reduction was demonstrated by Parkinson et al.,[109] where an indium-phosphide photoanode was used to reduce FDH from *Pseudomonas oxalaticus*, with methyl viologen acting as a mediator to couple the photogenerated electrons to the enzyme. Similar mediator-dependent strategies have been employed for CO_2 reduction to CO using CODH from *Moorella thermoacetica*,[110] and in

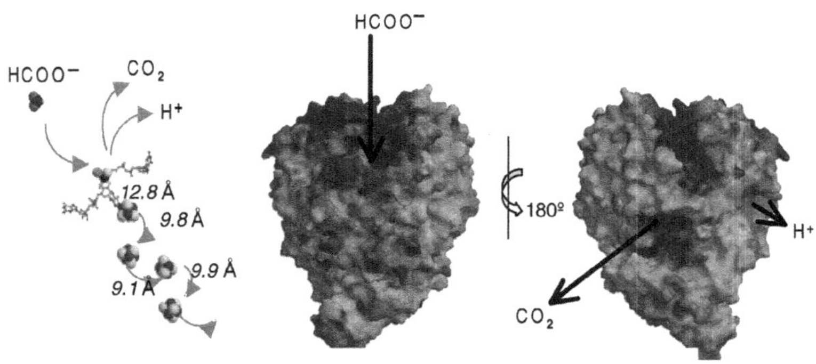

Figure 18.5
Diagram of *D. gigas* FDH metal cofactors (left, yellow and brown = FeS clusters; green = W/Se-pterin cofactor) and electrostatic-potential surface plot (right) contoured for positive charges (blue) and negative charges (red). The top cleft is lined with positively charged, basic amino acids and provides an electrostatically favorable surface for formate access. At the W/Se-pterin cofactor, the formate is split into CO_2, H^+, and two electrons.[107] *Reprinted with permission from Raaijmakers H, et al. Gene sequence and the 1.8 Å crystal structure of the tungsten-containing formate dehydrogenase from* Desulfovibrio gigas. Structure *2002;***10***:1261-72. Copyright 2002, Science Direct.*

a multienzyme systems of FDH and methanol dehydrogenase to produce formaldehyde and methanol from the reduction of CO_2.[111] Direct adsorption of both CODH and FDH on electrodes has also been demonstrated. Parkin et al. studied rapid and efficient electrocatalytic CO_2/CO interconversion by a Ni-CODH attached to a PGE electrode (Figure 18.6a).[112] Voltammograms revealed intense activity for this CODH on the electrode, with zero-current potentials that approximated to the formal potential for the CO_2/CO redox couple. Absorption of a W-containing FDH1 from *D. gigas* on PGE showed efficient electrochemical reduction of CO_2 to formate with small overpotentials required for both CO_2 reduction and formate oxidation (Figure 18.6b).[113] The formate oxidation reaction was significantly faster than CO_2 oxidations, with up to five times higher rates, but the catalysis of CO_2 reduction by FDH was more than two orders of magnitude faster than that of any known synthetic catalyst. To date, there have been no photoelectrochemical systems for CO_2 reduction by either CODH or FDH that do not rely on a mediator to shuttle electrons from the anode; however, the promising work on coupling both enzymes to electrodes will likely result in such a system in the near future.

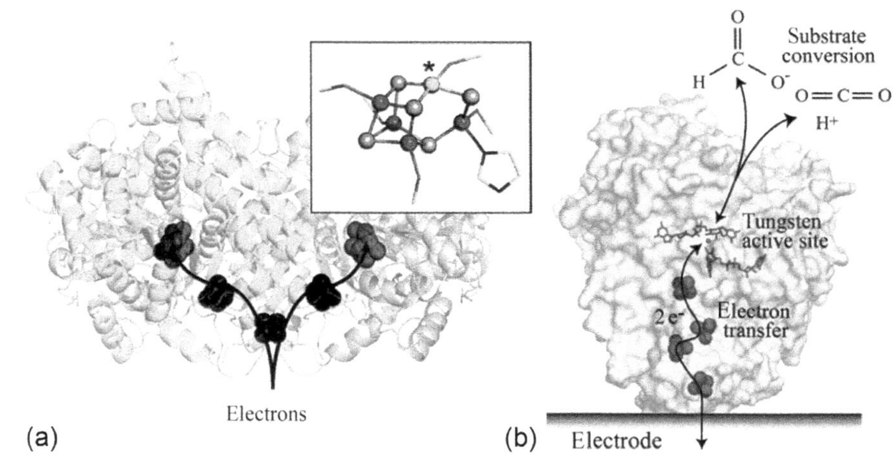

Figure 18.6

Electrochemical reduction of CO_2 by immobilized CODH and FDH. (a) X-ray structure of *Carboxydothermus hydrogenoformans* CODH; the two subunits are in blue and gray, electron-relay FeS clusters are shown in black, and the catalytic [Ni-Fe-S]-cluster (C-cluster) of the catalytic site is in red. Inset: the atomic structure of the C-cluster with the proposed Ni-binding site for substrate CO denoted by an asterisk (*).[112] *(Reprinted with permission from Parkin A, et al. Rapid and efficient electrocatalytic CO_2/CO interconversions by* Carboxydothermus hydrogenoformans *CO-dehydrogenase I on an electrode.* J Am Chem Soc *2007;***129***:10328-9. Copyright 2007, American Chemical Society.)* (b) Reversible electrocatalytic conversion of CO_2 and formate by *D. gigas* FDH. Electrons are exchanged between the electrode and the FDH catalytic site via a chain of FeS clusters in FDH. The relative rate of reduction of CO_2 to formate or oxidation of formate to CO_2 is controlled by the applied bias of the electrode potential.[113] *(Reprinted with permission from Reda T, et al. Reversible interconversion of carbon dioxide and formate by an electroactive enzyme.* Proc Natl Acad Sci U S A *2008;***105***:10654-8. Copyright 2008, National Academy of Sciences.)*

Nanomaterial-based CO_2 reduction systems

Reduction of CO_2 by direct adsorption of an enzyme to a light-harvesting nanomaterial has been demonstrated using the Ni-CODH from *Carboxydothermus hydrogenoformans*. This enzyme has been successfully assembled into complexes with several dye-sensitized metal oxide nanoparticles,[114,115] where use of TiO_2 showed the highest TOF at $0.14\,s^{-1}$. Higher TOFs were achieved with CdS nanorods, with CODH reaching a TOF of $1.23\,s^{-1}$.[116] While these TOFs are low compared with the maximum turnover rates of the enzyme in electrochemical cells,[112] they represent significant improvements compared to synthetic catalysts. Incorporation of FDH into a photocatalytic nanomaterial-based systems has also been achieved.[117] A novel graphene-based visible light photocatalyst with a covalently bonded chromophore was used to drive FDH-catalyzed CO_2 reduction to formate, using NADH as a mediator. This system led to high formate production rates for at least 2 h, showing the promise of both FDH-based photocatalysis and graphene-based systems for CO_2 photo reduction (Figure 18.7).

18.2.3 Enzymes for Fuel Consumption

While the focus of this chapter is on photobiohybrids for fuel production, a large body of work exists utilizing enzymes as catalysts in fuel-cell devices.[35,52,118,119,70,53] Enzymes have been used as both oxidation and reduction catalysts in fuel cells to produce electricity from a variety of chemicals, including sugars, alcohols, and H_2. Cathodes utilizing O_2 as an oxidant have been made using cytochrome oxidase, laccases,[120] and bilirubin oxidases[52] as catalysts. Anode catalysts have been constructed using a variety of enzymes, including glucose oxidase,[121] fructose dehydrogenase,[122] alcohol dehydrogenase,[123] and hydrogenases. As with the fuel-producing system discussed above, enzymes are unlikely to provide scalable solutions for fuel-cell technologies, but the substrate specificity, low overpotentials, and high reaction rates make them excellent model for catalyst development and provide benchmarks for what can be achieved in a well-engineered system.

18.3 Biological Antenna: Photoelectrochemical Biofuels Production Systems Based on Photosystem I

PSI is a ~500-kDa protein complex located in the thylakoid membranes of chloroplasts and participates in the light reactions photosynthesis. PSI performs several important functions during photosynthesis, including photon harvesting by chlorophylls oriented within its protein scaffolding, charge separation via a specialized reaction center, intraprotein electron transport, and finally, reduction of ferredoxin. PSI can perform these tasks with a quantum efficiency near unity,[124] and the iron sulfur complex at the terminus of its internal electron transfer chain achieves one of the most negative reduction potentials found in nature.[125] This rapid, efficient,

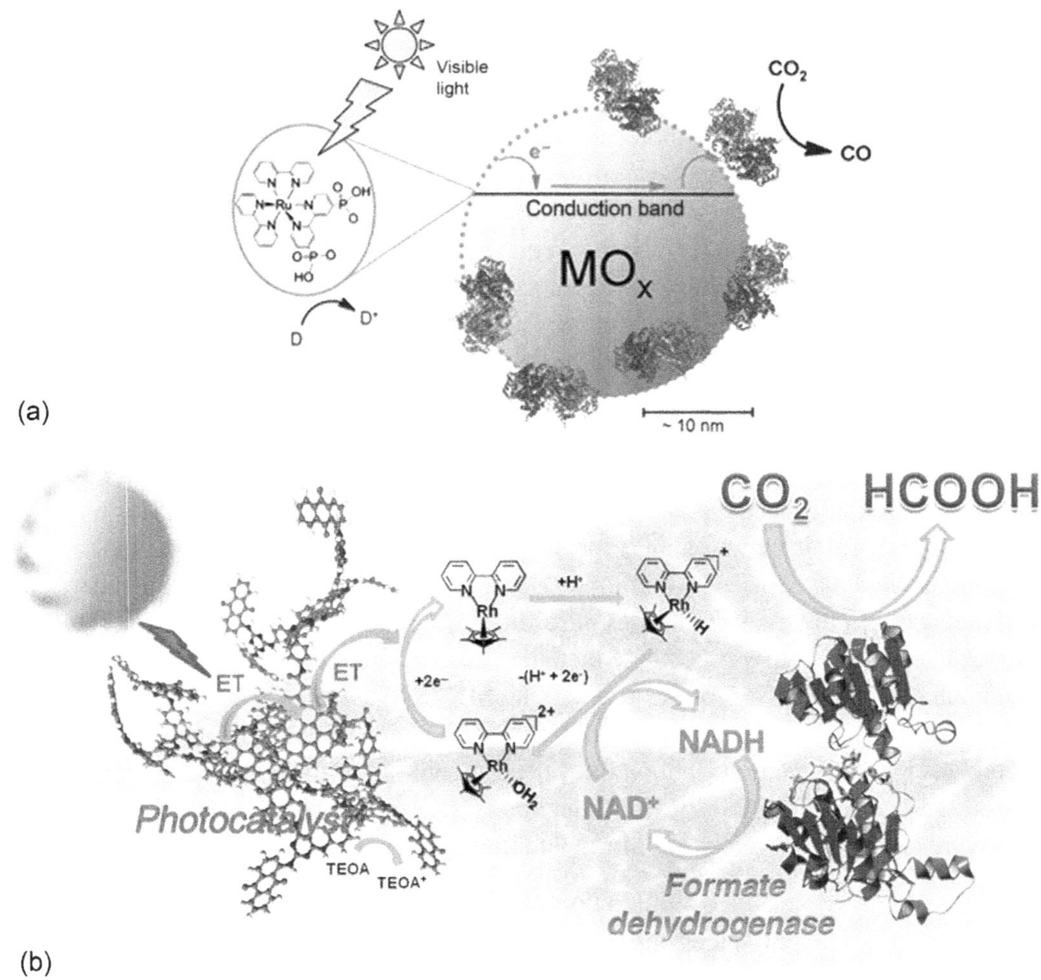

Figure 18.7

Nanomaterial photobiohybrids for light-driven CO_2 reduction. (a) A TiO_2 (MO_x) particle is photosensitized with ruthenium dyes (RuP) (orange circle). Under illumination, light absorption by RuP injects photoexcited electrons into TiO_2 which are transferred to the surface adsorbed CODH for reduction of $CO_2 \rightarrow CO$.[114] (Reprinted with permission from Woolerton TW, et al. CO_2 photoreduction at enzyme-modified metal oxide nanoparticles. *Energy Environ Sci* 2011;**4**:2393-9. Copyright 2011, Royal Society of Chemistry.) (b) Schematic of a graphene-based design that incorporates a dye molecule for light absorption with graphene as a electron acceptor/conductor material. Electrons are shuttled to NAD^+ to generate NADH which is used by FDH to reduce $CO_2 \rightarrow$ formate.[117] *(Reprinted with permission from Yadav RK, et al. A photocatalyst-enzyme coupled artificial photosynthesis system for solar energy in production of formic acid from CO_2.* J Am Chem Soc *2012;**134**:11455. Copyright 2012, American Chemical Society.)*

photoinduced charge transfer capability has seemingly obvious utility for photocatalysis and solar energy conversion systems. For these reasons, PSI has attracted much attention from researchers seeking to utilize its functionality for renewable energy applications including homogeneous and heterogeneous photocatalysis, specifically the production of H_2, as well as direct photoelectrochemical energy conversion.[126–128] Strictly in terms of thermodynamics, the remarkably negative reduction potential achieved by the F_B^- iron sulfur complex in PSI (~0.59 V vs. NHE) is more than sufficient to reduce protons to molecular hydrogen at neutral pH and ambient conditions. However, the electron-donating region of the PSI complex is specialized to interact with ferredoxin and does not facilitate the reduction of protons with appreciable kinetics. Therefore, researchers have developed strategies for the modification of PSI to improve its hydrogen production capabilities. The first of such strategies to prove successful was direct platinization by the photoinduced reduction of $Na_2[PtCl_6]$ to form metallic clusters at the electron-donating sites of PSI complexes. The origins of this approach stem from the work of Elias Greenbaum who first found that isolated choloroplasts may be platinized to yield a photoactive catalyst for the production of hydrogen.[129] Later, the same group demonstrated that isolated PSI complexes could be platinized directly[130] and also showed that several other metal chlorides, including $PtCl_4$, $OsCl_3$, $Na_2[OsCl_6]$, $(NH_4)_2RuCl_6$, and K_3RuCl_6, would similarly precipitate metallic clusters at the reducing site of PSI and provide photocatalytic activity for the production of H_2.[131] Eventually, researchers found that reducing site of PSI could be "wired" directly to metallic nanoparticles to achieve additional enhancements in H_2 production.[132] Soon thereafter, another interesting strategy for directing photoexcited electrons from PSI for H_2 production was reported, wherein various hydrogenase proteins were conjugated directly to PSI to yield impressive increases in H_2 yield.[12]

The systems discussed thus far all relied upon sacrificial electron donors, typically ascorbic acid, to continually resupply electrons to the oxidized P_{700}^+ site of PSI in order sustain H_2 production. In many cases, this strategy suffered from rate-limiting mass transport and, in all cases, required the replenishment of the sacrificial electron donor by some means in order to sustain H_2 production. While the modification described above centered around functionalizing the electron-donating sites of PSI, parallel efforts were underway focused on improving the rate at which electrons could be resupplied to the protein. Improvements to typical ascorbic acid system were offered by Millsaps and colleagues who found that incorporating plastocynanin, the copper-containing protein that transfers electrons to P_{700}^+ in natural photosynthesis, could improve electron replenishment to PSI. Electron transfer from plastocynanin to PSI was faster than that of ascorbic acid to PSI, which resulted in increased H_2 production rates.[133] This strategy was further refined by Evans et al. who covalently attached plastocynanin to PSI.[134] Meanwhile, direct electrochemical methods were investigated as another strategy to continually supply electrons to the system. Direct electrical connectivity between electrode surface and PSI was desirable because it facilitated rapid electron transfer without the requirement of an electrochemical mediator;

however, these strategies required the immobilization of PSI onto the electrode surface. The dominant strategy for PSI immobilization was the use of thiolate-terminated self-assembled monolayers (SAMs) with various terminal functional groups to direct the adsorption or attachment of PSI.

Preliminary studies of such SAM-based immobilization techniques were undertaken by Lee et al., who demonstrated that the orientation of PSI on the surface would affect the directionality of the electric field generated by photoinduced charge separation.[135] This finding immediately highlighted the importance of controlling the orientation of the protein complex in order to dictate which end of its electron transfer chain could access the electrode. The first attempt to confront this problem was to use a His-tagged PSI complex with a SAM containing Ni-NTA terminal groups.[136] "Molecular wire" type approaches, wherein SAMs with specialized functional groups designed to interact with specific regions of genetically modified PSI complexes were also developed.[137] Researchers also began to leverage emerging nanotechnologies to construct high surface area electrodes that could accommodate more PSI complexes than traditional planar electrodes and thus produce higher ensemble electron transfer rates.[138,139] Stacked, multilayer assemblies of PSI complexes were constructed on gold surfaces by various techniques, and these systems benefited from enhanced light absorption and increased photoactivity relative to monolayer films.[140,141] PSI complexes attached to substrates other than gold have also been the topic of many studies, and repots of PSI immobilized on Si,[142] GaAs,[143] TiO_2,[144] carbon nanotubes,[145] and graphene.[146]

Strategies have emerged that incorporate modifications to the reducing end of PSI to enhance its photocatalyic activity as well as take advantage of surface modification methods to facilitate rapid replenishment of electrons. In 2009, Krassen and colleagues reported an impressive system, wherein PSI was genetically fused to [Ni-Fe]-hydrogenase, and the entire assembly was immobilized on an electrode surface.[147] Later, LeBlanc et al. used a PSI multilayer assembly to reduce $Na_2[PtCl_6]$ to produce a photocatalytic film capable of H_2 production.[10] A simplified conceptual schematic of such a system is presented in Figure 18.8. This figure depicts a PSI complex in close proximity to an electrode surface from which its P_{700} site may be reduced by the electrode directly. Upon photoexitation, charge separation occurs at the P_{700} reaction center, and an excited electron travels through the electron transfer chain to eventually reduce the F_B iron sulfur complex. In case of the system depicted in Figure 18.8, the electron would be subsequently transferred to the Pt cluster where it becomes available for the reduction of adsorbed protons. Two such photo-driven reduction events occurring in succession would be required to produce one molecule of H_2. Systems such as these are likely the future of PSI-based photoelectrochemical catalysis, since they employ methods to improve the kinetics of electron replenishment to PSI as well as facilitate rapid reduction of desired substrates and thereby obtain the most out of the impressive photoconversion efficiency exhibited by this protein complex.

Figure 18.8
Idealized schematic of a PSI-based system used to photocatalyze production of H_2. A PSI complex immobilized an electrode surface harvests photons using chlorophylls oriented within its protein scaffolding. Inset: a zoom view showing the electron transfer chain (with the protein backbone hidden). Energy from photons is used for charge separation at the P_{700} site, resulting in an excited electron, which moves through an intraprotein electron transfer chain to Pt clusters precipitated at the electron donor site. The Pt cluster facilitates adsorption of protons and their subsequent reduction to form H_2. Electrons may be resupplied to PSI directly from the underlying electrode. Figure prepared by P. Ciesielski using the PDB structure file 2WSC contributed by Amunts et al.[148]

18.4 Perspectives

The hybrid systems reviewed here, incorporating both biological and inorganic materials, represent important benchmarks for the development of integrated synthetic systems for solar fuel production. The design principles established by studying photohybrid systems will be highly valuable in the advancement of fully artificial systems. While the enzymes and proteins used are not likely to be applicable in scalable commercial technologies, their integration with artificial components provides essential information for controlling the coupling between light capture antennas and fuel-forming catalyst, and the efficiencies achieved in photohybrid systems offer a target for artificial systems. Sunlight provides the energy necessary to both sustain current and meet future energy needs, and inventive technologies are needed in order to capture and convert that energy into useable and storable forms. Hybrid systems can supply a roadmap for the development of such systems, by combining the advantages of both materials chemistry and biology to achieve high specificity and quantum efficiencies for production of reduced chemicals from harvesting of solar energy.

Acknowledgments

The authors would like to thank support from the U.S. Department of Energy, Office of Science, Basic Energy Sciences, and NREL under U.S. Department of Energy under Contract No. DE-AC36-08-GO28308.

References

1. Lewis NS, Nocera DG. Powering the planet: chemical challenges in solar energy utilization. *Proc Natl Acad Sci U S A* 2006;**103**:15729–35. http://dx.doi.org/10.1073/pnas.0603395103.
2. Nocera DG. The artificial leaf. *Acc Chem Res* 2012;**45**:767–76. http://dx.doi.org/10.1021/ar2003013.
3. Artero V, Fontecave M. Solar fuels generation and molecular systems: is it homogeneous or heterogeneous catalysis? *Chem Soc Rev* 2013;**42**(6):2338–56.
4. Ghirardi ML, Posewitz MC, Maness PC, Dubini A, Yu J, Seibert M. Hydrogenases and hydrogen photoproduction in oxygenic photosynthetic organisms. *Annu Rev Plant Biol* 2007;**58**:71–91. http://dx.doi.org/10.1146/annurev.arplant.58.032806.103848.
5. Gust D, Moore TA, Moore AL. Mimicking photosynthetic solar energy transduction. *Acc Chem Res* 2000;**34**:40–8. http://dx.doi.org/10.1021/ar9801301.
6. Moore GF, Brudvig GW. Energy conversion in photosynthesis: a paradigm for solar fuel production. *Annu Rev Condens Matter Phys* 2011;**2**:303–27. http://dx.doi.org/10.1146/annurev-conmatphys-062910-140503.
7. Nocera DG. Chemistry of personalized solar energy. *Inorg Chem* 2009;**48**:10001.
8. Blankenship RE, et al. Comparing photosynthetic and photovoltaic efficiencies and recognizing the potential for improvement. *Science* 2011;**332**:805–9. http://dx.doi.org/10.1126/science.1200165.
9. Cogdell RJ, Gardiner AT, Molina PI, Cronin L. The use and misuse of photosynthesis in the quest for novel methods to harness solar energy to make fuel. *Philos Trans R Soc A Math Phys Eng Sci* 2013;**371**:http://dx.doi.org/10.1098/rsta.2011.0603.
10. LeBlanc G, Chen GP, Jennings GK, Cliffel DE. Photoreduction of catalytic platinum particles using immobilized multilayers of photosystem I. *Langmuir* 2012;**28**:7952–6. http://dx.doi.org/10.1021/La301019t.
11. Lee J-H, Lee JH, Lee YJ, Nam KT. Protein/peptide based nanomaterials for energy application. *Curr Opin Biotechnol* 2013;**24**:599–605. http://dx.doi.org/10.1016/j.copbio.2013.02.004.
12. Lubner CE, et al. Solar hydrogen-producing bionanodevice outperforms natural photosynthesis. *Proc Natl Acad Sci U S A* 2011;**108**:20988–91. http://dx.doi.org/10.1073/pnas.1114660108.
13. Rakowski DuBois M, DuBois DL. Development of molecular electrocatalysts for CO_2 reduction and H_2 production/oxidation. *Acc Chem Res* 1974;**42**:2009.
14. Armstrong FA, Hirst J. Reversibility and efficiency in electrocatalytic energy conversion and lessons from enzymes. *Proc Natl Acad Sci U S A* 2011;**108**:14049–54. http://dx.doi.org/10.1073/pnas.1103697108.
15. Armaroli N, Balzani V. The future of energy supply: challenges and opportunities. *Angew Chem Int Ed* 2007;**46**:52–66. http://dx.doi.org/10.1002/anie.200602373.
16. Gust D, Moore TA, Moore AL. Solar fuels via artificial photosynthesis. *Acc Chem Res* 2009;**42**:1890–8. http://dx.doi.org/10.1021/ar900209b.
17. Magnuson A, et al. Biomimetic and microbial approaches to solar fuel generation. *Acc Chem Res* 2009;**42**:1899–909. http://dx.doi.org/10.1021/ar900127h.
18. Casci JL, Lok CM, Shannon MD. Fischer-Tropsch catalysis: the basis for an emerging industry with origins in the early 20th century. *Catal Today* 2009;**145**:38–44. http://dx.doi.org/10.1016/j.cattod.2008.08.032.
19. Palo DR, Dagle RA, Holladay JD. Methanol steam reforming for hydrogen production. *Chem Rev* 2007;**107**:3992–4021. http://dx.doi.org/10.1021/cr050198b.
20. Reisner E. Solar hydrogen evolution with hydrogenases: from natural to hybrid systems. *Eur J Inorg Chem* 2011;**2011**:1005–16. http://dx.doi.org/10.1002/ejic.201000986.
21. Gordon RB, Bertram M, Graedel TE. Metal stocks and sustainability. *Proc Natl Acad Sci U S A* 2006;**103**:1209–14. http://dx.doi.org/10.1073/pnas.0509498103.

22. Jones AK, Sillery E, Albracht SPJ, Armstrong FA. Direct comparison of the electrocatalytic oxidation of hydrogen by an enzyme and a platinum catalyst. *Chem Commun* 2002;**866-867**. http://dx.doi.org/10.1039/b201337a.
23. Nicolet Y, Piras C, Legrand P, Hatchikian CE, Fontecilla-Camps JC. Desulfovibrio desulfuricans iron hydrogenase: the structure shows unusual coordination to an active site Fe binuclear center. *Structure* 1999;**7**:13–23. http://dx.doi.org/10.1016/S0969-2126(99)80005-7.
24. Volbeda A, et al. Structural differences between the ready and unready oxidized states of [NiFe] hydrogenases. *J Biol Inorg Chem* 2005;**10**:239–49. http://dx.doi.org/10.1007/s00775-005-0632-x.
25. Cohen J, Kim K, King P, Seibert M, Schulten K. Finding gas diffusion pathways in proteins: application to O_2 and H_2 transport in CpI [FeFe]-hydrogenase and the role of packing defects. *Structure* 2005;**13**:1321–9.
26. Cohen J, et al. Molecular dynamics and experimental investigation of H_2 and O_2 diffusion in [Fe]-hydrogenase. *Biochem Soc Trans* 2005;**33**:80.
27. Nicolet Y, Lemon BJ, Fontecilla-Camps JC, Peters JW. A novel FeS cluster in Fe-only hydrogenases. *Trends Biochem Sci* 2000;**25**:138–43.
28. Peters JW, Lanzilotta WN, Lemon BJ, Seefeldt LC. X-ray crystal structure of the Fe-only hydrogenase (CpI) from *Clostridium pasteurianum* to 1.8 Angstrom resolution. *Science* 1998;**282**:1853–8. http://dx.doi.org/10.1126/science.282.5395.1853.
29. Przybyla AE, Robbins J, Menon N, Peck Jr HD. Structure-function relationships among the nickel-containing hydrogenases. *FEMS Microbiol Lett* 1992;**88**:109–35.
30. Montet Y, et al. Gas access to the active site of Ni-Fe hydrogenases probed by X-ray crystallography and molecular dynamics. *Nat Struct Biol* 1997;**4**:523–6.
31. Dementin S, et al. A glutamate is the essential proton transfer gate during the catalytic cycle of the [NiFe] hydrogenase. *J Biol Chem* 2004;**279**:10508–13.
32. Bagley KA, et al. Infrared studies on the interaction of carbon monoxide with divalent nickel in hydrogenase from *Chromatium vinosum*. *Biochemistry* 1994;**33**:9229–36.
33. Happe RP, Roseboom W, Pierik AJ, Albracht S, Bagley KA. Biological activation of hydrogen. *Nature* 1997;**385**:126.
34. Léger C, Bertrand P. Direct electrochemistry of redox enzymes as a tool for mechanistic studies. *Chem Rev* 2008;**108**:2379–438. http://dx.doi.org/10.1021/cr0680742.
35. Woolerton TW, Sheard S, Chaudhary YS, Armstrong FA. Enzymes and bio-inspired electrocatalysts in solar fuel devices. *Energy Environ Sci* 2012;**5**:7470–90. http://dx.doi.org/10.1039/c2ee21471g.
36. Vincent KA, et al. Electrochemical definitions of O_2 sensitivity and oxidative inactivation in hydrogenases. *J Am Chem Soc* 2005;**127**:18179–89.
37. Armstrong FA, et al. Dynamic electrochemical investigations of hydrogen oxidation and production by enzymes and implications for future technology. *Chem Soc Rev* 2009;**38**:36–51. http://dx.doi.org/10.1039/b801144n.
38. Cracknell JA, Wait AF, Lenz O, Friedrich B, Armstrong FA. A kinetic and thermodynamic understanding of O_2 tolerance in [NiFe]-hydrogenases. *Proc Natl Acad Sci* 2009;**106**:20681–6. http://dx.doi.org/10.1073/pnas.0905959106.
39. Vignais PM, Billoud B. Occurrence, classification, and biological function of hydrogenases: an overview. *Chem Rev* 2007;**107**:4206–72. http://dx.doi.org/10.1021/cr050196r.
40. Vignais PM, Billoud B, Meyer J. Classification and phylogeny of hydrogenases. *FEMS Microbiol Rev* 2001;**25**:455–501.
41. Fontecilla-Camps JC, Volbeda A, Cavazza C, Nicolet Y. Structure/function relationships of [NiFe]- and [FeFe]-hydrogenases. *Chem Rev* 2007;**107**:4273–303. http://dx.doi.org/10.1021/cr050195z.
42. De Lacey AL, Fernández VM, Rousset M, Cammack R. Activation and inactivation of hydrogenase function and the catalytic cycle: spectroelectrochemical studies. *Chem Rev* 2007;**107**:4304–30. http://dx.doi.org/10.1021/cr0501947.
43. Lubitz W, Reijerse E, van Gastel M. [NiFe] and [FeFe] hydrogenases studied by advanced magnetic resonance techniques. *Chem Rev* 2007;**107**:4331–65. http://dx.doi.org/10.1021/cr050186q.
44. Ogata H, Lubitz W, Higuchi Y. [NiFe] hydrogenases: structural and spectroscopic studies of the reaction mechanism. *Dalton Trans* 2009;**37**:7577–87. http://dx.doi.org/10.1039/b903840j.

45. Heering HA, Canters GW. Activating redox enzymes through immobilisation and wiring. In: Davis JJ, editor. Engineering the bioelectronic interface. Royal Society of Chemistry, London 2009. p. 119–52.
46. Vincent KA, Parkin A, Armstrong FA. Investigating and exploiting the electrocatalytic properties of hydrogenases. *Chem Rev* 2007;**107**:4366–413. http://dx.doi.org/10.1021/cr050191u.
47. Tran PD, Artero V, Fontecave M. Water electrolysis and photoelectrolysis on electrodes engineered using biological and bio-inspired molecular systems. *Energy Environ Sci* 2010;**3**:727–47.
48. Baffert C, et al. Hydrogen-activating enzymes: activity does not correlate with oxygen sensitivity. *Angew Chem Int Ed* 2008;**47**:2052–4. http://dx.doi.org/10.1002/anie.200704313.
49. Baffert C, et al. Covalent attachment of FeFe hydrogenases to carbon electrodes for direct electron transfer. *Anal Chem* 2012;**84**:7999–8005. http://dx.doi.org/10.1021/ac301812s.
50. Lamle SE, Vincent KA, Halliwell LM, Albracht SPJ, Armstrong FA. Hydrogenase on an electrode: a remarkable heterogeneous catalyst. *Dalton Trans* 2003;**21**:4152–7. http://dx.doi.org/10.1039/b306234c.
51. Alonso-Lomillo MA, et al. Hydrogenase-coated carbon nanotubes for efficient H_2 oxidation. *Nano Lett* 2007;**7**:1603–8. http://dx.doi.org/10.1021/nl070519u.
52. Krishnan S, Armstrong FA. Order-of-magnitude enhancement of an enzymatic hydrogen-air fuel cell based on pyrenyl carbon nanostructures. *Chem Sci* 2012;**3**:1015–23. http://dx.doi.org/10.1039/c2sc01103d.
53. Lojou É, et al. Biocatalysts for fuel cells: efficient hydrogenase orientation for H_2 oxidation at electrodes modified with carbon nanotubes. *J Biol Inorg Chem* 2008;**13**:1157–67. http://dx.doi.org/10.1007/s00775-008-0401-8.
54. Svedruzic D, et al. High-performance hydrogen production and oxidation electrodes with hydrogenase supported on metallic single-wall carbon nanotube networks. *J Am Chem Soc* 2011;**133**:4299–306.
55. Joo H, Bae S, Kim C, Kim S, Yoon J. Hydrogen evolution in enzymatic photoelectrochemical cell using modified seawater electrolytes produced by membrane desalination process. *Sol Energy Mater Sol Cells* 2009;**93**:1555–61.
56. Morra S, et al. Direct electrochemistry of an [FeFe]-hydrogenase on a TiO_2 electrode. *Chem Commun* 2011;**47**:10566–8.
57. Krassen H, et al. Immobilization of the [FeFe]-hydrogenase CrHydA1 on a gold electrode: design of a catalytic surface for the production of molecular hydrogen. *J Biotechnol* 2009;**142**:3–9. http://dx.doi.org/10.1016/j.jbiotec.2009.01.018.
58. Millo D, et al. Spectroelectrochemical study of the [NiFe] hydrogenase from *Desulfovibrio vulgaris* Miyazaki F in solution and immobilized on biocompatible gold surfaces. *J Phys Chem B* 2009;**113**:15344–51. http://dx.doi.org/10.1021/jp906575r.
59. Wisitruangsakul N, et al. Monitoring catalysis of the membrane-bound hydrogenase from *Ralstonia eutropha* H16 by surface-enhanced IR absorption spectroscopy. *Angew Chem Int Ed* 2009;**48**:611–3. http://dx.doi.org/10.1002/anie.200802633.
60. Delamar M, Hitmi R, Pinson J, Saveant JM. Covalent modification of carbon surfaces by grafting of functionalized aryl radicals produced from electrochemical reduction of diazonium salts. *J Am Chem Soc* 1992;**114**:5883–4. http://dx.doi.org/10.1021/ja00040a074.
61. Bahr JL, et al. Functionalization of carbon nanotubes by electrochemical reduction of aryl diazonium salts: a bucky paper electrode. *J Am Chem Soc* 2001;**123**:6536–42. http://dx.doi.org/10.1021/ja010462s.
62. Kooi SE, Schlecht U, Burghard M, Kern K. Electrochemical modification of single carbon nanotubes. *Angew Chem Int Ed* 2002;**41**:1353–5.
63. Schmidt G, Gallon S, Esnouf S, Bourgoin J-P, Chenevier P. Mechanism of the coupling of diazonium to single-walled carbon nanotubes and its consequences. *Chem Eur J* 2009;**15**:2101–10. http://dx.doi.org/10.1002/chem.200801801.
64. Karyakin AA, et al. The limiting performance characteristics in bioelectrocatalysis of hydrogenase enzymes. *Angew Chem Int Ed* 2007;**46**:7244–6. http://dx.doi.org/10.1002/anie.200701096.
65. Morozov SV, et al. Bioelectrocatalytic hydrogen production by hydrogenase electrodes. *Int J Hydrogen Energy* 2002;**27**:1501–5. http://dx.doi.org/10.1016/S0360-3199(02)00091-5.
66. Morozov SV, et al. Tolerance to oxygen of hydrogen enzyme electrodes. *Electrochem Commun* 2006;**8**:851–4. http://dx.doi.org/10.1016/j.elecom.2006.03.007.

67. Qian D-J, et al. Electrochemical hydrogen evolution by use of a glass carbon electrode sandwiched with clay, poly(butylviologen) and hydrogenase. *Mater Lett* 2003;**57**:1130–4. http://dx.doi.org/10.1016/S0167-577X(02)00944-8.
68. Rüdiger O, Abad JM, Hatchikian EC, Fernandez VM, De Lacey AL. Oriented immobilization of *Desulfovibrio gigas* hydrogenase onto carbon electrodes by covalent bonds for nonmediated oxidation of H_2. *J Am Chem Soc* 2005;**127**:16008–9. http://dx.doi.org/10.1021/ja0554312.
69. Yoon J, Bae S, Shim E, Joo H. *Pyrococcus furiosus*-immobilized anodized tubular titania cathode in a hydrogen production system. *J Power Sources* 2009;**189**:1296–301. http://dx.doi.org/10.1016/j.jpowsour.2008.12.072.
70. Hambourger M, et al. [FeFe]-hydrogenase-catalyzed H_2 production in a photoelectrochemical biofuel cell. *J Am Chem Soc* 2008;**130**:2015–22. http://dx.doi.org/10.1021/ja077691k.
71. Nozik AJ. Nanoscience and nanostructures for photovoltaics and solar fuels. *Nano Lett* 2010;**10**:2735–41. http://dx.doi.org/10.1021/nl102122x.
72. Reisner E, Fontecilla-Camps JC, Armstrong FA. Catalytic electrochemistry of a [NiFeSe]-hydrogenase on TiO_2 and demonstration of its suitability for visible-light driven H_2 production. *Chem Commun* 2009;**5**:550–2. http://dx.doi.org/10.1039/b817371k.
73. Reisner E, Powell DJ, Cavazza C, Fontecilla-Camps JC, Armstrong FA. Visible light-driven H_2 production by hydrogenases attached to dye-sensitized TiO_2 nanoparticles. *J Am Chem Soc* 2009;**131**:18457–66. http://dx.doi.org/10.1021/ja907923r.
74. Brown KA, Dayal S, Ai X, Rumbles G, King PW. Controlled assembly of hydrogenase-CdTe nanocrystal hybrids for solar hydrogen production. *J Am Chem Soc* 2010;**132**:9672–80. http://dx.doi.org/10.1021/ja101031r.
75. Brown KA, Wilker MB, Boehm M, Dukovic G, King PW. Characterization of photochemical processes for H_2 production by CdS nanorod-[FeFe]-hydrogenase complexes. *J Am Chem Soc* 2012;**134**:5627–36. http://dx.doi.org/10.1021/ja2116348.
76. Wilker MB, et al. Electron transfer kinetics in CdS nanorod-[FeFe]-hydrogenase complexes and implications for photochemical H_2 generation. *J Am Chem Soc* 2014;**136**:4316–24.
77. Hamon C, et al. Synthesis and enzymatic photo-activity of an O_2 tolerant hydrogenase-CdSe@CdS quantum rod bioconjugate. *Chem Commun* 2014;**50**:4989–92.
78. Sullivan BP, Krist K, Guard H. Electrochemical and electrocatalytic reactions of carbon dioxide. Amsterdam: Elsevier; 1992.
79. Can M, Armstrong FA, Ragsdale SW. Structure, function, and mechanism of the nickel metalloenzymes, CO dehydrogenase, and acetyl-CoA synthase. *Chem Rev* 2014;**114**:g4149–74. http://dx.doi.org/10.1021/cr400461p.
80. Appel AM, et al. Frontiers, opportunities, and challenges in biochemical and chemical catalysis of CO_2 fixation. *Chem Rev* 2013;**113**:6621–58. http://dx.doi.org/10.1021/cr300463y.
81. Heo J, Staples CR, Ludden PW. Redox-dependent CO_2 reduction activity of CO dehydrogenase from *Rhodospirillum rubrum*. *Biochemistry* 2001;**40**:7604–11. http://dx.doi.org/10.1021/bi002554k.
82. Maynard EL, Lindahl PA. Evidence of a molecular tunnel connecting the active sites for CO_2 reduction and acetyl-CoA synthesis in acetyl-CoA synthase from *Clostridium thermoaceticum*. *J Am Chem Soc* 1999;**121**:9221–2. http://dx.doi.org/10.1021/ja992120g.
83. Meyer O, et al. The role of Se, Mo and Fe in the structure and function of carbon monoxide dehydrogenase. *Biol Chem* 2000;**38**:865–76.
84. Ermler U, Grabarse W, Shima S, Goubeaud M, Thauer RK. Active sites of transition-metal enzymes with a focus on nickel. *Curr Opin Struct Biol* 1998;**8**:749–58.
85. Ferry JG. CO dehydrogenase. *Annu Rev Microbiol* 1995;**49**:305–33.
86. Ragsdale SW, Kumar M. Nickel-containing carbon monoxide dehydrogenase/acetyl-CoA synthase. *Chem Rev* 1996;**96**:2515–40.
87. Svetlitchnyi V, Peschel C, Acker G, Meyer O. Two membrane-associated NiFeS-carbon monoxide dehydrogenases from the anaerobic carbon-monoxide-utilizing eubacterium *Carboxydothermus hydrogenoformans*. *J Bacteriol* 2001;**183**:5134–44.

88. Dobbek H, Gremer L, Meyer O, Huber R. Crystal structure and mechanism of CO dehydrogenase, a molybdo iron-sulfur flavoprotein containing S-selanylcysteine. *Proc Natl Acad Sci U S A* 1999;**96**:8884–9. http://dx.doi.org/10.1073/pnas.96.16.8884.
89. Hänzelmann P, Dobbek H, Gremer L, Huber R, Meyer O. The effect of intracellular molybdenum in *Hydrogenophaga pseudoflava* on the crystallographic structure of the seleno-molybdo-iron-sulfur flavoenzyme carbon monoxide dehydrogenase. *J Mol Biol* 2000;**301**:1221–35.
90. Dobbek H, Svetlitchnyi V, Gremer L, Huber R, Meyer O. Crystal structure of a carbon monoxide dehydrogenase reveals a [Ni-4Fe-5S] cluster. *Science* 2001;**293**:1281–5. http://dx.doi.org/10.1126/science.1061500.
91. Anderson ME, Lindahl PA. Organization of clusters and internal electron pathways in CO dehydrogenase from *Clostridium thermoaceticum*: relevance to the mechanism of catalysis and cyanide inhibition. *Biochemistry* 1994;**33**:8702–11. http://dx.doi.org/10.1021/bi00195a011.
92. Anderson ME, Lindahl PA. Spectroscopic states of the CO oxidation/CO_2 reduction active site of carbon monoxide dehydrogenase and mechanistic implications. *Biochemistry* 1996;**35**:8371–80. http://dx.doi.org/10.1021/bi952902w.
93. Chen J, et al. Infrared studies of carbon monoxide binding to carbon monoxide dehydrogenase/acetyl-CoA synthase from *Moorella thermoacetica*. *Biochemistry* 2003;**42**:14822–30. http://dx.doi.org/10.1021/bi0349470.
94. Gu W, Seravalli J, Ragsdale SW, Cramer SP. CO-induced structural rearrangement of the C cluster in *Carboxydothermus hydrogenoformans* CO dehydrogenase evidence from Ni K-edge X-ray absorption spectroscopy. *Biochemistry* 2004;**43**:9029–35. http://dx.doi.org/10.1021/bi036104e.
95. Hu Z, et al. Nature of the C-cluster in Ni-containing carbon monoxide dehydrogenases. *J Am Chem Soc* 1996;**118**:830–45. http://dx.doi.org/10.1021/ja9528386.
96. Lindahl P, Münck P, Ragsdale SW. CO dehydrogenase from *Clostridium thermoaceticum*. EPR and electrochemical studies in CO_2 and argon atmospheres. *J Biol Chem* 1990;**265**:3873–9.
97. Ragsdale S, Clark J, Ljungdahl L, Lundie L, Drake H. Properties of purified carbon monoxide dehydrogenase from *Clostridium thermoaceticum*, a nickel, iron-sulfur protein. *J Biol Chem* 1983;**258**:2364–9.
98. Ralston CY, et al. Characterization of heterogeneous nickel sites in CO dehydrogenases from *Clostridium thermoaceticum* and *Rhodospirillum rubrum* by nickel L-edge X-ray spectroscopy. *J Am Chem Soc* 2000;**122**:10553–60. http://dx.doi.org/10.1021/ja0009469.
99. Seravalli J, Kumar M, Lu W-P, Ragsdale SW. Mechanism of carbon monoxide oxidation by the carbon monoxide dehydrogenase/acetyl-CoA synthase from *Clostridium thermoaceticum*: kinetic characterization of the intermediates. *Biochemistry* 1997;**36**:11241–51. http://dx.doi.org/10.1021/bi970590m.
100. Wang P-h, Bruschi M, De Gioia L, Blumberger J. Uncovering a dynamically formed substrate access tunnel in carbon monoxide dehydrogenase/acetyl-CoA synthase. *J Am Chem Soc* 2013;**135**:9493–502. http://dx.doi.org/10.1021/ja403110s.
101. Wang W-N, Soulis J, Yang YJ, Biswas P. Comparison of CO_2 photoreduction systems: a review. *Aerosol Air Qual Res* 2014;**14**:533–49.
102. Gonzalez PJ, et al. Periplasmic nitrate reductases and formate dehydrogenases: biological control of the chemical properties of Mo and W for fine tuning of reactivity, substrate specificity and metabolic role. *Coord Chem Rev* 2013;**257**:315–31. http://dx.doi.org/10.1016/j.ccr.2012.05.020.
103. Lamzin VS, Dauter Z, Popov VO, Harutyunyan EH, Wilson KS. High resolution structures of holo and apo formate dehydrogenase. *J Mol Biol* 1994;**236**:759–85.
104. Richardson DJ. Bacterial respiration: a flexible process for a changing environment. *Microbiology* 2000;**146**:551–71.
105. Boyington JC, Gladyshev VN, Khangulov SV, Stadtman TC, Sun PD. Crystal structure of formate dehydrogenase H: catalysis involving Mo, molybdopterin, selenocysteine, and an Fe_4S_4 cluster. *Science* 1997;**275**:1305–8.
106. Jormakka M, Törnroth S, Byrne B, Iwata S. Molecular basis of proton motive force generation: structure of formate dehydrogenase-N. *Science* 2002;**295**:1863–8. http://dx.doi.org/10.1126/science.1068186.

107. Raaijmakers H, et al. Gene sequence and the 1.8 Å crystal structure of the tungsten-containing formate dehydrogenase from *Desulfovibrio gigas*. *Structure* 2002;**10**:1261–72.
108. Costa C, Teixeira M, LeGall J, Moura JJG, Moura I. Formate dehydrogenase from *Desulfovibrio desulfuricans* ATCC 27774: isolation and spectroscopic characterization of the active sites (heme, iron-sulfur centers and molybdenum). *J Biol Inorg Chem* 1997;**2**:198–208. http://dx.doi.org/10.1007/s007750050125.
109. Parkinson BA, Weaver PF. Photoelectrochemical pumping of enzymatic CO_2 reduction; 1984.
110. Shin W, Lee SH, Shin JW, Lee SP, Kim Y. Highly selective electrocatalytic conversion of CO_2 to CO at −0.57 V (NHE) by carbon monoxide dehydrogenase from *Moorella thermoacetica*. *J Am Chem Soc* 2003;**125**:14688–9.
111. Kuwabata S, Tsuda R, Nishida K, Yoneyama H. Electrochemical conversion of carbon dioxide to methanol with use of enzymes as biocatalysts. *Chem Lett* 1993;**22**:1631–4.
112. Parkin A, Seravalli J, Vincent KA, Ragsdale SW, Armstrong FA. Rapid and efficient electrocatalytic CO_2/CO interconversions by *Carboxydothermus hydrogenoformans* CO dehydrogenase I on an electrode. *J Am Chem Soc* 2007;**129**:10328–9.
113. Reda T, Plugge CM, Abram NJ, Hirst J. Reversible interconversion of carbon dioxide and formate by an electroactive enzyme. *Proc Natl Acad Sci U S A* 2008;**105**:10654–8.
114. Woolerton TW, Sheard S, Pierce E, Ragsdale SW, Armstrong FA. CO_2 photoreduction at enzyme-modified metal oxide nanoparticles. *Energy Environ Sci* 2011;**4**:2393–9.
115. Woolerton TW, et al. Efficient and clean photoreduction of CO_2 to CO by enzyme-modified TiO_2 nanoparticles using visible light. *J Am Chem Soc* 2010;**132**:2132–3.
116. Chaudhary YS, et al. Visible light-driven CO_2 reduction by enzyme coupled CdS nanocrystals. *Chem Commun* 2012;**48**:58–60.
117. Yadav RK, et al. A photocatalyst-enzyme coupled artificial photosynthesis system for solar energy in production of formic acid from CO_2. *J Am Chem Soc* 2012;**134**:11455–61. http://dx.doi.org/10.1021/ja3009902.
118. Bae S, Shim E, Yoon J, Joo H. Photoanodic and cathodic role of anodized tubular titania in light-sensitized enzymatic hydrogen production. *J Power Sources* 2008;**185**:439–44. http://dx.doi.org/10.1016/j.jpowsour.2008.06.094.
119. Cracknell JA, Vincent KA, Armstrong FA. Enzymes as working or inspirational electrocatalysts for fuel cells and electrolysis. *Chem Rev* 2008;**108**:2439–61. http://dx.doi.org/10.1021/cr0680639.
120. Gutiérrez-Sánchez C, Pita M, Vaz-Domínguez C, Shleev S, De Lacey AL. Gold nanoparticles as electronic bridges for laccase-based biocathodes. *J Am Chem Soc* 2012;**134**:17212–20. http://dx.doi.org/10.1021/ja307308j.
121. Katz E, Sheeney-Haj-Ichia L, Willner I. Electrical contacting of glucose oxidase in a redox-active rotaxane configuration. *Angew Chem Int Ed* 2004;**43**:3292–300. http://dx.doi.org/10.1002/anie.200353455.
122. Kamitaka Y, Tsujimura S, Setoyama N, Kajino T, Kano K. Fructose/dioxygen biofuel cell based on direct electron transfer-type bioelectrocatalysis. *Phys Chem Chem Phys* 2007;**9**:1793–801.
123. Yanagida S, et al. Semiconductor photocatalysis. I. Quantitative photoreduction of aliphatic ketones to alcohols using defect-free zinc sulfide quantum crystallites. *J Phys Chem* 1990;**94**:3104–11. http://dx.doi.org/10.1021/j100370a066.
124. He WZ, Malkin R. Photosystems I and II. In: Raghavendra AS, editor. Photosynthesis: a comprehensive treatise. Cambridge, UK: Cambridge University Press; 1998. p. 29–43.
125. Nelson N, Ben-Shem A. The structure of photosystem I and evolution of photosynthesis. *Bioessays* 2005;**27**:914–22. http://dx.doi.org/10.1002/Bies.20278.
126. Kargul J, Olmos JDJ, Krupnik T. Structure and function of photosystem I and its application in biomimetic solar-to-fuel systems. *J Plant Physiol* 2012;**169**:1639–53. http://dx.doi.org/10.1016/j.jplph.2012.05.018.
127. Yehezkeli O, Tel-Vered R, Michaeli D, Willner I, Nechushtai R. Photosynthetic reaction center-functionalized electrodes for photo-bioelectrochemical cells. *Photosynth Res* 2014;**120**:71–85. http://dx.doi.org/10.1007/s11120-013-9796-3.

128. Nguyen K, Bruce BD. Growing green electricity: progress and strategies for use of photosystem I for sustainable photovoltaic energy conversion. *Biochim Biophys Acta – Bioenergetics* 2014;**1837**:1553–66. http://dx.doi.org/10.1016/j.bbabio.2013.12.013.
129. Greenbaum E. Platinized chloroplasts—a novel photocatalytic material. *Science* 1985;**230**:1373–5. http://dx.doi.org/10.1126/science.230.4732.1373.
130. Lee JW, Greenbaum E. Bioelectronics and biometallocatalysis for production of fuels and chemicals by photosynthetic water-splitting. *Appl Biochem Biotechnol* 1995;**51-52**:295–305. http://dx.doi.org/10.1007/Bf02933433.
131. Lee JW, Collins RT, Greenbaum E. Molecular ionic probes: a new class of hill reagents and their potential for nanofabrication and biometallocatalysis. *J Phys Chem B* 1998;**102**:2095–100. http://dx.doi.org/10.1021/Jp973058h.
132. Grimme RA, Lubner CE, Bryant DA, Golbeck JH. Photosystem I/molecular wire/metal nanoparticle bioconjugates for the photocatalytic production of H-2. *J Am Chem Soc* 2008;**130**:6308–9. http://dx.doi.org/10.1021/Ja800923y.
133. Millsaps JF, Bruce BD, Lee JW, Greenbaum E. Nanoscale photosynthesis: photocatalytic production of hydrogen by platinized photosystem I reaction centers. *Photochem Photobiol* 2001;**73**:630–5. http://dx.doi.org/10.1562/0031-8655(2001)073<0630:Npppoh>2.0.Co;2.
134. Evans BR, O'Neill HM, Hutchens SA, Bruce BD, Greenbaum E. Enhanced photocatalytic hydrogen evolution by covalent attachment of plastocyanin to photosystem I. *Nano Lett* 2004;**4**:1815–9. http://dx.doi.org/10.1021/nl0493388.
135. Lee I, Lee JW, Greenbaum E. Biomolecular electronics: vectorial arrays of photosynthetic reaction centers. *Phys Rev Lett* 1997;**79**:3294–7. http://dx.doi.org/10.1103/PhysRevLett.79.3294.
136. Das R, et al. Integration of photosynthetic protein molecular complexes in solid-state electronic devices. *Nano Lett* 2004;**4**:1079–83. http://dx.doi.org/10.1021/Nl049579f.
137. Terasaki N, et al. Plugging a molecular wire into photosystem I: reconstitution of the photoelectric conversion system on a gold electrode. *Angew Chem Int Ed* 2009;**48**:1585–7. http://dx.doi.org/10.1002/anie.200805748.
138. Ciesielski PN, et al. Functionalized nanoporous gold leaf electrode films for the immobilization of photosystem I. *ACS Nano* 2008;**2**:2465–72. http://dx.doi.org/10.1021/Nn800389k.
139. Terasaki N, et al. Fabrication of novel photosystem I—gold nanoparticle hybrids and their photocurrent enhancement. *Thin Solid Films* 2006;**499**:153–6. http://dx.doi.org/10.1016/j.tsf.2005.07.050.
140. Frolov L, Wilner O, Carmeli C, Carmeli I. Fabrication of oriented multilayers of photosystem I proteins on solid surfaces by auto-metallization. *Adv Mater* 2008;**20**:263–6. http://dx.doi.org/10.1002/adma.200701.474.
141. Ciesielski PN, et al. Enhanced photocurrent production by photosystem I multilayer assemblies. *Adv Funct Mater* 2010;**20**:4048–54. http://dx.doi.org/10.1002/adfm.201001193.
142. LeBlanc G, Chen GP, Gizzie EA, Jennings GK, Cliffel DE. Enhanced photocurrents of photosystem I films on p-doped silicon. *Adv Mater* 2012;**24**:5959–62. http://dx.doi.org/10.1002/adma.201202794.
143. Frolov L, Rosenwaks Y, Richter S, Carmeli C, Carmeli I. Photoelectric junctions between GaAs and photosynthetic reaction center protein. *J Phys Chem C* 2008;**112**:13426–30. http://dx.doi.org/10.1021/Jp800586w.
144. Nikandrov VV, et al. Photochemical properties of photosystem 1 immobilized in a mesoporous semiconductor matrix. *High Energy Chem* 2012;**46**:200–5. http://dx.doi.org/10.1134/S0018143912030095.
145. Kaniber SM, Brandstetter M, Simmel FC, Carmeli I, Holleitner AW. On-chip functionalization of carbon nanotubes with photosystem I. *J Am Chem Soc* 2010;**132**:2872–3. http://dx.doi.org/10.1021/Ja910790x.
146. Gunther D, et al. Photosystem I on graphene as a highly transparent. *Photoactive Electrode Langmuir* 2013;**29**:4177–80. http://dx.doi.org/10.1021/La305020c.
147. Krassen H, et al. Photosynthetic hydrogen production by a hybrid complex of photosystem I and [NiFe]-hydrogenase. *ACS Nano* 2009;**3**:4055–61. http://dx.doi.org/10.1021/Nn900748j.
148. Amunts A, Toporik H, Borovikova A, Nelson N. Structure determination and improved model of plant photosystem I. *J Biol Chem* 2010;**285**:3478–86. http://dx.doi.org/10.1074/jbc.M109.072645.

CHAPTER 19

Scale-Up—Bioreactor Design and Culture Optimization

Joel L. Cuello, Takanori Hoshino, Sara Kuwahara, Cody L. Brown
Department of Agricultural and Biosystems Engineering, The University of Arizona, Tucson, Arizona, USA

19.1 Introduction

The scaled-up production of microbial biomass to produce biofuels constitutes an indispensable step in the commercialization process of microbial-based biofuels. In the case of microalgae, their large-scale commercial cultivation began in earnest, though not for biofuel production, in the 1960s with the culture of *Chlorella* in Japan[1] and the use of phytoplankton as a feedstuff for aquatic organisms.[2] By 1980, there were 46 large-scale facilities in Asia producing over 1000 kg of microalgae each month.[3] The global production of microalgal biomass was estimated to be over 5000 dry tons in 2005 with a value of over US$ 1.25 billion that excluded value of processed products.[4] About 3000 dry tons of *Spirulina* were produced in China, India, Myanmar, United States, and Japan; 2000 dry tons of *Chlorella* were produced in Taiwan, Germany, and Japan; and 1200 dry tons of *Dunaliella salina* were produced in Australia, Israel, United States, and China.[4] In 2008, the global production of microalgal biomass was estimated to have reached approximately 9000 dry tons per year.[5]

This chapter focuses on the scale-up of microalgae cultivation systems, specifically microalgae photobioreactors, whose basic principles of scale-up as elucidated here should generally apply to other types of photosynthetic microbial organisms.

19.2 General Scale-Up Process

The scale-up process for microalgae photobioreactors involves transitioning the volume of production from (1) small laboratory flasks to (2) laboratory-scale photobioreactors to (3) pilot-scale photobioreactors to (4) industrial-scale photobioreactors (Figure 19.1). Regardless of the type of the photobioreactor used, the principal goal of the scale-up process is to ensure that the desired level of microalgal productivity is maintained as the volume of production increases. This necessitates that the characteristic hydrodynamic conditions (e.g., degree of mixing) of the specific type of photobioreactor used as well as the accompanying environmental conditions (e.g., lighting) are maintained or appropriately adjusted with increasing production volume to maintain the desired level of microalgal productivity.

Biotechnology for Biofuel Production and Optimization. http://dx.doi.org/10.1016/B978-0-444-63475-7.00019-4

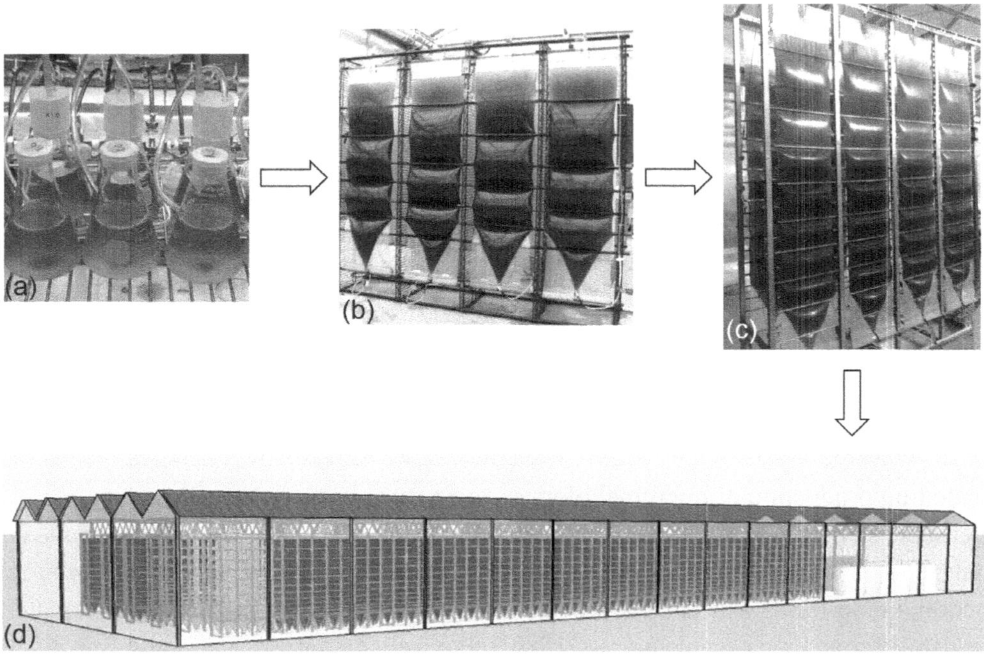

Figure 19.1
General scale-up process for microalgae photobioreactors including production volume transitions from small laboratory flasks (a) to laboratory-scale photobioreactors (b) to pilot-scale photobioreactors (c) to industrial-scale photobioreactors (d).[6]

19.3 Microalgal Photobioreactors

While microalgae can be grown in open ponds or raceways (Figure 19.2), the number of microalgal species that can be grown successfully in open-pond systems is significantly limited by the algae species that can be grown successfully in highly selective environments to ward off deleterious microbial contaminants that could cause culture crash.[9,10] For example, *Chlorella* is grown in a nutrient-rich medium, *Spirulina* at high pH

Figure 19.2
Open raceways for algae cultivation. Earthrise raceways (a) and Phyco raceways (b). *Source: Refs. 7,8*

and bicarbonate concentration, and *D. salina* at high salinity.[9,10] The mass cultivation of microalgal species that lack pronounced environmentally selective advantages, which is the case for most microalgal species, may thus necessitate the use of photobioreactors.[10] Photobioreactors are transparent containers or vessels designed to have reduced the light path to increase the amount of available light to the algal cells, and the cultures within are continuously mixed to enhance nutrient distribution and gas exchange. Photobioreactors for microalgae production have an optimal thickness of about 2-4 cm.[9] The tubular and the flat plate are the two most common general designs of microalgal photobioreactors (Figure 19.3).

Tubular photobioreactors have large surface-to-volume ratio. Because of their widespread availability, tubes have long been used as a basic photobioreactor material. The geometric configurations of tubular photobioreactors span a wide range from straight horizontal (Figures 19.3a and 19.4), straight vertical (Figures 19.3b and 19.5) to triangular

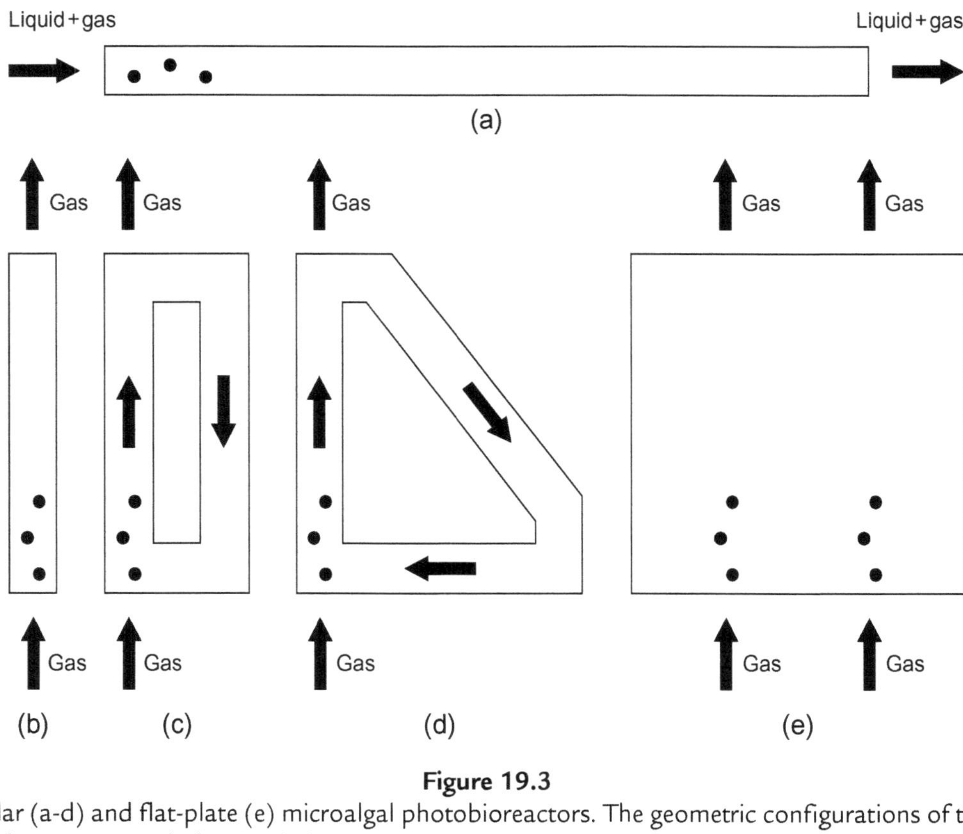

Figure 19.3
Tubular (a-d) and flat-plate (e) microalgal photobioreactors. The geometric configurations of tubular photobioreactors include straight horizontal (a), straight vertical (b), straight vertical with connecting arms, that is, external airlift (c) and triangular (d), among others. Photobioreactors (a), (c), and (d) involve bulk liquid flow, whereas (b) and (e) involve only liquid mixing through gas aeration.

Figure 19.4
Tubular photobioreactor consisting of straight horizontal tubes stacked in vertical fence-like arrays. *Source: Ref. 11*

Figure 19.5
Tubular photobioreactor consisting of straight vertical tubes (bubble columns). *Source: Ref. 12*

(Figure 19.3d and 19.6), among others. One of the world's largest photobioreactor facilities is in a greenhouse in Klotze, Germany. This facility consists of straight horizontal tubes stacked in vertical fence-like arrays (Figure 19.7). The facility has a total volume of 700 m^3, occupies a total land area of 10,000 m^2, and produces 35-41 g dry weight m^{-2} day^{-1} or 120-140 dry tons per year. Algae wall adhesion, biofouling, large pressure drop, and gradients in pH, dissolved oxygen (O_2), or carbon dioxide (CO_2) can occur along the tube

Figure 19.6
Triangular external airlift tubular photobioreactors. *Sources: Refs. 13,14*

Figure 19.7
Tubular photobioreactors of straight horizontal tubes stacked in vertical fence-like arrays housed in a greenhouse in Klotze, Germany. *Source: Ref. 15*

length. These factors are potential disadvantages of tubular photobioreactors,[16] which would need innovative engineering designs to resolve.

Flat-plate (or flat-panel) photobioreactors are transparent rectangular containers (usually vertical or inclined) with a light path of 1-30 cm (Figure 19.8). Flat-plate photobioreactors mix the liquid medium by vigorous air sparging from the bottom.

Productivities of algal biomass in photobioreactors vary with the type of geometric configuration used and on the algae species grown plus the operating and environmental conditions implemented (Table 19.1). Many novel production systems have been

Figure 19.8
Flat-plate photobioreactors. *Source: Refs. 17,18*

Table 19.1: Microalgae productivities in photobioreactors

Photobioreactor	Volume (L)	Microalgae species	Productivity (g DW L^{-1} day^{-1})	Reference
Airlift tubular	200	*Porphyridium cruentum*	1.5	19
Airlift tubular	200	*Phaeodactylon tricornutum*	1.2	20
Airlift tubular	200	*Phaeodactylon tricornutum*	1.9	21
Inclined tubular	6	*Chlorella sorokiniana*	1.47	22
Helical tubular	75	*Phaeodactylon tricornutum*	1.4	23
Parallel tubular (AGM)	25,000	*Haematococcus pluvialis*	0.05	24
Bubble column	55	*Haematococcus pluvialis*	0.06	25
Bubble column	2-200	*Chlorella ellipsoidea*	0.031-0.048	26
Flat plate	6-50	*Spirulina platensis*	0.3-4.3	27
Flat plate	30	*Chlorella vulgaris*	0.16-0.95	28
Flat plate	440	*Nannochloropsis* spp.	0.27	29
Flat plate	100	*Nannochloropsis* spp.	0.3	30
Accordion	60	*Monodus subterraneous*	0.4	6

designed and are currently being developed and tested. The new production systems aim to lower construction and maintenance costs close to those of open-pond systems and maintain the high, stable productivity and reduced contamination risk of closed photobioreactors. These systems include the Solix, Accordion, and Algenol photobioreactors.

The Solix photobioreactor is an elongated (low height-to-length ratio) flat-panel photobioreactor made of plastic.[31] It is designed to bridge the gap from flask to open-raceway pond by serving as a controlled environment test bed or as an algae inoculum scale-up device (Figure 19.9). The Solix photobioreactor allows for open-pond deployment by using the water as a thermal regulator for open-air field applications. Air sparging for aeration and mixing occurs along the full length of each panel.

Figure 19.9
Solix photobioreactor. *Source: Ref. 31*

The Accordion photobioreactor is an elongated plastic flat-panel photobioreactor with a wave or zig-zag geometry along the vertical axis, with aeration and mixing effected through gas sparging from the bottom (Figure 19.10). Alternatively, the Accordion photobioreactor may also assume a geometry of alternating vertical and angled continuous sections.[6] The waved or angled geometry is designed to improve the light incidence on surfaces and enhance the mixing and flow patterns inside the sections. For instance, a treatment for a 60-L Accordion photobioreactor with 45° plate angle and liquid flow rate of $14\,L\,min^{-1}$ resulted in algal productivity of $0.30\,g\,DW\,L^{-1}\,day^{-1}$ that was statistically indistinguishable from that for a 1-L shake-flask control, indicating successful maintenance of productivity from laboratory scale to an intermediate scale.[32] The Accordion photobioreactor is a modular design that can be scaled up by adding modules. This is equivalent to open-raceway ponds achieving scale-up by adding raceway units.

The Algenol photobioreactor is a low-rise (low height-to-length ratio) flat-panel photobioreactor made of plastic with discrete vertical tubular segments in the central section that are continuous with the lower and upper portions of the panel (Figure 19.11). Gas sparging provides aeration and mixing within the panel.

19.4 Design Performance of Microalgal Photobioreactors

Table 19.2 compares the general performance of the horizontal tubular and flat-plate photobioreactors, the two most common general designs of microalgal photobioreactors, based

Figure 19.10
Accordion photobioreactors.[6]

Figure 19.11
Algenol photobioreactors. *Source: Ref. 33*

Table 19.2: General comparison of tubular and flat-plate photobioreactors, with relative advantage and disadvantage designated by (+) and (−), respectively

Design factor	Horizontal tubular photobioreactor	Flat-panel photobioreactor
External light incidence (surface area/volume)	Greater (+)	Lower (−)
Material cost/volume (surface area/volume)	Greater (−)	Lower (+)
Liquid mixing (nutrients mixing; gas mass transfer)	Lower (−)	Greater (+)
Oxygen accumulation (photo-oxidation)	Greater (−)	Lower (+)
Maintenance (internal cleaning, etc.)	Greater (−)	Lower (+)

on specific design factors. The external light incidence on the tubular photobioreactor exceeds that on the flat-plate photobioreactor by virtue of the generally greater surface-to-volume ratio of the former. The greater surface-to-volume ratio of the tubular photobioreactor, however, serves as a disadvantage in terms of its associated greater material quantity per unit volume, and thus cost per unit volume, relative to that for the flat-plate photobioreactor. Liquid mixing is also generally better achieved in the flat-plate photobioreactor, with liquid mixing achieved more conveniently through gas sparging, than in the long horizontal tubular photobioreactor, wherein liquid mixing through convective liquid bulk flow generally requires more energy. Better liquid mixing in the flat-plate photobioreactor translates into better mixing of nutrients as well as better mass transfer of both CO_2 and O_2 in the liquid medium. Accumulation of photosynthetic oxygen when light is present, which could promote deleterious photo-oxidation of the microalgal culture, often is a concern in the horizontal tubular photobioreactor by virtue of the long residence time of the generated oxygen molecules within the tubes, giving them opportunity to accumulate to hazardous levels before they reach the gas exhaust point of the tube. Cleaning and maintenance of the tubular photobioreactor, particularly when algal adhesion to internal tubular walls occurs, are more onerous relative to those necessary for the flat-plate photobioreactor.

Table 19.3 compares the general design performance of photobioreactors with that of open raceways for photoautotrophic microalgal production. The low costs of construction and maintenance constitute one of the significant advantages of open raceways compared to photobioreactors. The conventional view is that the use of open raceway is more likely to achieve the goal of technoeconomic feasibility for producing microalgae for biofuels than the use of photobioreactors, though this remains questionable given the significant challenge open raceways face in regard to crop protection to avert contamination that could at any time cause crash of the algae culture. Scale-up of production also remains easier to implement with open

Table 19.3: General comparison of photobioreactors and raceways, with relative advantage and disadvantage designated by (+) and (−), respectively

Design factor	Raceways	Photobioreactors	Reference
Cost	Lower (+)	Higher (−)	34
Pumping energy	Lower (+)	Higher (−)	34,35
Ease of scale-up	Greater (+)	Lower (−)	34
Evaporative water loss	Up to $10\,L\,m^{-2}\,day^{-1}$ (−)	Negligible where wind cooling is sufficient (+) $1\text{-}2\,L\,m^{-2}\,day^{-1}$ when water spray cooling is used (+) Same as open system for photobioreactors submerged in ponds for cooling (−)	35
Land area required	Higher (−)	Lower (+)	9
Contamination risks	Higher (−)	Lower (+)	10,34
Productivity	Lower (−)	Higher (+)	See Tables 19.1 and 19.2
Productivity stability	More variable (−)	Less variable (+)	34
Sparged CO_2 loss	Higher (−)	Lower (+)	34,35

raceways than with photobioreactors, given that most photobioreactor configurations are scaled up, not only by multiplying units but also by increasing the unit volume. Increasing the unit volume of photobioreactors requires adjustments of physical variables to achieve appropriate or proportional flow dynamics within the new unit volume. Disadvantages of open raceways include losses of water to evaporation, risk of contamination by competing microorganisms, loss of algal biomass due to weather, and loss to the atmosphere of introduced CO_2.

Advantages of photobioreactors include significantly higher microalgal biomass productivity and greater production stability over time than in open raceways. For example, the volumetric productivity of *Spirulina platensis* in photobioreactors could exceed that in open raceways by as much as 24 times.[27,36]

The utilization efficiency of vital input resources—particularly water and land—are in general higher in closed photobioreactors than in open raceways. As shown in Table 19.3, evaporative water loss is of particular concern in open raceways. These losses could be as much as $10\,L\,m^{-2}\,day^{-1}$. Thus, a one hectare open raceway could lose 100,000 L of water per day or 36,500,000 L of water per year. Nonetheless, when the cooling of photobioreactors is achieved through water spray cooling or through submergence in open ponds, the evaporative water loss associated with photobioreactors can also be significant and can be just as much as in open systems.

The risk of biological contamination is significantly much greater in open raceways than in closed photobioreactors. With the exception of *Spirulina*, *D. salina* and perhaps *Chlorella*, which are cultivated in open systems under highly selective growing conditions, the lack of competitive advantages of many of the microalgae species is being tested for biofuel production in open ponds and their susceptibility to culture crashes are significant concerns.

The principal types of contaminants in algal cultures include other algal species, bacteria, zooplankton, fungi, insects, and viruses. The types and populations of contaminants in algal cultures depend on the local environmental conditions, the algal species cultivated, and the specific cultivation system in use.[35,37,38] Algal cultures that are contaminated with certain species of protozoans, particularly *Monas* spp., are likely to be completely destroyed in 12-18 h after the first detection.[37] The fungi chytrids have been detected on several algae species and often occur as epidemics, which sometimes result in the complete loss of cultures. The fungus *Chytridium* spp. is the most dangerous fungus for cultures of *Chlorophyceae*, and often appears together with the zooflagellate *Aphelidium* spp. Infections of *Scenedesmus* cultures by these organisms have been detected practically worldwide. Such infection is characterized by heavy flocculation of the algal suspension, a brown color of the culture medium, and decreased oxygen evolution.[35] By the time these symptoms appear in the algal culture, it is already too late to control the parasite. The only biological control for *Aphelidium* infection in an early stage is a manyfold dilution of the culture with fresh medium, enabling the algal population to remain in an exponential growth phase so that the algal population's multiplication rate exceeds that of the parasite. Infection of *Scenedesmus* spp. cultures with *Chytridium* spp. has been treated successfully in Israel by applying the fungicide Benomyl (methyl 1-butyl carbomoyl benzimidazolecarbamate) at a dosage of $1\,mg\,L^{-1}$.[35] Fungal contamination of *Chlamydomonas reinhardtii* can be controlled using the fungicides carbendazim ($1\,\mu g\,mL^{-1}$), thiophanate-methyl ($20\,\mu g\,mL^{-1}$), and benomyl ($20\,\mu g\,mL^{-1}$).[38] A combination of carbendazim and the antibiotics ampicillin and cefotaxime has also been shown to remove or reduce contamination of *C. reinhardtii* by three different bacteria and two different fungi tested.[39]

There have been occasional reports of contamination of algal cultures by the zooplanktons *Lycrymanis* spp., *Colpidium* spp. and *Vorticella* spp., though these organisms had negligible effects on algal growth.[35] Contamination by a group of rotifers called *Brachionus*, however, may impede the growth of algal cells and in extreme cases could spoil the entire algal culture. The most effective control has been to lower the pH of the culture to about 3.0 by adding acid and keeping the culture at that pH for 1-2 h before the pH is readjusted back to 7.5 with potassium hydroxide (KOH). The treatment effectively eliminates the rotifers without deleterious effects on the algal cultures.[35]

Thus, the low volumetric productivity and susceptibility to contamination could constitute a substantial risk to the economic sustainability of open raceways compared to closed photobioreactors.

19.5 Microalgal Culture Optimization

Since growth and productivity of microalgal cultures depend on essential production factors, including CO_2, nutrients (e.g., nitrogen, phosphorus, etc.), light intensity, light duration, temperature, liquid mixing, etc., microalgal growth and productivity may be significantly improved by determining the appropriate (or optimal) levels of the foregoing production factors. Various strategies for microalgal culture optimization are well documented in the literature. Table 19.4 lists some examples of optimization studies conducted for specific microalgal species based on selected production factors for various cultivation systems.

The culture optimization for any microalgae species should be tailored to the specific cultivation system or photobioreactor in which it is grown. This is because each photobioreactor has specific design characteristics pertaining to light path, liquid mixing, and aeration, which directly dictate the delivery or availability of light, nutrients, CO_2, and O_2 within the photobioreactor, influencing in turn the growth and productivity of the culture. For instance, an external photosynthetic photon flux (PPF) of 200 μmol m^{-2} s^{-1} would more than likely result in different average internal PPF levels in a tubular photobioreactor and in a flat-plate photobioreactor of comparable volumes, which could result in different productivity levels assuming all other factors to be the same. Indeed, the resulting average internal PPF level in each photobioreactor would also vary from laboratory scale to pilot scale. Consequently, the scale-up process of cultivating a specific algae species in a specific

Table 19.4: Microalgal culture optimization based on selected production factors in various cultivation systems

Production factors	Species	Cultivation method	Reference
Carbon dioxide	*Scenedesmus obliquus*	Flask	40
	Chlorella pyrenoidosa		
	Chlorella sp.	Bubble column	41
	Scenedesmus sp.	Helical tubular, Flat panel	42
	Chlorella sp.		
Nutrients (N, P, etc.)	*Scendesmus* sp.	Solid state sheet	43
	Arthrospira plantensis	Bubble column	44
Light	*Nannochloropsis* sp.	Flat panel	45
	Arthrospira plantensis	Tubular, Flat panel	46
	Nannochloropsis salina	Flat panel	47
	Dunaliella tertiolecta, Chlorella sorokiana	Flat panel	48
Temperature	*Monodus subterraneous*	Tubular	49
	Spirulina platensis	Tubular	50
	Scenedesmus obliquus	Chemostat	51
Hydrodynamic stress	*Arthrospira*	Tubular	52
	Phaeodactylum tricornutum	Bubble column, Airlift	53
	Dunaliella tertiolecta, Chlamydomonas reinhardtii	Bubble column	54

photobioreactor is in effect the stage-by-stage optimization or reoptimization of the culture as the photobioreactor volume goes from laboratory scale to pilot scale to industrial scale with the goal of ensuring that the desired level of microalgal productivity is maintained with the increase in scale (Figure 19.1).

19.6 Concluding Remarks

The scaled-up production of microbial biomass to produce biofuels constitutes an indispensable step in the commercialization process of microbial-based biofuels. Productivities of algal biomass in photobioreactors vary with the type of geometric configuration used (e.g., tubular versus flat panel) and on the algae species grown plus the operating and environmental conditions implemented. The scale-up process for microalgae photobioreactors involves transitioning the volume of production from (1) small laboratory flasks to (2) laboratory-scale photobioreactors to (3) pilot-scale photobioreactors to (4) industrial-scale photobioreactors (Figure 19.1). Regardless of the type of photobioreactor used, the principal goal of the scale-up process is to ensure that the desired level of microalgal productivity is maintained as the volume of production progressively rises. This necessitates that the characteristic hydrodynamic conditions (e.g., degree of mixing) and the accompanying environmental conditions (e.g., lighting) of the specific type of photobioreactor used are maintained or appropriately adjusted with increasing production volume to maintain the desired level of microalgal productivity. Higher microalgal productivity and significantly reduced incidence of deleterious culture contamination are two of the principal reasons why photobioreactors are considered favorable for the scaled-up production of microalgal biomass for biofuel production compared with open raceways.

References

1. Tsukuda O, Kawahara T, Miyachi S. Mass culture of *Chlorella* in Asian countries. In: Mitsui A, Miyachi S, San Pietro A, Tamura S, editors. *Biological solar energy conversion*. New York: Academic Press; 1977. p. 363–5.
2. Duerr EO, Molnar A, Sato V. Cultured microalgae as aquaculture feeds. *J Mar Biotechnol* 1998;**6**(2):65–70.
3. Kawaguchi K. Microalgae production systems in Asia. In: Shelef G, Soeder CJ, editors. *Algae biomass production and use*. Amsterdam, The Netherlands: Elsevier/North Holland Biomedical Press; 1980. p. 25–33.
4. Spolaore P, Joannis-Cassan C, Duran E, Isambert A. Commercial applications of microalgae. *J Biosci Bioeng* 2006;**101**(2):87–96.
5. Benemann JR. *Opportunities and challenges in algae biofuel production*; 2008. Available online at, http://www.fao.org/uploads/media/algae_positionpaper.pdf [accessed 21.10.11].
6. Cuello JL, Hoshino T, Kuwahara S, Brown CL. *Industrial algae production methods in photobioreactors and raceways*. In: Presentation to the global forum for innovations in agriculture on March 11, 2015, Abu Dhabi, UAE; 2015.
7. Earthrise Nutritional L. *Earthrise*; 2009. Available online at, http://www.earthrise.com/ [accessed 09.02.12].

8. Edwards MO. *Phyco biosciences super trough growing algae*; 2010. Available online at, http://www.algaeindustrymagazine.com/abundance-food/ [accessed 15.05.12].
9. Borowitzka MA. Commercial production of microalgae: ponds, tanks, tubes and fermenters. *J Biotechnol* 1999;**70**(1-3):313–21.
10. Milledge JJ. Commercial application of microalgae other than as biofuels: a brief review. *Rev Environ Sci Biotechnol* 2011;**10**(1):31–41.
11. AlgaePARC, Wageningen UR. Available online at, http://www.algaeparc.com/abo-ut.asp [accessed 15.03.15].
12. NanoVoltaix. *Photo-bioreactors*; 2012. Available online at, http://www.nanovoltaix.com/markets/algae.php [accessed 15.05.12].
13. Vunjak-Novakovic G, Kim Y, Wu X, Berzin I, Merchuk JC. Air-lift bioreactors for algal growth on flue gas: mathematical modeling and pilot-plant studies. *Ind Eng Chem Res* 2005;**44**(16):6154–63.
14. EOS. *Algal bioreactors*. Available online at, http://www.cambia.org/daisy/eos/4148.html [accessed 15.05.12].
15. IGV Biotech. *Photobioreactor*; 2012. Available online at, http://commons.wikimedia.org/wiki/File:Photobioreactor_PBR_4000_G_IGV_Biotech.jpg.
16. Chen P, Min M, Chen Y-F, Wang L, Li Y, Chen Q, et al. Review of the biological and engineering aspects of algae to fuels approach. *Int J Agric Biol Eng* 2009;**2**(4):1–30.
17. Algae Energy. *Photobioreactors*; 2012. Available online at, http://algae-energy.co.uk/biofuel_production/pbrs/ [accessed 15.05.12].
18. Arizona State University. *Flat panel algae photobioreactor*; 2012. Available online at, http://asulightworks.com/.
19. Camacho Rubio F, Acién Fernández FG, Sánchez Pérez JA, García Camacho F, Molina Grima E. Prediction of dissolved oxygen and carbon dioxide concentration profiles in tubular photobioreactors for microalgal culture. *Biotechnol Bioeng* 1999;**62**(1):71–86.
20. Acién Fernández FG, Fernández Sevilla JM, Sánchez Pérez JA, Molina Grima E, Chisti Y. Airlift-driven external-loop tubular photobioreactors for outdoor production of microalgae: assessment of design and performance. *Chem Eng Sci* 2001;**56**(8):2721–32.
21. Molina E, Fernández J, Acién FG, Chisti Y. Tubular photobioreactor design for algal cultures. *J Biotechnol* 2001;**92**(2):113–31.
22. Ugwu CU, Ogbonna JC, Tanaka H. Improvement of mass transfer characteristics and productivities of inclined tubular photobioreactors by installation of internal static mixers. *Appl Microbiol Biotechnol* 2002;**58**(5):600–7.
23. Hall DO, Acién Fernández FG, Guerrero EC, Rao KK, Grima EM. Outdoor helical tubular photobioreactors for microalgal production: modeling of fluid-dynamics and mass transfer and assessment of biomass productivity. *Biotechnol Bioeng* 2003;**82**(1):62–73.
24. Olaizola M. Commercial production of astaxanthin from *Haematococcus pluvialis* using 25,000-liter outdoor photobioreactors. *J Appl Phycol* 2000;**12**(3-5):499–506.
25. Garcia-Malea Lopez MC, Del Rio Sanchez E, Casas Lopez JL, Acien Fernandez FG, Fernandez Sevilla JM, Rivas J, et al. Comparative analysis of the outdoor culture of *Haematococcus pluvialis* in tubular and bubble column photobioreactors. *J Biotechnol* 2006;**123**(3):329–42.
26. Wang S-K, Hu Y-R, Wang F, Stiles AR, Liu C-Z. Scale-up cultivation of *Chlorella ellipsoidea* from indoor to outdoor in bubble column bioreactors. *Bioresour Technol* 2014;**156**:117–22.
27. Hu Q, Guterman H, Richmond A. A flat inclined modular photobioreactor for outdoor mass cultivation of photoautotrophs. *Biotechnol Bioeng* 1996;**51**:51–60.
28. Münkel R, Schmid-Staiger U, Werner A, Hirth T. Optimization of outdoor cultivation in flat panel airlift reactors for lipid production by *Chlorella vulgaris*. *Biotechnol Bioeng* 2013;**110**(11):2882–93.
29. Cheng-Wu Z, Zmora O, Kopel R, Richmond A. An industrial-size flat plate glass reactor for mass production of *Nannochloropsis* sp. (Eustigmatophyceae). *Aquaculture* 2001;**195**(1-2):35–49.
30. Rodolfi L, Zittelli GC, Bassi N, Padovani G, Biondi N, Bonini G, et al. Microalgae for oil: strain selection, induction of lipid synthesis and outdoor mass cultivation in a low-cost photobioreactor. *Biotechnol Bioeng* 2009;**102**(1):100–12.
31. Solix Biofuels. *The Lumian™ AGS4000: a high productivity algae growth system*; 2011. Available online at, http://www.solixbiofuels.com/content/products-/lumian-ags4000 [accessed 09.02.12].

32. Cuello J, Ley J. *ACCORDION photobioreactor for algae production of biofuels and biochemicals*. In: Presentation to the Conference EPA's GHG Emissions Standards for Power Plants and Oil Refineries on February 2, 2011; 2011.
33. Algenol Biofuels. *Photobioreactors*; 2015. Available online at, www.algen-ol.com [accessed 15.02.15].
34. Shen Y, Yuan W, Pei ZJ, Wu Q, Mao E. Microalgae mass production methods. *Trans ASABE* 2009;**52**(4):1275–87.
35. Becker EW. *Microalgae: biotechnology and microbiology*. Cambridge, UK: Cambridge University Press; 1994.
36. Richmond A, Lichtenberg E, Stahl B, Vonshak A. Quantitative assessment of the major limitations on productivity of *Spirulina platensis* in open raceways. *J Appl Phycol* 1990;**2**(3):195–206.
37. Baptist G, Meritt D, Webster D. *Growing microalgae to feed bivalve larvae*. North Dartmouth: University of Massachusetts Dartmouth; 1993.
38. Mahan KM, Odom OW, Herrin DL. Controlling fungal contamination in *Chlamydomonas reinhardtii* cultures. *Biotechniques* 2005;**39**(4):457–8.
39. Kan Y, Pan J. A one-shot solution to bacterial and fungal contamination in the green alga *Chlamydomonas reinhardtii* culture by using an antibiotic cocktail. *J Phycol* 2010;**46**(6):1356–8.
40. Tang EPY, Vincent WF, Proulx D, Lessard P, de la Noue J. Polar cyanobacteria versus green algae for tertiary waste-water treatment in cool climates. *J Appl Phycol* 1997;**9**(4):371–81.
41. Chiu S-Y, Kao C-Y, Chen C-H, Kuan T-C, Ong S-c, Lin C-S. Reduction of CO_2 by a high density culture of *Chlorella* sp. in a semicontinuous photobioreactor. *Bioresour Technol* 2008;**99**:3389–96.
42. Westerhoff P, Hu Q, Esparza-Soto M, Vermaas W. Growth parameters of microalgae tolerant to high levels of carbon dioxide in batch and continuous-flow photobioreactors. *Environ Technol* 2010;**31**(5):523–32.
43. Zheng P, Allen WB, Roesler K, Williams ME, Zhang S, Li J, et al. A phenylalanine in DGAT is a key determinant of oil content and composition in maize. *Nat Genet* 2008;**40**(3):367–72.
44. Markou G, Chatzipavlidis I, Georgakakis D. Effects of Phosphorus concentration and light intensity on the biomass composition of *Arthrospira* (*Spirulina*) *platensis*. *World J Microbiol Biotechnol* 2012;**28**:2661–70.
45. Richmond A, Cheng-Wu Z, Zarmi Y. Efficient use of strong light for high photosynthetic productivity interrelationships between the optical path, the optimal population density and cell-growth inhibition. *Biomol Eng* 2003;**20**:229–36.
46. Tredici MR, Zittelli GC. Efficiency of sunlight utilization: tubular versus flat photobioreactor. *Biotechnol Bioeng* 1998;**57**(2):187–97.
47. Sforza E, Simionato D, Giacometti GM, Bertucco A, Morosinotto T. Adjusted light and dark cycles can optimize photosynthetic efficiency in algae growing in photobioreactors. *PLoS ONE* 2012;**7**(6).
48. Zijffers J-WF, Schippers KJ, Zheng K, Janssen M, Tramper J, Wijffels RH. Maximum photosynthetic yield of green microalgae in photobioreactors. *Mar Biotechnol* 2010;**12**:708–18.
49. Vonshak A, Torzillo G, Masojidek J, Boussiba S. Sub-optimal morning temperature induces photoinhibition in dense outdoor cultures of the alga *Monodus subterraneus* (Eustigmatiphyte). *Plant Cell Environ* 2001;**24**:1113–8.
50. Torzillo G, Sacchi A, Materassi R, Richmond A. Effect of temperature on yield and night biomass loss in *Spirulina platensis* grown outdoor in tubular photobioreactors. *J Appl Phycol* 1991;**3**:103–9.
51. Hodaifa G, Martinez ME, Sanchez S. Influence of temperature on growth of *Scenedesmus obliquus* in diluted olive mill wastewater as culture medium. *Eng Life Sci* 2010;**10**(3):257–64.
52. Ferreira LS, Rodrigues MS, Converti A, Sato S. Kinetic and growth parameter of *Arthrospira* (*Spirulina*) *platensis* cultivated in tubular photobioreactors under different cell circulation systems. *Biotechnol Bioeng* 2012;**109**(2):444–50.
53. Sanchez-Miron A, Ceron Garcia MC, Contreras Gomez A, Garcia Camacho F, Molina Grima E, Chisti Y. Shear stress tolerance and biochemical characterization of *Phaeodactylum tricornutum* in quasi steady-state continuous culture in outdoor photobioreactors. *Biochem Eng J* 2003;**16**(3):287–97.
54. Barbosa MJ, Albrecht M, Wijffels RH. Hydrodynamic stress and lethal events in sparged microalgae cultures. *Biotechnol Bioeng* 2003;**83**(1):112–20.

CHAPTER 20

Scale-Up Considerations for Biofuels

David Humbird*, Qiang Fei†
**DWH Process Consulting, Centennial, Colorado, USA †National Bioenergy Center, National Renewable Energy Laboratory (NREL), Golden, Colorado, USA*

20.1 Introduction

Previous chapters discussed advances made at the laboratory scale to increase productivity and titers of various novel bioproducts—proposed liquid transportation fuels. Ultimately, production cost and scalability will draw the line between which can be considered a bioproduct versus a biofuel: specialty chemical produced at modest volume, designed and priced to fit a specific need; or, mass-produced, low-cost energy storage molecule, designed to be burned in an engine.

The incumbent fossil energy industry is extremely large and mature. In 2013, the United States consumed 130 billion gallons of gasoline.[1] By comparison, 13 billion gallons of corn ethanol were produced in 2013, blended into gasoline at 10%. The year 2014 saw the startup of two commercial-scale cellulosic ethanol plants in the United States[2,3] with a combined nameplate capacity of 50 million gallons per year—0.5% of corn ethanol capacity. Measurable displacements by advanced biofuels, such as those proposed here, will require significant advances in fermentation technology at scale.

For each new technology, it is instructive to develop a preliminary cost of production estimate using a very basic analysis of theoretical product yield (physical constraint) and feedstock cost (economic constraint). Consider the theoretical mass yields of various bioproducts from sugar (glucose) shown in Table 20.1.[5] Given a cost of sugar, one can compute a minimum, feedstock-only, cost of production. If this minimum cost is already unpalatable, additional work must be done to increase yield, negotiate a lower feedstock cost, or include external factors like mandates, credits, subsidies, and coproducts in this best-case calculation. For example, while this table assumes a base price of $0.17/lb conventional sugar, it has been shown that sugar from lignocellulosic biomass can cost less in theory,[4] though this depends of course on biomass feedstock cost, cost of cellulase enzymes, and whether or not a technology is able to use all C5 and C6 sugars produced in hydrolysis.

The conventional ethanol industry has been able to reach significant production volumes in part because the scale-up of anaerobic fermentation is rather straightforward. One can find

Biotechnology for Biofuel Production and Optimization. http://dx.doi.org/10.1016/B978-0-444-63475-7.00020-0

Table 20.1: Sugar feedstock cost and minimum O_2 requirements for various proposed biofuels and biofuel precursors

Product	Pathway	HHV (Btu/lb)	Theo. mass yield (%)	Glucose HHV retained (%)	Sugar cost in $/gal equivalent	Min. O_2 req. (lb O_2/lb product)
Gasoline replacements					$/GGE	
Ethanol	Fermentative	12,753	51	97	$3.25	0
Butanol	Fermentative	15,518	41	95	$3.32	0
Isobutanol	Nonfermentative	15,432	41	94	$3.34	0
Isoamyl alcohol	Nonfermentative	16,205	33	80	$3.95	0.54
Diesel replacements					$/GDE	
Farnesene	Mevalonate	19,690	25	73	$4.74	0.94
Farnesene	Deoxyxylulose	19,690	29	85	$4.09	0.39
Ethyl hexadecanoate	Fatty acid	16,936	35	88	$3.94	0.11
Pentadecane	Fatty acid	20,203	29	87	$3.99	0.20

After Rude and Schirmer.[4] Basis cost of sugar is $0.17/lb. Equivalent gasoline HHV = 124,340 Btu/GGE. Equivalent diesel HHV = 137,380 Btu/GDE.

ethanol fermentation vessels ranging from 1 gal in a homebrewer's basement to 1,000,000 gal at fuel ethanol plants in the United States and Brazil. Although extremely large, these fermenters are mechanically quite simple, generally comprising a sanitary tank and an external cooling loop for temperature control. For aerobically produced bioproducts like several of those proposed here, however, things are more complicated. The last column in Table 20.1 indicates a minimum oxygen requirement for some of the products. Regardless of whether the molecule produced contains O atoms, this is the minimum amount of oxygen that must be provided for the organism to enter and sustain the metabolic pathways that create these molecules. Providing the organism with this oxygen presents significant barriers to process scale-up in the form of mechanical complexity and practical equipment size limits. This chapter will focus primarily on some of the specific challenges associated with scale-up of aerobic bioprocesses.

20.2 Aeration in Large-Scale Cultures

20.2.1 Oxygen Mass Transfer and Mixing

Oxygen saturation and agitation are important for maintaining a desirable environment for cell growth and metabolite synthesis during biofuel production.[6,7] Except for obligate anaerobes that are harmed by the presence of oxygen, most microorganisms require oxygen as the terminal electron and/or energy supply for growth and the formation of products. As mentioned earlier, the metabolic pathways involved in producing the relevant molecules have a minimum sustaining oxygen requirement. Failure to deliver enough oxygen to the organism will cause it to shut down that pathway, leading to formation of undesirable side-products and possibly cell shock or death.

At small scale, aerobic bacteria can easily be grown in adequately mixed tubes and flasks using air supplied via the headspace. In large-scale submerged cultures, however, sufficient air/oxygen must be maintained for all living microorganisms by active sparging (bubbling). The stirred-tank reactor (STR) is the de facto aerobic bioreactor technology at larger scale. Specifications, metallurgies, control elements, and other accessories for aerobic STR designs vary depending on the bioproduct and type of cell culture, but the STR envisioned for biofuels production reflects the simplest (i.e., least capital-intensive) combination thereof: a stainless-steel batch reactor with compressor-driven gas sparging, motor-driven agitation by a rotating impeller, and forced-circulation cooling through an external heat exchanger. Figure 20.1 shows a single aerobic STR bioreactor with some of its associated support equipment, such as a cooler on the air compressor discharge and chilled-water system in turn connected to a cooling water system.

While the driving force for oxygen transfer between bubble and broth ultimately comes from the consumption of oxygen by the organism, liquid-gas mass transfer limitations can occur at several points in the transport path.[8] Gas molecules must travel across the bubble-liquid

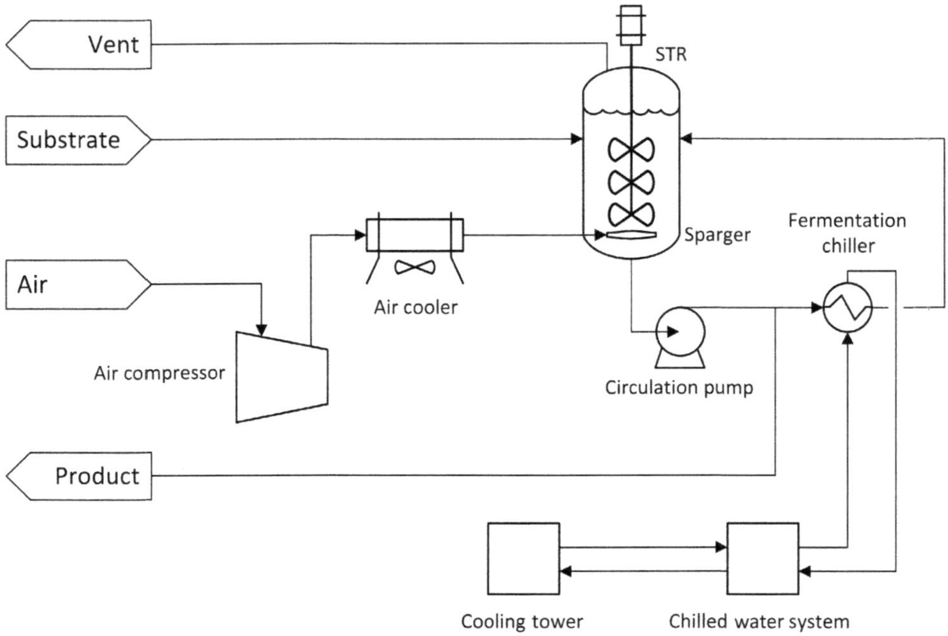

Figure 20.1
Flow diagram of a single STR and support equipment.

interface, through the aqueous phase (culture medium) and subsequently through the cell membrane into the cell body, where they can participate in metabolic reactions.[9] In order to improve the mass transfer in industrial-scale aerobic bioreactors, air is sparged into the STR and the bubbles are sliced and mixed by agitation, increasing the gas-liquid interfacial area.[10] In a perfectly mixed bioreactor, the size of bubbles should be uniform and dissolved oxygen (DO) should be the same anywhere in the bioreactor, including the exit stream. A general approach to generating smaller, uniform bubbles is to raise the agitation speed and/or modify the impeller design. These and other basic concepts for scaling up STRs have been discussed in many textbooks and review articles and will not be repeated here in detail.[11–13] Most of the literature, however, is primarily applicable to scale-up from lab scale to some production scale where "production" typically implies vessels with volumes in the tens of m^3—much smaller than we should realistically consider for a biofuel. A few concepts worth mentioning are

- *Hydrodynamic similarity*. To chemical and mechanical engineers, preservation of hydrodynamic similarity (i.e., constant Reynolds number) may be the first rule to come to mind. This approach is useful for single-phase, flow-only equipment like pipes and heat exchangers but breaks down in the case of the multiphase system (broth/cells/bubbles) where turbulence is aided by a rotating impeller.
- *Constant agitation power/volume ratio*. While generally successful for modest increases in scale, this method will lead to impractically large motor sizes at very large scale.

- *Constant impeller tip speed*. This approach is frequently used for shear-sensitive organisms, which may be damaged or inhibited by excessive agitation.[14] A constant impeller tip speed ensures a constant shear environment. While valid, it could be argued that shear-sensitive organisms in general are not good candidates for scale-up to biofuels.
- *Constant mass transfer coefficient*. Oxygen availability in submerged culture medium is measured in terms of the concentration of DO. Maintaining a homogeneous DO distribution is critical at large scale because oxygen limitation will inhibit cell growth and product formation. Ultimately, maintaining high and constant oxygen mass transfer is probably the most important consideration in scale-up.

As stated earlier, the organism ultimately drives oxygen transfer through metabolism. Oxygen uptake rate (OUR) from the supplied air or oxygen is used to characterize the organism's activity. Biologically, the OUR can be computed by:

$$\mathrm{OUR} = q_{\mathrm{O}_2} \cdot X \tag{20.1}$$

where q_{O_2} is the specific OUR of the cells (mmol/g h) and X is the cell density (g/L). Typical OUR values range from 40 to 200 mmol/g h in most large-scale systems. If the fermentation process is not substrate limited, it can be assumed that the organism's OUR is equal to the oxygen transfer rate (OTR):

$$\mathrm{OTR} = \mathrm{OUR} \tag{20.2}$$

In bioreactor design, OTR can be expressed as the product of a volumetric mass transfer coefficient $k_L a$ (h^{-1}) and a mean oxygen driving force:

$$\mathrm{OTR} = k_L a \left(C^* - C\right)_{\mathrm{MEAN}} \tag{20.3}$$

where C^* and C are, respectively, the equilibrium and the actual DO concentrations (mmol/L). Experimentally, $k_L a$ can be obtained by dynamic methods where, for example, oxygen sparging is stopped during fermentation and the DO response curve is measured.[12] Many correlations have also been proposed which relate $k_L a$ to fundamental design parameters. For STRs, the simplest of these correlations use the gassed power input to the bioreactor (P) per unit volume (V) and the gas superficial velocity within the reactor (u_S):

$$k_L a = K\left(P / V\right)^{\alpha}\left(u_S\right)^{\beta} \tag{20.4}$$

where the factor K and exponents α and β are adjustable for specific systems. More complex forms involving other design parameters and dimensionless groups that describe the reactor can also be found in the literature.

20.2.2 Aeration Process Economics

Put simply, the problem of how to attain "fuel-class" production scales biologically comes down to cost-effective delivery of oxygen. High mass transfer rates at low capital and operating costs are the keys to successful scale-up. We now present some simulations and calculations exploring the economies of scale in aerobic bioprocessing.

First, how many bioreactors do we need? Fortunately, the organism has already decided the total volume for us with its volumetric productivity. If we would like to produce 200,000,000 L/yr (50 MMgal/yr) of a bioproduct with 0.8 specific gravity, and our organism has a productivity of 1 g/L h, then a simple calculation will show that we need 18,000 m^3 of online reactor volume. The largest STRs currently in production are in the hundreds of cubic meters (≤500 m^3). It is postulated that the maximum practical STR size is 1000 m^3, owing to diminishing returns on oxygen transfer relative to volumetric power input, as well as practical limitations regarding the fabrication and maintenance of very large impellers, shafts, bearings, and motors. At biofuels scale, we will naturally want the largest vessel possible. Of course, by designing for the largest vessel possible, we lose economy of scale in the fermentation area—CAPEX must now scale linearly with capacity as we buy more and more identical bioreactors.

We set up a simulation in Aspen Plus to determine total system power demand for the model system shown in Figure 20.1 at different reactor volumes. We assumed STR volumes of 200, 500, and 1000 m^3, with a working volume of 80% and liquid H/D ratio of 2. The OUR was a dependent variable in the simulation, taken to be a volumetric consumption of substrate and oxygen (mmol O_2 consumed per liter per hour) by the combustion reaction:

$$C_6H_{12}O_6 + 6\,O_2 \rightarrow 6\,CO_2 + 6\,H_2O \tag{20.5}$$

The simulation was run at steady state with stoichiometric glucose feed and removal of CO_2 via the vent and removal of excess water via a liquid product stream. The simulated STR was run adiabatically with outlet temperature maintained at 32 °C by circulation through a heat exchanger with 20 °C outlet. All energy from the combustion reaction (approximately 110 kcal/mol O_2) was removed in this loop, along with an assumed 30% dissipation of the agitator power.

To compute the agitation and aeration rates, we used a correlation of the form shown in Equation (20.4). These correlations, as parameterized and published, are in most or all cases not intended for vessel volumes of >100 m^3. However, assuming that real life will not, in fact, be *better* than the correlations predict, we can use them to determine idealized scale-up scenarios and make cost predictions. The specific correlation we have used is[15]:

$$k_L a = 0.002\left(P/V\right)^{0.7}\left(u_S\right)^{0.2} \tag{20.6}$$

with $k_L a$ in s^{-1}, P/V in W/m^3, and u_S in m/s.

For a given OUR, an optimization routine was used to find the minimum total power for an equivalent OTR computed from Equations (20.3) and (20.6). Total power in this case included all contributors shown in Figure 20.1: air compressor, agitator, pump, chiller, and cooling tower (scaled by cooling duty).

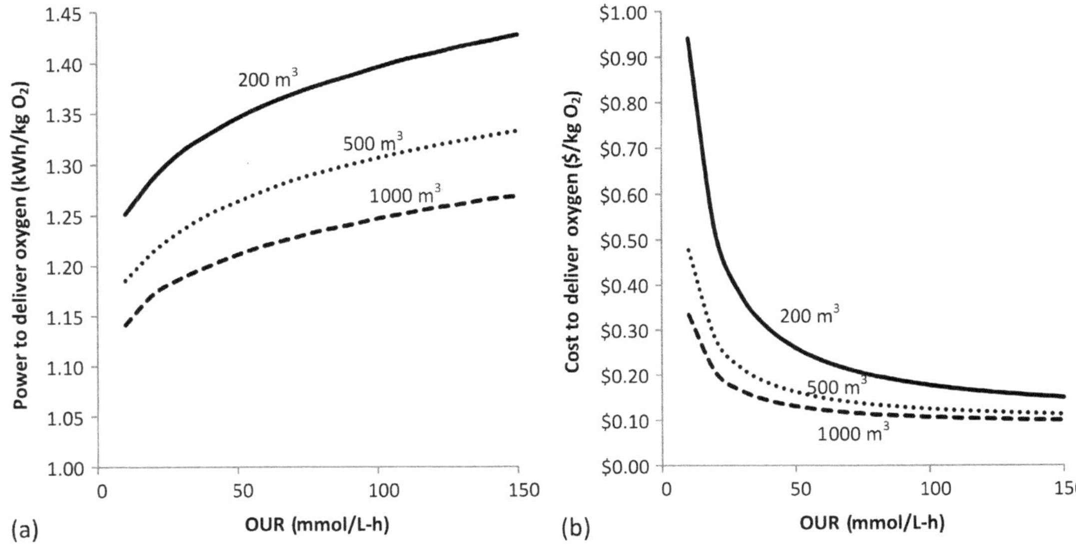

Figure 20.2

(a) Minimum total power demand and (b) aggregate capital and operating cost to deliver oxygen in a STR system at varying vessel volume and OUR.

Figure 20.2a shows the minimum power requirements to deliver a kilogram of oxygen at the three vessel sizes and OURs ranging from 10 to 150 mmol O_2/L h. Figure 20.2a indicates that the total power required to achieve a given OTR will not change significantly with vessel size (the power reduction from 500 to 1000 m^3 is only 5%). In other words, for a given OTR, the total system power will vary approximately linearly with liquid volume, which is determined by productivity as started earlier. This implies that any beneficial economies of scale realized at larger vessel sizes are more strongly a result of reduced capital costs for the vessels and agitators themselves. To investigate those, we placed capital and operating costs on the system shown in Figure 20.1 to arrive at Figure 20.2b, which shows the O_2 delivery cost in terms of \$/kg O_2 as a function of OUR and vessel size. Capital costs for all equipment were computed with Aspen Capital Cost Estimator. The bioreactor and agitator costs were additionally validated against vendor quotes. Operating costs included electricity at \$0.0572/kWh and maintenance as a percentage of the total capital. Note that the costs shown in Figure 20.2b represent the aggregate capital and operating expenses for just the system shown in Figure 20.1.

The curves in Figure 20.2b indicate that there is little difference in the O_2 delivery cost between bioreactor sizes of 500 and 1000 m^3, especially at higher OUR. This is good news considering that 1000 m^3 STRs do not actually exist at this time. Furthermore, it can be inferred from Figure 20.2b that there is probably no incentive to make a bioreactor larger than 1000 m^3 because the economy of scale to be gained will ultimately not pay for the development of such a large machine. We should point out that the aggregate cost in Figure 20.2b does not explicitly include piping, control systems, and manpower, which should naturally increase with higher number of (smaller) reactors. However, these are generally small effects.

20.2.3 Reducing Metabolic Oxygen Demand

Although much effort has gone into the design of bioreactors with high OTR, reducing the organism's demand for oxygen by manipulating metabolic pathways and carbon flux balancing can also play a role in reducing the cost of aerobic bioprocesses.

Microbial pathway modifications have been developed and used as an efficient approach of producing both current and promising biofuels from various carbon sources, including agriculturally prevalent sugars, lignocellulose, and methane.[16–19] In aerobic bioprocesses, one sugar molecule is first converted to two pyruvate molecules along with two ATP and four electrons (2 NADH) via the Embden-Meyerhof-Parnas (EMP) pathway, the most common type of glycolysis. In the presence of oxygen, pyruvate is oxidized to acetyl-CoA, where it can enter a number of metabolic and anabolic reactions. During this step, four electrons are generated from two pyruvate molecules.[20] Since oxygen molecules are required as electron acceptors at the end of the electron transport chain, any changes that result in fewer electrons generated or used in reactions might reduce the total oxygen demand in aerobic cultivation processes. Recently, a nonoxidative, cyclic pathway, termed nonoxidative glycolysis (NOG), was demonstrated both *in vivo* and *in vitro* in *Escherichia coli* for the production of stoichiometric amounts of two-carbon (C2) metabolites from hexose, pentose, and triose phosphates with a 100% carbon yield.[21] NOG enables not only complete carbon conversation from sugars to acetyl-CoA but also zero electron generation in the sugar catabolism compared to other glycolysis pathways such as EMP, Entner-Doudoroff, and phosphoketolase (Table 20.2). The NOG pathway therefore has the potential to decrease the amount of oxygen required in aerobic cultivation. However, the experimental result relative to oxygen utilization in NOG is not available yet.

Similarly, hypothetical pathways have been studied to reduce the oxygen demand in the so-called biological conversion of natural gas into liquid fuel (Bio-GTL) processes, which

Table 20.2: Comparison of electron generated from the conversion of glucose to acetyl-coA in various pathways

Pathway	Number of electrons (e^-) generated	Net ATP
Embden-Meyerhof-Parnas	8 e^-	2
Entner-Doudoroff	8 e^-	1
Phosphoketolase	4 e^-	−1
Nonoxidative glycolysis	0 e^-	−1

convert methane or biogas to liquid fuels.[9,22,23] Methanotrophic bacteria are capable of utilizing C1 compounds (methanol and/or methane) as their sole carbon and energy source to produce precursors of various biofuels.[9] Several routes have been postulated to improve the efficiency of aerobic methane bioconversion.[8,9] By engineering a dioxygenase-like enzyme to activate methane, the energy efficiency of methane to methanol can be improved by 20% and the oxygen demand for this step can also be reduced by 50%.[22] Another new concept to produce butanol from methane has been reported as well, which could improve both energy and carbon conversion efficiency.[9] These new proposed routes may potentially reduce the number of electrons required for the oxidation of methane to methanol from 2 to 0 by using benzene dioxygenase (BDIOX)-like enzymes instead of methane monooxygenases.

Although a number of new pathways and routes have been proposed and tested to reduce the oxygen demand in aerobic cultivation processes, experimental results are still limited at this stage. Therefore, more investigations are needed to prove these new concepts in the future.

20.3 Other Operating Concerns for Large-Scale Cultures

20.3.1 Nutrients

Microorganisms require a source of nutrients and energy for the biosynthesis of metabolic intermediates and desired products, cell maintenance, and reproduction. A simplified equation for biosynthesis, showing major nutrients for the cell growth and product formation, is

$$\begin{aligned}&\text{Carbon source} + \text{Nitrogen source} + O_2 + \text{Minerals} + \text{Trace elements} \rightarrow \\ &\text{Cellular material} + CO_2 + H_2O\end{aligned} \tag{20.7}$$

In addition to the metabolic energy, which can be obtained from a variety of sources, depending on the microorganism and the cultural conditions, microorganisms also need other nutrients for the synthesis of cellular materials and products. These nutrients can be classified as macronutrients including carbon, nitrogen, phosphorus, sulfur, potassium, and magnesium and micronutrients including vitamins and trace elements such as iron, zinc, manganese, copper, molybdenum, sodium, etc. Tables 20.3 and 20.4 itemize the physiological function of various macro- and micronutrients.

Traditional carbon sources such as carbohydrates, hydrocarbons, alcohols, and organic acids are excellent sources of carbon for energy and biosynthetic reactions. In aerobic cultivation, 50% of carbon substrate is used for cell mass, compared to 30% in anaerobic fermentations.[24] Carbohydrates used in industrial fermentations are simple sugars or short polysaccharides like molasses, starch, and corn syrup. Next-generation carbohydrate sources will include cellulose and sulfite waste liquor. Various biofuels (e.g., bioethanol, biobutanol, biodiesel, and jet fuel) have been produced from sugars derived from agricultural feedstocks,[25,26] lignocellulosic biomass,[27,28] and wastes[29] in order to reduce the production cost. Saturated and unsaturated hydrocarbons can also be used metabolically. Recently, methane has attracted considerable

Table 20.3: Physiological function of major macronutrients[24]

Macronutrients	Symbol	Function
Carbon	C	Constituent of organic cell materials
Oxygen	O	Constituent of cellular water and organic materials, as O_2, electron acceptor in respiration of aerobes
Nitrogen	N	Constituent of proteins, nucleic acids, and coenzymes
Phosphorus	P	Constituent of phospholipids, coenzymes, and nucleic acids
Sulfur	S	Constituent of cysteine, cystine, methionine, and proteins as well as some coenzymes as CoA and cocarboxylase
Magnesium	Mg	Important divalent cellular cation, inorganic cofactor for many enzymatic reactions, including those involving ATP; functions in binding enzymes to substrates and present in chlorophylls
Potassium	K	Principal intracellular cation, cofactor for some enzymes

Table 20.4: Physiological function of major micronutrients[24]

Macronutrients	Symbol	Function
Calcium	Ca	Important cellular cation, cofactor for enzymes as proteinases
Cobalt	Co	Constituent of vitamin B_{12} and its coenzyme derivatives
Copper	Cu	Inorganic constituents of special enzymes
Iron	Fe	Constituent of cytochromes and other heme or nonheme proteins, cofactor for a number of enzymes
Manganese	Mn	Inorganic cofactor cation, cofactor for enzymes as proteinases
Molybdenum	Mo	Inorganic constituents of special enzymes
Zinc	Zn	Inorganic constituents of special enzymes

attention as a carbon and energy source for the production of diesel fuel due to the surge of shale gas production.[9] Sugar alcohols like glycerol have been utilized as a low-cost carbon source for the production of biodiesel, bioethanol, and biohydrogen in fermentation bioprocesses.[30,31] Volatile fatty acids have also been explored as carbon sources for the biodiesel production in batch and fed-batch cultures.[32–34]

Following the carbon substrate, nitrogen source is the next most plentiful nutrient in cultivation media. Cell biomass is typically 10-14% nitrogen on a dry weight basis, in the form of proteins and nucleic acids in cell bodies.[35] The nitrogen source is always considered a key nutrient for the production of lipids and fatty acids, which are precursors for biodiesel production.[33,36–38] Ammonia and ammonium salts are the most widely used inorganic nitrogen sources in industrial fermentations due to their low price.[39] Urea is another cheap organic nitrogen source used by some microorganisms in synthetic media. More expensive organic nitrogen sources include yeast extract, peptone, amino acids, and purines. Nitrate salts are a favorable inorganic nitrogen source for the lipid production in the culture of algae,[40] but their high cost presents a significant obstacle at large scale. Some complex nitrogen sources derived from agricultural and food-processing by-products are also used in commercial cultivations to reduce production costs.[41]

Micronutrients are not only present in smaller concentrations but also have significant influence on cell growth rate and productivity.[42] The most commonly required trace elements during cell growth are iron, zinc, and manganese, which serve in coenzyme functions to catalyze many reactions, execration of primary metabolism, and regulation of secondary metabolism. Other trace elements required under specific growth conditions include copper, cobalt, and molybdenum, which contribute to both primary and secondary metabolite production and play roles in vitamin synthesis and cell wall transport. Apart from these mineral trace elements, chelating agents or ligands are also required in trace amounts to form soluble compounds with normally precipitating ions in culture media. The most widely used chelating agents in culture media include EDTA, vitamin B_{12}, citric acid, polyphosphates, tyrosine, and cysteine.

20.3.2 Sourcing Media at Scale

In biofuels, the life-cycle assessment (LCA) is typically thought of as a confirmation of sustainability, in the sense of carbon neutrality or energy efficiency. LCA can also be used to confirm that the demand for a required input does not exceed global supply. Individually, the nitrogen source, trace elements, and other media constituents described in the previous section are quite expensive and have limited production rates. At larger scale, one may wish to replace the defined media used in the lab with cheaper complex media that contain several of these constituents.

As a brief case study, we consider the proposed use of corn steep liquor (CSL) as inexpensive complex media for the cellulosic ethanol industry. CSL is a byproduct of the corn wet milling process, which produces sweeteners and starch. CSL is high in protein, vitamins, and minerals and is typically concentrated to 50% solids and mixed with other wet mill by-products to produce gluten feed for cattle.[43] A limited amount of CSL is sold outside the mill to pharmaceutical and enzyme production plants, as a nutrient source.

From wet mill technoeconomic analyses published elsewhere,[44] we can determine that 0.23 pounds of CSL are generated per pound of starch, of which roughly 50 billion pounds are produced annually,[45] meaning that there is a maximum CSL availability of about 11 billion pounds per year. Although outdated, one set of cellulosic ethanol production projections comes from the renewable fuels mandate in the Energy Independence and Security Act of 2007 (EISA),[46] which called for 16 MMgal/yr of cellulosic ethanol by 2022. Making some assumptions about the percentage of that ethanol produced by fermentation and about the addition rate of CSL in those fermentations, we can arrive at the range of CSL demand shown in Figure 20.3.

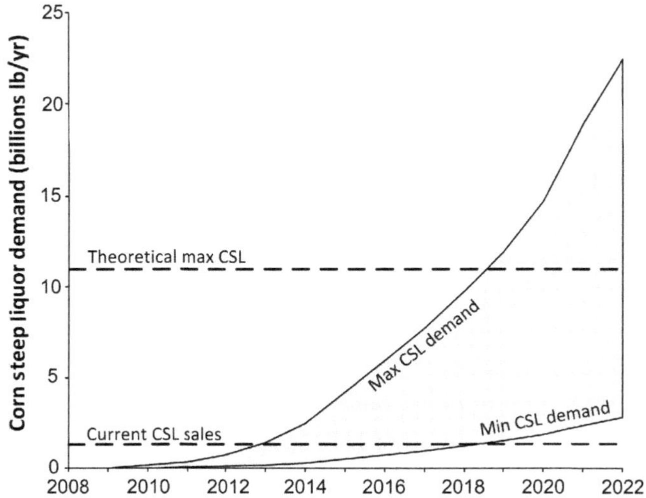

Figure 20.3
Min/max CSL demand predictions for biochemical ethanol.

Current CSL sales (~1.5 billion pounds per year)[45] and the theoretical maximum CSL production are also plotted on the demand curve in Figure 20.3. It becomes clear that while CSL may be a convenient and abundant resource in the short term, it is unlikely that CSL could scale to support cellulosic ethanol production at the mandated levels, without a simultaneous expansion in the starch business. It is also important to note that CSL is not technically a waste product, i.e., it already has a market (gluten feed). Even in the minimum demand case, a significant amount of CSL would have to be diverted from its current market outlet.

We further note that the trace minerals and nutrients obtained from complex media like CSL are not completely lost after fermentation—they are merely tied up in cell bodies. It is our opinion that nutrient recovery and recycle in fermentation is therefore a key area in which additional research can greatly enable biofuels scale-up.

20.3.3 Temperature

Culture temperature affects cell growth rate, productivity of desired products, oxygen solubility, and ultimately the production cost.[33,37] The minimum and maximum temperatures that define limits of growth and development of a particular microbe are called the cardinal temperatures. Between these exists an optimum temperature where the growth rate is highest. Although some microorganisms are able to grow at temperature below 0 °C or above 90 °C, the optimum temperature for most microorganisms is in the range of 20-40 °C.[47]

In process optimization, it is important to realize that temperature may influence cell growth rate and product synthesis rate independently. These two objective functions must be optimized accordingly during a bioprocess development. For instance, a temperature shift is used to induce the accumulation of lipids and/or fatty acids in the cultivation of various microbes (bacterial, yeasts, and microalgae) for the production of biofuels.[48–52]

In lab- and small-scale bioprocessing, reactor temperature is generally controlled with a jacket and circulating coolant. The ratio of heat transfer area to the reactor volume in a jacket becomes insufficient at medium scale, so internal coils are used instead.[53] In some reactor configurations, these coils double as agitation baffles. At the biofuel-scale, however, neither option will have enough surface area to adequately control the culture temperature, requiring a forced circulation loop through external heat exchangers, as shown in Figure 20.1.

Heat removal can be substantial for a large-scale operation,[9] and higher temperatures are usually preferred for commercial production to minimize cooling capital. Cooling water delivered through an evaporative cooling tower can nominally maintain a culture at 38 °C (100 °F); anything lower requires a refrigeration loop providing chilled water or glycol mixture, and a cooling tower of equal duty, as also shown in Figure 20.1.

20.3.4 pH

Regulation of culture pH is a significant and essential process for the growth and metabolism of microbes in bioprocessing.[54,55] Most microorganisms in cultures have an optimum pH at or close to neutral (6.5-7.5).[33,56] Yeasts prefer pH values of 4-5 but will tolerate a wide range of pH values, e.g., 3-8.5. Maintaining a relatively low pH in yeast cultures can prevent bacterial contamination, providing an advantage for yeast over bacteria in industrial fermentations. In any case, the pH of a culture generally has to be maintained within a narrow range to ensure optimum cell growth.

Monitoring of pH in industrial fermentations can provide qualitative and quantitative insight into microbial activities at a given point in the process. For example, the pH of a culture medium can change significantly during different growth periods due to differential uptake of nutrients (mainly nitrate or ammonium ions) and/or through H^+ fluxes.[57] It has also been accepted by many that uptake of NH_4^+ will generate H^+, causing a significant rapid decrease in pH, whereas NO_3^- metabolism will consume H^+, increasing the pH of a medium. The regulation of pH in large-scale bioprocesses is commonly performed by rapid replacement of the buffered medium (in continuous cultures) or the addition of a neutralizing agent to the medium (in batch and fed-batch cultures).

20.4 Process and Culture Optimization Strategies

Fermentation media and culture conditions can significantly affect the production cost, productivity, and quality of desired products.[9] Industrial scale-up must inevitably include some medium design efforts, where key culture conditions (temperature, pH, and agitation) are optimized for cell growth, product formation, and energy efficiency.

Methodologies and techniques that have been used for culture optimization include single-factor design, full factorial design, fractional factorial design, response surface methodology (RSM), artificial neural network (ANN), genetic algorithms (GAs), etc. There are two different strategies for improving fermentation performance, open and closed strategy.[58] A closed strategy normally considers a limited number of factors in optimization. An open strategy is more complex because it considers the best combination of all possible factors available. Most research starts with a closed strategy based on well-known medium recipes and culture conditions for selected microorganisms. Regardless of which optimization strategy is chosen, the commercial viability of medium components and the feasibility of scale-up based upon small-scale experimental data should also be considered during the fermentation optimization process. In this section, approaches to medium design and process optimization for biofuel production are discussed.

20.4.1 Conventional Design

One-factor-at-a-time method (OFAT), also known as single-factor design, is a conventional method of designing experiments by keeping all factors but one constant and changing a factor to be investigated over a desired range to determine its optimum level. The OFAT has been widely used for optimization problems of biofuel production due to its easy operation and analysis.[59] Based upon the knowledge of optimal nutritional requirements and/or metabolic pathways of a particular microorganism, a single factor can be optimized.[60,61] However, there are disadvantages to OFAT, which limit its application on optimizing biotechnological processes. First, OFAT experiments ignore interactions between factors, which may lead to false optimal conditions, especially when interactions among different factors are significant. Second, it involves numerous sequential experimental runs, a time-consuming strategy to study each factor's optimum, especially when a large number of factors need to be optimized. Taking into account the aforementioned drawbacks, OFAT is generally not a satisfactory approach to fermentation development.

20.4.2 Design of Experiments

An alternative to OFAT method is the design of experiments (DoE) proposed by Ronald A. Fisher,[62] which shows that changing two or more factors simultaneously can be more efficient than changing one at a time. Many biofuel production studies claim substantial improvement in culture conditions and media obtained by using DoE.[63–68] The development and optimization of culture media has attracted most attention in previous reports,[69,70] in which the most effective medium components are selected and their optimal levels are investigated. Nowadays, more process control parameters such as reaction time, pH, temperature, inoculum, agitation, and DO are integrated with medium factors in the DoE to provide a deeper understanding of bioprocesses.[71–74] Based upon the basic theory of the DoE, a number of approaches have been developed to find the optimum quickly and efficiently.

20.4.3 Statistical Experimental Design

Statistical experimental design plays an important role in the development of fermentation bioprocesses by screening main factors of interest and then optimizing these to improve industrial process performance.[75] Typical factors for industrial scale-up fermentations are listed in Table 20.5. A full factorial design can test every combination of each factor level, but this generally leads to an impractical number of experimental runs. A common two-level full factorial experimental design requires 2^n runs for n factors. Three or more levels require even more runs.[76,77] Therefore, prescreening a large number of factors such as medium components and control parameters is necessary. By eliminating certain factors that lack significant impact on fermentation performance, others are retained and studied more thoroughly. Fractional factorial design (e.g., Plackett-Burman, Taguchi, central composite, and Box-Behnken) has been widely used to accomplish this goal. Plackett-Burman design (PBD),[78] which is a two-level fractional factorial design, can examine up to 99 factors with 100 experimental runs. Taguchi design (based on PBD) can detect the effect of a few factors with two or more levels based on a response from a relatively small number of experimental runs and has been widely used in industry.[79] Both designs have been commonly employed to screen important factors for further investigation in biotechnology process studies.[64,68,80,81]

Once key factors are identified in initial experiments, it is necessary to determine the suitable ranges of the selected factors based on experience or using the steepest ascent method[82] for further optimization work. After identifying the significant parameters and their regions of optimal response, optimum culture conditions are obtained through more complex experimental designs such as central composite design (CCD) and Box-Behnken design (BBD). The use of CCD and BBD allows the determination of levels of various factors to be carried out with the interrelation between factors to fit a second-order polynomial model.[83] Kumar et al.,[84] for example, used a PBD to identify the olive oil, Tween 80 (surfactant), and KH_2PO_4 as key factors from 12 variables in a culture medium and used a CCD to study the effects of these three factors on lipase yield. The final optimized conditions provided a 3.25-fold increase in lipase production. In Saraphirom's study,[85] the optimal level of total sugar, initial pH, and $FeSO_4$ obtained from a PBD and a BBD provided fivefold higher biohydrogen production.

Table 20.5: Typical factors in a fermentation process

Medium component	Process control parameter
Carbon source	pH
Nitrogen source	Temperature
Carbon/nitrogen ratio	Agitation rate
Phosphate level	Dissolved oxygen level
Sulfur level	Back pressure
Magnesium level	Power input
Trace element	Inoculum

20.4.4 Response Surface Methodology

RSM, also known as Box-Wilson methodology,[86] is an optimization method applying statistical techniques based upon the factorial designs of CCD and BBD. RSM approach will not only determine optimum conditions from a minimal number of experiments but also give the information to estimate results in order to design a process. This has been successfully applied to optimize medium compositions and bioprocess control parameters in biofuel production.[31,71,80,87] RSM is a systematic and efficient procedure for studying interaction effects of various parameters by following the path of steepest ascent and locating a stationary point.[58] Analysis of variance (ANOVA) is always required to determine the significance and adequacy of mathematical models proposed by RSM. With correct models, different combinations of factors are able to lead to desired responses, which will be visually shown in three-dimensional surface plots. As shown in Figure 20.4, the sensitivity of responses over a region of key factor levels can be easily learned. Zabeti et al.[88] investigated the effect of precursor dosage, precursor type, and temperature on diesel production using RSM. Several response surface plots were generated to exhibit the effect of three factors on the biodiesel production yield. The results of an ANOVA indicated that temperature and precursor dosage had interactive effects on biodiesel production. In addition, with the help of the regression model, the highest biodiesel yield of 94% was obtained with the 100.54 wt.% of the precursor followed by calcination at 718 °C.

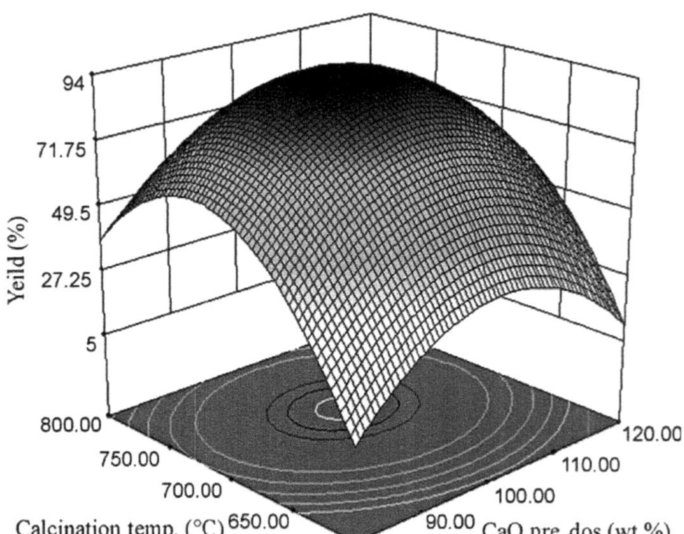

Figure 20.4
Response surface plot in the studied range of the variables for the predicted yields.[88]

RSM is an empirical modeling process derived from prior knowledge of the processes to achieve statistical model,[89] which is always complemented by a mechanistic model. Unlike empirical models developed based on the evaluation of the relationship of a set of controlled experimental factors and observed results (e.g., pH, DO, agitation, and temperature), mechanistic models are built on mass, heat, and momentum balance supplemented with mathematical formulation and equation of mechanisms (such as cell growth kinetics). During a mechanistic modeling process, prior knowledge of the processes is not mandatory. However, experimental validation is required for the most promising case from the simulations.[90] The mechanistic model has been extensively applied for bio-based processes to estimate performance of future production considering the scale of operation, the process design, and control parameters.[91–95] A recent biofuel production-related example illustrating the practical use of mathematical models demonstrated the prediction of composition of bio-oil from fast pyrolysis of biomass under different reaction conditions.[96] The model was developed based upon the fast pyrolysis reactions of cellulose, which can also be substituted for the fast pyrolysis of maltohexose, cellobiose, and glucose. It was illustrated in practice as well, by utilizing this model to optimize the timescale of the formation of major products and their qualities.

20.4.5 Other Approaches

In light of the rapid developments in computing power and inexpensive software, ANN, fuzzy logic (FL), and GA methods have also been applied for the optimization of bioprocesses. These approaches, all of which may be considered as examples of artificial intelligence techniques, are widely used not only for bioprocess control but also for optimization of culture media. ANN is a model that simulates the neurological function of a brain[97] and requires minimal knowledge of the biological system. It is able to describe interactive effects of various factors of a complicated process via a black box. The nonlinear regression and discrimination models derived from ANN are good at dealing with large amounts of data and complex pattern recognition, which makes them very suitable for medium design.[98–100] However, disadvantages of ANN, such as low tolerance of missing data, overtraining, black box process, and the requirement of large amount of data limit its development.[58,101] FL, which is a multivalued logic executing a series of rules, has also been explored for the regulation of fermentation process[102] and medium design.[58,103] FL is very tolerant of noisy and missing data, but it needs a very high mathematical expertise. A system integrating ANN and FL has exhibited considerable advantages by predicting tiny error results without attempting any experiments in a relatively short period of time.[104,105]

GA, which imitates natural evolution and natural selection to find solutions to a wide variety of problems, has established itself as a powerful optimization technique to solve problems with objective functions that are not continuous or differentiable. In recent decades, GA based on a neural network model has been applied successfully to optimize complicated bioprocesses,[79,80] suggesting it is a better optimization method than RSM. Using GA for medium optimization can be understood as follows.[106] Conditions for each experiment, such as medium composition, are coded in one "chromosome," where each medium component in the defined concentration represents one "gene." After completing the first-round generation of experiments, "chromosomes" with the highest productivity are chosen and replicated proportionally to the productivity. After replication, the crossover of "chromosomes" and the mutation of some randomly selected "genes" are performed. This process is kept running until a suitable result is achieved or no significant improvement is found anymore. It has also been experimentally verified that using a GA can improve the performance of bioprocess for both cell growth and product formation despite the large number of parameters under simultaneous consideration.[70,107–109] For instance, Na et al.[110] utilized different fed-batch culture strategies as chromosomes and different control parameters (feed rate, the amount of inoculum, and culture volume) as genes to improve cell growth rate and product titer. The advantage of GA is that it can handle a large number of components, and no new guessing is required at each round of experiments. However, the GA approach requires more domain-specific knowledge and more computational efforts to exactly evaluate a large population than a small one.[111,112]

Recently, a novel experimental design approach, named definitive screen design (DSD), was introduced by Jones and Nachtsheim,[113] which is more efficient and economical as compared to RSM. DSD can allow characterization and optimization of a bioprocess with a single experiment by screening important factors (continuous factors with three levels and nominal or categorical factors with two levels) and estimating the quadratic effect and two-way interaction simultaneously.[114] For example, a 6 factor, 17 run DSD was used to optimize a biomanufacturing process.[115] The resulting model from this DSD was successfully used to simulate the variability in the inputs expected on commercial-scale processes.

20.4.6 Optimization Tools

Software

Due to the requirement of programming skills and rapid analysis of data, commercial software has been used widely to find the best possible statistical models to predict the fermentation process behavior. These modern statistical packages have done a lot to make experimental design the status quo for the fermentation industry. Models derived from the software are used for the optimization and sensitivity analysis of fermentation processes. So far, several commercial packages including JMP (SAS Institute Inc., Cary, NC, USA), Design Expert (STAT-EASE Inc., MN, USA), MATLAB (The MathWorks, Inc. MA, USA), and Minitab (Minitab Inc., State College, PA, USA) have been applied to optimize the culture medium and bioprocess performance for biofuel production.[116–122] These packages are capable of conducting the above-mentioned optimization approaches (DoE, RSM, ANN, DSD, etc.)

and their analysis. The availability of commercial software with unique characters can let the users make choices based upon their needs.[123]

Hardware

The development of a biotechnological bioprocess is usually performed in bench-scale bioreactors with typical volumes of between 0.5 and 10 L to imitate large-scale cultivations. However, the relatively small number of experiments carried out in lab-scale bioreactors and the massive input of labor and raw materials limit the optimization of culture conditions. Therefore, tools for the scale-down design have been developed with requirements for cost-effective, rapid process development, and optimization as well as good understanding of critical process parameters. The parallelization of biological processes in a small-scale reactor system has attracted great attention in recent years.[124–126] Low-cost microbioreactors for high-throughput bioprocessing give them a leading position in the application of exiting scale-up methodologies.[127–130] This high-throughput system is able to provide geometric similarity to stirred-tank bioreactors, such as the control of process parameters (agitation, DO, and pH) during a culture, the capability of automated feeding, and the possibility of sampling.[131,132] However, it should be admitted that results from different scale bioprocesses could lead multiscale problems on each individual scale. The concept of high-throughput systems using microbioreactors is to provide an optimal process design solution in terms of optimum operation conditions and from economic point of view.

20.5 Concluding Remarks

In this chapter, we reviewed some technical and economic challenges facing the scale-up of such aerobic processes. It was shown that STRs exhibit limited economy of scale with respect to cost of oxygen delivery to cells, and there is probably a practical and economic maximum bioreactor size around 1000 m^3—a fraction of the volume of typical ethanol production fermentors. Various alternative bioreactor configurations have been proposed such as microbubble sparged reactors,[133] bubble column reactors,[134,135] trickle-bed reactors,[10] and membrane-based reactors.[136–138] In some cases, these may be less capital-intensive than STRs and may therefore provide more cost-effective oxygen delivery. Moreover, application of computational fluid dynamics (e.g., Euler-Lagrange approach) to scale-up and process design of all bioreactor configurations has shown promise toward precise predictions of flow and reaction parameters.[94,132,139,140]

Reducing an organism's oxygen demand through metabolic engineering can improve production economics at large scale. Advanced culture and process optimization strategies can also be beneficial. Ultimately, the potential market impact of the novel biotechnologies presented in previous chapters depends in large part on their ability to scale to huge production volumes. When many competing technologies are in play, prudent researchers should consider these scale-up and technoeconomic principles at increasingly early stages of development.

References

1. Energy Information Administration. *Annual energy outlook 2014*. US Department of Energy. Washington, DC; 2014. http://www.eia.gov/forecasts/aeo/.
2. Abengoa Bioenergy. Abengoa celebrates grand opening of its first commercial-scale next generation biofuels plant. Kansas: Hugoton; 2014.
3. POET-DSM. First commercial-scale cellulosic ethanol plant in the U.S. opens for business. Iowa: Emmetsburg; 2014.
4. Humbird D, Davis R, Tao L, Kinchin C, Hsu D, Aden A, et al. Process design and economics for biochemical conversion of lignocellulosic biomass to ethanol. 2011;**303**:275–3000.
5. Rude MA, Schirmer A. New microbial fuels: a biotech perspective. *Curr Opin Microbiol* 2009;**12**(3):274–81.
6. Chisti Y. Biodiesel from microalgae beats bioethanol. *Trends Biotechnol* 2008;**26**(3):126–31.
7. Li Q, Du W, Liu D. Perspectives of microbial oils for biodiesel production. *Appl Microbiol Biotechnol* 2008;**80**(5):749–56.
8. Munasinghe PC, Khanal SK. Syngas fermentation to biofuel: evaluation of carbon monoxide mass transfer and analytical modeling using a composite hollow fiber (CHF) membrane bioreactor. *Bioresour Technol* 2012;**122**:130–6.
9. Fei Q, Guarnieri MT, Tao L, Laurens LML, Dowe N, Pienkos PT. Bioconversion of natural gas to liquid fuel: opportunities and challenges. *Biotechnol Adv* 2014;**32**(3):596–614.
10. Bredwell MD, Srivastava P, Worden RM. Reactor design issues for synthesis-gas fermentations. *Biotechnol Prog* 1999;**15**(5):834–44.
11. Junker BH. Scale-up methodologies for Escherichia coli and yeast fermentation processes. *J Biosci Bioeng* 2004;**97**(6):347–64.
12. Garcia-Ochoa F, Gomez E. Bioreactor scale-up and oxygen transfer rate in microbial processes: an overview. *Biotechnol Adv* 2009;**27**(2):153–76.
13. Oldshue J. Fermentation mixing scale-up techniques. *Biotechnol Bioeng* 1966;**8**(1):3–24.
14. Kadic E. Survey of gas–liquid mass transfer in bioreactors. MS Thesis. Department of Mechanical Engineering, Iowa State University, IA, USA; 2010.
15. Van't Riet K. Review of measuring methods and results in nonviscous gas–liquid mass transfer in stirred vessels. *Ind Eng Chem Process Des Dev* 1979;**18**(3):357–64.
16. Fortman J, Chhabra S, Mukhopadhyay A, Chou H, Lee TS, Steen E, et al. Biofuel alternatives to ethanol: pumping the microbial well. *Trends Biotechnol* 2008;**26**(7):375–81.
17. Dugar D, Stephanopoulos G. Relative potential of biosynthetic pathways for biofuels and bio-based products. *Nat Biotechnol* 2011;**29**(12):1074–8.
18. Matsen JB, Yang S, Stein LY, Beck D, Kalyuzhnaya MG. Global molecular analyses of methane metabolism in methanotrophic alphaproteobacterium, Methylosinus trichosporium OB3b. Part I: transcriptomic study. *Front Microbiol* 2013;**4**(40):1–16.
19. Darzins A, Pienkos P, Edye L. Current status and potential for algal biofuels production. A report to IEA Bioenergy Task, 39; 2010.
20. Lee Y, Lafontaine Rivera JG, Liao JC. Ensemble modeling for robustness analysis in engineering non-native metabolic pathways. *Metab Eng* 2014;**25**:63–71.
21. Bogorad IW, Lin T-S, Liao JC. Synthetic non-oxidative glycolysis enables complete carbon conservation. *Nature* 2013;**502**(7473):693–7.
22. Conrado RJ, Gonzalez R. Envisioning the bioconversion of methane to liquid fuels. *Science* 2014;**343**(6171):621–3.
23. Haynes CA, Gonzalez R. Rethinking biological activation of methane and conversion to liquid fuels. *Nat Chem Biol* 2014;**10**:331–9.
24. Zabriske DW, Armiger WB, Phillips DH, Alabno PA. Traders guide to fermentation media formulation. Fort Worth, TX: Traders Protein Division, Traders Oil Mill; 1980.

25. Gao C, Zhai Y, Ding Y, Wu Q. Application of sweet sorghum for biodiesel production by heterotrophic microalga Chlorella protothecoides. *Appl Energy* 2010;**87**(3):756–61.
26. Pimentel D, Patzek TW. Ethanol production using corn, switchgrass, and wood; biodiesel production using soybean and sunflower. *Nat Resour Res* 2005;**14**(1):65–76.
27. Kurosawa K, Wewetzer SJ, Sinskey AJ. Engineering xylose metabolism in triacylglycerol-producing Rhodococcus opacus for lignocellulosic fuel production. *Biotechnol Biofuels* 2013;**6**(1):134.
28. Gírio F, Fonseca C, Carvalheiro F, Duarte L, Marques S, Bogel-Łukasik R. Hemicelluloses for fuel ethanol: a review. *Bioresour Technol* 2010;**101**(13):4775–800.
29. Yevich R, Logan JA. An assessment of biofuel use and burning of agricultural waste in the developing world. *Global Biogeochem Cycles* 2003;**17**(4):1–6.
30. Chatzifragkou A, Makri A, Belka A, Bellou S, Mavrou M, Mastoridou M, et al. Biotechnological conversions of biodiesel derived waste glycerol by yeast and fungal species. *Energy* 2011;**36**(2):1097–108.
31. Varrone C, Giussani B, Izzo G, Massini G, Marone A, Signorini A, et al. Statistical optimization of biohydrogen and ethanol production from crude glycerol by microbial mixed culture. *Int J Hydrog Energ* 2012;**37**(21):16479–88.
32. Fei Q, Chang HN, Shang L. Exploring low-cost carbon sources for microbial lipids production by fed-batch cultivation of Cryptococcus albidus. *Biotechnol Bioprocess Eng* 2011;**16**(3):482–7.
33. Fei Q, Chang HN, Shang L, Kim N, Kang J. The effect of volatile fatty acids as a sole carbon source on lipid accumulation by Cryptococcus albidus for biodiesel production. *Bioresour Technol* 2011;**102**(3):2695–701.
34. Fei Q, Smith H, Dowe N, Pienkos PT. *Effects of culture medium ingredients on cell growth in batch cultures of Methylomicrobium buryatense with methane as a sole carbon source*. Poster presented on Recent Advances in Fermentation Technology (RAFT) meeting. Marco Island, FL; November 3-6, 2013.
35. Shuler ML, Kargi F, Shuler ML, Kargi F. *Bioprocess engineering*. 2nd ed. Upper Saddle River, New Jersey: Prentice-Hall; 2002.
36. Fidalgo J, Cid A, Torres E, Sukenik A, Herrero C. Effects of nitrogen source and growth phase on proximate biochemical composition, lipid classes and fatty acid profile of the marine microalga Isochrysis galbana. *Aquaculture* 1998;**166**(1):105–16.
37. Converti A, Casazza AA, Ortiz EY, Perego P, Del Borghi M. Effect of temperature and nitrogen concentration on the growth and lipid content of Nannochloropsis oculata and Chlorella vulgaris for biodiesel production. *Chem Eng Process Process Intensif* 2009;**48**(6):1146–51.
38. Boyle NR, Page MD, Liu B, Blaby IK, Casero D, Kropat J, et al. Three acyltransferases and nitrogen-responsive regulator are implicated in nitrogen starvation-induced triacylglycerol accumulation in Chlamydomonas. *J Biol Chem* 2012;**287**(19):15811–25.
39. Davis R, Aden A, Pienkos PT. Techno-economic analysis of autotrophic microalgae for fuel production. *Appl Energy* 2011;**88**(10):3524–31.
40. Li Y, Horsman M, Wang B, Wu N, Lan CQ. Effects of nitrogen sources on cell growth and lipid accumulation of green alga Neochloris oleoabundans. *Appl Microbiol Biotechnol* 2008;**81**(4):629–36.
41. Júnior MM, Batistote M, Cilli EM, Ernandes JR. Sucrose fermentation by Brazilian ethanol production yeasts in media containing structurally complex nitrogen sources. *J Inst Brew* 2009;**115**(3):191–7.
42. Dou X, Lu X-H, Lu M-Z, Yu L-S, Xue R, Ji J-B. The effects of trace elements on the lipid productivity and fatty acid composition of Nannochloropis oculata. *J Renew Energy* 2013;**2013**(671545):1–6.
43. Ethanol-information.com. Ethanol co-products; 2007. http://ethanol-information.com/ethanol-coproducts.php
44. Ramirez EC, Johnston DB, McAloon AJ, Yee W, Singh V. Engineering process and cost model for a conventional corn wet milling facility. *Ind Crop Prod* 2008;**27**(1):91–7.
45. Corn Refiners Association, *Corn Annual*, Washington DC; 2001-2013.
46. Energy Independence and Security Act, H.R. 6, 110th United States Congress; 2007.
47. Farrell J, Rose A. Temperature effects on microorganisms. *Annu Rev Microbiol* 1967;**21**(1):101–20.
48. Zhu C, Lee Y, Chao T. Effects of temperature and growth phase on lipid and biochemical composition of Isochrysis galbana TK1. *J Appl Phycol* 1997;**9**(5):451–7.

49. Suutari M, Liukkonen K, Laakso S. Temperature adaptation in yeasts: the role of fatty acids. *J Gen Microbiol* 1990;**136**(8):1469–74.
50. Loffhagen N, Härtig C, Harms H. Impact of membrane fatty acid composition on the uncoupling sensitivity of the energy conservation of Comamonas testosteroni ATCC 17454. *Appl Microbiol Biotechnol* 2006;**70**(5):618–24.
51. Beales N. Adaptation of microorganisms to cold temperatures, weak acid preservatives, low pH, and osmotic stress: a review. *Compr Rev Food Sci Food Saf* 2004;**3**(1):1–20.
52. Berry ED, Foegeding PM. Cold temperature adaptation and growth of microorganisms. *J Food Prot* 1997;**60**(12):1583–94.
53. Gregory J, Green B. Heat transfer for huge-scale fermentation. *Chem Eng* 2013;**120**(11):44–6.
54. Li X, Xu H, Wu Q. Large-scale biodiesel production from microalga Chlorella protothecoides through heterotrophic cultivation in bioreactors. *Biotechnol Bioeng* 2007;**98**(4):764–71.
55. Zhu L, Zong M, Wu H. Efficient lipid production with Trichosporon fermentans and its use for biodiesel preparation. *Bioresour Technol* 2008;**99**(16):7881–5.
56. Minocha S. pH of the medium and the growth and metabolism of cells in culture. In: Bonga JM, Durzan D, editors. *Cell and tissue culture in forestry. Forestry sciences*, vols. 24-26. The Netherlands: Springer; 1987. p. 125–41.
57. Moolenaar WH. Effects of growth factors on intracellular pH regulation. *Annu Rev Physiol* 1986;**48**(1):363–76.
58. Kennedy M, Krouse D. Strategies for improving fermentation medium performance: a review. *J Ind Microbiol Biotechnol* 1999;**23**(6):456–75.
59. Wang J, Wan W. Experimental design methods for fermentative hydrogen production: a review. *Int J Hydrog Energ* 2009;**34**(1):235–44.
60. Pundle A, SivaRaman H. Medium optimization for the production of penicillin V acylase from Bacillus sphaericus. *Biotechnol Lett* 1994;**16**(10):1041–6.
61. Furuhashi K, Takagi M. Optimization of a medium for the production of 1, 2-epoxytetradecane by Nocardia corallina B-276. *Appl Microbiol Biotechnol* 1984;**20**(1):6–9.
62. Fisher RA. The arrangement of field experiments. In: Kotz S, Johnson NL, editors. *Breakthroughs in statistics*. New York: Springer; 1992. pp. 82–91.
63. Haaland PD. *Experimental design in biotechnology*, vol. 105. New York: CRC Press; 1989.
64. Azin M, Moravej R, Zareh D. Production of xylanase by Trichoderma longibrachiatum on a mixture of wheat bran and wheat straw: optimization of culture condition by Taguchi method. *Enzyme Microb Technol* 2007;**40**(4):801–5.
65. Silveira RG, Kakizono T, Takemoto S, Nishio N, Nagai S. Medium optimization by an orthogonal array design for the growth of Methanosarcina barkeri. *J Ferment Bioeng* 1991;**72**(1):20–5.
66. Vicente G, Coteron A, Martinez M, Aracil J. Application of the factorial design of experiments and response surface methodology to optimize biodiesel production. *Ind Crop Prod* 1998;**8**(1):29–35.
67. Ghose T, Tyagi R. Rapid ethanol fermentation of cellulose hydrolysate. II. Product and substrate inhibition and optimization of fermentor design. *Biotechnol Bioeng* 1979;**21**(8):1401–20.
68. Chi Z, Pyle D, Wen Z, Frear C, Chen S. A laboratory study of producing docosahexaenoic acid from biodiesel-waste glycerol by microalgal fermentation. *Process Biochem* 2007;**42**(11):1537–45.
69. Vinci VA, Parekh SR. *Handbook of industrial cell culture: mammalian, microbial, and plant cells*. Totowa, NJ: Humana Press; 2003.
70. Weuster-Botz D. Experimental design for fermentation media development: statistical design or global random search? *J Biosci Bioeng* 2000;**90**(5):473–83.
71. Khoramnia A, Lai OM, Ebrahimpour A, Tanduba CJ, Voon TS, Mukhlis S. Thermostable lipase from a newly isolated Staphylococcus xylosus strain; process optimization and characterization using RSM and ANN. *Electron J Biotechnol* 2010;**13**(5):15–6.
72. Ghadge SV, Raheman H. Process optimization for biodiesel production from mahua (Madhuca indica) oil using response surface methodology. *Bioresour Technol* 2006;**97**(3):379–84.

73. Ratnam B, Rao MN, Rao MD, Rao SS, Ayyanna C. Optimization of fermentation conditions for the production of ethanol from sago starch using response surface methodology. *World J Microbiol Biotechnol* 2003;**19**(5):523–6.
74. Prapulla S, Jacob Z, Chand N, Rajalakshmi D, Karanth N. Maximization of lipid production by Rhodotorula gracilis CFR-1 using response surface methodology. *Biotechnol Bioeng* 1992;**40**(8):965–70.
75. Mandenius CF, Brundin A. Bioprocess optimization using design-of-experiments methodology. *Biotechnol Prog* 2008;**24**(6):1191–203.
76. Thiel T, Bramble J, Rogers S. Optimum conditions for growth of cyanobacteria on solid media. *FEMS Microbiol Lett* 1989;**61**(1-2):27–32.
77. Valdez-Vazquez I, Ríos-Leal E, Muñoz-Páez KM, Carmona-Martínez A, Poggi-Varaldo HM. Effect of inhibition treatment, type of inocula, and incubation temperature on batch H2 production from organic solid waste. *Biotechnol Bioeng* 2006;**95**(3):342–9.
78. Plackett RL, Burman JP. The design of optimum multifactorial experiments. *Biometrika* 1946;**33**:305–25.
79. Stone R, Veevers A. The Taguchi influence on designed experiments. *J Chemometr* 1994;**8**(2):103–10.
80. Zhao C, Karakashev D, Lu W, Wang H, Angelidaki I. Xylose fermentation to biofuels (hydrogen and ethanol) by extreme thermophilic (70 C) mixed culture. *Int J Hydrog Energ* 2010;**35**(8):3415–22.
81. Chang M-Y, Tsai G-J, Houng J-Y. Optimization of the medium composition for the submerged culture of Ganoderma lucidum by Taguchi array design and steepest ascent method. *Enzyme Microb Technol* 2006;**38**(3):407–14.
82. Box GEP, Hunter JS, Hunter WG. Statistics for experimenters: design, innovation, and discovery. 2nd Edition. New York: Wiley-Interscience; 2005.
83. Bandaru VVR, Somalanka SR, Mendu DR, Madicherla NR, Chityala A. Optimization of fermentation conditions for the production of ethanol from sago starch by co-immobilized amyloglucosidase and cells of Zymomonas mobilis using response surface methodology. *Enzyme Microb Technol* 2006;**38**(1):209–14.
84. Kumar R, Mahajan S, Kumar A, Singh D. Identification of variables and value optimization for optimum lipase production by Bacillus pumilus RK31 using statistical methodology. *N Biotechnol* 2011;**28**(1):65–71.
85. Saraphirom P, Reungsang A. Optimization of biohydrogen production from sweet sorghum syrup using statistical methods. *Int J Hydrog Energ* 2010;**35**(24):13435–44.
86. Box GE, Wilson K. On the experimental attainment of optimum conditions. *J R Stat Soc Ser B Methodol* 1951;**13**(1):1–45.
87. Duarte SH, Ghiselli G, Maugeri F. Influence of culture conditions on lipid production by Candida sp. LEB-M3 using glycerol from biodiesel synthesis. *Biocatal Agric Biotechnol* 2013;**2**(4):339–43.
88. Zabeti M, Daud WMAW, Aroua MK. Optimization of the activity of CaO/Al2O3 catalyst for biodiesel production using response surface methodology. *Appl Catal Gen* 2009;**366**(1):154–9.
89. Sridevi V. Implementation of response surface methodology for phenol degradation using Pseudomonas putida (NCIM 2102). *J Bioremed Biodegrad* 2011;**2**:121.
90. Gernaey KV, Lantz AE, Tufvesson P, Woodley JM, Sin G. Application of mechanistic models to fermentation and biocatalysis for next-generation processes. *Trends Biotechnol* 2010;**28**(7):346–54.
91. Sin G, Ödman P, Petersen N, Lantz AE, Gernaey KV. Matrix notation for efficient development of first-principles models within PAT applications: integrated modeling of antibiotic production with Streptomyces coelicolor. *Biotechnol Bioeng* 2008;**101**(1):153–71.
92. Sin G, Gernaey KV, Lantz AE. Good modeling practice for PAT applications: propagation of input uncertainty and sensitivity analysis. *Biotechnol Prog* 2009;**25**(4):1043–53.
93. Sin G, Woodley JM, Gernaey KV. Application of modeling and simulation tools for the evaluation of biocatalytic processes: a future perspective. *Biotechnol Prog* 2009;**25**(6):1529–38.
94. Lapin A, Müller D, Reuss M. Dynamic behavior of microbial populations in stirred bioreactors simulated with Euler-Lagrange methods: traveling along the lifelines of single cells. *Ind Eng Chem Res* 2004;**43**(16):4647–56.

95. Armstrong RA, Lee C, Hedges JI, Honjo S, Wakeham SG. A new, mechanistic model for organic carbon fluxes in the ocean based on the quantitative association of POC with ballast minerals. *Deep-Sea Res II Top Stud Oceanogr* 2001;**49**(1):219–36.
96. Vinu R, Broadbelt LJ. A mechanistic model of fast pyrolysis of glucose-based carbohydrates to predict bio-oil composition. *Energy Environ Sci* 2012;**5**(12):9808–26.
97. Lippmann RP. An introduction to computing with neural nets. *IEEE ASSP Mag* 1987;**4**(2):4–22.
98. Nagata Y, Chu KH. Optimization of a fermentation medium using neural networks and genetic algorithms. *Biotechnol Lett* 2003;**25**(21):1837–42.
99. Sivapathasekaran C, Mukherjee S, Ray A, Gupta A, Sen R. Artificial neural network modeling and genetic algorithm based medium optimization for the improved production of marine biosurfactant. *Bioresour Technol* 2010;**101**(8):2884–7.
100. Yuste AJ, Dorado MP. A neural network approach to simulate biodiesel production from waste olive oil. *Energy Fuel* 2006;**20**(1):399–402.
101. Tu JV. Advantages and disadvantages of using artificial neural networks versus logistic regression for predicting medical outcomes. *J Clin Epidemiol* 1996;**49**(11):1225–31.
102. Hussain M, Ramachandran K. Design of a fuzzy logic controller for regulating substrate feed to fed-batch fermentation. *Food Bioprod Process* 2003;**81**(2):138–46.
103. Kennedy MJ, Prapulla S, Thakur M. Designing fermentation media: a comparison of neural networks to factorial design. *Biotechnol Tech* 1992;**6**(4):293–8.
104. Kuo RJ, Chen C, Hwang Y. An intelligent stock trading decision support system through integration of genetic algorithm based fuzzy neural network and artificial neural network. *Fuzzy Set Syst* 2001;**118**(1):21–45.
105. Topcu IB, Sarıdemir M. Prediction of compressive strength of concrete containing fly ash using artificial neural networks and fuzzy logic. *Comput Mater Sci* 2008;**41**(3):305–11.
106. Žužek M, Friedrich J, Cestnik B, Karalič A, Cimerman A. Optimization of fermentation medium by a modified method of genetic algorithms. *Biotechnol Tech* 1996;**10**(12):991–6.
107. Marteijn R, Jurrius O, Dhont J, De Gooijer C, Tramper J, Martens D. Optimization of a feed medium for fed-batch culture of insect cells using a genetic algorithm. *Biotechnol Bioeng* 2003;**81**(3):269–78.
108. Sarkar D, Modak JM. Optimisation of fed-batch bioreactors using genetic algorithms. *Chem Eng Sci* 2003;**58**(11):2283–96.
109. Chen L, Nguang SK, Chen XD, Li XM. Modelling and optimization of fed-batch fermentation processes using dynamic neural networks and genetic algorithms. *Biochem Eng J* 2004;**22**(1):51–61.
110. Na J-G, Chang Y, Chung B, Lim H. Adaptive optimization of fed-batch culture of yeast by using genetic algorithms. *Bioprocess Biosyst Eng* 2002;**24**(5):299–308.
111. Arroyo JM, Conejo AJ. A parallel repair genetic algorithm to solve the unit commitment problem. *IEEE Trans Power Syst* 2002;**17**(4):1216–24.
112. Li W. Using genetic algorithm for network intrusion detection. In: *Proceedings of the United States Department of Energy Cyber Security Training Conference, USA*, 2004, pp. 24–7.
113. Jones B, Nachtsheim CJ. A class of three-level designs for definitive screening in the presence of second-order effects. *J Qual Technol* 2011;**43**(1):1–15.
114. Xiao L, Lin DK, Bai F. Constructing definitive screening designs using conference matrices. *J Qual Technol* 2012;**44**(1):2–8.
115. Erler A, de Mas N, Ramsey P, Henderson G. Efficient biological process characterization by definitive-screening designs: the formaldehyde treatment of a therapeutic protein as a case study. *Biotechnol Lett* 2013;**35**(3):323–9.
116. Wong C-L, Huang C-C, Lu W-B, Chen W-M, Chang J-S. Producing 2, 3-butanediol from agricultural waste using an indigenous Klebsiella sp. Zmd30 strain. *Biochem Eng J* 2012;**69**:32–40.
117. Hameed B, Lai L, Chin L. Production of biodiesel from palm oil (Elaeis guineensis) using heterogeneous catalyst: an optimized process. *Fuel Process Technol* 2009;**90**(4):606–10.
118. Liu C-H, Lu W-B, Chang J-S. Optimizing lipase production of Burkholderia sp. by response surface methodology. *Process Biochem* 2006;**41**(9):1940–4.

119. Ong LK, Effendi C, Kurniawan A, Lin CX, Zhao XS, Ismadji S. Optimization of catalyst-free production of biodiesel from Ceiba pentandra (kapok) oil with high free fatty acid contents. *Energy* 2013;**57**:615–23.
120. Reaume SJ, Ellis N. Optimizing reaction conditions for the isomerization of fatty acids and fatty acid methyl esters to their branch chain products. *J Am Oil Chem Soc* 2011;**88**(5):661–71.
121. Zabeti M, Daud WMAW, Aroua MK. Biodiesel production using alumina-supported calcium oxide: an optimization study. *Fuel Process Technol* 2010;**91**(2):243–8.
122. Rajendra M, Jena PC, Raheman H. Prediction of optimized pretreatment process parameters for biodiesel production using ANN and GA. *Fuel* 2009;**88**(5):868–75.
123. Wikipedia. *Comparison of statistical packages*; 2014, Available from: http://en.wikipedia.org/wiki/Comparison_of_statistical_packages, References.
124. Hu W, Berdugo C, Chalmers JJ. The potential of hydrodynamic damage to animal cells of industrial relevance: current understanding. *Cytotechnology* 2011;**63**(5):445–60.
125. Betts JI, Doig SD, Baganz F. Characterization and application of a miniature 10 ml stirred-tank bioreactor, showing scale-down equivalence with a conventional 7 l reactor. *Biotechnol Prog* 2006;**22**(3):681–8.
126. Weuster-Botz D. Parallel reactor systems for bioprocess development. In: Kragl U, editor. *Technology transfer in biotechnology*, vol. 92. Berlin Heidelberg: Springer; 2005, pp. 125–43.
127. Zhang Z, Perozziello G, Boccazzi P, Sinskey AJ, Geschke O, Jensen KF. Microbioreactors for bioprocess development. *J Assoc Lab Autom* 2007;**12**(3):143–51.
128. Schäpper D, Alam MNHZ, Szita N, Lantz AE, Gernaey KV. Application of microbioreactors in fermentation process development: a review. *Anal Bioanal Chem* 2009;**395**(3):679–95.
129. Zhang S, Ye BC, Chu J, Zhuang Y, Guo M. From multi-scale methodology to systems biology: to integrate strain improvement and fermentation optimization. *J Chem Technol Biotechnol* 2006;**81**(5):734–45.
130. Enfors S-O, Jahic M, Rozkov A, Xu B, Hecker M, Jürgen B, et al. Physiological responses to mixing in large scale bioreactors. *J Biotechnol* 2001;**85**(2):175–85.
131. Formenti LR, Nørregaard A, Bolic A, Hernandez DQ, Hagemann T, Heins AL, et al. Challenges in industrial fermentation technology research. *Biotechnol J* 2014;**9**(6):727–38.
132. Neubauer P, Junne S. Scale-down simulators for metabolic analysis of large-scale bioprocesses. *Curr Opin Biotechnol* 2010;**21**(1):114–21.
133. Bredwell MD, Worden RM. Mass-transfer properties of microbubbles. 1. Experimental studies. *Biotechnol Prog* 1998;**14**(1):31–8.
134. Bouaifi M, Hebrard G, Bastoul D, Roustan M. A comparative study of gas hold-up, bubble size, interfacial area and mass transfer coefficients in stirred gas–liquid reactors and bubble columns. *Chem Eng Process Process Intensif* 2001;**40**(2):97–111.
135. Datar RP, Shenkman RM, Cateni BG, Huhnke RL, Lewis RS. Fermentation of biomass-generated producer gas to ethanol. *Biotechnol Bioeng* 2004;**86**(5):587–94.
136. Lee KC, Rittmann BE. Applying a novel autohydrogenotrophic hollow-fiber membrane biofilm reactor for denitrification of drinking water. *Water Res* 2002;**36**(8):2040–52.
137. Nerenberg R, Rittmann B. Hydrogen-based, hollow-fiber membrane biofilm reactor for reduction of perchlorate and other oxidized contaminants. *Water Sci Technol* 2004;**49**(11-12):223–30.
138. Tsai SP, Datta R, Basu R, Yoon SH, Tobey RE. Modular membrane supported bioreactor for conversion of syngas components to liquid products: US; 2011. US 8017384 B2.
139. Lencastre Fernandes R, Nierychlo M, Lundin L, Pedersen AE, Puentes Tellez P, Dutta A, et al. Experimental methods and modeling techniques for description of cell population heterogeneity. *Biotechnol Adv* 2011;**29**(6):575–99.
140. Lapin A, Schmid J, Reuss M. Modeling the dynamics of E. coli populations in the three-dimensional turbulent field of a stirred-tank bioreactor—a structured-segregated approach. *Chem Eng Sci* 2006;**61**(14):4783–97.

Index

Note: Page numbers followed by *b* indicate boxes, *f* indicate figures and *t* indicate tables.

F

G

H

N

S

Edwards Brothers Malloy
Ann Arbor MI. USA
July 25, 2016